Algal Ecology

Freshwater Benthic Ecosystems

AQUATIC ECOLOGY Series

Series Editor
James H. Thorp
Department of Biology
University of Louisville
Louisville, Kentucky

Other titles in the series:

Groundwater Ecology
Janine Gibert, Dan L. Danielopol, Jack A. Stanford

Algal Ecology

Freshwater Benthic Ecosystems

Edited by

R. Jan Stevenson
Department of Biology
University of Louisville
Louisville, Kentucky

Max L. Bothwell
Environmental Sciences
National Hydrological Research Institute
Saskatoon, Saskatchewan, Canada

Rex L. Lowe
Department of Biological Sciences
Bowling Green State University
Bowling Green, Ohio

ACADEMIC PRESS
San Diego New York Boston London Sydney Tokyo Toronto

Front cover photograph: The filamentous diatom *Aulacoseira* entangled among the stalked diatoms *Cymbella* and *Gomphonema*. (See Chapter 1, Figure 1 for more details.)

This book is printed on acid-free paper. ♾

Academic Press, Inc.
A Division of Harcourt Brace & Company
525 B Street, Suite 1900, San Diego, California 92101-4495

United Kingdom Edition published by
Academic Press Limited
24-28 Oval Road, London NW1 7DX

Library of Congress Cataloging-in-Publication Data

Algal ecology : freshwater benthic ecosystems / edited by R. Jan Stevenson, Max I. Bothwell, Rex L. Lowe.
p. cm. -- (Aquatic ecology series)
Includes bibliographical references and index.
ISBN 0-12-668450-2 (alk. paper)
1. Freshwater algae--Ecology. 2. Freshwater ecology.
I. Stevenson, R. Jan. II. Bothwell, M. L. (Max L.) III. Lowe, Rex L. IV. Series.
QK570.25.A43 1996
589.3'5--dc20 95-51737
CIP

PRINTED IN THE UNITED STATES OF AMERICA
96 97 98 99 00 01 MM 9 8 7 6 5 4 3 2 1

Jan and Rex graciously asked me to dedicate this book to my wife Carol R. Bothwell (deceased November 5, 1994) in remembrance of the selfless dedication she gave to me over the years so that my career might prosper. In this, she was not unlike many other wives, including Mariellyn Stevenson and Sheryn Lowe, who spend their lives nurturing their husbands and children so that they might grow and fulfill their own dreams.

Max L. Bothwell

Contents

4 Pattern in Wetlands

L. Gordon Goldsborough and Gordon G. C. Robinson

SECTION TWO

FACTORS AFFECTING BENTHIC ALGAE

5 Effects of Light

Walter R. Hill

7 Nutrients

Mark A. Borchardt

12 Effects of Grazers on Freshwater Benthic Algae

Alan D. Steinman

13 Response of Benthic Algal Communities to Natural Physical Disturbance

Christopher G. Peterson

14 *Ecotoxicology of Inorganic Chemical Stress to Algae*

Robert Brian Genter

15 Effects of Organic Toxic Substances

Kyle D. Hoaglund, Justin P. Carder, and Rebecca L. Spawn

16 Acidification Effects

D. Planas

SECTION THREE

THE NICHE OF BENTHIC ALGAE IN FRESHWATER ECOSYSTEMS

17 The Role of Periphyton in Benthic Food Webs

Gary A. Lamberti

Contributors

Numbers in parentheses indicate the pages on which the authors' contributions begin.

Barry J. F. Biggs (31), National Institute of Water and Atmospheric Research, Christchurch, New Zealand

Mark A. Borchardt (183), Marshfield Medical Research Foundation, Marshfield, Wisconsin 54449

Thomas L. Bott (573), Stroud Water Research Center, Academy of Natural Sciences, Avondale, Pennsylvania 19311

JoAnn M. Burkholder (253), Department of Botany, North Carolina State University, Raleigh, North Carolina 27695

Justin P. Carder (469), Department of Forestry, Fisheries, and Wildlife, University of Nebraska, Lincoln, Nebraska 68583

Dean M. DeNicola (149), Department of Biology, Slippery Rock University, Slippery Rock, Pennsylvania 16057

Robert Brian Genter (403), Department of Environment and Health Sciences, Johnson State College, Johnson, Vermont 05656

L. Gordon Goldsborough (77), Department of Botany, University of Manitoba, Winnipeg, Manitoba, Canada R3T 2N2

Stanley V. Gregory (669), Department of Fisheries and Wildlife, Oregon State University, Corvallis, Oregon 97331

Walter R. Hill (121), Environmental Sciences Division, Oak Ridge National Laboratory, Oak Ridge, Tennessee 37831

Kyle D. Hoagland (469), Department of Forestry, Fisheries, and Wildlife, University of Nebraska, Lincoln, Nebraska 68583

Gary A. Lamberti (533, 669), Department of Biological Sciences, University of Notre Dame, Notre Dame, Indiana 46556

Rex L. Lowe (57, 705), Department of Biological Sciences, Bowling Green State University, Bowling Green, Ohio 43403, and University of Michigan Biological Station, Pellston, Michigan 49769

Paul V. McCormick (229), Everglades Systems Research Division, South Florida Water Management District, West Palm Beach, Florida 33416

C. David McIntire (669), Department of Botany and Plant Pathology, Oregon State University, Corvallis, Oregon 97331

Patrick J. Mulholland (609), Environmental Sciences Division, Oak Ridge National Laboratory, Oak Ridge, Tennessee 37831

Yangdong Pan (705), Water Resources Laboratory, University of Louisville, Louisville, Kentucky 40292

Christopher G. Peterson (375), Department of Natural Sciences, Loyola University of Chicago, Chicago, Illinois 60626

D. Planas (497), Département des Sciences Biologiques, Université du Québec à Montréal, Montréal, Quebec, Canada H3C 3P8

Gordon G. C. Robinson (77), Department of Botany, University of Manitoba, Winnipcg, Manitoba, Canada R3T 2N2

Rebecca L. Spawn (469), Department of Forestry, Fisheries, and Wildlife, University of Nebraska, Lincoln, Nebraska 68583

Alan D. Steinman (341, 669), Department of Ecosystem Restoration, South Florida Water Management District, West Palm Beach, Florida 33416

R. Jan Stevenson (3, 321), Department of Biology, University of Louisville, Louisville, Kentucky 40292

Nancy C. Tuchman (299), Department of Biology, Loyola University of Chicago, Chicago, Illinois 60626

Robert G. Wetzel (641), Department of Biological Sciences, University of Alabama, Tuscaloosa, Alabama 35487

Preface

Benthic algae have been intensively studied, especially over the past two decades. This intensity has been stimulated by the widespread recognition that benthic algae are ideal indicators of the health of many, if not most, aquatic ecosystems. With this book we hope to synthesize this vital area of research and share its essence with our colleagues and students. We started with an outline of the myriad abiotic and biotic determinants of benthic algal ecology. Foremost was the realization that benthic algae are primary producers in food webs and are the fundamental components in biogeochemical cycles of aquatic ecosystems. We then chose experts to write and review chapters and most eagerly agreed to participate. These contributions are a tribute to the authors and to the reviewers.

Some necessary limitations had to be placed on the breadth of coverage of the book. Thus, these reviews are not encyclopedic, but are representative of the excitement we all share for our discipline. Of course, coverage was limited to freshwater benthic algal research, but often important observations and experimental results from planktonic and marine research were included to round out a chapter. We feel sure that many of the conceptual developments derived from studies of benthic algal ecology will also be interesting to our colleagues who concentrate on other aspects of algal and aquatic ecology.

Early research on energy sources in food webs of aquatic ecosystems focused on phytoplankton in lentic habitats, leaf litter in streams, and plants in wetlands. More recently, attention has focused on the algae on substrata. The cycle of science has taken benthic algal ecology from observation and experimentation to synthesis. We hope this book will generate a new wave of conceptual innovation for students and researchers of algal ecology.

R. Jan Stevenson
Max L. Bothwell
Rex L. Lowe

Acknowledgments

The development of this book was made possible by support for R. Jan Stevenson when on sabbatical at the Institute of Ecosystem Studies with funds from the Mellon Foundation to Gene Likens. In addition, the College of Arts and Sciences, through the Water Resources Laboratory at the University of Louisville, provided substantial support. Support from the National Hydrology Research Institute for Max Bothwell and from Bowling Green State University and the University of Michigan Biological Station for Rex Lowe is also acknowledged.

The following scientists contributed as reviewers: N. Aumen (South Florida Water Management District), B. Biggs (National Institute of Water and Atmospheric Research, New Zealand), D. Blinn (Northern Arizona University), R. Carlton (University of Notre Dame), W. Clements (Colorado State University), H. Carrick (State University of New York at Buffalo), A. Cattaneo (Université de Montréal), E. Cox (University of Sheffield), D. D'Angelo (Proctor and Gamble), J. Feminella (Auburn University), G. Goldsborough (Brandon University), S. Jensen (University of Nebraska), S. Kohler (Illinois Natural History Survey), M. Lewis (U.S. Environmental Protection Agency), P. Montagna (University of Texas Marine Science Institute), Y. Pan (University of Louisville), R. Pillsbury (Bowling Green State University), S. Porter (U.S. Geological Survey), A. Steinman (South Florida Water Management District), M. Sullivan (Mississippi State University), and M. Turner (Fisheries and Oceans, Canada).

Finally, all of us would like to thank our families. In particular, the editors would like to acknowledge the time that our families gave us to review manuscripts and consult with the authors. So thank you Mariellyn, Philip, and Virginia; Carol, Julie, and Peter; and Sheryn, Chris, and Terry.

R. Jan Stevenson
Max L. Bothwell
Rex L. Lowe

PATTERNS OF BENTHIC ALGAE IN AQUATIC ECOSYSTEMS

1

An Introduction to Algal Ecology in Freshwater Benthic Habitats

R. Jan Stevenson

Department of Biology
University of Louisville
Louisville, Kentucky 40292

I. THE DIVERSITY OF BENTHIC ALGAE AND THEIR HABITATS IN FRESH WATERS

A. The Taxonomic and Morphological Diversity of Benthic Algae

Algae are a highly diverse group of organisms that have important functions in aquatic habitats. Algae are an evolutionarily diverse group of photoautotrophic organisms with chlorophyll *a* and unicellular reproductive structures. By various taxonomic schemes, the number of algal divisions ranges from 4 to 13, with as many as 24 classes, and about 26,000 species (see Bold and Wynne, 1985; Raven and Johnson, 1992). The number of recognized species probably greatly underestimates the actual number of species because many habitats and regions have not been extensively sampled and many algae are very small and hard to distinguish from each other. Benthic algae are those that live on or in association with substrata. Phytoplankton are algae suspended in the water column. An individual alga may be benthic or planktonic at one time or another, but many species are characteristically found in just one habitat.

Most benthic algae in freshwater habitats are blue-green algae (Cyanophyta), green algae (Chlorophyta), diatoms (Bacillariophyta), or red algae (Rhodophyta). However, most other divisions of algae can occur in freshwater benthic habitats. The Chrysophyta, Xanthophyta, Cryptophyta, and Pyrrophyta have many species that usually occur in the phytoplankton, but they may also occur in physiologically active forms in some benthic habitats. In addition, resting cells of many algae can be found in the benthos, which may have originated there or may have settled from the water column (Sicko-Goad *et al.*, 1989). The latter divisions seldom constitute more algal biomass in a benthic habitat than blue-green algae, green algae, diatoms, or red algae.

The divisions of algae are distinguished by a variety of chemical and morphological differences (Bold and Wynne, 1985; Lee, 1989). All divisions have chlorophyll *a*, but different divisions can also have either chlorophylls *b*, *c*, or *d*. Distinctive accessory pigments, such as phycobilins and fucoxanthin, also are characteristic of different algal divisions. Accessory pigments may color the algae red, blue, or golden-brown, if they are not green with chlorophyll as the dominant pigment. The different divisions also have chemically different cell walls and storage products, or they have distinctive forms of motility or numbers of flagella. Ultrastructural features, such as the number of membranes around chloroplasts, also distinguish the different divisions and indicate that the algae have many ancestors and are an evolutionarily diverse group (Stewart and Mattox, 1980; Gibbs, 1981; Cavalier-Smith, 1986).

Even though these groups have great evolutionary, genetic, and chemical differences, they share many of the same growth forms (Table I). The blue-green algae, green algae, and diatoms have the greatest morphological diversity with unicellular, colonial, and filamentous forms. Many of the green algal filaments are individually macroscopic, whereas most of the other algae (except *Vaucheria*, a xanthophyte) are only macroscopically evident in multicellular masses. Each of the growth forms has a motile and nonmotile stage in one division or another. Blue-green algal filaments are motile because trichomes (a series of cells) can move through sheaths of mucilage. Unicellular green algae, chrysophytes, cryptomonads, and dinoflagellates are commonly found in benthic habitats, where they move by means of flagella. A few benthic algae, particularly chrysophytes, may be amoeboid. Few species of colonial green algae with flagella actually occur in motile form in benthic habitats, but *Gonium*, *Pandorina*, *Eudorina*, and even *Volvox* may occur incidentally in some benthic algal samples. Desmids, a family of green algae characteristic of acidic habitats, can

TABLE I Morphological Variability in the Divisions of Benthic Algae[a]

	Morphology						
	Unicellular		*Colonial*		*Filamentous*		
Taxon	*Mot.*	*N-M*	*Mot.*	*N-M*	*Mot.*	*N-M*	*Means of motility*
Cyanophyta (blue-green algae)		✓		✓	✓	✓	Sheaths
Chlorophyta (green algae)	✓	✓	✓	✓		✓	Flagella and pectin
Bacillariophyta (diatoms)	✓	✓		✓	1	✓	Raphe
Rhodophyta (red algae)						✓	
Chrysophyta (chrysophytes)	✓	✓	✓	✓		✓	Flagella and pseudopods
Xanthophyta (xanthophytes)						✓	
Euglenophyta (euglenoids)	✓						Flagella
Pyrrophyta (dinoflagellates)	✓	✓		✓			Flagella
Cryptophyta (cryptomonads)	✓						Flagella

[a]Mot., motile; N-M nonmotile.

move by extruding small bursts of pectin through the cell wall. Most motile diatoms are unicellular, are characteristically benthic, and move by means of a raphe (mucilage extruded through a long, narrow opening in the cell wall). But one genus of diatom, *Bacillaria,* is filamentous and motile and can occur in benthic habitats, even though it is characteristically planktonic.

Nonmotile forms of unicellular, colonial, and filamentous algae may be attached to substrata or entangled in the matrix of other organisms that are attached (Fig. 1). These organisms attach by specially adapted cells and mucilaginous secretions. Some taxa, such as the green filamentous alga *Stigeoclonium,* are heterotrichous, which means they have morphologically distinct basal and filamentous cells. The basal cells form broad horizontal expanses of cells across the substratum surface, and filaments develop vertically from the basal cells. Mucilaginous secretions can be amorphous for unicellular blue-green and green algae or organized into special pads, stalks, or tubes for diatoms (Fig. 1).

Benthic algal growth forms are hypothesized to confer competitive advantages during different stages of benthic algal community develop-

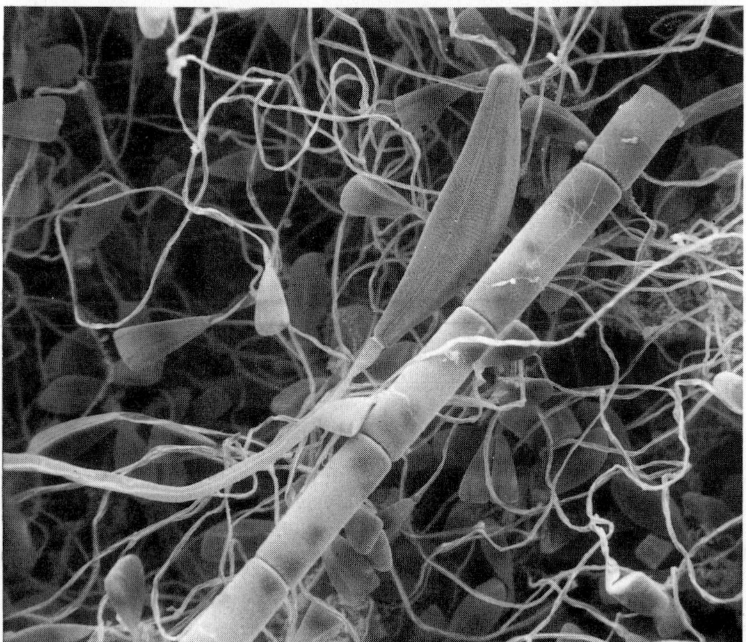

FIGURE 1 Growth forms of benthic algae. Upper left, adnate *Navicula* (×2400); upper right, apically attached *Synedra* (×2000); lower left, the filamentous diatom *Aulacoseira*

entangled among the stalked diatoms *Cymbella* and *Gomphonema* (×600); lower right, the filamentous green alga *Stigeoclonium* (×285). (Reprinted with permission from Hoagland *et al.*, 1982.)

ment (Patrick, 1976; Hoagland *et al.*, 1982; Hudon and Legendre, 1987). Adnate algae, mostly diatoms, that grow flat on substrata are close enough to substrata to lie within low-current-velocity boundary layers and thereby avoid the shear stress of severe currents. Adnate diatoms are also most resistant to grazing, however, they are easily overgrown by other benthic algae and may become nutrient- and light-limited by that overgrowth (McCormick and Stevenson, 1989). Apically attached algae, such as the diatom *Synedra* or the green alga *Characium,* stand erect on substrata in slow currents and are the first algae to overgrow adnate forms. Stalked diatoms and filamentous algae usually take longer to manifest true growth forms during community development. They then overgrow adnate and apically attached algae, thereby exploiting light and nutrients and outcompeting underlying forms for those resources (see McCormick, Chapter 8). Many motile algae, however, can still move through the filamentous and stalked algal overgrowth of substrata. Thus, the idealized succession of growth forms on inert substrata in habitats after communities are disturbed is from adnate forms that persist after disturbance, to fast-growing apically attached taxa, and finally to stalked, filamentous, or motile benthic algae.

B. The Habitats of Freshwater Benthic Algae

Freshwater benthic algae are found in the photic zones of streams and rivers, lakes, and wetlands. Many terms are used to distinguish the groups of benthic organisms that live in different aquatic habitats. In the following discussion, terms for benthic algae and associated organisms are defined according to common modern usage and to the widely read and respected limnology text by Robert Wetzel (1983a). My definition of terms is not meant to be a final definition of terms, because permanent definition impedes creativity. New definitions are required when new information is developed and new ideas evolve about how systems are organized, what their parts are, and how the parts arc linked. We (M. L. Bothwell, R. L. Lowe, and I) have tried to standardize use of most terms in the chapters because consistency does facilitate communication.

Benthos refers to organisms living on the "bottom" or associated with substrata. Wetzel (1983a) considers benthos to be the animals associated with any solid–liquid substratum. Common use of the term benthos includes most organisms associated with substrata in aquatic habitats: fish, macroscopic invertebrates and meiofauna, fungi, bacteria, and even hyporheic (below substratum surface) organisms. *Periphyton* and *aufwuchs* are also terms that are more or less synonymous with the term *benthic algae.* Aufwuchs is a German word that means "to grow upon" and is not often used in the modern literature. Periphyton is a commonly used term that refers to all the microflora on substrata (Wetzel, 1983a). Therefore, periphyton includes all the microscopic algae, bacteria, and fungi on (or associated with) substrata. According to this definition, macroscopic benthic algae would not be considered periphyton. Using this distinction between periphyton and benthic algae enables the use of periphyton to refer to the biofilms and thicker matrices of microscopic organisms in which flow, eddy diffusivity of nutrients, and exposure to other organisms are very different than among the filaments of macroscopic algae, such as *Cladophora, Spirogyra, Chara,* and *Vaucheria.*

The nature of the habitat in which these organisms are found depends on the habitat diversity of aquatic ecosystems and on the size of the organism. For example, macroalgae on substrata are in very different habitats than microalgae on the same substratum type, because they extend farther into the water column. However, one common criterion for habitat is the substratum type (see Burkholder, Chapter 9). *Epilithic* algae grow on hard, relatively inert substrata, such as gravel, pebble, cobble, and boulder, that are bigger than most algae (see Hynes, 1970, for substratum size definitions). *Epiphytic* algae grow on plants and larger algae, which provide relatively firm substrata that are bigger than the epiphytic algae, but can be highly active metabolically and can be a great source of nutrients (see Burkholder, Chapter 9). *Epipsammic* algae grow on sand, which is hard, relatively inert, and smaller than all but the smallest diatoms. Algae growing on inorganic or organic sediments that are smaller than most unicellular algae are called *epipelic.* Few large algae live among sand grains, because the sand is too unstable and may crush them; however, epipelic algae are characteristically large motile diatoms, motile filamentous blue-green algae, or larger motile flagellates like *Euglena.*

Metaphyton are the algae of the photic zone that are not directly attached to substrata, nor are they freely suspended in the water column. Metaphyton come in many forms and may have many origins. Metaphyton are usually clouds of filamentous green algae, like *Spirogyra, Mougeotia,* or *Zygnema,* that commonly do not attach directly to substrata, but become loosely aggregated and associated with substrata in areas that are protected from current or waves. They may become trapped between surface and substrata near shorelines or actually entangled with plants or other substrata. Denser assemblages of filamentous blue-green algae, diatoms, bacteria, and fungi can also form floating surface assemblages at the water surface that are usually entangled among or attached to substrata. The origin of metaphyton is usually algae from other substrata, typically epiphyton (see Goldsborough and Robinson, Chapter 4), but epipelon and epilithon can also form floating mats of microscopic organisms when they slough from submerged substrata. The persistence of these communities is highly dependent on their ability to withstand rain, mild currents, or animal disturbances. In the Everglades, metaphyton forms floating mats that are dominated by blue-green algae laden with calcareous depositions and that cover the surface of sloughs (Browder *et al.* 1994). My observations (unpublished) indicate that the calcareous deposition within the Everglades metaphyton increases its resistance to disturbance by winds and heavy rains.

Some algal ecologists would argue that *edaphic* (soil) algal communities are benthic in the sense that they are attached to substrata. Many of the same habitat characteristics apply to edaphic algae and periphytic algae (Metting, 1981; Starks *et al.*, 1981; Bell, 1993; Johansen, 1993): cells are packed closely on a substratum, provide habitat and nutrition for other organisms, stabilize substrata, and create their own microenvironment. Expanding the definition of periphyton to include edaphic microbial communities is an example of how definitions of terms can be revised to accommodate new information and conceptual approaches.

II. THE NICHE OF FRESHWATER BENTHIC ALGAE

A. The Role of Benthic Algae in Ecosystems

Benthic algae are important primary producers in streams, lakes, and wetlands. The main source of energy in streams was once thought to be detritus from terrestrial origin before Minshall's seminal paper (1978) that showed primary production by algae was important in many streams. Benthic algae are now predicted to be the primary energy source in many mid-sized (third to sixth order) streams (Vannote *et al.*, 1980). Benthic algae are also known to be important sources of energy for invertebrates in some headwater streams (Mayer and Likens, 1987). Wetzel (1964) has argued that benthic algae are important, and even dominant, primary producers in many shallow lakes and ponds. In wetlands, Goldsborough and Robinson (Chapter 4) have reported that algae may be significant primary producers because of their high turnover rate, even though macrophytes are the dominant photosynthetic biomass.

In addition to primary producers, benthic algae are chemical modulators in aquatic ecosystems (e.g., Lock *et al.*, 1984). They transform many inorganic chemicals into organic forms. The conversion of atmospheric N_2 to NH_3 and amino acids by blue-green algae and diatoms with endosymbionic blue-green algae may enable high primary productivity in low-nitrogen habitats (Fairchild *et al.*, 1985; Peterson and Grimm, 1992; DeYoe *et al.*, 1992). Benthic algae are primary harvesters of inorganic phosphorus and nitrogen in stream spiraling (see Mulholland, Chapter 19), in lake littoral modulation of influxes (see Wetzel, Chapter 20) and in wetlands (see Goldsborough and Robinson, Chapter 4). Effects of active nutrient uptake during the day explain the diurnal variation in nitrate concentrations in streams (Triska *et al.*, 1989). Benthic algae on surface sediments and plants are considered to be important sinks for nutrients before release into the water column (see Wetzel, Chapter 20). The uptake of nutrients in wetlands is often attributed to macrophytes, however, recent research shows that macrophytes actually pump nutrients out of the sediments. The benthic algal covering over macrophytes, however, traps nutrients before they reach the water column and returns them to the sediments when epiphytic algae settle to the bottom (Moeller *et al.*, 1988; Burkholder *et al.*, 1990; Wetzel, Chapter 20).

Benthic algae stabilize substrata in many aquatic habitats. Diatoms, filamentous blue-green algae, and *Vaucheria* can overgrow sands and sediments so that the substrata are less likely to move when current increases (see Biggs, Chapter 2). Many filamentous algae, particularly *Vaucheria* and *Chara,* can trap sediments. *Chara* can trap enough sand to form hummocks more than a meter in length in the sandy streams of many north-central states of the United States.

Benthic algae can also be important habitats for many other organisms. *Chara* hummocks can support a great diversity and density of aquatic invertebrates in streams where sand provides a poor habitat for most invertebrates. *Cladophora* and other filamentous algae that support epiphytes often also support great numbers of smaller invertebrates, such as chironomids, amphipods, and many smaller meiofauna (e.g., Chilton *et al.,* 1986; Holomuzki and Short, 1988; Power, 1990; see review by Dodds and Gudder, 1992). Even thicker periphyton matrices dominated by diatoms can shelter substantial numbers of chironomids and meiofauna.

B. How Are Algae in Benthos and Plankton Different?

Many species of algae are distinctly more common in benthic than in planktonic habitats. Some major taxonomic groups of algae can be characterized as "benthic" or "planktonic," such as orders of filamentous green algae and pennate diatoms that are generally benthic and the Volvocales and centric diatoms that tend to be planktonic. Other groups have been poorly characterized or have species that are found in either the benthos or plankton, such as many orders of blue-green algae and the diatom genera *Fragilaria, Synedra,* and even *Nitzschia.*

Obvious morphological characteristics, such as the diatom raphe, mucilage pads and stalks, or the holdfast mechanisms of filamentous algae, seem to confer a selective advantage for many algae to attach to substrata and remain there when disturbed by current. However, the same alga may utilize a morphological adaptation for attachment to substrata in the benthos and for attachment to other cells to form buoyant colonial growth forms in the plankton. The apically attached diatoms *Synedra* form hemispherical rosettes of cells on substrata and spherical colonies of radially arranged cells in plankton that are attached to each other at their apices.

The same habitat-specific growth forms are evident for blue-green algae with tapered trichomes in hemispherical rosettes of *Rivularia* on substrata and in spheres of *Gloeotrichia* in the plankton.

Many algal ecologists would hypothesize that algae characteristically found in benthic habitats may settle faster than algae usually found in plankton, because they have greater specific gravity. Few data are available to test that hypothesis. Stevenson and Peterson (1989) compared relative abundances of diatom species in the plankton and after settling on substrata, which should identify species with relatively high specific gravity, but most data were for pennate diatoms that are commonly found in the benthos. *Cyclotella meneghiniana,* a common diatom in stream plankton, was characterized in several studies and was found to be relatively more abundant in plankton than in immigration assemblages in two of the three studies (Stevenson and Peterson, 1989). Peterson and Hoagland (1990) found much higher relative abundances of planktonic taxa (*Asterionella* and two *Fragilaria* spp.) in settling traps than on substrata after short (3 d) incubations, which indicates that when planktonic taxa do have relatively high settling rates, they attach poorly or grow slowly when on substrata. Brown and Austin (1973) correlate decreases in abundances of taxa in plankton with their increases in the benthos, but do not actually compare settling rates or immigration rates of characteristically planktonic and benthic taxa.

One of the most important differences in benthic and planktonic habitats is the mode of nutrient delivery. The benthic habitats in an aquatic ecosystem probably have a greater diversity in nutrient conditions than planktonic habitats within the same habitat. Algae on substrata in substantial currents (which are often found in lakes and wetlands, as well as streams) may have greater supplies of nutrients than algae in the water column because currents reduce nutrient-poor and waste-rich boundary layers that develop around algae in still waters (see Borchardt, Chapter 7, and Stevenson, Chapter 11). However, algae on substrata in still waters may be in lower resource conditions than algae suspended in the water column because uptake by neighboring or overlying cells may create nutrient-poor regions within periphyton mats (see Borchardt, Chapter 7; Tuchman, Chapter 10; and Stevenson, Chapter 11). Therefore, algae on substrata in currents could have greater nutrient availability than algae in the water column, but the opposite may be the case in still waters. In addition, recycled nutrients are probably entrained within periphyton matrices more in still waters than in open-water planktonic habitats, such that interactions among organisms within periphyton are probably more tightly coupled than among freely suspended planktonic organisms. Exoenzymes secreted by organisms used to digest organic material may be more likely to confer selective advantage on the organism that produces them in relatively closed periphyton matrices than in open plankton habitats.

Since taxa characteristically found in plankton do not have specific morphological adaptations for attachment to substrata, they are not well adapted for maintaining position in currents. Therefore, planktonic taxa settling into benthic habitats would settle into still-water habitats, where nutrient availability may be lower and interactions among organisms may be higher than in the open water column. Since some benthic taxa can successfully grow in benthic habitats with slow currents and planktonic taxa generally cannot (Peterson and Hoagland, 1990), then the physiological difference between taxa may be that planktonic taxa cannot grow as fast as benthic taxa in low-resource, still-water benthic habitats. However, too little research has directly compared the physiological and morphological characteristics of algae that are more commonly found in benthic or planktonic habitats and determined whether those characteristics conferred effective fitness. The question "How are algae in benthos and plankton different?" remains to be answered.

III. METHODS FOR CHARACTERIZING FRESHWATER BENTHIC ALGAE

A. Population and Community Structure

Structural characteristics are measurements of system state. Algal population and community structure can be assessed for biomass, taxonomic composition, or chemical composition.

1. Biomass

Benthic algal biomass is the mass of algal organic matter per unit area of substratum. Many measurements are used to estimate algal biomass (Table II). Measurements of the area-specific masses of matter (such as chl *a*, C, N, or P, dry mass, and ash-free dry mass cm^{-2}) are relatively inexpensive methods for estimating algal biomass, however, their accuracy (susceptibility to bias) is lower than that of cell-counting methods. Chlorophyll *a* and other pigments are chemicals found only in algae in most benthic algal samples, unless plant or moss material also occurs in samples. Therefore, chl *a* and pigment densities more accurately indicate algal biomass than C, N, or P, which can be found in any living or nonliving organic matter. However, chromatic adaptation to low light or nutrient deficiency may alter the ratio between these chemicals and algal organic matter, so that pigment and chemical indicators of algal biomass may be biased when light and nutrient concentrations are ecological variables in the study. Dry mass (DM) and ash-free dry mass (AFDM) estimates of algal biomass (mg DM or AFDM cm^{-2}) may be biased, respectively, by inorganic matter and nonalgal organic matter (detritus, bacteria, fungi, etc.) in samples. So DM is a particularly poor indicator of benthic algal biomass when silt and inorganic deposition is great in samples, and AFDM is poor when detritus and heterotrophic organisms compose significant proportions of communities. Ash mass in samples, the portion of mass

TABLE II Advantages and Disadvantages of Using Different Measurements of Algal Community Biomass

Characteristic	*Pros*	*Cons*
chl *a* (μg cm^{-2})	Inexpensive, large literature base for comparison	Chromatic adaptation and nutrient limitation can bias estimates of biomass
C, N, or P (μg cm^{-2})	Also used to assess nutrient status of cells	Includes biomass of other living and nonliving matter
Dry mass (mg cm^{-2})	Inexpensive, can also be used to determine ash mass and entrapped silt or ash-free dry mass	Includes mass of all inorganic and organic matter in sample
Ash-free dry mass (mg cm^{-2})	Inexpensive, large literature base for comparison	Includes mass of all living and detrital matter
Cell density (cells cm^{-2})	Also used to assess species composition and biovolume, large literature base for comparison	Interspecific variation in cell sizes causes error in biomass estimates
Biovolume (μm^3 cm^{-2})	Accurately assesses algal biomass	Most time-consuming, must account for error due to cell vacuoles
Peak biomass	Good indicator of algal nuisance potential in a habitat	Requires monitoring algal community development

remaining after incineration during AFDM assessment, is a good indicator of silt and inorganic material that has accumulated in the periphyton.

Microscopic examination of cells is necessary to assess algal cell density and biovolume (Stevenson *et al.*, 1985). Random strewn-mounts of cells on microscope slides can be prepared in counting chambers (Sedgewick-Rafter and Palmer counting chambers or inverted microscope) or in different media on regular microscope slides [Palmer, 1962; Stevenson and Stoermer, 1981; Stevenson, 1984a; American Public Health Association (APHA), 1992]. Measurements of algal cell densities (cells cm^{-2}) are relatively good indicators of algal biomass if most of the algae are of similar sizes and the mass of all cells is assumed to be the same. Algal biovolumes (μm^3 cm^{-2}) are one of the best estimates of algal biomass if we assume that the mass of algal cytoplasm is the same among taxa. Biovolume corrects cell density estimates of biomass by accounting for size differences among species. However, algal biovolume estimates of algal biomass may be biased because of vacuole size variation among species, particularly among major taxonomic groups and algal growth forms. If vacuole size is estimated and subtracted from cell volume to estimate cytoplasm volume, then algal biovolume is a relatively accurate measure of algal biomass, particularly when great differences in cell sizes occur in the community. However, biovolume assays of algal communities take more time than the rest of the algal biomass assays. Thus the best approach for measuring algal biomasses varies among projects because of sample numbers and the personnel available.

Benthic algal biomass is temporally variable because of successive accumulation, autogenic sloughing, and disturbance that reset community development (Fig. 2). Peak biomass (PB) and time to peak biomass (T_{PB}) define two characteristics that may be useful in future research (see Biggs,

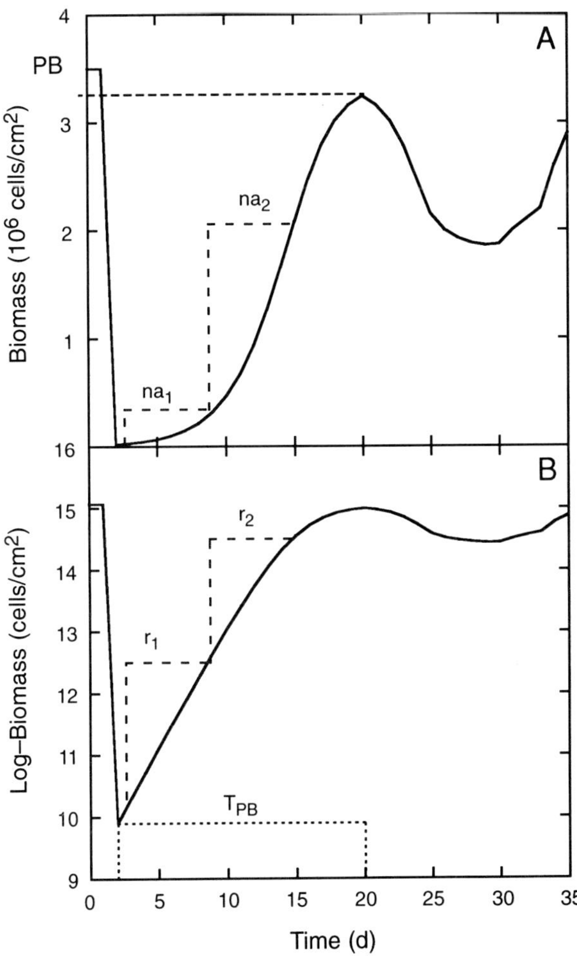

FIGURE 2 (A) Changes in biomass after a disturbance as algae accumulate on substrata to a peak biomass (PB), slough, and reaccumulate. Vertical lines indicate magnitude of net accrual rates (na_i, slopes of accumulation curve) at two times during community development (modified from Biggs, Chapter 2). (B) Changes in log-transformed biomass after disturbance. Vertical lines indicate growth rates (r_i, slopes of accumulation curve) at two times. T_{PB} = time to peak biomass.

Chapter 2, Fig. 2). Assessment of peak biomass and T_{PB} requires monitoring communities to ensure that the maximum biomass (Bothwell, 1989) is observed. Peak biomass better defines the potential of environments to support large standing crops of algae than biomass measured at specific times. Biomass at specific times can underestimate the ability of environments to support high algal standing crops because time to peak biomass is difficult to predict.

I recommend assay of as many different measures of algal biomass as can be afforded to best characterize algal biomass in samples. Specific assumptions are necessary for using each of the foregoing measures (i.e., chl *a*, AFDM, cell density, or cell biovolume) as measures of algal biomass. For example, chl:biomass or cell:biomass ratios must be assumed to be constant to use chl *a* or cell density as measures of biomass. This consistency does not occur often between samples within specific studies, and surely not between studies in different habitats. Ratios of biomass characteristics, such as chl *a*:AFDM or biovolume:AFDM, may be valuable indicators of the changes in algal proportions among samples. Therefore, the ability to characterize and compare algal biomass among samples and studies increases with the number of measures of algal biomass used.

I also recommend using standard methods (APHA, 1992) so that measurements are comparable as possible. Preferences for a nonstandard method may be great by individual investigators because of convenience and habitat-specific conditions, but effort should be made to use standard methods for consistency among studies.

2. Taxonomic Composition

Many approaches can be used to characterize the kinds of organisms in a benthic algal sample (Table III). Benthic algal species or taxonomic composition is typically assessed by identifying and counting cells microscopically (APHA, 1992). Ratios of species-specific cell densities or biovolumes to total density or biovolume are used with cell counts to define the proportions of communities (relative abundances) composed by different taxa. Another approach, but less taxonomically specific, is to assess different pigment types or other unique chemicals in different kinds of algae (APHA, 1992). Pigment ratios can be used to compare changes in proportions of green algae (chl *b*:chl *a*), diatoms (chl *c*:chl *a* or other chrysophytes), and red algae (chl *d*:chl *a*) among samples. Finally, the autotrophic indices [chl *a*:AFDM or biovolume:AFDM, modified from Weber's (1973) AFDM:chl *a* ratio] are also inexpensive and valuable characterizations of the algal proportion of biomass for comparison of communities.

Species composition is often summarized by the number of species in a community (species richness), by evenness of their abundances, by diversity indices that indicate both richness and evenness of communities, and by similarity of species among communities (Table III). Many diversity and

TABLE III Advantages and Disadvantages of Using Different Community Composition Assessments

Characteristic	*Pros*	*Cons*
Relative abundances	Less variable than relative biovolume	Poorly assesses shifts in biomass among taxa with different cell sizes, e.g., among divisions
Relative biovolume	Good assessment of taxonomic shifts in algal biomass	Can have high error variances
Pigment ratios	Relatively inexpensive assessment of shifts in divisional composition of assemblages	Relatively poor literature base for comparison, taxonomic resolution limited
Autotrophic indices	Distinguishes algal component of biomass from total community biomass	
Species diversity indices	Substantial literature base for comparison	High correlation among diversity indices using most common methods of algal enumeration, interpretation unclear
Nutrient status indices	May be good indicator of nutrient content of algal cells and habitat enrichment	Varies with algal density and may be biased by nonalgal sample components

evenness indices have been developed to characterize communities, but the most commonly used are Shannon diversity [Shannon (1948), also known as Shannon-Weaver and Shannon-Weiner] and Hurlbert's evenness (Hurlbert, 1971). Many richness, evenness, and diversity indices are highly correlated when using standard algal counting procedures (Archibald, 1972). Richness indices increase and evenness and diversity indices continue to change as more algae are counted in samples. Species numbers estimated by counting 200, 500, or 1000 cells of algae are positively correlated to the evenness of species abundances in communities. There are no examples of full community enumerations to assess assumptions about the relationship between species richness and evenness in communities, and theoretically the predictions for whether richness and evenness should be related are not clear. Analysis of a species frequency distribution curve based on log-transformed species abundances (e.g., Patrick *et al.*, 1954) indicates that as evenness of communities decreases, the probability of seeing new species in a count decreases. Therefore, the number of species in algal enumerations with specific numbers counted (e.g., 200 or 500) is really just an indicator of species evenness rather than the true number of species in the community. However, changes in diversity characteristics can be used as indicators of community change (Stevenson, 1984b).

3. Chemical Composition

Chemical composition of benthic algal samples has been used as an indicator of nutrient availability for benthic algae and community health. Resource ratios, such as N:C, P:C, and N:P of algal samples, have been used to indicate cellular nutrient status (Biggs and Close, 1989; Biggs, 1995; Humphrey and Stevenson, 1992; Peterson and Stevenson, 1992). One important source of covariation in N:C and P:C ratios with nutrient concentrations in the water column is benthic algal biomass. Whereas N:C and P:C ratios should increase with inorganic N and P concentrations in the water column, these ratios also decrease with algal density on substrata (Humphrey and Stevenson, 1992). Another chemical ratio used to indicate community conditions is the phaeophytin:chl *a* ratio, which can be used as an indicator of community senescence (e.g., Peterson and Stevenson, 1992).

B. Population and Community Function

Functional characteristics are measurements of the rate of change in a system state. These measurements include productivity, respiration, nutrient uptake rates, and enzyme activity.

1. Productivity

Community function can be measured as productivity, nutrient uptake and recycling, and enzyme activities. *Productivity* is defined by Wetzel (1983a) as the "rate of formation of organic material averaged over some defined period of time, such as a day or year." *Gross productivity* is defined by Wetzel (1983a) as the "change in biomass, plus all predatory and nonpredatory losses, divided by the time interval." *Net productivity* "is the gross accumulation or production of new organic matter, or stored energy, less losses, divided by the time interval" (Wetzel, 1983a). These definitions are adopted here because they are widely accepted by colleagues and these concepts will be the foundation of the following discussion of different methods of measuring productivity.

There are many ways of measuring productivity. For example, gross productivity can be measured as photosynthesis on an area- or biomass-specific basis with changes in dissolved oxygen concentrations (in chambers or over diurnal periods in the habitat) plus respiratory losses or with ^{14}C uptake rates (in chambers). Alternatively, gross productivity could be measured as reproduction rates of cells or changes in algal biovolume (a per capita accumulation rate) during a defined time interval, which requires accounting for losses from death (non-herbivory-related), emigration (drift or export of cells from substrata), and herbivory and gains from cell immigration (settling of cells from the water column onto substrata).

Photosynthesis and reproduction rates are two very different ways of measuring gross productivity because different resources may limit photosynthesis and accumulation rates. For example, biomass-specific photosynthesis is positively affected by light availability and negatively affected by the algal biomass present. Reproduction rates are limited not only by photosynthesis, but also by NO_3, NH_3, and PO_4 availability. Thus different determinants of algal ecology must be considered with different methods of measuring productivity.

The relationship between measurement of benthic algal community function and structure helps to link concepts and to define terms (Fig. 3). Dissolved inorganic nutrients (C, N, P, Si, etc.) are sequestered by benthic algae by various passive and active nutrient uptake mechanisms. Light, temperature, and current (among other factors) directly affect the uptake rate of nutrients by benthic algae. Light mediates uptake rates by regulating ATP and NADPH formation in the light reaction of photosynthesis, which then, based on dissolved inorganic carbon, nitrogen, and phosphorus availability, affects carbon fixation, production of amino acids, and photophosphorylation. Temperature affects the rate of metabolism and current affects nutrient transport and availability. Algal density itself negatively affects biomass-specific nutrient uptake rates because it negatively

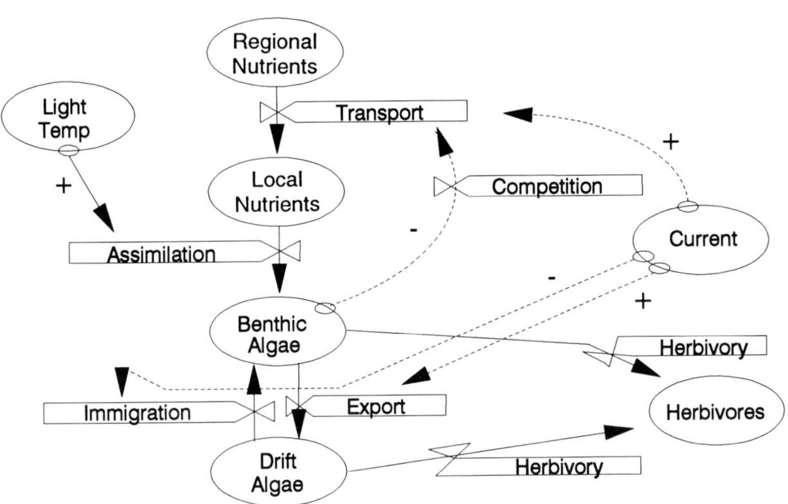

FIGURE 3 Components of community structure and function that are important determinants of benthic algal ecology. Matter (Nutes=C, N, P, Si, etc.) and energy flows are indicated by components and connected by arrows (solid lines) that indicate community processes [assimilation=biomass-specific gross productivity, biomass-specific (b-s) immigration rates, b-s emigration rates, b-s grazing rates, b-s death rates]. Many of the community structure parameters are modulators of community function. Abiotic and biotic modulators of community function are indicated by arrows (dashed lines) originating from small circles.

affects nutrient availability to cells. All of these factors then regulate benthic algal reproduction. So photosynthesis, or what is usually measured as gross primary production, is just one metabolic reaction among many that is required for the accumulation of organic material, which Wetzel (1983a) defines as gross production.

There are four basic approaches to measuring photosynthesis by benthic algae: by ^{14}C uptake in chambers or by dissolved oxygen, pH, or dissolved CO_2 changes in chambers or open waters of streams, lakes, or wetlands (Odum, 1956; Wright and Mills, 1967; Bott *et al.*, 1978, 1985). The first two methods involve placement of substrata into chambers. Changes in dissolved oxygen (DO) concentration in chambers are generally preferred over ^{14}C uptake because DO methods enable respiration measurement and do not involve radioisotope use. ^{14}C uptake is a useful measure of net primary production when algal biomass is low because it is more sensitive than changes in dissolved oxygen. Open water column measurements of DO are an advantage because they enable assessment of whole ecosystem net productivity, whereas chamber assays require measurement of individual habitat patches. However, oxygen change due to diffusion from the water surface can be a problem in turbulent waters because of reaeration (Bott *et al.*, 1978). All DO and ^{14}C methods are generally accepted and have value for productivity assessments at one spatial scale or another.

The terminology for the mechanisms of benthic algal accumulation has not been consistently defined in the literature. The value of this conceptual approach and its increasing use calls for some clarification of terms for different measurements of algal accumulation on substrata. First, area-specific and biomass-specific terms need to be distinguished for both gross and net productivity measurements that are based on biomass accumulation (Fig. 2). Biomass-specific or per capita changes in benthic algal organic material have been referred to as reproduction rates (Stevenson, 1984b, 1986), growth rates (Bothwell and Jasper, 1983; Stevenson, 1990), and relative specific growth rates (Biggs, 1990). Growth rate has been used to refer to both the net and gross productivity measurements of accumulation, but usually refers to reproduction minus loss (death) in the general ecological literature. Reproduction has only been used in reference to biomass-specific or per capita accumulation rates. Therefore biomass-specific *gross* accumulation (productivity) rates should be referred to as reproduction rates, which requires that immigration, emigration, death, and herbivory are taken into account (Fig. 3 and Table IV). One should refer to biomass-specific *net* accumulation rates as growth rates (Fig. 2). The term net with growth rate is redundant and is therefore an unnecessary adjective.

Area-specific accumulation rates (dN/dt, e.g., cells cm^{-2} d^{-1}) should be distinguished technically and conceptually from biomass-specific accumulation rates (dN/Ndt, e.g., cells cell^{-1} cm^{-2} d^{-1}, reproduction and growth rates), and therefore should have separately defined terms. Accrual rate has precedence as a term that could be used for area-specific accumulation rates (Biggs, 1988, 1990). Gross accrual and net accrual rates (biomass cm^{-2} d^{-1}) are appropriate terms for biomass changes that, respectively, do and do not take gains and losses into account (Fig. 2 and Table IV).

Determination of the various accumulation rates requires assaying algal biomass on successive dates, typically 3–4 d apart. Shorter times often do not have enough difference to be accurately measured. Longer times involve changes in algal biomass that are so great that accumulation rates change as a function of algal biomass. Reproduction and growth rates on a biomass-specific (per capita) basis are determined with natural-log-transformed algal biomass values and calculation of daily changes in log-transformed algal biomass (Fig. 2B and Table IV). Gross and net accrual rates are calculated with untransformed biomass (Fig. 2A).

Determination of gains and losses from accumulating algae requires additional measurements of the community. Generally, nongrowth gains and losses (except for massive sloughing) can be assumed to be less important in high-biomass communities than in low-biomass communities because daily accrual due to algal reproduction can be much greater in high- than in low-biomass communities. However, it is best to actually measure gains and losses to distinguish between gross and net productivity. Immigration rates can be determined by quantifying the algal biomass that accumulates on bare (clean) substrata without algae over a 24-h period (e.g., Stevenson, 1983). Death rates are difficult to determine for

TABLE IV Equations for Assessment of Reproduction, Growth, Gross Accrual, and Net Accrual Rates

Accumulation variable	*Formula*
Reproduction rate	$b=\frac{\ln(N_t+D_t-D_{t-1}+E_t-I_t+H_t)-\ln(N_{t-1})}{d_t-d_{t-1}}$
Growth rate	$r=\frac{\ln(N_t)-\ln(N_{t-1})}{d_t-d_{t-1}}$
Gross accrual rate	$ga=\frac{N_t+D_t-D_{t-1}+E_t-I_t+H_t}{d_t-d_{t-1}}$
Net accrual rate	$na=\frac{N_t-N_{t-1}}{d_t-d_{t-1}}$
Cell density on Days (d) t and t–1, respectively	N_t and N_{t-1}
Dead cell density on Days t and t–1, respectively	D_t and D_{t-1}
Number of cells exported, immigrating, and grazed by herbivores, respectively, during the time interval	E_t, I_t, and H_t

nondiatom algae, but for diatoms can be determined by the per capita increase in empty frustule accumulation ($D_t - D_{t-1}$) that occurs over a 3- to 4-d period (Stevenson, 1986; McCormick and Stevenson, 1991). Emigration rates are difficult to determine, but are best assayed as the differences in algal densities in the drift over a stream distance, after correcting for immigration losses (McCormick and Stevenson, 1991), or by assessing diurnal variation in drift algal abundances (Stevenson and Peterson, 1991). Herbivory rates are determined by assessing algal accumulation rate in grazer guts after grazers have been starved or with dual-isotope methods (Calow and Fletcher, 1972; Lamberti *et al.*, 1989). Alternatively, herbivory rates can be determined over short periods of time by assessing differences in algal accumulation on substrata with and without herbivores. Therefore, reproduction and gross accrual require subtraction of immigrating algae and addition of algae lost by death, emigration, or grazing, either on a biomass- or an area-specific basis, respectively (Table IV).

I recommend calculation of reproduction or growth rates, over accrual rates, because accrual patterns are often exponential (Fig. 2A) and conceptually they should fit an exponential growth curve. Accrual rates should vary much more than growth rates. Accrual rates increase with algal density on substrata and later decrease with density-dependent decreases in reproduction as biomass approaches peak levels. Growth (and reproduction) rates, however, decrease only with algal biomass (Fig. 2B). Yet accrual rates do provide a valuable measure of overall biomass accumulation of the mat and are a good indicator of the rate of change in the algal community's effect on the rest of the ecosystem.

2. *Respiration, Nutrient Uptake, and Enzyme Activity*

Many other measures of benthic algal activity can provide valuable insight into determinants of community structure and ecosystem process. Periphyton community respiration can be measured by changes in dissolved oxygen or pH in dark chambers or in the water column in natural habitats at night (Odum, 1956; Wright and Mills, 1967; Bott *et al.*, 1978, 1985). Phosphorus uptake can be measured as accumulation of ^{32}P in algae in chambers dosed with radiolabeled PO_4 (Lock and John, 1979; Riber and Wetzel, 1987). A convenient method of assessing nutrient uptake is to document loss of nutrients from overlying waters, either in chambers or in natural habitats (Triska *et al.*, 1989; Kim *et al.*, 1990, 1992). Assessments of spiraling distance are also indicators of algal nutrient uptake rates in streams (Newbold *et al.*, 1981). Assessments of enzyme concentrations can be valuable indicators of changes in benthic algal physiological activity in response to environmental change. Phosphatase activity is measured to assess phosphate limitation (Fitzgerald and Nelson, 1966; Biggs, 1990; Mulholland and Rosemond, 1992).

Distinguishing between biomass-specific and area-specific community function is also important because algal density has different effects. Many community functions expressed on a biomass-specific basis will decrease with algal biomass on substrata because resource availability (light and nutrient) usually decreases with algal density (e.g., Stevenson, 1990). Phosphatase activity should be the opposite, because phosphatase activity increases with PO_4 starvation, which theoretically increases with algal biomass. Conversely, most community functions expressed on an area-specific basis increase with algal biomass on substrata because there are more algae on substrata to perform function. Even though biomass-specific photosynthesis, growth rates, and nutrient uptake rates decrease with algal biomass, area-specific photosynthesis, accrual rates, and nutrient uptake rates usually increase with algal biomass on substrata.

IV. CONCEPTUAL FRAMEWORKS FOR BENTHIC ALGAL COMMUNITY ECOLOGY

Algal ecology has made significant contributions to the development of ecological theory. Lindeman's (1942) ideas of trophic dynamics in aquatic ecosystems were heavily influenced by the knowledge of phytoplankton ecology. Patrick's (1967) research with benthic algae tested MacArthur and Wilson's (1963) theories predicting that invasion rate and habitat area are major determinants of species numbers on islands. Hutchinson's (1961) exploration of the determinants of species richness led to his "Paradox of the Plankton." Tilman's (1976) development of resource-based competition theory started as one of many theories to address Hutchinson's paradox and has subsequently influenced understanding of plant community diversity and stability (Tilman and Downing, 1994). Thus, algae have played important roles in testing theories relating community diversity and stability. The main reasons for the success of algae in research applications are that algae live in ecosystems with relatively well-defined borders, they can be easily identified to species, they are spatially compact, and their short generation times enable testing hypotheses in relatively short amounts of time.

These contributions by algal research to the broader field of ecology demonstrate why algal communities are considered to be model ecosystems. Even though the results of hypothesis testing with algal communities have not been directly applied to the ecological understanding of terrestrial plant, microbial, and wildlife populations and communities, the results of algal studies often show how to approach ecological research with other organisms. Scaling up and down from the spatial and temporal scales of algal community ecology to appropriate scales for systems with other organisms can be informative (Allen, 1977). In addition, research with algal communities can expose processes or community states that can be important in community dynamics and that should be considered in research with other organisms.

The nonalgal ecological literature has also made significant contributions to the field of algal ecology. Contributions in terrestrial plant ecology (Pickett *et al.*, 1987; Pickett and McDonnell, 1989) may have great value as a general framework for conceptualizing community dynamics in many ecological settings. In particular, this framework, with some modifications, is useful for conceptualizing benthic algal community ecology, for it defines three causes of community dynamics: site availability, species availability, and differential species performance (Table V). Sites must be available for colonization to occur. In addition, a propagule pool and modes of dispersal must exist for organisms to immigrate to habitats. Species must occur in the region for the probability of immigration for taxa to be high enough for them to colonize. Dispersal barriers, such as a lack of surface water inputs into a system or wind barriers, may limit the routes of dispersal or immigration of species into a habitat, and therefore define the local species pool. Once organisms arrive, resources such as nutrients and light must be available so that species can successfully persist and colonize substrata. So community development processes, such as biomass accumulation, nutrient turnover, or energy flow through ecosystems, require resources (site) and species availability. However, changes in species composition (succession) will not occur unless species perform differently in the environment. Therefore, differential species performance causes changes in species composition as communities develop and environmental conditions autogenically and allogenically change. An organized set of null and alternative

TABLE V The "Causal Repertoire" of Vegetation Dynamics[a]

First-order cause	*Second-order cause*
Site availability	Change in environment that makes new resources available or old resources available to different species
Differential species availability	Dispersal modes
	Propagule pool
Differential species performance	Resource levels (nutrients, light)
	Life-history characteristics
	Ecophysiology
	Stress
	Allelopathy
	Competition
	Predation (herbivory)

[a]Modified from Pickett and McDonnell (1989).

hypotheses for community ecology emerges from this framework. Pickett and McDonnell (1989) refer to these alternative hypotheses as "causal repertoires."

Differential species performance drives community dynamics (i.e., succession). One of the basic causes of differential species performance is the effect that resource availability has on species. Different species of benthic algae have different minimal resource requirements and perform differently at different resource levels (e.g., Carrick *et al.,* 1988; Dodds and Gudder, 1992; Stevenson, 1996; Borchardt, Chapter 7). Resource availability sets performance levels of species that are then affected by species ecophysiological characteristics, such as their specific growth rates under specific resource levels. Nutrients, light, and space may be important resources that enable species colonization and cause differential species performance (Table V). For benthic algae, regional nutrient and light supplies may be modified by a number of factors to affect local nutrient and light availability for organisms (Fig. 3). These resource modulators, such as current and algal density, operate at a local scale. Nutrients and light are primary resources for all benthic algae, but for many algae space is a secondary resource. A primary resource in this context is one that algae cannot do without, whereas a secondary resource modulates availability of a primary resource. Thus algal density and space limitation decrease nutrient and light availability. For some algae, however, space is a primary resource because they cannot survive by attaching to other algae or moving through the benthic algal matrix. These algae must attach directly to the substratum and have space available for attachment.

In addition to resource availability, ecophysiology, life histories, stress, allelopathy, competition, and predation are causes of differential species performance. Differential life-history traits (strategies) for species can affect community composition as communities shift from those dominated by species with fast immigration rates to those with fast reproduction rates or low death rates (Stevenson *et al.,* 1991). Environmental stress is the deviation in environmental conditions from species optima, for example, in pH, salinity, or temperature. Environmental stress may affect species resource utilization efficiency and life-history characteristics. Allelopathy is a biotically based environmental stress about which little is known. Different algal species may have different competitive abilities that are important when algal density limits transport of nutrients from the regional to the local species pool, but little experimental evidence exists to support this hypothesis (McCormick and Stevenson, 1991; Stevenson, in press). Benthic algae have greatly different susceptibilities to predation (herbivory, e.g., Hunter, 1980; see Steinman, Chapter 12). Many algae probably also have differential susceptibility to parasitism or disease from fungal, bacterial, and viral infections (Steward, 1988; Peterson *et al.,* 1993).

Throughout this book, different chapters synthesize information about components of this conceptual framework. First, Biggs (Chapter 2), Lowe (Chapter 3), and Goldsborough and Robinson (Chapter 4) describe patterns of benthic algal community structure and function in streams, lakes, and wetlands. Then the "causal repertoire" for differential species performance is developed with chapters on the effects of light, nutrients, and heterotrophy (Hill, Chapter 5; Borchardt, Chapter 7; Tuchman, Chapter 10) describing resource availability for benthic algae. Other chapters describe how resource availability and algal metabolism are affected by stresses and resource modulators: temperature, pH, substrate, disturbance, current, and inorganic and organic toxic substances (DeNicola, Chapter 6; Burkholder, Chapter 9; Stevenson, Chapter 11; Peterson, Chapter 13; Genter, Chapter 14; Hoagland *et al.,* Chapter 15; Planas, Chapter 16). Other chapters describe how the biotic causes of differential species performance, competition and herbivory, affect benthic algal community dynamics (McCormick Chapter 8; Steinman, Chapter 12). The importance of benthic algal community-level dynamics is then related to trophic dynamics and ecosystem processes (Lamberti, Chapter 17; Bott, Chapter 18; Mulholland, Chapter 19; Wetzel, Chapter 20; McIntire *et al.,* Chapter 21). Finally, benthic algal ecology is related to water quality monitoring (Lowe and Pan, Chapter 22). After more than a decade of intensive experiments and field observations since the last compilation on the subject (Wetzel, 1983b), these chapters provide new syntheses and ideas for benthic algal ecologists to build upon and to alter during the years to come.

REFERENCES

Allen, T. F. H. (1977). Scale in microscopic algal ecology: A neglected dimension. *Phycologia* **16,** 253–257.

American Public Health Association (APHA). (1992). "Standard Methods for the Evaluation of Water and Wastewater," 18th ed. APHA, Washington, DC.

Archibald, R. E. M. (1972). Diversity in some South African diatom associations and its relation to water quality. *Water Res.* **6,** 1229–1238.

Bell, R. A. (1993). Cryptoendolithic algae of hot semiarid lands and deserts. *J. Phycol.* **29,** 122–139.

Biggs, B. J. F. (1988). Algal proliferations in New Zealand's shallow stony foothills-fed rivers: Toward a predictive model. *Verh—Int. Ver. Theor. Angew. Limnol.* **23,** 1405–1411.

Biggs, B. J. F. (1990). The use of relative specific growth rates of periphytic diatoms to assess enrichment of a stream. *N. Z. J. Mar. Freshwater Res.* **24,** 9–18.

Biggs, B. J. F. (1995). The contribution of flood disturbance, catchment geology and land use to the habitat template of periphyton in stream ecosystems. *Freshwater Biol.* **33,** 419–438.

Biggs, B. J. F., and Close, M. E. (1989). Periphyton biomass dynamics in gravel bed rivers: The relative effects of flows and nutrients. *Freshwater Biol.* **22,** 209–231.

Bold, H. C., and Wynne, M. J. (1985). "Introduction to the Algae Structure and Reproduction," 2nd ed. Prentice-Hall, Englewood Cliffs, NJ.

Bothwell, M. L., and Jasper, S. (1983). A light and dark trough methodology for measuring rates of lotic periphyton settling and net growth. *In* "Periphyton of Freshwater Ecosystems" (R. G. Wetzel, ed.), pp. 253–265. Dr. W. Junk Publishers, The Hague.

Bothwell, M. L. (1989). Phosphorus-limited growth dynamics of lotic periphytic diatom communities: areal biomass and cellular growth rate responses. *Can. J. Fish. Aquat. Sci.* 46, 1293–1301.

Bott, T. L., Brock, J. T., Cushing, C. E., Gregory, S. V., King, D., and Petersen, R. C. (1978). A comparison of methods for measuring primary productivity and community respiration in streams *Hydrobiologia* **60**, 3–12.

Bott, T. L., Brock, J. T., Dunn, C. S., Naiman, R. J., Ovink, R. W., and Petersen, R. C. (1985). Benthic community metabolism in four temperate stream ecosystems: An inter-biome comparison and evaluation of the river continuum concept. *Hydrobiologia* **123**, 3–45.

Browder, J. A., Gleason, P. J., and Swift, D. R. (1994). Periphyton in the Everglades: Spatial variation, environmental correlates, and ecological implications. *In* "Everglades: The Ecosystem and Its Restoration" (S. M. Davis and J. C. Ogden eds.), pp. 379–418. St. Lucie Press, Delray Beach, FL.

Brown, S.-D., and Austin, A. P. (1973). Diatom succession and interaction in littoral periphyton and plankton. *Hydrobiologia* **43**, 333–356.

Burkholder, J. M., Wetzel, R. G., and Klomparens, K. L. (1990). Direct comparison of phosphate uptake by adnate and loosely attached microalgae within an intact biofilm matrix. *Appl. Environ. Microbiol.* **56**, 2882–2890.

Calow, P., and Fletcher, C. R. (1972). A new radiotracer technique involving ^{14}C and ^{51}Cr for estimating the assimilation efficiencies of aquatic, primary consumers. *Oecologia* **9**, 155–170.

Carrick, H. J., Lowe, R. L., and Rotenberry, J. T. (1988). Functional associations of benthic algae along experimentally manipulated nutrient gradients: Relationships with algal community diversity. *J. North Am. Benthol. Soc.* **7**, 117–128.

Cavalier-Smith, T. (1986). The kingdom Chromist: Origin and systematics. *Prog. Phycol. Res.* **4**, 309–347.

Chilton, E. W., Lowe, R. L., and Schurr, K. M. (1986). Invertebrate communities associated with *Bangia atropurpurea* and *Cladophora glomerata* in western Lake Erie. *Great Lakes J. Res.* **12**, 149–153.

DeYoe, H. R., Lowe, R. L., and Marks, J. C. (1992). The effect of nitrogen and phosphorus on the endosymbiont load of *Rhopalodia gibba* and *Epithemia turgida* (Bacillariophyceae). *J. Phycol.* **28**, 773–777.

Dodds, W. K., and Gudder, D. A. (1992). The ecology of *Cladophora*. *J. Phycol.* **28**, 415–427.

Fairchild, G. W., Lowe, R. L., and Richardson, W. B. (1985). Nutrient-diffusing substrates in an *in situ* bioassay using periphyton: Algal growth responses to combinations of N and P. *Ecology* **66**, 465–472.

Fitzgerald, G. P., and Nelson, T. C. (1966). Extractive and enzymatic analysis for limiting or surplus phosphorus in algae. *J. Phycol.* **2**, 32–37.

Gibbs, S. P. (1981). The chloroplast endoplasmic reticulum, structure, function and evolutionary significance. *Int. Rev. Cyto.* **72**, 49–99.

Hoagland, K. D., Roemer, S. C., and Rosowski, J. R. (1982). Colonization and community structure of two periphyton assemblages, with emphasis on the diatoms (Bacillariophyceae). *Am. J. Bot.* **69**, 188–213.

Holomuzki, J. R., and Short, T. M. (1988). Habitat use and fish avoidance behaviors by the stream-dwelling isopod *Lirceus fontinalis*. *Oikos* **52**, 79–86.

Hudon, C., and Legendre, P. (1987). The ecological implications of growth forms in epibenthic diatoms. *J. Phycol.* **23**, 434–441.

Humphrey, K. P., and Stevenson, R. J. (1992). Responses of benthic algae to pulses of current and nutrients during simulations of subscouring spates. *J. North Am. Benthol.* **11**, 37–48.

Hunter, R. D. (1980). Effects of grazing on the quantity and quality of freshwater Aufwuchs. *Hydrobiologia* **69,** 251–259.

Hurlbert, S. H. (1971). The nonconcept of species diversity: A critique and alternative parameters. *Ecology* **52,** 577–586.

Hutchinson, G. E. (1961). The paradox of the plankton. *Am. Nat.* **95,** 137–145.

Hynes, H. B. N. (1970). "The Ecology of Running Waters." Liverpool Univ. Press, Liverpool, UK.

Johansen, J. R. (1993). Cryptogamic crusts of semiarid and arid lands of North America. *J. Phycol.* **29,** 140–147.

Kim, B. K. A., Jackman, A. P., and Triska, F. J. (1990). Modeling transient storage and nitrate uptake kinetics in a flume containing a natural periphyton community. *Water Resour. Res.* **26,** 505–515.

Kim, B. K. A., Jackman, A. P., and Triska, F. J. (1992). Modeling biotic uptake by periphyton in transient hyporrheic storage of nitrate in a natural stream. *Water Resour. Res.* **28,** 2743–2752.

Lamberti, G. A., Gregory, S. V., Ashkenas, L. R., Steinman, A. D., and McIntire, C. D. (1989). Productive capacity of periphyton as a determinant of plant–herbivore interactions in streams. *Ecology* **70,** 1840–1856.

Lee, R. E. (1989). "Phycology," 2nd ed. Cambridge Univ. Press, Cambridge, UK.

Lindeman, R. L. (1942). The trophic-dynamic aspect of ecology. *Ecology* **23,** 157–176.

Lock, M. A., and John, P. H. (1979). The effect of flow patterns on uptake of phosphorus by river periphyton. *Limnol. Oceanogr.* **24,** 376–383.

Lock, M. A., Wallace, R. R., Costerton, J. W., Ventullo, R. M., and Charlton, S. E. (1984). River epilithon: Toward a structural-functional model. *Oikos* **42,** 10–22.

MacArthur, R., and Wilson, E. O. (1963). An equilibrium theory of insular zoogeography. *Evolution (Lawrence, Kans.)* **17,** 373–387.

Mayer, M. S., and Likens, G. E. (1987). The importance of algae in a shaded headwater stream as food for an abundant caddisfly (Trichoptera). *J. North Am. Benthol. Soc.* **6,** 262–269.

McCormick, P. V., and Stevenson, R. J. (1989). Effects of snail grazing on benthic algal community structure in different nutrient environments. *J. North Am. Benthol. Soc.* **82,** 162–172.

McCormick, P. V., and Stevenson, R. J. (1991). Mechanisms of benthic algal succession in different flow environments. *Ecology* **72,** 1835–1848.

Metting, B. (1981). The systematics and ecology of soil algae. *Bot. Rev.* **47,** 195–312.

Minshall, G. W. (1978). Autotrophy in stream ecosystems. *BioScience* **28,** 767–771.

Moeller, R. E., Burkholder, J. M., and Wetzel, R. G. (1988). Significance of sedimentary phosphorus to a rooted submersed macrophyte (*Najas flexilis* (Willd.) Rostok. and Schmidt) and its algal epiphytes. *Aquat. Bot. 32,* 261–281.

Mulholland, P. J., and Rosemond, A. D. (1992). Periphyton response to longitudinal nutrient depletion in a woodland stream: Evidence of upstream–downstream linkage. *J. North Am. Benthol. Soc.* **11,** 405–419.

Newbold, J. D., Elwood, J. W., O'Neill, R. V., and Van Winkle, W. (1981). Measuring nutrient spiralling in streams. *Can. J. Fish. Aquat. Sci.* **38,** 860–863.

Odum, H. T. (1956). Primary production in flowing waters. *Limnol. Oceanogr.* **1,** 102–117.

Palmer, C. M. (1962). "Algae in Water Supplies." U. S. Dept. of Health, Education, and Welfare, Division of Water Supply and Pollution Control, Washington, DC.

Patrick, R. (1967). The effect of invasion rate, species pool, and size of area on the structure of the diatom community. *Proc. Natl. Acad. Sci. U.S.A.* **58,** 1335–1342.

Patrick, R. (1976). The formation and maintenance of benthic diatom communities. *Proc. Am Philos. Soc.* **120,** 475–484.

Patrick, R., Hohn, M. H., and Wallace, J. H. (1954). A new method for determining the pattern of the diatom flora. *Not. Nat. Acad. Nat. Sci. Philadelphia* **259,** 1–12.

Peterson, C. G., and Grimm, N. B. (1992). Temporal variation in enrichment effects during periphyton succession in a nitrogen-limited desert stream ecosystem. *J. North Am. Benthol. Soc.* **11**, 20–36.

Peterson, C. G., and Hoagland, K. D. (1990). Effects of wind-induced turbulence and algal mat development on epilithic diatom succession in a large reservoir. *Arch. Hydrobiol.* **188**, 47–68.

Peterson, C. G., and Stevenson, R. J. (1992). Resistance and recovery of lotic algal communities: Importance of disturbance timing, disturbance history, and current. *Ecology* **73**, 1445–1461.

Peterson, C. G., Dudley, T. L., Hoagland, K. D., and Johnson, L. M. (1993). Infection, growth, and community-level consequences of a diatom pathogen in a Sonoran desert stream. *J. Phycol.* **29**, 442–452.

Pickett, S. T. A., and McDonnell, M. J. (1989). Changing perspectives in community dynamics: A theory of successional forces. *Trends Ecol. Evol.* **4**, 241–245.

Pickett, S. T. A., Collins, S. L., and Armesto, J. J. (1987). Models, mechanisms, and pathways of succession. *Bot. Rev.* **53**, 335–371.

Power, M. E. (1990). Benthic turfs versus floating mats of algae in river food webs. *Oikos* **58**, 67–79.

Raven, P. H., and Johnson, G. B. (1992). "Biology," 3rd ed. Mosby-Year Book, St. Louis, MO.

Riber, H. H., and Wetzel, R. G. (1987). Boundary-layer and internal diffusion effects on phosphorus fluxes in lake periphyton. *Limnol. Oceanogr.* **32**, 1181–1194.

Shannon, C. F. (1948). A mathematical theory of communication. *Bell Syst. Tech. J.* **27**, 37–42.

Sicko-Goad, L., Stoermer, E. F., and Kociolek, J. P. (1989). Diatom resting cell rejuvenation and formation: Time course, species records, and distribution. *J. Plankton Res.* **11**, 375–389.

Starks, T. L., Schubert, L E., and Trainor, F. R. (1981). Ecology of soil algae. *Phycologia* **20**, 65–80.

Stevenson, R. J. (1983). Effects of current and conditions simulating autogenically changing microhabitats on benthic algal immigration. *Ecology* **64**, 1514–1524.

Stevenson, R. J. (1984a). Procedures for mounting algae in a syrup medium. *Trans. Am. Microsc. Soc.* **103**, 320–321.

Stevenson, R. J. (1984b). How currents on different sides of substrates in streams affect mechanisms of benthic algal accumulation. *Int. Rev. Gesamten Hydrobiol.* **69**, 241–262.

Stevenson, R. J. (1986). Mathematical model of epilithic diatom accumulation. *In* "Proceedings of the Eighth International Diatom Symposium" (M. Ricard, ed.), pp. 209–231. Koeltz Scientific Books, Koenigstein, Germany.

Stevenson, R. J. (1990). Benthic algal community dynamics in a stream during and after a spate. *J. North Am. Benthol. Soc.* **9**, 277–288.

Stevenson, R. J. Are evolutionary tradeoffs evident in responses of benthic diatoms to nutrients? *In* "Proceedings of the 13th International Diatom Symposium" (D. Marino, ed.). BioPress Ltd., Bristol, UK, in press.

Stevenson, R. J., and Peterson, C. G. (1989). Changes in benthic diatom (Bacillariophyceae) immigration with habitat characteristics and cell morphology. *J. Phycol.* **25**, 120–129.

Stevenson, R. J., and Peterson, C. G. (1991). Emigration and immigration can be important determinants of benthic diatom assemblages in streams. *Freshwater Biol.* **26**, 295–306.

Stevenson, R. J., and Stoermer, E. F. (1981). Quantitative differences between benthic algal communities along a depth gradient in Lake Michigan. *J. Phycol.* **17**(1), 29–36.

Stevenson, R. J., Singer, R., Roberts, D. A., and Boylen, C. W. (1985). Patterns of benthic algal abundance with depth, trophic status, and acidity in poorly buffered New Hampshire lakes. *Can. J. Fish. Aquat. Sci.* **42**, 1501–1512.

Stevenson, R. J., Peterson, C. G., Kirschtel, D. B., King, C. C., and Tuchman, N. C. (1991). Density-dependent growth, ecological strategies, and effects of nutrients and shading on benthic diatom succession in streams. *J. Phycol.* **27**, 59–69.

Stewart, A. J. (1988). A note on the occurrence of unusual patches of senescent periphyton in an Oklahoma stream. *J. Freshwater Ecol.* **4**, 395–399.

Stewart, K. D., and Mattox, K. (1980). Phylogeny of phytoflagellates. *In* "Phytoflagellates." (E. R. Cox, ed.), pp. 433–462. Elsevier/North-Holland, New York.

Tilman, D. (1976). Ecological competition between algae: Experimental confirmation of resource-based competition theory. *Science* **192**, 463–465.

Tilman, D., and Downing, J. A. (1994). Biodiversity and stability in grasslands. *Nature (London)* **367**, 363–365.

Triska, F. J., Kennedy, V. C., Avanzino, R. J., Zellweger, G. W., and Bencala, K. E. (1989). Retention and transport of nutrients in a third-order stream: Channel processes. *Ecology* **70**, 1877–1892.

Vannote, R. L., Minshall, G. W., Cummins, K. W., Sedell, J. R., and Cushing, C. E. (1980). The river continuum concept. *Can. J. Fish. Aquat. Sci.* **37**, 130–137.

Weber, C. I. (1973). Recent developments in the measurement of the response of plankton and periphyton to changes in their environment. *In* "Bioassay Techniques and Environmental Chemistry," (G. E. Glass, ed.), pp. 119–138. Ann Arbor Sci Pub. Ann Arbor, MI.

Wetzel, R. G. (1964). A comparative study of the primary productivity of higher aquatic plants, periphyton, and phytoplankton in a large, shallow lake. *Int. Rev. Gesamten Hydrobiol.* **49**, 1–61.

Wetzel, R. G. (1983a). "Limnology." 2nd ed. Saunders College Publishing, New York.

Wetzel, R. G., ed. (1983b). "Periphyton of Freshwater Ecosystems." Dr. W. Junk Publishers, The Hague.

Wright, J. C., and Mills, I. K. (1967). Productivity studies in the Madison River, Yellowstone National Park. *Limnol. Oceanogr.* **12**, 568–577.

2

Patterns in Benthic Algae of Streams

Barry J. F. Biggs

National Institute of Water and Atmospheric Research, Christchurch, New Zealand

I. INTRODUCTION

Benthic algae are the most successful primary producers to exploit streams as habitat. They are widely considered to be the main source of energy for higher trophic levels in many, if not most, unshaded temperate region streams (e.g., Minshall, 1978). Benthic algae also sequester inorganic nutrients (see Mulholland, Chapter 19) and labile organics (see Tuchman, Chapter 10), thereby helping to purify stream waters (Vymazal, 1988). However, in stable-flowing, enriched streams they can proliferate, causing water management problems (e.g., Trotter and Hendricks, 1979; Wharfe *et al.*, 1984, Biggs, 1985). Because of their rapid response to environmental change they are also useful as indicators of stream water quality (see Lowe and Pan, Chapter 22).

Algal Ecology

In the four previous reviews of the ecology of stream benthic algae (Blum, 1956; Hynes, 1970; Whitton, 1975; Lock, 1981), aspects of the distribution of common taxa and biomass are discussed in the context of environmental regulators such as current, substratum type, temperature, and light. Since these reviews, much has been learned about the effects of other factors, including nutrients (see Borchardt, Chapter 7), disturbance (see Peterson, Chapter 13), and invertebrate grazing (see Steinman, Chapter 12). The opportunities presented by new analytical technologies such as computer-based GIS and multivariate statistical systems have also enabled the definition of broader-scale relationships between benthic algal communities and their terrestrial settings of geology, land use, and climate (e.g., Biggs, 1990; Leland, 1995).

This review synthesizes patterns in biomass and taxonomic structure of stream benthic algae based on our more recent understandings. Temporal and spatial dimensions are addressed separately, although the two are partly interlinked in determining pattern. This review also considers a range of scales, from patterns occurring on individual substrata over periods of weeks, to patterns over periods of years that can be evident across continents. This synthesis will assist with understanding local, regional, and broadscale community structure and dynamics. It will also provide a conceptual framework for the placement of later chapters in this book, which provide more detailed reviews of specific processes.

A. Conceptual Overview: Disturbance-Resource Supply-Grazer Control of Stream Benthic Algae

The ability of benthic algae to grow and prosper in streams is the outcome of a complex series of interactions between hydrological, water quality, and biotic factors. The relative importance of proximate variables (i.e., variables directly controlling accrual and loss) largely reflects higher-scale environmental features of catchments such as their topography, slope, land use, and vegetation. In turn, these reflect the ultimate variables of the landscape: geology, climate, and humans (Fig. 1). Thus, broadscale patterns in benthic algae among streams over large geographic areas, and between years, largely reflect patterns in geology, climate, and human activity (Biggs, 1990, Biggs *et al.*, 1990).

The proximate factors controlling benthic algae can be categorized into those regulating processes of biomass accrual, and those regulating the counteracting processes of biomass loss (Fig. 2). The main factor that leads to accrual is the level of resources, particularly nutrients and light (components of "land use" and "water quality" in Fig. 1). Temperature interacts with these to influence rates of metabolism and growth. The main factor leading to loss is disturbance. This most often occurs through substratum instability and associated abrasion, high water velocities, and

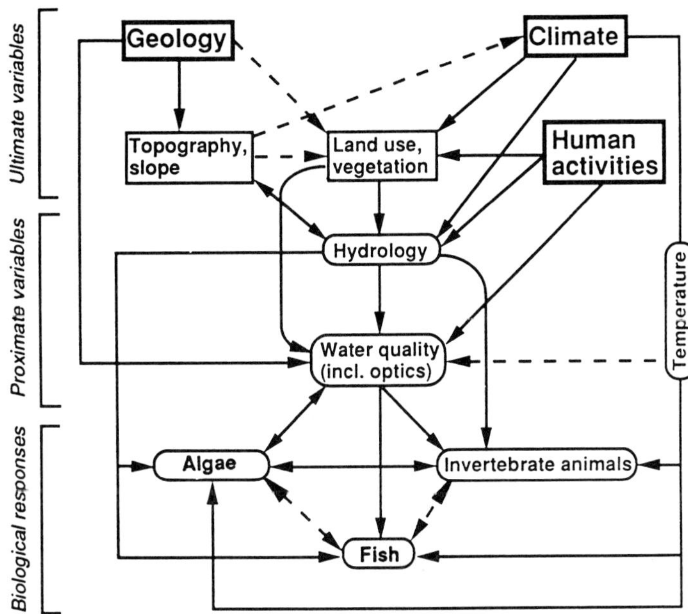

FIGURE 1 Diagram showing how landscape features link to control the fundamental physical variables of streams, which in turn control biological responses and interactions important to benthic algae. Strong causal interactions are shown as solid arrows and weaker interactions as dashed arrows. Double arrows indicate feedback relationships. Not all conceivable interactions are shown to avoid overcomplication. For example, land use affects benthic algae apart from through water quality, notably with regard to riparian shading, but this interaction is not shown. (Modified with permission from Biggs *et al.,* 1990.)

abrasion by suspended sediments (components of "hydrology" in Fig. 1). These factors can combine during floods to form more extreme, usually punctuated, disturbances that result in mass sloughing. Grazing by invertebrates and (to a much lesser extent) fish are biotic mechanisms by which biomass loss may also occur. Under medium to low flood disturbance frequency and grazing, accrual processes tend to dominate and where resource supply is medium to high, it results in communities that are dominated by erect, stalked diatoms and/or filamentous green algae (e.g., *Gomphoneis herculeana, Cladophora glomerata, Melosira varians*). Where disturbance frequency and resource supply is medium to low, the communities tend to be dominated by filamentous Cyanobacteria and red algae, with a limited number of diatoms (e.g., *Nostoc* sp., *Tolypothrix* sp., *Schizothrix* sp., *Phormidium* sp., *Audouinella hermanii, Epithemia* sp.). A large proportion of the taxa growing in these habitats are nitrogen fixers. With medium to high flood disturbance frequency, or heavy grazing, loss processes dominate and tend to result in communities dominated by low-

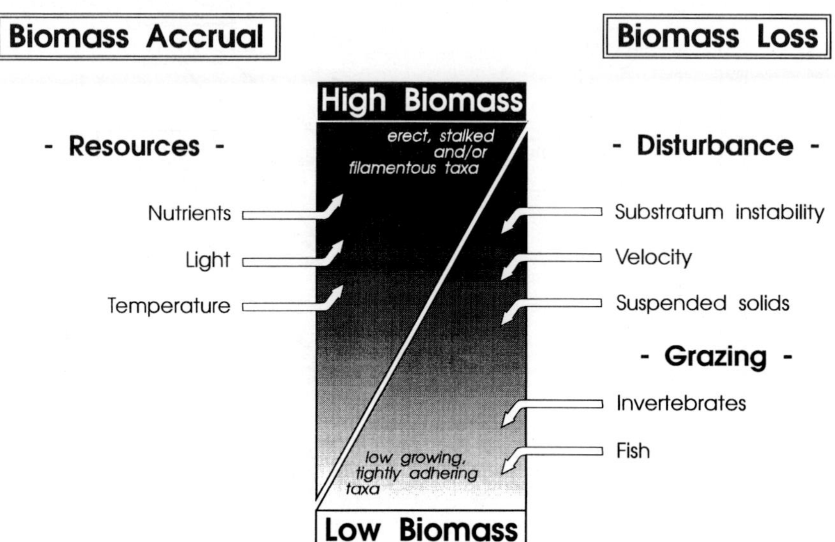

FIGURE 2 Summary of the disturbance-resource supply-grazer concept for the control of benthic algal development in streams. The relative balance of "biomass accrual" and "biomass loss" processes is depicted by the width of the triangles that make up the central rectangle. The physiognomy of the community likely to dominate each end of the gradient is also shown.

growing diatoms that adhere tightly to the substratum (e.g., *Achnanthidium minutissimum, Cocconeis placentula*). Other taxa capable of rapid colonization/replacement (particularly diatoms) can also become prominent (e.g., *Cymbella* spp., *Fragilaria vaucherie*).

In unshaded streams, the flood disturbance regime (reflecting "geology," "climate," and "topography" in Fig. 1) is perhaps the fundamental factor determining habitat suitability and pattern for benthic algae in streams. Apart from being a major loss mechanism, the frequency and intensity of floods can also influence a range of other variables that are important to benthic algal colonization and growth processes (Fig. 1). These include the availability of algal propagules (Hamilton and Duthie, 1987; Uehlinger, 1991), nutrient concentrations (e.g., Biggs and Close, 1989; Close and Davies-Colley, 1990; Humphrey and Stevenson, 1992), water clarity (Davies-Colley, 1990), stream geomorphology/baseflow velocities/substratum size (Jowett and Duncan, 1990), and density of grazing invertebrates (e.g., Sagar, 1986; Scrimgeour and Winterbourn, 1989; Quinn and Hickey, 1990). The frequency of floods also dictates the time available for benthic algal accrual. During the more physically benign, inter-flood periods, resource availability (light and nutrients), loss by grazing, spatial differences in water velocity and turbulence, and the growth strategies of individual species become important in determining net community development.

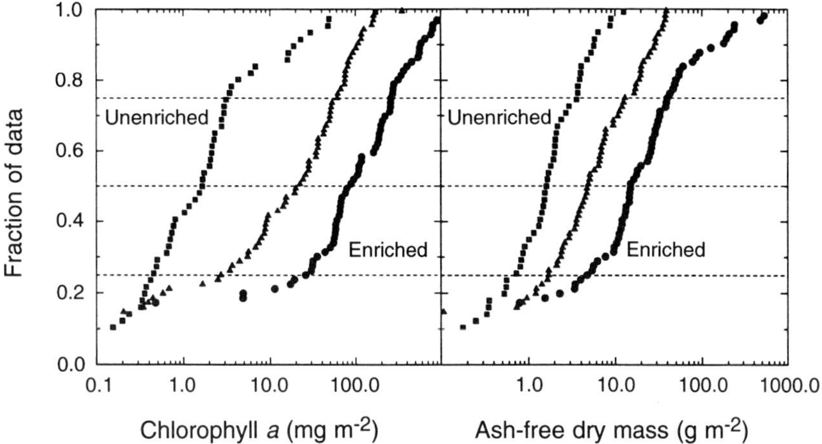

FIGURE 3 Cumulative frequency curves for chlorophyll *a* and ash-free dry mass from unenriched (squares), moderately enriched (triangles), and enriched (dots) stream sites in New Zealand, sampled every 4 weeks for a year [data pooled for groups of sites in each enrichment category with $N = 4$, 6, and 6 sites, respectively; see Biggs (1995) for sampling and site information]. The dashed lines denote the 25th, 50th, and 75th percentiles.

B. Typical Biomass Values for Streams

Lock (1981), Horner *et al.* (1983), Bott *et al.* (1985), and Biggs and Price (1987) summarize biomass data from a range of previous studies. However, it is difficult to determine from these "typical" algal biomass values for different stream types over a whole year because of differing sampling and analytical methodologies and the short duration of many of the studies. There has also been extensive use of artificial substrata, and such results are not necessarily typical of natural communities (Biggs, 1988a). Figure 3 shows chlorophyll *a* and ash-free dry mass (AFDM) cumulative frequency curves from 16 stream sites in New Zealand, which cover a range in disturbance frequency and enrichment status and were sampled monthly for a year. The sites are grouped according to level of enrichment (see Biggs, 1995, for details of methods and enrichment groupings).

These data permit the generalization that benthic chlorophyll *a* in streams can span four orders of magnitude and ash-free dry mass three orders over a year. Although the range in data was great, 75% of the values were <80 mg m^{-2} for chlorophyll *a* and <10.8 g m^{-2} for ash-free dry mass, with median values for the combined data of 20 mg m^{-2} chlorophyll *a* and 5 g m^{-2} ash-free dry mass. Some of the variation in values was caused by different degrees of enrichment. For unenriched streams that had forested catchments on hard metamorphic rocks (mean conductivity = 87 μS cm^{-1}), the upper and lower quartiles define that chlorophyll *a* was typically in the range of 0.5–3 mg m^{-2} (median of 1.7 mg m^{-2} chlorophyll *a* and 1.5 g m^{-2} AFDM).

For moderately enriched streams with catchments that were moderately developed for agriculture (mean conductivity = 106 μS cm^{-1}), chlorophyll *a* was typically in the range of 3–60 mg m^{-2} (median of 21 mg m^{-2} chlorophyll *a* and 4.8 g m^{-2} AFDM). Last, for enriched streams with catchments that were highly developed for agriculture and/or underlain by nutrient-rich rocks (mean conductivity = 271 μS cm^{-1}), chlorophyll *a* was typically in the range of 25–260 mg m^{-2} (median of 84 mg m^{-2} chlorophyll *a* and 15 g m^{-2} AFDM). High chlorophyll *a* values (i.e., >100 mg m^{-2}; see Section IV) occurred for approximately 40% of the year in these enriched streams. This compares with <1% of the year in the moderately enriched streams.

II. TEMPORAL PATTERNS

A. Short Term

A clear, and generally universal, pattern of short-term benthic algal biomass accrual is evident in streams (Fig. 4). It has been shown for communities grown on artificial substrata (Cattaneo *et al.*, 1975, Biggs, 1988a), at the commencement of experiments in artificial channels (e.g., Bothwell, 1989; Reiter, 1989; Peterson and Stevenson, 1990; Poff *et al.*, 1990, Mulholland *et al.*, 1991), and in natural streams following floods (e.g., Scrimgeour *et al.*, 1988; Biggs and Stokseth, 1996). This biomass pattern reflects a shift in importance among the primary processes (Fig. 2). Accrual through immigration/colonization and growth dominates early in

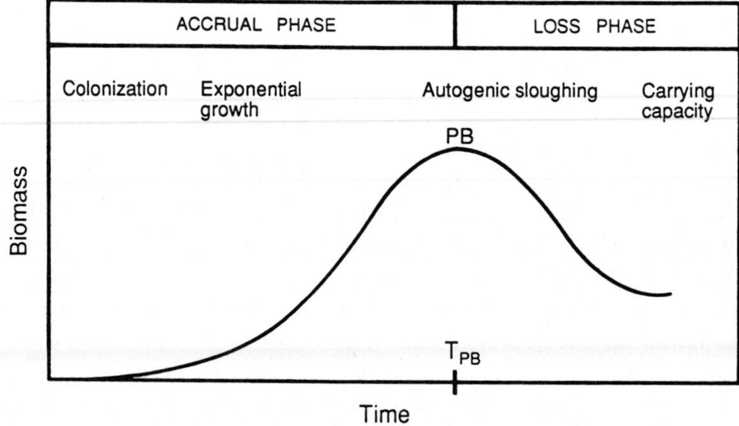

FIGURE 4 An idealized benthic algal accrual curve with different phases shown. PB (peak biomass) is the maximum accrual cycle biomass and T_{PB} is the time to PB from commencement of colonization.

the sequence (termed the "accrual phase"), but then there is a shift to dominance of loss processes through death, emigration, sloughing, and grazing later in the sequence (termed the "loss phase").

Immigration/colonization is a linear process dominated by passive settlement of cells. The settlement rate is governed by the size and type of the propagule pool (related to the abundance of cells in upstream refugia formed by bedrock, bryophytes, and seeps), immigration and dispersal properties (such as cell size/morphology), substratum texture, water velocity, and light intensity (Jones, 1978; Kaufman, 1979; Osborne, 1983; Stevenson, 1983; Bothwell *et al.*, 1989; Peterson and Stevenson, 1990; Stevenson and Peterson, 1989, 1991). A succession often occurs, beginning with the development of an organic matrix and bacterial flora followed by a transition from small adnate diatoms (Bacillariophyceae) to apically attached colonial diatoms, and finally to filamentous green algae (Chlorophyceae) (e.g., Hudon and Bourget, 1981; Korte and Blinn, 1983; Peterson and Stevenson, 1990; see also Tuchman, Chapter 10, and Peterson, Chapter 13). After settlement, the colonizing cells undergo exponential growth at a rate dictated by the availability of resources together with temperature and modes of reproduction (Fig. 2) (e.g., Bothwell 1988, 1989). In unshaded streams, the most important of these resources are nutrients (see Borchardt, Chapter 7). In heavily shaded streams, low light levels may primarily limit growth (Hill and Knight, 1988). Exponential cell division at this time results only in exponential development of mat biomass early in the accrual cycle. Later, kinetics become progressively linear (Fig. 4) as the newly synthesized biomass becomes small compared with that already accrued.

The proportion of new cells that can accumulate is generally determined by the degree of noncatastrophic losses through current-induced drag (see Stevenson, Chapter 11), emigration, and grazing (see Steinman, Chapter 12). The time taken to reach peak biomass (T_{PB}, Fig. 4) varies. Following low- to moderate-intensity floods, colonization can be rapid because of a high abundance of propagules from refugia. If this is also followed by rapid growth, peak biomass (PB) may be reached within two weeks (e.g., Stevenson, 1990). However, with more complete depopulation under severe floods, colonization may take many weeks. If this combines with low growth rates, T_{PB} can be as long as 70–100 days (e.g., Biggs and Stokseth, 1996). Intuitively one would expect T_{PB} to be shorter in enriched than in unenriched streams because specific growth rates should be higher (e.g., Horner *et al.*, 1983; Bothwell, 1989). However, this does not necessarily occur (e.g., Biggs, 1988a; Lohman *et al.*, 1992; Biggs *et al.*, 1996). This apparent paradox appears because of the earlier onset of nutrient limitation at the base of the accumulating benthic algal mat in nutrient-poor waters, resulting in mat degradation and sloughing early in the accrual cycle.

PB values (Fig. 4) vary asymptotically as a function of nutrient loading during the growing period (Horner *et al.*, 1983; Bothwell, 1989; Kjeldsen,

1994). Higher nutrient levels maintain higher diffusion gradients within algal mats, allowing them to develop into thicker structures. However, some variation in PB values can be expected for given nutrient loadings depending on other constraints such as water velocity (e.g., Horner, *et al.*, 1983, 1990; Biggs and Gerbeaux, 1993). Maximum PB values (termed PB_{max}, Bothwell, 1989) at high nutrient loadings vary for different communities and can range from 300 to 400 mg m^{-2} chlorophyll *a* for diatoms and cyanobacteria (Cyanophyceae) (Bothwell, 1989; Horner *et al.*, 1990) to >1200 mg m^{-2} chlorophyll *a* for filamentous green and chrysophyte algae (Chrysophyceae) (Kjeldsen, 1994, and personal communication; Biggs, 1995).

Biomass accrual slows when losses approach the rate of accrual. These losses can occur through a combination of death due to age, parasitism, disease, and removal by grazing. Very rapid loss often occurs as autogenic sloughing (e.g., Biggs and Close, 1989). This appears to be a function of resource stress of the underlying layers. Eventually loss processes dominate (i.e., the community enters the loss phase, Fig. 4). The carrying capacity is reached when rates of accrual and loss are in balance. The biomass at which this occurs appears to vary greatly. In some situations, carrying capacity approximately equals PB (e.g., Oemke and Burton, 1986), but often it is less (e.g., where communities are dominated by diatoms) (Biggs and Close, 1989). Factors probably affecting this balance include rates of metabolite mass transfer to the base of the mats, grazing, hydrodynamic shear stress, and mat tensile strength.

During the accrual cycle there is also a continual loss of cells through emigration. This process often peaks during the day time (Stevenson and Peterson, 1991; Barnese and Lowe, 1992). The concentration of these drifting emigrants in the stream waters is generally related to abundances of algae on the bed (at least for diatoms). Although the composition reflects benthic communities in general terms, there can be interspecific differences in drift activity (Stevenson and Peterson, 1991).

It should be stressed that these short-term biomass dynamics can be masked by variability if the spatial resolution of sampling is either too fine or too coarse. For example, quantifying dynamics at a small (say, cobble) scale will reveal much greater variability in biomass than the idealized accrual cycle in Fig. 4 would suggest. Patchy distribution occurs over individual substrata through variations in growth and loss due to spatial variations in hydrodynamic conditions (Biggs and Hickey, 1994; Biggs and Stokseth, 1996) and densities of grazers (Downes *et al.*, 1993). Conversely, when sampling at larger spatial scales that can incorporate several different habitat units (e.g., riffles, runs, and pools), growth rates will vary and accrual cycles may not be synchronized (Biggs and Stokseth, 1996). Thus, to determine the parameters for accrual dynamics such as depicted in Fig. 4, multiple (e.g., >10), whole-substratum samples are recommended that are stratified according to habitat units.

B. Long Term

Three main long-term temporal patterns in biomass can be distinguished among streams: (1) relatively constant, low biomass; (2) cycles of accrual and sloughing; and (3) seasonal cycles. These patterns are predominantly a result of interactions between processes of biomass accrual and loss (Fig. 2), with the disturbance regime being a fundamental determinant of the overall balance between these processes.

Relatively constant, low biomass can occur because of frequent disturbances in streams. Such disturbances can override the effects of other potentially limiting variables such as nutrients (Biggs, 1995). Disturbance can be a result of frequent floods or a streambed dominated by continuously unstable bed sediments. The latter condition is seen frequently in low-gradient streams with substrata dominated by sand. However, relatively constant, low biomass can also occur under conditions of very low nutrients (e.g., Biggs, 1995) and heavy riparian shading (e.g., Lyford and Gregory, 1975; Suren, 1992). Streams with a high disturbance frequency often are also very low in nutrients (e.g., Biggs and Close, 1989; Biggs, 1995). Thus, it may sometimes be difficult to separate the importance of these two constraints in determining long-term temporal patterns. Such conditions typically occur where streams drain highland/alpine and steep forested catchments (e.g., Biggs, 1995).

Conversely, perpetually low biomass may be evident where physical disturbance is very infrequent. Such conditions often allow dense invertebrate communities to develop. These can potentially heavily graze the periphytic algae throughout much of the growing season (e.g., Rosemond, 1994). This typically occurs in moderately enriched or unenriched spring-fed streams and those draining moderate-gradient forested catchments. The communities in such streams can be dominated for much of the year by disturbance- and grazing-resistant taxa, including species of *Achnanthidium, Cocconeis, Cymbella,* and *Synedra* and basal structures of *Stigeoclonium* (Biggs and Gerbeaux, 1993, 1994; Rosemond, 1994).

Cycles of sloughing and accrual can be found in streams that experience either a moderate frequency of, or seasonal, flood disturbances (Fig. 5). Extended periods of flow stability between floods (4–10 weeks) allows the accumulation of biomass (e.g., Douglas, 1958; Biggs, 1988b; Fisher and Grimm, 1988; Biggs and Close, 1989; Uehlinger, 1991; Lohman *et al.*, 1992) with the taxonomic structure sometimes proceeding through a succession from diatoms to large green filamentous algae and cyanobacteria (Fisher *et al.*, 1982). It should be noted that not every flood results in total biomass loss (Fig. 5). Incomplete removal may occur if the event has only a low intensity, the preflood algal biomass is low, or the resident taxa are highly scour resistant (Power and Stewart, 1987; Biggs and Close, 1989; Grimm and Fisher, 1989; Biggs and Thomsen, 1995). Under such circum-

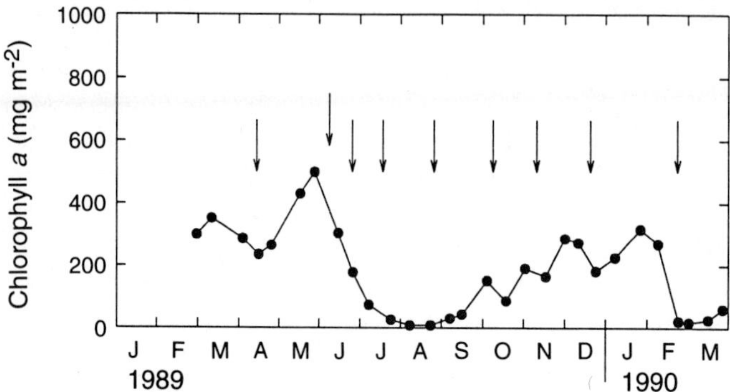

FIGURE 5 A pattern of accrual and sloughing in benthic algal communities of the Necker River, Switzerland (data averaged from 10 sites). (Modified with permission from Uehlinger, 1991.)

stances the remnant community can regenerate very rapidly, but be set back in successional stage (Biggs *et al.*, 1996).

Often in gravel/cobble bed streams there may be a window of time immediately after a major flood when invertebrate grazer densities are still low (probably due to slower invertebrate recolonization and reproduction). At such times benthic algal accumulation can proceed largely unconstrained by grazing, allowing high biomasses to develop (e.g., Power, 1992). However, with increasing time from the last disturbance, losses due to both grazing (Power, 1992) and autogenic sloughing (e.g., Biggs and Close, 1989) can become significant. Eventually grazing may be the predominant controller of accrual. Thus, there appears to be a shift from "abiotic" to "biotic" control of biomass until a further flood resets the sequence. Fisher and Grimm (1991) and Power (1992) proposed graphical models of these dynamics. A large number of streams in both temperate and desert climatic zones (commonly with low to moderate enrichment draining areas of moderate relief) appear to display the cycles of growth and sloughing (e.g., Douglas, 1958; Tett *et al.*, 1978; Jones *et al.*, 1984; Power and Stewart, 1987; Biggs, 1988b; Biggs and Close, 1989; Grimm and Fisher, 1989; Uehlinger, 1991; Lohman *et al.*, 1992). However, the shift from abiotic to biotic control of algal biomass late in the accrual cycle has yet to be demonstrated in many streams.

Strong seasonal patterns in community development (Fig. 6) appear to be mostly mediated by (1) seasonality in disturbance regimes (providing there are adequate nutrient resources), (2) seasonality in grazer activity (where flood disturbances are rare), or (3) seasonality in light regimes (where neither disturbance nor grazing is important). For dynamics controlled by seasonality in floods, accumulation of organic matter commences after the last late winter/early spring floods with the development

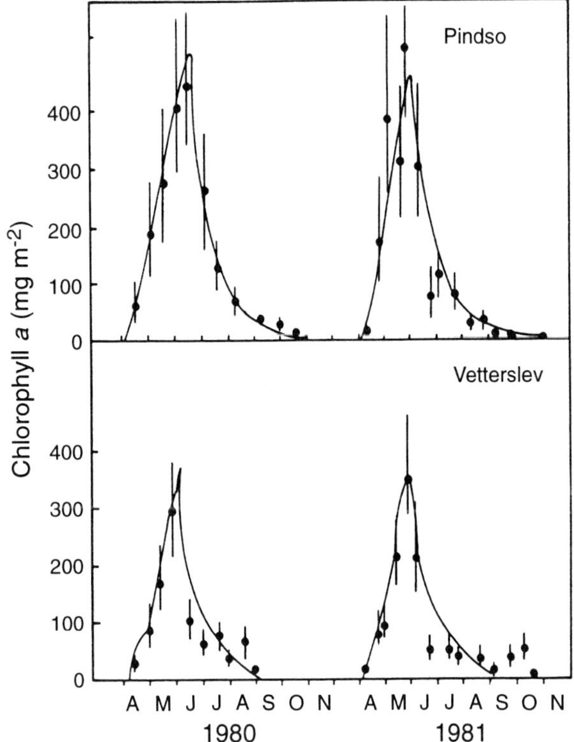

FIGURE 6 Chlorophyll *a* in two stable flowing streams in Denmark over 2 years showing strong seasonal growth responses. Observed chlorophyll biomass is represented by the dots (plus 95% confidence limits) and the solid line is the predicted biomass using an empirically derived model based on light (Reprinted with permission from Sand-Jensen et al., 1988.)

of diatom-dominated communities (e.g., *Diatoma, Synedra, Navicula*). As with short-term benthic algal dynamics, communities often become dominated by cyanobacteria progressing into summer (e.g., *Phormidium, Homeothrix*) and patchy growths of large filamentous algae such as *Vaucheria* and *Cladophora* in late summer (e.g., Marker, 1976; Power, 1992). By fall, much of the biomass in these streams may have degraded and been washed away (Power, 1992). Such seasonal dynamics typically occur in enriched spring-fed and lowland streams (e.g., Moore, 1977a; Sand-Jensen *et al.,* 1988), those in Mediterranean climatic zones (e.g., Power, 1992; Sabater and Sabater, 1992), and tropical/subtropical streams (e.g., Necchi and Pascoaloto, 1993).

In some stable streams, there is a seasonal biomass maxima in spring (e.g., Marker, 1976; Moore, 1977b; Sand-Jensen *et al.,* 1988; Shamsudin and Sleigh, 1994), late summer, or both (e.g., Biggs and Close, 1989).

Diatoms tend to be the dominant community in spring and filamentous cyanobacteria and/or green algae in late summer (Moore, 1977a). As in streams with cycles of accrual and sloughing, the communities can also experience a window of time during early spring when accumulation proceeds largely unchecked by herbivory because of low invertebrate activity in the cold waters (e.g., Power, 1992). Thus, it is possible that seasonality in grazing activity is also playing an important role in these seasonal dynamics (Shamsudin and Sleigh, 1994).

Some algae appear to have autogenic requirements that can limit their development to stable flow periods in specific seasons. These taxa include *Cladophora, Spirogyra, Drapernaldia, Batrachospermum, Gomphonema olivaceum,* and *Diatoma vulgare* (see Hynes, 1970, for further discussion).

III. SPATIAL PATTERNS

Spatial patterns in the distribution of stream benthic algae occur over a wide range of scales, from single grains of sand (e.g., Krejci and Lowe, 1986) to across continents (Sheath and Cole, 1992). As with temporal patterns, disturbance appears to be an important variable determining these spatial patterns.

A. Microscale: Substratum Patterns

At the finest level of resolution, algal taxa can show preferences for sand grains of different mineralogy and topography. Some stalked diatoms (e.g., *Fragilaria leptostauron*) appear more commonly on "hills" on the surface of stable grains. In contrast, some prostrate diatoms (e.g., *Achnanthidium lanceolata)* inhabit depressions and crevices on grains where they obtain protection from abrasion as grains roll (Krejci and Lowe, 1986).

A common pattern in algal distribution in streams during stable flows is a higher biomass on larger substrata (e.g., McConnell and Sigler, 1959; Tett *et al.,* 1978; Biggs and Shand, 1987), predominantly because of the higher stability of these during floods (Uehlinger, 1991). There may also be a zonation in community structure with increasing depth on boulders (Blum, 1956). Most taxa are able to colonize most substrata, but can develop into mature communities only if they are given a sufficiently long period of habitat stability. Small- and medium-sized floods (which can occur quite frequently in temperate region streams, Biggs and Close, 1989) often mobilize fine sediments that then abrade resident communities. However, some communities can survive on the larger, stable, substratum particles that protrude above the level of most of the tumbling particles (e.g., Uehlinger, 1991). Such communities can mature to form a high biomass, whereas those on mobile sands and gravels are more frequently set back to low-biomass, early-successional stages (see Peterson, Chapter 13). Such processes can result in extremely patchy distributions of algae during stable flows.

In some streams, regular movement of sand substrata maintains a low biomass. In such situations, isolated stable bedrock/boulders may support up to 15-fold higher biomass than the unconsolidated sand/silt (Biggs and Shand, 1987). If patches of silts and sands do stabilize for prolonged periods then a succession may occur. First, diatoms can colonize individual particles, but only rarely form a significant biomass (e.g., Stevenson and Hashim, 1989) because they increase the drag on the particles, resulting in destabilization of the particles. Eventually, cyanobacterial mats may develop over the diatoms. Unlike diatoms, these prostrate filaments can then bind the surficial layers of silt/sand particles and form a significant biomass (Biggs and Shand, 1987).

B. Mesoscale: Within Catchment Patterns

Differences in biomass and species composition can occur between pool, run, and riffle habitats in unshaded streams, and these reflect spatial differences in shear stress, nutrient mass transfer (see Stevenson, Chapter 11) and substratum type (size and disturbance history, see preceding section). There are few published data on patterns in algal accrual among these habitats for different streams, but it appears that opposite responses can exist depending on whether the waters are enriched or not (B. J. F. Biggs, unpublished observations). In enriched streams, higher-biomass communities often develop in low-velocity runs and pools. These communities are usually dominated by filamentous green algae (e.g., *Spirogyra, Oedogonium, Cladophora*). However, in riffles of these streams, higher shear stress restricts the maximum thickness of the algal mat (and thus biomass) compared with that in the pools. In these areas the communities are often dominated by low-growing diatoms (e.g., *Cocconeis, Cymbella, Nitzschia*) [e.g., see Peterson and Stevenson (1990), and Poff *et al.* (1990) for experimental data on community develoment as a function of velocity]. In contrast, in unenriched streams the highest benthic algal biomass (dominated by diatoms) is usually encountered in the higher-velocity riffles (e.g., Scarsbrook and Townsend, 1993; Biggs and Hickey, 1994; Biggs and Stokseth 1996). This response is probably due to increased mass transfer of metabolites with higher velocities and turbulence (see Borchardt, Chapter 7).

The River Continuum Concept (RCC, Vannote *et al.*, 1980) proposed that a downstream pattern of increasing benthic algal biomass should occur from headwater to midcatchment reaches as streams coalesce, the channel becomes wider, and riparian shading is reduced. Biomass is predicted to decrease again in the lower regions as a function of increasing light attenuation with higher water depths and increased turbidity. This proposed pattern is intuitively correct in catchments with forested headwaters and has been verified in a number of streams at certain times of the year (e.g., Minshall *et al.,* 1983). However, localized features of habitats can often interrupt such downstream trends. In nonforested catchments the RCC pattern can be reversed (Wiley *et al.,* 1990), whereas in catchments with frequent disturbances and/or localized nutrient inputs, benthic algal communities may display quite different patterns (e.g., Sabater and Sabater, 1992). In determining whether the distribution of benthic algae in a stream should conform to the RCC pattern, it is important to recognize the influence of downstream changes in channel geometry. The classic geomorphic model describes a pattern of decreasing bed gradient, widening of the active channel, and a decrease in substratum particle size moving downstream. This results in a progressive reduction in hydraulic stress (e.g., Statzner and Higler, 1986; Statzner *et al.,* 1988). However, in regions with complex geology and tectonic activity, major discontinuities in catchment gradient can occur that can impose strong, localized, hydraulic constraints on benthic algal development and thereby mask the simple downstream trends as proposed under the RCC.

In many catchments (particularly those with little forest cover), there is a downstream increase in nutrient loading that can lead to gradients in community composition and biomass. Headwater reaches are frequently dominated by diatoms (e.g., *Cymbella* spp., *Gomphoneis* sp., *Fragilaria* spp.), cyanobacteria (e.g., *Schizothrix* sp., *Phormidium* sp.), red algae (Rhodophyceae) (e.g., *Batrachospermum* sp.), and several low-biomass filamentous green algae (e.g., *Stigeoclonium* sp.). As enrichment increases down the catchment (e.g., with higher intensity land use), nutrient-demanding taxa such as the filamentous chrysophyte *Vaucheria* sp. and the filamentous green alga *Cladophora glomerata* become more prominent and can form a high biomass (e.g., Holmes and Whitton, 1981; Entwistle, 1989).

C. Broadscale: Intercatchment Patterns

An extremely wide range in average benthic algal biomass and community composition is possible among streams. This variability reflects spatial differences in broad environmental features of regions ("ultimate variables" in Fig. 1). A fundamental broadscale pattern can result from a varying frequency of flood events among streams. In accordance with the previous discussion, streams that experience frequent floods tend to have low average biomass (Fig. 7) (Biggs and Close, 1989; Biggs, 1995), with communities dominated by low-growing, highly shear-resistant taxa (Fig. 2; also see Peterson, Chapter 13). Conversely, more benign habitats (which have long time intervals between disturbances, little resource stress, and no heavy grazing) tend to have communities with higher biomass and greater architectural complexity.

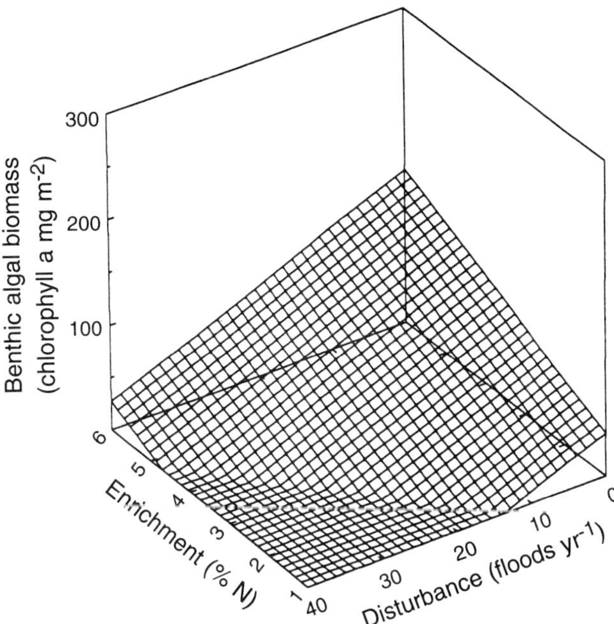

FIGURE 7 Mean monthly benthic chlorophyll *a* as a function of flood disturbance and enrichment (cellular nitrogen concentrations). The response surface was generated using data from 15 sites (multiple R^2 for chlorophyll $a = 0.87$). (Adapted with permission from Biggs, 1995.)

Differences in community composition and biomass among regions during periods of low flow may reflect regional differences in geology/land use and associated enrichment (e.g., Biggs, 1990; Biggs *et al.*, 1990; Leland, 1995). Green algae such as *Ulothrix zonata, Stigeoclonium* sp., and *Spirogyra* sp. and the cyanobacterium *Phormidium* sp. often form communities of low biomass in unenriched streams (Fig. 8). However, in enriched streams, algal communities are usually dominated by large filamentous taxa such as *Cladophora glomerata* and *Rhizoclonium* sp., which form a high biomass.

As could be anticipated from the earlier discussion and Fig. 2, there is also a strong interaction between disturbance and enrichment in determining broadscale spatial patterns. Biggs (1995) has shown that a combination of infrequent flood disturbances and high levels of enrichment are necessary to attain a high average monthly biomass of benthic algae in streams of New Zealand (Fig. 7). Conversely, if flood disturbances are very frequent, and/or levels of enrichment are very low, then there will be only a very low mean monthly biomass. High levels of enrichment result in some algal accrual at high frequencies of disturbance, because of more rapid interflood regeneration, and some biomass can also accrue with low levels of enrichment providing disturbance frequency is also low.

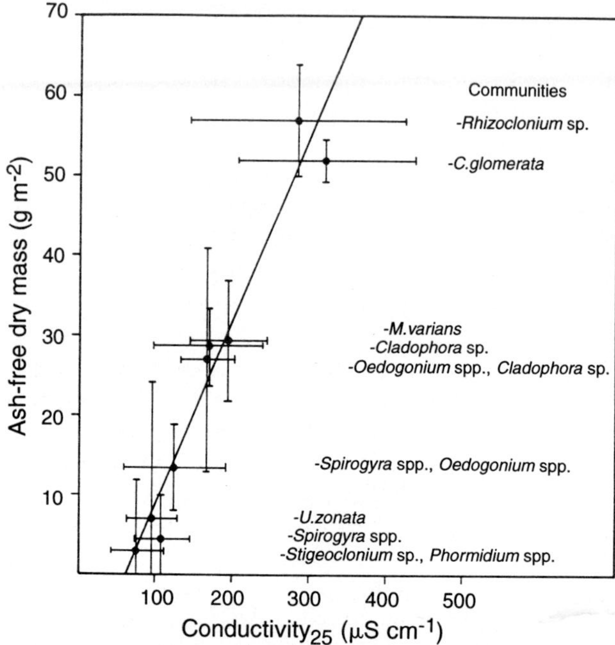

FIGURE 8 Relationship between mean AFDM biomass for nine different benthic algal communities and conductance of the water (a relative measure of enrichment) from a survey of over 400 New Zealand streams during summer low flows. The error bars are 1 S.D. *(C. glomerata = Cladophora glomerata, M. varians = Melosira varians, U. zonata = Ulothrix zonata).* (Modified with permission from Biggs and Price, 1987.)

Among streams on the North American continent, as latitude decreases there is a tendency for a decrease in filamentous chrysophytes (except in the tundra biome) and an increase in red algae (Sheath and Cole, 1992). No pattern is evident, however, for cyanobacteria. The boreal forest biome has the highest species richness and the tundra/desert–chaparral biomes have the lowest. In terms of biome similarities, macroalgal communities in the boreal and western conifer biomes show close similarity, with the eastern hemlock–hardwood forest and deciduous forest biomes also having floras with a moderately close similarity (Sheath and Cole, 1992).

IV. BENTHIC ALGAL PROLIFERATIONS

Excessive accumulations of benthic algae in streams at the peak of the accrual cycle are a widespread phenomenon (e.g., Wharfe *et al.*, 1984; Biggs, 1985; Biggs and Price, 1987; Welch *et al.*, 1989; Chessman *et al.*, 1992). They can potentially affect abstraction for water supply, aesthetic appeal, and instream recreation. They can also degrade ecosystem structure. Most severe cases occur under high nutrient loadings, which may originate from areas of nutrient-rich rocks, intensive agricultural development, and/or nutrient-rich waste discharges.

Filamentous green algae such as *Ulothrix* (in cold waters), *Cladophora,* and *Rhizoclonium* often present the greatest nuisance (Wharfe *et al.,* 1984; Biggs 1985; Welch *et al.,* 1989). In less enriched streams, stalked and tube-dwelling diatoms such as species of Gomphonemaceae (e.g., *Gomphoneis herculeana* and *Didymosphenia geminata*) and Cymbellaceae (e.g., *Cymbella affinis*) can also be a problem (e.g., Biggs and Hickey, 1994; Biggs, 1995). These algal groups form very different macroscopic communities. Green algae develop streaming beds of filaments that can be up to several meters long in low-velocity areas, and often detach and float near the surface (e.g., Power, 1990). These floating mats can then become entangled in branches or around cobbles protruding up from the bottom of the stream, where they often continue to proliferate. In contrast, diatoms tend to form thick gelatinous mats that consist predominantly of polysaccharides (Hoagland *et al.,* 1993). The mats may attain a thickness of a centimeter or more, completely smothering the substrata. In some streams they may strongly resemble whitish-gray sewage fungus.

There is some difficulty in defining what constitutes a nuisance growth because specific criteria need to be established for specific water uses. Figure 9 illustrates the visual appearance, and Table I gives cover and biomass values, for the range in community development of filamentous green algae that may occur in natural streams. As an initial guide for the control of proliferations, Horner *et al.* (1983) suggest that chlorophyll *a* >100–150 mg m^{-2}, or a cover >20% by filamentous algae, is unacceptable (cf. Fig. 9 and Table I). While surveying over 400 streams in New Zealand, Biggs and Price (1987)

TABLE I Summary of Percentage Cover, Chlorophyll *a,* and Ash-Free Dry Mass Data for Filamentous Green Algal Communities (Dominated by *Cladophora* sp. and *Oedogonium* sp.) for a Range of Development in a New Zealand Stream as Depicted in Fig. 2.9[a]

Photograph	*Percentage cover*	*Chlorophyll a* (mg m^{-2})	AFDM (g m^{-2})
A	20	80	25
B	30	120	35
C	40	160	40
D	55	300	50
E	70	900	200
F	95	640	90

[a]Note that the biomass at 95% cover is less than at 70% cover because the water is too shallow for the community to attain much vertical height.

FIGURE 9 Photos illustrating the visual appearance of a range in benthic algal cover that can occur in gravel/cobble bed streams (see Table I for cover and biomass data). (Reprinted with permission from Biggs, 1992.)

D

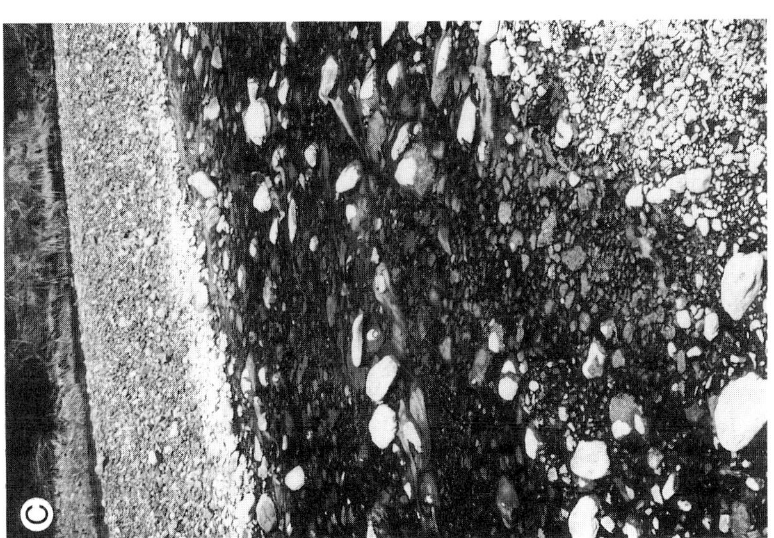
C

FIGURE 9 Continued

observed that when filamentous algae exceeded a cover of approximately 40%, the community became very conspicuous from the bank. They also observed that if biomass was >50 g m^{-2} AFDM (approximately 55% cover) it usually resulted in extensive smothering of the bed sediments (cf. Fig. 9 and Table 1). From a literature review of biomass values in various habitats, Nordin (1985) recommended the following benthic algal biomass criteria for streams of British Columbia, Canada: <50 mg m^{-2} chlorophyll *a* to protect recreational usage and <100 mg m^{-2} to protect other aquatic life. Similarly, in New Zealand the Ministry for the Environment (Zuur, 1992) have recommended to water management agencies that to protect contact recreational usage of streams, the seasonal maximum cover by filamentous algae should not exceed 40%, and/or biomass should not exceed 100 mg m^{-2} chlorophyll *a* or 40 g m^{-2} AFDM. The development of these guidelines demonstrates that the control of undesirable benthic algal growths in streams is becoming an important issue for water resources management.

V. CONCLUDING REMARKS

A complex array of factors and interactions govern the development of benthic algal communities in streams and the competitive success of any one species is currently difficult to predict. Benthic algae have high turnover and opportunistic life-history strategies that have enabled them to successfully exploit stream habitats, many of which are harsh environments. However, some patterns in biomass and community structure do exist. These patterns can be explained in general terms as interacting and separate functions of gradients in accrual (nutrient and light resources) and loss (disturbance and grazing) variables that operate over small to large spatial scales, and for periods from weeks to years.

ACKNOWLEDGMENTS

This chapter was prepared under the financial support of the New Zealand Foundation for Research, Science and Technology (NIWA River Ecosystems Programme, Output 32). I am grateful for review comments from Jan Stevenson, Rex Lowe, and Cathy Kilroy.

REFERENCES

Barnese, L. E., and Lowe, R. L. (1992). Effects of substrate, light, and benthic invertebrates on algal drift in small streams. *J. North Am. Benthol. Soc.* **11**, 49–59.

Biggs, B. J. F. (1985). Algae: A blooming nuisance in rivers. *Soil Water* **21**, 27–31.

Biggs, B. J. F. (1988a). Algal proliferations in New Zealand's shallow stony foothills-fed rivers: Toward a predictive model. *Verh—Int. Ver. Theor. Angew. Limnol.* **23**, 1405–1411.

Biggs, B. J. F. (1988b). Artificial substrate exposure times for periphyton biomass estimates in rivers. *N. Z. J. Mar. Freshwater Res.* **22,** 189–199.
Biggs, B. J. F. (1990). Periphyton communities and their environments in New Zealand rivers. *N. Z. J. M. Freshwater Res.* **24,** 367–386.
Biggs, B. J. F. (1992). Photographic illustration of filamentous algal growths in gravel-bed rivers. *In* "Water Quality Guidelines No. 1: Guidelines for the Control of Undesirable Biological Growths in Water" (B. Zuur, ed.), pp. 30–31. New Zealand Ministry for the Environment, Wellington.
Biggs, B. J. F. (1995). The contribution of disturbance, catchment geology and land use to the habitat template of periphyton in stream ecosystems. *Freshwater Biol.* **33,** 419–438.
Biggs, B. J. F., and Close, M. E. (1989). Periphyton biomass dynamics in gravel bed rivers: The relative effects of flows and nutrients. *Freshwater Biol.* **22,** 209–231.
Biggs, B. J. F., and Gerbeaux, P. (1993). Periphyton development in relation to macro-scale (geology) and micro-scale (velocity) limiters in two gravel-bed rivers, New Zealand. *N. Z. J. Mar. Freshwater Res.* **27,** 39–53.
Biggs, B. J. F., and Gerbeaux, P. J. (1994). Epilithic periphyton in temperate rivers: community composition in relation to differences in disturbance regimes. *NIWA-Christchurch Misc. Rep.* **177,** 1–33.
Biggs, B. J. F., and Hickey, C. W. (1994). Periphyton responses to a hydraulic gradient in a regulated river, New Zealand. *Freshwater Biol.* **32,** 49–59.
Biggs, B. J. F., and Price, G. M. (1987). A survey of filamentous algal proliferations in New Zealand rivers. *N. Z. J. Mar. Freshwater Res.* **21,** 175–191.
Biggs, B. J. F., and Shand, B. I. (1987). "Biological Communities and Their Potential Effects of Power Developments in the Lower Clutha River, Otago." Ministry of Works and Development, Christchurch, NZ.
Biggs, B. J. F., and Stokseth, S. (1996). Hydraulic habitat suitability for periphyton in rivers. *Regul. Rivers: Res.* 12 *(in press).*
Biggs, B. J. F., and Thomsen, H. A. (1995). Disturbance in stream periphyton by perturbations in shear stress: time to structural failure and differences in community resistance. *J. Phycol.* **31,** 233–241.
Biggs, B. J. F., Duncan, M. J., Jowett, I. G., Quinn, J. M., Hickey, C. W., Davies-Colley, R. J., and Close, M. E. (1990). Ecological characterisation, classification and modelling of New Zealand rivers: An introduction and synthesis. *N. Z. J. Mar. Freshwater Res.* **24,** 277–304.
Biggs, B. J. F., Tuchman, N. C., Lowe, R. L., and Stevenson, R. J. (1996). Disturbance of stream periphyton by hydrological scour: interactive effects of light and nutrients on resistance, resilience and accrual cycle dynamics. Submitted manuscript.
Blum, J. L. (1956). The ecology of river algae. *Bot. Rev.* **2,** 291–341.
Bothwell, M. L. (1988). Growth rate responses of lotic periphytic diatoms to experimental phosphorus enrichment: The influence of temperature and light. *Can. J. Fish. Aquat. Sci.* **45,** 261–270.
Bothwell, M. L. (1989). Phosphorus-limited growth dynamics of lotic periphytic diatom communities: Areal biomass and cellular growth rate responses. *Can. J. Fish. Aquat. Sci.* **46,** 1293–1301.
Bothwell, M. L., Suzuki, K. E., Bolin, M. K., and Hardy, F. J. (1989). Evidence of dark avoidance by phototrophic periphytic diatoms in lotic systems. *J. Phycol.* **25,** 85–94.
Bott, T. L., Brock, J. T., Dunn, C. S., Naiman, R. J., Ovink, R. W., and Petersen, R. C. (1985). Benthic community metabolism in four temperate stream systems: An inter-volume comparison and evaluation of the river continuum concept. *Hydrobiologia* **123,** 3–45.
Cattaneo, A., Ghittori, S., and Vendegna, V. (1975). The development of benthonic phytocoenosis on artificial substrates in the Ticino River. *Oecologia* **19,** 315–327.
Chessman, B. C., Hutton, P. E., and Burch, J. M. (1992). Limiting nutrients for periphyton growth in sub-alpine, forest, agricultural and urban streams. *Freshwater Biol.* **28,** 349–361.

Close, M. E., and Davies-Colley, R. J. (1990). Baseflow water chemistry in New Zealand rivers. 2. Influence of environmental factors. *N. Z. J. Mar. Freshwater Res.* **24**, 343–356.
Davies-Colley, R. J. (1990). Frequency distributions of visual water clarity in 12 New Zealand rivers. *N. Z. J. Mar. Freshwater Res.* **24**, 453–460.
Douglas, B. (1958). The ecology of the attached diatoms and other algae in a stony stream. *J. Ecol.* **46**, 295–322.
Downes, B. J., Lake, P. S., and Schreiber, E. S. G. (1993). Spatial variation in the distribution of stream invertebrates: Implications of patchiness for models of community organization. *Freshwater Biol.* **30**, 119–132.
Entwistle, T. J. (1989). Macroalgae in the Yarra River Basin: Flora and distribution. *Proc. R. So. Victoria* **101**, 1–76.
Fisher, S. G., and Grimm, N. B. (1988). Disturbance as a determinant of structure in a Sonoran Desert stream ecosystem. *Verh. Int. Ver. Theor. Angew. Limnol.* **23**, 1183–1190.
Fisher, S. G., and Grimm, N. B. (1991). Streams and disturbance: Are cross-ecosystem comparisons useful? *In* "Comparative Analyses of Ecosystems—Patterns, Mechanisms and Theories" (J. Cole, G. Lovett, and S. Findlay, eds.), pp. 196–221. Springer-Verlag, New York.
Fisher, S. G., Gray, L. J., Grimm, N. B., and Busch, D. E. (1982). Temporal succession in a desert stream ecosystem following flash flooding. *Ecol. Mono.* **52**, 93–110.
Grimm, N. B., and Fisher, S. G. (1989). Stability of periphyton and macroinvertebrates to disturbance by flash floods in a desert stream. *J. North Am. Benthol. Soc.* **8**, 292–307.
Hamilton, P. B., and Duthie, H. C. (1987). Relationship between algal drift, discharge and stream order in a boreal forest watershed. *Arch. Hydrobiol.* **110**, 275–289.
Hill, W. R., and Knight, A. W. (1988). Nutrient and light limitation of algae in two northern California streams. *J. Phycol.* **24**, 125–132.
Hoagland, K. D., Rosowski, J. R., Gretz, M. R., and Roemer, S. C. (1993). Diatom extracellular polymeric substances: Function, fine structure, chemistry, and physiology. *J. Phycol.* **29**, 537–566.
Holmes, N. T. H., and Whitton, B. A. (1981). Phytobenthos of the River Tees and its tributaries. *Freshwater Biol.* **11**, 139–168.
Horner, R. R., Welch, E. B., and Veenstra, R. B. (1983). Development of nuisance periphytic algae in laboratory streams in relation to enrichment and velocity. *In* "Periphyton of Freshwater Ecosystems" (R. G. Wetzel, ed.), pp. 121–134. Dr. W. Junk Publishers, The Hague.
Horner, R. R., Welch, E. B., Seeley, M. R., and Jacoby, J. M. (1990). Responses of periphyton to changes in current velocity, suspended sediment and phosphorus concentration. *Freshwater Biol.* **24**, 215–232.
Hudon, C., and Bourget, E. (1981). Initial colonization of artificial substrate: Community development and structure studied by scanning electron microscopy. *Can. J. Fish. Aquat. Sci.* **38**, 1371–1384.
Humphrey, K. P., and Stevenson, R. J. (1992). Responses of benthic algae to pulses in current and nutrients during simulations of subscouring spates. *J. North Am. Benthol. Soc.* **11**, 37–48.
Hynes, H. B. N. (1970). "The Ecology of Running Waters." Univ. of Toronto Press, Toronto.
Jones, J. R., Smart, M. M., and Burroughs, J. N. (1984). Factors related to algal biomass in Missouri Ozark streams. *Verh.—Int. Ver. Theor. Angew. Limnol.* **22**, 1867–1875.
Jones, R. C. (1978). Algal biomass dynamics during colonization of artificial islands: Experimental results and a model. *Hydrobiologia* **59**, 165–180.
Jowett, I. G., and Duncan, M. J. (1990). Flow variability in New Zealand rivers and its relationship to instream habitat and biota. *N. Z. J. Mar. Freshwater Res.* **24**, 305–318.
Kaufman, L. H. (1979). Stream aufwuchs accumulation processes: Effects of ecosystem depopulation. *Hydrobiologia* **70**, 75–81.
Kjeldsen, K. (1994). The relationship between phosphorus and peak biomass of benthic algae in small lowland streams. *Verh.—Int. Ver. Theor. Angew. Limnol.* **25**, 1530–1533.

Korte, V. L., and Blinn, D. W. (1983). Diatom colonization of artificial substrata in pool and riffle zones studied by light and scanning electron microscopy. *J. Phycol.* **19,** 332–341.

Krecji, M. E., and Lowe, R. L. (1986). Importance of sand grain mineralogy and topography in determining micro-spatial distribution of epipsammic diatoms. *J. North Am. Benthol. Soc.* **5,** 211–220.

Leland, H. (1995). Distribution of phytobenthos in the Yakima River Basin, Washington, in relation to geology, land use and other environmental factors. *Can. J. Fish. Aquat. Sci.* **52,** 1108–1129.

Lock, M. A. (1981). River epilithon—A light and organic energy transducer. *In* "Perspectives in Running Water Ecology" (M. A. Lock and D. D. Williams, eds.), pp. 3–40. Plenum, New York and London.

Lohman, K., Jones, J. R., and Perkins, B. D. (1992). Effects of nutrient enrichment and flood frequency on periphyton biomass in northern Ozark streams. *Can. J. Fish. Aquat. Sci.* **49,** 1198–1205.

Lyford, J. H., Jr., and Gregory, S. V. (1975). The dynamics and structure of periphyton communities in three Cascade Mountain streams. *Verh.—Int. Ver. Theor. Angew. Limnol.* **18,** 1610–1616.

Marker, A. F. H. (1976). The benthic algae of some streams in southern England. *J. Ecol.* **64,** 343–358.

McConnell, W. J., and Sigler, W. F. (1959). Chlorophyll and productivity in a mountain river. *Limnol. Oceanogr.* **4,** 335–351.

Minshall, G. W. (1978). Autotrophy in stream ecosystems. *BioScience* **28,** 767–771.

Minshall, G. W., Petersen, R. C., Cummins, K. W., Bott, T. L., Sedell, J. R., Cushing, C. E., and Vannote, R. L. (1983). Interbiome comparison of stream ecosystem dynamics. *Ecol. Monogr.* **53,** 1–25.

Moore, J. W. (1977a). Seasonal succession of algae in a eutrophic stream in southern England. *Hydrobiologia* **53,** 181–192.

Moore, J. W. (1977b). Some factors effecting algal densities in a eutrophic farmland stream. *Oecologia* **29,** 257–267.

Mulholland, P. J., Steinman, A. D., Palumbo, A. V., and DeAngelis, D. L. (1991). Influence of nutrients and grazing on the response of stream periphyton communities to a scour disturbance. *J. North Am. Benthol. Soc.* **10,** 127–142.

Necchi, O. J., and Pascoaloto, D. (1993). Seasonal dynamics of macroalgal communities in the Preto River Basin, Sao Paulo, southeastern Brazil. *Arch. Hydrobiol.* **129,** 231–252.

Nordin, R. N. (1985). "Water Quality Criteria for Nutrients and Algae (Technical Appendix)." Water Quality Unit, Resource Quality Section, Water Management Branch, British Columbia Ministry for the Environment, Victoria.

Oemke, M. P., and Burton, T.M. (1986). Diatom colonization dynamics in a lotic system. *Hydrobiologia* **139,** 153–166.

Osborne, L. L. (1983). Colonization and recovery of lotic epilithic communities: A metabolic approach. *Hydrobiologia* **99,** 29–36.

Peterson, C. G., and Stevenson, R. J. (1990). Post-spate development of epilithic algal communities in different current environments. *Can. J. Bot.* **68,** 2092–2102.

Poff, N. L., Voelz, N. J., and Ward, J. V. (1990). Algal colonization under four experimentally-controlled current regimes in a high mountain stream. *J. North Am. Benthol. Soc.* **9,** 303–318.

Power, M. E. (1990). Benthic turfs vs floating mats of algae in river food webs. *Oikos* **58,** 67–79.

Power, M. E. (1992). Hydrological and trophic controls of seasonal algal blooms in northern California rivers. *Arch. Hydrobiol.* **125,** 375–410.

Power, M. E., and Stewart, A. J. (1987). Disturbance and recovery of an algal assemblage following flooding in an Oklahoma stream. *Am. Mid. Nat.* **117,** 333–345.

Quinn, J. M., and Hickey, C. W. (1990). Magnitude of effects of substrate particle size, recent flooding and catchment development on benthic invertebrate communities in 88 New Zealand rivers. *N. Z. Mar. Freshwater Res.* **24,** 411–427.

Reiter, M. A. (1989). The effect of a developing algal assemblage on the hydrodynamics near substrates of different sizes. *Arch. Hydrobiol.* **115,** 221–244.

Rosemond, A. D. (1994). Multiple factors limit seasonal variation in periphyton in a forest stream. *J. North Am. Benthol. Soc.* **13,** 333–344.

Sabater, S., and Sabater, R. (1992). Longitudinal changes of benthic algal biomass in a Mediterranean river during two high production periods. *Arch. Hydrobiol.* **124,** 475–487.

Sagar, P. M. (1986). The effects of floods on the invertebrate fauna of a large, unstable braided river. *N. Z. J. Mar. Freshwater Res.* **20,** 37–46.

Sand-Jensen, K., Moller, J., and Olesen, B.H. (1988). Biomass regulation of microbenthic algae in Danish lowland streams. *Oikos* **53,** 332–340.

Scarsbrook, M. R., and Townsend, C. R. (1993). Stream community structure in relation to spatial and temporal variation: A habitat template study of two contrasting New Zealand streams. *Freshwater Biol.* **29,** 395–410.

Scrimgeour, G. J., and Winterbourn, M. J. (1989). Effects of floods on epilithon and benthic macroinvertebrate populations in an unstable New Zealand river. *Hydrobiologia* **171,** 33–44.

Scrimgeour, G. J., Davidson, R. J., and Davidson, J.M. (1988). Recovery of benthic macroinvertebrate and epilithic communities following a large flood, in an unstable, braided, New Zealand river. *N. Z. J. Mar. Freshwater Res.* **22,** 337–344.

Shamsudin, L., and Sleigh, M. A. (1994). Seasonal changes in composition and biomass of epilithic algal floras of a chalk stream and a soft water stream with estimates of production. *Hydrobiologia* **273,** 131–146.

Sheath, R. G., and Cole, K. M. (1992). Biogeography of stream macroalgae in North America. *J. Phycol.* **28,** 448–460.

Statzner, B., and Higler, B. (1986). Stream hydraulics as a major determinant of benthic invertebrate zonation patterns. *Freshwater Biol.* **16,** 127–139.

Statzner, B., Gore, J.A., and Vincent, H. R. (1988). Hydraulic stream ecology: Observed patterns and potential applications. *J. North Am. Benthol. Soc.* **7,** 307–360.

Stevenson, R. J. (1983). Effects of current and conditions simulating autogenically changing microhabitats on benthic diatom immigration. *Ecology* **64,** 1514–1524.

Stevenson, R. J. (1990). Benthic algal community dynamics in a stream during and after a spate. *J. North Am. Benthol. Soc.* **9,** 277–288.

Stevenson, R. J., and Hashim, S. (1989). Variation in diatom community structure among habitats in sandy streams. *J. Phycol.* **25,** 678–686.

Stevenson, R. J., and Peterson, C. G. (1989). Variation in benthic diatom (Bacillariophyceae) immigration with habitat characteristics and cell morphology. *J. Phycol.* **25,** 120–129.

Stevenson, R. J., and Peterson, C. G. (1991). Emigration and immigration can be important determinants of benthic diatom assemblages in streams. *Freshwater Biol.* **26,** 279–294.

Suren, A. M. (1992). Enhancement of invertebrate food resources by bryophytes in New Zealand alpine headwater streams. *N. Z. J. Mar. Freshwater Res.* **26,** 229–239.

Tett, P., Gallegos, C., Kelly, M. G., Hornberger, G. M., and Cosby, B. J. (1978). Relationships among substrate, flow, and benthic microalgal pigment density in the Mechums River, Virginia. *Limnol. Oceanogr.* **23,** 785–797.

Trotter, D. M., and Hendricks, A. C. (1979). Attached, filamentous algal communities. *In* "Methods and Measurements of Periphyton Communities: A Review" (R. L. Weitzel, ed.), pp. 58–69. Am. Soc. Tes. Mater., Philadelphia.

Uehlinger, U. (1991). Spatial and temporal variability of the periphyton biomass in a prealpine river (Necker, Switzerland). *Arch. Hydrobiol.* **123,** 219–237.

Vannote, R. L., Minshall, G. W., Cummins, K. W., Sedell, J. R., and Cushing, C. E. (1980). The river continuum concept. *Can. J. Fish. Aqua. Sci.* **37**, 130–137.

Vymazal, J. (1988). The use of periphyton communities for nutrient removal from polluted streams. *Hydrobiologia* **166**, 225–237.

Welch, E. B., Horner, R. R., and Patmont, C. R. (1989). Prediction of nuisance periphytic biomass: A management approach. *Water Res.* **23**, 401–405.

Wharfe, J. R., Taylor, K. S., and Montgomery, H. A. C. (1984). The growth of *Cladophora glomerata* in a river receiving sewage effluent. *Water Res.* **18**, 971–979.

Whitton, B. A. (1975). Algae. *In* "River Ecology" (B.A. Whitton, ed.), pp. 81–105. Univ. of California Press, Berkeley and Los Angeles.

Wiley, M. J., Osborne, L. L., and Larimore, R.W. (1990). Longitudinal structure of an agricultural prairie river system and its relationship to current stream ecosystem theory. *Can. J. Fish. Aquat. Sci.* **47**, 373–384.

Zuur, B., ed. (1992). Water Quality Guidelines No. 1: Guidelines for the Control of Undesirable Biological Growths in Water." New Zealand Ministry for the Environment, Wellington.

3

Periphyton Patterns in Lakes

Rex L. Lowe

Department of Biological Sciences
Bowling Green State University
Bowling Green, Ohio 43403
and University of Michigan Biological Station
Pellston, Michigan 49769

I. INTRODUCTION

Investigations of microbial producers in lentic ecosystems (lakes, ponds, deep-water wetlands, etc.) have focused largely on phytoplankton, the suspended component. This fact is understandable given the relative ease of plankton collection and the general perception that planktonic algae are the dominant primary producers of lentic ecosystems. The record of published research in leading aquatic journals *(Limnology and Oceanography, Hydrobiologia, Great Lakes Journal of Research, Freshwater Biology,* and *International Revue der Gesampten Hydrobiologie)* reflects this plankton research bias, with phytoplankton-based research papers outnumbering periphyton-based research papers by about 20 to 1 over the past five years.

Yet in most lakes where a large portion of the bottom area receives enough light to support photosynthesis (littoral zone, the area of substratum above the compensation level), periphyton can dominate carbon fixation (Wetzel, 1964; Hargrave, 1969; Wetzel *et al.*, 1972; Bjöjk-Ramberg and Änell, 1985; Wetzel, Chapter 20, this volume). Even in lakes with relatively narrow littoral zones, the role of periphyton in whole-lake carbon budgets can be substantial (Loeb and Reuter, 1981; Loeb *et al.*, 1983; Heath, 1988; Layne, 1990).

This chapter explores the nature of lentic periphyton. First, the lentic habitat and flora will be examined with the following questions in mind: Where are the periphyton microhabitats and what is their nature? Is there a distinct lentic subset of periphyton? Next, an overview of influential parameters affecting periphyton and particularly lentic periphyton will be presented, followed by a detailed examination of important biotic and abiotic factors that potentially regulate periphyton structure and function in lentic ecosystems.

II. THE LENTIC PERIPHYTON COMMUNITY

A. Habitat

Environmental variables influencing periphyton communities such as light, turbulence, water chemistry, and grazing pressure covary on both temporal and spatial scales in lentic habitats. The range of temporal variability within a lake is largely a function of local climate. For example, permanently frozen amictic lakes of Antarctica (Heath, 1988) provide periphyton habitats that are extremely different from those of a dimictic lake in the temperate zone (Kingston *et al.*, 1983). Polar amictic lakes experience 6-month intervals of light and dark and, except for occasional moat formation, are permanently under ice. Temperate dimictic lakes receive light every day but can experience great seasonal temperature changes. The range of seasonal temperature fluctuation varies among periphyton microhabitats along a depth gradient.

Basin morphometry can also strongly influence the quantity and quality of lentic periphyton. Deep, steep-sided lakes such as Lake Tahoe (Loeb *et al.*, 1983) provide a lower percentage of whole-lake periphyton habitat space than do shallow lakes with gradually sloping littoral zones (Wetzel, 1964).

Spatial variation in the distribution of the substratum of periphyton microhabitats in lentic ecosystems is largely a function of the sorting action of waves. The substratum near the lake margin or below shallow shoals is generally coarse and composed of rocks of a variety of sizes. Or, there may be a significant quantity of coarse sand that is regularly transported by waves. This is also the area where aquatic plants might provide additional microhabitat. In general this shallow (upper eulittoral) zone is more varied in microhabitats than are deeper areas of the lake, where the substratum tends to be more uniform. Below the zone of high-energy wave action, particle size of the substratum becomes finer and at the greatest depths consists of very fine silts and detritus. These varied substrata strongly influence the distribution and abundance of periphyton populations.

B. The Biota

Lentic periphyton communities are composed of both autotrophic (algae) and heterotrophic (fungi and bacteria) components that are often in tightly linked and highly structured communities. The algae are usually dominated by diatoms (Bacillariophyceae), green algae (Chlorophyceae), and blue-green algae (Cyanophyceae) with an occasional representation of red algae (Rhodophyceae). Although all of these algal classes occur in mixed communities, the upper eulittoral zone is often dominated by green algae and diatoms and the sublittoral by blue-green algae and diatoms (Loeb and Reuter, 1981) (Fig. 1).

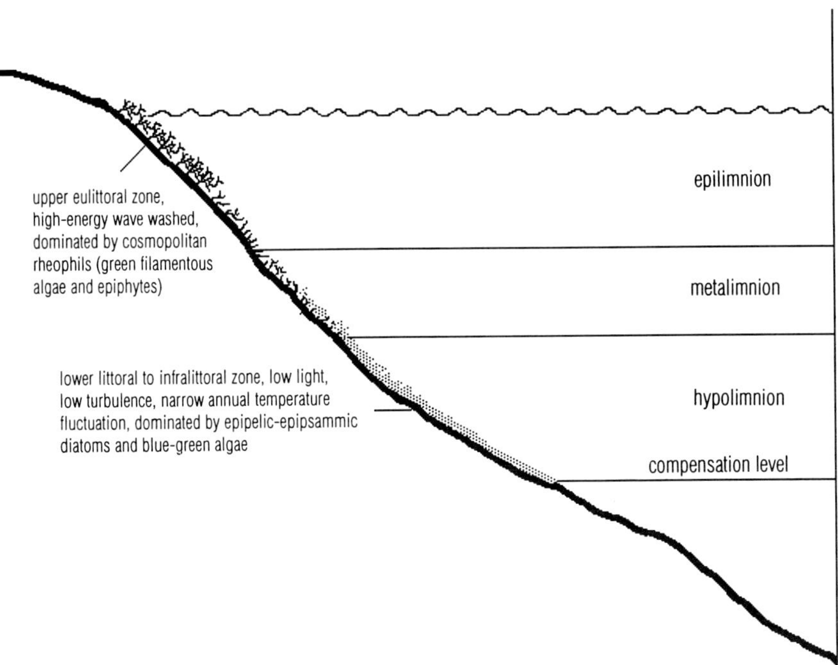

FIGURE 1 Periphyton distribution and habitat differences through depth in a thermally stratified lake.

The upper eulittoral zone, and especially that portion experiencing regular high-energy wave disturbance, is dominated by algae capable of attaching tightly to the substratum. This community often contains large numbers of rheophilous taxa and is somewhat analogous to periphyton of well-lighted streams. Dominant taxa change seasonally (Blum, 1982) as the epilimnion, at least in temperate climates, experiences strong seasonal temperature changes. In hard-water lakes, dominant filamentous genera include the green algae *Cladophora, Ulothrix,* and *Oedogonium* and occasionally the red alga *Bangia* (Lin and Blum, 1977; Auer *et al.,* 1983; Bohr *et al.,* 1983; Garwood, 1982; Millner and Sweeney, 1982; Millner *et al.,* 1982; Lorenz and Herdendorf, 1982). Filaments with coarse cellulose cell walls such as *Cladophora* support a complex community of epiphytes in the eulittoral zone (Fig. 2) (Sheath and Morison, 1982; Stevenson and Stoermer, 1982; Lowe *et al.,* 1982, Lowe *et al.,* 1984a).

The presence of *Cladophora* can increase the functional surface area of the littoral zone by a factor of >2000 (Taft and Kishler, 1973), thereby creating important microhabitat space for epiphytic microbes and the community of invertebrates that they support (Chilton *et al.,* 1986). The dominant algal epiphyton and epilithon of the eulittoral zone include diatoms *(Cocconeis pediculus, Diatoma vulgare, Gomphoneis olivacea, Cymbella prostrata* v. *auerswaldii, Rhoicosphenia curvata,* and species of *Gomphonema, Amphora, Navicula,* and *Achnanthidium),* the red alga *Chroodactylon ramosum,* and blue-green algae *(Chamaesiphon incrustans, Fischerella muscicola* and *Lyngbya diguetii)* (Obeng-Asamoa *et al.,* 1980; Stevenson and Stoermer 1981, 1982; Kingston *et al.,* 1983; Sheath and Morison, 1982).

In small soft-water lakes the eulittoral zone is normally dominated by desmids, filamentous members of the Zygnemataceae (Chlorophyceae), blue-green algae such as *Tolypothrix,* and the diatom genera *Eunotia, Frustulia,* and *Tabellaria* (Woelkerling and Gough, 1976; Brown, 1976; Bohr *et al.,* 1983).

In larger oligotrophic soft-water lakes such as Lake Tahoe (Aloi *et al.,* 1988) or the Canadian shield lakes (Stockner and Armstrong, 1971), the eulittoral zone may be dominated by diatoms rather than filamentous green algae. In Lake Tahoe (Aloi *et al.,* 1988), stalked diatoms such as *Gomphoneis herculeana* dominate the eulittoral zone, whereas in smaller lakes (Stockner and Armstrong, 1971) adnate diatoms such as *Achnanthidium* and *Anomoeoneis* (=*Brachysira*) dominate. Still, even in soft-water oligotrophic systems, the flora of the upper littoral zone displays remarkable taxonomic and physiognomic similarity to that of lotic ecosystems.

Below the upper eulittoral zone, where wave energy and light quantity decline and light quality shifts, the periphyton community structure also shifts (Stockner and Armstrong, 1971; Loeb and Reuter, 1981; Stevenson

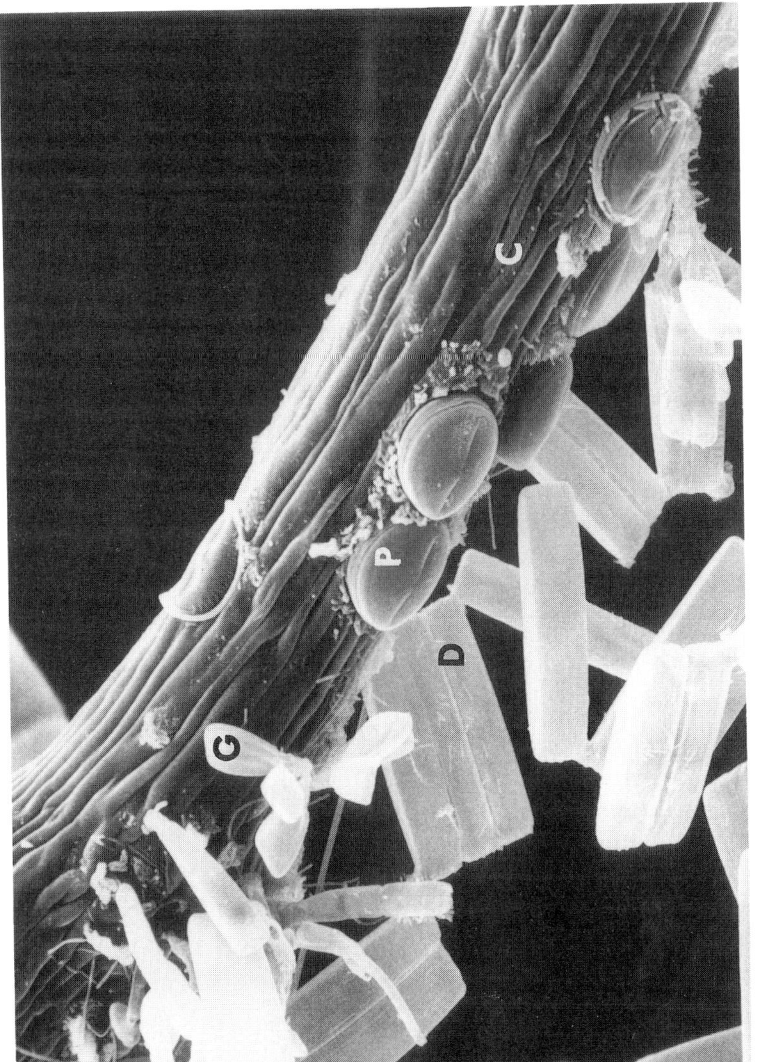

FIGURE 2 Scanning electron micrograph of *Cladophora glomerata* (C) and the associated epiphytes *Diatoma vulgare* (D), *Cocconeis pediculus* (P), and *Gomphoneis olivacea* (G). (Courtesy of John Kingston.)

and Stoermer, 1981; Kingston *et al.,* 1983). The deep littoral habitat below the maximum penetration of the epilimnion is characterized by dim light, relatively little wave-generated turbulence, and thermal stability. There have been relatively few investigations of periphyton in the deep littoral to infralittoral zones of lakes, but data suggest the presence of a highly diverse periphyton community. Rodriguez (1992) studied the regulation of this community through experiments in which periphyton communities were transplanted from the upper to the lower littoral zone of an oligotrophic Canadian lake. Even though his analyses involved a relatively low level of taxonomic resolution, he found that the upper littoral community converged in structure with lower littoral communities within about 11 weeks. This suggests that the deep-water periphyton communities experience very different patterns of environmental variables than the shallow littoral zone communities.

An investigation by Kingston *et al.* (1983) in Lake Michigan, U.S.A., found annual temperature fluctuations at a depth of 30 m of from 1° in winter to 8° during autumnal mixing. During the same period the eulittoral zone varied from 0 to 20°. These deep, thermally stable, lentic habitats support a rich species assemblage. In the northern part of Lake Tanganyika, Caljon and Cocquyt (1992) reported 277 diatom taxa from the surface sediments, and Kingston *et al.* (1983) reported 425 diatom taxa from deep benthic collections in Grand Traverse Bay, Lake Michigan. In both of these investigations, planktonic species, representing fallout from the water column, were rare and the majority of species were truly benthic, although Stevenson and Stoermer (1981) reported that benthic algal communities at depths of 22 to 27 m were dominated by fallout of planktonic species.

Roberts and Boylen (1988) and Loeb and Reuter (1981), on the other hand, reported the deep littoral zone to be dominated by filamentous and colonial blue-green algae such as *Hapalosiphon, Calothrix, Tolypothrix, Lyngbya,* and *Gloeocapsa,* in soft-water lakes. This unique and poorly studied deep-water periphyton habitat has been eliminated in many lakes, where increased plankton densities responding to nutrient loading prevent the euphotic zone from penetrating into the hypolimnion. Kuhn *et al.* (1981) presented a model of diatom life-form strategies in aquatic ecosystems that predicts maximum benthic algal development in habitats with low water column nutrient availability and low disturbance. Deep littoral and infralittoral zones of oligotrophic lakes exemplify such habitats.

It appears, then, that the unique species-rich community living below the maximum penetration of the summer thermocline is a distinct subset of lentic periphyton. The periphyton community occupying the wave-washed upper littoral zone, on the other hand, is not unique and contains many cosmopolitan species found in both streams and lakes. A summary of the distribution of lentic algal assemblages is presented in Table I.

TABLE I. Patterns of Algal Distribution in Lakes

Oligotrophic hard water[a]	*Oligotrophic soft water*[b]	*Eutrophic hard water*[c]
Upper eulittoral zone (high light, high turbulence, broad temperature fluctuation)		
Filaments and epiphytes	Filaments, desmids, and diatoms	Filaments and epiphytes
Ulothrix sp.	*Mougeotia* sp.	*Cladophora glomerata*
Ulothrix zonata	*Cosmarium* spp.	*Bangia atropurpurea*
Oedogonium spp.	*Staurastrum* spp.	*Ulothrix zonata*
Gomphonema spp.	*Gomphoneis herculeana*	*Rhoicosphenia curvata*
Cocconeis pediculus	*Synedra ulna* v. *spathulifera*	*Navicula tripunctata*
Rhoicosphenia curvata	*Eunotia pectinalis*	*Cymbella prostrata* v. *auerswaldii*
Diatoma vulgare	*Tabellaria fenestrata*	*Cocconeis pediculus*
Gomphoneis olivacea	*Tabellaria flocculosa*	*Rhoicosphenia curvata*
Lyngbya diguetii		
Deep littoral to infralittoral zone (low light, low turbulence, narrow temperature fluctuation)		
Epipelic diatoms	Blue-green and diatoms	Few taxa, below the photic zone
Achnanthidium inconspicua	*Calothrix* sp.	
Martyana ansata	*Tolypothrix* sp.	
Amphora ovalis	*Nostoc* sp.	
Diploneis petersenii	*Hapalosiphon pumilis*	
Fragilaria construens	*Navicula tenuicephala*	
Fragilaria brevistriata	*Eunotia* spp.	

[a]From Stevenson and Stoermer (1981) and Kingston *et al.* (1983).
[b]From Stockner and Armstrong (1971), Loeb and Reuter (1981), Roberts and Boylen (1988), and Aloi *et al.* (1988).
[c]From Stevenson and Stoermer (1982), Lowe *et al.* (1982), Sheath and Morison (1982), and Blum (1982).

III. IMPORTANT INFLUENTIAL FACTORS

A. Overview

The quantity and quality of periphyton in lentic ecosystems are influenced directly and indirectly by a broad array of abiotic and biotic factors. Because of the pivotal trophic position of periphyton at the interface of the abiotic and biotic components of the ecosystem, the interactions among these parameters can be quite complex. This section will address two related sets of these parameters individually, while maintaining an awareness that the parameters are not independent.

The first set of influential parameters falls under the heading of resources. Resources are defined as consumable requisites for maintenance and growth of a population (Tilman, 1982). Periphyton resources include light, nutrients (inorganic in most instances), and space. All of these resources are required and consumable, and competition for limiting resources can play a strong role in structuring lentic periphyton communities.

Disturbance constitutes the second set of parameters. Disturbance includes forms of abiotic (mechanical and chemical) phenomena such as turbulence, abrasion, and toxic chemicals, and biotic phenomena such as grazing by heterotrophic organisms.

B. Resources

1. Space

Space is not often considered a resource, but it is a requisite for periphyton growth and is consumable (see Burkholder, Chapter 9). Fisheries biologists have long been aware that space (structured surface area) can be a limiting factor to fish production. Periphyton too can be space-limited. Crowded rain forest-like communities exist in nutrient- and light-rich habitats with periphyton employing diverse physiognomies leading to architecturally complex communities. Hoagland *et al.* (1982) described the successional sequence of events in a space-limited habitat leading to "mature," structurally complex, three-dimensional communities.

In broad, shallow littoral zones with ample nutrients, such as the western basin of Lake Erie (U.S.A.), periphyton appear to be space-limited. Practically every suitable substratum (rocks) in the littoral zone is covered by *Cladophora glomerata* (Taft and Kishler, 1973) and, as mentioned earlier, the *Cladophora* itself serves as substratum space, increasing the functional littoral 2000 times! Thus, in lentic systems, space occasionally appears to be a limiting resource, particularly in the upper littoral zone.

2. Nutrients

Benthic algae require a large number of inorganic nutrients, including carbon, oxygen, hydrogen, nitrogen, phosphorus, silicon, potassium, sulfur, calcium, iron, manganese, copper, and several trace metals (Darley, 1982; Borchardt, Chapter 7). Some facultative and obligate heterotrophic taxa also utilize specific organic nutrients (Cooksey and Cooksey, 1988; Tuchman, Chapter 10). The abundance of nutrients in lentic ecosystems plays a strong role, both directly and indirectly, in determining the quantity, quality, and distribution (spatially and temporally) of periphyton.

Inorganic sources of nutrients in the water column of lakes may come either from atmospheric deposition, from surface or subsurface inflow, or from sediment release with subsequent turbulent mixing. Regardless of the source, water column nutrients are in more intimate contact with phytoplankton than with periphyton. In situations with relatively abundant nutrients, phytoplankton proliferations can shade periphyton and render them light-limited rather than nutrient-limited (Hansson, 1992). Bjöjk-Ramberg and Änell (1985) demonstrated this phenomenon elegantly in whole-lake fertilization experiments in subarctic Swedish lakes. The control lake, Lake Stugsjön, received no fertilizer and maintained relatively high periphyton densities that constituted 70–83% of the total lake primary production, whereas the contribution of benthic algae to primary production in the lake receiving phosphorus and nitrogen, Lake Hymenjaure, fell from 50% to 22% after fertilization. Periphyton in Lake Hymenjaure became shaded by phytoplankton following fertilization.

Cattaneo (1987) further demonstrated that the linkage between water column nutrients and algal biomass was much stronger for phytoplankton than for periphyton. However, periphyton may respond weakly to water column nutrients (Hansson, 1992; Fairchild and Sherman, 1992) and, in some instances, signal early stages of eutrophication before phytoplankton (Kann and Falter, 1989). Cattaneo (1987) found that only large or filamentous periphyton taxa were positively correlated with total water column phosphorus and that there was no communitywide response in an investigation of 10 lakes in Quebec, Canada.

The lack of a strong direct response from periphyton to increases in water column nutrients may be a function of the compactness of the periphyton community and steep resource gradients within the community. Bothwell (1988) demonstrated a two-fold reduction in water column nutrients within a lotic benthic algal mat. Gradients in lentic systems with little flow over the substratum should be considerably steeper. Other investigations have clearly demonstrated that periphyton mat chemistry is extremely different from water column chemistry (Revsbech and Jørgensen, 1986; Revsbech *et al.*, 1983), especially for periphyton in epipelic habitats (common in lentic systems). Riber and Wetzel (1987) and Carlton and Wetzel (1988) demonstrated that epipelic periphyton communities play significant roles in the release of sediment-associated phosphorus and in so doing generate phosphorus supplies for their own metabolic needs.

Phytoplanktonic species of algae can also exploit sedimentary phosphorus sources. Barbiero and Welch (1992) demonstrated that planktonic colonies of *Gloeotrichia echinulata* were heavily subsidized by benthic recruitment, which contributed a significant portion of the internal phosphorus loading to the lake.

Under circumstances of gross or chronic nutrient loading into littoral zones, periphyton may respond more strongly to water column nutrients. Goldman (1981) implicated nutrient loading into Lake Tahoe occurring over two decades in increases in benthic algal biomass. Hawes and Smith (1992) documented the same changes observed by Goldman (1981) (significant increases in *Gomphoneis herculeana* and *Synedra ulna*) in the littoral zone of Lake Taupo, New Zealand, following release of secondarily treated sewage effluent.

Periphyton experiencing resource-rich patches of water may also respond physiologically. Stevenson and Stoermer (1982) documented the accumulation of polyphosphate bodies by algal species epiphytic on *Cladophora* in a phosphorus plume in Lake Huron. Cells within a population nearest the source of the plume had relatively more polyphosphate bodies than cells farther from the plume.

Great progress in understanding periphyton–nutrient interactions has been made in the past decade following the adoption of two new techniques. First, the availability of microelectrodes (Revsbech and Jørgensen, 1983, 1986; Revsbech *et al.,* 1983) has facilitated the measurement of nutrient resource gradients over relatively short distances. Second, *in situ* nutrient manipulation experiments have facilitated a better understanding of nutrient resource competition within periphyton communities (Pringle and Bowers, 1984; Fairchild and Lowe, 1984; Fairchild *et al.,* 1985; Carrick and Lowe, 1988).

It is generally accepted that substratum or algal mat chemistry plays a much stronger role than water column chemistry in periphyton biology (Cattaneo, 1987; Bothwell, 1988; Hansson, 1988, 1992). These new methodologies have enabled researchers to measure and manipulate nutrient resource gradients on ecologically relevant scales and to manipulate the relative availability of those nutrients to periphyton biofilms. This has led to a greater appreciation of the algal mat as a true community of species, with each species responding differently to resource availability, rather than a homogeneous biofilm responding in unison to ambient environmental conditions.

Resource specialists, that is, species capable of securing resources that are present at concentrations limiting to sympatric species, within lentic periphyton communities have been identified. Fairchild *et al.* (1985) documented phosphorus specialists *(Achnanthidium minutissimum* and *Gomphonema tenellum),* nitrogen specialists *(Epithemia adnata, Rhopalodia gibba* and *Anabaena* sp.*)* and light–space specialists that require relatively high concentrations of both nitrate and phosphate *(Stigeoclonium tenue* and naviculoid diatoms). Carrick *et al.* (1988) described four species guilds of resource specialists in periphyton communities from northern Lake Michigan (U.S.A.). Resource specialists usually employ special mechanisms in resource exploitation. Space–light specialists, for example, are often filamentous or motile and can overgrow less competitive taxa, and nitrogen specialists fix elemental nitrogen (Lowe *et al.,* 1984b; DeYoe *et al.,* 1992). Lentic microhabitats exist where resource specialists exercise a competitive advantage. For example, *Epithemia adnata,* a nitrogen specialist, often dominates epiphytic microhabitats where the ratio of available nitrogen to phosphorus might be relatively low (Marks and Lowe, 1993).

While keeping in mind that nutrient resource limitation operates at the species level and that whole-community response to nutrient manipulation represents the sum of species responses, it is still informative to examine the response of periphyton biomass to nutrient manipulations. Nutrient manipulations in lentic ecosystems have revealed total periphyton biomass to be limited by phosphorus (Burkholder and Cuker, 1991; Fairchild and Sherman, 1992), nitrogen (Fairchild and Sherman, 1992; Marks and Lowe, 1993), carbon (Fairchild and Sherman, 1990, 1992, 1993), and nitrogen and phosphorus (Carrick and Lowe, 1988; Marks and Lowe, 1989). Carbon limitation appears to be most acute in lakes with alkalinity <100 μeq/liter (Fairchild and Sherman 1990, 1992, 1993). It must be remembered, though, that the response of periphyton to nutrient resources is influenced by many other variables in the lentic ecosystem.

3. Light

Light must be addressed on two spatial scales in lentic ecosystems. Light is attenuated with depth in the water column and light is also attenuated with depth as it enters the periphyton mat (see Hill, Chapter 5). Unlike phytoplankton, which may actively or passively experience changes in depth, lentic periphyton are more likely to display light-mediated differences in community structure and function along a depth gradient. Attenuation of light with depth is strongly affected by suspended solids, including phytoplankton.

In oligotrophic lakes, the photic zone may extend well below the summer thermocline. For example, in Lake Tahoe (Nevada/California border), approximately 1% of the light penetrates to 60 m (Loeb *et al.*, 1983). As mentioned earlier, the deep-water periphyton community occupying this habitat has been poorly studied, however, the role of light in the upper littoral zone has received considerable attention.

In the littoral zone of Lake 239 in the Experimental Lakes Area of northwestern Ontario, Canada, Turner *et al.* (1983) found photosynthesis to be light-saturated at about 200 $\mu E\ m^{-2}\ s^{-1}$ and Lorenz *et al.* (1991) found that average daily light intensities below 28 $\mu E\ m^{-2}\ s^{-1}$ were limiting for *Cladophora* colonization in Lake Erie.

Although many investigators have implicated light as an important variable affecting the community composition of periphyton through depth gradients, the evidence has largely been correlative in nature (Kingston *et al.*, 1983; Stevenson and Stoermer, 1982). *In situ* experiments where light quantity and/or quality are manipulated in lentic habitats while holding other variables constant are rare (Hudon and Bourget, 1983; Marks and Lowe, 1993; van Dijk, 1993). This fact undoubtedly attests to the reality that such experiments are logistically difficult to execute and usually require SCUBA-assisted experiments if done *in situ*.

Marks and Lowe (1993) manipulated light with shade cloth at a depth of 3 m in the littoral zone of oligotrophic Flathead Lake (Montana, U.S.A.). Their most intensive shade treatment reduced light levels to only 270 $\mu E\ m^{-2}$, well above levels found to be saturating by Turner *et al.* (1983) (200 $\mu E\ m^{-2}\ s^{-1}$), and resulted in no significant changes in the periphyton community. Hoagland and Peterson (1990) manipulated periphyton communities in a large reservoir by shifting the depth of substrata in a reciprocal transplant experiment. Although they identified sets of algal taxa that exhibited definite depth preferences, the causative mechanism was obscured by the fact that light and disturbance covaried along the depth gradient. In a combination shading–transplant experiment, Hudon and Bourget (1983) concluded that periphyton community structure, physiognomy, and density were dependent on both depth and light intensity.

Finally, periphyton communities themselves attenuate light very strongly (Losee and Wetzel, 1983), and in epiphytic communities have been shown to severely modify light quality reaching the host plant (Sand-Jensen and Søndergaard, 1981; Bulthuis and Woelkerling, 1983; van Dijk, 1993). Fiber-optic microprobes have made it possible to measure clines of light within periphyton mats at a resolution scale of 100 μm (Jørgensen and Des Marais, 1986; Jørgensen and Nelson, 1988; Dodds, 1989). Dodds (1992) has employed a modified light probe that measured spherically integrated irradiance to examine light penetration in two structurally dissimilar periphyton communities. Light was shown to penetrate to greater depths in a *Ulothrix*-dominated periphyton mat than in a mat dominated by diatoms, although it must be remembered that periphyton communities are patchy and spatial variability within the mat leads to variability in the distribution of light. Losee and Wetzel (1983), for example, showed that calcium carbonate crystals, weakly pigmented algal cells, and open space can generate light gaps in periphyton communities.

C. Disturbance

Disturbance is a relative term. What might constitute disturbance to one population could act neutrally or positively to others (see Peterson, Chapter 13). In examining the role of disturbance in lentic periphyton communities, I will address events that generally disrupt the community by exceeding the bounds of natural environmental variability within the habitat. Routine high-energy wave activity in the upper portions of the littoral zone would not constitute disturbance under this definition. Periphyton communities in this habitat develop under the influence of frequent turbulence. Turbulent water movement over periphyton communities in the hypolimnion or excessive wave action associated with storm events (Young, 1945; Fox *et al.*, 1969; Kairesalo, 1983), however, would constitute disturbance. The following discussion of the role of disturbance in lentic periphyton communities includes abiotic, mechanical, and chemical disturbance (turbulence, abrasion, toxic chemicals, etc.) and biotic disturbance (grazing and movement by heterotrophs).

1. Abiotic, Mechanical

Water turbulence (wave action) in lakes has been shown to have a positive relationship with periphyton biomass, especially on filamentous algae and long-stalked diatoms (Cattaneo, 1990). However, when periphyton communities that have developed in relatively nonturbulent habitats are exposed to wave action, some populations are affected adversely. Hoagland (1983) examined the impact of an episodic storm event and subsequent increase in turbulence on periphyton community structure in a eutrophic reservoir and found that older communities lost a greater percentage of their biomass than younger ones, but that community diversity was not significantly affected. In a later study, Hoagland and Peterson (1990) identified two diatom taxa, *Achnanthidium minutissimum* and *Cymbella mexicana*, that appear to be tolerant of turbulence disturbance. They noted, as did Cattaneo (1990), that stalked diatoms appear to tolerate wave turbulence better than do nonstalked forms. Stalks attach diatoms tightly to substrata yet remain flexible enough to tolerate turbulence.

2. Abiotic, Chemical

Chemical disturbances (chemical extremes beyond normal aquatic chemical variability) are usually anthropogenic in nature. Chemical disturbance to periphyton in lentic systems has traditionally been less severe and less well studied than in lotic systems. This phenomenon probably reflects the generally greater volume to substratum ratio of lakes and thus their ability to dilute quantities of chemicals that would exceed the tolerance of periphyton in lotic ecosystems. The implications of excessive nutrient loading have already been discussed (phytoplankton stimulation and subsequent shading of the periphyton community), thus chemical disturbance of lentic periphyton communities that is nonnutrient in nature will be considered here. See Genter (Chapter 14) and Hoagland *et al.* (Chapter 15) for more detailed treatment of the impact of toxic substances on periphyton communities.

The focus of this chapter is the pattern of periphyton distribution in lakes and there do not appear to be circumstances that make lentic periphyton differ from lotic periphyton in vulnerability to toxic substances. Studies on the impact of toxic substances on natural periphyton communities are rare, although there are scattered examples in the literature (Austin, 1983; Tuchman *et al.*, 1984). However, the abundant literature on the impact of toxic substances on periphyton that has been generated in microcosms or mesocosms can be cautiously extrapolated to natural lentic ecosystems (Genter *et al.*, 1987; Belanger *et al.*, 1994).

Arguably, the most well-studied and well-documented class of chemical disturbance in lakes is the atmospheric deposition of substances leading to acidification (see Smol *et al.,* 1986, for a general review). Acidification of poorly buffered lakes appears to impact the phytoplankton community more severely than the periphyton community (Pillsbury and Kingston, 1990). In fact, a general phenomenon observed in acidified lakes has been a decrease in phytoplankton abundance, leading to increased light penetration and a deepening of the photic zone, resulting in increased periphyton biomass (Stevenson *et al.,* 1985).

Diatom frustules and other autochthonous microfossils have been employed extensively in paleolimnological investigations and a generally observed phenomenon in cores from lakes that have become acidified is a relative decrease in planktonic species and a relative increase in periphyton species (Charles, 1985). This phenomenon may be a function of the difference between the higher hydrogen ion concentration and the associated chemistry of the water column compared to the modified chemistry of benthic periphyton mats.

3. Biotic

Biotic disturbance is usually associated with the activity of grazers in the periphyton community (see Lamberti, Chapter 17). Periphyton grazers in lentic ecosystems range in size from protozoa <100 μm in length to bottom-feeding fish several decimeters long. Most attention, however, has been focused on grazing by relatively small invertebrate taxa. Grazer–periphyton interaction has been the focus of a large number of recent investigations that are addressed thoroughly in Chapter 11 of this book. Biotic disturbance seems to lack aspects that are unique to lentic systems, but a brief consideration of the phenomenon is appropriate here.

Grazers impact periphyton communities both by direct consumption of algal cells and by dislodgment of cells from the substratum. The intensity of grazing pressure in lentic habitats is, in some instances, controlled secondarily by fish. Brönmark *et al.* (1992) demonstrated that in mesotrophic lakes of Wisconsin, periphyton biomass and community structure is regulated by top-down interactions of sunfish and snails. In the consideration of grazing activity as a disturbance, it is necessary to return to the earlier definition of disturbance as a phenomenon that exceeds the bounds of natural environmental variability within the habitat. With this in mind, the regular reduction of a periphyton community to a monolayer of *Cocconeis pediculus* (Medlin, 1981) by the continuous grazing activity of a dense population of gastropods would not constitute disturbance. However, a large gastropod infrequently passing through a mature, physiognomically complex periphyton community and removing algae in the upper portion of the community would constitute disturbance (Lowe and Hunter, 1988).

An example of biotic disturbance to lentic periphyton communities that is occurring on a relatively large scale is the habitat modifications generated by the recent invasion of the exotic zebra mussels *(Dreissena polymorpha)* into North American lakes (Nalepa and Schloesser, 1993). Since their introduction into North American waters in the late 1980s, zebra mussels have proliferated and now maintain population densities in the tens of thousands per square meter, where they modify periphyton habitat in the littoral zone. In portions of Lakes Erie and Huron, they have not only changed the nature of substratum from rock and rubble to zebra mussel shells, but their filtering activity has dramatically changed the light environment (Lowe and Pillsbury, 1995). These changes are associated with a dramatic shift from a diatom-dominated epilithic community to an epizooic community dominated by green filamentous algae (Zygnematales).

IV. CONCLUSIONS AND MISSING PIECES

In spite of several recent experimental investigations, our knowledge of lentic periphyton communities lags far behind our knowledge of lotic periphyton and even farther behind our knowledge of phytoplankton biology. Lentic periphyton is strongly influenced by its position along a depth gradient. The community structure shifts somewhat predictably through depth from epiphytic and epilithic rheophils to epipelic and epipsammic limnobionts. Because of covariation of many potentially important parameters through depth (nature of the substratum, quantity and quality of dissolved chemicals, light quantity and quality, temperature, turbulence, and grazers), the mechanisms regulating periphyton community structure through depth have been poorly elucidated. Few *in situ* research investigations have attempted to sort out these covarying parameters with the exception of several nutrient manipulation experiments. This lack of *in situ* experimentation on periphyton is probably a function of difficulties and expense encountered in working on lake bottoms (securing ship time, SCUBA, etc.), but may also be a function of difference in the quantities of lakes relative to streams in densely populated areas and, therefore, the accessibility of research sites.

Several basic but important pieces of information on lentic periphyton are still in need of much research. For example, the role of periphyton in lentic systems has been inadequately investigated. Although rates of photosynthesis and its contribution to whole-lake carbon budgets have been developed for a few lakes, the sinks and pathways for periphyton-fixed carbon are poorly known. Some grazing investigations have been undertaken but information about detailed pathways of energy transfer up the food web is rare. How important is periphyton carbon fixation to the life cycles of heterotrophs? Which taxa or assemblages of periphyton are most important in this process and to which heterotrophs?

Most of the data on lentic periphyton have been generated in the upper littoral zone, with the lower littoral to infralittoral zone remaining poorly understood. This is the zone of maximum periphyton diversity with hundreds of species packed into a seemingly uniform habitat. Why is this community so species-rich? Where are all of the niches in this relatively uniform habitat? What is the role of the deep-water periphyton in whole-lake biology? Is this a community of cosmopolitan distribution in cold, dimly lighted habitats? It is obvious that there are many critical experiments to perform and many fascinating revelations yet to be unshrouded regarding lentic periphyton communities. It is hoped that this brief review will stimulate interest in their discovery.

ACKNOWLEDGMENT

I thank John Kingston for providing the scanning electron micrograph of *Cladophora* with epiphytes.

REFERENCES

Aloi, J. E., Loeb, S. E., and Goldman, C. R. (1988). Temporal and spatial variability of the eulittoral epilithic periphyton, Lake Tahoe, California–Nevada. *J. Freshwater Ecol.* **4**, 401–410.

Auer, M. T., Graham, J. M., Graham, L. E., and Kranzfelder, J. A. (1983). Factors regulating the spatial and temporal distribution of *Cladophora* and *Ulothrix* in the Laurentian Great Lakes. *In* "Periphyton of Freshwater Ecosystems" (R. G. Wetzel, ed.), pp. 135–144. Dr. W. Junk Publishers, The Hague.

Austin, A. P. (1983). Evaluation of changes in a large oligotrophic wilderness park lake exposed to mine tailings effluent for 14 years: The periphyton. *Nat. Can. (Que.) (Rev. Écol. Syst.)* **110**, 119–134.

Barbiero, R. P., and Welch, E. B. (1992). Contribution of benthic blue-green algal recruitment to lake populations and phosphorus translocation. *Freshwater Biol.* **27**, 249–260.

Belanger, S. E., Barnum, J. B., Woltering, D. M., Bowling, J. W., Ventullo, R. M., Schermerhorn, S. D., and Lowe, R. L. (1994). Algal periphyton structure and function in response to consumer chemicals in experimental stream mesocosms. *In* "Utilization of Simulated Ecosystems in Ecological Risk Assessment" (R. L. Granney, J. H. Kennedy, and J. H. Rogers, eds.). Pergamon, New York.

Bjöjk-Ramberg, S., and Änell, C. (1985). Production and chlorophyll concentration of epipelic and epilithic algae in fertilized and nonfertilized subarctic lakes. *Hydrobiologia* **126**, 213–219.

Blum, J. L. (1982). Colonization and growth of attached algae at the Lake Michigan water line. *J. Great Lakes Res.* **8**, 10–15.

Bohr, R., Luscinska, M., and Oleksowicz, A. S. (1983). Phytosociological associations of algal periphyton. *In* "Periphyton of Freshwater Ecosystems" (R. G. Wetzel, ed.), pp. 23–29. Dr. W. Junk Publishers, The Hague.

Bothwell, M. L. (1988). Growth rate response of lotic periphytic diatoms to experimental phosphorus enrichment: The influence of temperature and light. *Can. J. Fish. Aquat. Sci.* **45**, 261–270.

Brönmark, C., Klosiewski, S. P., and Stein, R. A. (1992). Indirect effects of predation in a freshwater benthic food chain. *Ecology* **73**, 1662–1674.

Brown, H. D. (1976). A comparison of the attached algal communities of a natural and artificial substrate. *J. Phycol.* **12**, 301–306.

Bulthuis, D. A., and Woelkerling, W. J. (1983). Biomass accumulation and shading effects of epiphytes on the leaves of *Heterozostera tasmanica* in Victoria, Australia. *Aquat. Bot.* **16**, 137–140.

Burkholder, J. M., and Cuker, B. E. (1991). Response of periphyton communities to clay and phosphate loading in a shallow reservoir. *J. Phycol.* **27**, 373–384.

Caljon, A. G., and Cocquyt, C. Z. (1992). Diatoms from the surface sediments of the northern part of Lake Tanganyika. *Hydrobiologia* **230**, 135–156.

Carlton, R. G., and Wetzel, R. G. (1988). Phosphorus flux from lake sediments: Effect of epipelic oxygen production. *Limnol. Oceanogr.* **33**, 562–570.

Carrick, H. J., and Lowe, R. L. (1988). Response of Lake Michigan benthic algae to *in situ* enrichment with Si, N and P. *Can. J. Fish. Aquat. Sci.* **45**, 271–279.

Carrick, H. J., Lowe, R. L., and Rotenberry, J. T. (1988). Guilds of benthic algae along nutrient-gradients: Relationships to algal community diversity. *J. North Am. Benthol. Soc.* **7**, 117–128.

Cattaneo, A. (1987). Periphyton in lakes of different trophy. *Can. J. Fish. Aquat. Sci.* **44**, 296–303.

Cattaneo, A. (1990). The effect of fetch on periphyton spatial variation. *Hydrobiologia* **206**, 1–10.

Charles, D. F. (1985). Relationships between surface sediment diatom assemblages and lakewater characteristics in Adirondack lakes. *Ecology* **66**, 994–1011.

Chilton, E. W., Lowe, R. L., and Schurr, K. M. (1986). Invertebrate communities associated with *Bangia atropurpurea* and *Cladophora glomerata* in western Lake Erie. *J. Great Lakes Res.* **12**, 149–153.

Cooksey, B., and Cooksey, K. E. (1988). Chemical signal response in diatoms of the genus *Amphora*. *J. Cell Sci.* **91**, 523–529.

Darley, W. M. (1982). "Algal Biology: A Physiological Approach," Basic Biol. Vol. 9. Blackwell, Oxford.

De Yoe, H. R., Lowe, R. L., and Marks, J. C. (1992). Effects of nitrogen and phosphorus on the endosymbiont load of *Rhopalodia gibba* and *Epithemia turgida* (Bacillariophyceae). *J. Phycol.* 28, 773–777.

Dodds, W. K. (1989). Microscale vertical profiles of N_2 fixation, photosynthesis, O_2, chlorophyll *a* and light in a cyanobacterial assemblage. *Appl. Environ. Microbiol.* **55**, 882–886.

Dodds, W. K. (1992). A modified fiber-optic microprobe to measure spherically integrated photosynthetic photon flux density: Characterization of periphyton photosynthesis–irradiance patterns. *Limnol. Oceanogr.* **37**, 871–878.

Fairchild, G. W., and Lowe, R. L. (1984). Artificial substrates which release nutrients: Effects upon periphyton and invertebrate succession. *Hydrobiologia* **114**, 29–37.

Fairchild, G. W., and Sherman, J. W. (1990). Effects of liming on nutrient limitation of epilithic algae in an acidic lake. *Water, Air, Soil Pollut.* **52**, 133–147.

Fairchild, G. W., and Sherman, J. W. (1992). Linkage between epilithic algal growth and water column nutrients in softwater lakes. *Can. J. Fish. Aquat. Sci.* **49**, 1641–1649.

Fairchild, G. W., and Sherman, J. W. (1993). Algal periphyton response to acidity and nutrients in softwater lakes: Lake comparison vs. nutrient enrichment approaches. *J. North Am. Benthol. Soc.* **12**, 157–167.

Fairchild, G. W., Lowe, R. L., and Richardson, W. T. (1985). Algal periphyton growth on nutrient-diffusing substrates: An *in situ* bioassay. *Ecology* **66**, 465–472.

Fox, J. L., Olaug , T. O., and Olson, T. A. (1969). The ecology of periphyton in western Lake Superior. *Res. Cent. Univ. Minn.* **14**, 1–97.

Garwood, P. E. (1982). Ecological interactions among *Bangia, Cladophora* and *Ulothrix* along the Lake Erie shoreline. *J. Great Lakes Res.* **8**, 54–60.

Genter, R. B., Cherry, D. S., Smith, E. P. and Cairns, J. (1987). Algal periphyton population and community changes from zinc stress in stream mesocosms. *Hydrobiologia* **153**, 261–275.

Goldman, C. R. (1981). Lake Tahoe: Two decades of change in a nitrogen deficient oligotrophic lake. *Verh.—Int. Ver. Theor. Angew. Limnol.* **24**, 411–415.

Hansson, L. A. (1988). Effects of competitive interactions on the biomass development of planktonic and periphytic algae in lakes. *Limnol. Oceanogr.* **37**, 322–328.

Hansson, L. A. (1992). Factors regulating periphytic algal biomass. *Limnol. Oceanogr.* **33**, 121–128.

Hargrave, B. T. (1969). Epibenthic algal production and community respiration in the sediments of Marion Lake. *J. Fish. Res. Board Can.* **26**, 2003–2026.

Hawes, I. H., and Smith, R. (1992). Effect of localised nutrient enrichment on the shallow epilithic periphyton of oligotrophic Lake Taupo. *N. Z. J. Mar. Freshwater Res.* **27**, 365–372.

Heath, C. W. (1988). Primary productivity of an Antarctic continental lake: Phytoplankton and benthic algal mat production strategies. *Hydrobiologia* **165**, 77–87.

Hoagland, K. D. (1983). Short-term standing crop and diversity of periphytic diatoms in a eutrophic reservoir. *J. Phycol.* **19**, 30–38.

Hoagland, K. D., and Peterson, C. G. (1990). Effects of light and wave disturbance on vertical zonation of attached microalgae in a large reservoir. *J. Phycol.* **26**, 450–457.

Hoagland, K. D., Roemer, S. C., and Rosowski, J. R. (1982). Colonization and community structure of two periphyton assemblages, with emphasis on the diatoms (Bacillariophyceae). *Am. J. Bot.* **69**, 188–213.

Hudon, C., and Bourget, E. (1983). The effect of light on the vertical structure of epibenthic algal communities. *Bot. Mar.* **26**, 317–330.

Jørgensen, B. B., and Des Marais, D. S. (1986). A simple fiber optic microprobe for high resolution light measurements: Application in marine sediment. *Limnol. Oceanogr.* **31**, 1376–1383.

Jørgensen, B. B., and Nelson, D. C. (1988). Bacterial zonation, photosynthesis and spectral light distribution in hot spring microbial mats of Iceland. *Microb. Ecol.* **16**, 133–147.

Kairesalo, T. (1983). Dynamics of epiphytic communities on *Equisetum fluviatile* L.: Response to short-term variation in environmental conditions. *In* "Periphyton of Freshwater Ecosystems" (R. G. Wetzel, ed.), pp. 153–160. Dr. W. Junk Publishers, The Hague.

Kann, J., and Falter, C. M. (1989). Periphyton indicators of enrichment in Lake Pend Oreille, Idaho. *Lake Reserv. Manage.* **5**, 39–48.

Kingston, J. C., Lowe, R. L., Stoermer, E. F., and Ludewski, T. (1983). Spatial and temporal distribution of benthic diatoms in northern Lake Michigan. Ecology **64**, 1566–1580.

Kuhn, D., J., Plafkin, J., Cairns, J., and Lowe, R. L. (1981). Qualitative characterization of environmental conditions using diatom life-form strategies. *Trans. Am. Microsc. Soc.* **100**, 165–182.

Layne, C. D. (1990). The algal mat of Douglas Lake, Michigan: Its composition, role in lake ecology, and response to chemical perturbations. Thesis, Bowling Green State University, Bowling Green, OH.

Lin, C. K., and Blum, J. L. (1977). Recent invasion of a red alga *(Bangia atropurpurea)* in Lake Michigan. *J. Fish. Res. Board Can.* **34**, 2413–2416.

Loeb, S. L., and Reuter, J. E. (1981). The epilithic periphyton community: A five lake comparative study of community productivity, nitrogen metabolism and depth distribution of standing crop. *Verh.—Int. Ver. Theor. Angew. Limnol.* **21**, 346–352.

Loeb, S. L., Reuter, J. E., and Goldman, C. R. (1983). Littoral zone production of oligotrophic lakes: The contributions of phytoplankton and periphyton. *In* "Periphyton of Freshwater Ecosystems" (R. G. Wetzel ed.), pp. 161–167. Dr. W. Junk Publishers, The Hague.

Lorenz, R. C., and Herdendorf, C. E. (1982). Growth dynamics of *Cladophora glomerata* in relation to some environmental factors. *J. Great Lakes Res.* **8**, 42–53.

Lorenz, R. C., Monaco, M. E., and Herdendorf, C. E. (1991). Minimum light requirements for substrate colonization by *Cladophora glomerata. J. Great Lakes Res.* **17**, 536–542.

Losee, R. F. and Wetzel, R. G. (1983). Selective light attenuation by the periphyton complex. *In* "Periphyton of Freshwater Ecosystems" (R. G. Wetzel, ed.), pp. 89–96. Dr. W. Junk Publishers, The Hague.

Lowe, R. L., and Hunter, R. D. (1988). The effect of grazing by *Physa integra* on periphyton community structure. *J. North Am. Benthol. Soc.* **7**, 29–36.

Lowe, R. L., and Pillsbury, R. W. (1995). The impact of zebra mussels *(Dreissena polymorpha)* on littoral periphyton communities in Saginaw Bay, Lake Huron. *J. Great Lakes Res.* **34**, 558–567.

Lowe, R. L., Rosen, B. H., and Kingston, J. C. (1982). A comparison of epiphytes on *Bangia atropurpurea* (Rhodophyta) and *Cladophora glomerata* (Chlorophyta) from Grand Traverse Bay, Lake Michigan. *J. Great Lakes Res.* **8**, 164–168.

Lowe, R. L., Rosen, B. H., and Larson, T. M. (1984a). Seasonal dynamics of periphyton epiphytic on *Cladophora* in the Western Basin of Lake Erie. *Bull. North Am. Benthol. Soc.* **32**, 96.

Lowe, R. L., Rosen, B. H., and Fairchild, G. W. (1984b). Endosymbiotic blue-green algae in freshwater diatoms: An advantage in nitrogen-poor habitats. *J. Phycol., Suppl.* **20**, 24.

Marks, J. C., and Lowe, R. L. (1989). The independent and interactive effects of snail grazing and nutrient enrichment on structuring periphyton communities. *Hydrobiologia* **185**, 9–17.

Marks, J. C., and Lowe, R. L. (1993). Interactive effects of nutrient availability and light levels on the periphyton composition of a large oligotrophic lake. *Can J. Fish. Aquat. Sci.* **50**, 1270–1278.

Medlin, L. K. (1981). Effect of grazers on epiphytic diatom communities. *In* "Proceedings of the Sixth Symposium on Recent and Fossil Diatoms" (R. Ross, ed.), pp. 399–412. Koeltz Scientific Books, Koenigstein, Germany.

Millner, G. C., and Sweeney, R. A. (1982). Lake Erie *Cladophora* in perspective. *J. Great Lakes Res.* **8**, 27–29.

Millner, G. C., Sweeney, R. A., and Frederick, V. R. (1982). Biomass and distribution of *Cladophora glomerata* in relation to some physical-chemical variables at two sites in Lake Erie. *J. Great Lakes Res.* **8**, 35–41.

Nalepa, T. F., and Schloesser, D. W. (1993). "Zebra Mussels: Biology, Impacts and Control." Lewis, Boca Raton, Fl.

Obeng-Asamoa, E. K., John, D. M., and Appler, H. N. (1980). Periphyton in the Volta Lake. I. Seasonal changes on the trunks of flooded trees. *Hydrobiologia* **76**, 191–200.

Pillsbury, R. W., and Kingston, J. C. (1990). The pH-independent effect of aluminum on cultures of phytoplankton from an acidic Wisconsin lake. *Hydrobiologia* **194**, 225–233.

Pringle, C. M., and Bowers, J. A. (1984). An *in situ* substratum fertilization technique: Diatom colonization on nutrient enriched substrata. *Can J. Fish. Aquat. Sci.* **41**, 1247–1251.

Revsbech, N. P., and Jørgensen, B. B. (1983). Photosynthesis of benthic microflora measured with high spatial resolution by the oxygen microprofile method: Capabilities and limitations of the method. *Limnol. Oceanogr.* **28**, 749–756.

Revsbech, N. P., and Jørgensen, B. B. (1986). Microelectrodes: Their use in microbial ecology. *Adv. Microb. Ecol.* **9**, 749–756.

Revsbech, N. P., Jørgensen, B. B., Blackburn, T. H., and Cohen, Y. (1983). Microelectrode studies of photosynthesis and O_2, H_2S and pH profiles of a microbial mat. *Limnol. Oceanogr.* **28**, 1062–1074.

Riber, H. H., and Wetzel, R. G. (1987). Boundary layer and internal diffusion effects on phosphorus fluxes in lake periphyton. *Limnol. Oceanogr.* **32**, 1181–1194.

Roberts, D. A., and Boylen, C. W. (1988). Patterns of epipelic algal distribution in an Adirondack lake. *J. Phycol.* **24**, 146–152.

Rodriguez, M. A. (1992). An empirical analysis of community structure regulation in periphyton. *Oikos* **65**, 419–427.

Sand-Jensen, K., and Søndergaard, M. (1981). Phytoplankton and epiphytic development and their shading effect on submerged macrophytes in lakes of different nutrient status. *Int. Rev. Gesamten Hydrobiol.* **66**, 529–552.

Sheath, R. G., and Morison, M. O. (1982). Epiphytes on *Cladophora glomerata* in the Great Lakes and St. Lawrence Seaway with particular reference to the red alga *Chroodactylon ramosum* (= *Asterocytis smargdina*). *J. Phycol.* **18**, 385–391.

Smol, J. P., Battarbee, R. W., Davis, R. B., and Meriläinen, J. (1986). "Diatoms and Lake Acidity." Junk Publ., Dordrecht, The Netherlands.

Stevenson, R. J., and Stoermer, E. F. (1981). Quantitative differences between benthic algal communities along a depth gradient in Lake Michigan. *J. Phycol.* **17**, 29–36.

Stevenson, R. J., and Stoermer, E. F. (1982). Seasonal abundance patterns of diatoms on *Cladophora* in Lake Huron. *J. Great Lakes Res.* **8**, 169–183

Stevenson, R. J., Singer, R., Roberts, D. A., and Boylen, C. W. (1985). Patterns of benthic algal abundance with depth, trophic status, and acidity in poorly buffered New Hampshire lakes. *Can J. Fish. Aquat. Sci.* **42**, 1501–1512.

Stockner, J. G., and Armstrong, F. A. J. (1971). Periphyton of the experimental lakes area, northwestern Ontario. *J. Fish. Res. Board Can.* **28**, 215–229.

Taft, C. E., and Kishler, W. J. (1973). "*Cladophora* as Related to Pollution and Eutrophication in Western Lake Erie." Center for Lake Erie Area Research, Ohio State University, Columbus.

Tilman, D. (1982). "Resource Competition and Community Structure." Monogr. Popul. Biol., No. 17. Princeton Univ. Press, Princeton, NJ.

Tuchman, M. L., Stoermer, E. F., and Carney, H. J. (1984). Effects of increased salinity on the diatom assemblage of Fonda Lake, Michigan. *Hydrobiologia* **109**, 179–188.

Turner, M. A., Schindler, D. W., and Graham, R. W. (1983). Photosynthesis-irradiance relationships of epilithic algae measured in the laboratory and *in situ*. *In* "Periphyton of Freshwater Ecosystems" (R. G. Wetzel, ed.), pp. 73–88. Dr. W. Junk Publishers, The Hague.

van Dijk, G. M. (1993). Dynamics and attenuation characteristics of periphyton upon artificial substratum under various light conditions and some additional observations on periphyton upon *Potamogeton pectinalis* L. *Hydrobiologia* **252**, 143–161.

Wetzel, R. G. (1964). A comparative study of the primary productivity of higher aquatic plants, periphyton and phytoplankton in a large shallow lake. *Int. Rev. Gesamten Hydrobiol.* **48**, 1–61.

Wetzel, R. G., Rich, P. H., Miller, M. C., and Allen, H. L. (1972). Metabolism of dissolved and particulate detrital carbon in a temperate hard-water lake. *Mem. Ist. Ital. Idrobiol., Suppl.* **29**, 185–243.

Woelkerling, W. J., and Gough, S. B. (1976). Wisconsin desmids. III. Desmid community composition and distribution in relation to lake type and water chemistry. *Hydrobiologia* **51**, 3–32.

Young, O. W. (1945). A limnological investigation of the periphyton in Douglas Lake, Michigan. *Trans. Am. Microsc. Soc.* **64**, 1–20.

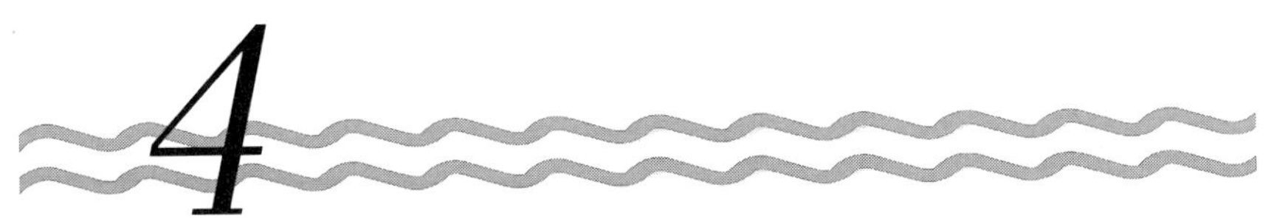

4 Pattern in Wetlands*

L. Gordon Goldsborough
and Gordon G. C. Robinson

*Department of Botany,
University of Manitoba
Winnipeg, Manitoba, Canada R3T 2N2*

*Contribution 248 from the University Field Station (Delta Marsh), University of Manitoba, Winnipeg.

I. INTRODUCTION

Wetlands are ecosystems in which the soil, despite periodic fluctuations in water level, is more or less continuously waterlogged. Most wetlands generally have a water depth <2 m and, by this definition, comprise as much as 6% of the global land area (Mitsch and Gosselink, 1993). Regionally, wetlands may be even more abundant; for example, about 14% of the land area of Canada is wetlands (Zoltai, 1988), much in the form of northern peatlands. Wetlands are increasingly recognized for their high gross primary production and their role as nutrient sinks (Dolan *et al.*, 1981; Grimshaw *et al.*, 1993; Cronk and Mitsch, 1994a; Reeder, 1994), flood control buffers, and breeding grounds for waterfowl (e.g., Sheehan *et al.*, 1987; Batt *et al.*, 1989). Given that wetlands provide abundant colonizable substrata (principally macrophytes and sediments), they would appear to offer ideal habitat for benthic algae.

Wetlands occur on every continent except Antarctica (Mitsch and Gosselink, 1993). Despite the global abundance of inland wetlands, there are few data on their benthic algae (Crumpton, 1989). Regions where research on wetland algae is especially active include the North American Great Plains (e.g., Barica *et al.*, 1980; Hanson and Butler, 1994; Murkin *et al.*, 1994), southern Florida (Browder *et al.*, 1994), eastern and southern coastal areas of North America (Sullivan and Moncreiff, 1990; Pinckney and Zingmark, 1991), eastern England (Moss, 1983), eastern Europe (Eiseltová and Pokorný, 1994) and southern Africa (Haines *et al.*, 1987). Within these boundaries are studies in wetlands as diverse as those established in kettle holes left in the till plains of extinct glaciers ("prairie potholes"; Haertel, 1976; Barica, 1975), marshes bordering the Great Lakes (Millie and Lowe, 1983) and other interior large lakes (Hooper and Robinson, 1976), beaver impoundments (Weaks, 1988), flooded medieval peat excavation pits (Moss, 1983) and tundra ponds (Stanley, 1976), among others (Table I).

II. ALGAL ASSEMBLAGES IN WETLANDS

Just as the term "wetland" eludes precise definition, algae in wetlands comprise a series of structurally overlapping, intimately interacting assemblages that defy traditional categories. For the purposes of discussion, we consider the algae of wetlands to include the following:

1. **Epipelon** includes motile algae inhabiting soft sediments. A related algal type is **plocon,** which includes nonmigratory algal crusts, typically composed of cyanobacteria and diatoms, that form on the surface of exposed or submersed sediments. These crusts occasionally detach owing

TABLE I Summary of Studies on Freshwater and Saltwater Wetland Algae

Wetland type	*Location*	*Source*
Prairie pothole	Manitoba	Barica *et al.*, (1980); Barica (1990); Shamess *et al.*, (1985)
	Michigan	Henebry and Cairns (1984)
	Minnesota	Hanson and Butler (1990, 1994)
	South Dakota	Haertel (1976)
Northern marsh	Manitoba (Interlake)	Campeau *et al.* (1994); Garbor *et al.* (1994); Murkin *et al.* (1991, 1994)
Lakeshore marsh	Manitoba (Delta Marsh)	Goldsborough and Robinson (1983, 1985, 1986); Gurney and Robinson (1988, 1989); Hann (1991); Hooper and Robinson (1976); Hooper-Reid and Robinson (1978a,b); Hosseini and van der Valk (1989a,b); Kotak and Robinson (1991); E.J. Murkin *et al.* (1992); Neill and Cornwell (1992)
	Minnesota (Lake Superior)	Keough *et al.* (1993)
	Ohio (Old Woman Creek Estuary)	Klarer and Millie (1992); Mitsch and Reeder (1991); Reeder (1994)
	Ohio (Lake Erie marshes)	Millie and Lowe (1983)
	Ontario (Holland Marsh)	Nicholls (1976)
Constructed riverine marsh	Illinois	Cronk and Mitsch (1994a,b)
Constructed shallow fish ponds	South Bohemia (Trebon Biosphere Reserve)	Eiseltová and Pokorný (1994)
Flooded peat pits	England (Norfolk Broads)	Eminson and Moss (1980); Leah *et al.* (1978); Mason and Bryant (1975); Phillips *et al.* (1978)
Dystrophic wetland	South Carolina (Carolina Bays)	Schalles and Shure (1989)
Cypress swamp	Virginia (Great Dismal Swamp)	Atchue *et al.* (1983)
Coastal salt marsh	California (Tijuana Estuary)	Zedler (1980)
	Massachusetts (Great Sippewissett Marsh)	van Raalte *et al.* (1976)
	Mississippi (Graveline Bay Marsh)	Sullivan and Moncreiff (1988, 1990)
	Netherlands	de Jonge and Colijn (1994)
	New Brunswick	Tracy and South (1989)
	South Carolina (North Inlet Estuary)	Pinckney and Zingmark (1991, 1993a,b,c)
	Texas (East Galveston Bay)	Hall and Fisher (1985)

(continues)

TABLE I (continued)

Wetland type	*Location*	*Source*
Subtropical freshwater marsh	Florida (Everglades)	Browder *et al.* (1994); Grimshaw *et al.* (1993); Steward and Ornes (1975); van Meter-Kasanof (1973)
Subtropical swamp	Australia (Murrumbidgil Swamp)	Briggs *et al.* (1993)
Beaver impoundment	Tennessee	Weaks (1988)
Alpine wetland	Colorado	Mihuc and Toetz (1994)
Peat bog/fen	Georgia (Okefenokee Swamp)	Bosserman (1983); Scherer (1988); Schoenberg and Oliver (1988)
	Michigan	Henerbry and Cairns (1984); Richardson and Schwegler (1986)
	Minnesota	Kingston (1982)
	North Wales	Duthie (1965)
	Ontario	Flensburg and Sparling (1973)
	Wisconsin	Woelkerling (1976)
Tundra pond	Alaska	Stanley (1976); Stanley and Daley (1976)

to buoyancy of accumulated gas bubbles (Moss, 1968; Durako *et al.*, 1982) and float at the water surface. Plocon may correspond to "epibenthic" mats or "edaphic algae" studied in estuarine marshes (e.g., Sage and Sullivan, 1978; Hall and Fisher, 1985; Vernberg, 1993) where they can become several centimeters thick. The assemblage has been studied very rarely in freshwater wetlands so, consequently, virtually nothing is known of its production and ecology.

2. **Epiphyton** is composed of prostrate, erect, and heterotrichous algae growing on the external surfaces of submersed and emergent vascular and nonvascular plants that may, under some circumstances, be differentiated into "loosely" and "firmly" attached components (Haines *et al.*, 1987).

3. **Metaphyton,** also known as "flab" (Hillebrand, 1983), forms cohesive floating and subsurface mats, usually composed of filamentous green algae. The mats originate as epiphyton but are detached by water turbulence and may float owing to trapped gases within them. In its early stages of development, metaphyton may be synonymous with "loosely" attached epiphyton.

4. **Phytoplankton** includes algae entrained in the water column that may or may not be motile, often arising from detachment of epipelic or epiphytic forms.

Development of a true phytoplankton assemblage occurs most often in wetlands connected to large lakes, which serve as sources of inocula and nutrients (Klarer and Millie, 1992). For example, phytoplankton may form periodic, short-lived blooms, dominated by genera such as *Aphanizomenon, Stephanodiscus* (Brown, 1972), and *Volvox* (L. G. Goldsborough, personal observation), in Delta Marsh, a large freshwater wetland bordering Lake Manitoba, Canada. However, the positive correlation of chlorophyll concentration in the marsh water column with indices measuring wind stress (Kotak, 1990) suggests that much of what is assumed to be phytoplankton is, in reality, derived from the detachment and suspension of benthic forms (Ganf, 1973). Common diatom taxa in the phytoplankton of an isolated pond within the marsh include *Achnanthes, Cocconeis, Cymbella, Gomphonema,* and *Rhoicosphenia* (Brown, 1972), which are also major constituents of the epiphyton (e.g., Goldsborough and Robinson, 1986). As such, this assemblage may be more accurately termed "tychoplankton" (*sensu* Round, 1981). Lack of habitat fidelity between phytoplankton and periphyton has been noted by others working in wetlands, including van Meter-Kasanof (1973), who coined the term "waif periphyton" to describe detached, entrained cells. De Jonge and van Beusekom (1992) estimated that entrained epipelon made up 22% of the total phytoplankton in one Dutch estuary and about 60% in another; over 30% of all chlorophyll in the water column was derived from epipelic sources.

Deposition of phytoplanktonic cells into the epiphyton (pseudoperiphyton") also occurs; Moss (1981) reported that the diatom epiphyton of a Norfolk Broad contained some genera in common with the phytoplankton (*Diatoma, Synedra*) and others exclusive to the epiphyton (*Achnanthes, Cocconeis, Epithemia, Gomphonema, Mastogloia,* and *Rhopalodia*). Another case of the indistinct separation of algal assemblages in wetlands is the gradual isolation of metaphyton mats, which often lack any connection to a substratum (and therefore are not truly benthic), from the epiphyton or epipelon over time. Epipelon and plocon share a common substratum but they are architecturally and structurally distinct. Epipelon is typically dominated by motile diatoms such as *Amphora, Navicula,* and *Nitzschia* (e.g., Sullivan, 1975; Shamess *et al.,* 1985; de Jonge and Colijn, 1994), whereas plocon often comprises cyanobacteria bound in a gelatinous matrix. Indeed, the two assemblages likely intergrade in wetlands that experience alternating periods of drawdown and flooding (cf. van der Valk and Davis, 1978) due to the changing availability and exposure of sediment.

III. THE ROLE OF ALGAE IN WETLANDS

Algae are fundamental players in the physical, chemical, and biological processes that characterize wetland ecosystems. Most obvious is their role as primary producers and, consequently, their place in the wetland food web. Measurements of total annual C production by epiphyton in

Delta Marsh (1 to 49 g m^{-2} yr^{-1}—Hooper and Robinson, 1976; 3.6 to 548 g m^{-2} yr^{-1}—Robinson *et al.*, 1996b) are comparable to those of benthic algae in other wetlands (Table II) and lakes (Wetzel, 1983). It is difficult to generalize on the quantitative contribution made by algae to gross primary production in wetlands because few studies have been sufficiently inclusive to measure all potential producers. The proportionate contribution of epiphytes to total primary production in constructed wetlands in Illinois ranged from 1 to 65% (Cronk and Mitsch, 1994a). Pinckney and Zingmark (1993b) calculated that epipelic algae in a South Carolina estuary contributed 22–38% of total primary production, as compared to 15–27% for phytoplankton, 10–17% for macroalgae, and 30–59% for vascular plants. Other data from the same salt marsh site indicated that epipelon contributed 20% of annual net primary production; phytoplankton and epiphyton plus neuston each contributed 10% to total net production and the remainder (60%) was due to macroalgae and vascular plants (Vernberg, 1993). Epipelic algal production was 76 to 140% of vascular plant production in a California salt marsh studies by Zedler (1980). In an Alaskan tundra pond, annual primary production by epipelon was estimated as 4–10g C m^{-2}, as compared to 1 g C m^{-2} for phytoplankton and >15 g C m^{-2} for the emergent macrophyte *Carex aquatilis* (Stanley and Daley, 1976). In a shallow dystrophic wetland in the Carolina Bays area of South Carolina, Schalles and Shure (1989) estimated that "littoral algae" (including epiphyton, phytoplankton, and photosynthetic bacteria) contributed slightly over one-third of net primary production, whereas macrophytes made up the remainder.

The importance of wetland algae as a resource to herbivores is due, in part, to their availability throughout the ice-free period as compared to other foods such as macrophytes and insects. Algae may be more important to herbivores than macroscopic plants because, as single or small clusters of cells, they are easily assimilated (see Lamberti, Chapter 17, this volume). Few aquatic herbivores have the capacity to ingest living macrophyte tissue and, consequently, macrophytes are accessible only when in detrital form. Epipelon and phytoplankton are available to grazers throughout the growning season, unlike higher plants, which die back periodically through the year. Moreover, carnivores and omnivores that normally specialize on invertebrates may consume algae during periods of scarcity. For example, benthic algae are a critical food for fish in the Florida Everglades during the winter, when insect abundance decreases (Browder *et al.*, 1994). These authors also noted that crayfish acquire as much as 50% of their diet from algae. Studies by Mason and Bryant (1975), Hann (1991), and Campeau *et al.*, (1994) have shown that wetland benthic algae are a major food resource for cladocerans, copepods, chironomid larvae, amphipods, oligochaetes, and planorbid snails, affecting their growth, development, survivorship, and reproduction. Confirmation of the importance of benthic

TABLE II Summary of Data on Algal Production in Freshwater and Saltwater Wetlands[a]

Algal type	*Wetland*	*Chlorophyll a concentration*	*Dry weight*	*C fixation rate*	*Reference*
Epipelon	Freshwater marsh	1–2 mg m^{-2}			Campeau *et al.* (1994)
		7.5 mg m^{-2}			Gabor *et al.* (1994)
				0.2–0.5 μg cm^{-2} h^{-1}	Gurney and Robinson (1989)
		0.4–4.8 mg m^{-2}			Murkin *et al.* (1991)
		0.4–4.2 mg m^{-2}			Murkin *et al.* (1994)
				<0.1–29 g m^{-2} yr^{-1}	Robinson *et al.* (1996a)
		13–435 mg m^{-2}			Shamess (1980)
	Salt marsh	29–247 mg m^{-2}			de Jonge and Colijn (1994)
				80 g m^{-2} yr^{-1}	Gallagher and Daiber (1974)
		20–150 mg m^{-2}			Pinckney and Zingmark (1991)
		60–102 mg m^{-2}			Pinckney and Zingmark (1993a)
				1.5–6.1 μg cm^{-2} h^{-1}	Pinckney and Zingmark (1993b)
		34–85 mg m^{-2}			Pinckney and Zingmark (1993c)
		57–160 mg m^{-2}		28–151 g m^{-2} yr^{-1}	Sullivan and Moncreiff (1988)
		231 mg m^{-2}		71 g m^{-2} yr^{-1}	Hall and Fisher (1985)
				106 g m^{-2} yr^{-1}	van Raalte *et al.* (1976)
				200 g m^{-2} yr^{-1}	Vernberg (1993)
				185–341 g m^{-2} yr^{-1}	Zedler (1980)
	Tundra ponds			4.1–10.1 g m^{-2} yr^{-1}	Stanley (1976)
Epiphyton	Freshwater marsh		0.2–268 mg cm^{-2}	−65–201 μg cm^{-2} d^{-1}	Browder *et al.* (1994)
		0.1–0.2 μg cm^{-2}			Campeau *et al.* (1994)
		0.2–0.4 μg cm^{-2}	1.5–2.3 mg cm^{-2}	2–85 g m^{-2} yr^{-1}	Cronk and Mitsch (1994a)
		0.5–0.8 μg cm^{-2}			Gabor *et al.* (1994)
			4–45 mg cm^{-2}		Gleason and Spackman (1974, cited in Browder *et al.*, 1994)
		7–650 mg m^{-2}			Goldsborough (1993)
		0–3.7 μg cm^{-2}		0.2–4 μg cm^{-2} h^{-1}	Goldsborough and Robinson (1983)

(continues)

TABLE II (continued)

Algal type	*Wetland*	*Chlorophyll a concentration*	*Dry weight*	*C fixation rate*	*Reference*
Epiphyton	Freshwater marsh	0–7.9 μg cm^{-2}		0.2–11 μg cm^{-2} h^{-1}	Goldsborough *et al.* (1986)
		0–0.4 μg cm^{-2}		0.3–4.3 μg cm^{-2} h^{-1}	Gurney and Robinson (1989)
		<0.1 μg cm^{-2}	0.1–0.3 mg cm^{-2}	0.04–0.37 μg cm^{-2} h^{-1}	Haines *et al.* (1987)
		2.1–3.0 μg cm^{-2}	0.1–2.2 mg cm^{-2}	0.24–0.27 μg cm^{-2} h^{-1}	Haines *et al.* (1987)
		0.2–2.5 μg cm^{-2}	0.2 mg cm^{-2}		Hann (1991)
			0–1.4 mg cm^{-2}	1–49 **g m^{-2} yr^{-1}**	Hooper and Robinson (1976)
		0–6.30 μg cm^{-2}			Hooper-Reid and Robinson (1978a)
		1.7–3.3 μg cm^{-2}			Hosseini and van der Valk (1989a)
		0–4 μg cm^{-2}			Kotak and Robinson (1991)
			0–1.8 mg cm^{-2}	17 μg cm^{-2} d^{-1}	Mason and Bryant (1975)
		0.1–2.4 μg cm^{-2}			Murkin *et al.* (1991)
		0.3–4.3 μg cm^{-2}			Murkin *et al.* (1992)
		0.2–0.8 μg cm^{-2}			Murkin *et al.* (1994)
				0.2–3 μg cm^{-2} h^{-1}	Robinson and Pip (1983)
				3.6–548 **g m^{-2} yr^{-1}**	Robinson *et al.* (1996a)
		0–0.88 μg cm^{-2}			Shamess (1980)
			1.2–35 mg cm^{-2}		van Meter-Kasanof (1973)
		4–37 μg cm^{-2}			Weaks (1988)
			0–596 mg cm^{-2}		Wood and Maynard (1974, cited in Browder *et al.*, 1994)
	Peat bog	2–238 **mg m^{-2}**			Goldsborough (1993)
		0.04–0.4 μg cm^{-2}			Schoenberg and Oliver (1988)
	Salt marsh			100 **g m^{-2} yr^{-1}**	Vernberg (1993)
Metaphyton	Freshwater marsh		0.3–6.6 mg cm^{-2}		Hosseini and van der Valk (1989b)
			0.5 mg cm^{-2}		Richardson and Schwegler (1986)
				12.6–1119 **g m^{-2} yr^{-1}**	Robinson *et al.* (1996a)
			80.8 mg cm^{-2}		van der Valk (1986)
	Shallow fish ponds		1–49 mg cm^{-2}		Eiseltová and Pokorný (1994)
	Peat bog		0–6 mg cm^{-2}		Schoenberg and Oliver (1988)
	Cypress swamp		1–3 mg cm^{-2}		Atchue *et al.* (1983)

Phytoplankton	Freshwater marsh	4–335 μg liter^{-1}			Barica (1978)
		3–10 μg liter^{-1}			Campeau *et al.* (1994)
		84–94 μg liter^{-1}			Carper and Bachmann (1984)
		20–58 μg liter^{-1}			Haertel (1976)
		11–45 μg liter^{-1}			Hanson and Butler (1994)
		<1–18 μg liter^{-1}		3–16 μg liter^{-1} h^{-1}	Henebry and Cairns (1984)
		3–64 μg liter^{-1}			Hosseini and van der Valk (1989a)
		<10–53 μg liter^{-1}			Kotak and Robinson (1991)
		42–211 μg liter^{-1}			Leah *et al.* (1978)
		5–75 μg liter^{-1}			Mitsch and Reeder (1991)
		2–60 μg liter^{-1}			Murkin *et al.* (1991)
		1–4 μg liter^{-1}			Murkin *et al.* (1994)
		0–358 μg liter^{-1}			Nicholls (1976)
		44–104 **mg m^{-2}**			Reeder (1994)
				368 **g m^{-2} yr^{-1}**	Reeder and Mitsch (1989, cited in Klarer and Millie, 1992)
				1.1–381 **g m^{-2} yr^{-1}**	Robinson *et al.* (1996a)
		3–280 μg liter^{-1}			Shamess(1980)
		2–260 μg liter^{-1}			Timms and Moss (1984)
		0–16 μg liter^{-1}			Vörös and Padisák (1991)
		80–200 μg liter^{-1}			Watson and Osborne (1979)
		4–200 μg liter^{-1}			Ytow *et al.* (1994)
	Cypress swamp		3.4 **g m^{-2}**		Atchue *et al.* (1983)
	Peat bogs	3–13 μg liter^{-1}		5–32 μg liter^{-1} h^{-1}	Henebry and Cairns (1984)
		<1–1000 μg liter^{-1}			Schoenberg and Oliver (1988)
	Salt marsh			100 **g m^{-2} yr^{-1}**	Vernberg (1983)
	Tundra ponds			0.6–0.9 **g m^{-2} yr^{-1}**	Stanley (1976)
Phytoplankton + epiphyton	Freshwater marsh (dystrophic)	15–36 **mg m^{-2}**		81 **g m^{-2} yr^{-1}**	Schalles and Shure (1989)

[a]These data should be interpreted cautiously because of variation arising from geographic and seasonal factors, differences in analytical methods (e.g., dry weight or ash-free dry weight; spectrophotometric or HPLC determination of pigments; measurement of photosynthesis by O_2 evolution or C assimilation), varying contributions by other pigments (e.g., phaeopigments and bacteriochlorophylls), and inconsistent bases on which data are derived (wetland surface area, substratum surface area, or volume). Units involving m^2 (in bold) represent production per m^2 of wetland surface area, which is equal to substratum surface area only for epipelon.

algae in the diets of these herbivores has come from the similarity of the respective ^{13}C, ^{15}N, and ^{35}S composition (Neill and Cornwell, 1992; Mihuc and Toetz, 1994). Analyses of the stable isotopic composition in algae and invertebrates from a lakeshore wetland on Lake Superior (Keough *et al.,* 1993) indicate that entrained algae are an important food resource for planktonic herbivores (mostly copepods and cladocerans). However, zooplankters were 3–4‰ ^{15}N-enriched with respect to wetland phytoplankton, suggesting that other food sources, possibly benthic in origin, were also being used. Sullivan and Moncreiff (1990), in their study of C, N, and S stable isotopes in a salt marsh ecosystem on the Gulf of Mexico, concluded that the "ultimate food source for this marsh's invertebrate and fish fauna are the benthic and planktonic algae; direct contributions from vascular plants appear to be minor."

Benthic algal metabolism causes changes in water chemistry that indirectly affect other wetland organisms. These include oxygenation of the water column (van Meter-Kasanof, 1973; Hillebrand, 1983; Browder *et al.,* 1994), increases in pH, and decreased concentrations of CO_2 and bicarbonate (Browder *et al.,* 1994). Nocturnal respiration and decomposition of algal blooms causes deoxygenation, leading to hypoxia and death of fish and other organisms (Barica, 1975). Diurnal fluctuations in oxygen concentration among dense macrophytes and algal cover (Cronk and Mitsch, 1994b) can affect the spatial and temporal distribution of fish and invertebrates in wetlands (e.g., Suthers and Gee, 1986).

Algae contribute to wetland nutrient cycles as sources of dissolved organic matter (Hall and Fisher, 1985; Briggs *et al.,* 1993) and N, the latter fixed by heterocystous cyanobacteria (Granhall and Selander, 1973; Hooper-Reid and Robinson, 1978b; Bazely and Jefferies, 1989). They also serve as short-term sinks for P (Howard-Williams, 1981; Grimshaw *et al.,* 1993), N (Moraghan, 1993), and metals (Bosserman, 1983). Epipelic algae serve another important, but often overlooked, role in nutrient cycling. Wetland sediments typically contain higher levels of N and P as compared to those in the overlying water (e.g., Kadlec, 1986), which suggests that the sediments are a major source of nutrients to macrophytes and algae. Epipelic algae mediate the nutrient efflux rate from sediment interstitial water (Jansson, 1980; Hansson, 1989; Wetzel, Chapter 20, this volume). This role may be direct, in that algal assimilation of sediment nutrients precludes their release to the overlying water, and indirect, as a result of an aerobic microzone at the sediment/water interface, arising from epipelic photosynthesis, that prevents efflux of reduced N and P species. Carlton and Wetzel (1988) demonstrated, using oxygen-specific microelectrodes and a ^{32}P tracer, that P release from sediments varies diurnally, with maximum efflux occurring during the night, when algal photosynthesis stops. Evidence in support of this hypothesis derives from toxicological studies (e.g., Gurney and Robinson, 1989), which showed that photosynthetic activity by epipelon in marsh enclosures treated with photosynthetic inhibitors decreased as compared to untreated controls. Ammonia and P concentration in the water column increased following treatment in direct proportion to the inhibitor concentration (Goldsborough and Robinson, 1985).

Benthic algae serve several physical roles in wetlands. In alkaline systems such as the Florida Everglades, periphyton contribute to sediment formation via their role in the precipitation of calcite (Browder *et al.*, (1994). The mucilaginous film produced by benthic algal mats in estuaries stabilizes the substratum and reduces erosion (e.g., Holland *et al.*, 1974; Grant *et al.*, 1986); the same likely holds true for cyanobacterial mats in fresh water and salt marshes. Thick floating mats of metaphyton offer refuge from predation to invertebrates and habitat for frogs (L. G. Goldsborough, personal observation). The epiphyton matrix on macrophytes provides habitat for invertebrates, fungi, and bacteria. Browder *et al.* (1994) reported that shaded microhabitats under floating "rafts" of detached epiphyton in the Everglades are used by largemouth bass (*Micropterus salmoides*) and other aquatic animals. Laboratory experiments have demonstrated the extent to which epiphytic films on macrophytes attenuate light at the substratum surface (Losee and Wetzel, 1983), which contributes to the decline of macrophytes when epiphyton biomass is high (Phillips *et al.*, 1978). Epiphytic coverage of eelgrass can reduce macrophyte photosynthesis by as much as 58% (Sand-Jensen, 1977), with attenuation of incident light by epiphytes ranging up to 82% (Sand-Jensen and Borum, 1984).

Finally, there is growing awareness of the role played by algae in the wetland detrital food web, as producers of organic matter, consumers of organic substrates (Saks *et al.*, 1976), and facilitators of microbial decomposition. In the latter case, Neely (1994) found that breakdown and mineralization of *Typha* stems in a freshwater marsh occurs more quickly in the presence of epiphytic algae.

IV. SPECIES COMPOSITION OF WETLAND ALGAE

All major algal groups are represented in the wetland flora, although the most commonly reported groups are diatoms, green algae, and cyanobacteria. Determinants of the dominant species in any given wetland include the nature of the assemblage (epipelon, epiphyton, metaphyton, or phytoplankton), the geographic location of the wetland (as it determines solar insolation and water temperature), season, water depth and hydroperiod, macrophyte types present, and nutrient and mineral concentrations.

The epipelic algal flora in wetlands depends on the nature of the habitat. In exposed sites subject to desiccation and drought (such as tidal and freshwater mudflats), cyanobacterial mats (plocon) of *Lyngbya, Microcoleus, Nostoc, Oscillatoria, Schizothrix,* and other taxa predominate (Granhall and Selander, 1973; Sage and Sullivan, 1978; Zedler, 1980; Bazely and Jeffries, 1989; Tracy and South, 1989; Browder *et al.,* 1994). *Nostoc* mats covered the sediments of a small freshwater marsh where macrophytes were absent (Brown, 1972), whereas in a nearby prairie pothole, pennate diatoms (*Amphora, Cocconeis, Cymbella, Epithemia, Nitzschia, Pleurosigma,* and *Surirella*) comprised >90% of total algal biovolume (Shamess *et al.,* 1985). In shallow tundra ponds, where midsummer irradiance is high and benthic production predominates, the epipelic algal flora consisted of cyanobacteria (*Aphanizomenon, Microcystis*), green algae (*Ankistrodesmus, Chlamydomonas,* and *Closterium*), and diatoms (Stanley, 1976). Desmids were common in the sediments of a Welsh *Sphagnum* bog, where the distribution of specific taxa reflected microvariation in macrophyte species cover (Duthie, 1965). Conversely, chlamydomonads were more abundant than desmids, diatoms, euglenids, or cyanobacteria in acidic Priddy Pool, England (Happey-Wood, 1980).

Dominant algal groups in the epiphyton of nutrient-rich temperate freshwater wetlands include pennate diatoms (typically genera such as *Achnanthes, Cocconeis, Cymbella, Epithemia, Fragilaria, Gomphonema, Navicula,* and *Nitzschia*), filamentous green algae (such as *Cladophora, Coleochaete, Oedogonium,* and *Stigeoclonium*), and cyanobacteria (*Anabaena, Gloeotrichia, Oscillatoria,* and others) (Hooper-Reid and Robinson, 1978a; Pip and Robinson, 1984; Shamess *et al.,* 1985; Hann, 1991; Cronk and Mitsch, 1994a). Unicellar green algae, chrysophytes, cryptomonads, euglenoids, dinoflagellates, and desmids are occasionally reported. Typically, a thin film of diatoms initiates the successional sequence on the macrophyte, followed later by green algae. Cyanobacteria appear in late summer. However, intensive study of early substratum colonization in Delta Marsh by Kruszynski (1989) has shown that species composition of the immigrant pool is seasonally variable. In some wetlands, algal species distribution reflects spatial environmental gradients. Browder *et al.,* (1994) recognized that epiphytes were distributed in the Everglades according to hydroperiod and gradients of P and calcium concentration. Green algae (*Oedogonium, Stigeoclonium*) and diatoms were favored at sites having moderate calcium and P concentration, whereas desmids were common at sites with low $CaCO_3$ and low nutrient concentration. Cyanobacteria (especially *Microcoleus*) were abundant at sites where nutrients and conductivity were high.

An interesting special case is the epiphyton in dystrophic wetlands that have highly reduced, poorly oxygenated water near the sediments. In these cases, green and purple photosynthetic bacteria can contribute significantly to apparent "algal" biomass (Goldsborough and Brown, 1991) and primary production (Schalles and Shure, 1989).

Metaphyton mats in wetlands are typically composed of filamentous green algae such as *Chaetophora, Cladophora, Enteromorpha, Oedogonium,* and *Spirogyra* (Brown, 1972; Richardson and Schwegler, 1986; Neill and Cornwell, 1992; Eiseltová and Pokorný, 1994). Given the epiphytic origin of metaphyton, it is not surprising that these taxa are also commonly reported as epiphytes in wetlands. A "secondary" epiphyton of smaller algae (e.g., diatoms, chlorophytes, and cyanobacteria) is found on the submersed surfaces of the mat (Moss, 1976; L. G. Goldsborough, personal observation).

The algal flora of bogs and fens is unique from that of other wetlands because it generally has low diversity and is dominated by desmids and, to a lesser extent, diatoms, cyanobacteria, euglenoids, chrysophytes, or dinoflagellates (Duthie, 1965; Flensburg and Sparling, 1973; Woelkerling, 1976; Kingston, 1982; Scherer, 1988). There are few data on the ecological requirements of bog and fen algae. Species are assumed to be distributed on the basis of vegetation type (Woelkerling, 1976), proximity of the water table, and nutrient status (Kingston, 1982). The flora is particularly impoverished in ombrotrophic peat bogs (Flensburg and Sparling, 1973), where precipitation is the sole source of nutrient input. The scarcity of nutrients in the water column invites the possibility that nutrients provided by macrophyte substrata may be particularly important in such oligotrophic environments, with the result that host specificity may develop most strongly here (Eminson and Moss, 1980).

V. ALGAL PRODUCTION IN WETLANDS

Epipelon biomass in freshwater wetlands (<10 mg m^{-2} of chlorophyll *a;* Table II) is generally lower than that in other freshwater habitats tabulated by Moss (1968). This may be due to the fact that marsh epipelon typically forms under dense canopies of submersed and emergent macrophytes, their associated epiphytes, floating metaphyton mats, or phytoplankton blooms, all of which reduce substantially the irradiance reaching the sediment surface throughout much of the growing season. Several workers have noted that epipelon biomass is higher in unvegetated areas than in ones containing abundant macrophytes (e.g., van Raalte *et al.,* 1976; Murkin *et al.,* 1991). The markedly higher epipelic biomass in salt marshes (>20 mg m^{-2} of chlorophyll *a;* Table II) may be due, in part, to procedural differences in sampling. Most studies of freshwater marsh epipelon have used the method of Eaton and Moss (1966), which, because it relies on the migration of algal cells into lens paper placed onto the exposed sediment surface, samples only motile components of the total flora. In contrast, studies in salt marshes have generally used intact sediment cores, which include both motile and nonmotile species, as well as detrital pigments. By one estimate, degraded chlorophyll accounted for 85% of total chlorophyll (Sullivan and Moncreiff, 1988), so that living algal biomass was severely overestimated. An indication of the importance of procedural differences comes from a study of epipelon in freshwater prairie marsh in which whole-sediment samples were analyzed (Shamess, 1980); the range of chlorophyll *a* values (13–435 mg m^{-2}) was considerably higher than estimates made using the lens paper technique and are more comparable to those in salt marshes (Table II). Another possible basis for the difference is that salt marshes typically have daily periods of tidal drawdown, during which time irradiance at the sediment surface is high and conditions are conducive for epipelon and plocon growth.

Epiphyton biomass is rarely expressed in terms of wetland surface area because of the difficulty of accurately quantifying aquatic plant surface area. Consequently, data are commonly based on measurements using morphologically simpler artificial substrata, with values in freshwater marshes generally <5 μg cm^{-2} of chlorophyll *a* (Table II). Where area-based values exist, epiphyton biomass (up to 650 mg m^{-2} of chlorophyll *a;* Table II) is comparable to that reported for other shallow waters (e.g., Moss, 1968; Wetzel, 1983).

The few data on metaphyton biomass in wetlands are expressed in terms of dry weight (Table II), making it difficult to compare them with data sets based on chlorophyll *a* concentration (e.g., Moss, 1968). Biomass is generally <10 mg cm^{-2} except in cases where metaphyton has formed dense surface mats (van der Valk, 1986).

Levels of phytoplankton chlorophyll *a* in wetlands commonly exceed 50 μg $liter^{-1}$ (Table II); small, eutrophic wetlands often have chlorophyll *a* values >200 μg $liter^{-1}$ during periodic cyanobacterial blooms (Barica, 1975). For comparative purposes, a phytoplankton chlorophyll *a* concentration of 100 μg $liter^{-1}$ equals 100 mg m^{-2} when expressed in terms of wetland surface area, assuming an average water depth of 1 m. Therefore, it is clear from Table II that when phytoplankton are abundant, they may be as productive as epiphyton or metaphyton in other, less turbid wetlands.

VI. FACTORS AFFECTING ALGAL PRODUCTION

A. Hydrodynamics

Arguably, the single most important variable that defines a wetland, and therefore its algae, is water depth (Kadlec, 1979), because it affects, directly or indirectly, all the factors whose individual effects we consider in the following discussion. The characteristic shallowness of wetlands can result in extensive reworking and resuspension of sediments by wind (e.g., Carper and Bachmann, 1984), high turbidity, and a thoroughly mixed water column. Prolonged thermal stratification rarely develops and, consequently, dissolved and particulate nutrients are distributed readily throughout the water column. Subsurface irradiance is generally high, although it varies spatially and temporally with turbidity. These effects require that wind velocity is sufficiently high in shallow water that Langmuir cells reach the bottom; dense macrophyte or metaphyton cover may be mitigating factors that reduce mixing. The role of water depth is further complicated in that most wetlands experience regular or periodic fluctuations in water level. Good *et al.* (1978) argued that prairie wetlands are maintained by a cycle of flooding and drought. Flooding reduces macrophyte abundance and releases nutrients locked up in standing dead vegetation, litter, and sediments. Drought and exposure of the sediment surface is needed to allow recruitment of plants from the seed bank (van der Valk, 1986).

Fluctuations in water level undoubtedly affect the availability of substratum and, therefore, the potential for development of specific algal assemblages. For instance, sites in the Florida Everglades that undergo frequent desiccation are dominated by benthic cyanobacterial mats whereas diatoms and green algae are common at sites with frequent, persistent water; desmids occur only at sites having continuous standing water (Browder *et al.*, 1994)

The influence of wind on wetland algal production is generally negative. On one hand, increased turbidity and decreased subsurface irradiance reduces photosynthesis (Hellström, 1991). For example, C fixation by algae colonizing vertically positioned artificial substrata in Delta Marsh decreased by 75% over a depth of 40 cm, closely paralleling the extinction profile of photosynthetically available radiation (Goldsborough *et al.*, 1986). Schalles and Shure (1989) noted that diel changes in dissolved oxygen in a dystrophic South Carolina wetland were pronounced at the water surface but were absent at 40 cm depth. Wind physically disturbs floating mats, epiphyton, and sediments. At the same time, it is reasonable to assume, based on empirical correlations between nutrient levels and wave action in shallow lakes (e.g., Hamilton and Mitchell, 1988), that wind increases the nutrient supply in the water column of wetlands, benefiting epiphytes and phytoplankters. Water motion may also eliminate nutrient-depleted boundary layers around colonized substrata, thereby stimulating growth.

B. Nutrients

Sources of nutrients in wetlands are varied. Efflux from the sediments is a major source of nutrients to the water column (the foregoing). In lakeshore and riverine wetlands, the significance of water input from the adjacent lake, river or stream as a source of nutrients (and algal inoculum) may be high. Cronk and Mitsch (1994a) found that high-flow constructed wetlands in Illinois supported higher algal biomass on artificial substrata than did low-flow wetlands. They attributed this finding to increased nutrient import and waste export with increased water throughput.

It is widely assumed that most freshwater lakes are P-limited, but there are too few data to assess whether this generalization applies to freshwater wetlands as a whole. For wetlands in north-temperate North America, it is supported equivocally. Here, the N : P ratio in the water column is mostly >10 (Table III). In Delta Marsh, estimates of the N : P ratio range from 11 in a *Lemna*-covered channel to 25 in the open marsh (Table III), with the latter value most closely matching the value for sediment interstitial water (21). The nature of nutrient deficiency in northern wetlands has also been assessed based on the biochemistry and physiology of their phytoplankton and benthic algae. Hooper-Reed and Robinson (1978b) concluded, based on measurements of silicon uptake rates, alkaline phosphatase and nitrogenase activities, cellular P and chlorophyll content, and the ratio of protein to carbohydrate plus lipid, that benthic algae at two sites in a small pond in Delta Marsh experienced short intervals of N, P, or Si nutrient deficiency periodically through the season. The significance of external nutrient loading in alleviating nutrient deficiency of wetland phytoplankton was demonstrated by Murkin *et al.* (1991), in which measurements of alkaline phosphatase activity, ammonia uptake rate, and P/C, N/P, N/C and chlorophyll/C composition ratios were consistent in showing severe N and P deficiency in an uncontaminated marsh throughout the

TABLE III Nitrogen : Phosphorus Ratio (Total N : P) of Water in Temperate Freshwater Wetlands[a]

N:P	*Location*	*Reference*
1–23	Michigan (various bogs, fens, marshes and swamps)	Henebry and Cairns (1984)
7	Minnesota (prairie pothole)	Haertel (1976)
9–27	Manitoba (Erickson potholes)	Barica (1990)
11	Manitoba (Delta Marsh)	L.G. Goldsborough (unpublished data)
11	Manitoba (Delta Marsh)	Kadlec (1986)
13	Iowa (prairie pothole)	Carper and Bachmann (1984)
18	Manitoba (Delta Marsh; sediment interstitial water)	Kadlec (1986)
18–34	Minnesota (prairie pothole)	Hanson and Butler (1994)
21	Manitoba (Delta Marsh; sediment interstitial water)	Kadlec (1989)
25	Manitoba (Delta Marsh)	L. G. Goldsborough (unpublished data)
27	Manitoba (Cruise Marsh)	Murkin *et al.* (1991)
33	Manitoba (boreal forest ponds)	L. G. Goldsborough and D. J. Brown (unpublished data)
126	Manitoba (Narcisse Marsh)	Murkin *et al.* (1991)

[a]After Barica (1990).

summer. A nearby marsh that was enriched by cattle feedlot effluent showed few symptoms of deficiency. Campeau *et al.* (1994) found that additions of N and P to enclosures positioned in a nutrient-poor marsh stimulated biomass of phytoplankton, epiphyton, and metaphyton but not epipelon. Other studies have been less conclusive on the role of nutrients. Murkin *et al.* (1994) found that phytoplankton and periphyton in the same marsh remained nutrient-limited, based on measurements of alkaline phosphatase activity and N/C, P/C, and N/P cell composition ratios, despite N and P enrichment. The varying response by phytoplankton to nutrient enrichment of the water column is especially interesting, as it indicates that the abundance of algae may be regulated, in part, by the type and abundance of macrophytes and their ability to compete effectively with phytoplankton for nutrients.

Some shallow systems are dominated by phytoplankton (e.g., Haertel, 1976; Barica *et al.*, 1980), whereas others have abundant macrophytes and associated epiphytes but few phytoplankton (e.g., Hooper and Robinson, 1976). One explanation is that dominance is determined by the absolute level of P loading, with phytoplankton flourishing with progressive eutrophication (e.g., Phillips *et al.*, 1978). Despite this view, and contrary to the contention that macrophytes meet their P requirement from the sediments (Barko *et al.*, 1991), nutrient enrichment studies have shown that macrophytes can assimilate nutrients added to the water column (Steward and Ornes, 1975; Howard-Williams, 1981), effectively outcompeting associated epiphyton and leading to its elimination (Steward and Ornes, 1975).

The role of fluctuating water depth on wetland nutrient status is unresolved. It is assumed that exposure of the sediments during droughts promotes the decomposition of organic matter and liberation of nutrients during subsequent reflooding (Kadlec, 1979; Schoenberg and Oliver, 1988). In late stages of flooding, macrophytes are killed by flood stress (van der Valk, 1994) and their decomposition releases N and P (Murkin *et al.*, 1989). Surprisingly, Kadlec (1986) reported that experimental flooding of a region of Delta Marsh did not change dissolved and suspended nutrient levels. This may be due to the removal of nutrients by remaining plants and benthic algae, particularly given that thick, floating metaphyton mats proliferated in flooded areas (van der Valk, 1986). Hosseini and van der Valk (1989b) observed that although the biomass of algae colonizing artificial substrata in flooded areas decreased, total algal N content increased, resulting in higher overall N content per unit of algal biomass.

Wetland algae and plants release a diverse suite of soluble organic substances into their environment (e.g., Hall and Fisher, 1985; Briggs *et al.*, 1993) during growth and decomposition. The resulting load of dissolved organic C in many wetlands (Thurman, 1985; Briggs *et al.*, 1993), and the abundance of organic N and P species as compared to inorganic forms, argues that benthic algae inhabiting such environments may be facultatively heterotrophic. Saks *et al.* (1976) cultured several epiphytic diatoms and green algae from a coastal salt marsh on media containing amino acids, sugars, or vitamins. Results were inconsistent in that individual substrates stimulated growth of some taxa while inhibiting others. Although they showed clearly that heterotrophy may be a viable means of supplementary nutrition for some species, it is noteworthy that all but one of cultures tested were obligately phototrophic and were unable to grow in the dark. In a direct demonstration (using microautoradiography) of heterotrophic utilization of organic substrates, Pip and Robinson (1982) found that most epiphytes in a shallow lake, including diatoms, green algae, and cyanobacteria, were capable of assimilating a radiolabeled mixture of glucose, fructose, and sucrose. However, uptake rates about one-hundredth for inorganic C, suggesting that heterotrophy is not the primary method of nutrition for these algae.

C. Light

Numerous studies of lakeshore wetlands and shallow lakes have demonstrated that sediment resuspension by winds increases the turbity of the water column (e.g., Klarer and Millie, 1992) and, therefore, decreases significantly the irradiance penetrating the water column. Coupled with periodic disruption of the bottom by such benthivorous fishes as carp (*Cyprinus carpio*) (Meijer *et al.*, 1990), the subsurface light environment of wetlands changes frequently with velocity and direction of prevailing winds and the abundance of macrophytes to stabilize bottom sediments, slow wind, and dampen turbulence of water movement (e.g., Brix, 1994). During periods of maximum *Phragmites* growth, up to 95% of incident light can be extinguished by the emergent canopy (Roos and Meulemans, 1987; Brix 1994). Transmittance of light through floating *Lemna* mats may be ≈0.1% of surface irradiance (Goldsborough, 1993).

Most studies of factors affecting benthic algal accrual on substrata have emphasized the importance of nutrient supply. However, E. J. Murkin *et al.*, (1992) noted that epiphyton biomass was low within a dense *Typha* stand in Delta Marsh despite high nutrient concentrations from decomposing *Typha* litter. They concluded that epiphyton were light-limited, not nutrient-limited.

There is considerable variation in the photosynthesis–irradiance relationship measured for benthic algae from wetlands (Table IV), which likely reflects differences in analytical methodology and the physiological state of organisms collected from diverse habitats. A portion of the variability, at least for assemblages such as epiphyton with discrete three-dimensional physiognomy, may be attributed to the accumulated biomass of the assemblage, for there is an inverse correlation between biomass (expressed as chlorophyll *a* concentration) and photosynthetic efficiency (Fig. 1). It may

TABLE IV Parameters of the Photosynthesis–Irradiance Curves for Wetland Algae[a]

Algal type	α [μg C (μg Chl a)$^{-1}$h^{-1} (μE m^{-2}s^{-1})$^{-1}$]	I_k (μE m^{-2}s^{-1})	SP_{max} [μg C (μg Chl a)$^{-1}$h^{-1}]	Reference
Epipelon	0.002–0.003*	579–834	1–3*	Pinckney and Zingmark (1993c)
		237–563	0.8–4	Robinson (1988)
	0.006	407	2	Robinson *et al.* (1995b)
Epiphyton	0.016	430	3–21	Kotak (1990)
		198–659	0.5–4	Robinson (1988)
	0.008	292	2	Robinson *et al.* (1995b)
Metaphyton		36–306	0.2–0.7	Gurney and Robinson (1988)
	0.007*	344	2*	Richmond (1992)
	0.003	399	1	Robinson *et al.* (1995b)
Phytoplankton		253–446	2–3	Gurney and Robinson (1988)
	0.092	303	3–79	Kotak (1990)
		235–392	6–14	Robinson (1988)
	0.025	282	7	Robinson *et al.* (1995b)

[a]α is a measure of photosynthetic efficiency under light-limited conditions, I_k is the saturing irradiance at which maximum photosynthesis occurs and SP_{max} is the maximum rate of photosynthesis, standardized for algal biomass (asterisk indicates converted to μg C from μmole O_2 with an assumed photosynthetic quotient of 1).

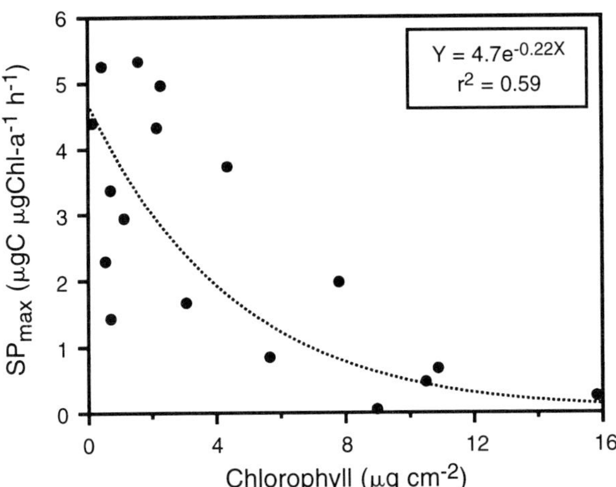

FIGURE 1 Relationship between maximum photosynthetic rate [μg C (μg Chl *a*)$^{-1}$ h^{-1}] and biomass (μg Chl *a* cm^{-2}) of epiphytic algae in Delta Marsh. (From G. G. C. Robinson, unpublished data.)

be that as biomass accumulates, self-shading and diffusive availability of nutrients limit those cells mostly deeply buried within the algal matrix (see Hill, Chapter 5, and McCormick, Chapter 8, this volume). Given the few data, generalizations on the photosynthetic efficiency of various algal assemblages are unwarranted except to note that values for I_k, the irradiance at which maximum rate of photosynthesis occurs, are similar in freshwater marsh assemblages (ca. 300–400 μE m^{-2} s^{-1}). Values for epipelon in a tidal mudflat, where irradiance is likely to be higher than in a well-vegetated marsh, are notable higher (Table IV; Pickney and Zingmark, 1993c). These I_k values indicate that algal photosynthesis is likely light-saturated at the water surface. However, irradiance in turbid waters and below a macrophyte canopy can easily fall below I_k. In an extreme case, subsurface irradiance below a dense floating mat of duckweeds *Lemna minor*) was often <5 μE m^{-2} s^{-1} (Goldsborough, 1993). Few studies have reported evidence of photoinhibition in surface waters, even at full solar irradiance, an exception being the study of Robinson and Pip (1983) of species-specific C assimilation by benthic algae in Delta Marsh.

D. Temperature

Based on the characteristic shallowness of wetlands, it is usually assumed that differences in temperature with depth are minor and, given wind-induced mixing, transitory. Consequently, temperature effects on spatial differences in algal growth are probably negligible. However, temperature profiles with depth may be prolonged when the water surface is occluded by vegetation. For instance, thick floating mats of duckweed may develop among emergent macrophytes at the periphery of wetlands and in secluded channels away from direct influence of winds. In such mats, which achieve biomass >300 g m^{-2} in a layer a few centimeters thick, rapid light extinction results in a thermal profile of up to 15°C in 25 cm depth developing diurnally (Goldsborough, 1993). Floating metaphyton mats have been shown to do likewise (Hillebrand, 1983). Van Meter-Kasanof (1973) observed a temperature change of 6°C within 35 cm of the surface in an area of the Florida Everglades that was covered by "coalesced floating heavy periphyton." Associated epiphytes within such mats must be adapted to wide daily fluctuations in temperature.

Wetlands are less buffered from seasonal changes in water temperature than deeper bodies of water. In temperature lakeshore wetlands, for example, initiation of the spring ice-off and autumnal freezeup typically precedes that of the adjacent lakes. Seasonal trends in primary production presumably track such differences.

E. Macrophytes

Macrophytes are the dominant element of most freshwater wetlands, contributing much of total ecosystem biomass. For example, compared to algal biomass in Delta Marsh (<100 g m^{-2} dry weight for epiphyton; L. G. Goldsborough, unpublished data), macrophyte biomass is high. Emergent macrophyte aboveground biomass is ca. 400 g m^{-2} (van der Valk, 1994) and submersed macrophytes contribute up to 500 g m^{-2} (Pip and Stewart, 1976; Wrubleski, 1991); an equal or greater biomass can occur in belowground parts (Shay and Shay, 1986). It is through this dominance that macrophytes influence, directly or indirectly, the biotic and abiotic environment of the wetland. Effects pertinent to wetland algae include provision of colonizable substratum, modification of the aquatic light regime, serving as both sources and sinks for nutrients and allelochemicals, causing physical abrasion of algal assemblages, and providing habitat for grazers.

Direct measurements of macrophyte surface area available to wetland epiphyton are rare. Surface area varies with plant growth, and between plants of differing morphology. For example, data from Pip and Stewart (1976) show that the surface area of two *Potamogeton* species per square meter of bottom area in Delta Marsh increased from 0.2m^2 m^{-2} in mid-May to a maximum of 15.6 m^2 m^{-2} in late July, then decreased with plant senescence through August. At any given time, the surface area of filiform leaves (*P. pectinatus*) was 120% of that for an equivalent mass of laminate leaves (*P. richardsonii*). Hooper and Robinson (1976) estimated that the seasonal mean surface area in the marsh ranged from 0.2 m^2 m^{-2} in emergent vegetation (*Typha glauca* and *Phragmites australis*) to 5.6 m^2 m^{-2} among dense *Potamogeton pectinatus* stands. The former value is within the range reported by Cronk and Mitsch (1994a) for emergent macrophytes (*Typha* spp., *Phalaris arundinacea*) in constructed wetlands (0.1–0.2 m^2 m^{-2}). To our knowledge, comparable data exist only for the littoral zone of Lawrence Lake, Michigan, where values range from 0.2–0.5 m^2 m^{-2} for emergent *Scirpus acutus,* to 6–12 m^2 m^{-2} for submerged *Najas* and *Chara* (Allen, 1971), to 24 m^2 m^{-2} for submerged *Scirpus subterminalis* (Burkholder and Wetzel, 1989).

Macrophytes negatively alter the environment for wetland algae in several ways. Wrubleski (1991) showed experimentally that epipelon biomass in Delta Marsh increased when he removed a dense *Potamogeton pectinatus* overstory. The biomass of phytoplankton and algae on artificial substrata also increased after macrophyte harvesting. Murkin *et al.* (1994) reported that epipelon biomass in areas of a marsh vegetated by emergent macrophytes (*Scirpus acutus*) supported significantly less epipelon biomass than areas devoid of macrophytes. Similar findings have been made in salt marshes (van Raalte *et al.,* 1976). There may be several bases for this observation, among which is the possibility that macrophyte cover reduced irradiance at the sediment surface sufficiently to reduce algal colonization. It is also possible that macrophytes outcompete epipelon for sedimentary nutrients or release chemicals that selectively inhibit the growth of epipelic algae.

Some wetland macrophytes have been shown to produce allelochemicals (Elakovich and Wooten, 1989), presumably as a defense against epiphytes on their surfaces or phytoplankton in their vicinity (Wium-Anderson *et al.,* 1982). Dodds (1991) noted that an extract from *Cladophora glomerata* slightly reduced the photosynthetic rate of a pure *Nitzschia fonticola* culture established from its epiphyton. Weaks (1988) suggested allelopathy to explain why algal species diversity and biomass on artificial substrata were lowest for sites near a stand of dead, decaying macrophytes in a newly formed beaver impoundment, as compared to sites farther away. The benefit of such allelochemicals to the plant, if released only after its death, are unclear.

Positive effects of macrophyte abundance in wetlands accrue to algae from the stabilizing influence that the plants have on bottom sediments. In the absence of plant cover, wind activity is often sufficient to cause extensive sediment resuspension and turbidity. Conversely, stabilization of the sediments by plant roots and reduction of water flow by abundant, highly dissected leaves of submersed macrophytes reduce turbidity in areas of dense growth (Dieter, 1990). Water clarity in channels of Delta Marsh is noticeably high in summer following development of profuse macrophyte stands. In addition, macrophytes release inorganic nutrients and photosynthate during growth (Wetzel and Manny, 1972) and senescence (Landers, 1982), which are subsequently assimilated by their epiphytes (e.g., McRoy and Goering, 1974). Pakulski (1992) observed that foliar release of P by *Spartina alterniflora* varied seasonally, with highest rates occurring in midsummer when the plant essentially acted as a "pump" to transfer P from marsh sediments to the water column. In return, N fixed by heterocystous cyanobacteria can be assimilated by macrophytes (Jones, 1974).

The observation that different macrophyte species at the same site can support different epiphyte assemblages, owing to variations in their physiology and morphology, is an important theme of epiphyton studies (e.g., Eminson and Moss, 1980; Pip and Robinson, 1984). For instance, Cronk and Mitsch (1994a) found that *Typha* spp. in Illinois wetlands supported significantly less epiphyton biomass per unit of surface area than *Polygonum* spp. at the same sites. In a clear demonstration of the host specificity of epiphyton in a eutrophic prairie wetland, Shamess *et al.* (1985) compared algal community structure on *Typha* stems and acetate strips. Both smooth and artificially roughened artificial substrata supported high algal biomass dominated by green algae and cyanobacteria. *Typha* epiphyton, on the other hand, had lower biomass and consisted mostly of diatoms with periodic peaks of green algae and cyanobacteria. The maintenance of a distinct epiphyton on living substrata as compared to inert artificial ones in a highly eutrophic wetland runs counter to the hypothesis of Phillips *et al.* (1978) that host specificity should be most pronounced in oligotrophic systems, where the importance of host-derived exudates, and therefore the stimulus for host specificity, should be greatest.

F. Herbivory and Other Faunal Influences

Wetlands typically support a diverse aquatic insect fauna, of which chironomids are probably the most abundant, often comprising >50% of all emerging insects (Wrubleski, 1987). Variation in water level may affect the number and diversity of grazing insects. Wrubleski (1987) noted that drawdown, particularly in temporary wetlands, can significantly reduce the survival of chironomids. Various other invertebrates are abundant in wetlands (Pip and Stewart, 1976; Hann, 1991), although less is known of their ecology, particularly their impact on wetland algae. Wetlands provide spawning and feeding grounds for numerous vertebrates, including fish, waterfowl, muskrats, and beaver. These animals fulfill a number of roles in the wetland, serving as planktivores, herbivores, and detritivores, promoting resuspension of sediments and destruction of submersed and emergent macrophytes, and providing sources and sinks for nutrients.

In one of the few quantitative studies of the impact of herbivory on benthic algae in a wetland, Hann (1991) observed that grazers altered the biomass, species composition, and physiognomy of epiphyton. Under normal grazing pressure, algal biomass was maintained at a low, constant level by cladocerans, copepods, and ostracods. When the composition of grazers was selectively altered using rotenone, leading to dominance by chironomids and oligochaetes and reductions in microcrustacea, chlorophyll *a* levels increased by as much as two to three times. This suggests that chironomids and oligochaetes are less effective grazers than microcrustacea and, more importantly, that algal standing crop is, in large part, regulated by grazing pressure (i.e., top-down control). Community structure was also affected by herbivory; under unperturbed conditions, a two-dimensional assemblage was dominated by *Stigeoclonium* with a few other filamentous green algae, cyanobacteria, and prostrate diatoms as subdominants. As grazing pressure was alleviated using rotenone, a more diverse three-dimensional assemblage developed (Fig. 2).

Mason and Bryant (1975) demonstrated how the nature of a substratum and the specific grazer can affect epiphyton biomass and community structure in a wetland. They observed that algal biomass was usually much higher on glass rods than on adjacent *Typha* stems, presumably because the artificial substratum excluded burrowing chironomid larvae that were the predominant epiphyton grazers. Examination of the gut contents of the

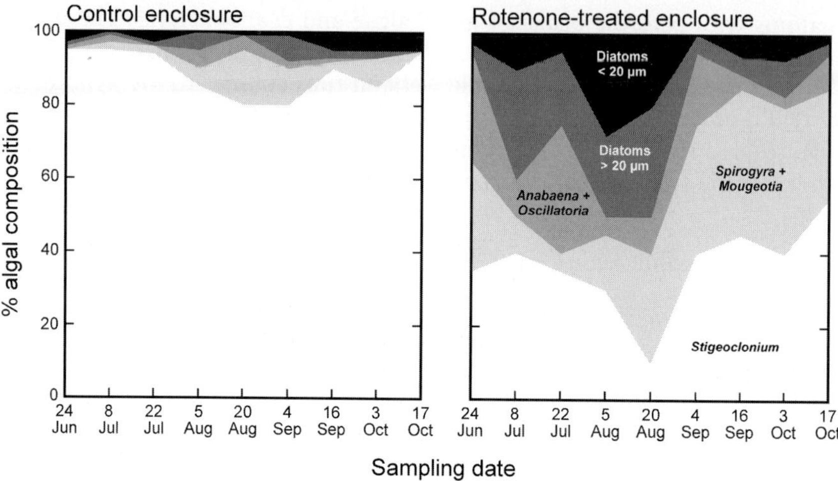

FIGURE 2 Periphytic algal community composition in unperturbed (left) and rotenone-treated (right) enclosures in Delta Marsh (see text). [Figure is redrawn from Hann (1991) by permission of Blackwell Scientific Publications Ltd.]

chironomids *Glyptotendipes* and *Cricotopus* from *Typha* stems showed that they were actively consuming diatoms and filamentous algae, whereas *Chironomus tentans* was eating rotting *Typha* detritus but no algae. Grazer selectivity was also reported by Browder *et al.* (1994), who noted that fish and macroinvertebrates in the Everglades selected diatoms and green algae preferentially to cyanobacteria. They attributed selectivity to morphological and nutritional differences between the algal groups.

Resistance to grazing pressure has been proposed as a mechanism by which the filamentous green algae comprising metaphyton mats flourish; only when the mats decompose are they consumed by detritivores (Dodds and Gudder, 1992). Despite anecdotal observations that metaphyton mats can harbor large invertebrate populations (e.g., Hillebrand, 1983) and may be grazed by dabbling ducks (Murkin, 1989), there are few direct data on the herbivory of metaphyton mats. Supporting evidence comes from dietary analysis of grazers, via direct gut examination and analysis of the stable isotopes of C, N, and S of grazer and food resource. One such study, done in an alpine wetland, found that the isotopic signature of metaphyton did not match that of the predominant chironomid herbivores, suggesting that it was not actively grazed (Mihuc and Toetz, 1994). Similarly, the ^{13}C and ^{15}N composition of metaphyton from Delta Marsh, composed of *Chaetophora* sp. and *Enteromorpha intestinalis,* did not resemble that of any potential consumers, including muskrats (*Ondatra zibethicus*), minnows (*Pimephales promelas*), *Daphnia pulex,* chironomids (*Chironomus* sp. and *Glyptotendipes* sp.), amphipods (*Gammarus lacustris, Hyalella azteca*), snails (*Lymnaea stagnalis*), corixids (Corixidae), odonates (Odonata), and dytiscids (Dytiscidae) (Neill and Cornwell, 1992).

Animals also affect wetland algae indirectly via their impact on sediment efflux of nutrients through bioturbation (Gallepp, 1979), through the nutrients released in their excretions and the decomposition of their carcasses, and by the effects of predation on herbivore abundance (Hanson and Butler, 1994). The importance of terrestrial animals, particularly of waterfowl, to the nutrient loading of wetlands has rarely been quantified, despite the fact that many wetlands serve as breeding and staging grounds for thousands of migratory birds each season. Moss and Leah (1982) studied an English wetland, described as being "guanotrophic," for which excreta of black-headed gulls (*Larus ridibundus*) accounted for 53–72% of the total P input. Snow geese (*Chen caerulescens caerulescens*) feces are known to make significant contributions to the N budget of subarctic salt marshes during their summer breeding period (Ruess *et al.*, 1989). Schoenberg and Oliver (1988) ascribed high phytoplankton biomass at a wetland site near an ibis (*Eudocimus albus*) rookery to nutrients released from bird guano. Flocks of Canada geese (*Branta canadensis*) and mallard ducks (*Anas platyrhyunchos*) contribute as much as 70% of the P budget of Wintergreen Lake, Michigan (Manny *et al.*, 1994); these same species are common on shallow northern wetlands, where their nutrient inputs are at least as important. Decomposition of waterfowl carcasses is also an important contributor of nutrients; Parmenter and Lamarra (1991) found that pintail ducks (*Anas acutas*) lost 65% of their total N and 30% of their P over a period of 10 months.

The contributions of aquatic animals to wetland nutrient loading are not well documented. The dead bodies of fish and mammals presumably contribute nutrients to the water column; decomposing rainbow trout (*Oncorhynchus mykiss*) carcasses have been shown to release 95% of their N and 60% of their P over 10 months (Parmenter and Lamarra, 1991). In addition, bioturbation of sediments by burrowing chironomid larvae (Gallepp, 1979; Phillips *et al.*, 1990) and by benthivorous fish such as carp (*Cyprinus carpio;* Meyer, *et al.*, 1990) may release nutrients to the overlying water, although its quantitative significance is unknown.

G. Anthropogenic Factors

The number of wetlands in North America has declined since the turn of the century (Mitsch and Gosselink, 1993) as many were drained to be brought into agricultural production. Those that remain are often closely encroached upon by active farmland, with the result that waterfowl populations are disrupted (Batt *et al.*, 1989) and wetlands are increasingly subject to contamination by fertilizers and pesticides (Neely and Baker, 1989). The impacts of these additions on algae are largely unknown (see

Hoagland *et al.*, Chapter 15), as there have been few studies of the magnitude of the contamination, the nature of the contaminants, and their environmental chemistry in the wetland habitat. Nevertheless, studies of the impact of agricultural herbicides on marsh epiphyton (e.g., Goldsborough and Robinson, 1983, 1986; Gurney and Robinson, 1989) have shown that these can have pronounced, long-lasting effects on the productivity, species composition, and physiognomy of the algal assemblage. For example, in a study of the short-term response by algae colonizing artificial substrata in small enclosures treated with the photosynthetic inhibitor simazine, Goldsborough and Robinson (1986) observed recolonization of substrata following removal of the inhibitor, resulting in a profound change in the species composition and physiognomy of the assemblage. While control samples supported a dense, three-dimensional community dominated by *Cocconeis placentula* and *Stigeoclonium tenue,* herbicide-exposed substrata were colonized by a nearly unispecific, two-dimensional carpet of *C. placentula* (Fig. 3).

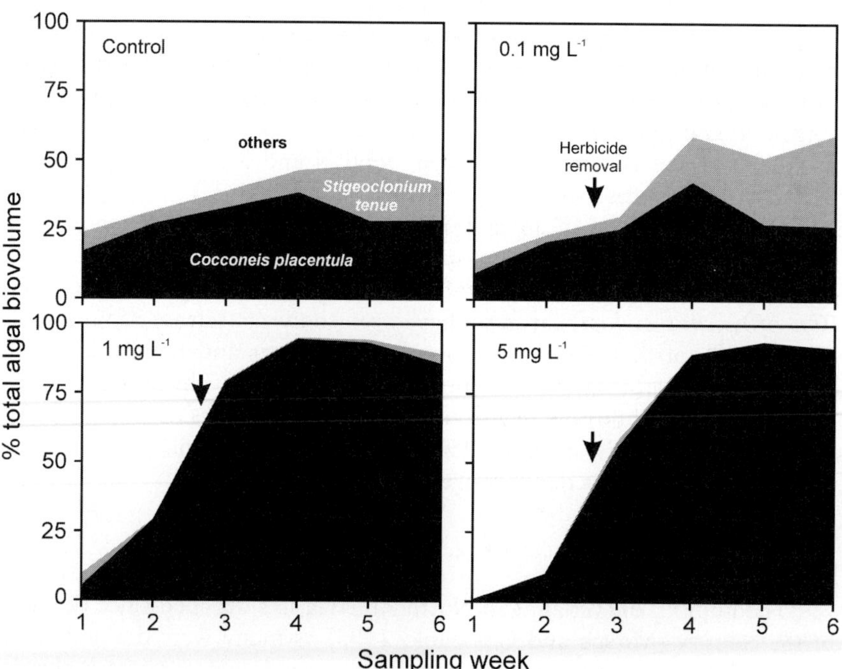

FIGURE 3 Relative abundance of *Cocconeis placentual* and *Stigeoclonium tenue* in the periphyton of an unperturbed enclosure, and in enclosures treated with three concentrations of simazine, a photosynthetic inhibitor. The vertical arrows represent a flooding event that displaced the inhibitor from all enclosures between Weeks 2 and 3. [Figure is modified from Goldsborough and Robinson (1986); reprinted by permission of Kluwer Academic Publishers.]

VII. CONCEPTUAL MODEL OF WETLAND ALGAE

Any attempt to integrate the foregoing observations into a conceptual model must recognize that wetlands contain four major algal assemblages (Section II) whose proportionate abundances vary spatially between wetlands and temporally at different stages in wetland development. Our experience with wetlands in western Canada suggests that each one has a characteristically dominant algal assemblage that, without major environmental changes such as water level fluctuation, can persist for decades. For instance, a shallow isolated pond within Delta Marsh has supported dense metaphyton mats over much its surface intermittently since at least the 1970s (Brown, 1972), whereas the main channels of the marsh have supported dense epiphyton on submersed macrophytes, with no perceptible metaphyton. Yet, epiphyton/metaphyton dominance is not the only stable end point in wetlands. In the Old Woman Creek Estuary on Lake Erie, for instance, macrophytes are sparse so phytoplankters are the dominant contributors to total primary production (Reeder, 1994). Many of the shallow "prairie potholes" of the North American prairies, which were hypereutrophic with periodic phytoplankton blooms in the 1970s (e.g., Barica, 1975; Shamess *et al.*, 1985), remain so today. Biomanipulative experiments involving removal of planktivorous fish from wetlands in Minnesota (Hanson and Butler, 1994) emphasize that replacement of an epiphyte/macrophyte-dominated system by phytoplankton tends to be, in the absence of human intervention, a persistent condition. The same conclusion has been reached for the Norfolk Broads, a collection of culturally eutrophicated wetlands in eastern England (e.g., Moss, 1983).

A reasonable approach to modeling may be to recognize four quasi-stable states in wetlands dominated, alternatively, by epipelon, epiphyton, metaphyton, or phytoplankton (Fig. 4; Robinson *et al.*, 1996a). Recognizing that wetlands are increasingly subject to management, such a scheme, unlike other models of wetlands ontogeny (e.g., van der Valk and Davis, 1978), should not assume cyclic passage of a wetland through all four states; it should enforce no directionality on shifts from one state to another. Instead, the dominant algal assemblage is determined by natural grazing pressure and water column stability, and by anthropogenic nutrient loading, water level control, and biomanipulation. These combined, interacting factors determine the duration that a specific state will persist. We characterize each state (Table V), and the conditions leading to its wax and wane, in the following sections.

A. Dry State

The Dry State exists when or where very low water levels occur following a period of drought or drawdown in an existing wetland or in the early flooding of a new wetland (Fig. 4). The exposed sediments must

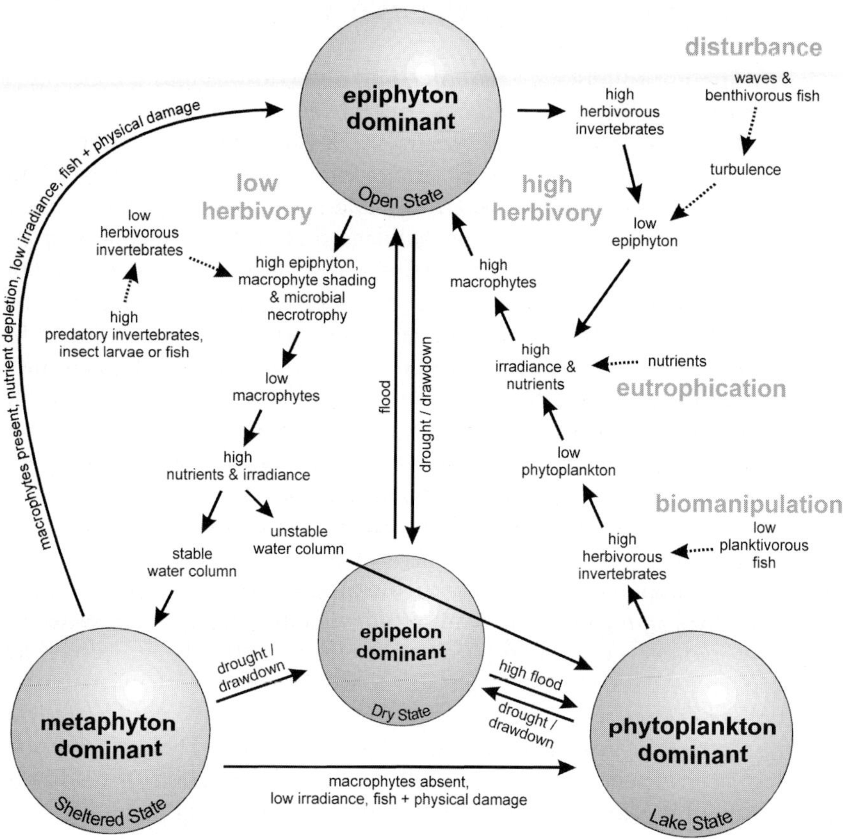

FIGURE 4 Predicted trophic interactions occurring in freshwater wetlands, leading to one of four stable states dominated, alternately, by epipelon, epiphyton, metaphyton, or phytoplankton (see text).

TABLE V Predicted Properties of Four Quasi-stable States in the Ontogeny of a Freshwater Wetland Depicted in Fig. 4

Property	*Dry State*	*Open State*	*Sheltered State*	*Lake State*
Dominant algae	Epipelon	Epiphyton	Metaphyton	Phytoplankton
Algal primary production	Low	Medium	High	Variable
Water level	Low	Medium	Medium	High
Water column disturbance	Rare	Frequent	Rare	Frequent
Water column transparency	High	Variable	High	Low
Nutrients	High (?)	Medium	High	High
Aquatic macrophytes	Few	Abundant	Medium	Few
Herbivory	Low	High	Low	Low
Secondary production	Low	High	Low	Variable

remain sufficiently moist to prevent desiccation of algal cells, although it may be possible for cells with mucilaginous sheaths similar to those in terrestrial soils (Round, 1981) to persist through short dry periods. Biomass of aquatic macrophytes is low, although they may be replaced by "wet meadows" of grasses and sedges that do not provide habitat for epiphytic algae or their grazers. Accumulation of litter with restricted water flow gives rise to a nutrient-rich environment. Irradiance at the sediment surface should be high owing to shallowness of the water column but it may be reduced substantially by vegetative cover (Brix, 1994). Consequently, conditions are suitable for surficial plocon crusts and migratory epipelic algae inhabiting the surface layers of illuminated sediments. There have been few studies of algae in dry wetlands so their quantitative contribution to total primary production is unknown, but is probably low.

Following flooding, dry wetlands are transformed to a state dependent on the quantity and timing of water input. If water is sufficiently shallow that aquatic macrophytes develop, the increase in substratum availability may lead to an Open State wetland in which epiphytes are the predominant algal assemblage (Fig. 4). On the other hand, if water input is too rapid for macrophytes to colonize, or if vegetation is excluded by excessively deep water, phytoplankton predominate in a Lake State wetland.

B. Open State

An Open State wetland is characterized by a turbulent water column populated by profuse submersed macrophytes in deep areas and emergent macrophytes in shallower areas. Such a wetland may arise in at least two ways: it may occur as a dry wetland gradually fills with water and macrophytes develop, or it may occur because of the biomanipulation of a phytoplankton-dominated Lake State wetland (Fig. 4). Epiphyton biomass is high owing to the abundance of colonizable surface area on macrophytes and the increased nutrient supply from newly flooded soil (Kadlec, 1979; Schoenberg and Oliver, 1988). Epipelon abundance is reduced from the dry wetland because of the combined shading of the sediment surface by macrophytes and their epiphytes. Phytoplankton is outcompeted for nutrients by other primary producers so its abundance is also low. Metaphyton may develop under appropriate conditions of high nutrient status, high irradiance, low grazing pressure, and a stable water column, leading to the alternate Sheltered State wetland (Fig. 4).

Maintenance of epiphyton dominance depends on periodic physical disturbance of the assemblage by benthivorous fish and other fauna, wave action, and high grazing pressure (Table V). Such disturbance prevents accumulation of epiphyton biomass to the point that it causes macrophyte decline due to the effects of shading (Losee and Wetzel, 1983) and enhancement of bacterial necrotrophy by epiphytes (Neely, 1994). This

Open State definition applies to many temperate freshwater wetlands, including most areas of Delta Marsh (Table VI) where profuse macrophytes, coupled with water column turbulence caused by high wind exposure, periodic floods from a large lake to which it connects, and the activity of carp, give rise to abundant epiphyton (96% of total algal biomass; Table VI). Epipelon and phytoplankton are minor contributors to total production, and metaphyton is largely absent.

Where epiphyton is not kept in check by these factors, or if excess nutrient loading results in greatly increased epiphyton biomass, progressive shading of macrophytes by dense epiphyton and, eventually, phytoplankton leads to the replacement of macrophytes by phytoplankton (the Lake State; Fig. 4). This view is consistent with the "habitat fidelity" model proposed by Phillips *et al.* (1978) to explain the shift in algal dominance of Norfolk Broads, from epiphyton to phytoplankton, upon eutrophication. However, transition of an epiphyton-dominated system to one overwhelmed by phytoplankton is not an inevitable result of nutrient loading; blooms of mat-forming metaphyton may succeed epiphyton if the water column remains reasonably stable (e.g., if sufficient macrophytes persist to reduce turbulence that would otherwise disaggregate metaphyton mats, and to provide substratum for the initial development of epiphytes that give rise to metaphyton). It is also noteworthy that epiphyte-dominant wetlands can result from "cascading" biomanipulations in which removal of planktivorous fish from phytoplankton-dominated shallow lakes increases herbivorous invertebrates, which in turn consume phytoplankton (e.g., Schindler and Comita, 1972; Hanson and Butler, 1990, 1994). An underlying assumption is that macrophytes grow sufficiently quickly fol-

TABLE VI Total Biomass (Chlorophyll *a* Concentration per m^2 of Wetland Area) of Epipelon, Epiphyton, Metaphyton, and Phytoplankton in a Natural Area (Open State) of Delta Marsh, as Compared to an Area in Which Water Level Was Experimentally Drawn Down and Raised (Sheltered State)

	Open State[a]		*Sheltered State*[b]	
Algal type	*mg m^{-2}*	%	*mg m^{-2}*	%
Epipelon	3	1	4	1
Epiphyton	250	96	67	11
Metaphyton	0	0	530	87
Phytoplankton	7	3	7	1
Total	260		608	

[a]L. G. Goldsborough (unpublished data).
[b]Robinson *et al.* (1996a).

lowing the biomanipulation, providing substratum for epiphytes, to prevent recurrence of phytoplankton dominance.

C. Sheltered State

Where nutrient loading is high and the water column is stable as a consequence of shelter provided by macrophytes or by bordering vegetation, epiphyton may develop to the extent that it detaches from macrophytes, forming dense mats of metaphyton that carpet the bottom and water surface. Such mats may eventually lead to the decline, but not necessarily the elimination, of macrophytes from which they arose. They persist in the absence of physical disturbance (by winds or animals). These conditions typify a Sheltered State wetland (Fig. 4).

Anecdotal observations indicate that metaphyton abundance may be symptomatic of water input to a dry wetland. For example, van der Valk and Davis (1978) noted that metaphyton contributed 29% of the cover in a portion of a "regenerating marsh" in Iowa that was free of emergent macrophytes and <1% in an area with abundant macrophytes. Experimentally flooded marshes develop massive floating meadows of metaphyton that persist for several years (e.g., Gerney and Robinson, 1988; Hosseini and van der Valk, 1989a; Robinson *et al.,* 1996a). Data from the Marsh Ecosystem Research Program (MERP) in Delta Marsh, covering a 5-year period after an experimental reflooding, show that metaphyton contributed 87% of total algal biomass, with the remainder made up primarily by epiphyton (11%); epipelon and phytoplankton each contributed <1% of total biomass (Table VI). Metaphyton biomass was highest among emergent macrophyte litter, as compared to more open sites (Gurney and Robinson, 1988; Hosseini and van der Valk, 1989b). During a 7-year study of an Iowa wetland that developed from dryness to a shallow lake state (Weller and Fredrickson, 1973), metaphyton was abundant only in the dry wetland and during early stages of its reflooding.

Conditions that promote metaphyton development have not been fully resolved. In a comprehensive review on the ecology of *Cladophora,* a common constituent of metaphyton mats, Whitton (1970) listed high nutrient status, high irradiance, alkaline pH, high calcium, and low N:P ratios as contributors to its abundance, to which Dodds (1991) added grazer resistance and high conductivity. Fong and Zedler (1993) noted that development of floating masses of *Cladophora, Ulva* and *Enteromorpha* in an estuarine lagoon was seasonal, with maximum development occurring in summer when irradiance was highest, a connection that was supported by laboratory microcosm experiments. Evidence from MERP wetlands supports the importance of irradiance in mat proliferation (Hosseini and van der Valk, 1989b). However, light is clearly not the sole determining factor, as open-water areas of manipulated MERP marshes did not develop extensive metaphyton mats (Gurney and Robinson, 1988). Evidence from controlled experiments, in which abundant, floating masses of filamentous green algae developed in response to nutrient enrichment (Phillips *et al.*, 1978; Howard-Williams, 1981), supports the view that high nutrient status is a necessary prerequisite for mat development. Nutrient-poor wetlands typically do not support metaphyton mats, but they do develop abundantly in enclosures enriched with N and P in such wetlands (Murkin *et al.*, 1994; Gabor *et al.*, 1994).

Of particular interest in the context of wetland development are the consequent effects of metaphyton mats. For instance, decreased light and temperature beneath the mat (Eiseltová and Pokorný, 1994) may reduce the growth of macrophytes and algae and, therefore, reduce competition for resources. Coupled with the speculation that metaphyton is resistant to grazing is the implication that metaphyton mats would be self-stabilizing. Conversely, nutrient-enrichment experiments by Portielje and Lijklema (1994) in macrophyte-free shallow channels, in which a transitory bloom of *Cladophora* gave way to phytoplankton, indicate that metaphyton dominance can be short-lived. Conditions determining the persistence of metaphyton, and the wetland state to which transition may pass (epiphyton or phytoplankton), include the presence or absence of macrophytes, irradiance level, and the extent of damage to metaphyton caused by fish and physical factors. As macrophytes are necessary to stabilize the water column and to provide inoculum for metaphyton, their absence may lead to phytoplankton dominance, whereas if sufficient macrophytes persist to provide algal substratum, epiphyton may resume abundance (Fig. 4).

D. Lake State

The Lake State is distinguished by high water and nutrient levels, and a turbid water column containing abundant phytoplankton (Fig. 4 and Table V). Opportunities for development of epiphyton, metaphyton, and epipelon are poor owing to low irradiance and limited macrophyte substratum. The Lake State may be the consequence of rapid water input to other wetland states that leads to loss of macrophytes that otherwise promote dominance by epiphyton or metaphyton. As discussed earlier, the Lake State persists if water levels remain high and grazing pressure on phytoplankton is low.

As the plant canopy opens with loss of emergent macrophytes from deeper portions of the wetland, metaphyton is restricted to peripheral, calm areas. Progressive loss of macrophytes provides decreasing habitat for epiphyton, whereas increased wind action leads to elimination of metaphyton. Consequently, phytoplankton dominates a Lake State wetland because of reduced competition for nutrients from epiphytes, metaphyton, and macrophytes (Fig. 4).

VIII. CONCLUSION

Despite the abundance of wetland habitat globally, phycological research has focused on rivers, lakes, and oceans. Consequently, information on the algal assemblages of wetlands remains fragmentary. As Klarer and Millie (1992) contend, "the role of algae in the wetlands ecosystem is largely unknown and represents a major void in our understanding of wetland ecology." The characteristic shallowness of wetlands, and the high surface area of submersed and emergent macrophytes, provides abundant habitat for the development of epiphytic and, in some cases, metaphytic algae. In areas of lower macrophyte density, epipelic algae and plocon crusts can develop on illuminated sediments. In addition, detached benthic algae can contribute significantly to the phytoplankton assemblage in the water column. Major factors affecting the abundance and predominant type of wetland algae are the hydrodynamics of the system, the relative stability of its water column, the supply of colonizable substratum and nutrients, and herbivory. Future research should expand our knowledge of the environmental controls on algal biomass, productivity, and species composition in wetlands, with particular emphasis in areas for which knowledge is especially incomplete. Included among these may be a detailed evaluation of the proportionate contributions by epipelon, epiphyton, metaphyton, and phytoplankton to food web dynamics in each of the four wetland states; further study of the epipelon, plocon, and metaphyton of freshwater wetlands; and research on wetlands in less intensively studied subarctic and tropical areas.

REFERENCES

Allen, H. L. (1971). Primary productivity, chemo-organotrophy, and nutritional interactions of epiphytic algae and bacteria on macrophytes in the littoral of a lake. *Ecol. Monogr.* **41**, 97–127.

Atchue, J. A., III, Day, F. P., Jr., and Marshall, H. G. (1983). Algal dynamics and nitrogen and phosphorous cycling in a cypress stand in the seasonally flooded Great Dismal Swamp. *Hydrobiologia* **106**, 115–122.

Barica, J. (1975). Collapses of algal blooms in prairie pothole lakes: Their mechanism and ecological impact. *Verh.—Int. Theor. Angew. Limnol.* **19**, 606–615.

Barica, J. (1978). Variability in ionic composition and phytoplankton biomass of saline eutrophic prairie lakes within a small geographic area. *Arch. Hydrobiol.* **81**, 304–326.

Barica, J. (1990). Seasonal variability of N : P ratios in eutrophic lakes. *Hydrobiologia* **191**, 97–103.

Barica, J., Kling, H., and Gibson, J. (1980). Experimental manipulation of algal bloom composition by nitrogen addition. *Can. J. Fish. Aquat. Sci.* **37**, 1175–1183.

Barko, J. W., Gunnison, D., and Carpenter, S. R. (1991). Sediment interactions with submersed macrophyte growth and community dynamics. *Aquat. Bot.* **41**, 41–65.

Batt, B. D. J., Anderson, M. G., Anderson, C. D., and Caswell, F. D. (1989). The use of prairie potholes by North American ducks. *In* "Northern Prairie Wetlands" (A. van der Valk, ed.), pp. 204–227. Iowa State Univ. Press, Ames.

Bazely, D. R., and Jefferies, R. L. (1989). Lesser snow geese and the nitrogen economy of a grazed salt marsh. *J. Ecol.* **77,** 24–34.
Bosserman, R. W. (1983). Elemental composition of *Utricularia*–periphyton ecosystems from Okefenokee Swamp. *Ecology* **Ecology 64,** 1637–1645.
Briggs, S. V., Maher, M. T., and Tongway, D. J. (1993). Dissolved and particulate organic carbon in two wetlands in southwestern New South Wales, Australia. *Hydrobiologia* **264,** 13–19.
Brix, H. (1994). Functions of macrophytes in constructed wetlands. *Water Sci. Technol.* **29,** 71–78.
Browder, J. A., Gleason, P. J., and Swift, D. R. (1994). Periphyton in the Everglades: Spatial variation, environmental correlates, and ecological implications. *In* "Everglades: The Ecosystem and Its Restoration" (S. M. Davis and J. C. Ogden, eds.), pp. 379–418. St. Lucie Press, Delray Beach, FL.
Brown, D. J. (1972). Primary production and seasonal succession of the phytoplankton component of Crescent Pond, Delta Marsh, Manitoba. MSc. Thesis, University of Manitoba.
Burkholder, J. M., and Wetzel, R. G. (1989). Epiphytic microalgae on natural substrata in a hardwater lake: Seasonal dynamics of community structure, biomass and ATP content. *Arch. Hydrobiol., Suppl.* **83,** 1–56.
Campeau, S., Murkin, H. R., and Titman, R. D. (1994). Relative importance of algae and emergent plant litter to freshwater marsh invertebrates. *Can. J. Fish. Aquat. Sci.* **51,** 681–692.
Carlton, R. G., and Wetael, R. G. (1988). Phosphorus flux from lake sediments: Effects of epipelic algal oxygen production. *Limnol. Oceanogr.* **33,** 562–570.
Carper, G. L., and Bachmann, R. W. (1984). Wind resuspension of sediments in a prairie lake. *Can. J. Fish. Aquat. Sci.* **41,** 1763–1767.
Cronk, J. K., and Mitsch, W. J. (1994a). Periphyton productivity on artificial and natural surfaces in constructetd freshwater wetlands under different hydrologic regimes. *Aquat. Bot.* **48,** 325–341.
Cronk, J. K., and Mitsch, W. J. (1994b). Water column primary productivity in four newly constructed freshwater wetlands with different hydrologic inputs. *Ecol. Eng.* **3,** 449–468.
Crumpton, W. G. (1989). Algae in northern prairie wetlands. *In* "Northern Prairie Wetlands" (A. van der Valk, ed.), pp. 188–203. Iowa State Univ. Press, Ames.
de Jonge, V. N., and Colijn, F. (1994). Dynamics of microphytobenthos biomass in the Ems estuary. *Mar. Ecol.: Prog. Ser.* **104,** 185–196.
de Jonge, V. N., and van Beusekom, J. E. E. (1992). Contribution of resuspended microphytobenthos to total phytoplankton in the Ems estuary and its possible role for grazers. *Neth. J. Sea Res.* **30,** 91–105.
Dieter, C. D. (1990). The importance of emergent vegetation in reducing sediment resuspension in wetlands. *J. Freshwater Ecol.* **5,** 467–473.
Dodds, W. K. (1991). Community interactions between the filamentous alga *Cladophora glomerata* (L.) Kützing, its epiphytes, and epiphytic grazers. *Oecologia* **85,** 572–580.
Dodds, W. K., and Gudder, D. A. (1992). The ecology of *Cladophora. J. Phycol.* **28,** 415–427.
Dolan, T. J., Bayley, S. E., Zoltek, J., Jr., and Hermann, A. J. (1981). Phosphorus dynamics of a Florida freshwater marsh receiving treated wastewater. *J. Appl. Ecol.* **18,** 205–219.
Durako, M. J., Medlyn, R. A., and Moffler, M. D. (1982). Particulate matter resuspension via metabolically produced gas bubbles from benthic estuarine microalgae communities. *Limnol. Oceanogr.* **27,** 752–756.
Duthie, H. C. (1965). A study of the distribution and periodicity of some algae in a bog pool. *J. Ecol.* **53,** 343–359.
Eaton, J. W., and Moss, B. (1966). The estimation of numbers of pigment content in epipelic algal populations. *Limnol. Oceanogr.* **11,** 584–585.

Eiseltová, and Pokorný, J. (1994). Filamentous algae in fish ponds of the Trebon Biosphere Reserve—Ecophysiological study. *Vegetatio* **113,** 155–170.

Elakovich, S. D., and Wooten, J. W. (1989). Allelophathic potential of sixteen aquatic and wetland plants. *J. Aquat. Plant Manage.* **27,** 78–84.

Eminson, D., and Moss, B. (1980). The composition and ecology of periphyton communities in freshwaters. 1. The influence of host type and external environment on community composition. *Br. Phycol. J.* **15,** 429–446.

Flensburg, T., and Sparling, J. H. (1973). The algal microflora of a string mire in relation to the chemical composition of the water. *Can. J. Bot.* **51,** 743–749.

Fong, P., and Zedler, J. B. (1993). Temperature and light effects on the seasonal succession of algal communities in shallow coastal lagoons. *J. Exp. Mar. Biol. Ecol.* **171,** 259–272.

Gabor, T. S., Murkin, H. R., Stainton, M. P., Boughen, J. A., and Titman, R. D. (1994). Nutrient additions to wetlands in the interlake region of Manitoba, Canada: Effects of a single pulse addition in spring. *Hydrobiologia* **280,** 497–510.

Gallagher, J. L., and Daiber, F. C. (1974). Primary production of edaphic algal communities in a Delaware salt marsh. *Limnol. Oceanogr.* **19,** 390–395.

Gallepp, G. W. (1979). Chironomid influence on phosphorus release in sediment–water microcosms. *Ecology* **60,** 547–556.

Ganf, G. G. (1973). Incident solar irradiance and underwater light penetration as factors controlling the chlorophyll *a* content of a shallow equatorial lake (Lake George, Uganda). *J. Ecol.* **62,** 593–609.

Goldsborough, L. G. (1993). Diatom ecology in the phyllosphere of the common duckweed (*Lemna minor* L.). *Hydrobiologia* **269/279,** 463–471.

Goldsborough, L. G., and Brown, D. J. (1991). Periphyton production in a small, dystrophic pond on the Canadian Precambrian Shield. *Verh.—Int. Ver. Theor. Angew. Limnol.* **24,** 1497–1502.

Goldsborough, L. G., and Robinson, G. G. C. (1983). The effect of two triazine herbicides on the productivity of freshwater marsh periphyton. *Aquat. Toxicol.* **4,** 95–112.

Goldsborough, L. G., and Robinson, G. G. C. (1985). Effect of an aquatic herbicide on sediment nutrient flux in a freshwater marsh. *Hydrobiologia* **122,** 121–128.

Goldsborough, L. G., and Robinson, G. G. C. (1986). Changes in periphytic algal community structure as a consequence of short herbicide exposures. *Hydrobiologia* **139,** 177–192.

Goldsborough, L. G., Robinson, G. G. C., and Gurney, S. E. (1986). An enclosure/substratum system for *in situ* ecological studies of periphyton. *Arch. Hydrobiol.* **106,** 373–393.

Good, R. E., Whigham, D. F., and Simpson, R. L. (1978). "Freshwater Wetlands: Ecological Processes and Management Potential." Academic Press, New York.

Granhall, U., and Selander, H. (1973). Nitrogen fixation in a subarctic mire. *Oikos* **24,** 8–15.

Grant, J., Mills, E. L., and Hopper, C. M. (1986). A chlorophyll budget of the sediment–water interface and the effect of stabilizing biofilms on particle fluxes. *Ophelia* **26,** 207–219.

Grimshaw, H. J., Rosen, M., Swift, D. R., Rodberg, K., and Noel, J. M. (1993). Marsh phosphorus concentrations, phosphorus content and species composition of Everglades periphyton communities. *Arch. Hydrobiol.* **128,** 257–276.

Gurney, S. E., and Robinson, G. G. C. (1988). The influence of water level manipulation on metaphyton production in a temperate freshwater marsh. *Verh.—Int. Ver. Theor. Angew. Limnol.* **23,** 1932–1040.

Gurney, S. E., and Robinson, G. G. C. (1989). The influence of two triazine herbicides on the productivity, biomass and community composition of freshwater marsh periphyton. *Aquat. Bot.* **36,** 1–22.

Haertel, L. (1976). Nutrient limitation of algal standing crops in shallow prairie lakes. *Ecology* **57,** 664–678.

Haines, D. W., Rogers, K. H., and Rogers, F. E. J. (1987). Loose and firmly attached epiphyton, their relative contributions to algal and bacterial carbon productivity in a *Phragmites* marsh. *Aquat. Bot.* **29,** 169–176.

Hall, S. L., and Fisher, F. M. (1985). Annual productivity and extracellular release of dissolved organic compounds by the epibenthic algal community of a brackish marsh. *J. Phycol.* **21,** 277–281.

Hamilton, D. P., and Mitchell, S. F. (1988). Effects of wind on nitrogen, phosphorous, and chlorophyll in a shallow New Zealand lake. *Verh.—Int. Ver. Theor. Angew. Limnol.* **23,** 624–628.

Hann, B. J. (1991). Invertebrate grazer–periphyton interactions in a eutrophic marsh pond. *Freshwater Biol.* **26,** 87–96.

Hanson, M. A., and Butler, M. G. (1990). Early responses of plankton and turbidity to biomanipulation in a shallow prairie lake. *Hydrobiologia* **200/201,** 317–327.

Hanson, M. A., and Butler, M. G. (1994). Responses to food web manipulation in a shallow waterfowl lake. *Hydrobiologia* **279/280,** 457–466.

Hansson, L.-A. (1989). The influence of a periphytic biolayer on phosphorus exchange between substrate and water. *Arch. Hydrobiol.* **115,** 21–26.

Happey-Wood, C. M. (1980). Periodicity of epipelic unicellular Volvocales (Chlorophyceae) in a shallow acid pool. *J. Phycol.* **16,** 116–128.

Hellström, T. (1991). The effect of resuspension on algal production in a shallow lake. *Hydrobiologia* **213,** 183–190.

Henebry, M. S., and Cairna, J., Jr. (1984). Protozoan colonization rates and trophic status of some freshwater wetland lakes. *J. Protozool.* **31,** 456–467.

Hillebrand, H. (1983). Development and dynamics of floating clusters of filamentous algae. *In* "Periphyton of Freshwater Ecosystems" (R. G. Wetzel, ed.), pp. 31–39. Dr. W. Junk Publishers, The Hague.

Holland, A. F., Zingmark, R. G., and Dean, J. M. (1974). Quantitative evidence concerning the stabilization of sediments by marine benthic diatoms. *Mar. Biol.* **27,** 191–196.

Hooper, N. M., and Robinson, G. G. C. (1976). Primary production of epiphytic algae in a marsh pond. *Can. J. Bot.* **54,** 2810–2815.

Hooper-Reid, N. M., and Robinson, G. G. C. (1978a). Seasonal dynamics of epiphytic algal growth in a marsh pond: Productivity, standing crop, and community composition. *Can. J. Bot.* **56,** 2434–2440.

Hooper-Reid, N. M., and Robinson, G. G. C. (1978b). Seasonal dynamics of epiphytic algal growth in a marsh pond: Composition, metabolism, and nutrient availability. *Can. J. Bot.* **56,** 2441–2448.

Hosseini, S. M., and van der Valk, A. G. (1989a). Primary productivity and biomass of periphyton and phytoplankton in flooded freshwater marshes. *In* "Freshwater Wetlands and Wildlife" (R. R. Sharitz and J. W. Gibbons, eds.). DOE Symp. Ser. No. 61, pp. 303–315. USDOE Office of Scientific and Technical Information, Oak Ridge, TN.

Hosseini, S. M., and van der Valk, A. G. (1989b). The impact of prolonged, above-normal flooding on metaphyton in a freshwater marsh. *In* "Freshwater Wetlands and Wildlife" (R. R. Shritz and J. W. Gibbons, eds.). DOE Sym. Ser. No. 61, pp. 317–324. USDOE Office of Scientific and Technical Information, Oak Ridge, TN.

Howard-Williams, C. (1981). Studies on the ability of a *Potamogeton pectinatus* community to remove dissolved nitrogen and phosphorus compounds from lake water. *J. Appl. Ecol.* **18,** 619–637.

Jansson, M. (1980). Role of benthic algae in transport of nitrogen from sediment to lake water in a shallow clearwater lake. *Arch. Hydrobiol.* **89,** 101–109.

Jones, K. (1974). Nitrogen fixation in a salt marsh. *J. Ecol.* **62,** 553–565.

Kadlec, J. A. (1979). Nitrogen and phosphorus dynamics in inland freshwater wetlands. *In* "Waterfowl and Wetlands: An Integrated Review" (T. A. Bookhout, ed.), pp. 17–41. The Wildlife Society, Madison, WI.

Kadlec, J. A. (1986). Effects of flooding on dissolved and suspended nutrients in small diked marshes. *Can. J. Fish. Aquat. Sci.* **43,** 1999–2008.

Kadlec, J. A. (1989). Effects of deep flooding and drawdown on freshwater marsh sediments. *In* "Freshwater Wetlands and Wildlife" (R. R. Sharitz and J. W. Gibbons, eds.), DOE Symp. Ser. No. 61, pp. 127–143. USDOE Office of Scientific and Technical Information, Oak Ridge, TN.

Keough, J. R., Sierszen, M. E., and Hagley, C. A. (1993). Analysis of a Great Lakes coastal wetland food web using stable isotope ratios. *Bull. Ecol. Soc. Am.* **74**(Suppl.2), 304–305.

Kingston, J. C. (1982). Association and distribution of common diatoms in surface samples from northern Minnesota peatlands. *Nova Hedwigia, Beih.* **73,** 333–346.

Klarer, D. M., and Millie, D. F. (1992). Aquatic macrophytes and algae at Old Woman Creek estuary and other Great Lakes coastal wetlands. *J. Great Lakes Res.* **18,** 622–633.

Kotak, B. G. (1990). The effects of water turbulence on the limnology of a shallow, prairie wetland. M.Sc. Thesis, University of Manitoba.

Kotak, B. G., and Robinson, G. G. C. (1991). Artificially-induced water turbulence and the physical and biological features within small enclosures. *Arch. Hydrobiol.* **122,** 335–349.

Kruszynski, G. M. (1989). Investigations into the existence of asociations within benthic diatom communities. M.Sc. Thesis, University of Manitoba.

Landers, D. H. (1982). Effects of naturally senescing aquatic macrophytes on nutrient chemistry and chlorophyll *a* of surrounding waters. *Limnol. Oceanogr.* **27,** 428–439.

Leah, R. T., Moss, B., and Forrest, D. E. (1978). Experiments with large enclosures in a fertile, shallow, brackish lake, Hickling Broad, Norfolk, United Kingdom. *Int. Rev. Gesamten Hydrobiol.* **63,** 291–310.

Losee, R. F., and Wetzel, R. G. (1983). Selective light attenuation by the periphyton complex. *In* "Periphyton of Freshwater Ecosystems" (R. G. Wetzel, ed.), pp. 89–96. Dr. W. Junk Publishers, The Hague.

Manny, B. A., Johnson, W. C., and Wetzel, R. G. (1994). Nutrient additions by waterfowl to lakes and reservoirs—Predicting their effects on productivity and water quality. *Hydrobiologia* **280,** 121–132.

Mason, C. F., and Bryant, R. J. (1975). Periphyton production and grazing by chironomids in Alderfen Broad, Norfolk. *Freshwater Biol.* **5,** 271–277.

McRoy, C. P., and Goering, J. J. (1974). Nutrient transfer between the eelgrass *Zostera marina* and its epiphytes. *Nature (London)* **248,** 173–174.

Meijer, M. A., deHaan, M. W., Breukelaar, A. W., and Buiteveld, H. (1990). Is reduction of the benthivorous fish as important cause of high transparency following biomanipulation in shallow lakes? *Hydrobiologia* **200/201,** 303–315.

Mihuc, T., and Toetz, D. (1994). Determination of diets of alpine aquatic insects using stable isotopes and gut analysis. *Am. Midl. Nat.* **131,** 146–155.

Millie, D. F., and Lowe, R. L. (1983). Studies on Lake Erie's littoral algae: Host specificity and temporal periodicity of epiphytic diatoms. *Hydrobiologia* **99,** 7–18.

Mitsch, W. J., and Gosselink, J. G. (1993). "Wetlands," 2nd ed. Van Nostrand-Reinhold, New York.

Mitsch, W. J., and Reeder, B. C. (1991). Modelling nutrient retention of a freshwater coastal wetland: Estimating the roles of primary productivity, sedimentation, resuspension and hydrology. *Ecol. Modell.* **54,** 151–187.

Moraghan, J. T. (1993). Loss and assimilation of ^{15}N-nitrate added to a North Dakota cattail marsh. *Aquat. Bot.* **46,** 225–234.

Moss, B. (1968). The chlorophyll *a* content of some benthic algal communities. *Arch. Hydrobiol.* **65,** 51–62.

Moss, B. (1976). The effects of fertilization and fish on community structure and biomass of aquatic macrophytes and epiphytic algal populations: An ecosystem experiment. *J. Ecol.* **64,** 313–342.

Moss, B. (1981). The composition and ecology of periphyton communities in freshwaters. II. Inter-relationships between water chemistry, phytoplankton populations and periphyton populations in a shallow lake and associated experimental reservoirs ('Lund tubes'). *Br. Phycol. J.* **16,** 59–76.

Moss, B. (1983). The Norfolk Broadland: Experiments in the restoration of a complex wetland. *Biol. Rev. Cambridge Philos. Soc.* **58,** 521–561.

Moss, B., and Leah, R. T. (1982). Changes in the ecosystem of a guanotrophic and brackish shallow lake in eastern England: Potential problems in its restoration. *Int. Rev. Gesamten Hydrobiol.* **67,** 625–659.

Murkin, E. J., Murkin, H. R., and Titman, R. D. (1992). Nektonic invertebrate abundance and distribution at the emergent vegetation–open water interface in the Delta Marsh, Manitoba, Canada. *Wetlands* **12,** 45–52.

Murkin, H. R. (1989). The basis for food chains in prairie wetlands. *In* "Northern Prairie Wetlands" (A. van der Valk, ed.), pp. 316–338. Iowa State Univ. Press, Ames.

Murkin, H. R., van der Walk, A. G., and Davis, C. B. (1989). Decomposition of four dominant macrophytes in the Delta Marsh, Manitoba. *Wildl. Soc. Bull.* **17,** 215–221.

Murkin, H. R., Stainton, M. P., Boughen, J. A., Pollard, J. B., and Titman, R. D. (1991). Nutrient status of wetlands in the interlake region of Manitoba, Canada. *Wetlands* **11,** 105–122.

Murkin, H. R., Pollard, J. B., Stainton, M. P., Boughen, J. A., and Titman, R. D. (1994). Nutrient additions to wetlands in the interlake region of Manitoba, Canada. *Hydrobiologia* **280,** 483–495.

Neely, R. K. (1994). Evidence for positive interactions between epiphytic algae and heterotrophic decomposers during the decomposition of *Typha latifolia. Arch. Hydrobiol.* **129,** 443–457.

Neely, R. K., and Baker, J. L. (1989). Nitrogen and phosphorus dynamics and the fate of agricultural runoff. *In* "Northern Prairie Wetland" (A. van der Valk, ed.), pp. 92–131. Iowa State Univ. Press, Ames.

Neill, C., and Cornwell, J. C. (1992). Stable carbon, nitrogen, and sulfur isotopes in a prairie marsh food web. *Wetlands* **12,** 217–224.

Nicholls, K. H. (1976). Nutrient–phytoplankton relationships in the Holland Marsh, Ontario. *Ecol. Monogr.* **46,** 179–199.

Pakulski, J. D. (1992). Foliar release of soluble reactive phosphorus from *Spartina alterniflora* in a Georgia (USA) salt marsh. *Mar. Ecol.: Prog. Ser.* **90,** 53–60.

Parmenter, R. R., and Lamarra, V. A. (1991). Nutrient cycling in a freshwater marsh: The decomposition of fish and waterfowl carrion. *Limnol. Oceanogr.* 976–987.

Phillips, G. L., Eminson, D., and Moss, B. (1978). A mechanism to account for macrophyte decline in progressively eutrophicated freshwaters. *Aquat. Bot.* **4,** 103–126.

Phillips, G., Jackson, R., Bennett, C., and Chilvers, A. (1994). The importance of sediment phosphorus release in the restoration of very shallow lakes (The Norfolk Broads, England) and implications for biomanipulation. *Hydrobiologia* **275/276,** 445–456.

Pinckney, J., and Zingmark, R. G. (1991). Effects of tidal stage and sun angles on intertidal benthic microalgal productivity. *Mar. Ecol.: Prog. Ser.* **76,** 81–89.

Pinckney, J., and Zingmark, R. G. (1993a). Biomass and production of benthic microalgal communities in estuaring habitats. *Estuaries* **16,** 887–897.

Pinckney, J., and Zingmark, R. G. (1993b). Modeling the annual production of intertidal benthic microalgae in estuarine ecosystems. *J. Phycol.* **29,** 396–407.

Pinckney, J., and Zingmark, R. G. (1993c). Photophysiological responses of intertidal benthic microalgal communities to *in situ* light environments: Methodological considerations. *Limnol. Oceanogr.* **38,** 1373–1383.

Pip, E., and Robinson, G. G. C. (1982). A study of the seasonal dynamics of three phycoperiphytic communities using nuclear track autoradiography. II. Organic carbon uptake. *Arch. Hydrobiol.* **96,** 47–64.

Pip, E., and Robinson, G. G. C. (1984). A comparison of algal periphyton composition on eleven species of submerged macrophytes. *Hydrobiol. Bull.* **18**, 109–118.

Pip, E., and Stewart, J. M. (1976). The dynamics of two aquatic plant-snail associations. *Can. J. Zool.* **54**, 1192–1205.

Portielje, R., and Lijklema, L. (1994). Kinetics of luzury uptake of phosphate by algae-dominated benthic communities. *Hydrobiologia* **275/276**, 349–358.

Reeder, B. C. (1994). Estimating the role of autotrophs in nonpoint source phosphorus retention in a Laurentian Great Lakes coastal wetland. *Ecol. Eng.* **3**, 161–169.

Richardson, C. J., and Schwegler, B. R. (1986). Algal bioassay and gross productivity experiments using sewage effluent in a Michigan wetland. *Water Resour. Bull.* **22**, 111–120.

Richmond, K.-A. (1992). A comparison of photosynthesis of metaphyton in eutrophic littoral waters with that of an acidified lake. B.Sc. Thesis, University of Manitoba.

Robinson, G. G. C. (1988). "Productivity–irradiance Relationships of the Algal Communities in the Delta Marsh: A Preliminary Report," Annu. Rep. No. 23, pp. 100–110. University Field Station (Delta Marsh), University of Manitoba, Winnipeg.

Robinson, G. G. C., and Pip, E. (1983). The application of a nuclear track autoradiographic technique to the study of periphyton photosynthesis. *In* "Periphyton of Freshwater Ecosystems" (R. G. Wetzel, ed.), pp. 267–273. Dr. W. Junk Publishers, The Hague.

Robinson, G. G. C., Gurney, S. E., and Goldsborough, L. G. (1996a). The comparative biomass of algal associations of a prairie wetland under controlled water-level regimes. *Wetlands* (in review).

Robinson, G. G. C., Gurney, S. E., and Goldsborough, L. G. (1996b). The primary production of algal associations of a prairie wetland under controlled water-level regimes. *Wetlands* (in review).

Roos, P. J., and Meulemans, J. T. (1987). Under water light regime in a reedstand—Short-term, daily, and seasonal. *Arch. Hydrobiol.* **111**, 161–169.

Round, F. (1981). "The Ecology of Algae." Cambridge Univ. Press, Cambridge, UK.

Ruess, R. W., Kik, D. S., and Jefferies, R. L. (1989). The role of lesser snow geese as nitrogen processors in a sub-arctic salt marsh. *Oecologia* **79**, 23–29.

Sage, W. W., and Sullivan, M. J. (1978). Distribution of bluegreen algae in a Mississippi gulf coast salt marsh. *J. Phycol.* **14**, 333–337.

Saks, N. M., Stone, R. J., and Lee, J. J. (1976). Autotrophic and heterotrophic nutritional budget of salt marsh epiphytic algae. *J. Phycol.* **124**, 443–448.

Sand-Jensen, K. (1977). Effects of epiphytes on eelgrass photosynthesis. *Aquat. Bot.* **3**, 55–63.

Sand-Jensen, K., and Borum, J. (1984). Epiphyte shading and its effect on photosynthesis and diel metabolism of *Lobelia dortmanna* L. during the spring bloom in a Danish lake. *Aquat. Bot.* **20**, 109–119.

Schalles, J. F., and Shure, D. J. (1989). Hydrology, community structure, and productivity patterns of a dystrophic Carolina Bay wetland. *Ecol. Monogr.* **59**, 365–385.

Scherer, R. P. (1988). Freshwater diatom assemblages and ecology palaeoecology of the Okefenokee Swamp marsh complex, southern Georgia, U.S.A. *Diatom Res.* **3**, 129–157.

Schindler, D. W., and Comita, G. W. (1972). The dependence of primary production upon physical and chemical factors in a small, senescing lake, including the effects of complete water oxygen depletion. *Arch. Hydrobiol.* **69**, 423–451.

Schoenberg, S. A., and Oliver, J. D. (1988). Temporal dynamics and spatial variation of algae in relation to hydrology and sediment characteristics in the Okefenokee Swamp, Georgia. *Hydrobiolobia* **162**, 123–133.

Shamess, J. J. (1980). A description of the epiphytic, epipelic and planktonic algal communities in two shallow eutrophic lakes in southwestern Manitoba. M.Sc. Thesis, University of Manitoba.

Shamess, J. J., Robinson, G. G. C., and Goldsborough, L. G. (1985). The structure and comparison of periphytic and planktonic algal communities in two eutrophic prairie lakes. *Arch. Hydrobiol.* **103,** 99–116.

Shay, J. M., and Shay, C. T. (1986). Prairie marshes in western Canada, with specific reference to the ecology of five emergent macrophytes. *Can. J. Bot.* **64,** 443–454.

Sheehan, P. J., Baril, A., Mineau, P., Smith, D. K., Harfenist, A., and Marshall, W. K. (1987). "The Impact of Pesticides on the Ecology of Prairie Nesting Ducks," Can. Wildl. Serv. Tech. Rep. Ser. No. 19. Environment Canada, Ottawa.

Stanley, D. W. (1976). Productivity of epipelic algae in tundra ponds and a lake near Barrow, Alaska. *Ecology* **57,** 1015–1024.

Stanley, D. W., and Daley, R. J. (1976). Environmental control of primary productivity in Alaskan tundra ponds. *Ecology* **57,** 1025–1033.

Steward, K. K., and Ornes, W. H. (1975). Assessing a marsh environment for wastewater renovation. *J. Water Pollut. Control Fed.* **47,** 1880–1891.

Sullivan, M. J. (1975). Diatom communities from a Delaware salt marsh. *J. Phycol.* **11,** 384–390.

Sullivan, M. J., and Moncreiff, C. A. (1988). Primary production of adaphic algal communities in a Mississippi salt marsh. *J. Phycol.* **24,** 49–58.

Sullivan, M. J., and Moncreiff, C. A. (1990). Edaphic algae are an important component of salt marsh food-webs: Evidence from multiple stable isotope analyses. *Mar. Ecol.: Prog. Ser.* **62,** 149–159.

Suthers, I. M., and Gee, J. H. (1986). Role of hypoxia in limiting diel spring and summer distribution of juvenile yellow perch (*Perca flavescens*) in a prairie marsh. *Can. J. Fish. Aquat. Sci.* **43,** 1562–1570.

Thurman, E. M. (1985). "Organic Geochemistry of Natural Waters." Martinus Nijhoff/Dr. W. Junk Publishers, Dordrecht, The Netherlands.

Timms, R. M., and Moss, B. (1984). Prevention of growth of potentially dense phytoplankton populations by zooplankton grazing in the presence of zooplanktivorous fish in a shallow wetland ecosystem. *Limnol. Oceanogr.* **29,** 472–486.

Tracy, E. J., and South, G. R. (1989). Composition and seasonality of micro-algal mats on a salt marsh in New Brunswick, Canada. *Br. Phycol. J.* **24,** 285–291.

van der Valk, A. G. (1986). The impact of litter and annual plants on recruitment from the seed bank of a lacustrine wetland. *Aquat. Bot.* **24,** 13–26.

van der Valk, A. G. (1994). Effects of prolonged flooding on the distribution and biomass of emergent species along a freshwater wetland coenocline. *Vegetatio* **110,** 185–196.

van der Valk, A. G., and Davis, C. B. (1978). The role of seed banks in the vegetation dynamics of prairie glacial marshes. *Ecology* **59,** 322–335.

van Meter-Kasanof, N. (1973). Ecology of the microalgae of the Florida Everglades. Part 1. Environment and some aspects of freshwater periphyton, 1959 to 1963. *Nova Hedwigia* **24,** 619–664.

van Raalte, C. D., Valiela, I., and Teal, J. M. (1976). Production of epibenthic salt marsh algae: Light and nutrient limitations. *Limnol. Oceanogr.* **21,** 862–872.

Vernberg, F. J. (1993). Salt-marsh processes: A review. *Environ. Toxicol. Chem.* **12,** 2167–2165.

Vörös, L., and Padisák, J. (1991). Phytoplankton biomass and chlorophyll-*a* in some shallow lakes in central Europe. *Hydrobiologia* **215,** 111–119.

Watson, R. A., and Osborne, P. L. (1979). An algal pigment ratio as an indicator of the nitrogen supply to phytoplankton in three Norfolk broads. *Freshwater Biol.* **9,** 585–594.

Weaks, T. E. (1988). Allelopathic interference as a factor influencing the periphyton community of a freshwater marsh. *Arch. Hydrolbiol.* **111,** 369–382.

Weller, M. W., and Fredrickson, L. H. (1973). Avian ecology of a managed glacial marsh. *Living Bird* **12,** 269–291.

Wetzel, R. G. (1983). "Limnology," 2nd ed. Saunders College Publishing, New York.

Wetzel, R. G., and Manny, B. A. (1972). Secretion of dissolved organic carbon and nitrogen by aquatic macrophytes. *Verh.—Int. Ver. Theor. Agnew. Limnol.* **18,** 162–170.
Whitton, B. (1970). Biology of *Cladophira* in freshwaters. *Water Res.* **4,** 457–476.
Wium-Andersen, S., Anthoni, U., Christophersen, C., and Houen, G. (1982). Allelophathic effects on phytoplankton by substances isolated from aquatic macrophytes (Charales). *Oikos* **39,** 187–190.
Woelkerling, W. J. (1976). Wisconsin desmids. I. Aufwuchs and plankton communities of selected acid bogs, alkaline bogs, and closed bogs. *Hydrobiologia* **48,** 209–232.
Wrubleski, D. A. (1987). Chironomidae (Diptera) of peatlands and marshes in Canada. Mem. Entomol. Soc. Can. **140,** 141–161.
Wrubleski, D. A. (1991). Chironomidae (Diptera) community development following experimental manipulation of water levels and aquatic vegetation. Ph.D. Thesis, University of Alberta.
Ytow, N., Seki, H., Mizusaki, T., Kojima, S., and Batomalaque, A. E. (1994). Trophodynamic structure of a swampy bog at the climax stage of limnological succession. I. Primary production dynamics. *Water, Air, Soil Pollut.* **76,** 467–479.
Zedler, J. B. (1980). Algal mat productivity: Comparisons in a salt marsh. *Estuaries* **3,** 122–131.
Zoltai, S. C. (1988). Wetland environments and classification. *In* "Wetlands of Canada, Ecological Land Classification Series, No. 24" (National Wetlands Working Group, ed.), pp. 1–26. Environment Canada, Ottawa.

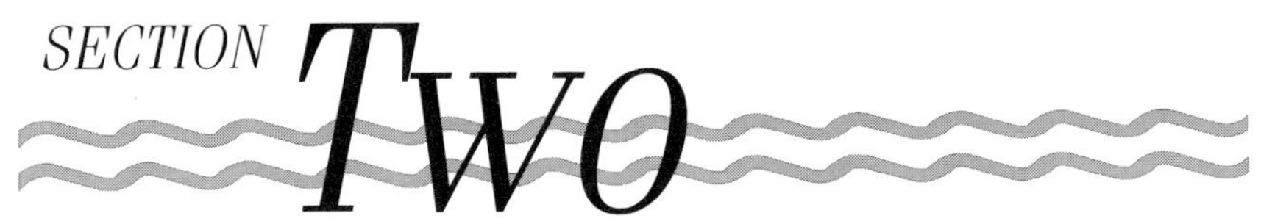

SECTION Two

FACTORS AFFECTING BENTHIC ALGAE

Effects of Light

Walter Hill

Environmental Sciences Division
*Oak Ridge National Laboratory**
Oak Ridge, Tennessee 37831

I. INTRODUCTION

Light is a fundamental variable for benthic algae, allowing these organisms to photosynthesize inorganic compounds into living biomass.

*Managed by Lockheed Martin Energy Research, under contract DE-AC05-96OR22464 with the Department of Energy.

Because photosynthesis responds quantitatively to changes in light, environmental variation in its quantity and quality potentially accounts for much of the variation in the physiology, population growth, and community structure of benthic algae. Other environmental factors such as substrate stability, temperature, nutrients, and grazers may influence the distribution and abundance of benthic algae, but adequate light is clearly a prerequisite for a phototrophic existence.

Despite light's crucial role in photosynthesis and growth, its effect on freshwater benthic algae remains understudied. Compared to the large number of publications dealing with nutrient effects, relatively few papers have been published on light effects on benthic algae. Research on algal–light interactions in the freshwater benthos also lags considerably behind that in the water column. For example, fewer than 20 photosynthesis–irradiance (P–I) studies on benthic algae have been published for freshwater benthic algae, whereas hundreds of phytoplankton P–I responses are in the literature (e.g., Richardson *et al.*, 1983). This paucity of effort on the part of freshwater benthic phycologists is certainly not due to an absence of interesting research topics. Benthic algae are subject to extreme light regimes, and they face a variety of special problems: (1) existence in a physically compressed community that promotes severe self-shading; (2) shading by macrophytes (terrestrial and aquatic); (3) light attenuation by phytoplankton, inorganic particles, and dissolved substances in the overlying water column; (4) accentuated temporal and spatial variability in light regimes caused by adjacent terrestrial features; and (5) growth in a relatively fixed position under potentially inhibiting visible and ultraviolet light.

II. PHOTOSYNTHETIC PROCESSES

Physiologists separate photosynthesis into two processes: *light reactions* and *dark reactions.* Light reactions involve the capture of light energy by photosynthetic pigments and the transfer of this energy to ATP and NADPH. Water is split as part of these reactions, making available electrons that reduce NADP. The light reactions occur in the thylakoids, where the photosynthetic pigments are located. Because these reactions are photochemical in nature and not enzymatically driven, they are not influenced by temperature. Dark reactions are so named because they can occur in the absence of light, but they actually follow the light reactions by only a few hundredths of a second. Dark reactions utilize the chemical energy and reducing power of ATP and NADPH to reduce CO_2 to hexose. The dark reactions occur in the spaces between thylakoids (the stroma in eukaryote chloroplasts), and because they are principally chemical reactions, are temperature sensitive. Both light and dark reactions have ecophysiological manifestations, particularly in photosynthesis–irradiance responses. However, light reactions and the cellular structures supporting them are generally more relevant to the discussion of light effects on benthic algae, so they will be treated in greater detail.

There are two sets of light reactions in photosynthesis: *photosystem I* (PS I) is associated with the reduction of NADP to NADPH, and *photosystem II* (PS II) is responsible for the splitting of water (and the consequent evolution of oxygen). Each photosystem consists of a reaction center, where a special core molecule of chlorophyll *a* mediates electron transfer, and an assembly of light-absorbing molecules that function like antennae to absorb and funnel light energy to the reaction centers. Antennae pigments can be more molecules of chlorophyll *a*, or a variety of *accessory* pigments such as chlorophyll *b*, chlorophyll *c*, carotenoids, phycocyanins, and phycoerythrins, depending on the particular taxon of algae. Although both photosystems are required for complete photosynthesis, they are physically distinct and have independent requirements for photons (e.g., Hoober, 1984). The two photosystems even appear to have different types of antenna pigments in some taxa. For example, in rhodophytes and cyanobacteria, the phycoerythrins appear to be connected primarily to PS II, whereas most of the chlorophyll *a* molecules funnel energy to PS I reaction centers (Kirk, 1983).

Algae and other phototrophs do not use all wavelengths of solar radiation. Photosynthesis is restricted to wavelengths of 400–700 nm, a range of wavelengths termed *photosynthetically active radiation (PAR)*, and different wavelengths of PAR are used with different efficiencies. At subsaturating intensities, the action spectrum of photosynthesis resembles the absorption spectrum of the principal light-harvesting (antennae) pigments. Consequently, photosynthesis by taxa with antenna pigments of mostly chlorophyll *a* and *b* (green algae) peaks at around 440 and 680 nm, whereas photosynthesis by taxa with phycoerythrin (red and some cyanobacteria) often shows peaks at 570 nm (Kirk, 1983). Observations on the taxonomic differences in absorption and photosynthesis maxima in algae gave rise to the notion of *chromatic adaptation*, in which the depth distribution of algae is explained by differences in the spectral composition of light at different depths. However, the experimental support for chromatic adaptation is weak (e.g., Ramus, 1983).

III. BENTHIC LIGHT ENVIRONMENTS

Benthic light environments are quite variable. Maximum light intensities to which benthic algae are exposed range from near zero in deep or turbid lakes to full sunlight in shallow, clear waters. There is considerable temporal variability in benthic environments as well, ranging from short term (e.g., sunflecks in forested streams) to long term (e.g., seasonal cycles in day length and sun angle). Before light reaches algal cells on the bottom of freshwater ecosystems, it must pass through a variety of obstacles whose integrated effect may leave few photons available for photosynthesis.

A. Terrestrial Vegetation

The shading effects of terrestrial vegetation are most pronounced in streams, where the leaf canopy of adjacent trees can completely overhang the streambed. Lakes and rivers are less affected by the shade of terrestrial vegetation because these bodies of water are large enough to prevent canopy overhang, but benthic areas close to the water's edge may still be shaded for at least part of the day. In small streams, leaf canopies can intercept 95% or more of incident radiation, reducing maximum photon flux densities (PFDs) to less than 40 μmol m^{-2} s^{-1} (Hill *et al.,* 1995). The effects of streamside vegetation diminish downstream as the streambed widens, resulting in a longitudinal continuum of light reaching the water surface from headwaters to estuaries (Vannote *et al.,* 1980).

Streamside vegetation not only affects light regimes by reducing the total amount of light, but it also creates considerable heterogeneity. Canopy cover is rarely uniform even in undisturbed forests. Localized areas of the streambed can receive direct sunlight that is several orders of magnitude more intense than the diffuse irradiation received by surrounding shaded areas. These areas range in size from sunflecks of centimeter-scale dimensions to sunpatches meters in scale. Sunflecks are transient: they move across the streambed as the sun moves across the sky, and the high light intensity of a sunfleck on any particular point on the streambed may last for as little as 2 s. Although their duration may be brief, sunflecks can contribute from 10 to 85% of total daily irradiance because of their high PFDs, which approach those of full sunlight (Chazdon and Pearcy, 1991). Seasonal variation in light regime is enhanced by terrestrial vegetation if the vegetation is deciduous. Leaf emergence in spring and leaf-fall in autumn cause orders-of-magnitude changes in photon flux to streams located in deciduous forests.

As sunlight filters through vegetation, its spectral distribution is altered by photosynthetic pigments. Absorption of red and blue wavelengths by leaf chlorophyll results in a light environment in forests that is weighted toward green wavelengths (Federer and Tanner, 1966). Large gaps in the vegetative canopy that are not directly illuminated by the sun produce spectral distributions that are weighted toward blue wavelengths because the major light source is diffuse light from the blue sky (Endler, 1993). The contribution of blue wavelengths to the light environment beneath large gaps or relatively open stream sites increases late in the day because the proportion of unfiltered diffuse skylight (blue sky) increases as sun angle decreases (DeNicola and Hoagland, 1992). When clouds block the sun, light in gaps of almost any size is nearly always white (Endler, 1993). Although the relative contributions of different wavelengths of PAR may increase with vegetation cover over streams, it is important to remember that shade decreases the absolute quantities of all wavelengths.

B. Attenuation by the Water Column

Light is attenuated exponentially as it penetrates the water column. Attenuation occurs because water, dissolved organic matter, suspended inorganic particles, and phytoplankton scatter and absorb light. In clear waters with little dissolved or particulate matter, 1% of surface intensity (the traditional limit of the photic zone) can be measured deeper than 50 m. In highly productive or turbid waters, light is attenuated to 1% of surface intensity in less than a meter of depth. Attenuation of light by the water column is usually of greater importance to benthic algae in lakes than in running waters, but eroded silts and clays seriously reduce light penetration in some streams (e. g., Davies-Colley *et al.*, 1992). Water column effects in running waters increase with increasing stream size because of the greater depths in large streams and rivers, and because these bodies are more likely to develop phytoplankton communities and carry larger loads of suspended inorganic particles. Attenuation is selective within the PAR range, and different components of the water column have different effects on the color of light reaching benthic algae. Pure water absorbs long wavelengths (red) best, but both ends of the spectrum are preferentially absorbed by water in clear lakes, resulting in the predominance of green light as depth increases. Dissolved humic substances strongly absorb short wavelengths (including ultraviolet light), transmitting red light to depth (Kirk, 1983; Watras and Baker, 1988). Phytoplankton are expected to absorb according to their dominant photosynthetic pigments, so that red (665 nm) and blue (450 nm) wavelengths should be attenuated most in lakes with well-developed phytoplankton communities containing large proportions of chlorophyll, but the light absorption properties of phytoplankton populations *in situ* have not been well studied (Kirk, 1983). Despite changes in their relative proportions, all wavelengths decrease with depth. Water column effects on light transmission have been described in detail in a variety of texts (e.g., Kirk, 1983; Wetzel, 1983).

C. Matrix Effects

Once light reaches the benthic community, it is further attenuated by the vertical matrix of algal cells and inorganic particles. These components effectively absorb and scatter light, attenuating downwelling light very quickly. Attenuation within the matrix is generally logarithmic, as it is in the water column, but it occurs on a much smaller scale. Photic zones are usually restricted to the upper few millimeters, even when surface intensities are relatively high (Meulemans, 1987; Dodds, 1992). One percent light levels may penetrate as deeply as 5–6 mm into epiphytic communities that contain relatively few inorganic particles (Meulemans, 1987). Although sand and clay particles are obviously important attenuators of light in sediment/epipelic communities, inorganic constituents may be responsible for only 20% or less of light attenuation in some epilithic or epiphytic communities (van Dijk, 1993). Attenuation generally decreases with decreasing organic matter (e.g., algal cells) and increasing particle size (Kuhl and Jorgensen, 1994), suggesting that light penetration should be deeper in coarse sand than in substrata dominated by small organic particles.

The relative contributions of absorption and scattering are difficult to determine if only downwelling light is measured, but measurements of upwelling light in marine microbenthic communities indicate that scattering is very significant for sedimentary substrates (Kuhl and Jorgensen, 1994). Backscattering by these communities actually increases the total (scalar) irradiance experienced by algal cells located in surface and subsurface layers to levels greater than incident light alone, especially at near-infrared wavelengths (Ploug *et al.*, 1993; Kuhl and Jorgensen, 1994). Backscattering appears to be more important in communities with high proportions of inorganic particles (Kuhl and Jorgensen, 1994), with absorption likely to be more important in organic matrices; algal cells have obviously evolved to be efficient absorbers.

The spectrum of light penetrating freshwater benthic algal matrices is undoubtedly altered by selective absorption and scattering. In particular, absorption by photosynthetic pigments in overlying algal cells should result in the selective attenuation of photosynthetically useful wavelengths, lowering the quality of light transmitted to lower layers. Relatively little work has been done to examine the spectral distribution of light within freshwater benthic algal communities, but in marine sediment microalgal communities, selective absorption by chlorophyll *a* results in subsurface light regimes depauperate in blue (430 nm) and red (675 nm) wavelengths (Ploug *et al.*, 1993). Accessory pigments also modify the light field: carotenoids in diatom communities reduce the penetration of 450 to 500-nm wavelengths, and phycocyanin in cyanobacteria reduce the penetration of 620-nm wavelengths to underlying cells. Selective scattering by inorganic particles also theoretically affects light quality, but sediments such as sand that have low densities of photosynthetic organisms exhibit relatively low spectral variability (Kuhl and Jorgensen, 1994). Little spectral variation was observed by Meulemans (1987) in one of the few examinations of light transmission in freshwater periphyton, but he attributed the lack of specific attenuation peaks to the predominance of structural material (diatom stalks) within the periphyton matrix.

IV. PHOTOSYNTHESIS–IRRADIANCE RELATIONS

Photosynthesis-irradiance relationships are measured for two basic reasons: (1) to evaluate ecophysiological responses to light, and (2) to predict *in situ* photosynthesis. Ecophysiological interest appears to have been the proximate stimulus for many phytoplankton P–I measurements, but numerous predictive models of primary production have evolved. These models range from simple static equations with fixed P–I parameters (Ryther, 1956) to more complex dynamic models (Pahl-Wostl and Imboden, 1990). Photosynthesis–irradiance studies on freshwater benthic algae have been primarily ecophysiologically oriented; there have been little use of P–I data to predict or estimate *in situ* primary production by benthic algae.

Photosynthesis by most freshwater benthic algae (and the vast majority of other plants) is a nonlinear function of light intensity (Fig. 1). At low irradiances, photosynthetic rate increases linearly with increasing light, and appears to be limited primarily by the number of photons captured by photosynthetic pigments. The slope of this linear portion of the P–I curve is termed α. At midlevel irradiances, photosynthesis begins to level off as light becomes saturating. Photosynthesis at this point appears to be limited by dark reactions such as ribulose 1,5-bisphosphate carboxylase activity (Rivkin, 1990). With further increases in irradiance, photosynthetic rate approaches an asymptotic maximum or decreases because of photoinhibition. The maximum rate of photosynthesis, whether reached asymptotically (no photoinhibition) or as a peak (photoinhibition), is referred to as P_{max}.

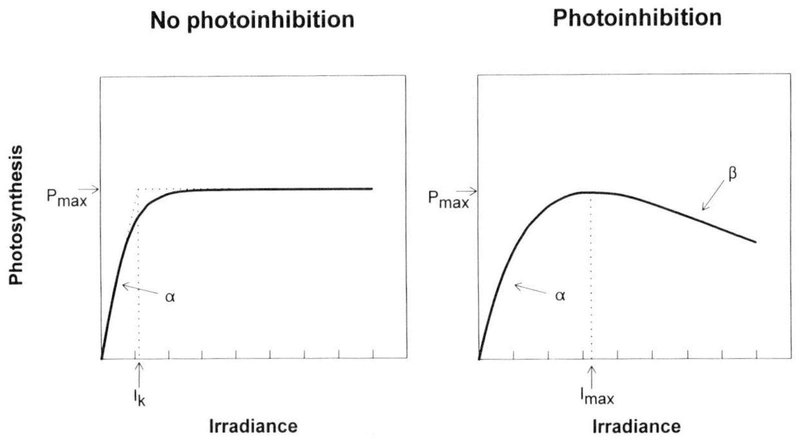

FIGURE 1 Typical photosynthesis–irradiance curves and associated parameters.

Photosynthetic responses by freshwater benthic algae have been described mathematically with a number of equations originally developed for phytoplankton. In the absence of photoinhibition, the most popular equation has been the hyperbolic tangent equation of Jassby and Platt (1976). In this model, $P = P_{max} \tanh(\alpha I/P_{max})$, where P = photosynthetic rate for any irradiance I, P_{max} is the asymptotic maximum rate of photosynthesis, and α is the initial slope of the P–I curve. Note that only two parameters are estimated with this equation: P_{max} and α. When photoinhibition is obvious in P–I data, the three-parameter model of Platt *et al.* (1980) is appropriate. In this model, $P = P_s (1 - e^{-a})e^{-b}$, where P_s is the theoretical maximum photosynthetic rate in the absence of photoinhibition, $a = \alpha I/P_s$, and $b = \beta I/P_s$. Three parameters are fit in this equation: P_s, α, and β, the photoinhibition parameter. Although a model proposed by Neale and Richerson (1987) that incorporates a saturation plateau prior to photoinhibition appears to be a more realistic model, the Platt *et al.* (1980) equation fits most photoinhibition data fairly well. Several other meaningful P–I parameters are calculated from P–I models, including I_k, the irradiance representing the onset of photosaturation, and I_{max}, the irradiance of maximum photosynthesis (P_{max}). If net photosynthesis is measured, then it is possible to calculate the compensation point I_c, the irradiance at which respiratory losses equal photosynthetic gains.

A. Exposure Diversity

A fundamental problem in the construction and interpretation of P–I curves for benthic algae is determining actual exposure intensity. Natural substrata tend to be three dimensional, so algal cells on the sides of these substrata receive significantly less direct light than cells on the top when the light source is directly overhead. Algal cells situated in crevices also receive less than the measured irradiance. Even if the substrate is uniform (e.g., glass slides, ceramic tiles), cells on the surface shade those below them, causing subsurface cells in the vertical matrix to receive less light than is measured by a sensor outside the matrix. Photosynthesis measured with whole substrata is necessarily the sum of photosynthesis of many individual algal cells: some cells may be receiving saturating irradiances and are photosynthesizing at P_{max}, whereas others are receiving subsaturating irradiances and are in the α range of their P–I response. The problem of diverse exposures is compounded by the possibility that cells occupying different light microhabitats are photoadapted to those particular microhabitats, and so have different photosynthetic responses as well as different exposures. Another potentially large source of heterogeneity are interspecific differences in P–I responses; even the simplest benthic algal community is composed of more than one species.

The diversity of exposures and potential responses suggests that P–I curves for intact algal communities on whole substrata are the integration of many separate P–I curves, despite the investigator's best attempts to control the exposure regimen. The effect of this heterogeneity is probably to extend the gradual transition between the linear and horizontal parts of an uninhibited P–I curve (McIntire and Phinney, 1965) and to increase the apparent intensity at which photoinhibition takes place in photoinhibited P–I curves (Hill and Boston, 1991). It seems likely that transition areas (linear to saturation and saturation to photoinhibition) should be considerably broader for natural than for artificial substrata, for whole-stream measurements than for individual substrata measurements, and for thick than for thin communities. This hypothesis has not been tested.

The effect of self-shading and vertical heterogeneity on P–I responses in freshwater benthic algae has been the focus of a number of studies. Sand-Jensen and Revsbech (1987) examined gross photosynthesis by 2-mm-thick epiphytes of *Potamogeton* with oxygen microelectrodes, and observed profound differences in photosynthesis between different matrix depths. Photosynthesis occurred only in the upper 0.5 mm of the matrix at low irradiances (13 and 49 μmol m^{-2} s^{-1}), but it increased substantially at lower depths as irradiance increased, presumably because more light penetrated to the epiphyte understory. Since light was reduced by about 50% in 1 mm, Sand-Jensen and Revsbech calculated that self-shading alone would reduce I_k and I_c twofold. Self-shading also appears to have dramatic effects on the slope of biomass-specific P–I curves: whole-community measurements of *Phragmites* epiphytes (Meulemans, 1988) and stream periphyton (Hill and Boston, 1991) show very strong declines in $\alpha_{chlorophyll}$ (chlorophyll-specific alpha) as algal biomass accumulates vertically during community development. Biomass-specific P_{max} also declines with vertical development (Hill and Boston, 1991).

There is some evidence that algae at different depths in the benthic algal matrix have different P–I responses because of photoacclimation. Paul and Duthie (1989) obtained separate P–I curves for understory and overstory components of stream periphyton with track autoradiography, reporting lower "light compensation points" (presumably saturation levels) for the understory. Dodds (1992) used a fiber-optic light microprobe and an oxygen microelectrode to measure separate P–I responses at 250 = μm depth intervals inside *Ulothrix*-dominated periphyton, finding that α increased with depth. Decreased saturation irradiances and increased alpha are attributes of shade adaptation.

B. Shade Adaptation

Shade adaptation or acclimation (Falkowski and LaRoche, 1991) has been little documented in freshwater benthic algae, despite the fact that light levels are extremely low in many benthic habitats. Shade adaptation is a common response of phytoplankton and macrophytes in habitats characterized by low ambient irradiances. It is usually manifest in increased photosynthetic efficiency at low irradiances (increased α), lower photosynthetic rates at saturating irradiances (lower P_{max}), and decreased saturation parameters (lower I_k or I_{max}). These functional changes typically reflect increased quantities of antenna pigments and decreased quantities of dark reaction enzymes. Shade adaptation theoretically buffers the photosynthesis-limiting effects of low irradiances; its effects will be positive for primary production if respiratory costs are not too high.

Only a few published studies have documented community-level shade adaptation by freshwater benthic algae. McIntire and Phinney (1965) measured photosynthetic responses of periphyton grown under 40 and 110 μmol m^{-2} s^{-1} in laboratory streams; the community grown under 40 μmol m^{-2} s^{-1} appeared to have slightly higher alpha, a lower I_k, and somewhat lower P_{max} than the community grown under 110 μmol m^{-2} s^{-1}. More dramatic adaptation was reported by Hill *et al.* (1995), who compared P–I curves of stream periphyton from shaded and open sites. Maximum ambient irradiances at the shaded sites averaged 50 μmol m^{-2} s^{-1} whereas those at the open sites averaged 1100 μmol m^{-2} s^{-1}. At low light intensities, photosynthesis by periphyton from the shaded sites was twice that of periphyton from the open sites (two times greater α of shaded periphyton), and photosynthesis by shaded periphyton saturated at lower intensities (Fig. 2). Interestingly, shaded periphyton also had the highest P_{max} in these laboratory-obtained P–I curves, but it is important to note that the shaded periphyton would rarely have achieved P_{max} *in situ* because of the relatively low (maximum of 50 μmol m^{-2} s^{-1}) irradiances in the shade. Low ambient irradiances at shaded sites restricted daily photosynthesis at these sites to less than 25% of that at open sites, despite the high photosynthetic efficiency of shaded periphyton at low irradiances (Hill *et al.*, 1995).

Photoinhibition was evident in the P–I response of shaded periphyton in Fig. 2, whereas the P–I response of periphyton from open sites did not indicate significant photoinhibition. When benthic algae from shaded habitats are exposed to relatively high irradiances (e.g., >1000 μmol m^{-2} s^{-1}) in the laboratory, they often experience much more photoinhibition than algae from unshaded habitats (Boston and Hill, 1991; Hill and Boston, 1991). Hill *et al.* (1995) hypothesized that some of the sensitivity of shaded algae to photoinhibition in the laboratory was due to low levels of light-shielding accessory pigments (e.g., carotenoids) that ordinarily protect unshaded algae from inhibitory PAR or UV intensities. Carotenoids that accumulate in algae exposed to high irradiances may diminish photoinhibition by absorbing excess photons, but they may also reduce α (and perhaps P_{max} as well) by intercepting photons that would be photosynthetically useful at low intensities. Some phytoplankton acclimated to high

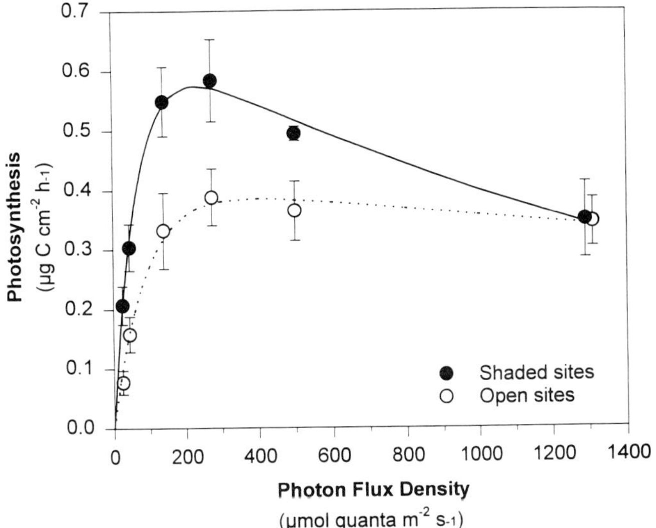

FIGURE 2 Photosynthesis–irradiance responses by periphyton in upper White Oak Creek. (Reprinted with permission from Hill *et al.*, 1995.)

light accumulate carotenoids and experience reduced quantum yields (Falkowski and LaRoche, 1991; Dubinsky *et al.*, 1986). Differences between shade-adapted and light-adapted algae may therefore result from both adaptation by shaded algae to enhance the efficiency of photon use and adaptation by unshaded algae to minimize photoinhibitory damage to sensitive light-harvesting complexes. In a study of Antarctic freshwater benthic mats, Howard-Williams and Vincent (1989) attributed the relatively high saturation intensities of *Nostoc* communities in high-light environments to the shielding effect of accessory pigments.

C. Photoinhibition

Inhibition of short-term photosynthesis by high irradiances is often observed in P–I curves for freshwater benthic algae. However, does this mean that high irradiances actually inhibit short-term photosynthesis in nature? When photoinhibition has been reported, it has usually occurred at irradiances greater than 600 μmol m^{-2} s^{-1} (Table I), a moderately high intensity. It is not at all clear that benthic algae exhibiting photoinhibition in their P–I curves actually experience these inhibitory irradiances *in situ*. In a relatively large P–I survey of stream periphyton, Boston and Hill (1991) and Hill and Boston (1991) found that only periphyton from highly shaded sites exhibited photoinhibition (at 1100 μmol m^{-2} s^{-1}); periphyton

TABLE I Photosynthesis–Irradiance Studies on Freshwater Benthic Algae

Source	*Community*	*Saturation*	I_k	*Inhibition (maximum I)*	I_c
McIntire and Phinney (1965)	Lotic periphyton	200–400		No (450)	10,20
Hunding (1971)	Lentic periphyton	200	60–195	Yes, above 590	
Kelly *et al.* (1974)	Lotic periphyton	None		No	
Hornberger *et al.* (1976)	Lotic periphyton	None		No	
Kremer (1983)	Lotic rhodophytes	200–400		No (500)	
Turner *et al.* (1983)	Lentic periphyton	200		No (600)	12
Jasper and Bothwell (1986)	Lotic periphyton	ND		Yes, various *I*	
Sand-Jensen and Revsbech (1987)	Lentic epiphytes	300–400	200	Yes, 330	22,37
Meulemans (1988)	Lentic epiphytes	150–300		Yes	20–40?
Paul and Duthie (1989)	Lotic diatoms	200	95–160	Yes above 250	
Howard-Williams and Vincent (1989)	Lotic cyanobacteria	100–400	20–105	No (700)	
Kaczmarczyk and Sheath (1991)	*Batrachospermum*	250			
Boston and Hill (1991)	Lotic periphyton	200–400	100–400	Yes, above 400	
Hill and Boston (1991)	Lotic periphyton	200–400	148–254	Yes, above 400	
Davies-Colley *et al.* (1992)	Lotic periphyton	200–900	100–400	No (1200)	30–350
Dodds (1992)	Ditch algae	None, 600		No	
Hill *et al.* (1992)	Lotic periphyton	200–400	66–95	Yes, above 400	
Wootton and Power (1993)	Lotic periphyton	None		No	
Graham *et al.* 1995	*Spirogyra*	375–1500		Yes, above 1500	7–23
Hill *et al.* (1995)	Lotic periphyton	200–400		Yes, above 350	

from open sites exposed to full sunlight did not show any evidence of photoinhibition in the laboratory. Photosynthesis–irradiance responses of benthic algae inhabiting high-light environments (such as open sites in clear streams) should be under considerable selective pressure to develop mechanisms that reduce the potentially damaging effects of high irradiances, whether it be accessory pigments as suggested earlier or sheath pigments such as scytonemin (Garcia-Pichel and Castenholz, 1991).

Absence of photoinhibition in laboratory P–I measurements does not provide unequivocal evidence that photoinhibition does not occur. First, localized photoinhibition may be hidden in whole-community incubations. High irradiances at the surface of benthic algal mats may inhibit photosynthesis by surface cells while shaded subsurface cells receive only saturating or subsaturating irradiances. If the matrix is deep enough, then photosynthesis by underlayers at high surface irradiances may compensate for inhibition in surface layers, so there is no evidence of photoinhibition at the community level (Hill and Boston, 1991). This is apparent in the microelectrode study of Sand-Jensen and Revsbech (1987), in which surface layers were inhibited by 487 μmol m^{-2} s^{-1} but vertically integrated photosynthesis was not. Second, photoinhibition is time-dependent (Falkowski, 1984), and most P–I measurements may have been too brief to measure photoinhibition. Third, photoinhibition is strongly wavelength-dependent as well, with greatest effects occurring at short wavelengths (e.g., Smith *et al.*, 1980; Vincent and Roy, 1993). Lamps used in laboratory photosynthesis–irradiance studies may not provide enough short-wavelength radiation to induce photoinhibition, and photosynthetic incubations employing chambers made of ultraviolet-absorbing glass or acrylic may underrepresent photoinhibition.

D. Saturation

Despite the varied light regimes of benthic algae, most P–I studies indicate that photosaturation occurs in a relatively narrow range of PFDs, from 100 to 400 μmol m^{-2} s^{-1}. Saturation irradiances are often well beyond the typical irradiances found *in situ*, implying that *in situ* photosynthesis is very strongly light-limited. For example, photosynthesis by attached algae in heavily shaded forest streams saturates between 100 and 200 μmol m^{-2} s^{-1}, yet maximum irradiances during summer (when shade is greatest) rarely exceed 30 μmol m^{-2} s^{-1} (Hill *et al.*, 1995). The relatively high saturation intensities are somewhat puzzling, given the phylogenetic and/or ontogenetic flexibility other microalgae exhibit in adapting to dark habitats. Microalgae living under the ice in the Arctic and Antarctic exhibit saturation at less than 10 μmol m^{-2} s^{-1} (Cota, 1985; Palmisano *et al.*, 1985). The photosynthetic machinery of the freshwater species studied to date may be geared to harness higher irradiances that occur infrequently (e.g., sunflecks) or seasonally (e.g., winter and spring maxima for streams shaded by deciduous trees). Further work in consistently dark freshwater habitats such as lake benthos may reveal saturation values under 50 μmol m^{-2} s^{-1}.

Saturation is not apparent in the results of all freshwater benthic P–I studies. Kelly *et al.* (1974) reported that photosynthesis in an open stream was a linear function of light intensity, even at very high intensities. They suggest that nonlinear P–I curves result from nutrient depletion or other container effects. However, other open-stream P–I responses show typical nonlinear curves (Erich Marzolf, unpublished data). It is difficult to give much credence to linear P–I relationships, given the tremendous volume of literature reporting photosaturation in C_3 plants and algae, but two other benthic algal P–I studies also reported linear relationships between light and photosynthesis (Dodds, 1992; Wootton and Power, 1993). However, the data in both of these studies were variable enough that nonlinear functions could also be fit to the data.

E. Compensation Point

The compensation point (I_c) is the irradiance at which photosynthesis equals respiration. It is a useful estimate of the lowest irradiance at which algae can maintain an autotrophic existence. The compensation point has been reported infrequently in benthic P–I studies, partly because respiration cannot be directly measured with the ^{14}C technique. In the relatively few studies where I_c was calculated (McIntire and Phinney, 1965; Turner *et al.,* 1983; Sand-Jensen and Revsbech, 1987), it ranged from 10 to 40 $\mu mol\ m^{-2}\ s^{-1}$. Like other parameters of P–I curves, I_c is likely to be affected by photoadaptation/photoacclimation. Some species of phytoplankton and benthic marine algae have I_cs less than 2 $\mu mol\ m^{-2}\ s^{-1}$ (Richardson *et al.,* 1983; Cota, 1985), and are clearly well adapted to habitats where photons are scarce. Further research on freshwater benthic algae may uncover lower I_c values, but respiration by the abundant heterotrophs that are usually associated with benthic algae causes I_c values to to be higher than those measured in communities with fewer heterotrophs (McIntire and Phinney, 1965).

F. Estimating *in Situ* Primary Production

Photosynthesis–irradiance equations such as the simple hyperbolic tangent model can be used to estimate primary production *in situ* if ambient irradiances are measured continuously. These estimates of primary production are much superior to those based on photosynthesis measured at only one or two irradiances because of the extreme temporal variability of light in most habitats. To date, P–I models coupled with *in situ* irradiances have been little used to estimate primary production *in situ* (but see Hill *et al.,* 1995). Dynamic P–I models, in which photosynthesis does not simply depend on actual instantaneous light, but also on the prior light history (e.g., Pahl -Wostl and Imboden, 1990), could provide even better estimates of *in situ* photosynthesis. These models allow for diurnal changes in the P–I relationship that may occur because of time-dependent photoinhibition or other diurnal physiological changes within algal cells. Although dynamic photosynthetic responses by phytoplankton are known (e.g., Harris and Piccinin, 1977), the effect of previous exposure on P–I responses has not been investigated in freshwater benthic algae.

V. ECOLOGICAL EFFECTS OF LIGHT INTENSITY

Photosynthesis is obviously closely related to light intensity in short-term measurements, but what are the longer-term, ecological consequences of spatial and temporal variation in light? Benthic algae are subject to a variety of biotic and abiotic constraints in addition to light, and these factors may interact with the effects of light or even override them. For example, maximum photosynthetic rates may be several times higher in unshaded habitats than in shaded habitats, but these differences may not be expressed in algal standing crops because of heavy grazing pressure (Hill *et al.,* 1995). Community responses to light can also be modified or obscured by indirect effects of light. Temperature frequently correlates with light intensity, so it is often difficult to parcel out the effects of the two factors in observational studies. Indirect effects of light on community architecture are also likely to influence community composition: increased vertical structure caused by light-enhanced primary production may select for species that are capable of moving to or extending above the community surface. Manipulative, *in situ* experiments can help distinguish the effect of light on important ecological parameters such as biomass, production, and community composition; unfortunately, very few have been attempted.

A. Biomass and Productivity

Because light acts primarily on growth or turnover rates, its influence is best assessed with dynamic measures such as *in situ* primary production. However, estimations of periphyton biomass [dry mass, ash-free dry mass (AFDM), or chlorophyll *a*] are rather easily obtained, and are consequently the parameters most often measured in studies of light effects. Cell counts and associated biovolumes, parameters requiring more time and expertise, are also used as standing crop estimators. All of these static parameters suffer from the disadvantage that they are not necessarily representative of growth rates, particularly where loss rates (mortality and emigration) are high. Biomass accrual is the integration of both gains and losses, and losses from grazing, physical disturbance, pathogens, and so on can be significant (Hill *et al.,* 1992).

In streams, algal biomass often correlates with the amount of streamside vegetation (e.g., Hill and Harvey, 1990). Chlorophyll *a* or AFDM can be four to five times higher at open sites than at sites with full tree canopy development (Lowe *et al.,* 1986; Hill and Knight, 1988), particularly when grazing pressure is low. When grazing pressure is high, however, biomass may not be correlated with light. Feminella *et al.* (1989), Steinman (1992), and Hill *et al.* (1995) observed biomass responses to increased light in shaded streams, but only after grazers were experimentally removed.

Light-enhanced primary production that is not expressed in increased periphyton biomass when grazing is heavy is instead reflected in increased grazer growth or densities (Hill *et al.*, 1995). Other loss factors also modify the effects of canopy cover: frequent substrate disturbance (such as would occur during spates) significantly reduces biomass differences between open and closed-canopy sites (Robinson and Minshall, 1986). Interactive and overriding effects notwithstanding, almost all experimental and observational studies support the idea that benthic algal production is strongly constrained by the shade from well-developed streamside vegetation (but see Rosemond, 1993).

In lakes, light is attenuated primarily by absorption and scattering in the water column. Light attenuation obviously establishes a lower limit to the depth at which benthic algae can photosynthesize. The maximum depth at which *Cladophora glomerata* grows in Lake Erie appears to be a simple function of light intensity, which must exceed 28 μmol m^{-2} s^{-1} for this macroscopic alga to survive (Lorenz *et al.*, 1991). However, light intensity alone does not appear to determine the depth distribution of benthic algal biomass in lakes. If it did, benthic algal biomass should decline exponentially with depth, mirroring the exponential decrease in light penetration. The few studies that have examined the abundance of benthic algae at different depths do not indicate an exponential decline. For example, Stevenson and Stoermer (1981) observed that sediment-associated benthic algae in Lake Michigan were relatively scarce at shallow depths (<9 m), increased at intermediate depths (9–15 m), then declined at greater depths (>23 m) where light intensity was <6% of surface irradiances. Disturbance by wave action appears to be an important constraint on algal development at shallow depths (e.g., Evans and Stockner, 1972), but photoinhibition may also restrict biomass accrual in clear, shallow waters. Other depth-associated factors, such as temperature, nutrient concentration, and substrate type, have the potential to confound the effect of light on the depth distribution of benthic algae.

Turbidity from silt and other inorganic particles substantially reduces light penetration in many rivers, reservoirs, and wave-swept lakes. It can severely limit benthic algal development in lowland rivers and estuaries (e.g., Vannote *et al.*, 1980), where Secchi depths are often less than 50 cm. The effect of turbidity is undoubtedly increased by human activities: land-use practices affecting erosion provide nonpoint sources of acute and chronic stress, and activities such as mining provide point sources of inorganic particles. Clay particles from placer gold-mining in New Zealand coastal streams reduce PFDs from 340 μmol m^{-2} s^{-1} to 80 μmol m^{-2} s^{-1} at 0.3 m depth, diminishing algal biomass and primary production manyfold (Davies-Colley *et al.*, 1992). Benthic algal development in shallow reservoirs receiving particle-rich inflows is also affected by turbidity, especially after large rain events. Burkholder and Cuker (1991) suggest that the light-reducing effect of inorganic particles may in some instances be mitigated by (1) clay particles providing a source of phosphorus for benthic algae at shallow depths where light is relatively high, and (2) clay particles coflocculating overlying phytoplankton.

Light is a key component in the competitive interaction between phytoplankton and benthic algae in lakes. Because phytoplankton are superimposed over benthic algae, and because they strongly absorb PAR, phytoplankton abundance can control the quantity and quality of light that reaches benthic algae in lakes. An inverse relationship is therefore expected between phytoplankton biomass and benthic algal biomass (Hansson, 1988). In a survey of epipelic algal biomass along a productivity gradient of Swedish and Antarctic lakes, Hansson (1992) demonstrated a negative correlation between phytoplankton biomass (measured as light extinction in the water column) and benthic algal biomass, but only for extinction coefficients greater than 1.5 m^{-1} (Fig. 3). At extinction coefficients less than 1.5 m^{-1}, light reaching benthic algae at the 0.75-m sampling depth appeared to be nonlimiting. Benthic biomass was positively correlated with light extinction in oligotrophic lakes (extinction coefficients <0.5 m^{-1}). Hansson argued that benthic algae were nutrient-limited in the relatively clear waters of oligotrophic lakes, and increasing phytoplankton biomass (as indicated by increasing light extinction) reflected increasing nutrient supply rates to both benthic algae and phytoplankton. Maximum benthic

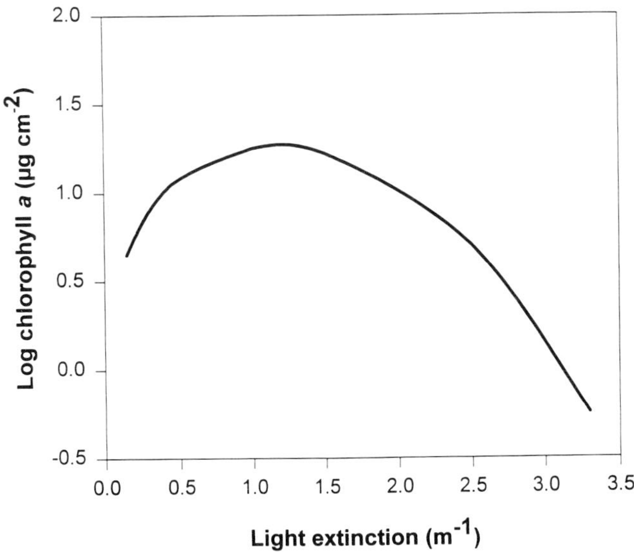

FIGURE 3 Benthic algal biomass versus light extinction in lakes. (After Hansson, 1992 with permission.)

algal biomass occurred in lakes with extinction coefficients ranging from 0.5 to 1.5 m^{-1}, presumably owing to the optimal combination of light and nutrients in these lakes. Although photoinhibition is an alternative explanation for lower benthic algal biomass in oligotrophic lakes, Hannson's hypothesis of switching controls (nutrient to light) on benthic primary production is provocative and merits experimental attention.

B. Taxonomic Responses

Autecological light requirements by individual benthic species are essentially unknown because unialgal growth versus irradiance measurements have simply not been made. Inherent differences between species in basic parameters such as compensation point and maximum growth irradiance are likely to exist, and these differences may be important determinants of community composition. Unfortunately, the magnitude of these differences and their ecological effect are unmeasured. Differences in community composition under different light regimes have been reported in a few observational and experimental studies, but the causal relationship between light and community composition in these studies is clouded by potential indirect effects such as temperature or competition for space in a rapidly growing algal matrix. Major physiological responses to light within species can also obfuscate differences between species, especially when light conditions for growth are heterogeneous. Nonetheless, there are indications that benthic community composition does change with different light intensities.

Differences in light response by major taxonomic categories (i.e., divisions) of algae have been suggested because taxonomic categories are in part defined by differences in light-capturing pigments and membranes. Although there is considerable interspecific variability within large taxonomic groups, some evidence from phytoplankton indicates that there are overall differences between major divisions. Richardson *et al.* (1983) compared published minimum (I_c) and maximum (I_{max}) growth irradiances of major phytoplankton groups and concluded that chlorophytes did not tolerate low light intensities as well as cyanobacteria or diatoms. The latter two groups had average I_c values of 5–6 μmol m^{-2} s^{-1}, whereas chlorophytes had average I_c values of 21 μmol m^{-2} s^{-1}. Langdon (1988) also concluded that chlorophytes were less adapted to grow at low irradiances than were diatoms. An inherent requirement for relatively high light intensity by chlorophytes is consistent with observations in clear-cut forest streams, where Zygnematales and other filamentous chlorophytes are unusually common (e.g., Lyford and Gregory, 1975; Shortreed and Stockner, 1983). In the more controlled environment of laboratory streams, Steinman and McIntire (1987) observed that filaments of the chlorophytes *Stigeoclonium* and *Ulothrix* became abundant only at growth irradiances of 150 μmol m^{-2} s^{-1} and above. Also consistent with the notion that chlorophytes require high light was the unprecedented appearance of the desmid *Hyalotheca dissiliens* that occurred when gypsy moth larvae defoliated the tree canopy overhanging a Rhode Island stream (Sheath *et al.*, 1986). Chlorophytes do not always require high light, however. *Stigeoclonium* basal cells dominate the sparse periphyton in highly shaded forest streams in east Tennessee (Steinman, 1992; Hill *et al.*, 1995). However, the abundance of *Stigeoclonium* basal cells in these streams may result more from their resistance to grazing by the abundant snail *Elimia clavaeformis* than from an affinity for low light (Steinman, 1992). Generalizing from their results of light manipulations in laboratory streams, Steinman *et al.* (1989) predicted that diatoms will dominate moderately grazed benthic algal communities at <50 μmol m^{-2} s^{-1}, diatoms plus some cyanobacteria and chlorophytes will constitute the community at 50–100 μmol m^{-2} s^{-1}, and chlorophytes will dominate at irradiances >100 μmol m^{-2} s^{-1}.

Freshwater Rhodophyta have been characterized as being adapted to low light because some species are abundant in highly shaded environments (Hynes, 1970). More recent summaries of rhodophyte distributions indicate that freshwater red algae exhibit a wide variety of light requirements (Sheath, 1984). The abundance of some species (e.g., *Audouinella violacea*) is positively correlated with light (Korch and Sheath, 1989), some species (e.g., *Batrachospermum macrosporum*) deteriorate under high light intensities (Dillard, 1966), and others (e.g., *Batrachospermum moniliforme*) are euryphotic (Sheath, 1984). Photosynthesis–irradiance responses of four taxa of freshwater rhodophytes (Kremer, 1983) do not suggest strong shade adaptation: no photoinhibition was evident at moderately high irradiances and the range of saturation intensities (200–400 μmol m^{-2} s^{-1}) is quite similar to that exhibited by other species of benthic algae (Table I).

Although diatoms, cyanobacteria, and rhodophytes may fare better than chlorophytes under low light, most taxa probably grow best under moderately high irradiances (e.g. 200-800 μmol m^{-2} s^{-1}. Filamentous chlorophytes may become more abundant at these irradiances not because other taxa grow suboptimally, but because the growth form of filamentous chlorophytes enables them to overgrow prostrate taxa such as diatoms and cyanobacteria. It would be interesting to compare photosynthesis–irradiance responses of filamentous chlorophytes to prostrate forms to determine if chlorophytes generally exhibit higher P_{max} at higher irradiances than do prostrate forms; it would also be interesting to see if prostrate forms are inherently more efficient at converting light into chemical energy at low irradiances. Unfortunately, data are currently unavailable to make this comparison. Large growth forms, regardless of their taxonomic affinity, have a competitive advantage for light, as their higher profile allows them to intercept photons before the photons reach forms lower in the community (Hudon and Bourget, 1983).

Nonmotile, prostrate taxa are at a distinct disadvantage in vertically developing benthic communities. Prostrate forms (e.g., *Achnanthes*) that dominate underdeveloped communites on bare or repeatedly disturbed substrates will suffer reduced access to light as the community develops vertically, and probably will decrease in relative abundance. If a community's vertical development is due to increased light (e.g., increased water clarity or loss of terrestrial shading), then light's apparent effect on community composition (decline in the relative abundance of prostrate forms) will be indirect, essentially identical to the effects of other factors (e.g., nutrient addition or grazer removal) that allow vertical expansion of periphyton. For example, Steinman (1992) found that prostrate basal cells of *Stigeoclonium* decreased in relative abundance as light increased in a shaded stream, but only after the removal of grazers allowed vertical development of the periphyton community. Some prostrate forms may have physiological mechanisms that allow them to persist under low-light conditions at the bottom of benthic algal mats. *Achnanthes rostrata* can survive extended dark periods (≥ 30 days) in a quiescent or heterotrophic state, resuming growth when reintroduced to the light (Tuchman *et al.*, 1994). *Stigeoclonium* basal cells can survive >92 days of darkness with only moderate losses of chlorophyll, and begin photosynthesizing almost immediately (although at reduced rates) upon reexposure to light (Steinman *et al.*, 1990). The mechanism of *Stigeoclonium* survival is not known, but reduced metabolism is probably the primary means by which most algae survive dark or highly shaded periods (Dehning and Tilzer, 1989).

Facultative heterotrophy could be advantageous under conditions of low light, whether the light reduction is caused by overlying cells in a benthic mat, extensive terrestrial shading, or water column attenuation. Admiraal and Peletier (1979) demonstrated that the growth of the diatoms *Navicula* and *Nitzschia* was supplemented by heterotrophy at low light levels, leading Stevenson and Stoermer (1981) to suggest that the increased relative abundance of these taxa with depth in Lake Michigan is due to facultative heterotrophy. The ability to grow heterotrophically on dissolved organic matter will depend on the type and concentration of the organic compounds, and it is unclear if adequate concentrations of labile organics commonly occur in or on benthic substrata. Heterotrophy's contribution to algal growth is likely to be small, but it may help some taxa in organically rich media survive transient dark conditions (Palmisano and Sullivan, 1982; see Tuchman, Chapter 10, this volume).

Motile benthic taxa such as raphe-bearing diatoms may regulate their light environment through phototaxis, allowing them to escape the extreme ends of a light gradient if the scale of the gradient is on the order of millimeters. Epipelic diatoms migrate to the surface of estuarine sediments when the tide is out and return to subsurface layers when the tide returns (e.g., Faure-Fremiet, 1951). Movement of these diatoms to the surface during low tide presumably maximizes their light exposure while their retreat during flooding high tide presumably minimizes the potential of being washed away. Diurnal patterns of vertical migration are also seen in freshwater benthic algae, where *Navicula* and *Nitzschia* species move to surface of pond sediments during the day and return to lower layers at night (Round and Eaton, 1966). Interestingly, the upward movement begins before sunrise and the downward movement begins before sunset, and the rhythm persists for some time in continuous light or dark, implying an internal clock. Diurnal vertical movements have not been reported on other substrata, but such migrations in thick epilithic communities would clearly benefit those taxa that can move. Mobility may allow a fine-tuning of exposure regimes within sediments or algal mats, so that inhibitory and limiting irradiances are avoided, enabling motile algae to experience I_{max} as often as possible. Diurnal migration may also allow mobile taxa to elude surface grazers or exploit nutrient gradients during dark hours.

VI. ECOLOGICAL EFFECTS OF LIGHT QUALITY

Selective absorption of different wavelengths of PAR by photosynthetic pigments, coupled with heterogeneity in the spectral distribution of light in aquatic habitats, suggests the potential for light-quality effects. Considerable discussion and experimentation have revolved around photosynthetic effects of light quality on phytoplankton and marine macrophytes. Few light-quality experiments have been performed on freshwater benthic algae, whose spectral environment is strongly affected by surrounding vegetation, water, dissolved compounds, suspended particles, sediments, and overlying benthic algae. Agents that reduce light intensity tend to attenuate certain wavelengths more than others, so decreases in intensity are confounded by potential effects of light quality. For example, leaves on trees overhanging forest streams strongly absorb red and blue wavelengths in addition to reducing total PAR (e.g., Federer and Tanner, 1966), resulting in green light that is poorly absorbed by chlorophyll *a* and *b*. Sheath and Burkholder (1985) speculated that the scarcity of chlorophytes in shaded streams is due to their lack of accessory pigments that efficiently absorb green light.

The few experiments that manipulated light quality for freshwater benthic algae do not provide much evidence for significant light-quality effects. Antoine and Benson-Evans (1983a) observed species-specific responses to red, blue, and green filters in laboratory experiments, but the treatments were confounded by differences in intensity. Kaczmarczyk and Sheath (1991) controlled light intensity independently while using color filters to provide different light regimes for *Batrachospermum boryanum*.

They reported a broad photosynthetic response by *B. boryanum*, indicating that the rhodophyte was equally productive over most of the visible spectrum. Pigment composition in *B. boryanum* seemed to be controlled by light intensity, not light quality. The paucity of data makes it difficult to generalize about the overall significance of light quality to freshwater benthic algae, but investigators in other aquatic habitats have concluded that variation in light intensity is much more important than spectral quality (Dring, 1981; Ramus, 1983; Morel *et al.*, 1987; Falkowski and LaRoche, 1991; Ploug *et al.*, 1993).

VII. ULTRAVIOLET RADIATION

Most UV radiation arriving at the water's surface is in the UV-A range (320–400 nm), and a small amount is in the UV-B range (280–320 nm). In very clear waters, UV-A wavelengths penetrate deeper than visible wavelengths, so that benthic algae even several meters deep may receive significant doses (Vincent and Roy, 1993). Significant amounts of UV-B can also penetrate to depth: Karentz and Lutze (1990) measured biologically effective UV-B as deep as 20 m in Antarctic waters. The flux of UV radiation is important because: (1) UV radiation is a potent inhibitor of photosynthesis (PS II) in unprotected phototrophs, and (2) UV-B is strongly absorbed by DNA, causing genetic damage (Vincent and Roy, 1993). Ultraviolet radiation is of special concern because decreasing ozone concentrations in the stratosphere are predicted to allow increasing amounts of UV-B to fall upon the earth's surface (e.g., Frederick, 1993).

The effects of current levels of UV radiation on microalgae are somewhat unclear. UV inhibits phytoplankton photosynthesis in static bottle incubations (Smith *et al.*, 1980), but phytoplankton chronically exposed to high light intensities show considerable resistance to UV (Helbling *et al.*, 1992). Protective mechanisms in microalgae appear to mitigate at least some of the harmful effects of UV; these mechanisms include mycosporine-like amino acids that absorb strongly in the UV-B region, sheath pigments (scytonemin) that absorb strongly in the UV-A region and whose production is induced in filamentous cyanobacteria by UV-A, and carotenoids that may act to quench UV-produced toxic intermediates (Garcia-Pichel and Castenholz, 1991; Vincent and Roy, 1993). Repair mechanisms for UV-damaged DNA are also operable in microalgae, but vary from species to species, and small taxa may be more sensitive to DNA damage than large taxa (Karentz *et al.*, 1991).

What is the ecological impact of UV radiation on freshwater benthic algae? Little UV work has been done on this group of organisms, but research suggests that UV effects may be complex. Bothwell *et al.* (1993) shielded developing diatom communities from UV in riverside flumes in

British Columbia, and observed that the communities shielded from UV grew 30–40% faster than those exposed to UV. However, after 5 weeks, the effect was reversed, and algal accumulation was two to four times greater in communities exposed to UV. Mean cell size in communities exposed to UV was twice that of UV-shielded communities, consistent with predictions based on phytoplankton data (Karentz *et al.*, 1991). Species succession under UV resulted in a doubling of the mean cell size. Subsequent experiments demonstrated that the initial inhibition was due principally to direct effects of UV-A, whereas the enhancement that occurred after 5 weeks resulted from indirect UV-B effects. Algal accumulation after 5 weeks increased because UV-B reduced the densities of grazing chironomids, which limited algal accumulation in communities shielded from UV-B (Bothwell *et al.*, 1994). The magnitude of these indirect effects in natural settings is unknown, but direct effects of UV on periphyton in the artificial streams of Bothwell *et al.* were clearly insignificant compared to grazing effects. Because benthic algae persist in exposed habitats, it is likely that they successfully employ at least some of the protective mechanisms mentioned previously. The cost of protective mechanisms is unknown. Further research on UV effects is needed.

VIII. SUMMARY

Benthic light environments are diverse, ranging from full sunlight in shallow, clear waters, to near darkness in deep or turbid habitats. Microenvironments within benthic communities also vary tremendously with depth in the matrix; self-shading and scattering by inorganic particles can reduce light penetration to a few millimeters or less. Responses by benthic algae to spatial and temporal light variation are not particularly well documented, and need much more attention. Results obtained to date suggest that light effects are very important to freshwater benthic algal ecology. Community development and the self-shading that accompanies vertical expansion appear to be very important in benthic algae, altering photosynthesis–irradiance responses and ultimately limiting the community development. Photoadaptation by stream algae in response to shading by terrestrial vegetation can be very significant, involving substantial increases in photosynthetic efficiency. However, published saturation intensities for freshwater benthic algae do not vary greatly, ranging from 100 to 400 μmol m^{-2} s^{-1}, despite great variability in benthic light regimes. Shading by terrestrial vegetation is a major constraint upon benthic primary production in streams, although biomass accrual is not always closely related to canopy cover because loss factors such as grazing can obscure shade effects. Attenuation of light by the water column should be a major determinant of benthic algal distributions in lakes, but little light-related research has been done in these ecosystems. The depth distribution of benthic algal biomass does not appear to precisely follow the exponential decrease in light intensity with depth; other depth-related factors such as wave disturbance, nutrients, or temperature may interact with light to govern algal depth distribution. Light does appear to be a very important part of the phytoplankton–benthic algae interaction in lakes though, and benthic algal biomass is negatively correlated with the increasing light extinction that accompanies increasing phytoplankton biomass. The effects of light on benthic algal community structure are not clear-cut, but in streams, filamentous green algae generally prosper in high-light environments. Spectral variation in PAR has not yet been implicated as an important effector of community structure or *in situ* primary production. Recent research suggests that freshwater benthic algae may be susceptible to ultraviolet radiation, but it is unclear that ultraviolet effects are widespread and indirect effects may be more important than direct effects.

Much is left to be learned about light effects on benthic algae. We know little about minimal light requirements or saturation irradiances *in situ*. Photoinhibition is still an unknown factor: benthic algae may be particularly susceptible to high fluxes of visible and ultraviolet radiation because of their fixed position, yet shielding and repair mechanisms may prevent or mitigate direct effects. The respiratory costs of potential photoinhibition–protection mechanisms are unknown, as are the costs of shade adaptation. No information exists on the time scale of photoadaptation in benthic algae, and it is unclear if static P–I models are sufficient for estimating *in situ* primary production. In streams, the importance of sunflecks to daily photosynthesis is a mystery. Sunflecks may be too brief to contribute much to primary production, but if stream algae are like some shade-adapted understory plants on the forest floor, then their photosynthetic machinery may be geared to harvesting high-intensity light over short periods. The roles that facultative heterotrophy and respiratory quiescence play in allowing benthic algae to persist during extended dark periods need investigating. Last, the interaction of other potential limiting factors (e.g., nutrients) with light may be as important as the effect of light alone, and our knowledge of benthic algal ecology would profit from multifactor experiments that explore these interactions.

REFERENCES

Admiraal, W., and Peletier, H. (1979). Influence of organic compounds and light limitation on the growth rate of estuarine benthic diatoms. *Br. Phycol. J.* **14,** 197–206.

Antoine, S. E., and Benson-Evans, K. (1983a). The effect of light intensity and quality on the growth of benthic algae. *I.* Phytopigment variations. *Arch. Hydrobiol.* **98,** 299–306.

Antoine, S. E., and Benson-Evans, K. (1983b). The effect of light intensity and quality on the growth of benthic algae. *II*. Population dynamics. *Arch. Hydrobiol.* **99**, 118–128.

Boston, H. L., and Hill, W. R. (1991). Photosynthesis–light relations of stream periphyton communities. *Limnol. Oceanogr.* **36**, 644–656.

Bothwell, M. L., Sherbot, D., Roberge, A. C., and Daley, R. J. (1993). Influence of natural ultraviolet radiation on lotic periphytic diatom community growth, biomass accrual, and species composition: Short-term versus long-term effects. *J. Phycol.* **29**, 24–35.

Bothwell, M. L., Sherbot, D. M. J., and Pollock, C. M. (1994). Ecosystem response to solar ultraviolet-B radiation: Influence of trophic-level interactions. *Science* **265**, 97–100.

Burkholder, J. M., and Cuker, B. E. (1991). Response of periphyton communities to clay and phosphate loading in a shallow reservoir. *J. Phycol.* **27**, 373–384.

Chazdon, R. L., and Pearcy, R. W. (1991). The importance of sunflecks for forest understory plants. *BioScience* **41**, 760–766.

Cota, G. F. (1985). Photoadaptation of high Arctic ice algae. *Nature (London)* **315**, 219–221.

Davies-Colley, R. J., Hickey, C. W., Quinn, J. M., and Ryan, P. A. (1992). Effects of clay discharges on streams. 1. Optical properties and epilithon. *Hydrobiologia* **248**, 215–234.

Dehning, I., and Tilzer, M. M. (1989). Survival of *Scenedesmus acuminatus* (Chlorophyceae) in darkness. *J. Phycol.* **25**, 509–515.

DeNicola, D. M., and Hoagland, K. D. (1992). Influences of canopy cover on spectral irradiance and periphyton assemblages in a prairie stream. *J. North Am. Benthol. Soc.* **11**, 391–404.

Dillard, G. E. (1966). The seasonal periodicity of *Batrochospermum macrosporum* Mont. and *Audouinella violacea* (Kuetz.) Hamel in Turkey Creek, Moore County, North Carolina. *J. Elisha Mitchell Sci. Soc.* **82**, 204–207.

Dodds, W. K. (1992). A modified fiber-optic light microprobe to measure spherically integrated photosynthetic photon flux density: Characterization of periphyton photosynthesis–irradiance patterns. *Limnol. Oceanogr.* **37**, 871–878.

Dring, M. J. (1981). Chromatic adaptation of photosynthesis in benthic marine algae: An examination of its ecological significance using a theoretical model. *Limnol. Oceanogr.* **26**, 271–284.

Dubinsky, A., Falkowski, P. G., and Wyman, K. (1986). Light harvesting and utilization by phytoplankton. *Plant Cell Physiol.* **27**, 1335–1349.

Endler, J. A. (1993). The color of light in forests and its implications. *Ecol. Monogr.* **63**, 1–27.

Evans, D., and Stockner, J. G. (1972). Attached algae on artificial and natural substrates in Lake Winnipeg, Manitoba. *J. Fish. Res. Board Can.* **29**, 31–44.

Falkowski, P. G. (1984). Physiological responses of phytoplankton to natural light regimes. *J. Plankton Res.* **6**, 295–307.

Falkowski, P. G., and LaRoche, J. (1991). Acclimation to spectral irradiance in algae. *J. Phycol.* **27**, 8–14.

Falkowski, P. G., Owens, T. G., Ley, A. C., and Mauzerall, C. C. (1981). Effects of growth irradiance levels on the ratio of reaction centers in two species of marine phytoplankton. *Plant Physiol.* **68**, 969–973.

Faure-Fremiet, E. (1951). The tidal rhythm of the diatom *Hantzschia amphioxys*. *Biol. Bull. (Woods Hole, Mass.)* **100**, 173–177.

Federer, C. A., and Tanner, C. B. (1966). Spectral distribution of light in the forest. *Ecology* **47**, 555–560.

Feminella, J. W., Power, M. E., and Resh, V. H. (1989). Periphyton responses to invertebrate grazing and riparian canopy in three northern California coastal streams. *Freshwater Biol.* **22**, 445–457.

Frederick, J. E. (1993). Ultraviolet sunlight reaching the earth's surface: A review of recent research. *Photochem. Photobiol.* **57**, 175–178.

Garcia-Pichel, F., and Castenholz, R. W. (1991). Characterization and biological implications of scytonemin, a cyanobacterial sheath pigment. *J. Phycol.* **27,** 395–409.

Graham, J. M., Kranzfelder, J. A., and Auer, M. T. (1985). Light and temperature as factor regulating seasonal growth and distribution of *Ulothrix zonata* (Ulvophyceae). *J. Phycol.* **21,** 228–234.

Graham, J. M., Lembi, C. A., Adrian, H. L., and Spencer, D. F. (1995). Physiological responses to temperature and irradiance in *Spirogyra* (Zygnematales, Charophyceae). *J. Phycol.* **31,** 531–540.

Hansson, L.-A. (1988). Effects of competitive interactions on the biomass development of planktonic and periphytic algae in lakes. *Limnol. Oceanogr.* **33,** 121–128.

Hansson, L.-A. (1992). Factors regulating periphytic algal biomass. *Limnol. Oceanogr.* **37,** 322–328.

Harris, G. P., and Piccinin, B. B. (1977). Photosynthesis by natural phytoplankton populations. *Arch. Hydrobiol.* **80,** 405–456.

Helbling, E. W., Villafane, V., Ferrario, M., and Holm-Hansen, O. (1992). Impact of natural ultraviolet radiation on rates of photosynthesis and on specific marine phytoplankton species. *Mar. Ecol.: Prog. Ser.* **80,** 89–100.

Hill, W. R., and Boston, H. L. (1991). Community development alters photosynthesis–irradiance relations in stream periphyton. *Limnol. Oceanogr.* **36,** 1375–1389.

Hill, W. R., and Harvey, B. C. (1990.) Periphyton responses to higher trophic levels and light in a shaded stream. *Can. J. Fish. Aquat. Sci.* **12,** 2307–2314.

Hill, W. R., and Knight, A. W. (1988). Nutrient and light limitation of algae in two northern California streams. *J. Phycol.* **24,** 125–132.

Hill, W. R., Boston, H. L., and Steinman, A. D. (1992). Grazers and nutrients simultaneously limit lotic primary productivity. *Can. J. Fish. Aquat. Sci.* **49,** 504–512.

Hill, W. R., Ryon, M. G., and Schilling, E. M. (1995). Light limitation in a stream ecosystem: Responses by primary producers and consumers. *Ecology* **76,** 1297–1309.

Hoober, J. K. (1984). "Chloroplasts." Plenum, New York.

Hornberger, G. M., Kelly, M. G., and Eller, R. M. (1976). The relationship between light and photosynthetic rate in a river community and implications for water quality modeling. *Water Resour. Res.* 12, 723–730.

Howard-Williams, C., and Vincent, W. F. (1989). Microbial communities in southern Victoria Land streams (Antarctica). I. Photosynthesis. *Hydrobiologia* **172,** 27–38.

Hudon, C., and Bourget, E. (1983). The effect of light on the vertical structure of epibenthic diatom communities. *Bot. Mar.* **26,** 317–330.

Hunding, C. (1971). Production of benthic microalgae in the littoral zone of a eutrophic lake. *Oikos* 22, 389–397.

Hynes, H. B. N. (1970). "The Ecology of Running Waters." Univ. of Toronto Press, Toronto.

Jasper, J., and Bothwell, M. L., (1986). Photosynthetic characteristics of lotic periphyton. *Can. J. Fish. Aquat. Sci.* **43,** 1960–1969.

Jassby, A. D., and Platt, T. (1976). Mathematical formulation of the relationship between photosynthesis and light for phytoplankton. *Limnol. Oceanogr.* **21,** 540–547.

Kaczmarczyk, D., and Sheath, R. G. (1991). The effect of light regime on the photosynthetic apparatus of the freshwater red alga *Batrachospermum boryanum*. *Cryptogam.: Algol.* **12,** 249–263.

Karentz, D., and Lutze, L. H. (1990). Evaluation of biologically harmful ultraviolet radiation in Antarctica with a biological dosimeter designed for aquatic environments. *Limnol. Oceanogr.* **35,** 549–561.

Karentz, D., Cleaver, J., and Mitchell, D. L. (1991). Cell survival characteristics and molecular responses of Antarctic phytoplankton to ultraviolet-B radiation. *J. Phycol.* **27,** 326–341.

Kelly, M. G., Hornberger, G. M., and Cosby, B. J. (1974). Continuous automated measurement of rates of photosynthesis and respiration in an undisturbed river community. *Limnol. Oceanogr.* **19,** 305–312.

Kirk, J. T. O. (1983). "Light and Photosynthesis in Aquatic Ecosystems." Cambridge Univ. Press, Cambridge, UK.

Korch, J. E., and Sheath, R. G. (1989). The phenology of *Audouinella violacea* (Acrochaetiacieae, Rhodophyta) in a Rhode Island stream (USA). *Phycologia* **28**, 228–236.

Kremer, B. P. (1983). Untersuchungen zur Okophysiologie eininger Susswasserrotalgen. *Decheniana.* **136**, 31–42.

Kuhl, M., and Jorgensen, B. B. (1994). The light field of microbenthic communities: Radiance distribution and microscale optics of sandy coastal sediments. *Limnol. Oceanogr.* **39**, 1368–1398.

Langdon, C. (1988). On the causes of interspecific differences in the growth–irradiance relationship for phytoplankton. II. A general review. *J. Plankton Res.* **10**, 1291–1312.

Lorenz, R. C., Monaco, M. E., and Herdendorf, C. E. (1991). Minimum light requirements for substrate colonization by *Cladophora glomerata. J. Great Lakes Res.* **17**, 536–542.

Lowe, R. L., Golladay, S. W., and Webster, J. R. (1986). Periphyton response to nutrient manipulation in streams draining clearcut and forested watersheds. *J. North Am. Benthol. Soc.* **5**, 221–229.

Lyford, J. H., and Gregory, S. V. (1975). The dynamics and structure of periphyton communities in three Cascade Mountain streams. *Verh.—Int. Ver. Theor. Angew. Limnol.* **19**, 1610–1616.

McIntire, C. D., and Phinney, H. K. (1965). Laboratory studies of periphyton production and community metabolism in lotic environments. *Ecol. Monogr.* **35**, 237–258.

Meulemans, J. T. (1987). A method for measuring selective light attenuation within a periphytic community. *Arch. Hydrobiol.* **109**, 139–145.

Meulemans, J. T. (1988). Seasonal changes in biomass and production of periphyton growing upon reed in Lake Maarsseveen I. *Arch. Hydrobiol.* **112**, 21–42.

Morel, A., Lazzara, L. and Gostan, J. (1987). Growth rate and quantum yield time response for a diatom to changing irradiances (energy and color). *Limnol. Oceanogr.* **32**, 1066–1084.

Neale, P. J., and Richerson, P. J. (1987). Photoinhibition and the diurnal variation of phytoplankton photosynthesis. I. Development of a photosynthesis–irradiance model from studies of *in situ* responses. *J. Plankton Res.* **9**, 167–193.

Pahl-Wostl, C., and Imboden, D. M. (1990) DYPHORA—A dynamic model for the rate of photosynthesis of algae. *J. Plankton Res.* **12**, 1207–1221.

Palmisano, A. C., and Sullivan, C. W. (1982) Physiology of sea ice diatoms. 1. Response of three ploar diatoms to a simulated summer–winter transition. *J. Phycol.* **18**, 489–498.

Palmisano, A. C., SooHoo, J. B., and Sullivan, C. W. (1985). Photosynthesis–irradiance relationships in sea ice microalgae from McMurdo Sound, Antarctica. *J. Phycol.* **21**, 341–346.

Paul, B. J., and Duthie, H. C. (1989). Nutrient cycling in the epilithon of running waters. *Can. J. Bot.* **67**, 2302–2309.

Platt, T., Gallegos, C. L., and Harrison, W. G. (1980). Photoinhibition of photosynthesis in natural assemblages of marine phytoplankton. *J. Mar. Res.* **38**, 687–701.

Ploug, H., Lassen, C., and Jorgensen, B. B. (1993). Action spectra of microalgal photosynthesis and depth distribution of spectral scalar irradiance in a coastal marine sediment of Limfjorden, Denmark. *FEMS Microbiol. Ecol.* **102**, 261–270.

Ramus, J. (1983). A physiological test of the theory of complementary chromatic adaptation. II. Brown, green and red seaweeds. *J. Phycol.* **19**, 173–178.

Richardson, K., Beardall, J., and Raven, J. A. (1983). Adaptation of unicellular algae to irradiance: An analysis of strategies. *New Phytol.* **93**, 157–191.

Rivkin, R. B. (1990). Photoadaptation in marine phytoplankton: Variations in ribulose 1,5-bisphosphate activity. *Mar. Ecol.: Prog. Ser.* **62**, 61–72.

Robinson, C. T., and Minshall, G. W. (1986) Effects of disturbance frequency on stream benthic community structure in relation to canopy cover and season. *J. North Am. Benthol. Soc.* **5**, 237–248.

Rosemond, A. D. (1993). Interactions among irradiance, nutrients, and herbivores constrain a stream algal community. *Oecologia* **94,** 585–594.
Round, F. E., and Eaton, J. W. (1966). Persistent, vertical-migration rhythms in benthic microflora. III. The rhythm of epipelic algae in a freshwater pond. *J. Ecol.* **54,** 609–615.
Ryther, J. H. (1956). Photosynthesis in the ocean as a function of light intensity. *Limnol. Oceanogr.* **1,** 61–70.
Sand-Jensen, K., and Revsbech, N. P. (1987). Photosynthesis and light adaptation in epiphyte–macrophyte associations measured by oxygen microelectrodes. *Limnol. Oceanogr.* **32,** 452–457.
Sheath, R. G. (1984). The biology of freshwater red algae. *Prog. Phycol. Res.* **3,** 89–157.
Sheath, R. G., and Burkholder, J. M. (1985). Characteristics of softwater streams in Rhode Island, II. Composition and seasonal dynamics of macroalgal communities. *Hydrobiologia* **128,** 109–118.
Sheath, R. G., Burkholder, J. M., Morison, M. O., Steinman, A. D., and VanAlstyne, K. L. (1986). Effect of tree canopy removal by gypsy moth larvae on the macroalgae of a Rhode Island headwater stream. *J. Phycol.* **22,** 567–570.
Shortreed, K. S., and Stockner, J. G. (1983). Periphyton biomass and species composition in a coastal rainforest stream in British Columbia: Effects of environmental changes caused by logging. *Can. J. Fish. Aquat. Sci.* **40,** 1887–1895.
Smith, R. C., Baker, K. S., Holm-Hansen, O., and Olson, R. (1980). Photoinhibition of photosynthesis in natural waters. *Photochem. Photobiol.* **31,** 585–592.
Steinman, A. D. (1992). Does an increase in irradiance influence periphyton in a heavily-grazed woodland stream? *Oecologia* **91,** 163–170.
Steinman, A. D., and McIntire, C. D. (1987). Effects of irradiance on the community structure and biomass of algal assemblages in laboratory streams. *Can. J. Fish. Aquat. Sci.* **44,** 1640–1648.
Steinman, A. D., McIntire, C. D., Gregory, S. V., and Lamberti, G. A. (1989). Effects of irradiance and grazing on lotic algal assemblages. *J. Phycol.* **25,** 478–485.
Steinman, A. D., Mulholland, P. J., Palumbo, A. V., Flum, T. F., Elwood, J. W., and DeAngelis, D. L. (1990). Resistance of lotic ecosystems to a light elimination disturbance: A laboratory stream study. *Oikos* **58,** 80–90.
Stevenson, R. J., and Stoermer, E. F. (1981). Quantitative differences between benthic algal communities along a depth gradient in Lake Michigan. *J. Phycol.* **17,** 29–36.
Tuchman, N. C., Panella, J., Donovan, P., and Smarrelli, J., Jr. (1994). Facultative heterotrophy as a survival mechanism for light-limited benthic diatoms. *Bull. North Am. Benthol. Soc.* **11,** 129.
Turner, M. A., Schindler, D. W., and Graham, R. W. (1983). Photosynthesis–irradiance relationships of epilithic algae measured in the laboratory and *in situ*. "Periphyton of Freshwater Ecosystems" *In* (R. G. Wetzel, ed.) pp. 73–87. Dr. W. Junk Publishers, The Hague.
van Dijk, G. M. (1993). Dynamics and attenuation characteristics of periphyton upon artificial substratum under various light conditions and some additional observations on periphyton upon *Potamogeton pectinatus* L. *Hydrobiologia* **252,** 143–161.
Vannote, R. L., Minshall, G. W., Cummins, K. W., Sedell, J. R., and Cushing, C. E. (1980). The river continuum concept. *Can. J. Fish. Aqua. Sci.* **37,** 130–137.
Vincent, W. F., and Roy, S. (1993). Solar ultraviolet-B radiation and aquatic primary production: Damage, protection, and recovery. *Environ. Rev.* **1,** 1–12.
Watras, C. J., and Baker, A. L. (1988). The spectral distribution of downwelling light in northern Wisconsin Lakes. *Arch. Hydrobiol.* **112,** 481–494.
Wetzel, R. G. (1983). "Limnology," 2 ed. Saunders, Philadelphia.
Wootton, J. T., and Power, M. E. (1993). Productivity, consumers, and the structure of a river food chain. *Proc. Nat. Acad. Sci.* **90,** 1384–1387.

6

Periphyton Responses to Temperature at Different Ecological Levels

Dean M. DeNicola

Department of Biology
Slippery Rock University
Slippery Rock, Pennsylvania 16057

I. INTRODUCTION

A. Temporal and Spatial Variation of Temperature in Benthic Habitats

The effect of temperature on biochemical reactions makes it one of the most important environmental factors affecting freshwater lentic and lotic periphyton communities. The temperature of cells in periphyton may be influenced directly by solar radiation, particularly in clear, shallow water, where infrared wavelengths are transmitted. However, cell temperatures are primarily determined by conduction and convection transfer of heat from the surrounding water. Therefore, spatial and temporal variations of temperature in the aquatic habitat strongly influence biological responses of periphyton.

The temperature variation in lakes and streams is influenced by factors that operate at different temporal and spatial scales, with temperature changes at each level constrained by the level above. In most freshwater periphyton habitats, water temperature is determined primarily by direct solar radiation, and at a large spatial scale, factors such as latitude, elevation, continentality, aspect, and morphometry determine the overall temperature regime (Hutchinson, 1957). Over long time scales, temperature varies with long-term climatic cycles, with recent trends possibly influenced by anthropogenic changes in global climate (Table I).

On a smaller scale, local input from groundwater can have significant effects on temperature in stream or lake systems. In large lotic watersheds, water temperature often increases downstream with discharge in summer owing to decreases in elevation and in the relative magnitude of groundwater inputs. This temperature pattern may reverse in the winter when groundwater is warmer than surface runoff (Hynes, 1970). In addition, temperatures near the maximum for periphyton growth are created by geothermal groundwater inputs, with distinct temperature gradients usually occurring downstream from hot springs. Thermal input from surface runoff (tributaries) also may alter the temperature along the main stream in a watershed or at the inlet of a lake (Ward, 1992). Seasonal fluctuations in air and water temperature created by changes in sun angle and duration determine the degree of thermal stratification in lentic habitats. Stratification and overturn events play a major role in determining seasonal succession in freshwater phytoplankton, and may influence lentic periphyton community structure through associated effects on disturbance, dissolved oxygen and nutrient cycling within the mat (Round, 1972; Wetzel, 1983; Riber and Wetzel, 1987). Lotic habitats in deciduous forests may have temperature changes created by variation in thermal radiation due to changes in canopy cover either seasonally or spatially in a watershed. In general, the presence of canopy cover tends to result in lower mean and maximum summer temperatures, less diurnal variation, and higher mean and minimum winter temperatures in streams (Lynch *et al.*, 1984; Weatherley and Ormerod, 1990). In a heat budget for low-order streams in the Oregon Coast Range, Brown (1969) estimated that direct solar radiation was the predominant (>90%) source of heat for the streams during the day. Moreover, 25% of midday heat was absorbed by bedrock substrata and reradiated to the water at night. Seasonal input from snowmelt also can change temperature in lotic and lentic periphyton habitats in cold climates, and anchor ice formation in streams can scour substrata (Table I). Overall, the range of mean monthly temperature in rivers and lakes usually decreases with water volume, and often is much less than that of the land surface (Hynes, 1970). In addition, shallow regions of deep lakes and streams generally have a greater seasonal temperature variation than deep-water areas (Ward, 1992).

At a smaller spatial scale, pools may have higher temperatures in summer than adjacent riffles in streams (Ward, 1992). Diurnal temperature variations occur in both lentic and lotic habitats through variations in daily solar radiation and rainfall (Table I). In large lakes and rivers, diurnal fluctuations are primarily restricted to shallow depths, whereas diurnal changes may significantly affect the entire water column in a shallow lake

TABLE I Temporal and Spatial Scales of Temperature Variation in Freshwater Periphyton Habitats

Habitat level	*Processes affecting temperature*	*Linear spatial scale (km)*	*Temporal scale (y)*	*Potential temperature variation (°C)*
Latitude and elevation Long-term climate	Plate tectonics, Milankovitch cycles, volcanic activity, greenhouse gas levels, meteor/comet impacts, El Niño, geomorphic processes	10^2–10^5	10^1–10^8	10^1–10^2
Watershed Seasonal climate	Day length, sun angle, canopy cover, groundwater, snow melt, stratification, ice formation, discharge	10^{-1}–10^2	10^0	10^1–10^2
Microhabitat Diurnal climate	Sun angle, sunflecks, rainfall, current velocity	$<10^{-1}$	10^{-3}	10^{-1}–10^1

or stream (Hynes, 1970; Wetzel, 1983). In temperate watersheds, diurnal variation in stream temperature usually is greatest in midorder streams, as groundwater and a large water volume buffer diurnal changes in headwater and high-order streams, respectively (Vannote and Sweeney, 1980; Ward, 1992). For example, Brown (1969) found that diurnal fluctuations in a forested third-order stream were as large as 9°C. Finally, diurnal temperature changes are often smaller for interstitial water than for surface water, although this depends on the degree of upwelling groundwater at the site (Shepherd *et al.*, 1986).

B. Examination of Periphyton Responses to Temperature

The effects of temperature on freshwater periphyton can be examined at different ecological levels; autecological or physiological, population, community, and ecosystem. In this review, information concerning periphyton–temperature interactions is examined for each level in this hierarchical scheme (Table II). Responses of periphyton at a given ecological level are determined by synthesizing interactions for the lower levels within the constraints and properties of higher levels. A hierarchical approach may lead to a more complete understanding of the influence of temperature on periphyton physiology, structure, function, and distribution at different spatiotemporal scales.

There has been considerably more research examining the effects of temperature on phytoplankton than on periphyton at all ecological levels, but especially for cellular responses. Epply (1972) and Li (1980) have done extensive reviews of the influence of temperature on physiology and growth in marine phytoplankton. Partial reviews of temperature effects on

TABLE II Responses of Freshwater Periphyton to Temperature at Different Ecological Levels

Ecological level	*Periphyton response*
Autecological	Concentrations of photosynthetic and respiratory enzymes; changes in cell quota and nutrient uptake; alteration of fatty acids; heat shock proteins?; timing of life history and cell cycle.
Population	Potential maximum growth rate; minimum, maximum, and optimal temperatures for growth.
Community	Dominance of major algal classes; species composition and diversity; seasonal succession; geographic distribution; competitive and trophic interactions.
Ecosystem	Potential maximum areal primary production; biomass
Global	Species composition of fossil assemblages for paleoreconstruction of temperature.

freshwater phytoplankton can be found in Hutchinson (1967), Fogg (1975), and Reynolds (1984). The mechanisms of physiological effects of temperature on phytoplankton and periphyton cells are probably similar in many cases, and given the paucity of studies on periphytic species, several studies on phytoplanktonic species are used as examples of likely physiological responses.

Ecological effects of temperature at the population, community, and ecosystem levels can differ substantially between phytoplankton and periphyton. These differences result primarily from the manner in which temperature interacts with other environmental factors in the two habitats. The more variable physicochemical environment of the benthos relative to the plankton usually results in more complex spatiotemporal patterns in species distribution (Round, 1972; Wetzel, 1983). Studies of community- and ecosystem-level responses of periphyton to temperature employing direct experimental manipulation of temperature independent of other factors are relatively few compared to controlled studies of nutrient, light, and grazing effects. Most of the studies that controlled temperature involved the use of mesocosms. More commonly, semicontrolled studies of temperature effects have been carried out in habitats subjected to thermal effluent, where periphyton responses are examined along thermal gradients. Much of the research concerning temperature and periphyton involves studies of seasonal succession, but interactions between temperature and other variables, such as nutrients, gases, metabolites, developmental stage, trophic interactions, and especially light, are not clearly separated. One cannot assume that such multifactor interactions are simply additive, and conclusions based on seasonal correlations between temperature and periphyton are dubious for all but the most obvious cases. The most information concerning temperature effects on phytoplankton has been obtained from studies that combine field observations with controlled multifactorial experiments, yet very few such studies have been conducted so far for freshwater periphyton.

II. AUTECOLOGICAL RESPONSES

Physiological effects of temperature on microalgae can be considered in terms of capacity and resistance adaptations (Li, 1980). Capacity adaptations occur within the temperature tolerances for a taxon, and cell responses are dependent on the temperature experienced during growth. These adaptations are related to temperature limits on enzymatic rates, which determine the accumulation of various metabolites in the cell, and will be discussed in this section. Resistance adaptations to temperature refer to physiological mechanisms that determine the upper and lower temperature extremes for growth. Such mechanisms usually are related to denaturation of proteins at high temperature and freezing of cell contents at low temperature (Li, 1980). Resistance adaptations will be discussed in the section on population growth.

Most of the research concerning capacity-related physiological responses of algae to temperature has involved laboratory studies with marine phytoplankton species (e.g., Steemann-Nielsen and Jørgensen, 1968; Mortain-Bertrand *et al.*, 1988; Levasseur *et al.*, 1990; Thompson *et al.*, 1992a), with few freshwater taxa having been examined (Mosser and Brock, 1976; Rhee and Gotham, 1981; Schlesinger and Shuter, 1981). Many of the phytoplankton species studied are within the same family as common periphyton taxa, and physiological responses and adaptations to temperature for both groups are presumed to be similar.

A. Photosynthesis and Respiration

Temperature primarily affects algal photosynthetic metabolism through its control of enzyme reaction rates. As temperature increases, the increased kinetic energy of reacting molecules results in higher reaction rates until the point where denaturation rate exceeds kinetic effects. As a result, enzymatic rates and the processes they determine usually have a Q_{10} of approximately 2, up to an optimal temperature (Salisbury and Ross, 1985). Independent temperature effects on the activation energy, maximum enzymatic rate (V_{max}), and half-saturation constant (K_m) determine the specific Q_{10} value for an enzyme. Mantal (1974) determined the Q_{10} for photosynthesis in the common periphytic green alga *Cladophora glomerata* to be about 2, which is a typical value for most algal species (Davison, 1991). The short-term responses (i.e., no acclimation) of photosynthesis to temperature depend on whether light is saturating or not. Short-term, light-saturated responses are determined primarily by carbon fixation by RUBISCO and by the enzymes controlling the rate of phosphate transport into chloroplasts for ATP regeneration. In addition, photorespiration may increase with increasing temperature owing to differential changes in the affinity of RUBISCO for O_2 versus CO_2 (Davison, 1991). At subsaturating light intensity, the short-term photosynthetic rate is controlled primarily by physical photochemical reactions that have been usually considered to be independent of temperature (Steemann-Nielsen and Jørgensen, 1968). However, in some studies, rates of enzymatic reactions involved in photophosphorylation and electron transport have been found to be temperature-dependent, with light-harvesting efficiency (alpha, or the initial slope of the photosynthesis–irradiance curve) decreasing with temperature. Because respiration rate increases with temperature, the amount of light necessary to reach the photosynthetic compensation point may increase as temperature increases for periphyton in light-limiting environments (Davison, 1991).

Most species of algae have been shown to acclimate photosynthetic metabolism to temperature by altering their enzyme and chlorophyll *a* concentrations (Mortain-Bertrand *et al.*, 1988; Thompson *et al.*, 1992a). Concentrations of RUBISCO and other Calvin cycle enzymes increase as temperature decreases to compensate for the lower activity of individual molecules (Davison, 1991). For example, in a study involving a freshwater species *(Scenedesmus obliquus),* Rhee and Gotham (1981) found that net photosynthesis varied little from 10° to 20° C because of the compensation in cell enzyme content. Similarly, maintenance of respiration rate at different temperatures involves changes in cell protein concentrations (Harris, 1986). Enzymatic adaptations to temperature by the use of isozymes and/or allosteric effects have been shown to occur in some animals (Cossins and Bowler, 1987), but have not been examined for algae.

In addition to altering photosynthetic enzyme concentrations, several studies have found that algal cells acclimate light-harvesting efficiency by increasing chlorophyll *a* concentration (both increases in size and number of photosynthetic units) at higher temperatures to compensate for increased CO_2 fixation (Yoder, 1979; Thompson *et al.*, 1992a). An increase in assimilation number (Chl *a*/C) with increasing temperature occurs for many marine phytoplankton species (Epply, 1972), and in the few freshwater taxa examined (Schlesinger and Shuter, 1981). Similarly, *Dunaliella* sp. was found to decrease light harvesting (turning off photosynthetic units) at low temperatures to compensate for reductions in CO_2 fixation (Levasseur *et al.*, 1990). Taken together, these results suggest that cells acclimate their photosynthetic physiology to temperature by balancing photon capture (chlorophyll *a* concentration for light reactions) with carbon fixation (concentrations of Calvin cycle enzymes). Most studies of photosynthetic adaptations of algae to temperature have used cultures grown at an acclimation temperature for several weeks. The few studies that have addressed the rate of acclimation suggest that photosynthetic adaptations begin within a few hours following a temperature change (Davison, 1991).

In a study of temperature–photosynthesis relationships in freshwater periphyton, Mosser and Brock (1976) found that the cyanobacteria *Protococcus, Oscillatoria,* and *Phormidium* spp., and the green alga *Zygnema* sp. from a cold mountain stream, had optimal short-term photosynthetic rates at 20° to 30°C despite being obtained from sites that ranged from 1° to 12°C. Similarly, optimal temperatures for photosynthesis in snow algae were 10° to 20°C higher than ambient temperatures (Mosser *et al.*, 1977). Many other studies have shown that algae typically have short-term temperature optima for photosynthesis several degrees higher than the temperature at which they were growing. The reasons for this are not well known. Li (1980) suggested that the relationships depend on whether net or gross photosynthetic rates are measured because of temperature effects on respiration. Davison (1991) presents evidence, however, that many of these studies involved short-term measurements of photosynthesis, but when cells are allowed to acclimate, optimal temperatures for photosynthesis and growth are more similar. Since acclimation of photosynthesis to temperature involves altering enzyme concentrations, the amount of organic matter and nutrient uptake needed for growth and reproduction should also change (Steemann-Nielsen and Jørgensen, 1968).

B. Cell Composition

Temperature is known to influence the biochemical composition of algae (Jørgensen, 1968; Yoder, 1979; Goldman and Mann, 1980; Thompson *et al.*, 1992a), which is important in determining its nutritional and caloric value for herbivores. As with photosynthesis, studies examining the influence of temperature on cell composition of algae have been primarily restricted to marine phytoplankton species. Among many of these species, an increase in cell biomass as temperature decreases has been measured (Epply, 1972). As noted in the previous section, increases in cellular protein with decreasing temperature can result from increased enzyme production as an adaptive mechanism for maintaining rates of photosynthesis and respiration. The pattern of cellular carbon and nitrogen quotas (as well as cell volume) has been widely assumed to be a U-shaped response, with minimal concentrations (and cell size) at the optimal temperature for growth (Goldman and Mann, 1980; Rhee, 1982; Harris, 1986). In other words, it requires more carbon and nutrients to produce a cell (at the same growth rate) at a nonoptimal temperature (Darley, 1982). However, in a study involving eight species of marine phytoplankton, Thompson *et al.* (1992a) found no general response in carbon and nitrogen quotas, or cell volume to temperature, although the foregoing relationship held for cell protein concentration and temperature for most of the species that were examined. Species in which cell quotas do not change with temperature should be better adapted to varying temperature during nutrient-limited growth (Darley, 1982). Studies of the effects of temperature on cell quotas for other nutrients and biochemical products are few, but some marine diatoms become more highly silicified at lower temperatures (Rhee, 1982).

Cell quotas for nutrients also are determined by effects of temperature on K_s and V_{max} values for enzymes involved in nutrient uptake. For seaweeds, nitrogen uptake rates are correlated with short-term temperature variation, and it is thought that long-term temperature acclimation can occur by changes in the concentrations of enzymes involved in uptake (Duke *et al.*, 1989).

In many organisms, it is known that the degree of saturation in fatty acids of cellular membranes changes in response to temperature, presumably to maintain stability and fluidity of the membranes. Several studies have found that decreasing temperature caused a general increase in the degree of unsaturation of fatty acids in marine phytoplankton cells (e.g., Lynch and Thompson, 1982). Thompson *et al.* (1992b) also found a moderate but significant relationship between temperature and the percentage of polyunsaturated fats in cells. However, they point out that such relationships are complicated by temperature-associated changes in light, growth rate, and physiological state of the cells, and membrane changes are probably not based solely on maintaining fluidity and stability at a given temperature. In addition, other changes in membranes, such as protein : lipid and cholesterol : lipid ratios, or variation in fatty acid chain length, can alter fluidity at different temperatures (Thompson *et al.,* 1992b).

C. Heat Shock Proteins

Heat shock genes are activated in a wide range of prokaryotic and eukaryotic organisms in response to sudden elevation of temperature. The resulting proteins (heat shock proteins) help the cell/organism to survive the stress period by protecting essential cellular components from heat damage, allowing normal metabolism to resume quickly during recovery (Nover, 1984). The function of heat shock proteins appears to usually involve folding, transport, or regulation of proteins. The role of heat shock proteins in photosynthetic cells is largely unstudied at the present, although it is known that some heat shock proteins are associated with the thylakoid membranes, where they may repair proteins that have been partially denatured by heat (Lehel *et al.,* 1992). A few studies have demonstrated the expression of heat shock genes in algae, but only the green alga *Chlamydomonas* and a few cyanobacterial strains have been examined so far (Mannan *et al.,* 1986; Bhagwat and Apte, 1989). Lehel *et al.* (1992) induced the production of four heat shock proteins in the cyanobacterium *Synechocystis* by elevating the temperature of cells grown at 30°C to 42.5°C. Production of heat shock proteins reached a maximum within 0.5–2.0 h and was accompanied by a sharp reduction in total protein synthesis. The presence and role of heat shock proteins in periphyton taxa have not been examined to date.

D. Life Cycles

Temperature is an environmental cue for sexual reproduction in many algal species. In diatoms, cell size is of importance in determining sexual reproduction, however, temperature and light have been found to influence sexuality and auxospore formation. Similarly, formation and germination of resting spores of diatoms are affected by temperature (Dring, 1974; Drebes, 1977).

Temperature has been found to be an important factor in determining zoosporogenesis in the benthic green algae *Coleochaete, Ulothrix,* and *Cladophora.* In some cases, temperature thresholds for zoospore formation are correlated with the seasonal appearance of these taxa in lakes and streams, and may be related to the geographic distribution of some species (Hoffmann and Graham, 1984; Graham *et al.*, 1985, 1986). Temperature limits for zoospore production in species of *Oedogonium, Tetraspora, Bumilleria,* and *Vaucheria* are reported by Patrick (1974).

Timing of the cell division cycle for many microalgal species is synchronized or phased to some degree, often entrained by the light/dark cycle. In some species, this timing of cell division within a population has been found to be shifted by temperature (Chisholm, 1981).

III. POPULATION RESPONSES

A. Temperature and Maximum Growth Rates

Growth is the integrated expression of metabolism leading to cell synthesis, and the optimal temperatures for growth of a population reflect the evolution of physiological adaptations. Temperature is assumed to set the upper limit of growth rates and temperature effects should be most apparent when all other factors are optimized (Epply, 1972; Li, 1980). Using data from a large number of studies, Epply fitted a curve to describe the maximum specific growth rate of phytoplankton as a function of temperature (up to 40°C). The expected maximum growth rate was determined as

$$\mu = 0.851(1.066)^{T}$$

where μ = maximum growth rate in doublings per day and T = temperature in degrees centigrade. This relationship has a Q_{10} of 1.88. Goldman and Carpenter (1974) used a theoretical approach to describe the relationship between maximum specific growth rate and temperature by applying the Arrenius equation to a large data set and obtaining the equation

$$\mu = (5.35 \times 10^{9})e^{-6472/T}$$

where μ = maximum growth rate (base e) and T = temperature on the Kelvin scale. This relationship has a Q_{10} of 2.08. The implication of both equations is that although each species has its own temperature optimum for growth, the temperature-specific maximum growth rate for a species approaches a theoretical, absolute maximum. Combining all species, these maxima increase exponentially with temperature with a Q_{10} of approximately 2.0.

In a study of temperature effects on growth rates in lotic periphyton, Bothwell (1988) found that P-saturated growth rates (to obtain maximum

μ) for periphyton in laboratory stream channels followed a linear rather than exponential relationship with temperature. However, he suggested that nitrogen limitation in the summer may have reduced growth to prevent the relationship from becoming exponential at higher temperatures. The Q_{10} values for maximum growth rates in this study were also about 2.0.

B. Temperature Tolerance Ranges and Optima for Growth

Freshwater periphyton have been shown to grow actively from temperatures of about 0°C up to 75°C (Fogg, 1969). Hustedt's (1927–1959) classification for diatom species based on their preferred temperature and the width of their tolerance range can be applied to all periphyton species (Table III). Temperature limits (i.e., resistance adaptations, *sensu* Li, 1980) of algae are considered to be mainly determined by genetically fixed traits, and attempts to "train" populations to significantly expand minimum or maximum temperature tolerances by subculturing at progressively higher or lower temperatures is generally not possible (Fogg, 1969). The temperature range for the growth of a particular periphyton species may be related to the absolute range of temperature and the stability of temperature in its habitat (Seaburg *et al.*, 1981; Seaburg and Parker, 1983). This is tentatively supported by the tendency toward greater stenothermy in polar phytoplankton (relatively low temperature fluctuations) relative to polar soil algae and periphyton (Seaburg *et al.*, 1981).

Periphyton communities in hot springs are primarily cyanobacteria, which can survive in temperatures up to about 75°C (Brock, 1967). The

TABLE III Thermal Tolerance Classification of Diatoms[a]

Classification	*Temperature range (°C)*
Stenotherms: can withstand <10°C variation	
Cold-water stenotherms	15
Temperate stenotherms	15–25
Warm-water stenotherms	>25
Meso-stenotherms: can withstand 10°C variation	
Tropical cold-water forms	10–20
Temperate forms	15–25 and 20–30
Warm-water forms	25–35 and 30–40
Meso-eurytherms: can withstand 15°C variation	
Cold-water to temperate forms	10–25
Temperate forms	15–30
Temperate to warm-water forms	30–45
Eu-eurytherms: can live in >20°C variation	

[a]After Patrick (1977) and Hustedt (1927–1959) with permission of Blackwell Scientific Publications Ltd.

most widespread thermophilic periphyton taxon is the cyanobacterium *Mastigocladium laminosces,* probably because it can also tolerate freezing and desiccation. The unicellular red alga *Cyanidium caldarium* often dominates in low-pH hot springs, where cyanobacteria are absent (Darley, 1982). Diversity of cyanobacteria increases and some species of diatoms and strains of the green algal *Chlorella* can occur in temperatures of 30–50°C (Wallace, 1955; Fogg, 1969; Patrick, 1974).

It appears that heat tolerance of algae in hot springs is primarily a result of special protein configurations that depend less on hydrogen bonding for their spatial structure (Marrè, 1962). In addition, the cell membranes of thermophilic algae contain lipids with shorter chains, higher degrees of saturation, and higher melting points (Soeder and Stengel, 1974). Thermophilic species are well adapted to their environments and optimum temperatures for growth are close to ambient (Brock, 1967). Many hot springs algae can be grown experimentally at lower temperatures, though in natural habitats they may be outcompeted and overgrown at these temperatures (Marrè, 1962).

Cryophilic periphyton species are generally found in habitats at high altitudes or near the polar regions. Many different species can survive temperatures around freezing and no major class of algae found in the periphyton is entirely excluded from cold water (Fogg, 1969). Survival of freeze-drying at temperatures well below 0°C has been reported for several species of cyanobacteria (Fogg, 1969), and dry, frozen cyanobacterial mats that have overwintered in Antarctic streams resume metabolism within hours of rehydration (Vincent and Howard-Williams, 1986). Unlike thermophilic periphyton species, optimal growth temperatures for species in very cold habitats are usually well above ambient temperatures (Vincent and Howard-Williams, 1989). In general, microorganisms have optimal temperatures for growth (in culture) closer to their maximum temperature than their minimum temperature (Seaburg *et al.,* 1981). Seaburg *et al.* (1981) measured temperature tolerance for growth in culture for several periphyton isolates from Antarctic streams and found that most isolates were eurythermal, growing from 2° to 20°C. Optimal temperatures for growth were usually at the high end of this range, well above ambient, and approximately one-third of the isolates did not grow below 2°C. Similarly, periphyton collected in a cold mountain stream at 1–12°C grew optimally at 20–30°C (Mosser and Brock, 1976). This is in contrast to snow algae and phytoplankton from polar waters, which tend to be more stenothermal and have optima closer to their cold ambient temperatures.

Physiological resistance adaptations to cold in algae have received less study than adaptations to heat. Species that live in very cold environments are able to tolerate freezing and thawing processes, perhaps through changes in osmolarity (Marrè, 1962). *In situ* growth rates of periphyton that grow in near-freezing environments are very low, and temperature-related adaptations involving supercooling through the synthesis of oligosaccharides, specialized protein structure/function, or other mechanisms that may allow taxa to tolerate very cold conditions have not been examined.

Within the minimum and maximum temperature for growth, growth curves for individual species typically show a gradual decline in μ from the optimum to the minimum temperature, and a more abrupt decline in μ at temperatures above the optimum (Epply, 1972). Seaburg *et al.* (1981) suggested that ecologically important points in this curve are the lower and upper temperature limits, the range of temperature for active growth and the optimal temperature (Fig. 1). Batch cultures have been used for many years to examine the relationship between growth and temperature in algae. The temperature optimum for growth rate is a reproducible parameter that can be used to characterize species or clones (Epply, 1977), although the majority of these measurements have been for phytoplankton species. Temperature–growth curves for periphyton species have been determined primarily for either thermophilic taxa, cryophilic taxa (e.g., Brock, 1970; Seaburg *et al.*, 1981), or filamentous green algae (DeVries and Hillebrand, 1986). Results of several laboratory studies determining the optimum temperature for growth for freshwater periphyton taxa are summarized by Wallace (1955) and Patrick (1969).

A few studies have examined effects of short-term, temperature shock on algal growth. In most cases, small heat shocks (5–10°C) for short peri-

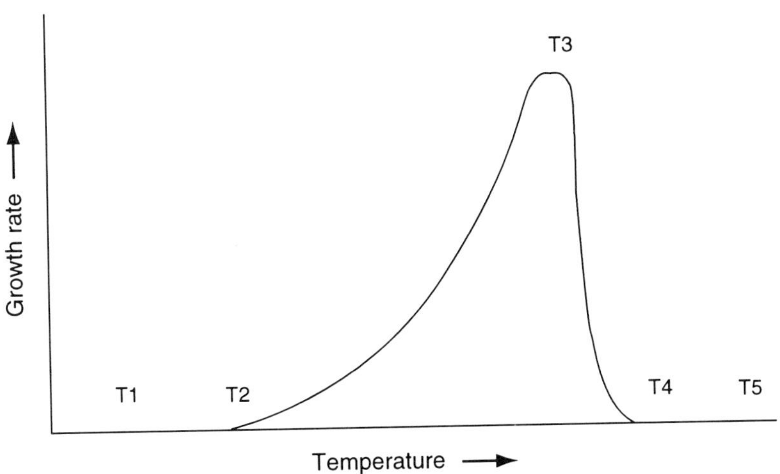

FIGURE 1 Hypothetical temperature ranges for the survival and growth of algae: T1 = minimum temperature for survival, T2 = minimum temperature for growth, T3 = optimum temperature for growth, T4 = maximum temperature for growth, T5 = maximum temperature for survival. (After Seaburg *et al.*, 1981 with permission.)

ods (<1 h) appear to have only temporary adverse effects on some species. However, the temperature preceding shock and the phase of the cell cycle that experiences shock are important factors in determining the overall effect on growth (Lanza and Cairns, 1972; Patrick, 1974). For example, Lanza and Cairns (1972) found severe effects of short-term heat stress on the growth and physiology of cultures of *Navicula seminulum* only when populations were initially growing near their thermal maximum.

C. Interactive Effects of Temperature and Nutrient Limitation on Growth Rates

Many growth models for microalgae are based on characteristics of nutrient-limited growth. The two major models for steady-state growth conditions describe growth either as a function of external nutrient concentration (Monod model) or as a function of internal cell concentrations (i.e., cell quota, Droop model). Although it is known that maximum growth rates are set by temperature, it was thought that temperature was unimportant in determining phytoplankton growth in natural environments because cells are usually limited by nutrients (Epply, 1972). However, other studies have shown that temperature interacts with nutrient limitation in a complex manner to affect growth. For example, studies have found that temperature significantly affects the half-saturation constants for growth in phytoplankton (K_s) (Rhee, 1982). In addition, the optimal temperature for nutrient uptake and growth is usually not the same, and the degree of difference seems to be species specific (Rhee, 1982). As described earlier, cell quota also is affected by temperature, which in turn alters nutrient uptake (Droop model). As a result, combined effects of nutrient limitation and temperature stress are greater than the sum of their individual effects, and simple insertion of a temperature factor into nutrient-based growth rate equations (e.g., Goldman and Carpenter, 1974) is not sufficient (Goldman, 1977; Rhee, 1982). Finally, in non-steady-state conditions there is a further uncoupling of nutrient uptake and growth rates. In more complex growth models, the effect of temperature on nutrient-limited growth is to alter fluxes between compartments of cell components (Shuter, 1979).

Bothwell (1988) examined the effects of phosphorus limitation on periphyton growth during different seasons (temperatures) in outdoor stream channels. Similar to phytoplankton studies, temperature exerted the dominant effect on periphyton growth rates when phosphorus was not limiting (i.e., when growth rates approach maximum values). Under natural, seasonally varying nutrient limitation, the relationship between temperature and growth was weaker, but still significant. Growth response to P addition also varied seasonally with changes in temperature, light, and ambient phosphate concentrations. However, relative specific growth rates (μ/μ_{max}) as a function of external phosphorus concentrations were constant throughout the year and growth saturation always occurred around an ambient P level of around 0.6 μg/Liter. In contrast to phytoplankton studies, this suggests that temperature effects on nutrient kinetics were small.

IV. COMMUNITY STRUCTURE

A. Lotic Periphyton

1. Controlled and Semicontrolled Studies

Controlled and semicontrolled studies of the effects of temperature on species composition of lotic periphyton communities have been primarily restricted to epilithic assemblages. Three main types of studies have been conducted: (1) experimental studies in which temperature was directly manipulated either in the laboratory or *in situ*, (2) examination of periphyton community structure in thermal gradients downstream from hot springs, and (3) examination of communities downstream from power stations releasing heated effluent. Studies correlating seasonal changes in lotic periphyton composition to temperature are reviewed in the next section.

At the highest taxonomic level, temperature has been suggested to cause shifts in lotic periphyton composition based on the average temperature ranges for the major classes of algae. Although each species has its own temperature optimum and range, several studies have shown that Bacillariophyceae (diatoms) tend to dominate at approximately 5–20°C, Chlorophyceae (green algae) and Xanthophyceae (yellow-green algae) at 15–30°C, and Cyanophyceae at >30°C (Patrick, 1969; Wilde and Tilly, 1981; Lamberti and Resh, 1985). Although absolute temperature values for each group vary between studies, the overall trend often holds. For example, Cairns (1956) and Patrick *et al.* (1969) manipulated temperature in the laboratory and *in situ*, respectively, and observed that diatoms dominated in the lotic periphyton communities on glass slides at temperatures of 20–28°C, with green algae dominating at 30–35°C and cyanobacteria at 35–40°C. Gradual lowering of temperature reversed these trends. Wilde and Tilly (1981) artificially altered temperature in 3-year-old streamside channel communities (a blackwater stream) for 3 years and found that 60% of the most abundant taxa were significantly affected by the thermal treatment. The yellow-green alga, *Vaucheria,* and diatoms dominated in control streams (5–20°C). The cyanobacterium, *Schizothrix,* increased in all heated streams, and dominated in streams heated the most (12°C above ambient, 20–35°C). The red algae, *Rhodochorton* and *Tuomeya,* were absent from the most heated treatment. An examination of postthermal recovery of these communities showed that the species composition of the heated streams resembled that of the control streams in less than 5 months (Wilde, 1982). In a study of thermal pollution downstream from a power plant on the Provo River (Utah), Squires *et al.* (1979) observed alteration in epilithic community structure only in the most heated sections (ca. 10°C above ambient to a maximum of 24°C). In these areas, *Cladophora* (Chlorophyceae) and *Vaucheria* abundance increased relative to that of most diatom species, and *Hydrurus* (Chrysophyceae) was eliminated. In a thermal gradient surrounding a hot spring, Stockner (1967) found that *Mougeotia* (Chlorophyceae) replaced *Schizothrix* below 47°C, and diatom diversity and relative abundance increased as temperature decreased. Several other studies have documented a replacement of a diatom-dominated flora by a cyanobacterial flora at temperatures <35°C (e.g., Kullberg, 1971; Patrick, 1974; Vinson and Rushforth, 1989). McIntire (1968a) reported that the increase in abundance of Chlorophyceae and Xanthophyceae in warmer months in laboratory streams depended on light and current conditions.

Changes in community structure with temperature also are determined by differences at lower taxonomic levels. Although many species of diatoms are eurythermic (Patrick, 1969), Vinson and Rushforth (1989) found that species within certain diatom genera tended to have similar temperature distributions in a thermal gradient downstream from a hot spring. In cooler temperatures, species of *Achnanthes* (<14°C) and *Navicula* (8–12°C) were most abundant, whereas species of *Cocconeis* (>25°C) and *Nitzschia* (22–39°C) were more common in warm water. Species of *Cocconeis* also have been found to increase in warmer water by Patrick (1971), Squires *et al.* (1979), and Klarer and Hickman (1975). Although generalizations concerning dominance at high taxonomic levels are fairly well described, there is much variation at the species level. Lowe (1974) lists thermal preferences for many freshwater periphytic diatom species, based on field distributions.

Changes in taxonomic composition of periphyton with temperature are reflected by species diversity. In general, species richness and diversity have been found to increase with temperature up to 25–30°C, then decrease above 30°C, as species-rich diatom floras are replaced by fewer species of green algae or cyanobacteria (Patrick, 1969; Patrick *et al.*, 1969; Kullberg, 1971; Vinson and Rushforth, 1989). This means that the effect of thermal stress on periphyton community structure in natural environments depends on the ambient (i.e., seasonal) temperature. Patrick (1971) controlled temperature *in situ* in a small temperate stream and examined changes in species composition on glass slides for different seasons. Small rises (<10°C) in temperature increased diatom diversity in the winter when ambient temperature was near the lower tolerance range for diatoms. When the temperature was artificially raised at the upper end of the tolerance range (i.e., in summer), diversity decreased as cyanobacterial species dominated. Similarly, Squires *et al.* (1979) and Wilde and Tilly (1981) found that the effect of artificial heating on periphyton diversity depended on the seasonal temperature.

Few studies have examined the effect of temperature on secondary successional sequences following scour of the substrata. Kullberg (1982) suggested that timing of succession following scour was influenced by temperature for some species along a thermal gradient in a hot spring.

2. Geographic and Seasonal Distributions

Although experimental manipulation of temperature has demonstrated that temperature can affect species composition of periphyton, the importance of temperature to the temporal and spatial distribution of periphytic species in natural environments is less certain. Studies attempting to correlate seasonal or spatial differences in natural lotic periphyton composition to temperature are complicated by cyclical and short-term changes in other environmental factors.

Many studies have examined the seasonal changes in periphyton composition in streams. In most instances, seasonal patterns were observed, however, relationships to temperature were obscured by simultaneous changes in light, flow, and nutrient concentrations (e.g., Moore, 1977a,b; Cox, 1990). For example, in a principal components analysis of seasonal changes in periphyton composition in a mountain stream, Wehr (1981) found significant correlations of species composition with temperature, cation concentration, current, and depth. As in other studies, most species were present year-round but tended to wax and wane in abundance. Patrick (1969) suggests that changes in composition occur by some species forming resting stages during periods of adverse temperatures, and resuming growth when temperatures become favorable. In spatial classification schemes for lotic periphyton communities, temperature is usually considered as a secondary factor (Margalef, 1960; Blum, 1960). In some cases the restriction of a few species to very narrow temperature environments (e.g., cold headwater streams) is obvious, but the relative importance of temperature on geographic distribution is unknown.

A more successful approach to understanding the role of temperature in determining temporal and spatial trends in periphyton composition in natural environments has been to combine field observations with autecological studies. In a study of seasonal differences in temperature tolerances of periphyton in a Virginia stream with an annual temperature range of 0–27°C, Seaburg and Parker (1983) found that most isolates collected from different seasons grew well between 7.5° and 20°C. Twelve percent of all periphyton taxa collected were theoretically perennial, and there were more obligate warm-water species than cold-water taxa. This suggests that a larger percentage of taxa from the total periphyton community would become metabolically inactive during winter, compared to those that would become inactive at late summer temperatures. In other words, the minimum temperature for growth was more important in determining seasonal succession changes in these streams than the maximum temperature for growth, with most warm-water taxa not being able to grow below 7.5°C. The authors suggest that a 5°C increase in annual temperature would tend to suppress seasonal succession of periphyton in these streams by allowing warm-water taxa to grow during the winter months.

Cox (1993) examined the growth of several common species of periphytic diatoms at different combinations of temperature, light, nutrients, and pH in the laboratory, and results successfully explained natural distributions of these taxa in the field. DeVries and Hillebrand (1986) found the temporal distribution of *Tribonema minus* and *Spirogyra singularis* could be explained by differences in their optimum temperatures for growth. However, several studies have found that known temperature conditions determined in the laboratory for a given species do not correspond to its distribution in a natural community, and concluded that biotic factors (competition) are also important (Patrick, 1971). Temperature has been shown to strongly affect the outcome of competition in mass-cultured phytoplankton communities (Goldman and Ryther, 1976). In a removal experiment with periphyton, Stockner (1967) demonstrated that *Schizothrix,* even though it could grow over a wider range of temperatures, was restricted to hot-water areas in a thermal spring because it was overgrown by *Oscillatoria* in cool-water areas. Further experimental studies are needed to determine relative roles of autecological versus synecological effects of temperature on periphyton community composition.

In phytoplankton studies, geographical and seasonal distribution of species can be related to differences in temperature optima for growth (e.g., Braarud, 1961). However, these relationships are complicated by differences in temperature responses between clonal isolates of the same species. The demonstration of a large amount of genetic diversity in marine (Wood and Leatham, 1992) and freshwater algae (Soudek and Robinson, 1983; Hoagland *et al.,* in press) is making it difficult to determine if variation between species (morphologically defined) is greater than that within a species. Several studies have shown that strains of the same phytoplankton species isolated from regions of cold or warm water differ in their extreme and optimal temperatures for growth (e.g., Braarud, 1961; Hulbert and Guillard, 1968). Similar effects may occur for periphyton species. In the previously mentioned study by Seaburg and Parker (1983), of six species of periphyton in which separate strains were isolated, all had differences in thermal tolerance between strains.

B. Lentic Periphyton

1. Thermal Effluent Studies

The majority of studies examining the effects of temperature on lentic periphyton have involved epipelic and epiphytic communities in lakes receiving thermal effluent. Hickman (1974) measured the response of epipelic communities in a shallow eutrophic lake in Alberta that received thermal discharge from a power plant. Water at heated sites in the lake was 18°C above ambient (control sites) in the winter and 7°C above in the summer. At all sites, dissolved oxygen was high and stratification did not occur. Control sites were dominated by diatoms year-round. *Oscillatoria* occurred only at heated sites in the lake and was especially abundant in the most heated area, the discharge canal. The diatom species were similar at all sites except for the discharge canal, where the number of species was reduced and *Craticula cuspidata* dominated in summer (31°C). In the same study, Hickman (1974) found no effect of thermal discharge on the epipsammic flora of the lake. Freshwater epipsammic communities are usually a specialized flora of tightly attached diatoms (Krejci and Lowe, 1986). Hickman (1974) suggested that the inability of *Oscillatoria* to attach to the sand may have prevented its growth at elevated temperatures in his study.

A detailed study of the effect of thermal discharge on epiphytic communities of *Scirpus* in a shallow moderately eutrophic pond in Alberta was carried out from 1971 to 1978 (Klarer and Hickman, 1975; Hickman, 1982). Temperatures in the lake varied from 0 to 24°C, with heated surface water 8–9°C higher in the summer and 14–19°C higher in the winter. There was little difference in species composition between heated and control sites in the spring, with the flora consisting mostly of diatoms. However, differences between the sites were pronounced when ambient temperatures were higher in the summer and fall. Relative to control sites, heated sites showed a shift in dominance from diatoms to the filamentous green algae *Oedogonium, Spirogyra,* and *Cladophora.* Diatoms at the control sites were primarily *Achnanthes minutissima* and *Gomphonema gracile,* but were almost entirely *Cocconeis placentula* at the heated sites. Overall, diversity was lower at heated sites than at control sites. After removal of thermal discharge at one of the heated sites in 1975, the flora become similar to that of unheated areas within one year, as filamentous green taxa diminished in importance and several previously absent diatom species increased (Hickman, 1982).

It appears from these studies that some effects of temperature on lentic epipelic and epiphytic communities are similar to those described for lotic periphyton. At temperatures >30°C diversity is reduced, as a few species of green algae, cyanobacteria, and diatoms become dominant (however, see Eloranta, 1982). Also, as in streams, effects of thermal effluent depend on seasonal water temperatures. In contrast to streams, increased temperature does not seem to have as great an effect on community composition in the winter. Perhaps very low light levels in benthic lake habitats under the ice mask temperature effects on lentic periphyton. Because of their larger water volume, lakes also require more heat to raise overall temperatures, and the development of a flora dominated by thermophilic cyanobacteria may be less common than in streams. Heating in shallow areas along lake margins may be significant, however.

2. Seasonal Studies

Many studies have examined seasonal succession in lentic periphyton communities. As with stream studies, correlations between seasonal changes in temperature and other factors such as light, grazing, and disturbance complicate interpretations. In addition, settling of phytoplankton cells into periphyton communities and seasonal effects of overturn on water chemistry are related to seasonal temperature changes in lakes.

Brown and Austin (1973) and Castenholtz (1960) found distinct seasonal differences in epilithic (artificial substrata of glass) communities in lakes. Hoagland *et al.* (1982) noted seasonal differences in colonization and timing of periphyton succession on Plexiglas substrata. All three studies emphasized that seasonal changes in a complex of physicochemical and biological factors were probably responsible for their observations. Moore (1978) demonstrated that the seasonal change from a *Ulothrix*- to a *Cladophora*-dominated epilithic community in Lake Erie occurs above 10°C and is primarily determined by temperature. He observed that growth of *Cladophora* in an area of thermal effluent (10–14°C above ambient) occurred almost 3 months before it appeared in nonheated areas, which were similar in all other environmental factors. The decline of *Cladophora* in midsummer (ca. 25°C) was not a direct effect of temperature because rates of photosynthesis for *Cladophora* in the laboratory increased with temperature up to a maximum at 25°C, with inhibition not occurring until 33°C. Increased epiphyte load or decreased nutrients in midsummer was suggested as the cause of the decline in *Cladophora.*

For lentic epiphytic communities, Klarer and Hickman (1975) found that the relationship between seasonal temperature changes and species composition in their study did not correspond to other studies, suggesting that temperature played a secondary role. Gons (1982) found that seasonal changes in epiphyte communities in dense macrophyte beds appeared to be related more to light and phytoplankton sedimentation than to temperature.

In epipelic communities, Round (1972) describes four growth periods during the year, which differ in species composition. He suggests that temperature is not the major factor affecting seasonal succession in the epipelon because major growth periods correspond to different temperatures in lakes of the English Lake District, and blooms can occur at very cold temperatures when other factors are favorable (Round, 1960). However, Hickman (1978a) found that cyanobacteria and green algae increased abundance during summer in the epipelon of some northern shallow, eutrophic ponds. Also Gruendling (1971) observed cyanobacteria in the epipelon at temperatures >15°C for a shallow oligotrophic pond.

Few studies have attempted experimentally to examine the role of temperature on seasonal succession in epipelic periphyton. Admiraal *et al.* (1984) demonstrated that the species of epipelic diatoms that dominated in summer in an intertidal mudflat required higher temperatures for rapid cell division than the species that dominated in cold seasons. By manipulating combinations of temperature, light, and desiccation in the laboratory, they were able to reproduce seasonal communities observed in the natural environment. Similar types of studies would be useful in understanding the importance of temperature to seasonal succession in freshwater periphyton.

V. ECOSYSTEM RESPONSE

A. Lotic Periphyton Biomass and Primary Production

Few generalizations can be made concerning seasonal changes in periphyton biomass in streams and their relationships to temperature. In temperate streams, there is generally a biomass minimum in winter, with a spring maximum, followed by unpredictable fluctuations in biomass during summer and fall (e.g., Marker, 1976a; Moore, 1977a,b; Sumner and Fisher, 1979; Cox, 1990). In most cases, spring growth can be related to increases in irradiance and fluctuations are due to flood events. In areas where spring floods are common and in subarctic streams, periphyton biomass is often highest in summer (e.g., Gumtow, 1955; Moore, 1977c). Bothwell (1988) found that growth rates of epilithon in streamside channels were correlated with seasonal changes in temperature but not light.

Studies in areas receiving thermal effluent found that biomass in the epilithon increased in heated areas (Squires *et al.*, 1979; Descy and Mouvet, 1984). Similarly, Lamberti and Resh (1983) artificially raised temperature from 20° to 30°C in streamside channels and found a 40-fold increase in chlorophyll *a*. Conversely, cooling of the geothermally heated water reduced algal biomass. Patrick (1971) manipulated temperatures of temperate periphyton communities on glass slides *in situ* and found that moderate temperature increases (<14°C) increased biomass, particularly at low ambient temperature. However, temperature increases near the upper end of the range of tolerance (ca. 32°C) caused biomass to decrease. She concluded that at lower temperatures, small rises in temperature increase biomass, but when the optimum temperature is exceeded, biomass decreases. Brock (1970) postulated that increases in temperature should increase periphyton biomass accumulation for habitats at <25°C but not in habitats >25°C, where thermophilic taxa should already be close to their optimum for growth.

Thermal discharges also can indirectly change algal biomass and composition by affecting the growth, emergence, biomass, and composition of grazers (Lamberti and Resh, 1983, 1985). At temperatures of 40–50°C, the biomass of periphyton can be quite high. Lamberti and Resh (1985) found that periphyton biomass decreased almost 200 times along a temperature gradient from 52° to 23°C, downstream from a hot spring. The absence of grazers at high temperatures (>40–55°C) allows cyanobacterial mats to build up high standing crops (Stockner, 1967; Lamberti and Resh, 1985). At temperatures of 55–75°C, biomass of cyanobacterial mats decreases as cells reach their thermal limits (Brock, 1970). The productivity of cyanobacterial mats in hot springs is lowest when biomass is highest (ca. 55°C), possibly because of diffusion limitations within the thick mat (Brock, 1970).

In general, temperature appears to play a secondary role in determining seasonal changes in primary productivity of epilithon, particularly in forested or low-nutrient streams (e.g., Marker, 1976b; Stockner and Shortreed, 1976; Hornick *et al.,* 1981; Jasper and Bothwell, 1986). However, interactions of temperature with seasonal changes in other factors may be significant. For example, Jasper and Bothwell (1986) found that the degree of light inhibition for photosynthesis was inversely correlated with water temperature, possibly due to enhanced recovery of photoinhibition at increased temperature. In addition, Sumner and Fisher (1979) found that primary production per unit biomass in a Massachusetts river was lower in seasons when algal biomass was high, particularly at temperatures >5°C. R.L. Fuller (unpublished data) raised temperatures from 5° to 12°C for periphyton communities on clay tiles and observed an increase in biomass-specific net primary production from onefold to ninefold, with the magnitude of the increase differing among seasons.

A few studies have experimentally examined direct effects of temperature on rates of periphyton community photosynthesis and respiration using laboratory streams. Kevern and Ball (1965) found little change in net primary production by increasing temperature from 20° to 25°C, and suggested that respiration rates increased as much as gross photosynthetic rates. Also using laboratory streams, Phinney and McIntire (1965) and McIntire (1968b) demonstrated that rates of photosynthesis in epilithic algal assemblages had a mean Q_{10} of 1.49 when light was saturating. There was no effect of temperature on photosynthetic rates when light was limiting. For a similar system, McIntire (1966) reported a Q_{10} for respiration of around 2.0. However, temperature had a greater effect on respiration rate when the temperature was raised from 8° to 13°C than between 18° and 23°C. At temperatures >23°C, respiration rates became erratic, suggesting that these assemblages were near their thermal tolerance. Effects of temperature on periphyton respiration also showed interactions with current velocity and dissolved gas concentrations (CO_2 and O_2), suggesting that at high levels of periphyton biomass, temperature affects community metabolism indirectly by altering the rate of gas diffusion through the algal mat (McIntire, 1966). Such an effect could explain the interaction between biomass-specific primary production and temperature observed by Sumner and Fisher (1979) and R.L. Fuller (unpublished data) in the field.

I tested the hypothesis that temperature sets the upper limit of areal primary productivity of periphyton in streams when other factors are not limiting. Ninety-four measurements (most done *in situ*) of primary production for epilithic communities were recorded at various temperatures from 14 studies that employed a wide variety of methods (Fig. 2). Units were converted following Hornick *et al.* (1981) and values are best considered to be net primary production as large heterotrophic organisms were not present in most measurements. Despite the large variability in methods and types of lotic systems represented, maximum areal net primary production (i.e., not corrected for biomass) increases exponentially with temperature for temperatures <30°C. This relationship is described by the equation $P_{max} = e^{0.1056T + 2.9602}$, where P_{max} is the maximum, areal net primary production (mg C m^{-2} h^{-1}) and T is temperature (degrees centigrade). The Q_{10} value for this equation is 2.9. This is similar to the relationship obtained by Epply (1972) for phytoplankton growth rates and temperature. Measurements of lotic net primary productivity at tempera-

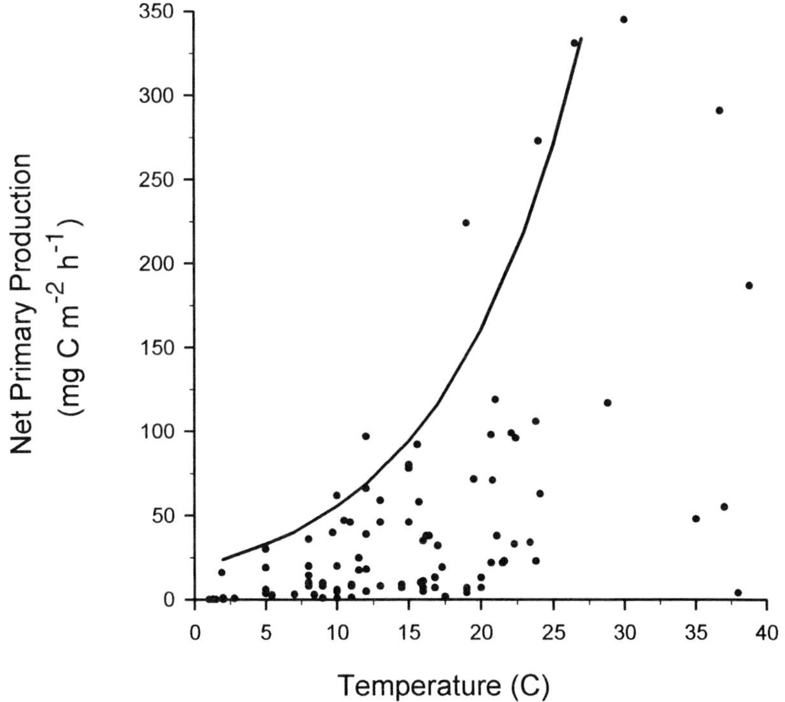

FIGURE 2 Variation in areal net primary production in lotic epilithon assemblages with temperature based on 94 measurements in 14 studies. The curve was determined by regression of maximum rates for various temperatures, r=0.96. See text for further explanation.

tures >30°C are from desert streams and hot springs, and show a general decline in productivity from 30° to 40°C (Fig. 2). It is clear that for most of the studies in this data set, areal primary production was limited by factors other than temperature, but temperature can be used to roughly predict maximum productivity of lotic epilithon in a system when other factors are not limiting.

B. Lentic Periphyton Biomass and Primary Production

Seasonal changes in lentic periphytic biomass are usually characterized by maxima in spring and late summer/fall for epilithic (e.g., Castenholtz, 1960; Eloranta, 1982), epipelic (e.g., Round, 1960; Hickman, 1978b), and epiphytic communities (e.g., Klarer and Hickman, 1975) in temperate lakes. In arctic lakes there is often one maximum in late summer (Moore, 1974). Seasonal changes in primary production have been found to be highly correlated to those for biomass for epiphytic (Hickman and Round, 1970) and epipelic (Gruendling, 1971) communities. In a multifactor correlation analysis, Gruendling (1971) found that seasonal fluctuations in biomass and productivity in the epipelon of a shallow lake in southern British Columbia were most related to temperature and light. The temperatures that correspond to seasonal changes in periphyton biomass vary widely between studies, indicating that factors other than temperature (particularly light) play an important role (Klarer and Hickman, 1975).

Studies in areas receiving thermal effluent have generally shown an increase in lentic periphyton biomass with elevated temperatures for epilithic (Eloranta, 1982) and epipelic (Hickman, 1974) communities. However, Hickman (1974) found no effect of heated effluent on epipsammic assemblages. The annual biomass accumulation of epiphytes on *Scirpus* was greatly increased in areas of a shallow eutrophic lake receiving thermal effluent (Klarer and Hickman, 1975). The primary effects of heating were to prevent ice cover in that area, allowing increased light for winter growth. Consequently, the size of the spring maximum was greatly increased. Biomass values became similar to unheated sites within a year of the thermal effluent being diverted from the lake (Hickman, 1982).

Few studies have examined the direct effect of temperature on primary production of lentic periphyton. In the preceding study on epiphyton, mean primary production was greater at the heated sites (22°C) than at nonheated sites (14°C). However, the assimilation efficiency of photosynthesis was lower at heated than at nonheated sites, indicating that the overall increase in productivity at the heated site was due to its greater amount of biomass (Hickman and Klarer, 1975). The authors suggested that increased respiration at the higher temperature sites (*sensu* Kevern and Ball, 1965) lowered assimilation efficiency.

Stanley and Daley (1976) measured primary production in epipelic communities in small tundra ponds at various combinations of light and temperature. Epipelic communities had a higher temperature optimum (>20°C) than the phytoplankton community (14°C), and the Q_{10} for the epipelon was 2.2 compared to 3.0 for the phytoplankton. This implied that the epipelic community was well adapted for the sediment habitat, where temperatures were higher and more stable than for the water column. During the growing season (summer), photosynthesis in the epipelic community was usually limited by light, with temperature setting the potential maximum rate.

A lack of data prevented examination of primary productivity in lentic periphyton as a function of temperature, as was done in the foregoing for lotic periphyton. Presumably, temperature sets the maximum rate of areal primary production for lentic periphyton as well, although maximum rates may seldom be achieved owing to limitations by other factors, especially given the low light levels often found in benthic lake habitats.

VI. GLOBAL TEMPERATURE CHANGES

A report from the Experimental Lakes Area in northwest Ontario indicates that increases in annual air temperatures over the past 20 years have resulted in an increase in mean and maximum temperatures of lakes in that region (Schindler and Beaty, 1990). Byron and Goldman (1990) also predict dramatic increases in freshwater algal productivity and shifts in species composition using a model based on long-term limnological data from Castle Lake, California. Similarly, stream temperatures are predicted to rise with projected global climatic changes based on loss of shading riparian vegetation (Stefan and Sinokrot, 1993).

Attempts to use siliceous algae (Bacillariophyceae and Chrysophyceae) preserved in lake sediments to infer past changes in global climate are just beginning. Using canonical correspondence analysis, Pienitz *et al.* (1994) found that water depth and temperature were the two environmental variables that explained most of the variance in diatom assemblages preserved in surface sediments from 59 lakes along a latitudinal gradient in northern Canada. They developed calibration models from this data set to infer paleoclimatic changes in freshwater habitats. Several studies have also examined the relationship between salinity and fossil assemblages in saline lakes to reconstruct long-term climatic changes (Fritz *et al.*, 1991, 1993). Yang (1994) found that Holocene diatom-inferred temperature changes from cores in Lake Ontario embayments were correlated with temperature signals from oxygen isotope ratios. Periphytic diatoms can be abundantly represented in sediment from lakes with large littoral zones and a poorly developed phytoplanktonic diatom flora (Bjork-Ramberg, 1984; DeNicola, 1986; Jones and Flower, 1986). Therefore, understanding effects of temperature on lentic periphyton communities is of importance for reconstructing and predicting changes in global temperature.

VII. SYNTHESIS

Most of the research concerning the effect of temperature on freshwater periphyton has been at the community and ecosystem levels. The approach of the majority of these studies was to examine correlations between temperature and seasonal changes in community structure, biomass, and primary production. However, interactions of temperature with other factors are complex, making it difficult to separate specific effects. Though studies examining community and ecosystem responses of periphyton in areas of thermal effluent are somewhat more controlled, they are also complicated by interactions of temperature with other factors. The most information has been gained by examining periphyton in environments where temperature and other factors were artificially controlled (e.g., Kevern and Ball, 1965; McIntire, 1966; Patrick, 1971; Wilde and Tilly, 1981; Lamberti and Resh, 1983). Given this, several broad generalizations concerning periphyton responses to temperature changes at the community and ecosystem level can be made. (1) As temperature increases, there is a shift in the dominance of algal classes from Bacillariophyceae (<20°C) to Chlorophyceae (15–30°C) to Cyanophyceae (>30°C). (2) Species diversity increases from approximately 0° to 25°C and decreases at temperatures >30°C, when a few cyanobacterial taxa dominate. (3) The degree to which community composition changes with thermal input depends on the initial ambient temperature. Increases in temperature in environments near 25–30°C usually cause greater changes in community structure than in environments <25°C. (4) Community structure usually recovers rapidly (<1 year) when temperature stress is discontinued. (5) Biomass increases with temperature from approximately 0–30°C, and decreases at 30–40°C. The biomass of cyanobacterial mats in hot springs peaks at approximately 50–55°C owing to a lack of grazers, but biomass-specific primary productivity is low. (6) Community structure and biomass of epipsammic assemblages seem to be less affected by temperature changes than other periphyton communities, possibly because the epipsammon is a specialized flora. (7) In many natural communities, temperature does not usually limit biomass and primary productivity, but it does set an upper limit for production when other factors are optimal. Maximum areal productivity of lotic periphyton increases exponentially with temperature for temperatures <30°C. (8) The degree to which primary productivity is limited by factors such as light, nutrients, and grazing depends on temperature.

Many of these generalizations were known 15 years ago and recent studies of temperature effects on periphyton are few. In fact, field experiments examining changes in periphyton community structure and productivity in response to light, nutrients, grazing, and so on often do not report the temperature regime of the habitat. The next phase of research is to determine the degree to which periphyton responses to temperature at the community and ecosystem levels are influenced by responses at autecological and population levels (Table II). More studies such as Seaburg and Parker (1983), DeVries and Hillebrand (1986), and Cox (1993) are needed, where species population-level responses are used to explain shifts in community composition or biomass with temperature. Tentatively, these studies suggest that temperature sets limits for the growth of each population in a community within which they can compete for light and nutrients, or respond to grazing pressure and disturbance.

Most of the autecological responses of algae to temperature were obtained from research on marine phytoplankton. Whether freshwater periphytic species also respond to temperature by adjusting rates of light versus dark reactions of photosynthesis, changing concentrations of enzymes, altering cell quota for nutrients, modifying fatty acids, and employing heat shock proteins needs to be investigated. These responses can maintain high growth rates for small temperature variations within the tolerance limits for the cell. Selection pressure on the physiological apparatus of algal cells may result in their adaptations becoming genetically fixed, and could explain thermal strains within a species, as well as differences between species with different temperature requirements and responses (Davison, 1991). Adaptations that set minimum, maximum, and optimal temperatures for growth of a species help to determine the presence and abundance of its population in a community. A fuller understanding of the relationship between physiological/population responses and community/ecosystem-level responses requires experiments to determine the plasticity of physiological adaptations and growth rates for a cell, ecotype, or species. The relative importance of these autecological characteristics to synecological interactions, such as competition, then needs to be determined.

A more complete understanding of the effects of temperature on periphyton can be obtained by defining the scale of temperature variation (Table I) and conducting a combination of controlled field and laboratory experiments to determine the integrated response of periphyton at different ecological levels (Table II). Although lakes and streams contribute <1% to the earth's carbon budget (Woodwell *et al.*, 1978), they represent a finite resource critical to the maintenance of terrestrial ecosystems. Given the predicted temperature changes for lakes and streams resulting from global climate change, more accurate predictions of periphyton response in terms of changes in species composition and productivity can be made using an approach that integrates effects at different scales.

ACKNOWLEDGMENTS

Thanks to Al Steinman and Kyle Hoagland for suggesting references, and to Dennis Anderson for checking several reference sources. Comments by the editors and two anonymous reviewers improved the manuscript. Rosemary Grgurina kindly assisted in the preparation of the manuscript.

REFERENCES

Admiraal, W., Peletier, H., and Brouwer, T. (1984). The seasonal succession patterns of diatom species on an intertidal mudflat: An experimental analysis. *Oikos* **42,** 30–40.

Bhagwat, A. A., and Apte, S. K. (1989). Comparative analysis of proteins induced by heat shock, salinity and osmotic stress in the nitrogen-fixing cyanobacterium *Anabaena* sp. strain L-31. *J. Bacteriol.* **171,** 5187–5180.

Bjork-Ramberg, S. (1984). Species composition and biomass of an epipelic algal community in a subarctic lake before and during lake fertilization. *Holarctic Ecol.* **7,** 195–201.

Blum, J. T. (1960). Algal populations in flowing waters. *In* "The Pymatuning Symposium in Ecology" (C. A. Tryon, Jr. and R. T. Hartman, eds.), Pymatuning Lab. Field Biol., Spec. Publ. No. 2, pp. 11–21. Univ. of Pittsburgh Press, Pittsburgh, PA.

Bothwell, M. L. (1988). Growth rate responses of lotic periphytic diatoms to experimental phosphorus enrichment. *Can. J. Fish. Aquat. Sci.* **45,** 261–270.

Braarud, T. (1961). Cultivation of marine organisms as a means of understanding environmental influences on populations. *In* "Oceanography" (M. Sears, ed.), Publ. No. 67, pp. 271-298. Am. Assoc. Adv. Sci., Washington, DC.

Brock, T. D. (1967). Relationships between standing crop and primary productivity along a hot spring thermal gradient. *Ecology* **48,** 566–571.

Brock, T. D. (1970). High temperature systems. *Annu. Rev. Ecol. Syst.* **1,** 191–220.

Brown, G. W. (1969). Predicting temperatures of small streams. *Water Resour. Res.* **5,** 66–75.

Brown, S. D., and Austin, A. P. (1973). Spatial and temporal variation in periphyton and physico-chemical conditions in the littoral of a lake. *Hydrobiologia* **71,** 183–232.

Byron, E. R., and Goldman, C. R. (1990). The potential effects of global warming on the primary productivity of a subalpine lake. *Wat. Res. Bull.* **26,** 983–989.

Cairns, J. (1956). Effects of increased temperatures on aquatic organisms. *Ind. Water Wastes* **1,** 150–152.

Castenholtz, R. W. (1960). Seasonal changes in the attached algae of freshwater and saline lakes in the Lower Grand Coulee, Washington. *Limnol. Oceanogr.* **5,** 1–28.

Chisholm, S. W. (1981). Temporal patterns of cell division in unicellular algae. *Can. Bull. Fish. Aquat. Sci.* **210,** 151–181.

Cossins, A. R., and Bowler, K. (1987). "Temperature Biology of Animals." Chapman & Hall, New York.

Cox, E. J. (1990). Studies on the algae of a small softwater stream. I. Occurrence and distribution with particular reference to the diatoms. *Arch. Hydrobiol., Suppl.* **83,** 525–552.

Cox, E. J. (1993). Freshwater diatom ecology: Developing an experimental approach as an aid to interpreting field data. *Hydrobiologia* **269/270,** 447–452.

Darley, W. M. (1982). "Algal Biology: A Physiological Approach." Blackwell, London.

Davison, I. R. (1991). Environmental effects on algal photosynthesis: Temperature. *J. Phycol.* **27,** 2–8.

DeNicola, D. M. (1986). The representation of living diatom communities in deep-water sedimentary diatom assemblages in two Maine (USA) lakes. *In* "Diatoms and Lake Acidity" (J. P. Smol, R. W. Battarbee, R. B. Davis, and J. Merilainen, eds.), pp. 73–85. Junk Pub., Dordrecht, The Netherlands.

Descy, J. P., and Mouvet, C. (1984). Impact of the Tihange nuclear power plant on the periphyton and phytoplankton of the Meuse River (Belgium). *Hydrobiologia* **119,** 119–128.

DeVries, P. J. R., and Hillebrand, H. (1986). Growth control of *Tribonema minus* (Wille) Hazen and *Spirogyra singularis* Nordstedt by light and temperature. *Acta Bot. Neerl.* **35,** 65–70.

Drebes, G. (1977). Sexuality. *In* "The Biology of Diatoms" (D. Werner, ed.), pp. 250–283. Univ. of California Press, Berkeley.

Dring, M. J. (1974). Reproduction. *In* "Algal Physiology and Biochemistry" (W. D. P. Stewart, ed.), pp. 814–838. Univ. of California Press, Berkeley.

Duke, C. S., Litaker, W., and Ramus, J. (1989). Effects of temperature, nitrogen supply, and tissue nitrogen on ammonium uptake rates of the chlorophyte seaweeds *Ulva curvata* and *Codium decorticatum*. *J. Phycol.* **25**, 113–120.

Eloranta, P. V. (1982). Periphyton growth and diatom community structure in a cooling water pond. *Hydrobiologia* **96**, 253–265.

Eppley, R. W. (1972). Temperatue and phytoplankton growth in the sea. *Fish. Bull.* **70**, 1063–1085.

Eppley, R. W. (1977). The growth and culture of diatoms. *In* "The Biology of Diatoms" (D. Werner, ed.), pp. 24–64. Univ. of California Press, Berkeley.

Fogg, G.E. (1969). Survival of algae under adverse conditions. *Symp. Soc. Exp. Biol.* **23**, 123–142.

Fogg, G. E. (1975). "Algal Cultures and Phytoplankton Ecology." Univ. of Wisconsin Press, Madison.

Fritz, S. C., Juggins, S., and Battarbee, R. W. (1991). Reconstruction of past changes in salinity and climate using a diatom-based transfer function. *Nature (London)* **352**, 706–708.

Fritz, S. C., Juggins, S., and Battarbee, R. W. (1993). Diatom assemblages and ionic characterization of lakes of the northern Great Plains, North America: A tool of reconstructing past salinity and climate fluctuations. *Can. J. Fish. Aquat. Sci.* **50**, 1844–1856.

Goldman, J. C. (1977). Temperature effects on phytoplankton growth in continuous culture. *Limnol. Oceanogr.* **22**, 932–936.

Goldman, J. C., and Carpenter, E. J. (1974). A kinetic approach to the effect of temperature on algal growth. *Limnol. Oceanog.* **19**, 756–766.

Goldman, J. C., and Mann, R. (1980). Temperature influenced variations in speciation and the chemical composition of marine phytoplankton in outdoor mass cultures. *J. Exp. Mar. Biol. Ecol.* **46**, 29–40.

Goldman, J. C., and Ryther, J. H. (1976). Temperature-influenced species competition in mass cultures of marine phytoplankton. *Biotechnol. Bioeng.* **18**, 1125–1144.

Gons, H. J. (1982). Structural and functional characteristics of epiphyton and epipelon in relation to their distribution in Lake Vechten. *Hydrobiologia* **95**, 79–114.

Graham, L. E., Graham, J. M., and Kranzfelder, J. A. (1985). Light and temperature as factors regulating seasonal growth and distribution of *Ulothrix zonata* (Ulvophyceae). *J. Phycol.* **21**, 228–34.

Graham, L. E., Graham, J. M., and Kranzfelder, J. A. (1986). Irradiance, daylength and temperature effects on zoosporogenesis in *Coleochaete scutata* (Charophyceae). *J. Phycol.* **22**, 5–39.

Gruendling, G. K. (1971). Ecology of the epipelic algal communities in Marion Lake, British Columbia. *J. Phycol.* **7**, 239–249.

Gumtow, R. B. (1955). An investigation of the periphyton in a riffle of the West Gallatin River, Montana. *Trans. Am. Microsc. Soc.* **74**, 278–292.

Harris, G. P. (1986). "Phytoplankton Ecology." Chapman & Hall, New York.

Hickman, M. (1974). Effects of discharge of thermal effluent from a power station on Lake Wabamun, Alberta, Canada—The epipelic and epipsammic algal communities. *Hydrobiologia* **45**, 199–215.

Hickman, M. (1978a). Studies on the epipelic algal community—Seasonal changes and standing crops at shallow littoral stations in four lakes. *Arch. Protistenkd.* **120**, 1–15.

Hickman, M. (1978b). Ecological studies on the epipelic algal community in five prairie—parkland lakes in central Alberta. *Can. J. Bot.* **56**, 991–1009.

Hickman, M. (1982). The removal of a heated water discharge from a lake and the effect upon an epiphytic algal community. *Hydrobiologia* **87**, 21–32.

Hickman, M., and Klarer, D. M. (1975). The effect of the discharge of thermal effluent from a power station on the primary productivity of an epiphytic algal community. *Br. Phycol. J.* **10**, 81–91.

Hickman, M., and Round, F. E. (1970). Primary production and standing crops of epipsammic and epipelic algae. *Br. Phycol. J.* **5**, 247–255.

Hoagland, K. D., Roemer, S. C., and Rosowski, J. R. (1982). Colonization and community structure of two periphyton assemblages, with emphasis on the diatoms (Bacillariophyceae). *Am. J. Bot.* **69**, 188–213.

Hoagland, K. D., Ernst, S. G., Jensen, S. I., Miller, V. I., and DeNicola, D. M. Genetic variation in *Fragilaria capucina* clones along a latitudinal gradient across North America: A baseline for detecting global climate change. *Diatom Research* (in press).

Hoffmann, J. P., and Graham, L. E. (1984). Effects of selected physicochemical factors on growth and zoosporogenesis of *Cladophora glomerata* (Chlorophyta). *J. Phycol.* **20**, 1–7.

Hornick, L. E., Webster, J. R., and Benfield, E. F. (1981). Periphyton production in an Appalachian mountain trout stream. *Am. Midl. Nat.* **106**, 22–37.

Hulbert, E. M., and Guillard, R. R. L. (1968). The relationship of the distribution of the diatom *Skeletonema tropicum* to temperature. *Ecology* **49**, 337–339.

Hustedt, F. (1927–1959). Die Kieselalgen Deutschlands, Österreichs und der Schweiz mit Berücksichtigung der übrigen Länder Europas sowie der angrenzenden Meeresgebiete. *In* "Kryptogamen" (L. Rabenhorst, ed.), Band VII. Akademische Verlagsgesellschoft, Leipzig.

Hutchinson, G. E. (1957). "A Treatise on Limnology," Vol. 1. Wiley, New York.

Hutchinson, G. E. (1967). "A Treatise on Limnology," Vol. 2. Wiley, New York.

Hynes, H. B. N. (1970). "The Ecology of Running Waters." Liverpool Univ. Press, Liverpool, UK.

Jasper, S., and Bothwell, M. L. (1986). Photosynthetic characteristics of lotic periphyton. *Can. J. Fish. Aquat. Sci.* **43**, 1960–1969.

Jones, V. J., and Flower, R. J. (1986). Spatial and temporal variability in periphytic diatom communities: Paleoecological significance in an acidified lake. *In* "Diatoms and Lake Acidity" (J. P. Smol, R. W. Battarbee, R. B. Davis, and J. Merilainen, eds.), pp. 87–94. Junk Publ., Dordrecht, The Netherlands.

Jørgensen, E.G. (1968). The adaptation of plankton algae. 2. Aspects of the temperature adaptation of *Skeletonema costatum. Physiol. Plant.* **21**, 423–427.

Kevern, N. R., and Ball, R. C. (1965). Primary productivity and energy relationships in artificial streams. *Limnol. Oceanogr.* **10**, 74–87.

Klarer, D. M., and Hickman, M. (1975). The effect of thermal effluent upon the standing crop of an epiphytic algal community. *Hydrobiologia* **60**, 17–62.

Krejci, M. E., and Lowe, R. L. (1986). Importance of sand grain mineralogy and topography in determining micro-spatial distribution of epipsammic diatoms. *J. North Am. Benthol. Soc.* **5**, 211–220.

Kullberg, R. G. (1971). Algal distribution in six thermal spring effluents. *Trans. Am. Microsc. Soc.* **90**, 412–434.

Kullberg, R. G. (1982). Algal succession in a hot spring community. *Am. Midl. Nat.* **108**, 224–244.

Lamberti, G. A., and Resh, V. H. (1983). Geothermal effects on stream benthos: Separate influences of thermal and chemical components on periphyton and macroinvertebrates. *Can. J. Fish. Aquat. Sci.* **40**, 1995–2009.

Lamberti, G. A., and Resh, V. H. (1985). Distribution of benthic algae and macroinvertebrates along a thermal stream gradient. *Hydrobiologia* **128**, 13–21.

Lanza, G. R., and Cairns, J. (1972). Physio-morphological effects of abrupt thermal stress on diatoms. *Trans. Am. Microsc. Soc.* **91**, 276–298.

Lehel, C., Wada, H., Kovacs, E., Torok, Z., Gombos, S., Horvath, I., Murata, N., and Vigh, L. (1992). Heat shock protein synthesis of the cyanobacterium *Synechocystis* PCC 6803: Purification of the GroEL-related chaperonin. *Plant Mol. Biol.* **18,** 327–336.

Levasseur, M., Morissette, J., Popovic, R., and Harrison, P. J. (1990). Effects of long term exposure to low temperature on the photosynthetic apparatus of *Dunaliella tertiolecta* (Chlorophyceae). *J. Phycol.* **26,** 479–484.

Li, W. K. W. (1980). Temperature adaptation in phytoplankton: Cellular and photosynthetic characteristics. *In* "Primary Productivity of the Sea" (P.G. Falkowski, ed.), Vol. 19, pp. 259–279. Plenum, New York.

Lowe, R. L. (1974). "Environmental Requirements and Pollution Tolerance of Freshwater Diatoms," USEPA 670/4-74-005. USEPA. Cincinnati, OH.

Lynch, D. V., and Thompson, G. A. (1982). Low temperature-induced alterations in the chloroplast and microsomal membranes of *Dunaliella salina. Plant Physiol.* **69,** 1369–1375.

Lynch, J. A., Rishel, G. B., and Corbett, E. S. (1984). Thermal alteration of streams draining clearcut watersheds: Quantification and biological implications. *Hydrobiologia* **111,** 161–169.

Mannan, R. M., Krishan, M., and Gnaham, A. (1986). Heat shock proteins of cyanobacterium *Anacystis nidulans. Plant Cell Physiol.* **27,** 377–381.

Mantal, K. M. (1974). Some aspects of photosynthesis in *Cladophora glomerata. J. Phycol.* **10,** 288–291.

Margalef, R. (1960). Ideas for a synthetic approach to the ecology of running waters. *Hydrobiologia* **45,** 133–153.

Marker, A. F. H. (1976a). The benthic algae of some streams in southern England. I. Biomass of the epilithon in some small streams. *J. Ecol.* **64,** 343–358.

Marker, A. F. H. (1976b). The benthic algae of some streams in southern England. II. The primary production of the epilithon in a small chalk-stream. *J. Ecol.* **64,** 359–373.

Marrè, E. (1962). Temperature. *In* "Physiology and Biochemistry of Algae" (R. A. Lewin, ed.), pp. 541–550. Academic Press, New York.

McIntire, C. D. (1966). Some factors affecting respiration of periphyton communities in lotic environments. *Ecology* **47,** 918–929.

McIntire, C. D. (1968a). Structural characteristics of benthic algal communities in laboratory streams. *Ecology* **49,** 520–537.

McIntire, C. D. (1968b). Physiological-ecological studies of benthic algae in laboratory streams. *J.—Water Pollut. Control Fed.* **40,** 1940–1952.

Moore, J. W. (1974). Benthic algae of southern Baffin Island. II. The epipelic communities in ponds. *J. Ecol.* **62,** 809–819.

Moore, J.W. (1977a). Seasonal succession of algae in a eutrophic stream in southern England. *Hydrobiologia* **53,** 181–192.

Moore, J. W. (1977b). Seasonal succession of algae in rivers. II. Examples from Highland Water, a small woodland stream. *Hydrobiologia* **80,** 160–171.

Moore, J. W. (1977c). Ecology of algae in a subarctic stream. *Can. J. Bot.* **55,** 1838–1847.

Moore, L. F. (1978). Attached algae at thermal generating stations—The effect of temperature on *Cladophora. Verh.—Int. Theor. Angew. Limnol.* **20,** 1727–1733.

Mortain-Bertrand, A., Descolas-Gros, C., and Jupin, H. (1988). Growth, photosynthesis and carbon metabolism in the temperate marine diatom *Skeletonema costatum* adapted to low temperature and low photon-flux density. *Mar. Biol.* **100,** 135–141.

Mosser, J. L., and Brock, T. D. (1976). Temperature optima for algae inhabiting cold mountain streams. *Arct. Alp. Res.* **8,** 111–114.

Mosser, J. L., Mosser, A. G., and Brock, T. D. (1977). Photosynthesis in the snow: The alga *Chlamydomonas nivalis* (Chlorophyceae). *J. Phycol.* **13,** 22–27.

Nover, L. (1984). "Heat Shock Response of Eukaryotic Cells." Springer-Verlag, New York.

Patrick, R. (1969). Some effects of temperature on freshwater algae. *In* "Biological Aspects of Thermal Pollution" (P. A. Krendel and F. L. Parker, eds.), pp. 161–198. Vanderbilt Univ. Press, Knoxville, TN.

Patrick, R. (1971). The effects of increasing light and temperature on the structure of diatom communities. *Limnol. Oceanogr.* **16,** 405–421.

Patrick, R. (1974). Effects of abnormal temperatures on algal communities. *In* "Thermal Ecology" (W. J. Gibbons and R. R. Sharitz, eds.), pp. 335–370. U.S. Atomic Energy Commission, Washington, DC.

Patrick, R. (1977). Ecology of freshwater diatoms—Diatom communities. *In* "The Biology of Diatoms" (D. Werner, ed.), pp. 284–332. Univ. of California Press, Berkeley.

Patrick, R., Crum, B., and Coles, J. (1969). Temperature and manganese as determining factors in the presence of diatom or blue-green algal floras in streams. *Proc. Natl. Acad. Sci. U.S.A.* **64,** 472–478.

Phinney, H. K., and McIntire, C. D. (1965). Effect of temperature on metabolism of periphyton communities developed in laboratory streams. *Limnol. Oceanogr.* **10,** 341–344.

Pienitz, R., Reinhard, J., Smol, J. P., and Birks, H. J. B. (1994). Assessment of freshwater diatoms as quantitative indicators of past climatic change in the Yukon and Northwest Territories, Canada. *J. Paleolimnol.*

Reynolds, C. S. (1984). "The Ecology of Freshwater Phytoplankton." Cambridge Univ. Press, Cambridge, UK.

Rhee, G.-Y. (1982). Effects of environmental factors and their interactions on phytoplankton growth. *Adv. Microb. Ecol.* **6,** 33–74.

Rhee, G.-Y., and Gotham, I. J. (1981). The effect of environmental factors on phytoplankton growth: Temperature and the interactions of temperature with nutrient limitation. *Limnol. Oceanogr.* **26,** 635–648.

Riber, H. H., and Wetzel, R. G. (1987). Boundary-layer and internal diffusion effects on phosphorus fluxes in lake periphyton. *Limnol. Oceanogr.* **32,** 1181–1194.

Round, F. E. (1960). Studies on bottom-living algae in some lakes of the English lake district. *J. Ecol.* **48,** 529–47.

Round, F. E. (1972). Patterns of seasonal succession of freshwater epipelic algae. *Br. Phycol. J.* **7,** 213–220.

Salisbury, F. B., and Ross, C. W. (1985). "Plant Physiology," 3rd ed. Wadsworth, Belmont, CA.

Schlesinger, D. A., and Shuter, B. J. (1981). Patterns of growth and cell composition of freshwater algae in light-limited continuous cultures. *J. Phycol.* **17,** 250–256.

Schindler, D. W., and Beaty, K. G. (1990). Effects of climatic warming on lakes of the central boreal forest. *Science* **250,** 967–970.

Seaburg, K. G., and Parker, B. C. (1983). Seasonal differences in the temperature ranges of growth of Virginia algae. *J. Phycol.* **19,** 380–386.

Seaburg, K. G., Parker, B. C., Wharton, R. A., and Simmons, G. M. (1981). Temperature–growth responses of algal isolates from Antarctic oases. *J. Phycol.* **17,** 353–360.

Shepherd, B. G., Hartman, G. F., and Wilson, W. J. (1986). Relationships between stream and intragravel temperatures in coastal drainages, and some implications for fisheries workers. *Can. J. Fish. Aquat. Sci.* **43,** 1818–1822.

Shuter, B. (1979). A model of physiological adaptation in unicellular algae. *J. Theor. Biol.* **78,** 519–552.

Soeder, C. J., and Stengel, E. (1974). Physico-chemical factors affecting metabolism and growth rate. *In* "Algal Physiology and Biochemistry" (W. P. D. Stewart, ed.), pp. 714–740. Univ. of California Press, Berkeley.

Soudek, D., and Robinson, G. G. C. (1983). Electrophoretic analysis of the species and population structure of *Asterionella formosa. Can. J. Bot.* **61,** 418–433.

Squires, L. E., Rushforth, S. R., and Brotherson, D. J. (1979). Algal response to a thermal effluent: Study of a power station on the Provo River, Utah, USA. *Hydrobiologia* **63**, 1011–1017.

Stanley, D. W., and Daley, R. J. (1976). Environmental control of primary productivity in Alaskan tundra ponds. *Ecology* **57**, 1025–1033.

Steemann-Nielsen, E., and Jørgensen, E. G. (1968). The adaptation of plankton algae. I. General part. *Physiol. Plant.* **21**, 401–413.

Stefan, H. G., and Sinokrot, B. A. (1993). Projected global climate change impact on water temperature in five north central U.S. streams. *Clim. Change* **24**, 353–381.

Stockner, J.G. (1967). Observations of thermophilic algal communities in Mt. Rainer and Yellowstone National Parks. *Limnol. Oceanogr.* **12**, 13–17.

Stockner, J. G., and Shortreed, K. R. S. (1976). Autotrophic production in Carnation Creek, a coastal rainforest stream on Vancouver Island, British Columbia. *J. Fish. Res. Board Can.* **33**, 1553–1563.

Sumner, W. T., and Fisher, S. G. (1979). Periphyton production in Fort River, Massachusetts. *Freshwater Biol.* **9**, 205–212.

Thompson, P. A., Guo, M., and Harrison, P. J. (1992a). Effects of variation in temperature. I. On the biochemical composition of eight species of marine phytoplankton. *J. Phycol.* **28**, 481–488.

Thompson, P. A., Guo, M., Harrison, P. J., and Whyte, J. N. C. (1992b). Effects of variation in temperature. II. On the fatty acid composition of eight species of marine phytoplankton. *J. Phycol.* **28**, 488–497.

Vannote, R. L., and Sweeney, B. W. (1980). Geographic analysis of thermal equilibria: A conceptual model for evaluating the effect of natural and modified thermal regimes on aquatic insect communities. *Am. Nat.* **115**, 667–695.

Vincent, W. F., and Howard-Williams, C. (1986). Antarctic stream ecosystems: Physiological ecology of a blue-green algal epilithon. *Freshwater Biol.* **16**, 219–233.

Vincent, W. F., and Howard-Williams, C. (1989). Microbial communities in southern Victoria Land streams (Antarctica). II. The effects of temperature. *Hydrobiologia* **172**, 39–49.

Vinson, D. K., and Rushforth, S. R. (1989). Diatom species composition along a thermal gradient in the Portneuf River, Idaho, USA. *Hydrobiologia* **185**, 41–54.

Wallace, N. M. (1955). The effect of temperature on the growth of some fresh-water diatoms. *Not. Nat. Acad. Nat. Sci. Philadelphia.* **280**, 1–11.

Ward, J. V. (1992). "Aquatic Insect Ecology." Wiley, New York.

Weatherly, N. S., and Ormerod, S. J. (1990). Forests and the temperature of upland streams in Wales: A modelling exploration of the biological effects. *Freshwater Biol.* **24**, 109–122.

Wehr, J. D. (1981). Analysis of seasonal succession of attached algae in a mountain stream, the North Alouette River. *Can. J. Bot.* **59**, 1465–1474.

Wetzel, R. G. (1983). "Limnology," 2nd ed. Saunders, New York.

Wilde, E. W. (1982). Responses of attached algal communities to termination of thermal pollution. *Hydrobiologia* **94**, 135–138.

Wilde, W. E., and Tilly, L. J. (1981). Structural characteristics of algal communities in thermally altered artificial streams. *Hydrobiologia* **76**, 57–63.

Wood, A. M., and Leatham, T. (1992). The species concept in phytoplankton ecology. *J. Phycol.* **28**, 723–729.

Woodwell, G. M., Whittaker, R. H., Reiners, W. A., Likens, G. E., Delwiche, C. C., and Botkin, D. B. (1978). The biota and the world carbon budget. *Science* **199**, 141–146.

Yang, J. (1994). Reconstruction of paleo-environmental conditions in Hamilton Harbour and East Lake, Ontario. Ph.D. Dissertation, University of Waterloo, Waterloo, Ontario.

Yoder, J. A. (1979). Effect of temperature on light-limited growth and chemical composition of *Skeletonema costatum* (Bacillariophyceae). *J. Phycol.* **15**, 362–370.

7

Nutrients

Mark A. Borchardt
Marshfield Medical Research Foundation
Marshfield, Wisconsin 54449

I. INTRODUCTION

Ask any layperson why there is floating green scum on a pond or why a lake is the color of pea soup and the response will likely refer, either obliquely or directly, to the amount of nutrients in the water. Such has been the impact of the debate over the causes of eutrophication in the 1960s and 1970s. The linkage between nutrients and algae has been firmly imprinted in the public's mind—and in the minds of aquatic scientists. Certainly, as this book illustrates, algae need more than nutrients to thrive, and their abundance and species composition are determined by biotic interactions, such as grazing and parasitism, as well as other physical and chemical factors. But the importance of nutrients in algal ecology is difficult to overstate. Surprisingly then, of the more than 6000 citations that contain the key words "algae" and "nutrients" in *Biosis* since 1969, only a handful have been concerned with benthic algae. Does this discrepancy matter, or is the free association between nutrients and algae in our consciousness justified for benthic algae as well?

Historically, two perspectives have been taken to empirically describe the relationship between nutrient levels and algal growth: (1) An ecosystem approach, where annual nutrient inputs into aquatic systems are related to algal biomass and primary productivity (Sakamoto, 1966; Dillon and Rigler, 1974; Vollenweider, 1976; Schindler, 1977); and (2) an autecological approach, where either ambient or intracellular nutrient levels are related to the growth of specific algal taxa (Caperon, 1968; Droop, 1968; Rhee, 1973; Goldman, 1977; Kilham, 1978). Both perspectives have been fruitful, yielding insights into nutrient cycling, anthropogenic eutrophication, algal community structure, and interactions among algae and other organisms. Both perspectives have also focused on planktonic algae.

A key tool in the development of these relationships for planktonic algae has been the chemostat. Algae grown in chemostats are in physiological and numerical equilibrium. This condition allows precise and relatively rapid studies of many aspects of nutrient-limited growth: coupling between growth and nutrient uptake, ambient nutrient concentration, intracellular nutrient pools, and nonlimiting nutrients; optimum nutrient ratios; interactions with temperature and light; and the effects of nutrient perturbations and other nonequilibrious conditions. Defining the relationships between nutrients and the growth of benthic algae may be more problematic. Benthic algae, by virtue of their attached growth habit, are not amenable to chemostat growth, which makes some problems less tractable and more laborious.

Extending nutrient kinetic theory from phytoplankton to benthic algae is confounded by fundamental differences between the two groups. Unlike phytoplankton, which, when actively growing, are entrained in the motion of water, benthic algae are relatively fixed in position and subject to flow velocities 10 to 100,000 times greater than the sinking rates of planktonic forms. Water motion alters the physicochemical environment near the algal cell surface, most notably the movement of dissolved nutrients (Schumacher and Whitford, 1965) and gases (e.g., O_2, Dodds, 1991) between the bulkwater and the water adjacent to the algae. Moreover, there is some evidence that water motion can be physiologically costly, the costs manifest as an increased requirement for growth-limiting nutrients (Borchardt, 1994). These consequences of a benthic life may be more acute in streams and rivers, where water motion is unrelenting and unidirectional flow presents unique nutrient conditions.

Another important difference between planktonic and benthic algae is the spatial organization of the algal community. Phytoplankton, for the most part, are separate entities suspended in the water column with nutrients available from any direction. Benthic algae, on the other hand, often create mats on the substratum that are many cells thick. The development of a mat alters the hydrodynamic environment (Reiter, 1989) and establishes a microenvironment whose physicochemical properties may vary primarily in a direction perpendicular to the mat surface (Dodds, 1989; Jørgensen and Des Marais, 1990; Vincent *et al.*, 1993). Nutrients must enter the mat from either below in upwelling regions or above from the overlying water. A vertical nutrient gradient, combined with the normal presence of an unmixed "dead zone" of water above the mat surface, creates the potential for algae within the mat to be separated from the bulkwater nutrient source.

The scope of the literature on phytoplankton and nutrients has justified numerous review articles and chapters, not all of which can be cited here. Reviews by Droop (1983), Button (1985), Zevenboom (1986), and Turpin (1988) focus on the physiological aspects of nutrient-limited growth and the consequences for algal competition and community structure. Hecky and Kilham (1988) compare the type and degree of nutrient limitation between marine and freshwater systems. Harlin and Wheeler (1985) provide a summary of methodology, whereas Goldman and Glibert (1983) and Harrison *et al.* (1989) critique those methods commonly employed. Benthic algae received attention from Wheeler (1988), who reviewed the hydrodynamic theory relevant to nutrient acquisition by large marine algae, and from Raven (1992), who examined aspects of light and nutrient acquisition by lotic algae.

This chapter will focus specifically on the relationship between nutrients and the biomass and growth rate of benthic algae using autecological and community-level perspectives. I intend to demonstrate that nutrient kinetic theory is applicable to benthic algae and, moreover, that it offers a valuable conceptual framework hitherto underutilized in benthic algal ecology. I will further argue that aspects of the benthic growth habit, namely, exposure to greater water motion and the formation of mats, have ramifications for nutrient uptake, utilization, recycling, and competition that are unique to benthic algae.

II. CONCEPTUAL FRAMEWORK: NUTRIENT UPTAKE AND GROWTH KINETICS

A. Empirical Models

The first stone laid in the foundation of what is now called algal nutrient kinetics is often attributed to the seminal work of Dugdale (1967). He proposed that algal nutrient uptake could be described by a rectangular hyperbola function mathematically equivalent to the Michaelis–Menten model of enzyme kinetics (Fig. 1A).

$$V = V_{\text{max}} \frac{S}{K_s + S} \tag{7.1}$$

where V = specific nutrient uptake rate (time^{-1}), V_{max} = maximum specific uptake rate, S = substrate concentration (μmoles Liter^{-1}), and K_s = the substrate concentration that yields half the maximum rate.

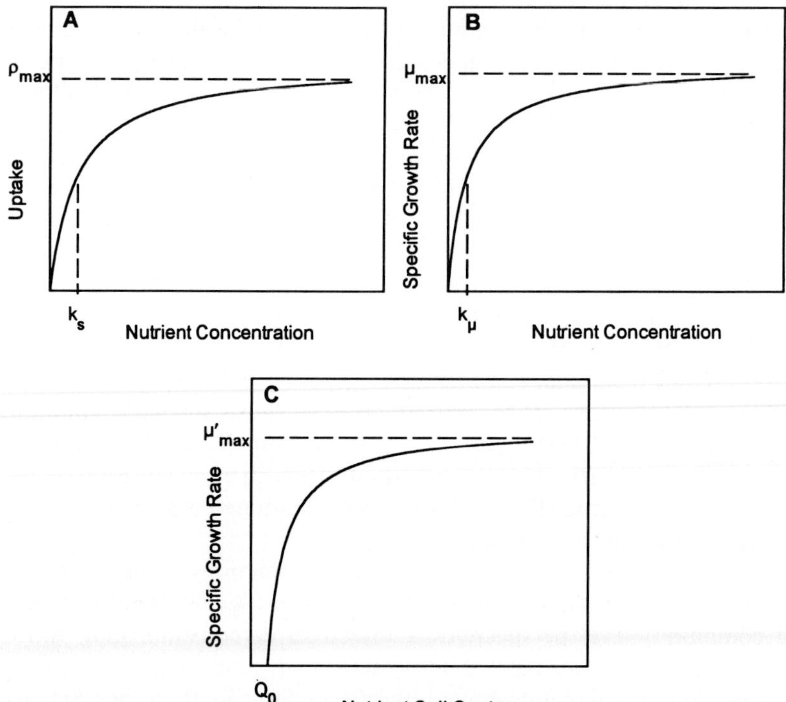

FIGURE 1 Graphical representation of algal nutrient uptake and nutrient-limited growth as described by (A) Michaelis–Menten, (B) Monod, and (C) Droop models. See text for description of parameters.

Nutrient uptake can also be expressed as an absolute rate, ρ (mass/cell/time), with the same relationship to external nutrient concentration as in Eq. (7.1):

$$\rho = \rho_{max} \frac{S}{K_s + S} \tag{7.2}$$

Absolute uptake is related to specific uptake by $V = \rho/Q$, where Q is the nutrient cell quota (i.e., intracellular nutrient concentration).

The initial appeal of Michaelis–Menten kinetics arose partly from the idea that only two parameters were needed to characterize the nutrient acquisition and growth potential of an alga. The half-saturation constant was a measure of the uptake affinity for a particular nutrient, and the specific uptake rate reflected specific growth rate because these two parameters share the same inverse time dimension. Moreover, it appeared that V_{max} and K_s, presumably unique for each species, could help explain algal succession and the outcome of competitive interactions (Dugdale, 1967; Eppley *et al.*, 1969). The "paradox of the plankton" (Hutchinson, 1961) had been realized a few years earlier and Dugdale's elegant linkage of algal physiology with fundamental questions in algal ecology was immediately acknowledged.

Evidence quickly accumulated that Michaelis–Menten kinetics were applicable for algal uptake of a variety of nutrients: nitrogen (Eppley *et al.*, 1969), vitamin B_{12} (Droop, 1968), silicate (Goering *et al.*, 1973; Paasche, 1973), and phosphorus (Rhee, 1973). However, the prospects of one model for both growth and uptake were short-lived. Caperon (1968) and Droop (1968) pointed out that nutrient uptake and growth were coupled only under steady-state conditions as in a chemostat. Algae have the capacity to store some nutrients so that uptake and growth may be temporally separated.

Droop (1961) and Caperon (1967) proposed that algal nutrient-limited growth could be modeled by the Monod equation (Monod, 1950), which was empirically derived for bacterial growth and is analogous in form to the Michaelis–Menten equation (Fig. 1B):

$$\mu = \mu_{max} \frac{S}{K_\mu + S} \tag{7.3}$$

where μ is the specific growth rate ($time^{-1}$), μ_{max} is the maximum specific growth rate, and K_μ is the half-saturation constant for growth for the limiting nutrient. How K_μ and K_s relate depends on how much the algal cell can vary the intracellular stores of the limiting nutrient. When the range is great, as with phosphorus, then $K_\mu << K_s$. When the range is small, as with carbon, then K_μ approaches the value of K_s (Turpin, 1988). The ratio $K_\mu : K_s$ reflects the ability of an alga to fluctuate its intracellular nutrient stores in response to nutrient limitation and may be adaptive (Turpin, 1988).

Early applications of the Monod model to algal growth were disappointing. It appeared to work well for carbon (Goldman *et al.*, 1974; although cf. Turpin *et al.*, 1985), but for the other major nutrients, many investigators reported deviations at low growth rates (e.g., Droop, 1968; Paasche, 1973). Many hypotheses for the source of error were proposed, the most likely that the external nutrient concentrations at low growth rates were near the limit of analytical detection (Rhee, 1980; Goldman and Glibert, 1983). Sometimes multiphasic growth kinetics restrict use of the model to a narrow range of growth rates rather than the entire range of possible rates (Turpin, *et al.*, 1985; Olsen, 1989). On the other hand, the Monod model has been successfully applied to data obtained using short-term batch culture techniques (Tilman and Kilham, 1976; Grover, 1989; van Donk and Kilham, 1990) and, indeed, was a keystone model in the development of resource ratio partitioning theory (Tilman, 1977, 1981; Tilman *et al.*, 1982). Based on K_μ and μ_{max}, or parameters derived from these, the outcome of competitive interactions among algae for growth-limiting nutrients can be predicted (Kilham, 1984; van Donk and Kilham, 1990; cf. Grover, 1990).

Given the capacity of algae to store nutrients and the apparent problems of the Monod model, alternative models of nutrient-limited algal growth were developed. Droop (1968), Caperon (1968), and Fuhs (1969) proposed that growth was a function of the intracellular concentration of the limiting nutrient. The mathematical forms of their empirical models differ, but the relationship they describe is essentially the same. The Droop model has become the accepted formulation (Fig. 1C):

$$\mu = \mu'_{max}\,(1 - Q_0/Q) \qquad (7.4)$$

where Q is the cell quota of the growth-limiting nutrient and Q_0 is the minimum amount of nutrient required for growth (Q when $\mu = 0$). Note that μ'_{max} is the maximum growth rate at infinite Q and is greater than μ_{max} of the Monod model by a factor characteristic for each nutrient. The ratio $\mu'_{max} : \mu_{max}$, like $K_\mu : K_s$, indicates the range in cell quota for a particular nutrient and alga (Turpin, 1988).

A critical assumption of the Droop model is that the total cellular pool of limiting nutrient, as given by Q, is proportional to that cellular pool most closely coupled to growth processes. Other models have compartmentalized the intracellular pool (e.g., Grenney *et al.*, 1973; Shuter, 1979) or the Droop model has been applied to specific nutrient pools such as polyphosphates and free amino acids (Rhee, 1973, 1978). Some questions may require that the active nutrient pool be identified, if possible, but generally the total nutrient pool may be used. The Droop model has described algal growth limited by vitamin B_{12} (Droop, 1968), phosphorus (Rhee,

1973; Droop, 1974; Goldman, 1977; Senft, 1978; Borchardt, 1994), nitrogen (Goldman and Peavey, 1979; Borchardt, 1994), silicate (Davis *et al.*, 1978), and iron (Davies, 1970).

These three simple empirical models, Michaelis–Menten, Monod, and Droop, form the mathematical core for most nutrient kinetic studies with algae. Combining the Michaelis–Menten and Droop equations creates a two-step model, uptake and utilization, called Variable Internal Stores (VIS). Written as a set of differential equations with respect to time, the VIS model can describe the dynamics of algal density, cell quota, and external nutrient concentration (Burmaster and Chisholm, 1979; Grover, 1991a). At steady state the Monod and VIS models are equivalent (Droop, 1973; Kilham, 1978; Burmaster, 1979).

B. Concept of Threshold Nutrient Limitation and Optimum Nutrient Ratio

An important concept underpinning algal nutrient kinetics is single-nutrient limitation. Growth of an algal species can be limited by only one nutrient at a time. Two or more nutrients can be simultaneously *near* growth-limiting concentrations, but only one will actually be limiting. Simultaneous limitation with some combinations of nutrients was tested with different approaches by Droop (1974) and Rhee (1974). Both investigations showed definitively that algae responded to nutrient limitation in a threshold manner, that is, with sharp transitions between limitation by one or another nutrient. The transition occurs at that proportion of nutrients within the cell that perfectly matches the growth requirements of the alga. For example, if physiological processes require 17 parts nitrogen to 1 part phosphorus, then at a cellular N:P ratio of 18:1 the alga would be limited by P, whereas if the ratio were 16:1, the alga would be limited by N. The cellular ratio at which neither nutrient is growth-limiting is termed the optimum ratio. The optimum ratios of various nutrient pairs differ among algal species (Rhee and Gotham, 1980; Tilman *et al.*, 1982) and are the cornerstone of resource ratio partitioning theory (Tilman *et al.*, 1982, see the following). The concept of single-nutrient limitation does not apply to algal communities where different species may be limited by different nutrients.

C. Differential Limitation of Growth Rate and Biomass

When considering nutrient limitation, it is important to distinguish between the rate of biomass production (growth rate) and the amount of biomass produced (yield). The rate of growth is determined by the supply rate of limiting nutrient, whereas the potential yield is determined by the total quantity of nutrients available. A classic example of this distinction is acidic lakes, where the inorganic carbon concentration is low and algal growth rate is set by the influx of atmospheric CO_2 (Schindler and Fee, 1973). However, given the large pool of CO_2 in the atmosphere, algal yield in these lakes will never be carbon-limited, but instead will be phosphorus-limited (Schindler, 1977). Hence, it is possible for growth rate and yield to be limited by different nutrients or the same nutrient. Another possible situation is growth rate limitation without yield limitation. For benthic algae in streams and rivers, downstream solute transport presents a never-ending supply of nutrients to be utilized in the production of biomass. But if the concentration (i.e., supply rate) of nutrients is low, growth rate limitation can occur (Grimm and Fisher, 1986).

D. Temperature and Light Interactions

Nutrient uptake and growth, and hence the model parameters that encapsulate these processes, may vary with light and temperature. Uptake of a nutrient, particularly those that are frequently growth-limiting, like phosphorus, is usually counter to its free-energy gradient between the inside and the outside of the algal cell. Thus any environmental conditions that limit cellular energetics, such as low light, would be expected to also limit nutrient uptake. Light-dependent uptake has been shown for phosphate (e.g., Falkner, *et al.*, 1980; Wynne and Rhee, 1988). However, the effects of abiotic factors on nutrient uptake are rarely unequivocal or consistent and are too diverse to even summarize briefly. In addition, uptake may depend on the environmental conditions or form of the limiting nutrient *prior* to uptake (e.g., Dortch *et al.*, 1991; Borchardt *et al.*, 1994). For an introduction to the complex interactions among abiotic factors on P uptake, the reader should consult Cembella *et al.* (1984) and Ahlgren (1988).

Algae often require more nutrients when light and temperature conditions are less than optimum for growth. This would be evident as an increase in the half-saturation constant for growth (K_μ) or minimum cell quota (Q_0) for the growth-limiting nutrient. Temperature was inversely related to the minimum cell quotas for nitrogen, phosphorus, and silicate for several algal species (Goldman, 1979; Rhee and Gotham, 1981a; Ahlgren, 1987; van Donk and Kilham, 1990), although for some taxa Q_0^P and Q_0^{Si} were unaffected (van Donk and Kilham, 1990). Ahlgren (1987) reviewed several data sets and concluded that K_μ is independent of temperature, but van Donk and Kilham (1990) showed that for some diatom taxa a 5°C change in temperature can have a significant effect on K_μ. The effects of light intensity are similarly taxon- and nutrient-specific. Subsaturating light intensities have been consistently shown to increase the minimum nitrogen cell quotas of algae (Zevenboom *et al.*, 1980; Rhee and Gotham, 1981b; Healey, 1985; Wynne and Rhee, 1986), but its effects on Q_0^P vary by taxon (Healey, 1985; Wynne and Rhee, 1986). Light wavelength can also alter Q_0^N and to a lesser extent Q_0^P (Wynne and Rhee, 1986).

E. Steady-State versus Non-Steady-State Growth

Steady-state growth, the condition under which Eq. (7.1)–(7.4) were developed, occurs when nutrient uptake matches the amount of nutrients required by the algal cell to maintain its current growth rate. Mathematically, this can be expressed as $V = \mu$ or $\rho = \mu Q$. In other words, the amount of nutrient transported into the cell is equal to the amount of nutrient diluted by growth. This relationship holds if nutrient efflux from the cells is negligible. Deviation from steady state occurs under variable nutrient conditions when uptake is less than or greater than the rate of nutrient dilution by growth and the cell quota will, if possible, decrease or increase, respectively. The degree to which uptake can become uncoupled from growth depends on the storage capacity of the alga for a nutrient. For example, with carbon there is little cellular capacity for storage, so uptake and growth are more closely coupled (Turpin, *et al.*, 1985) than for a nutrient like phosphorus, where algal storage capacity is great and P uptake can be 5 to 50 times greater than necessary to maintain maximum growth (Cembella, *et al.*, 1984).

When algae are exposed to a nutrient pulse, the uptake rate usually varies with the level of cellular nutrient stores. Uptake of ammonium (e.g., Eppley and Renger, 1974; Goldman and Glibert, 1982), bicarbonate (Miller *et al.*, 1984), and phosphate (Gotham and Rhee, 1981; Auer and Canale, 1982a; Riegman and Mur, 1984; Olsen, 1989; Borchardt, 1991) have been shown to decrease nonlinearly with increasing cell quota. Feedback from the increasing intracellular nutrient stores to plasmalemma transporters during uptake may be rapid, on the order of 2–3 minutes, causing uptake to become nonlinear over the course of an incubation (Fig. 2). Use of the nonlinear portion of a depletion time course can lead to substantial errors in uptake rate measurements (Goldman and Glibert, 1983; Harrison *et al.*, 1989). For other algae feedback is slower and nutrient uptake is sustained for many minutes before diminishing (Olsen, 1989; Borchardt *et al.*, 1994). The relationship between ρ and Q is not definitive as there are reports of uptake being independent of cell quota, for example, with phosphorus (Perry, 1976; Nyholm, 1977; Burmaster and Chisholm, 1979; Olsen, 1989; Grover, 1991b) and nitrate (Dortch *et al.*, 1982). Recent theory, however, considers that uptake is inversely related to cell quota and that there is not one ρ versus S uptake curve, but a family of curves, each one unique for the growth rate or level of nutrient sufficiency of the alga at the time of the nutrient perturbation (Morel, 1987; Turpin, 1988). There is no clear evidence that K_s varies with Q, perhaps because of the methodological problems and large variance associated with measuring K_s. Turpin (1988) thus argues that the principal physiological adaptation of algae to nutrient limitation is flexibility in ρ_{max}, each change in this parameter generating a different uptake curve. A clear picture of how

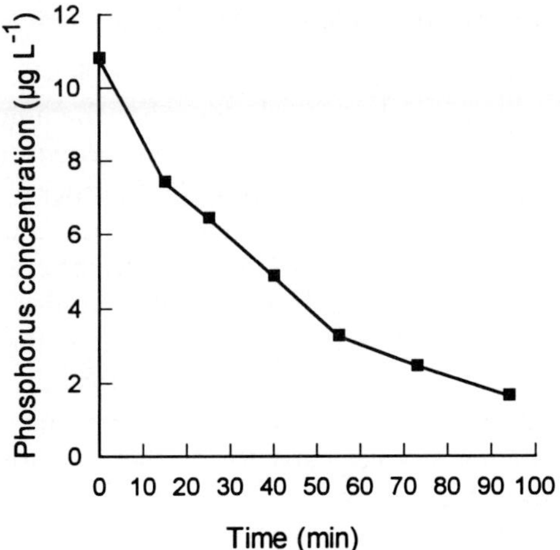

FIGURE 2 Nonlinear phosphorus uptake by *Spirogyra fluviatilis.* Initial PO_4 pulse was 10.8 μg PO_4-P $Liter^{-1}$. Data from Borchardt (1991).

nutrient uptake varies with cellular nutrient level is unlikely to emerge fully until the mechanisms of uptake regulation are elucidated.

Whether models derived from algae in physiological equilibrium are appropriate for understanding algae living in variable natural systems is still debatable. Clearly, the body of theory outlined here has advanced our understanding of nutrient-limited algal growth, nutrient competition among algae, and the consequences of these interactions for algal community structure (e.g., Rhee and Gotham, 1980; Tilman *et al.,* 1982; Turpin, 1988; Olsen, 1989). The next step is to understand how algal growth and community structure are affected by variable nutrient conditions. The models described in Eqs. (7.1)–(7.4) will likely remain the best approach for several reasons: (1) Nutrient supply rates in natural systems may not be as variable as conjectured and the algae are therefore growing near steady state. (2) The Monod model and, particularly, the Droop model are applicable to nutrient conditions that change gradually (Grover, 1991b). (3) Deviation from steady state and the nutrient uptake that immediately follows can be modeled. One must assume that the rectangular hyperbolae of Eqs. (7.1)–(7.4) are an accurate depiction of uptake and growth processes and that ρ_{max} is a linear decreasing function of Q, similar to much of the experimental evidence (Morel, 1987). (4) The competitive abilities of algae for growth-limiting nutrients determined at steady state may be the same as their competitive abilities at non-steady state. Then, insofar as algal community structure is the result of nutrient competition, predictions based on steady state would be valid for all conditions. Numerical simulations using equations from Morel (1987) and the VIS model were inconclusive in predicting whether different taxa would dominate at equilibrium or pulsed nutrient regimes (Grover, 1991a). The simulations suggested that if algae have a physiological trade-off between the highest ρ_{max} and the maximum Q, then with increasing frequency of nutrient pulses the algal community composition will shift. Whether that trade-off exists is unknown. Experimental evidence shows that nonequilibrium nutrient conditions can alter algal community composition (e.g., Turpin and Harrison, 1979; Sommer, 1985; Olsen *et al.,* 1989). The advances in algal nutrient kinetics since Dugdale's paper notwithstanding, the paradox of the plankton remains and is no less paradoxical for benthic algal communities.

III. CONCEPTUAL FRAMEWORK: MICROHYDRODYNAMIC ENVIRONMENT

A. Boundary Layer and Sublayer

To understand the nutrient-limited growth of benthic algae one must also understand the hydrodynamic environment in which they live. This environment is often complex in that it can vary in space and time, but certain characteristics are common to all objects in moving fluids. For the sake of simplicity, consider a flat plate with its long axis parallel to the direction of water moving at a constant velocity. On a line running perpendicular to the surface of the plate, there are points at which the velocity does not vary with changes in vertical position. This is the free-stream velocity of the water. Moving along the line toward the plate, friction slows the water velocity, at first asymptotically and then linearly, with diminishing vertical distance (Fig. 3). At the plate surface, the frictional force is greater than the force of the moving water and the velocity is zero. The boundary layer is defined by hydrologists as the region where the velocity gradient ranges from 0 to 99% of the free-stream velocity. For biological purposes an outer limit of 90% free-stream velocity has been suggested (Vogel, 1981).

A boundary layer, then, is where a velocity gradient exists. It is not a "dead zone" with a velocity of zero. The layer is "dead" only in the sense that at relatively low free-stream velocities over small objects the flow in the boundary layer is laminar and mixing is absent. But at faster velocities, or if the object is large enough, at some point downstream from the leading edge, the boundary layer becomes turbulent and materials within it are mixed. Even with a turbulent boundary layer there may remain, beneath the turbulence, a sublayer where the flow is essentially laminar and mixing is again rare. It is through these layers of laminar flow, either boundary layer or sublayer, that nutrient movement toward benthic algae is predominantly by molecular diffusion.

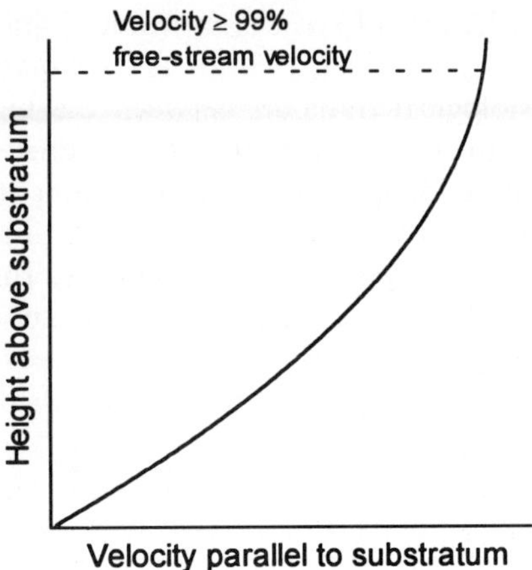

FIGURE 3 Water velocity gradient in the boundary layer above a substratum. The boundary layer begins at the height where the velocity is ≤99% of the free-stream velocity.

B. Molecular and Eddy Diffusion

The rate of molecular diffusion is given by Fick's first law:

$$J_{MD} = D(dC/dz) \tag{7.5}$$

where, considering solute movement only in one direction, the molecular diffusive flux, J_{MD}, is proportional to the concentration gradient, that is, the change in concentration (dC) relative to the change in distance (dz). The proportionality constant (D) is the diffusion coefficient, a temperature- and concentration-dependent quantity unique for each kind of diffusing molecule. The greater the concentration difference between the surface of an alga and the mixed water or the shorter the diffusion distance, the faster the diffusion rate.

The diffusion distance, which roughly corresponds to the thickness of the laminar boundary layer or sublayer, ranges from microns to millimeters (Gundersen and Jørgensen, 1990; Jørgensen and Des Marais, 1990). It is inversely related to the free-stream velocity; the faster the velocity, the thinner the layers and the shorter the diffusion distance. The thickness also depends on the specific location on an object and the shape of the object (Silvester and Sleigh, 1985).

Solute transport also occurs by eddy diffusion, whereby a "packet" of moving water carries the solutes dissolved within it. Eddy diffusion is described by an equation analogous to Eq. (7.5):

$$J_{ED} = K(dC/dz) \tag{7.6}$$

where the eddy diffusive flux, J_{ED}, is proportional to the concentration gradient (dC/dz). The eddy diffusion coefficient, K, unlike the molecular diffusion coefficient, varies proportionally with flow velocity (Nobel, 1983). Thus, in a turbulent boundary layer where advection is occurring in the z direction, K will be larger than D and eddy diffusion will dominate solute transport. In the laminar sublayer there is little velocity in the z direction, D will be larger than K, and molecular diffusion will predominate. The contribution of eddy diffusivity to solute transport varies with height above the substratum or algal mat. Therefore, in a laminar layer, especially near its juncture with a turbulent layer, some degree of mixing still occurs. At half the height of the diffusive boundary layer, eddy diffusion may be 15% of molecular diffusion (Boudreau, 1988, cited in Jørgensen and Des Marais, 1990). The lesson for algal ecologists is that even though a benthic algal community may live within a laminar layer, the diffusion environment may differ among individuals or taxa. Those algae in the upper region of the laminar layer will receive more of their nutrients by eddy diffusivity than those closer to the substratum. The upper-layer algae may thus procure more nutrients than their lower-layer counterparts (Burkholder *et al.*, 1990).

In assessing the microhydrodynamic environment of benthic algae, one must also consider which velocity gradient, of possibly several, is most relevant. For example, a solitary alga on a rock in a river could exist within three velocity gradients: the riverbed gradient, the rock gradient, and the gradient around the alga. A number of situations are possible. The boundary layer over the rock may be laminar and present a long diffusion distance, or it could be turbulent with the alga protruding into the turbulence; diffusion would then occur through the layer immediately adjacent to the alga. Flow can penetrate and mix the water among loosely scattered elements on a substratum; in this case the relevant gradient is around an individual element and the diffusion distance is probably short (Mierle, 1985). When the elements are more tightly packed, as with dense algal mats, then the flow lines remain above the surface of the elements (Nowell and Jumars, 1984) and the diffusion distance to the elements/algae increases. Roughness can enhance the total diffusive flux by increasing surface area (Jørgensen and Des Marais, 1990).

C. Diffusion Limitation

Diffusion limitation occurs when the nutrient concentration is so low or the diffusion distance so great that the diffusion rate is slower than the cellular uptake rate. In other words, diffusion cannot "keep up" in replacing nutrients at the cell surface that have been removed by active transport, and a nutrient-depleted layer develops around the cell (Whitford, 1960). Diffusion limitation is evident as a deviation in Michaelis–Menten kinetics because the measured uptake rate is slower than the rate predicted by the model (Winne, 1973; Pasciak and Gavis, 1975; Mierle, 1985). The Hill–Whittingham equation, where the nutrient concentration at the plasmalemma surface is solved from Fick's first law and substituted into the

Michaelis–Menten equation, incorporates diffusion in the description of nutrient uptake (Hill and Whittingham, 1955). Modeling nutrient uptake with both the Michaelis–Menten and Hill–Whittingham equations and comparing the goodness of fit will help determine whether diffusion or active transport is the rate-limiting step (Smith and Walker, 1980; Borchardt, 1991).

D. Boundary Layer Instability

The establishment of a boundary layer or sublayer and the corresponding diffusion distance should not be viewed as temporally constant. The hydrodynamic environment will vary as algal growth waxes and wanes and the physiognomy of the algal mat changes (Reiter, 1989). On shorter time scales, advective currents can stochastically penetrate the laminar sublayer, quickly transporting nutrients from the bulkwater into an algal mat (e.g., Dade, 1993, Gundersen and Jørgensen, 1990). This "flushing" of the laminar sublayer can dramatically alter the transport rate and concentration of solutes near the algal surface (Gundersen and Jørgensen, 1990). The importance of these so-called "splat" events to benthic algae is unknown. Suffice to note that benthic algae do not necessarily grow in a stable laminar layer where nutrient movement is solely by diffusion. Both diffusion and advection contribute to nutrient transport, the relative contribution of each dependent on the degree of turbulence and the height above the substratum.

For a more thorough discussion of the hydrodynamic environment and hydraulic parameters relevant to benthic algal ecology, the reader should consult Vogel (1981), Nowell and Jumars (1984), Silvester and Sleigh (1985), Statzner *et al.* (1988), Wheeler (1988), and Carling (1992).

IV. NUTRIENT LIMITATION OF BENTHIC ALGAE

A. Patterns

Several approaches have been used by benthic algal ecologists to elucidate the relationship between nutrient levels and benthic algal abundance and growth rate: (1) nutrient enrichment of water and/or artificial substrata *in situ* or in artificial streams; (2) comparisons of habitat or seasonal variation in biomass with nutrient levels and other factors using multiple regression or multivariate statistics; and (3) characterization of nutrient uptake and growth kinetics of benthic algal species or assemblages. Table I summarizes some studies that are representative primarily of the first two approaches.

Nearly all studies on nutrient limitation of benthic algae have been conducted in temperate climes, with the majority of studies conducted in running-water habitats in the United States and Canada (Table I). There is insufficient information to make generalizations for benthic algae in the tropics, the Arctic, or the Antarctic, or whether there are differences in the type or degree of nutrient limitation of benthic algae growing in lentic versus lotic habitats. Moreover, the emphasis has been on the benthic algal community; there is much less information on the nutrient requirements of individual species. Hence, given a substantial change in the concentrations of nutrients, it is now possible to broadly predict the changes in benthic algal biomass, but there is much less confidence in predicting changes in community structure.

Phosphorus and nitrogen are the most commonly investigated nutrients, and, of the benthic habitats examined, these two nutrients are the most likely to be growth-limiting (Table I). Which of the two is the primary growth-limiting nutrient appears to vary geographically. Studies conducted in roughly the northern half of the U.S.A. have found phosphorus to be in short supply, whereas in the Missouri Ozarks (Lohman *et al.*, 1991) and Southwest (Grimm and Fisher, 1986; Hill and Knight, 1988; Peterson and Grimm, 1992), nitrogen is limiting. In the Pacific Northwest, phosphorus (Horner and Welch, 1981) and nitrogen (Triska *et al.*, 1983) are both reported to be growth-limiting. The regional pattern of nitrogen limitation in the American Southwest and Pacific Northwest appears to coincide with soils of volcanic origin, which are abundant in phosphate-rich minerals (Dillon and Kirchner, 1975). Nitrogen limitation in streams may also occur during periods of low discharge when denitrification may reduce the N supply (Lohman *et al.*, 1991). When P limitation is mitigated by P enrichment, nitrogen is often found to be secondarily limiting (Stockner and Shortreed, 1978; Pringle and Bowers, 1984; Fairchild *et al.*, 1985; Winterbourn, 1990; Marks and Lowe, 1993).

Other nutrients receiving some attention are carbon, which can be growth-limiting in soft-water lakes (Fairchild *et al.*, 1989; Fairchild and Sherman, 1993), trace metals (Wuhrmann and Eichenberger, 1975; Winterbourn, 1990), organic phosphorus (Pringle, 1987), and vitamin B_{12} (Hoffmann, 1990). Surprisingly, because diatoms frequently dominate benthic algal communities, little attention has been given to silicate (an exception is Duncan and Blinn, 1989). Silicate can be undetectable in the water column of some lakes (Sommer, 1988) and therefore may be growth-limiting for benthic diatoms in those habitats.

B. Growth-Limiting Concentrations and Ratios of N and P

Whereas many studies have identified N and P as the nutrients in short supply for benthic algal communities, fewer studies have quantified the N and P concentrations that are growth-limiting (Table I). Bothwell (1988) showed definitively that 0.3–0.6 μg PO_4-P liter^{-1} saturated the growth rate

TABLE I Representative Studies on Nutrient Limitation of Benthic Algae[a]

Habitat and location	*Nutrients[b] manipulated*	*Nutrients[b] limiting*	*Method*	*Results*	*Reference*
Upper Rawdon River, Nova Scotia	N, P, K		125 lb bag of 4-12-6 fertilizer placed intact into pool of river	This may have been the first nutrient enrichment experiment in a river. A year following fertilization, upstream from the bag the substratum was barren. Downstream, thick mats of filamentous green algae had developed.	Huntsman (1948)
Outdoor artificial streams	N, P, and trace metals	Trace metals	Enrichment of groundwater used in artificial streams.	N and P enrichment did not enhance periphyton growth because ambient conc. of ca. 8 μg P liter^{-1} and 500–700 μg N liter^{-1} was already growth-saturating. However, sewage or trace metal mixture did increase biomass.	Wuhrmann and Eichenberger (1975)
Six rivers, Ontario			Surveyed total P, total N, and algal tissue N and P and related to photosynthesis by *Cladophora glomerata*.	*C. glomerata* became N- or P-limited when tissue conc. fell below 1.5% and 0.16%, respectively. Tissue P of 0.16% was obtained when ambient total P = 60 μg liter^{-1}.	Wong and Clark (1976)
Outdoor artificial streams, British Columbia	N, P	P and N+P[c]	Enrichment of streamwater.	N addition from 3.6 to 11 μg liter^{-1} had no effect. P addition from 0.1 to 0.3 μg liter^{-1} increased chl *a* 4.7×. N+P (same conc. changes as when added separately) increased chl *a* 7.9×. Shifts in N:P ratio did not alter relative abundance of major algal groups.	Stockner and Shortreed (1978)
Outdoor artificial streams, Norway			Oligotrophic lake water amended with sewage effluent from different types of treatment plants.	0.5% effluent from sewage treated with $AlSO_4$ and stored in a pond for 10 d (0.7 μg P liter^{-1}) increased gross primary productivity (GPP) 1.35×. 5% effluent after primary treatment (200 μg P liter^{-1}) increased GPP 6×.	Traaen (1978)
Walker Branch, Tennessee	P		Two adjacent reaches enriched to 60 and	Periphyton was P-limited as indicated by increase in ash-free dry mass with enrichment. But P	Elwood *et al.* (1981)

			450 μg P $liter^{-1}$ for 95 days. Upstream control was ca. 4 μg $liter^{-1}$.	limitation of algal periphyton (chl *a*) was equivocal, perhaps because of increased selective grazing on algae in enriched reaches.	
Six streams, Washington	P, NO_3, NH_4	P	Monitored spatial and temporal variation in ambient nutrients in one summer.	Chl *a* accrual rate was greatest at water velocities between 20–50 cm s^{-1} and 40–50 μg P $liter^{-1}$.	Horner and Welch (1981)
Indoor artificial streams	P	P	Nitrogen enrichment to ensure P limitation. Six [P] from 2 to 75 μg $liter^{-1}$.	Ch *a* (biomass primarily from *Mougeotia*) was greatest at 25 μg P $liter^{-1}$.	Horner *et al.* (1983)
Carp Creek, Michigan	N, P	P and N+P	Substratum of agar and sand in petri plates enriched with NO_3 and PO_4. N:P=25:1.	Diatom biovolume increased 2 and 4× with P and N+P enrichment, respectively, compared to the control. N addition alone had no effect.	Pringle and Bowers (1984)
Outdoor artificial streams, Thompson River, British Columbia	P	P	P limitation of periphyton (primarily diatoms) assessed by cellular nutrient ratios, physiological indicators, and growth rates.	Relative growth rate (μ:μ_{max}) was directly related to soluble reactive phosphorus (SRP). At 4–9°C, growth was near maximum when SRP was 3–4 μg $liter^{-1}$.	Bothwell (1985)
Douglas Lake, Michigan	N, P	P, N+P	Clay flowerpots filled with nutrient-enriched agar.	P in flowerpots increased chl *a* 6–10× compared to unenriched control. N enrichment did not have a significant effect. Biomass was greatest with N+P combined. The response to nutrient treatments was species-specific.	Fairchild *et al.* (1985)
Agricultural and forested streams, New York		P	P limitation assessed by alkaline phosphatase activity of microorganisms in stream sediments.	Degree of P limitation was controlled by abiotic sorption by sediments and not the P conc. of the streamwater.	Klotz (1985)

(continues)

TABLE I (continued)

Habitat and location	*Nutrients[b] manipulated*	*Nutrients[b] limiting*	*Method*	*Results*	*Reference*
Kuparuk River, Alaska	P	P	Increased P by 10 μg $liter^{-1}$ for 6 weeks by continuously adding PO_4 to stream.	Chl *a* increased 10× compared to upstream unenriched control. Diatom species richness decreased with enrichment.	Peterson *et al.* (1985)
Sycamore Creek, Arizona	P, NO_3, NO_4	N	Clay flowerpots filled with nutrient-enriched agar, and continuous enrichment of streamwater flowing through artificial channels placed on stream bottom.	In both flowerpot and streamwater enrichment experiments, N stimulated growth whereas P had no effect singly or in combination with N. At the highest ambient NO_3-N conc. (55 μg $liter^{-1}$) algal periphyton were still N-limited.	Grimm and Fisher (1986)
Big Hurricane Branch and Hugh White Creek, North Carolina	N, P, Ca		Clay flowerpots filled with nutrient-enriched agar and placed in streams in clear-cut (high light) and forested (shaded) regions.	Light had more effect on algal biovolume and chl *a* than did nutrients.	Lowe *et al.* (1986)
Indoor artificial streams	P			Filamentous algal growth was not P-limited when SRP was >7 μg $liter^{-1}$.	Seeley (1986), cited in Welch *et al.* (1988)
Keogh River, British Columbia	N+P		Enriched stream with barley and inorganic fertilizer.	Chl *a* accrual rate increased ca. 10× with 200 μg N $liter^{-1}$ and 15 μg P $liter^{-1}$. Barley did not increase benthic algal biomass.	Perrin *et al.* (1987)
Carp Creek, Michigan	N, P, and β-glycerophosphate	P	Nutrient-enriched substrata located in clear cylinders	Benthic algae derived nutrients from both the substratum and overlying water. N+P combined produced the greatest chl *a* increase; N alone had	Pringle (1987)

			submerged in stream and held parallel to water flow. Water flowing through some cylinders was nutrient-enriched.	no effect. β-glycerophosphate enrichment enhanced algal biomass as much as PO_4 enrichment.	
Outdoor artificial streams, Thompson River, British Columbia	P	P	Enrichment with 0–5 μg P $liter^{-1}$ at temperatures ranging from 4° to 18°C. Diatoms dominant.	Growth became saturated at all temperatures when PO_4-P was 0.3–0.6 μg $liter^{-1}$. Discrepancy with Bothwell (1985) attributed to difference between SRP and PO_4.	Bothwell (1988)
Barnwell Creek and Fox Creek, California	N, P	N	Clay saucers filled with nutrient-enriched agar, placed in streams under varying light regimes.	Ambient total inorganic nitrogen and SRP were 0.3–14 and 25–42 μg $liter^{-1}$ respectively. Diatoms dominant. In unshaded reaches (PAR 4.1–20.7 E m^{-2} d^{-1}), N enrichment increased chl *a* 4× compared to control. In natural shade (PAR 0.8–1.5 E m^{-2} d^{-1}), nutrient enrichment had no effect on algal biomass.	Hill and Knight (1988)
Streams in northern U.S.A. and Sweden			Surveyed SRP, total P, NH_4, NO_3, and NO_2 and related to algal biomass.	Benthic algal biomass was not related to N or P conc. Differences in algal biomass among streams attributed to water velocity, light, and grazing.	Welch *et al.* (1988)
Nine rivers in Canterbury, New Zealand			Multiple regression model of algal biomass as a function of hydrogeological variables and N and P levels over a 13-month period.	SRP explained 53% of the variance in chl *a* among rivers. Nearly the same proportion of variance was explained by percentage of time a river was in flood. Cellular nutrient levels and alkaline phosphatase activity suggested algae in some rivers were P-limited.	Biggs and Close (1989)
Outdoor artificial streams, Thompson River, British Columbia	P	P	Enrichment with 0–100 μg P $liter^{-1}$. Diatoms dominant.	Maximum area diatom biomass was attained with 25–50 μg P $liter^{-1}$. 70% of maximum biomass occurred with 1 μg P $liter^{-1}$.	Bothwell (1989)

(continues)

TABLE I (continued)

Habitat and location	*Nutrients[b] manipulated*	*Nutrients[b] limiting*	*Method*	*Results*	*Reference*
Oak Creek, Arizona			Multiple regression model of benthic algal abundance as a function of physicochemical variables over a 14-month period.	Seasonal abundance of diatom species related to light, temperature, and discharge, but not pH or nutrients. NO_3-N, PO_4-P, and SiO_2 averaged 50, 10, and 9400 μg liter^{-1}, respectively, during the study period. Abundance of most chlorophyte species was not related to any of the physicochemical variables.	Duncan and Blinn (1989)
Lake Lacawac, Pennsylvania	C, NH_4, NO_3	C	Clay flowerpots filled with nutrient-enriched agar. P was added in excess to all treatments.	In this soft-water lake, addition of HCO_3 increased periphytic chl *a* >20× and altered species composition. Other manipulations (nitrogen form, grazer density, water depth) had no effect.	Fairchild *et al.* (1989)
Douglas Lake, Michigan	N, P		Clay flowerpots filled with nutrient-enriched agar.	N+P combined increased algal biovolume ca. 5×. Nutrients had a greater effect on community composition (by algal class) than did grazing. Chlorophytes dominated with high N+P.	Marks and Lowe (1989)
Wilson Creek, Kentucky	N, P		Clay dishes filled with nutrient-enriched agar placed in enclosures with varying snail densities.	With grazing absent, P enrichment significantly reduced algal density and biovolume. However, with intermediate snail densities (40–80 m^{-2}), these parameters were greatly enhanced by P compared to control and N treatments.	McCormick and Stevenson (1989)
Indoor artificial streams	P		Streamwater enriched with NO_3 and SRP levels ranging from 5 to ca. 45 μg liter^{-1}. Experiments conducted at 6 water velocities, 10–80 cm s^{-1}.	For all water velocities, maximum algal biomass occurred when SRP was ca. 8 μg liter^{-1}.	Horner *et al.* (1990)

Guadalupe River, Texas	N, P	P	Clay flowerpots filled with nutrient-enriched agar, and enrichment of river water in riverside flow-through troughs.	P increased chl *a*, N had no effect, and N+P combined yielded the greatest chl *a*. The effect of nutrient enrichment varied with season and river site.	Stanley *et al.* (1990)
Middle Bush stream, South Island, New Zealand	N, P, micronutrients	N, P	Plastic cups filled with nutrient-enriched agar, covered with plankton netting to provide artificial substratum.	Both N and P added singly stimulated biomass (chl *a*). N+P combined had a significantly greater effect than either nutrient added singly. The response to nutrient enrichment was usually greater in open-canopy sites.	Winterbourn (1990)
Ford and Peshekee rivers, Upper Peninsula, Michigan	N, P	P, N+P	Glass jars filled with nutrient-enriched agar, capped with 10 μM mesh to provide substratum.	Chl *a* increased with N+P treatment in both hard-water and soft-water rivers, although lesser response in latter suggested carbon limitation. Response to nutrient enrichment was species-specific.	Burton *et al.* (1991)
In situ flow-through troughs, Saline Creek, Missouri	N, P	N	Water in trough was continuously enriched with N, P, or N+P for 30 days.	N and N+P enrichment increased chl *a* 4–6× compared to the control. P alone had only a minor effect. Suggest that [N] ≤100 μg $liter^{-1}$ is growth-limiting. Ca. 50% of the sites in 10 streams in the Ozarks had TN:TP ratios ≤20:1, suggesting N limitation.	Lohman *et al.* (1991)
Outdoor artificial streams, Harts Run, Kentucky	N, P	Not determined	Diatom community development followed for 30 d in control (6 μg N and 4 μg $Pliter^{-1}$ and N+P-enriched (1060 μg N and 286 μg P $liter^{-1}$) streams.	Effect of nutrient enrichment on diatom abundance and community growth rate depended on the successional stage of communities, the effect greatest in early stages. Succession rate was faster with nutrient enrichment.	Stevenson *et al.* (1991)
16 lakes, Sweden, and 18 lakes, Antarctica			Comparison of benthic algal biomass among lakes differing in productivity, light attenuation, temperature, substrata, and food web composition.	In low-productivity lakes (i.e., high light penetration), biomass was directly related to total [P] in the sediment interstitial water. In high-productivity lakes, nutrients were unimportant and light limited biomass. Maximum biomass occurred in intermediate productivity lakes that apparently had the optimum combination of light and nutrients.	Hansson (1992)

(continues)

TABLE I *(continued)*

Habitat and location	*Nutrients[b] manipulated*	*Nutrients[b] limiting*	*Method*	*Results*	*Reference*
Outdoor artificial streams, Wilson Creek, Kentucky	N, P		Enriched water in streams from 3.5 to 30 μg P liter^{-1} and from 110 to 600 μg N liter^{-1} for 12 h at beginning of 6-day experiment.	Short-term nutrient enrichment did not enhance algal abundance or growth rate.	Humphrey and Stevenson (1992)
Walker Branch, Tennessee			Quantified algal biomass, productivity, and nutrient content along longitudinal gradients of SRP and NO_2 + NO_3 in the stream on 7 dates.	Longitudinal SRP gradient (1–3 μg over 140 m) was not reflected by a gradient of biomass or productivity on any particular date. Suspected that intense snail herbivory masked effects of SRP. When all sample dates were combined, productivity was related to SRP (range 1–5 μg liter^{-1}).	Mulholland and Rosemond (1992)
Sycamore Creek, Arizona	N	N	Clay saucers filled with nutrient-enriched agar.	N enrichment altered algal species composition, successional dynamics, and enhanced biomass. Biomass between control and enriched substrata became more similar as succession proceeded.	Peterson and Grimm (1992)
Twelve lakes, Pennsylvania	N, P, C	Varied by lake and algal species	Clay flowerpots filled with nutrient-enriched agar, and comparison of biomass and ambient nutrients among lakes.	Among lakes, pH and alkalinity explained more variation in algal species composition than nutrients. The growth-limiting nutrient varied by species and, for an individual species, by lake. More species were C-limited in acidic lakes.	Fairchild and Sherman (1993)
Flathead Lake, Montana	N, P	N, P, N+P	Clay flowerpots filled with nutrient-enriched agar.	N had a greater effect on algal biovolume and cell density than P, the response varying by species. Bauman (cited in Marks and Lowe, 1993) found that P was limiting periphyton growth in this lake the previous year. Annual variation in nutrient limitation may occur.	Marks and Lowe (1993)

Kuparuk River, Alaska	NH_4, P	P	Increased P by 10 μg $liter^{-1}$ continuously for 6–7 weeks in four consecutive summers. In one summer increased N by 100 μg $liter^{-1}$ continuously for 7 weeks.	In the first two years, P enrichment increased chl *a* ≥10× and algal production doubled the second year. In years 3 and 4, chl *a* differed little between control and enriched reaches because of greater grazing in the enriched reach. Diatom community structure was altered only slightly by P enrichment.	Peterson *et al.* (1993)
Artificial streams and *in situ* experiments, Walker Branch, Tennessee	NO_3, NH_4, P		Ambient and nutrient-enriched streamwater and ambient and reduced grazers.	When grazers were removed and N+P were added (final conc. ca. 40 and 250 μg $liter^{-1}$ respectively), chl *a* and productivity increased ca. 5×. These parameters only doubled if nutrients were added and grazers were not removed. Grazing had a greater effect than nutrients on species composition.	Rosemond *et al.* (1993)
Walker Branch, Tennessee			Multiple regression of snail density and abiotic factors (including SRP, NO_3, and NH_4) to 2-year seasonal variation in algal periphyton biomass and productivity.	Area-specific productivity was related to total inorganic nitrogen. However, algal biomass was not related to nutrients or other abiotic factors. Previous experimental evidence suggested that snail grazing masked effects from seasonal variation of abiotic factors.	Rosemond (1994)

[a] All concentrations refer to μg-atoms of nutrient. Note that this is not a comprehensive list of studies, and only the results pertaining to nutrient limitation of benthic algae are summarized. Some studies had much broader objectives and the reader should not conclude that the results listed were the only findings.

[b] N = nitrogen from NO_3, P = phosphorus from PO_4, C = carbon from HCO_3.

[c] + sign between nutrients (e.g., N+P) means both nutrients added together.

of benthic diatoms in the Thompson River, British Columbia, the same order of magnitude concentration for the growth saturation of planktonic diatoms. Benthic filamentous chlorophytes may require more P than diatoms to saturate growth (Wong and Clark, 1976; Rosemarin, 1982; Seeley, 1986, cited in Welch *et al.*, 1988). Because a developing benthic mat will impede molecular and eddy diffusion of nutrients, a higher P concentration, 25–50 μg PO_4-P liter^{-1}, is needed to attain maximum benthic algal biomass (Horner *et al.*, 1983; Bothwell, 1989). In one study, 8 μg PO_4-P liter^{-1} was sufficient to achieve maximum biomass (Horner *et al.*, 1990).

Nitrogen limitation of benthic algae has been reported when ambient concentrations were 55 μg NO_3-N liter^{-1} in Sycamore Creek, Arizona (Grimm and Fisher, 1986) and 100 μg NO_3-N liter^{-1} in Saline Creek, Missouri (Lohman *et al.*, 1991). These are minimal concentrations, in other words, if the ambient concentration had been higher, a response to N enrichment may have still occurred. Investigations similar to those conducted for phosphorus are needed to determine the nitrogen levels that yield maximum growth rate and biomass.

Whether the growth-limiting nutrient is N or P can be gauged roughly from the ambient N:P atomic ratio (often dissolved inorganic nitrogen to soluble reactive phosphorus). Redfield (1958) proposed for oceanic phytoplankton that the physiological processes of growth were balanced when the cellular carbon, nitrogen, and phosphorus atomic ratio was 106:16:1. The Redfield ratio can be considered a community wide optimum nutrient ratio, and like species-specific optimum ratios, it is growth-rate-dependent (Goldman *et al.*, 1979). Nonetheless, the Redfield ratio provides a benchmark for assessing nutrient limitation, most commonly between N and P. In studies of benthic algae, a range of ambient or cellular N:P ratios has been used as the transition between N and P limitation. Ambient N:P ratios greater than 20:1 are considered P-limited, less than 10:1 are N-limited, and between 10 and 20 to 1 the distinction is equivocal (Schanz and Juon, 1983). Nutrient enrichment studies have corroborated that in broad terms these ratios represent transitions between N and P limitation (e.g., Grimm and Fisher, 1986; Peterson *et al.*, 1993). Nutrient ratios are useful for assessing limitation insofar as ambient concentrations are near growth-limiting levels. When nutrients are in excess, the supply ratio becomes irrelevant. Moreover, because the uptake rates of two nutrients may differ, the ambient ratio may not reflect the cellular ratio relevant to the physiological processes of growth. Hence, cellular ratios should be used when feasible.

C. Overriding Effects of Light, Disturbance, and Grazing

Benthic algal biomass and growth rate do not always relate to nutrient levels. Other factors, such as light, disturbance, and grazing, may be the primary determinants of biomass and growth while nutrients are either replete or secondarily limiting (Table I). Light can be well below saturation intensities under macrophytes and under riparian canopies. If there is insufficient light, nutrient enrichment will have little or no effect on growth (Hill and Knight, 1988; Winterbourn, 1990; Triska *et al.*, 1983). In some habitats, the spatial and seasonal variation in benthic algal biomass is due, in part, to variation in light, not nutrients (Welch *et al.*, 1988; Duncan and Blinn, 1989; Hansson, 1992). Disturbance from the shearing forces of fast-moving water is a recurring event for many benthic algal communities. In multiple regression models, parameters that reflect the degree and frequency of disturbance, such as water velocity, discharge, and number and duration of floods, explain as much or more of the variation in benthic algal biomass as compared to nutrients (Welch *et al.*, 1988; Biggs and Close, 1989; Duncan and Blinn, 1989). Grazing can be the major factor controlling accumulation of benthic algae in some systems (top-down control). If nutrients are limiting and enrichment occurs, grazing can offset or lessen the increase in biomass (Mundie *et al.*, 1991; Hill *et al.*, 1992; Rosemond *et al.*, 1993), sometimes years later after grazer densities have increased in response to the greater availability of algal food resources (Peterson *et al.*, 1993). Hence, nutrient limitation, if it is judged by changes in biomass or areal-specific productivity, can be masked by grazing (Rosemond, 1994). Specific growth rate (μ) may be the least sensitive growth parameter to the direct effects of grazing, although indirect effects like nutrient regeneration and release of inhibitory compounds (Hill *et al.*, 1992) will affect it as well.

A strong relationship exists between nutrient levels, particularly phosphorus, and phytoplankton biomass (e.g., Vollenweider, 1976). The relationship between nutrients and benthic algal biomass is weaker because of the effects of light, disturbance, and grazing (Cattaneo, 1987). With regard to nutrients, this perhaps is the most significant difference between planktonic and benthic algae.

D. Nutrients and Species Composition

Despite myriad nutrient enrichment studies, the relationship of nutrients to benthic algal community structure is not well understood. Sometimes a change in nutrient levels does not alter algal species composition (Shortreed *et al.*, 1984; Lohman *et al.*, 1991; Peterson *et al.*, 1993), other times it does (e.g., Burton *et al.*, 1991). The response of the same species to seemingly similar nutrient enrichment conditions can differ among studies (cf. *Nitzschia palea* in the studies by Hill and Knight, 1988, and Marks and Lowe, 1993), suggesting that factors other than nutrients are more important in determining species composition. Despite the conflicting information, a few generalizations can be made. Filamentous chlorophytes, like *Cladophora glomerata* and *Stigeoclonium tenue,* become abundant when N and P levels are relatively high and there is sufficient sunlight. The latter taxon is frequently found in hypereutrophic habitats (Davis *et al.*, 1990). *Achnanthes minutissima,* a diatom, appears to prefer a nitrogen- and/or phosphorus-enriched environment (Pringle and Bowers, 1984; Fairchild *et al.*, 1985; McCormick and Stevenson, 1989; Stevenson *et al.*, 1991; Peterson and Grimm, 1992). Nitrogen limitation tends to favor diatoms in the Epithemiaceae (e.g., *Rhopalodia gibba, Epithemia* spp.) (Fairchild *et al.*, 1985; Hill and Knight, 1988; Peterson and Grimm, 1992), which have cyanobacterial endosymbionts capable, presumably, of fixing atmospheric nitrogen (DeYoe *et al.*, 1992, and citations therein). With phytoplankton, phosphorus enrichment predictably leads to a community dominated by nitrogen-fixing cyanobacteria. However, in benthic algal communities the effects of P enrichment on species composition are less predictable; sometimes it leads to dominance by cyanobacteria (Elwood *et al.*, 1981; Fairchild *et al.*, 1985), and other times it does not (Stockner and Shortreed, 1978; Hill and Knight, 1988; Lohman *et al.*, 1991).

The method commonly used to examine the relationship of nutrients to benthic algal community structure, nutrient-enriched agar in clay flowerpots or petri plates, has some serious shortcomings. The method is fine for identifying nutrients that are growth-limiting for the entire algal community. But the nutrients are provided in such great excess and at unknown concentrations at the juncture of mat and substratum that it is difficult to partition species along nutrient supply ratios. Moreover, ascribing nutrient limitation to a particular taxon is equivocal, because interactions among species (e.g., competition, commensalism) can increase or decrease a species abundance, regardless of whether it was nutrient-limited (Fairchild and Sherman, 1993). To achieve greater understanding on the role of nutrients in structuring benthic algal communities, it will be necessary to move on to other approaches. Nutrient kinetics has worked well for phytoplankton ecologists and will likely be fruitful for benthic algal ecologists. Multivariate statistical techniques are equally promising, as demonstrated in the studies by Carrick *et al.* (1988) and Fairchild and Sherman (1993).

V. NUTRIENT KINETICS OF BENTHIC ALGAE

Only a handful of studies have used the nutrient kinetics conceptual framework for examining the nutrient requirements of benthic algae. Of these, several will be briefly reviewed to emphasize the applicability and advantages of using a nutrient kinetic approach with benthic algae.

Auer and Canale (1982a,b) successfully developed a model for predicting the effectiveness of various effluent management programs in reducing the growth of *Cladophora glomerata* in Lake Huron. Environmental factors (light, temperature, and phosphorus) were related to the growth response of the alga. Phosphorus uptake was described by the Michaelis–Menten equation (Auer and Canale, 1982a) and the resultant intracellular phosphorus pool was translated to specific growth rate with the Droop equation (Auer and Canale, 1982b). Knowledge of these relationships enhanced the predictive power of the general model. Because the maximum P cell quota was approximately four times greater than the Q^P at which growth limitation occurred, for those field populations of *Cladophora* near Q^P_{max}, small reductions in P loading were predicted to have no effect on the alga's growth.

Rosemarin (1982) measured K_s, ρ_{max}, and Q_0 of phosphorus for *Cladophora* and *Stigeoclonium tenue* and then, assuming steady state, calculated K_μ for each alga. From the derived Monod curves, *Stigeoclonium* was predicted to outcompete *Cladophora* for P, which may be true, but did not correspond to the abundance patterns of the algae in the littoral of the Great Lakes. Hoffmann (1990) found that photosynthesis and yield of *Cladophora* were related to cellular vitamin B_{12} by the Droop equation; dependence of these processes on external B_{12} was modeled by a hyperbolic tangent rather than the Monod equation. Lohman and Priscu (1992) inferred that *Cladophora* in the Clark Fork of the Columbia River, Montana, was nitrogen-limited based on seasonal patterns of K_s, and cellular N:P ratios.

Bothwell (1985, 1988, 1989) used a nutrient kinetic approach to examine phosphorus limitation of benthic diatoms in the Thompson River system of British Columbia. Before these studies, the reported soluble reactive phosphorus (SRP) values for saturating lotic diatom growth were 10 to 20 times higher than those required by planktonic diatoms. However, when lotic diatoms were dispersed and suspended in water and P uptake was measured, the Michaelis–Menten parameters indicated that the affinity of lotic diatoms for P was similar to that of phytoplankton (Bothwell, 1985). Considering growth kinetics next, a series of flow-through troughs were built adjacent to the river at three sites with different levels of SRP. The measured specific growth rates did not always correspond to SRP. However, if μ was normalized by μ_{max} as estimated by the Droop equation (for *P*, μ'_{max} and μ_{max} are similar), this compensated for any differences in physical factors among sites (Goldman, 1980) and $\mu:\mu_{max}$ was correlated to SRP. Combined with other lines of kinetic-based evidence (e.g., cellular nutrient ratios), Bothwell (1985) convincingly argued that benthic diatoms in the Thompson River were P-limited. In a later set of experiments in which the river water was amended with P, μ_{max} was estimated with the Monod equation and the relative specific growth rates ($\mu:\mu_{max}$) were used to find the inorganic phosphorus concentration that saturated growth, independent of temperature (Bothwell, 1988). Similar to planktonic diatoms, growth of lotic diatoms was saturated at less than 1 μg P liter^{-1}, although to achieve maximum areal biomass, 50 μg P liter^{-1} was required (Bothwell, 1989).

Biggs (1990) used relative specific growth rates, which controlled for seasonal temperature-dependent differences in μ, to compare benthic diatom growth upstream and downstream of a wastewater effluent pipe in the South Brook, Canterbury, New Zealand. Based on $\mu:\mu_{max}$ ratios near 1, high P cell quotas, low P uptake rates, and low alkaline phosphatase activity for the algae at the upstream site, Biggs (1990) concluded that the algae in the stream were already nutrient-saturated and the effluent from the wastewater treatment plant would not appreciably alter algal growth.

Borchardt *et al.* (1994) elucidated the effects of flow velocity on the P uptake kinetics of *Spirogyra fluviatilis.* The alga was cultured in laboratory streams with flow velocities ranging from 3 to 30 cm s^{-1}. When the alga was P-replete ($Q_P \geq 0.21\%$), flow velocity had no effect on P uptake. When the alga was P-deficient ($0.06\% \leq Q^P \leq 0.18\%$), P uptake was enhanced by flow velocities up to 15 cm s^{-1}, but at higher velocities, 22 and 30 cm s^{-1}, uptake was reduced by 12 to 26% relative to uptake at the optimum flow velocities. This occurred for every P pulse concentration tested. Maximum uptake (ρ_{max}) was similarly affected.

In a companion study, Borchardt (1994) used the Droop equation to describe nitrogen- and phosphorus-limited growth by *Spirogyra fluviatilis*

TABLE II Nutrient Uptake Parameters for Freshwater Benthic Algae following Michaelis–Menten-Type Kinetics

Species	*Nutrient*	K_s *(μM atoms nutrient)*	ρ_{max} *(μg atoms nutrient g dry wt*$^{-1}$ *hr*$^{-1}$*)*	*Reference*	*Note*[a]
Spirogyra fluviatilis	PO_4	0.3–1.5	322–803	Borchardt *et al.* (1994)	A, B
Stigeoclonium tenue	PO_4	3.0	4320	Rosemarin (1982)	C
Periphytic diatoms	PO_4	0.02–0.2	408–4082	Bothwell (1985)	A, D
Cladophora glomerata	PO_4	1–8.1	167–1875	Auer and Canale (1982a)	A
Cladophora glomerata	PO_4	1.0	210	Rosemarin (1982)	C
Cladophora glomerata	PO_4	0.5–2.8	62–166	Lohman and Priscu (1992)	A
Cladophora glomerata	NO_3	7.3–15.2	0–984	Lohman and Priscu (1992)	A
Cladophora glomerata	NH_4	17.4–41.9	1137–6991	Lohman and Priscu (1992)	A
Cladophora glomerata	B_{12}	—	0.1	Hoffman (1990)	E

[a](A) Range reported for different degrees of nutrient limitation. (B) Biphasic uptake. (C) Converted to dry weight from mass carbon assuming carbon content of 36%. (D) Converted to dry weight from mass chl *a* using average C : chl *a* ratio (w/w) of 97 (Bothwell, 1985) and assuming carbon content of 36%. (E) Empirically derived maximum uptake rate.

in water flowing at 3, 12, and 30 cm s^{-1}. For both nutrients, the minimum cell quotas required for growth (Q_0) generally increased with flow velocity, and μ'_{max} appeared to be greatest at the intermediate velocity of 12 cm s^{-1}. When the alga was either N- or P-limited, net photosynthesis was inversely related to flow velocity. The Droop model controlled for differences in N and P uptake at the different velocities and made apparent the metabolic costs *Spirogyra fluviatilis* must incur when growing in flowing water. Unless the extra requirement for N and P can be offset with enhanced nutrient uptake, which occurred under some nutrient and flow conditions (Borchardt *et al.*, 1994), the alga's growth rate would diminish the faster the flow velocity.

Thus far in the studies of the nutrient and growth kinetics of benthic algae there have not been any inconsistencies with the theory developed using phytoplankton. The uptake and growth parameters reported for benthic algae (Tables II and III), as few as there are, are similar to the range of values reported for phytoplankton (cf. Cembella *et al.*, 1984; Button, 1985). It is likely that the parameters depend more on cell size and shape and phy-

TABLE III Growth Kinetic Parameters for Freshwater Benthic Algae

Species	*Nutrient*	*Q_0 (%)*	*Maximum μ (day^{-1})*	*Reference*	*Notes*[a]
Spirogyra fluviatilis	N	1.41–1.81	3.1–3.8 (μ'_{max})	Borchardt (1994)	A, B
Cladophora glomerata	B_{12}	7×10^{-6}	0.4 (μ'_{max})	Hoffman (1990)	C
Cladophora glomerata	P	0.06	0.714 (μ'_{max})	Auer and Canale (1982b)	—
Cladophora glomerata	P	0.4	0.8	Rosemarin (1982)	D, E
Stigeoclonium tenue	P	0.3	2.0	Rosemarin (1982)	D, E
Spirogyra fluviatilis	P	0.062–0.077	2.6–2.7 (μ'_{max})	Borchardt (1994)	A, B
Periphytic diatoms	P	—	0.12–0.47 (μ'_{max})	Bothwell (1988)	F

[a](A) Net O_2 evolution rate converted to specific growth rate assuming a photosynthetic quotient of 1 and carbon content of 36%. (B) Range in μ'_{max} measured at flow velocities of 3, 12, and 30 cm s^{-1}. (C) Carbon fixation converted to specific growth rate assuming carbon content of 36%. (D) Cell quota normalized by mass carbon converted to percentage by assuming carbon content of 36%. (E) Maximum growth rate empirically derived from nutrient-replete batch cultures. (F) Seasonal range in maximum growth rate calculated from the Monod equation and empirically derived by nutrient enrichment. K_m ranged from 0.15 to 0.3 μg PO_4-P liter^{-1}.

logenetic affinities than on whether the alga is planktonic or benthic. The effects of light and temperature on nutrient uptake and nutrient-limited growth, as reflected in the parameters, will probably be similarly varied for benthic algae. Information on this is lacking but may be crucial for understanding benthic algal community structure. The principal caveat in applying nutrient kinetic theory to benthic algae is that solutions from algebraic equations (e.g., $\rho = \mu Q$) are only valid if steady state is achieved. As demonstrated by Grover (1991b), and by inference from the excellent fit of the models used in the benthic algal studies cited earlier, growth with gradual changes in substrate concentration or cell quota can be modeled by the Monod or Droop equations. But one must not be too quick to assume steady state and derive relationships that, because conditions were nonequilibrous, are spurious. With that caveat in mind, nutrient kinetic models should provide a well-founded conceptual framework for taking the next steps in understanding the role of nutrients in benthic algal ecology.

VI. EFFECTS OF WATER MOTION ON BENTHIC ALGAL NUTRIENT UPTAKE AND NUTRIENT-LIMITED GROWTH

A. Positive Effects of Water Motion

The first quantitative studies on the effects of water motion on algal nutrient uptake were conducted by Whitford and Schumacher (1961) over 30 years ago. They showed that moving water could greatly enhance P uptake for a number of algal taxa. Remarkably, compared to still water, a flow velocity of 18 cm s^{-1} enhanced P uptake 150 to 4242% depending on the alga tested (Schumacher and Whitford, 1965). Subsequent studies with other algae differing widely in phylogeny, habitat, morphology, and composition of the algal community have confirmed that moving water can enhance algal uptake of phosphorus (Sperling and Grunewald, 1969; Canelli and Fuhs, 1976; Lock and John, 1979; Riber and Wetzel, 1987; Horner *et al.,* 1990), nitrogen (Parker, 1981; Gerard, 1982), and carbon (Sperling and Hale, 1973; Rodgers and Harvey, 1976; Thirb and Benson-Evans, 1982; Dodds, 1989). The mechanism for this phenomenon, first suggested for the algae by Munk and Riley (1952), is that moving water decreases the laminar layer around the algae, thereby increasing nutrient diffusion. As it pertains to the algae, the mechanism has received some theoretical and experimental attention (Pasciak and Gavis, 1974, 1975; Gavis, 1976; Riber and Wetzel, 1987; Wheeler, 1988). For benthic algae, perhaps a greater frequency of splat events with faster flow velocities is as important in contributing to enhanced nutrient uptake.

Flowing water has an effect on algal nutrient uptake only to the extent that a nutrient concentration gradient exists between the periphyton mat and bulkwater. A gradient exists because nutrients are removed from solution via periphyton uptake. Thus, whatever factors affect uptake ultimately affect the impact of flowing water. Areal nutrient uptake rates and the steepness of the concentration gradient are directly related to algal density, and Stevenson and Glover (1993) found that flowing water had a greater enhancement effect on NO_3 flux through periphyton as periphyton density increased. In contrast, chloride is not removed by uptake and its flux through periphyton decreased at high periphyton densities, probably because of the tortuosity of thick periphyton mats (Stevenson and Glover, 1993). Because nutrient uptake rates generally vary inversely with nutrient cell quota, flowing water will have little effect on uptake when benthic algae are nutrient-replete (Borchardt *et al.*, 1994).

Enhanced nutrient uptake is the most likely explanation for the increase in algal production often noted in habitats with moving water (waste removal is an alternative hypothesis that remains untested, although see Gonen *et al.* 1995). This observation has a long history, beginning with Ruttner's metaphor in 1926 that water is "physiologically richer" when flowing than when quiescent (Ruttner, 1953). Experiments have confirmed that benthic algal communities grow faster and can accumulate more biomass the faster the water velocity, unless the force of moving water is too great and algae are sheared from the substratum (McIntire, 1966; Pfeifer and McDiffett, 1975; Horner and Welch, 1981; Horner *et al.*, 1990; Carpenter *et al.*, 1991). That water motion, through enhanced nutrient uptake, is beneficial to benthic algal growth is considered by many to be axiomatic.

B. Negative Effects of Water Motion

Many of the studies on benthic algal nutrient uptake in flowing water have been at the community level. The autecological studies by Whitford and Schumacher, novel and informative for their time, did not control for feedback effects via intracellular nutrient stores and compared nutrient uptake under the worst conditions, still water, to uptake at a modest flow velocity. Though it probably is axiomatic that flowing water is beneficial to the nutrient-limited growth of benthic algal communities, too few benthic algal species have been examined, often with a small range of flow velocities and an incomplete characterization of nutrient uptake and growth, for the generalization to be made across taxa. Indeed for some taxa, above a certain threshold flow velocity, nutrient uptake and growth are impaired. Ammonium uptake by *Ulva lactuca* was reduced at some irradiances when flow velocities were equal or greater than 15 cm s^{-1} (Parker, 1981). Phosphorus uptake by *Spirogyra fluviatilis* growing at flow velocities of 22 and 30 cm s^{-1} was reduced by 12 to 26% relative to uptake at 12 and 15 cm s^{-1} (Borchardt *et al.*, 1994). When *Spirogyra fluviatilis* was nitrogen- or phosphorus-limited, net photosynthesis, as described by the Droop equation, was highest at 3 cm s^{-1} and lowest at 30 cm s^{-1} (Borchardt,

1994). Similar deleterious effects of water motion on algal growth and carbon fixation have been reported (Gerard and Mann, 1979; Antoine and Benson-Evans, 1982; Dodds, 1991; Berdalet, 1992; Humphrey and Stevenson, 1992).

From the perspective of a benthic algal mat, and for those individuals embedded within, the influx of nutrients carried in by flowing water is advantageous. But from the perspective of particular benthic algal species, the velocity of water may not be optimum for the procurement and utilization of nutrients and, perhaps, ultimately the success of the species in that habitat. If benthic algal species differ in their nutrient uptake and growth response to flow, this could be one reason why some taxa are partitioned along a continuum of velocities in lotic habitats (Traaen and Lindstrøm, 1983; Lindstrøm and Traaen, 1984).

C. Interaction of Flow Velocity and Nutrient Concentration

The interaction of flow velocity and nutrient concentration on benthic algal growth is often examined by treating these parameters separately. A simple alternative approach is the concept of mass transport. Nutrient mass transport is the product of discharge (volume/time) and nutrient concentration (mass/volume). It indicates the mass of nutrient moving past a given point per unit time and, at a convective scale, encapsulates in one term the interaction of flow velocity and nutrient concentration. Two sets of data suggest that the concept may apply to benthic algae in unidirectional flow: (1) At a very low, relatively invariate phosphorus concentration that was identical among flow velocity treatments, P uptake by *Spirogyra fluviatilis* was approximately proportional to discharge (Borchardt *et al.*, 1994). (2) Benthic diatoms growing in stream microcosms with different concentrations of phosphorus (0.1, 0.2, 0.4, 1.2, and 6 μg P liter^{-1}) had similar growth rates because the SRP hourly mass transport was manipulated so that it was constant among streams (M. Bothwell, unpublished data). Under the conditions used in these studies, mass transport rather than P concentration was a better predictor of uptake or growth. The usefulness of the mass-transport concept for exploring the mechanisms and consequences of flow-mediated nutrient uptake needs to be considered.

VII. NUTRIENT COMPETITION

A. Characterizing Nutrient Competitive Ability

Given that in many aquatic systems algal growth is restricted by a shortage of one or more nutrients, when two or more algal species are limited by the same nutrient, competition ensues. The species with the greatest ability to acquire and utilize the nutrient for growth will produce the greatest biomass and may, given enough time, eliminate its competitors from the system. Competitive ability for nutrients stems from several species-specific attributes that range from enzyme affinity to changes in morphology or life cycle (Turpin, 1988). Physiological attributes are used frequently to characterize competitive ability and have been used successfully to predict the outcome of competitive interactions between algae for a nutrient. The attributes can be generally categorized as nutrient uptake, efflux, storage, and utilization for growth.

Nutrient uptake ability is often summarized by the parameter affinity, which, for uptake that follows Michaelis–Menten kinetics, is the initial slope of the ρ versus S curve calculated by ρ_{max}/K_m (Healey, 1980). For general transport kinetics, regardless of the model followed, Button (1985) presents a thorough historical and theoretical treatment of the affinity concept. Suffice to note that algae with high nutrient affinity will procure more nutrients per unit time than algae with low affinity.

However, high affinity is of little benefit if the nutrient is not retained in the cell. Algae differ in their nutrient efflux rates (Borchardt *et al.*, 1994, and citations therein); some species readily lose some nutrients from intracellular pools. A competitively inferior alga in terms of nutrient affinity may, nevertheless, be the competitive dominant, if by its low nutrient efflux rate it sequesters more growth-limiting nutrient than its "leaky" competitors (e.g., Olsen, 1989, Olsen *et al.*, 1989).

Another physiological attribute that may lead to competitive success is the ability to store a growth-limiting nutrient. All other things being equal, an alga with a high storage capacity (Q_{max}/Q_0) for a growth-limiting nutrient has a greater chance of survival between periods when the nutrient is unavailable than an alga with lower capacity. Large algal cells with high storage capacities may be superior competitors in habitats with infrequent but large pulses of nutrients (Malone, 1980).

In comparing the efficiency of nutrient utilization for growth among algae, two basic approaches are used depending on whether growth is expressed as a function of external or intracellular nutrients. When based on external nutrients, as with the Monod model, for a given growth rate the alga that requires the least amount of nutrient will have the lowest steady-state ambient nutrient concentration (Tilman, 1981). This concentration is denoted R^* and is obtained by solving for S in the Monod equation. Alternatively, the initial slope of the Monod curve, as calculated by μ_{max}/K_μ, can be used to compare competitive abilities among species (Healey, 1980). When based on intracellular nutrients, as with the Droop model, comparisons can be made using the yield coefficient, calculated as $1/Q$ and interpreted as the biomass produced per unit nutrient. For a given growth rate, the alga with the highest yield coefficient has the greatest growth efficiency.

Competitive ability for a nutrient cannot be summarized or predicted from a single kinetic parameter. All steps of the process, namely, uptake, assimilation, and growth must be characterized. Moreover, the set of kinetic parameters considered competitively superior depends on the nutrient supply regime. For example, under a low and constant nutrient supply, rapid uptake would be advantageous, whereas with infrequent but large nutrient perturbations, high storage capacity would be advantageous. Moreover, there is no one set of optimum physiological traits that will allow one species to always outcompete all others under all nutrient supply regimes; there appear to be physiological trade-offs. Maximum uptake (ρ_{max}) may be inversely related to Q_{max} (Grover, 1991a) and/or the rate of assimilation (Turpin, 1988). Nor does the ability to procure one nutrient relate to the ability to procure a different nutrient. For example, freshwater diatoms that are excellent competitors for silicate are poor competitors for phosphorus and vice versa (Sommer, 1988).

B. Predicting Competition with Optimum Ratios

Optimum nutrient ratios indicate which nutrient is growth-limiting and hence, insofar as steady-state kinetics are applicable, can be used to predict whether two algal species will compete for the same nutrient or whether their growth is limited by different nutrients and coexistence is possible. This concept is succinctly portrayed in Fig. 4. Two hypothetical algal species, A and B, have different optimum ratios (R_c) for resources R_i and R_j. When μ'_{max} calculated from the Droop model for nutrient i is greater than μ'_{max} for nutrient j, it has been shown that R_c (Q_i/Q_j, each cell quota based on the same growth rate) is inversely related to growth rate (Terry *et al.*, 1985; Turpin, 1986; Borchardt, 1994). This relationship can be superimposed on a R_i/R_j supply ratio continuum (Fig. 4). If the relationship between the optimum ratio and growth rate differs between species A and B, the R_c curves will cross and zones of growth limitation by different nutrients will be established. In zone 1, the R_i/R_j supply ratio is higher than the R_i/R_j optimum ratio for both species; in other words, more R_i is supplied than what is needed by the algae, and both species are limited by R_j. In zone 2, the R_i/R_j supply ratio is greater than R_c for species A but less than R_c for species B, so that the algae are limited by different nutrients, R_j and R_i, respectively. If the supply ratio was very low, both species would be limited by R_i (zone 3). Zone 4 is the converse of zone 2; the supply ratio is less than optimum for species A (R_i limiting) and greater than optimum for species B (R_j limiting). If the R_i/R_j supply ratio was identical to R_c for one of the algae, the growth requirements for R_i and R_j would be perfectly balanced. In the zone where each species is limited by the nutrient for which it is the superior competitor (either zone 2 or 4), stable coexistence is possible (Tilman, 1982). Along the continuum of possi-

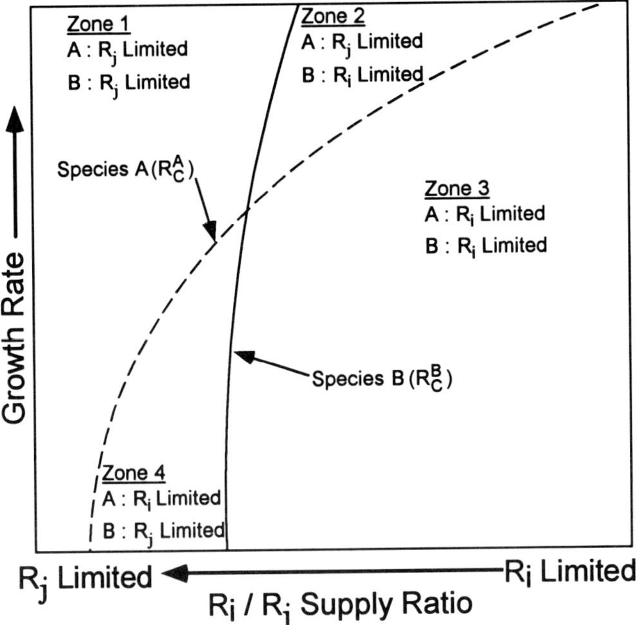

FIGURE 4 Optimum ratios (R_c) and zones of nutrient limitation for hypothetical species A and B for two resources, R_i and R_j. The optimum ratios curve because they are growth-rate-dependent. The zones of limitation are established along the R_i/R_j supply ratio, and the resource limiting each species is given for each of the four zones. Modified from Turpin, D. H. (1988) Physiological mechanisms in phytoplankton resource competition. *In* "Growth and Reproductive Strategies of Freshwater Phytoplankton" (C. Sandgren, ed.), pp. 316–368. Reprinted with permission of Cambridge University Press.

ble supply ratios of two nutrients, there is thought to be a corresponding continuum of algal species each with a species-specific optimum ratio matching some point on the supply ratio continuum [Rhee and Gotham (1980); Tilman, *et al.*, (1982); for an optimum ratio continuum see Sommer (1988)]. For thorough coverage of the theory of resource ratio partitioning, the reader should consult Tilman (1982) and, as it applies specifically to algae, Turpin (1988).

C. Effects of Water Motion on Nutrient Competition

One potentially important factor in nutrient competition among benthic algae may be water motion. Optimum nutrient ratios of algae are known to vary with irradiance, light quality, and temperature (Healey, 1985; Wynne and Rhee, 1986; van Donk and Kilham, 1990). Presumably, insofar as these abiotic factors vary on a spatial and temporal scale relevant to the size, movement, and generation time of algae, greater partitioning of nutrient resources among species may occur than what is predicted by invariant optimum ratios alone. Given that flowing water alters rates of nutrient uptake and growth, it may not be unreasonable to expect that optimum nutrient ratios are also affected. Borchardt (1994) examined the optimum N:P ratio of *Spirogyra fluviatilis* for net photosynthesis (PS_{net}) at three flow velocities: 3, 12, and 30 cm s^{-1}. The optimum ratio was similar among velocities at high photosynthetic rates, but when PS_{net} was less than half of maximum, $R_c^{N:P}$ was higher at 12 cm s^{-1} than at the other two velocities. The maximum increase in $R_c^{N:P}$ was 16% between 3 and 12 cm s^{-1}. Compared to 20–100% changes in the optimum N:P ratio attributed to irradiance and light quality (Healey, 1985; Wynne and Rhee, 1986), flow velocity had a minor effect. However, at present, the magnitude of change in an optimum ratio that is ecologically significant is unknown. Combined with its effects on growth and nutrient uptake, water motion may be important in mediating nutrient partitioning among benthic algae.

D. Nutrient Competition among Benthic Algae

For all of the emphasis given to nutrient competition in structuring phytoplankton communities, what do we know about the nutrient competitive abilities and optimum ratios of benthic algae? Very little. Perhaps this reflects the habitat in which many benthic algae have been studied—streams—and the importance of light, grazing, and disturbance placed on structuring lotic algal communities. There is some evidence that benthic algae can be partitioned along a nutrient ratio continuum (Fairchild *et al.*, 1985; Carrick *et al.*, 1988; Stevenson *et al.*, 1991). Certainly, a nutrient kinetic approach is feasible when studying nutrient competition among benthic algae. The characterization of competitive ability must also include the type of substratum, whether the principal source of nutrients is the overlying water or substratum beneath the algae, and the potential for motility that would allow an alga to move closer to an area with more nutrients (Pringle, 1990). With these caveats in mind, application of nutrient kinetic theory to benthic algae should provide the least ambiguous assessment of the importance of nutrient competition in structuring benthic algal communities.

REFERENCES

Ahlgren, G. (1987). Temperature functions in biology and their application to algal growth constants. *Oikos* **49**, 177–190.

Ahlgren, G. (1988). Phosphorus as growth-regulating factor relative to other environmental factors in cultured algae. *Hydrobiologia* **170**, 191–210.

Antoine, S. E., and Benson-Evans, K. (1982). The effect of current velocity on the rate of growth of benthic algal communities. *Int. Rev. Gesamten Hydrobiol.* **67**, 575–583.

Auer, M. T., and Canale, R. P. (1982a). Ecological studies and mathematical modeling of

Cladophora in Lake Huron. 2. Phosphorus uptake kinetics. *J. Great Lakes Res.* **8**, 84–92.

Auer, M. T., and Canale, R. P. (1982b). Ecological studies and mathematical modeling of *Cladophora* in Lake Huron. 3. The dependence of growth rates on internal phosphorus pool size. *J. Great Lakes Res.* **8**, 93–99.

Berdalet, E. (1992). Effects of turbulence on the marine dinoflagellate *Gymnodinium nelsonii. J. Phycol.* **28**, 267–272.

Biggs, B. J. F. (1990). Use of relative specific growth rates of periphytic diatoms to assess enrichment of a stream. *N. Z. J. Mar. Freshwater Res.* **24**, 9–18.

Biggs, B. J. F., and Close, M. E. (1989). Periphyton biomass dynamics in gravel bed rivers: The relative effects of flows and nutrients. *Freshwater Biol.* **22**, 209–231.

Borchardt, M. A. (1991). Phosphorus uptake kinetics of *Spirogyra fluviatilis* Hilse (Charophyceae) in flowing water. Ph.D. Thesis, University of Vermont.

Borchardt, M. A. (1994). Effects of flowing water on nitrogen- and phosphorus-limited photosynthesis and optimum N:P ratios by *Spirogyra fluviatilis* (Charophyceae). *J. Phycol.* **30**, 418–430.

Borchardt, M. A., Hoffmann, J. P., and Cook, P. W. (1994). Phosphorus uptake kinetics of *Spirogyra fluviatilis* (Charophyceae) in flowing water. *J. Phycol.* **30**, 403–417.

Bothwell, M. L. (1985). Phosphorus limitation of lotic periphyton growth rates: An intersite comparison using continuous-flow troughs (Thompson River system, British Columbia). *Limnol. Oceanogr.* **30**, 527–542.

Bothwell, M. L. (1988). Growth rate responses of lotic periphytic diatoms to experimental phosphorus enrichment: The influence of temperature and light. *Can. J. Fish. Aquat. Sci.* **45**, 261–270.

Bothwell, M. L. (1989). Phosphorus-limited growth dynamics of lotic periphytic diatom communities: Areal biomass and cellular growth rate responses. *Can. J. Fish. Aquat. Sci.* **46**, 1293–1301.

Boudreau, B. P. (1988). Mass-transport constraints on the growth of discoidal ferromanganese nodules. *Am. J. Sci.* **288**, 777–797.

Burkholder, J. M., Wetzel, R. G., and Klomparens, K. L. (1990). Direct comparison of phosphate uptake by adnate and loosely attached microalgae within an intact biofilm matrix. *Appl. Environ. Microbiol.* **56**, 2882–2890.

Burmaster, D. (1979). The continuous culture of phytoplankton: Mathematical equivalence among three steady-state models. *Am. Nat.* **113**, 123–134.

Burmaster, D. E., and Chisholm, S. W. (1979). A comparison of two methods for measuring phosphate uptake by *Monochrysis lutheri* Droop grown in continuous culture. *J. Exp. Mar. Biol. Ecol.* **39**, 187–202.

Burton, T. M., Oemke, M. P., and Molloy, J. M. (1991). Contrasting effects of nitrogen and phosphorus additions on epilithic algae in a hardwater and a softwater stream in northern Michigan. *Verh.—Int. Ver. Theor. Angew. Limnol.* **24**, 1644–1653.

Button, D. K. (1985). Kinetics of nutrient-limited transport and microbial growth. *Microbiol. Rev.* **49**, 270–297.

Canelli, E., and Fuhs, G. W. (1976). Effect of the sinking rate of two diatoms (*Thalassiosira* spp.) on uptake from low concentrations of phosphate. *J. Phycol.* **12**, 93–99.

Caperon, J. (1967). Population growth in micro-organisms limited by food supply. *Ecology* **48**, 715–722.

Caperon, J. (1968). Population growth response of *Isochrysis galbana* to nitrate variation at limiting concentrations. *Ecology* **49**, 866–872.

Carling, P. A. (1992). The nature of the fluid boundary layer and the selection of parameters for benthic ecology. *Freshwater Biol.* **28**, 273–284.

Carpenter, R. C., Hackney, J. M., and Adey, W. H. (1991). Measurements of primary productivity and nitrogenase activity of coral reef algae in a chamber incorporating oscillatory flow. *Limnol. Oceanogr.* **36**, 40–49.

Carrick, H. J., Lowe, R. L., and Rotenberry, J. T. (1988). Guilds of benthic algae along nutrient gradients: Relationships to algal community diversity. *J. North Am. Benthol. Soc.* **7,** 117–128.

Cattaneo, A. (1987). Periphyton in lakes of different trophy. *Can. J. Fish. Aquat. Sci.* **44,** 296–303.

Cembella, A. D., Antia, N. J., and Harrison, P. J. (1984). The utilization of inorganic and organic phosphorous compounds as nutrients by eukaryotic microalgae: A multidisciplinary perspective. Part 1. *CRC Crit. Rev. Microbiol.* **10,** 317–391.

Dade, W. B. (1993). Near-bed turbulence and hydrodynamic control of diffusional mass transfer at the sea floor. *Limnol. Oceanogr.* **38,** 52–69.

Davies, A. G. (1970). Iron, chelation and the growth of marine phytoplankton. I. Growth kinetics and chlorophyll production in cultures of the euryhaline flagellate *Dunalliela tertiolecta* under iron-limiting conditions. *J. Mar. Biol. Assoc. U. K.* **50,** 65–68.

Davis, C. O., Breitner, N. F., and Harrison, P. J. (1978). Continuous culture of marine diatoms under silicon limitation. 3. A model of Si-limited diatom growth. *Limnol. Oceanogr.* **23,** 41–52.

Davis, L. S., Hoffmann, J. P., and Cook, P. W. (1990). Seasonal succession of algal periphyton from a wastewater treatment facility. *J. Phycol.* **26,** 611–617.

DeYoe, H. R., Lowe, R. L., and Marks, J. C. (1992). Effects of nitrogen and phosphorus on the endosymbiont load of *Rhopalodia gibba* and *Epithemia turgida* (Bacillariophyceae). *J. Phycol.* **28,** 773–777.

Dillon, P. J., and Kirchner, W. B. (1975). The effects of geology and land use on the export of phosphorus from watersheds. *Water Res.* **9,** 135–148.

Dillon, P. J., and Rigler, F. H. (1974). The phosphorus–chlorophyll relationship in lakes. *Limnol. Oceanogr.* **19,** 767–773.

Dodds, W. K. (1989). Photosynthesis of two morphologies of *Nostoc parmelioides* (Cyanobacteria) as related to current velocities. *J. Phycol.* **25,** 258–262.

Dodds, W. K. (1991). Micro-environmental characteristics of filamentous algal communities in flowing freshwaters. *Freshwater Biol.* **25,** 199–209.

Dortch, Q., Clayton, J. R., Jr., Thoresen, S. S., Bressler, S. L., and Ahmed, S. I. (1982). Response of marine phytoplankton to nitrogen deficiency: Decreased nitrate uptake vs. enhanced ammonium uptake. *Mar. Biol. (Berlin)* **70,** 13–19.

Dortch, Q., Thompson, P. A., and Harrison, P. J. (1991). Variability in nitrate uptake kinetics in *Thalassiosira pseudonana* (Bacillariophyceae). *J. Phycol.* **27,** 35–39.

Droop, M. R. (1961). Vitamin B12 and marine ecology: The response of *Monochrysis lutheri. J. Mar. Biol. Assoc. U. K.* **41,** 69–76.

Droop, M. R. (1968). Vitamin B12 and marine ecology. IV. The kinetics of uptake, growth and inhibition in *Monochrysis lutheri. J. Mar. Biol. Assoc. U. K.* **48,** 689–733.

Droop, M. R. (1973). Some thoughts on nutrient limitation in algae. *J. Phycol.* **9,** 264–272.

Droop, M. R. (1974). The nutrient status of algal cells in continuous culture. *J. Mar. Biol. Assoc. U. K.* **54,** 825–855.

Droop, M. R. (1983). Twenty-five years of algal growth kinetics. A personal view. *Bot. Mar.* **26,** 99–112.

Dugdale, R. C. (1967). Nutrient limitation in the sea: Dynamics, identification, and significance. *Limnol. Oceanogr.* **12,** 685–695.

Duncan, S. W., and Blinn, D. W. (1989). Importance of physical variables on the seasonal dynamics of epilithic algae in a highly shaded canyon stream. *J. Phycol.* **25,** 455–461.

Elwood, J. W., Newbold, J. D., Trimble, A. F., and Stark, R. W. (1981). The limiting role of phosphorus in a woodland stream ecosystem: Effects of P enrichment on leaf decomposition and primary producers. *Ecology* **62,** 146–158.

Eppley, R. W., and Renger, E. H. (1974). Nitrogen assimilation of an oceanic diatom in nitrogen-limited continuous culture. *J. Phycol.* **10,** 15–23.

Eppley, R. W., Rogers, J. N., and McCarthy, J. J. (1969). Half-saturation constants for uptake of nitrate and ammonium by marine phytoplankton. *Limnol. Oceanogr.* **14,** 912–920.

Fairchild, G. W., and Sherman, J. W. (1993). Algal periphyton response to acidity and nutrients in softwater lakes: Lake comparison vs. nutrient enrichment approaches. *J. North Am. Benthol. Soc.* **12,** 157–167.

Fairchild, G. W., Lowe, R. L., and Richardson, W. B. (1985). Algal periphyton growth on nutrient-diffusing substrates: An *in situ* bioassay. *Ecology* **66,** 465–472.

Fairchild, G. W., Sherman, J. W., and Acker, F. W. (1989). Effects of nutrient (N, P, C) enrichment, grazing and depth upon littoral periphyton of a softwater lake. *Hydrobiologia* **173,** 69–83.

Falkner, G., Horner, F., and Simonis, W. (1980). The regulation of the energy-dependent phosphate uptake by the blue-green alga *Anacystis nidulans*. *Planta* **149,** 138 143.

Fuhs, G. W. (1969). Phosphorus content and rate of growth in the diatoms *Cyclotella nana* and *Thalassiosira fluviatilis*. *J. Phycol.* **5,** 312–321.

Gavis, J. (1976). Munk and Riley revisited: Nutrient diffusion transport and rates of phytoplankton growth. *J. Mar. Res.* **34,** 161–179.

Gerard, V. A. (1982). *In situ* water motion and nutrient uptake by the giant kelp *Macrocystis pyrifera*. *Mar. Biol. (Berlin)* **69,** 51–54.

Gerard, V. A., and Mann, K. H. (1979). Growth and production of *Laminaria longicruris* (Phaeophyta) populations exposed to different intensities of water movement. *J. Phycol.* **15,** 33–41.

Goering, J. J., Nelson, D. M., and Carter, J. A. (1973). Silicic acid uptake by natural populations of marine phytoplankton. *Deep-Sea Res.* **20,** 777–789.

Goldman, J. C. (1977). Steady state growth of phytoplankton in continuous culture: Comparison of internal and external nutrient equations. *J. Phycol.* **13,** 251–258.

Goldman, J. C. (1979). Temperature effects on steady-state growth, phosphorus uptake, and the chemical composition of a marine phytoplankter. *Microb. Ecol.* **5,** 153–166.

Goldman, J. C. (1980). Physiological processes, nutrient availability, and the concept of relative growth rate in marine phytoplankton ecology. *Brookhaven Symp. Biol.* **31,** 179–194.

Goldman, J. C., and Glibert, P. M. (1982). Comparative rapid ammonium uptake by four marine phytoplankton species. *Limnol. Oceanogr.* **27,** 814–827.

Goldman, J. C., and Glibert, P. M. (1983). Kinetics of inorganic nitrogen uptake by phytoplankton. *In* "Nitrogen in the Marine Environment" (E. J. Carpenter and D. G. Capone, eds.), pp. 233– 274. Academic Press, New York.

Goldman, J. C., and Peavey, D. G. (1979). Steady-state growth and chemical composition of the marine chlorophyte *Dunaliella tertiolecta* in nitrogen-limited continuous cultures. *Appl. Environ. Microbiol.* **38,** 894–901.

Goldman, J. C., Oswald, W. J., and Jenkins, D. (1974). The kinetics of inorganic carbon limited algal growth. *J. Water Pollut. Control Fed.* **46,** 554–574.

Goldman, J. C., McCarthy, J. J., and Peavy, D. G. (1979). Growth rate influence on the chemical composition of phytoplankton in oceanic waters. *Nature (London)* **279,** 210–215.

Gonen, Y., Kimmel, E., and Friedlander, M. (1995). Diffusion boundary layer transport in *Gracilaria conferta* (Rhodophyta). *J. Phycol.* 31, 768–773.

Gotham, I. J., and Rhee, G.-Y. (1981). Comparative kinetic studies of phosphate-limited growth and phosphate uptake in phytoplankton in continuous culture. *J. Phycol.* **17,** 257–265.

Grenney, W. J., Bella, D. A., and Curl, H. C., Jr. (1973). A theoretical approach to interspecific competition in phytoplankton communities. *Am. Nat.* **107,** 405–425.

Grimm, N. B., and Fisher, S. G. (1986). Nitrogen limitation in a Sonoran Desert stream. *J. North Am. Benthol. Soc.* **5,** 2–15.

Grover, J. P. (1989). Influence of cell shape and size on algal competitive ability. *J. Phycol.* **25,** 402–405.

Grover, J. P. (1990). Resource competition in a variable environment: Phytoplankton growing according to Monod's model. *Am. Nat.* **136,** 771–789.

Grover, J. P. (1991a). Resource competition in a variable environment: Phytoplankton growing according to the variable-internal-stores model. *Am. Nat.* **138,** 811–835.

Grover, J. P. (1991b). Non-steady state dynamics of algal population growth: Experiments with two chlorophytes. *J. Phycol.* **27,** 70–79.

Gundersen, J. K., and Jørgensen, B. B. (1990). Microstructure of diffusive boundary layers and the oxygen uptake of the sea floor. *Nature (London)* **345,** 604–607.

Hansson, L.-A. (1992). Factors regulating periphytic algal biomass. *Limnol. Oceanogr.* **37,** 322–328.

Harlin, M. M., and Wheeler, P. A. (1985). Nutrient uptake. *In* "Phycological Methods, Ecological Field Methods: Macroalgae" (M. M. Littler and D. S. Littler, eds.), pp. 493–508. Cambridge Univ. Press, London.

Harrison, P. J., Parslow, J. S., and Conway, H. L. (1989). Determination of nutrient uptake kinetic parameters: A comparison of methods. *Mar. Ecol.: Prog. Ser.* **52,** 301–312.

Healey, F. P. (1980). Slope of the Monod equation as an indicator of advantage in nutrient competition. *Microbiol. Ecol.* **5,** 281–286.

Healey, F. P. (1985). Interacting effects of light and nutrient limitation on the growth rate of *Synechococcus linearis* (Cyanophyceae). *J. Phycol.* **21,** 134–146.

Hecky, R. E., and Kilham, P. (1988). Nutrient limitation of phytoplankton in freshwater and marine environments: A review of recent evidence on the effects of enrichment. *Limnol. Oceanogr.* **33,** 796–822.

Hill, R., and Whittingham, C. P. (1955). "Photosynthesis." Methuen, London.

Hill, W. R., and Knight, A. W. (1988). Nutrient and light limitation of algae in two northern California streams. *J. Phycol.* **24,** 125–132.

Hill, W. R., Boston, H. L., and Steinman, A. D. (1992). Grazers and nutrients simultaneously limit lotic primary productivity. *Can. J. Fish. Aquat. Sci.* **49,** 504–512.

Hoffmann, J. P. (1990). Dependence of photosynthesis and vitamin B_{12} uptake on cellular vitamin B_{12} concentration in the multicellular alga *Cladophora glomerata* (Chlorophyta). *Limnol. Oceanogr.* **35,** 100–108.

Horner, R. R., and Welch, E. B. (1981). Stream periphyton development in relation to current velocity and nutrients. *Can. J. Fish. Aquat. Sci.* **38,** 449–457.

Horner, R. R., Welch, E. B., and Veenstra, R. B. (1983). Development of nuisance periphytic algae in laboratory streams in relation to enrichment and velocity. *In* "Periphyton of Freshwater Ecosystems" (R. G. Wetzel, ed.), pp. 121–134. Dr. W. Junk Publishers, The Hague.

Horner, R. R., Welch, E. B., Seeley, M. R., and Jacoby, J. M. (1990). Responses of periphyton to changes in current velocity, suspended sediment and phosphorus concentration. *Freshwater Biol.* **24,** 215–232.

Humphrey, K. P., and Stevenson, R. J. (1992). Responses of benthic algae to pulses in current and nutrients during simulations of subscouring spates. *J. North Am. Benthol. Soc.* **11,** 37–48.

Huntsman, A. G. (1948). Fertility and fertilization of streams. *J. Fish. Res. Board Can.* **7,** 248–253.

Hutchinson, G. E. (1961). The paradox of the plankton. *Am. Nat.* **95,** 137–146.

Jørgensen, B. B., and Des Marais, D. J. (1990). The diffusive boundary layer of sediments: Oxygen microgradients over a microbial mat. *Limnol. Oceanogr.* **35,** 1343–1355.

Kilham, S. S. (1978). Nutrient kinetics of freshwater planktonic algae using batch and semicontinuous methods. *Mitt.—Int. Ver. Theor. Angew. Limnol.* **21,** 147–157.

Kilham, S. S. (1984). Silicon and phosphorus growth kinetics and competitive interactions between *Stephanodiscus minutus* and *Synedra* sp. *Verh.—Int. Ver. Theor. Angew. Limnol.* **22,** 435–439.

Klotz, R. L. (1985). Factors controlling phosphorus limitation in stream sediments. *Limnol. Oceanogr.* **30,** 543–553.

Lindstrøm, E.-A., and Traaen, T. S. (1984). Influence of current velocity on periphyton distribution and succession in a Norwegian soft water river. *Verh.—Int. Ver. Theor. Angew. Limnol.* **22,** 1965–1972.

Lock, M. A., and John, P. H. (1979). The effect of flow patterns on uptake of phosphorous by river periphyton. *Limnol. Oceanogr.* **24,** 376–383.

Lohman, K., and Priscu, J. C. (1992). Physiological indicators of nutrient deficiency in Cladophora (Chlorophyta) in the Clark Fork of the Columbia River, Montana. *J. Phycol.* **28,** 443–448.

Lohman, K., Jones, J. R., and Baysinger-Daniel, C. (1991). Experimental evidence for nitrogen limitation in a northern Ozark stream. *J. North Am. Benthol. Soc.* **19,** 14–23.

Lowe, R. L., Golladay, S. W., and Webster, J. R. (1986). Periphyton response to nutrient manipulation in streams draining clearcut and forested watersheds. *J. North Am. Benthol. Soc.* **5,** 221–229.

Malone, T. C. (1980). Algal size. *In* "The Physiological Ecology of Phytoplankton" (I. Morris, ed.), pp. 433–463. Blackwell, Oxford.

Marks, J. C., and Lowe, R. L. (1989). The independent and interactive effects of snail grazing and nutrient enrichment on structuring periphyton communities. *Hydrobiologia* **185,** 9–17.

Marks, J. C., and Lowe, R. L. (1993). Interactive effects of nutrient availability and light levels on the periphyton composition of a large oligotrophic lake. *Can. J. Fish. Aquat. Sci.* **50,** 1270–1278.

McCormick, P. V., and Stevenson, R. J. (1989). Effects of snail grazing on benthic algal community structure in different nutrient environments. *J. North Am. Benthol. Soc.* **8,** 162–172.

McIntire, C. D. (1966). Some effects of current velocity on periphyton communities in laboratory streams. *Hydrobiologia* **27,** 559–570.

Mierle, G. (1985). Kinetics of phosphate transport by *Synechococcus leopoliensis* (Cyanophyta): Evidence for diffusion limitation of phosphate uptake. *J. Phycol.* **21,** 177–181.

Miller, A. G., Turpin, D. H., and Canvin, D. T. (1984). Growth and photosynthesis of the cyanobacterium *Synechococcus leopoliensis* in HCO_3^--limited chemostats. *Plant Physiol.* **75,** 1064–1070.

Monod, J. (1950). La technique de culture continue, théorie et applications. *Ann. Inst. Pasteur, Paris* **79,** 390–410.

Morel, F. M. M. (1987). Kinetics of nutrient uptake and growth in phytoplankton. *J. Phycol.* **23,** 137–150.

Mulholland, P. J., and Rosemond, A. D. (1992). Periphyton response to longitudinal nutrient depletion in a woodland stream: Evidence of upstream–downstream linkage. *J. North Am. Benthol. Soc.* **11,** 405–419.

Mundie, J. H., Simpson, K. S., and Perrin, C. J. (1991). Responses of stream periphyton and benthic insects to increases in dissolved inorganic phosphorus in a mesocosm. *Can. J. Fish. Aquat. Sci.* **48,** 2061–2072.

Munk, W. H., and Riley, G. A. (1952). Absorption of nutrients by aquatic plants. *J. Mar. Res.* **11,** 215–240.

Nobel, P. S. (1983). "Biophysical Plant Physiology and Ecology." Freeman, New York.

Nowell, A. R. M., and Jumars, P. A. (1984). Flow environments of aquatic benthos. *Annu. Rev. Ecol. Syst.* **15,** 303–328.

Nyholm, N. (1977). Kinetics of phosphate limited algal growth. *Biotechnol. Bioeng.* **19,** 467–492.

Olsen, Y. (1989). Evaluation of competitive ability of *Staurastrum luetkemuellerii* (Chlorophyceae) and *Microcystis aeruginosa* (Cyanophyceae) under P limitation. *J. Phycol.* **25,** 486–499.

Olsen, Y., Vadstein, O., Andersen, T., and Jensen, A. (1989). Competition between *Staurastrum luetkemuellerii* (Chlorophyceae) and *Microcystis aeruginosa* (Cyanophyceae) under varying modes of phosphate supply. *J. Phycol.* **25**, 499–508.

Paasche, E. (1973). Silicon and the ecology of marine plankton diatoms. *Thalassiosira pseudonana (Cyclotella nana)* grown in a chemostat with silicate as the limiting nutrient. *Mar. Biol. (Berlin)* **19**, 117–126.

Parker, H. S. (1981). Influence of relative water motion on the growth, ammonium uptake, and carbon and nitrogen composition of *Ulva lactuca* (Chlorophyta). *Mar. Biol. (Berlin)* **63**, 309–318.

Pasciak, W. J., and Gavis, J. (1974). Transport limitation of nutrient uptake in phytoplankton. *Limnol. Oceanogr.* **19**, 881–888.

Pasciak, W. J., and Gavis, J. (1975). Transport limited nutrient uptake rates in *Ditylum brightwellii. Limnol. Oceanogr.* **20**, 604– 617.

Perrin, C. J., Bothwell, M. L., and Slaney, P. A. (1987). Experimental enrichment of a coastal stream in British Columbia: Effects of organic and inorganic additions on autotrophic periphyton production. *Can. J. Fish. Aquat. Sci.* **44**, 1247–1256.

Perry, M. J. (1976). Phosphate utilization by an oceanic diatom in phosphorus-limited chemostat culture and in the oligotrophic waters of the central North Pacific. *Limnol. Oceanogr.* **21**, 88–107.

Peterson, B. J., Hobbie, J. E., Hershey, A. E., Lock, M. A., Ford, T. E., Vestal, J. R., McKinley, V. L., Hullar, M. A. J., Miller, M. C., Ventullo, R. M., and Volk, G. S. (1985). Transformation of a tundra river from heterotrophy to autotrophy by addition of phosphorus. *Science* **229**, 1383–1386.

Peterson, B. J., Deegan, L., Helfrich, J., Hobbie, J. E., Hullar, M., Moller, B., Ford, T. E., Hershey, A., Hiltner, A., Kipphut, G., Lock, M. A., Fiebig, D. M., McKinley, V., Miller, M. C., Vestal, J. R., Ventullo, R., and Volk, G. (1993). Biological responses of a tundra river to fertilization. *Ecology* **74**, 653–672.

Peterson, C. G., and Grimm, N. B. (1992). Temporal variation in enrichment effects during periphyton succession in a nitrogen-limited desert stream ecosystem. *J. North Am. Benthol. Soc.* **11**, 20–36.

Pfeifer, R. F., and McDiffett, W. F. (1975). Some factors affecting primary productivity of stream riffle communities. *Arch. Hydrobiol.* **75**, 306–317.

Pringle, C. M. (1987). Effects of water and substratum nutrient supplies on lotic periphyton growth: An integrated bioassay. *Can. J. Fish. Aquat. Sci.* **44**, 619–629.

Pringle, C. M. (1990). Nutrient spatial heterogeneity: Effects on community structure, physiognomy, and diversity of stream algae. *Ecology* **71**, 905–920.

Pringle, C. M., and Bowers, J. A. (1984). An *in situ* substratum fertilization technique: Diatom colonization on nutrient-enriched, sand substrata. *Can. J. Fish. Aquat. Sci.* **41**, 1247–1251.

Raven, J. A. (1992). How benthic macroalgae cope with flowing freshwater: Resource acquisition and retention. *J. Phycol.* **28**, 133–146.

Redfield, A. C. (1958). The biological control of chemical factors in the environment. *Am. Sci.* **46**, 205–222.

Reiter, M. A. (1989). The effect of a developing algal assemblage on the hydrodynamics near an uneven substrate. *Arch. Hydrobiol.* **20**, 1–24.

Rhee, G.-Y. (1973). A continuous culture study of phosphate uptake, growth rate and polyphosphate in *Scenedesmus* sp. *J. Phycol.* **9**, 495–506.

Rhee, G.-Y. (1974). Phosphate uptake under nitrate limitation by *Scenedesmus* and its ecological implications. *J. Phycol.* **10**, 470–475.

Rhee, G.-Y. (1978). Effects of N/P atomic ratios and nitrate limitation on algal growth, cell composition, and nitrate uptake: A study of dual nutrient limitation. *Limnol. Oceanogr.* **23**, 10–25.

Rhee, G.-Y. (1980). Continuous culture in phytoplankton ecology. *In* "Advances in Aquatic Microbiology 2" (M. R. Droop and H. W. Jannasch, eds.), pp. 151–203. Academic Press, New York.

Rhee, G.-Y., and Gotham, I. J. (1980). Optimum N:P ratios and the coexistence of planktonic algae. *J. Phycol.* **16,** 486–489.

Rhee, G.-Y., and Gotham, I. J. (1981a). The effect of environmental factors on phytoplankton growth: Temperature and the interactions of temperature with nutrient limitation. *Limnol. Oceanogr.* **26,** 635–648.

Rhee, G.-Y., and Gotham, I. J. (1981b). The effect of environmental factors on phytoplankton growth: Light and the interactions of light with nitrate limitation. *Limnol. Oceanogr.* **26,** 649–659.

Riber, H. H., and Wetzel, R. G. (1987). Boundary layer and internal-diffusion effects on phosphorus fluxes in lake periphyton. *Limnol. Oceanogr.* **32,** 1181–1194.

Riegman, R., and Mur, L. R. (1984). Regulation of phosphate uptake kinetics in *Oscillatoria agardhii. Arch. Microbiol.* **139,** 28–32.

Rodgers, J. H., and Harvey, R. S. (1976). The effect of current on periphytic productivity as determined using carbon-14. *Water Resour. Bull.* **12,** 1109–1118.

Rosemarin, A. S. (1982). Phosphorus nutrition of two potentially competing filamentous algae, *Cladophora glomerata* (L.) Kutz. and *Stigeoclonium tenue* (Agardh) Kutz. from Lake Ontario. *J. Great Lakes Res.* **8,** 66–72.

Rosemond, A. D. (1994). Multiple factors limit seasonal variation in periphyton in a forest stream. *J. North Am. Benthol. Soc.* **13,** 333–344.

Rosemond, A. D., Mulholland, P. J., and Elwood, J. W. (1993). Top-down and bottom-up control of stream periphyton: Effects of nutrients and herbivores. *Ecology* **74,** 1264–1280.

Ruttner, F. (1953). "Fundamentals of Limnology," p. 199. Univ. of Toronto Press, Toronto.

Sakamoto, M. (1966). Primary production by phytoplankton community in some Japanese lakes and its dependence on lake depth. *Arch. Hydrobiol.* **62,** 1–28.

Schanz, F., and Juon, H. (1983). Two different methods of evaluating nutrient limitations of periphyton bioassays using water from the River Rhine and eight of its tributaries. *Hydrobiologia* **102,** 187–195.

Schindler, D. W. (1977). Evolution of phosphorus limitation in lakes. *Science* **195,** 260–262.

Schindler, D. W., and Fee, E. J. (1973). Diurnal variation of dissolved inorganic carbon and its use in estimating primary production and CO_2 invasion in Lake 227. *J. Fish. Res. Board Can.* **30,** 1501–1510.

Schumacher, G. J., and Whitford, L. A. (1965). Respiration and P-32 uptake in various species of freshwater algae as affected by a current. *J. Phycol.* **1,** 78–80.

Senft, W. H. (1978). Dependence of light-saturated rates of algal photosynthesis on intracellular concentrations of phosphorus. *Limnol. Oceanogr.* **23,** 709–718.

Shortreed, K. S., Costella, A. C., and Stockner, J. G. (1984). Periphyton biomass and species composition in 21 British Columbia lakes: Seasonal abundance and response to whole-lake nutrient additions. *Can. J. Bot.* **62,** 1022–1031.

Shuter, B. (1979). A model of physiological adaptation in unicellular algae. *J. Theor. Biol.* **78,** 519.

Silvester, N. R., and Sleigh, M. A. (1985). The forces on microorganisms at surfaces in flowing water. *Freshwater Biol.* **15,** 433–448.

Smith, F. A., and Walker, N. A. (1980). Photosynthesis by aquatic plants: Effects of unstirred layers in relation to assimilation of CO_2 and HCO_3^- and to carbon isotopic discrimination. *New Phytol.* **86,** 245–259.

Sommer, U. (1985). Comparison between steady state and non-steady state competition: Experiments with natural phytoplankton. *Limnol. Oceanogr.* **30,** 335–346.

Sommer, U. (1988). Growth and survival strategies of planktonic diatoms. *In* "Growth and Reproductive Strategies of Freshwater Phytoplankton" (C. Sandgren, ed.), pp. 227–260. Cambridge Univ. Press, New York.

Sperling, J. A., and Grunewald, R. (1969). Batch culturing of thermophilic benthic algae and phosphorus uptake in a laboratory stream model. *Limnol. Oceanogr.* **14,** 944–949.

Sperling, J. A., and Hale, G. M. (1973). Patterns of radiocarbon uptake by a thermophilic blue-green alga under varying conditions of incubation. *Limnol. Oceanogr.* **18,** 658–662.

Stanley, E. H., Short, R. A., Harrison, J. W., Hall, R., and Wiedenfeld, R. C. (1990). Variation in nutrient limitation of lotic and lentic algal communities in a Texas (USA) river. *Hydrobiologia* **206,** 61–71.

Statzner, B., Gore, J. A., and Resh, V. H. (1988). Hydraulic stream ecology: Observed patterns and potential applications. *J. North Am. Benthol. Soc.* **7,** 307–360.

Stevenson, R. J., and Glover, R. (1993). Effects of algal density and current on ion transport through periphyton communities. *Limnol. Oceanogr.* **38,** 1276–1281.

Stevenson, R. J., Peterson, C. G., Kirschtel, D. B., King, C. C., and Tuchman, N. C. (1991). Density-dependent growth, ecological strategies, and effects of nutrients and shading on benthic diatom succession in streams. *J. Phycol.* **27,** 59–69.

Stockner, J. G., and Shortreed, K. R. S. (1978). Enhancement of autotrophic production by nutrient addition in a coastal rainforest stream on Vancouver Island. *J. Fish. Res. Board Can.* **35,** 28–34.

Terry, K. L., Laws, E. A., and Burns, D. J. (1985). Growth rate variation in the N:P requirement ratio of phytoplankton. *J. Phycol.* **21,** 323–329.

Thirb, H. H., and Benson-Evans, K. (1982). The effect of different current velocities on the red alga *Lemanea* in a laboratory stream. *Arch. Hydrobiol.* **96,** 65–72.

Tilman, D. (1977). Resource competition between planktonic algae: An experimental and theoretical approach. *Ecology* **58,** 338–348.

Tilman, D. (1981). Tests of resource competition theory using four species of Lake Michigan algae. *Ecology* **62,** 802–815.

Tilman, D. (1982). "Resource Competition and Community Structure." Princeton Univ. Press, Princeton, NJ.

Tilman, D., and Kilham, S. S. (1976). Phosphate and silicate growth and uptake kinetics of the diatoms *Asterionella formosa* and *Cyclotella meneghiniana* in batch and semicontinuous culture. *J. Phycol.* **12,** 375–383.

Tilman, D., Kilham, S. S., and Kilham, P. (1982). Phytoplankton community ecology: The role of limiting nutrients. *Annu. Rev. Ecol. Syst.* **13,** 349–372.

Traaen, T. S. (1978). Effects of effluents from a variety of sewage treatment methods on primary productivity, respiration and algal communities in artificial stream channels. *Verh.—Int. Ver. Theor. Angew. Limnol.* **20,** 1767–1771.

Traaen, T. S., and Lindstrøm, E.-A. (1983). Influence of current velocity on periphyton distribution. *In* "Periphyton of Freshwater Ecosystems" (R. G. Wetzel, ed.), pp. 97–99. Dr. W. Junk Publishers, The Hague.

Triska, F. J., Kennedy, V. C., Avanzino, R. J., and Reilly, B. N. (1983). Effect of simulated canopy cover on regulation of nitrate uptake and primary production by natural periphyton communities. *In* "Dynamics of Lotic Ecosystems" (T. D. Fontaine and S. M. Bartell, eds.), pp 129–159. Ann Arbor Sci. Publ., Ann Arbor, MI.

Turpin, D. H. (1986). Growth rate dependent optimum ratios in *Selenastrum minutum* (Chlorophyta): Implications for competition, coexistence and stability in phytoplankton communities. *J. Phycol.* **22,** 94–102.

Turpin, D. H. (1988). Physiological mechanisms in phytoplankton resource competition. *In* "Growth and Reproductive Strategies of Freshwater Phytoplankton" (C. Sandgren, ed.), pp. 316–368. Cambridge Univ. Press, New York.

Turpin, D. H., and Harrison, P. J. (1979). Limiting nutrient patchiness and its role in phytoplankton ecology. *J. Exp. Mar. Biol. Ecol.* **39,** 151–166.

Turpin, D. H., Miller, A. G., Parslow, J. S., Elrifi, I. R., and Canvin, D. T. (1985). Predicting the kinetics of dissolved inorganic carbon limited growth from the short-term kinetics of photosynthesis in *Synechococcus leopoliensis* (Cyanophyta). *J. Phycol.* **21,** 409–418.

van Donk, E., and Kilham, S. S. (1990). Temperature effects on silicon- and phosphorus-limited growth and competitive interactions among three diatoms. *J. Phycol.* **26,** 40–50.

Vincent, W. F., Castenholz, R. W., Downes, M. T., and Howard-Williams, C. (1993). Antarctic cyanobacteria: Light, nutrients, and photosynthesis in the microbial mat environment. *J. Phycol.* **29,** 745–755.

Vogel, S. (1981). "Life in Moving Fluids. The Physical Biology of Flow." Princeton Univ. Press, Princeton, NJ.

Vollenweider, R. A. (1976). Advances in defining critical loading levels for phosphorus in lake eutrophication. *Mem. Ist. Ital. Idrobiol. Dott. Marco de Marchi* **33,** 53–83.

Welch, E. B., Jacoby, J. M., Horner, R. R., and Seeley, M. R. (1988). Nuisance biomass levels of periphytic algae in streams. *Hydrobiologia* **157,** 161–168.

Wheeler, W. N. (1988). Algal productivity and hydrodynamics—A synthesis. *Prog. Phycol. Res.* **6,** 23–58.

Whitford, L. A. (1960). The current effect and growth of fresh-water algae. *Trans. Am. Microsc. Soc.* **79,** 302–309.

Whitford, L. A., and Schumacher, G. J. (1961). Effect of current on mineral uptake and respiration by a freshwater alga. *Limnol. Oceanogr.* **6,** 423–425.

Winne, D. (1973). Unstirred layer, source of biased Michaelis constant in membrane transport. *Biochim. Biophys. Acta* **298,** 27–31.

Winterbourn, M. J. (1990). Interactions among nutrients algae and invertebrates in a New Zealand mountain stream. *Freshwater Biol.* **23,** 463–474.

Wong, S. L., and Clark, B. (1976). Field determination of the critical nutrient concentrations for *Cladophora* in streams. *J. Fish. Res. Board Can.* **33,** 85–92.

Wuhrmann, K., and Eichenberger, E. (1975). Experiments on the effects of inorganic enrichment of rivers on periphyton primary production. *Verh.—Int. Ver. Theor. Angew. Limnol.* **19,** 2028–2034.

Wynne, D. and Rhee, G.-Y. (1986). Effects of light intensity and quality on the relative N and P requirement (the optimum N:P ratio) of marine planktonic algae. *J. Plankton Res.* **8,** 91–103.

Wynne, D., and Rhee, G.-Y. (1988). Changes in alkaline phosphatase activity and phosphate uptake in P-limited phytoplankton, induced by light intensity and spectral quality. *Hydrobiologia* **160,** 173–178.

Zevenboom, W. (1986). Ecophysiology of nutrient uptake, photosynthesis, and growth. *Can. Bull. Fish. Aquat. Sci.* **214,** 391–422.

Zevenboom, W., DeGroot, G. J., and Mur, L. R. (1980). Effects of light on nitrate-limited *Oscillatoria agardhii* in chemostat cultures. *Arch. Microbiol.* **125,** 59–65.

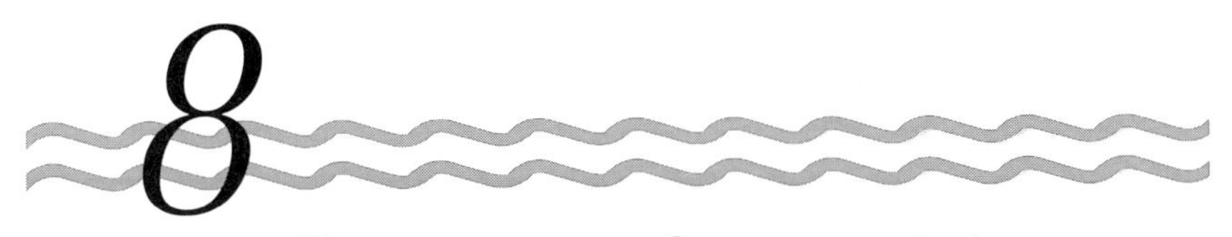

Resource Competition and Species Coexistence in Freshwater Benthic Algal Assemblages

Paul V. McCormick

Everglades Systems Research Division
South Florida Water Management District
West Palm Beach, Florida 33416

I. INTRODUCTORY REMARKS

Competition as a force determining population distributions and the taxonomic composition of ecological communities has been one of the most widely studied and vigorously debated topics in ecology. Competition may arise when organisms rely on the same consumable resource and that resource is present in limited supply. Under such conditions, consumption of the resource by one individual effectively reduces the performance (e.g., survivorship, reproductive capacity) of others. When competition occurs among species, the result is often a reduction in the density and, perhaps, the local extinction of one or more populations. Correlative studies that link reductions in species performance to increased densities of other populations provide circumstantial evidence for the prevalence and importance of competition. Competition became a cornerstone of classical ecological theory largely on the basis of such observations and a limited number of simple laboratory experiments (e.g., Gause, 1934). Although available evidence from field experiments indicates that interspecific competitive interactions may be widespread in nature (Connell, 1983; Schoener, 1983), critical tests of the importance of this process in determining community composition are lacking and viable alternative views (e.g., regulation by density-independent forces) have been presented (e.g., Strong *et al.,* 1979; Price, 1984). Contemporary views of community organization treat competition as but one of many possible structuring forces and acknowledge that the importance of this process may vary greatly among different types of communities (e.g., Dayton, 1984).

There is considerable circumstantial evidence to suggest that negative interspecific interactions such as competition are an important determinant of species composition in benthic algal assemblages. As a group, algae possess exceptional powers of dispersal (Round, 1981). Consequently, an extremely large species pool should be available to colonize most habitats. Algal species typically exhibit broad environmental tolerances as well and, thus, should be capable of successfully colonizing a wide range of habitats. The fact that relatively few algal species dominate most habitats compared to the potentially large pool of colonists suggests that biotic interactions are the principal determinant of the species composition of algal assemblages. Benthic algae typically grow as dense "mats" of cells. Resources become limiting within benthic mats as indicated by a pronounced reduction in light and nutrient penetration as the mat develops (Meulemans, 1987; Riber and Wetzel, 1987; Stevenson and Glover, 1993) and a corresponding reduction in biomass-specific photosynthetic activity (Hill and Boston, 1991). Reduced resource availability, coupled with an extensive overlap in nutritional requirements among species, favors the development of intense competition for resources.

In a comprehensive review of experimental studies of competition, Schoener (1983) noted the paucity of field studies of competition for freshwater producers including algae. Although important experimental studies of competitive interactions and species coexistence in benthic algal assemblages have been performed in a few habitats, most notably the marine intertidal (e.g., Dayton, 1975; Sousa, 1979; Lubchenco, 1978), direct tests of the importance of competition are lacking for most aquatic habitats and progress toward a conceptual model of community organization has been slow. This is in sharp contrast to the preeminence of competition theory in terrestrial plant ecology. Given the many physiological and ecological traits that are shared by all photoautotrophs, existing evidence from terrestrial communities should provide a useful basis for predicting the nature and importance of competitive interactions in benthic algal assemblages. The reliance of both plants and algae on essential (i.e., nonsubstitutable) resources distinguishes them from heterotrophs, which generally rely on complementary or switching resources. This distinction has important implications when evaluating alternative competitive models (e.g., Tilman, 1982). Similarities in the physiognomic structure (e.g., vertical development and stratification) of many plant and benthic algal assemblages (e.g., Hoagland *et al.*, 1982) suggest the existence of analogous types of competitive interactions and adaptive traits related to light limitation within both types of assemblages (Fig. 1). Interestingly, because nutrients are supplied to benthic algal mats from the overlying water column as well as from the sediments, traits that enhance an individual's ability to compete for light (e.g., height) also confer an advantage for nutrient procurement. The simplicity of algal life histories reduces some of the need for complex models of resource competition developed for plants (e.g., allocation to below- versus above-ground biomass, cf. Tilman, 1989). However, certain ecological strategies for avoiding competition (e.g., facultative heterotrophy, see Tuchman, Chapter 10, this volume) may be unique to algae. Though these environmental and physiological differences are not trivial,

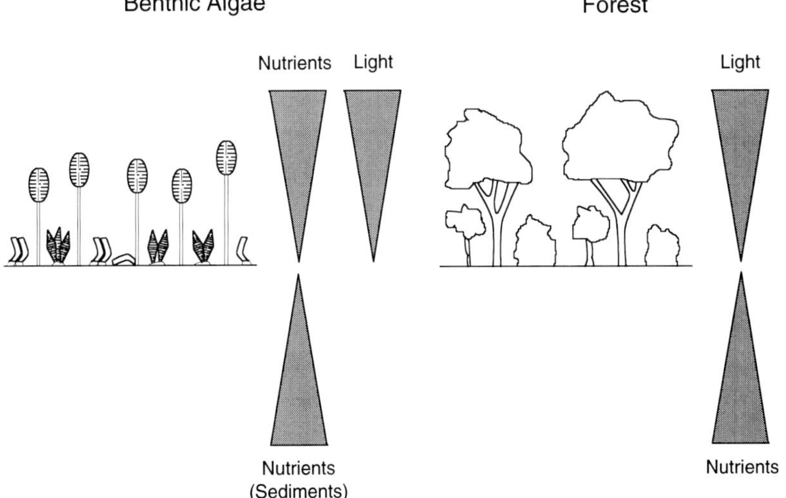

FIGURE 1 Idealized physiognomic structure and resource supply directions in benthic algal and terrestrial plant assemblages. Shaded triangles indicate the direction of resource attenuation. Note that in benthic assemblages growing on hard substrata, both light and nutrients enter the assemblage from above, suggesting similar adaptive traits for the acquisition of both resources. In contrast, different adaptive traits and patterns of energy allocation are required for the acquisition of light and nutrients by terrestrial plants.

they do not prevent the useful exchange of competition theory between plant and algal assemblages.

Rather than attempting an exhaustive review of the literature on algal and plant competition, I will focus on the basic principles of competition and species coexistence as they relate to benthic algal assemblages and propose one explanation of patterns in these assemblages based on existing models for plant communities. Discussion will be limited to *exploitative competition,* which involves only those interactions and adaptations relating directly to resource procurement. *Interference competition* (e.g., allelopathic interactions) undoubtedly occurs in algal assemblages as well; however, such interactions have rarely been investigated and no assessment of their general significance can be made. A heavy reliance on theoretical and empirical evidence from plant and phytoplankton assemblages is necessitated by the relative absence of direct tests of competitive interactions in benthic algal assemblages. The aim is to present a conceptual framework that provides falsifiable predictions concerning important ecological strategies related to population persistence in competitively structured assemblages and species distributions under different environmental conditions. A brief discussion of methodology relevant to testing the hypotheses presented herein is provided at the end.

II. BASIC THEORY AND PREDICTIONS: THE LOTKA–VOLTERRA EQUATIONS AND THE COMPETITIVE EXCLUSION PRINCIPLE

The Lotka–Volterra competition model provides a simple, generic description of competitive interactions among species and allows for basic predictions of the outcomes of competition to be made. Competitive interactions between two species are described mathematically by the following equations:

$$\frac{dN_1}{dt} = r_1N_1 \frac{(K_1 - N_1 - \alpha_{12}N_2)}{K_1}$$

$$\frac{dN_2}{dt} = r_2N_2 \frac{(K_2 - N_2 - \alpha_{21}N_1)}{K_2}$$

where r_i = the instantaneous rate of increase, N_i = the density, and K_i = the carrying capacity of the ith competing population (i=1,2). These equations describe two populations undergoing logistic growth (i.e., intraspecific competition implied) while competing with each other for one or more limiting resources. The intensity of competition between the two species is indicated by the magnitude of the competition coefficients, α_{12} and α_{21}. The term α_{ij} represents the inhibitory effect that population j exerts on population i. In the absence of any competing population, a population grows logistically and eventually attains a certain carrying capacity where intraspecific competition for resources limits population reproduction to a level approximately equal to the rate of death (assuming a closed system). When two competitors occupy the same habitat, two outcomes are possible: 1) the inferior competitor is driven to extinction or 2) the two species coexist. Conceptually, it is a rather simple matter to extend the two-species scenario to a multispecies assemblage; assuming equilibrium conditions, the species that is the best competitor either excludes all other species or may coexist with one or more of the other competing populations.

The Lotka–Volterra equations are closely linked with the concept of competitive exclusion, which predicts that overlapping requirements for a limiting resource between two or more species will result in the local elimination of all but the best competitor for that resource. However, even these simple equations predict coexistence as an outcome of competition in at least two circumstances. Coexistence can occur if each population inhibits the growth of other populations more than its own. However, this equilibrium condition is unstable in that slight changes in the abundance of even one species can result in the extinction of one or more of the other populations. Transient coexistence is also predicted for assemblages of populations with identical competitive abilities, a situation termed symmetric competition. Stable coexistence is predicted if the magnitude of interspecific competition is less than that of intraspecific competition within each population, that is, if each population inhibits its own growth more than that of other species. The effect of interspecific competition under such circumstances is to maintain competing populations at densities below their respective carrying capacities. Though interspecific competition is, thus, predicted to affect community structure under all conditions, the strongest influence (i.e., complete exclusion of populations from a habitat) is predicted in circumstances where interspecific competition is asymmetric and intense relative to intraspecific interactions.

Asymmetric competition is common in terrestrial and wetland plant communities (e.g., Weiner and Thomas, 1986; Keddy and Shipley, 1989) and individual species within the community can be ranked based on their position within a competitive hierarchy. The existence of competitive hierarchies is favored under conditions where several co-occurring species are limited by the same essential resource because species inevitably vary in their ability to acquire a particular resource as a result of various physiological and morphological adaptations and constraints. Asymmetry in competitive ability for light, for example, is often an inevitable consequence of size; as Keddy (1989) notes, "Short plants simply do not shade tall ones to the same extent that tall ones shade short ones." Competitive hierarchies have been inferred by experimentally measuring species performance (e.g., biomass yield) in the presence and absence of potential competitors or a common reference species or "phytometer." Populations that exert a strong negative effect on other species (i.e., compete better for the limiting resource) but are not themselves greatly affected by the presence of other populations are ranked high in the hierarchy, whereas those that are suppressed by other populations are ranked low. Information on relative competitive abilities of populations gained in this way allows for predictions of species dominance within a community to be made. Failure to validate these predictions would indicate that processes other than competition are more important in determining community structure.

The existence of competitive hierarchies is likely in benthic algal assemblages since the growth of many species is typically limited by the same resource under any set of circumstances and individual species inevitably vary in their ability to acquire the resource as a result of specific physiological and/or morphological adaptations. Interspecific variation in nutrient uptake kinetics (Darley, 1982) exemplifies how evolutionary adaptations constrain the ability of a species to compete successfully with other populations under different resource regimes. Morphological adaptations contribute to differential light-capturing ability among species in benthic mats since cells that are stalked or otherwise extended disproportionately reduce the availability of light to other cells within the mat. Although the existence of competitive hierarchies has been more thoroughly documented for phytoplankton assemblages (Sommer, 1990), direct evidence to support the existence of asymmetric competitive abilities among species has been provided for benthic assemblages in intertidal habitats (Dayton, 1975) and at least one freshwater system (McCormick and Stevenson, 1991).

An alternative view, that algal species can maintain an unstable coexistence as a result of symmetric competitive abilities for a common limiting resource, has been forwarded by a few investigators. Hutchinson (1967) suggested that symmetric competition was a logical outcome of species evolution in similar environments, but noted at the time that there was little empirical evidence to support this hypothesis. This alternative view, termed the "coexistence principle" by den Boer (1980), has subsequently gained some support from studies of phytoplankton assemblages (Lewis, 1977; Ghilarov, 1984). However, theory already discussed predicts that such coexistence is inherently unstable. More importantly, this principle contradicts a growing body of empirical evidence indicating that differential abilities to acquire resources determine the outcome of competitive interactions in phytoplankton assemblages (Tilman, 1982; Sommer, 1990).

Extensive overlap in resource requirements among algal species suggests that the magnitude of intra- and interspecific interactions within benthic assemblages should be similar. However, the intensity of competition among individuals is determined not only by overlap in resource utilization but by location within the mat and the proximity to other cells of the same and different species. Intraspecific interactions should be intense in sessile populations that grow largely as a result of clonal division or other forms of asexual reproduction that produce clumped distributions. Examination of algal mats by means of scanning electron microscopy (SEM) supports the tendency toward clumped cell distributions for many stalked and apically attached species (Hoagland *et al.*, 1982; Steinman and McIntire, 1986). Conversely, vertical stratification of the mat increases the potential for interspecific resource competition, as in the case of stalked species reducing penetration of light and nutrients to populations with prostrate growth forms. Thus, in vertically structured communities, which describe many benthic algal assemblages, it is likely that the growth of competitively dominant populations (i.e., ranked high in the hierarchy) may be limited by intraspecific competition, whereas competitively inferior species (i.e., ranked low in the hierarchy) are limited as a result of resource preemption by competitive dominants. At least one experimental investigation supports this prediction for stream algal assemblages (McCormick and Stevenson, 1991). This experiment, which manipulated diatom species densities in natural assemblages, found that while the growth of stalked species tended to be negatively correlated with their own abundance, growth rates of apically attached species were negatively correlated with the abundance of one or more stalked species.

III. LIMITATIONS TO CLASSICAL THEORY: EQUILIBRIUM AND NONEQUILIBRIUM COEXISTENCE IN COMPETITIVE HIERARCHIES

Arguments presented in the foregoing suggest that interspecific competition is a widespread and important process in algal assemblages. The suppression, even exclusion, of populations is predicted because species differ in their ability to compete for limiting nonsubstitutable resources. Indeed, species exclusion generally results when two or more species are maintained under constant laboratory conditions (e.g., Tilman, 1977; Holm and Armstrong, 1981; Healey and Hendzel, 1988). In contrast, the relatively high species richness of most algal assemblages in nature indicates that many populations with similar resource requirements can be maintained in the same habitat. This raises questions as to how species coexistence is possible given the potential for asymmetric density-dependent interactions. Whereas resource overlap in heterotroph guilds can be minimized by partitioning or switching usage of resources in response to competitive pressures (e.g., MacArthur, 1958; Pianka, 1969), opportunities for resource partitioning within plant and algal assemblages are strictly limited owing to a reliance of constituent populations on the same nonsubstitutable resources. Although theory predicts that algal species limited by different resources (i.e., inorganic nutrients, vitamins, and light) will coexist, no more than a couple dozen resources or so could conceivably limit algal growth. In fact, coexisting populations of benthic algae often appear to be limited by the same resource as evidenced by growth response to artificial enrichment (e.g., Fairchild *et al.*, 1985; Carrick *et al.*, 1988; McCormick, 1990). Thus, resource partitioning alone cannot explain population coexistence in species-rich algal assemblages.

Predictions of the Lotka–Volterra equations are contingent on the assumption that populations interact in a spatially and temporally homogeneous environment. Under such circumstances competitive interactions are allowed to proceed uninterrupted with competitive exclusion being the eventual outcome. In fact, stable, homogeneous conditions are rarely encountered in nature and environmental fluctuations can mediate the intensity of competitive interactions in numerous ways, leading to outcomes not anticipated by classical theory. In his essay on the "paradox of the plankton," Hutchinson (1961) predicted that fluctuations in resource supply and other environmental conditions should allow for species coexistence provided that the time scale of such fluctuations was similar to that of the generation times of competing species. Implicit in this theory is the assumption that fluctuations in environmental conditions alter competitive relationships by switching the relative positions of species within the competitive hierarchy or at least by periodically ameliorating the intensity of interspecific competition. Experimental studies with phytoplankton assemblages have supported (Sommer, 1985) and rejected (Grover, 1991) the prediction that introduction of non-steady-state conditions fosters species coexistence. These contrary findings serve to emphasize that environmental variation per se is not always sufficient to allow coexistence. The frequency and magnitude of environmental fluctuations must be sufficient to alter the competitive balance and, in general, the more asymmetric the interaction between two species, the more likely the outcome will be relatively independent of environmental conditions (Keddy, 1989).

The most pronounced environmental changes in many benthic habitats and, perhaps, the most important in terms of effects on the intensity of density-dependent interactions are associated with disturbance. Algal mats are susceptible to disruption or complete removal by spates in streams or storm-enhanced wave action in the littoral zone of lakes. Disturbances temporarily reduce the intensity of density-dependent interactions, including interspecific competition, by removing biomass from the substrate and increasing the penetration of light and nutrients to remaining cells. Periodic senescence and sloughing of algal mats produce similar effects even in the absence of abiotic disturbance. Disturbance typically results in disproportionate losses in populations of those species that, by virtue of their high vertical profile, are competitively dominant within the mat. Selective removal of competitive dominants enhances the potential for species coexistence and, consequently, habitats exposed to periodic disturbance of moderate intensity tend to exhibit higher species richness than comparable habitats that are undisturbed (e.g., Dayton, 1975). Whereas intermediate levels of disturbance can promote species richness by preventing competitive displacement, frequent and intense disturbance typically has the opposite effect by effectively excluding from the community all species other than those most resistant to removal. These opposing effects of disturbance, a reduction in the intensity of competition and selection for disturbance-resistant species, underlie the "humped" relationship between species richness and disturbance intensity (Figure 2), which has been recognized for different types of ecological communities, both terrestrial and aquatic (Connell, 1978; Sousa, 1985).

Equilibrium models of resource competition have been proposed that predict species coexistence in spatially heterogeneous habitats in the absence of temporal variation (Tilman, 1982). Even in the absence of physical disturbance, all habitats exhibit a degree of spatial heterogeneity in resource availability and other environmental parameters that may influence the relative competitive ability of different species. Tilman's model of resource competition predicts that spatial variation in the relative availability of as few as two potentially limiting resources can allow for the coexistence of an almost limitless number of plant or algal species under certain conditions. The key assumption of this model is that competing species perform optimally at different ratios of these resources and that trade-offs exist in a species' ability to tolerate depletion of these different resources. For example, diatom species that are relatively good at maintaining growth under conditions of low phosphorus availability compete rather poorly under conditions of silica depletion and vice versa (Tilman *et*

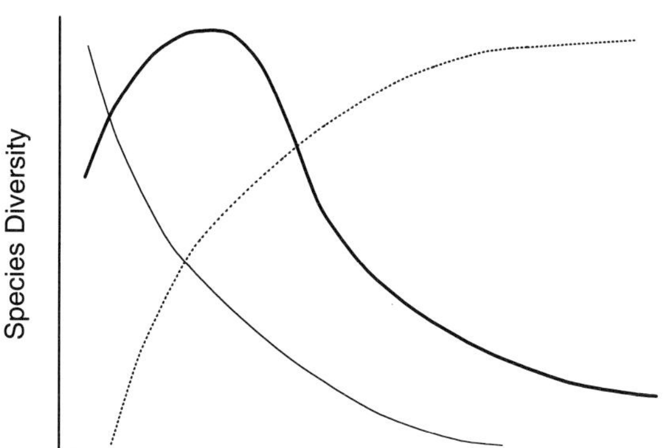

FIGURE 2 Idealized relationship between species diversity and the degree of disturbance (thick solid line). The thin solid line illustrates the decreasing importance of competitive interactions (density-dependent factors) in determining species composition with increasing disturbance, whereas the dashed line shows the increasing importance of adaptations for coping with disturbance (density-independent factors).

al., 1982). The tenets and predictions of this model have been verified experimentally for several phytoplankton assemblages (see review in Sommer, 1990), although no critical tests of the model have been performed using benthic algal assemblages. Tolerance to suboptimal conditions unrelated to resource supply (e.g., extreme current, desiccation) may allow competitively inferior species to survive in certain microsites that are unsuitable for the growth of other species (Grime, 1979). Rather than representing a mechanism whereby species maximize their ability to compete for resources, tolerance to "nonresource" environmental stress provides a means of avoiding competition.

Trophic interactions represent a third principal means by which competing species can coexist. Herbivory is important in controlling algal biomass and productivity in most benthic environments (see Steinman, Chapter 12). The effect of herbivory on competition intensity and the potential for species coexistence is quite similar to the effects of disturbance. Removal of biomass by herbivores relaxes density-dependent interactions for limiting resources. Coexistence is further enhanced if herbivores selectively consume competitively dominant species; this is often the case in both macroalgal (Lubchenco, 1978) and microalgal (McCormick and Stevenson, 1989) assemblages since the competitive dominant tends to be large and/or loosely attached to the substrate. Intense grazing results in the loss of all but the most grazer-resistant species from the assemblage and relatively low species diversity. The predicted shape of the relationship between species diversity and herbivore intensity (Paine and Vadas, 1969; Lubchenco, 1978; McCormick and Stevenson, 1989) is similar to that for diversity and disturbance (Figure 2) as a result of the similar mechanisms underlying the two relationships. Exceptions to this relationship may occur if the competitively dominant species is also grazer-resistant. In such cases, grazing may only reduce the growth of inferior competitors further, thus increasing the probability of exclusion (Porter, 1977).

IV. ADAPTIVE TRAITS CONFERRING COMPETITIVE DOMINANCE AND AVOIDANCE

The various forms of environmental heterogeneity just discussed provide the template for the evolution of different ecological strategies that allow individual species to maximize performance under different environmental conditions (Southwood, 1977). Given the variable conditions encountered in most benthic habitats, it is untenable to assume that the intensity of competitive relationships remains constant through time and space. This suggests two general types of strategies that confer ecological success under different circumstances: those that maximize a species' ability to compete and those that allow species to avoid or minimize competitive interactions. Dominance by species adapted to compete effectively for resources is predicted where environmental conditions favor the development of large standing crops of benthic algae and density-dependent growth is realized. Under circumstances where competitive interactions are weak compared with other environmental influences (e.g., frequently disturbed sites), alternative ecological strategies unrelated to resource competition are favored. Various adaptive traits associated with each of these strategies have been identified for both plant and algal assemblages and are discussed in the following.

Any ecological strategy necessarily involves evolutionary trade-offs in a species' ability to maximize performance under different circumstances. In the absence of such trade-offs, species could evolve to competitively displace all other populations with overlapping resource requirements under any set of conditions, resulting in a progressive loss of global biodiversity through evolutionary time. Tilman (1982) notes that the fossil record refutes such a trend and shows how genetic evolution in heterogeneous environments precludes the formation of such a "superspecies." In many cases, the ecological trade-offs associated with an adaptive trait are clear, for example, a growth form that positions a species high within benthic algal mats increases access to light and nutrients but also increases susceptibility to removal by spates. In other cases, trade-offs have been inferred by documenting species performance under different conditions. For example, laboratory studies with phytoplankton have used this approach to show that species that compete well for one limiting resource are poor competitors for another limiting resource (Tilman, 1982; Sommer, 1990). A general trade-off that has been repeatedly emphasized by plant ecologists, and will be the focus of discussion here, is the inverse correlation between a species' competitive ability and its capacity to maintain populations under suboptimal (e.g., low resource) or adverse conditions (Grime, 1979; Chapin, 1980; Keddy, 1990). An understanding of important adaptive traits and attendant trade-offs associated with alternative ecological strategies allows for predictions to be made concerning changes in population abundances in response to changes in competitive pressures and other environmental influences.

A. Traits Conferring Competitive Dominance

To identify the adaptive traits that enhance competitive ability within benthic algal assemblages, a precise definition of competitiveness and the mechanisms whereby competitive exclusion occurs must be established. Two very different definitions of competitiveness have been debated in the literature on plant competition (see summary in Grace, 1990) and both can be applied with some success to algal assemblages. Tilman (1982) proposed a theory of resource competition that essentially defines competitiveness on the basis of a species' ability to tolerate low levels of a limiting resource. His model assumes that when two or more species utilize the same limiting resource, levels of that resource in the surrounding environment decline. Species are eliminated as resource levels drop below the minimum level required for the population to maintain itself. The species with the lowest maintenance requirement for the resource (termed R^*) effectively outcompetes all other species by continued resource depletion. Species coexistence is predicted on the basis of trade-offs in abilities to compete successfully for different resources in a spatially and temporally heterogeneous environment.

Grime (1979) posited a distinctly different definition of competitiveness, which emphasized a species' ability to effectively sequester resources and, in so doing, to suppress the growth of other populations. According to this view, species that exhibit a high capacity to capture available resources and convert them into biomass can effectively preempt the growth of other populations. Grime termed these species *competitors* and predicted that they would dominate in habitats with high resource supply rates that favor fast-growing populations. Thus, whereas Tilman predicts that the species with the lowest resource requirement will be competitively dominant, Grime equates competitive success with a species' intrinsic rate of growth. Indeed, Grime's theory proposes that tolerance of low resource levels, rather than conferring competitive success, is an adaptation that allows inferior competitors (by his definition) to avoid competition by maintaining populations under suboptimal environmental conditions. Species capable of withstanding low resource levels or other forms of environment "stress," which Grime termed *stress-tolerant,* are capable of maintaining populations in locations where other species are unable to grow. Grime proposed a third category of species, termed *ruderals,* which exhibit adaptations that allow them to maintain populations on highly disturbed sites, where loss rates are high and recolonization mechanisms are important. Whereas the range of strategies proposed by Grime explicitly considers the influence of disturbance and environmental stress on species performance and competition intensity, in Tilman's model these forces alter the competitive balance only to the extent that they affect the relative availability of different resources at a site.

Both of the theories just described have received support from experiments on plant and phytoplankton assemblages, although neither has been directly tested using benthic algae. The tenets of Tilman's model are consistent with adaptive traits, trade-offs, and outcomes of species competition in phytoplankton assemblages (reviewed in Sommer, 1990) and, thus, might seem the most logical starting point for developing a mechanistic theory of competition in benthic algal assemblages, given taxonomic and physiological similarities between benthic and planktonic algae. However, the ecological strategies that allow species to compete effectively in benthic and planktonic habitats are not necessarily similar given obvious differences in the physical structure of algal assemblages in these two environments. Vertical gradients of light and nutrients occur within benthic algal mats (Meulemans, 1987; Stevenson and Glover, 1993; see Fig. 1). Substrate-associated populations vary in their ability to actively seek these resources by growing or moving upward through the mat. It follows that species that exhibit extended growth forms can disproportionately reduce resource supply rates to populations deeper (e.g., prostrate, attached forms) in the mat while continuing to experience a relatively favorable resource environment. If this interpretation is correct, then knowledge of the R^* of individual species would be insufficient for predicting competitive outcomes since different populations are experiencing different resource environments. Competitive relationships in benthic algal assemblages would seem to be more similar to those in vertically structured plant communities, where Grime's definition of competitive success has gained support (e.g., Givnish, 1982; Grace, 1988; Keddy and Shipley, 1989), than those in the phytoplankton.

As indicated earlier, both physiological and morphological characteristics determine the competitiveness of species in benthic algal mats. Competitive dominants should be those species that have intrinsic rates of increase high enough to allow them to proliferate under favorable environmental conditions. Although critical studies of algal growth rates in nature are lacking, observational evidence suggests that algal species differ dramatically in their intrinsic rate of increase and, thus, in their ability to capitalize on increasingly higher rates of resource supply. Many algal species never appear to exhibit high rates of growth and are perpetually rare wherever they are found. A relatively small number of species are responsible for most of the algal "blooms" reported in benthic habitats. These species, which have the capacity to exhibit high rates of growth under favorable conditions, appear to fit the classification of *competitors* as defined by Grime (1979).

Growth form is often, although not always, linked to a species' ability to dominate benthic assemblages. Stature within the mat is important since it increases the probability that an individual will be able to acquire light and nutrients supplied from above. As in plant communities (Grace, 1988; Keddy and Shipley, 1989), the height of individuals of a species is positively correlated with its rank in competitive hierarchies in benthic algal assemblages. Competitively dominant species in stream diatom assemblages were typically those capable of producing mucilaginous stalks (McCormick, 1989; McCormick and Stevenson, 1991). As mat development proceeds and biomass accrues on the substrate, the advantage of height is evident in the increasing dominance of species exhibiting extended growth forms (e.g., stalks, filaments) (Hoagland *et al.*, 1982; Hudon and Bourget, 1983; Korte and Blinn, 1983).

Species that do not exhibit a filamentous or stalked growth form can maintain themselves in the upper sections of the mat in other ways. Motility, an adaptation not found in plants, provides the capacity for species to position themselves within favorable resource environments (e.g., positive phototaxis). Small prostrate diatoms (e.g., *Cocconeis*) often maintain themselves high in the mat by growing epiphytically on macroalgae (e.g., *Cladophora*). Sheer capacity for growth alone can confer competitive dominance to a population in some cases. The diatom *Achnanthes minutissima* was found to dominate the epilithon of a stream in Kentucky (U.S.A.) for several weeks, apparently by its ability to divide more quickly than other species in the assemblage, including stalked and filamentous species, and to form a mucilaginous mat several millimeters thick (P. V. McCormick, unpublished data). Observations such as this indicate that a high intrinsic rate of growth may be the only prerequisite for competitive success in benthic mats when the resource environment is favorable for growth.

B. Tolerance to Stress

Dominance by species capable of maintaining positive growth rates under suboptimal conditions are favored in habitats where resource availability is low or where other environmental factors (e.g., pH, salinity) inhibit the growth of other populations. The ability to tolerate adverse environmental conditions is an important adaptation that allows certain plant populations to dominate marginal sites in habitats where they might otherwise be excluded and a similar strategy can be envisioned for algal populations. Discussion here is limited to physiological adaptations to cope with shortages of essential resources, although other natural and anthropogenic environmental stressors are certainly important in many benthic habitats.

Both laboratory experimentation and field observation support the prediction that the ability of an algal population to capitalize on high resource availability is inversely related to the ability of the same population to maintain itself under low-resource conditions. Studies of algal nutrient kinetics suggest a trade-off between a species' physiological capacity to respond to increases in resource availability (i.e., high maximal uptake velocity) and its ability to obtain nutrients at low environmental concentrations (i.e., low half-saturation constant) (see Darley, 1982). Different patterns of response to changes in nutrient supply are predicted for species that are good at capturing nutrients when concentrations are high and those that exhibit efficient uptake mechanisms at low concentrations (Fig. 3). Equally important is the ability to convert resources into biomass: good competitors (*sensu* Grime) are predicted to have a greater capacity to increase biomass production at high supply rates, whereas tolerant species are able to maintain positive population growth rates at low resource levels. Physiological trade-offs in the utilization of light by algal cells parallel those documented for nutrient utilization.

The dichotomy between competitiveness and tolerance just presented contrasts sharply with the predictions of Tilman's model of resource competition discussed earlier, which defines competitiveness based on a species' ability to tolerate low resource levels. In fact, Tilman's view may accurately describe the types of species interactions occurring deep within the mat, where resources are most scarce. To the extent that light and nutrient supplies are depleted as a result of localized uptake, resource competition does occur within the mat and depletion of resources can be identified as the mechanism whereby species are outcompeted by populations that are more efficient at acquiring resources (i.e., possess a lower R^*). However, resource supply rates into the mat are also reduced by species growing near the top of the mat. To the extent that light and nutrient availability is reduced within the mat, tolerance of low resource levels by species within the mat is more accurately viewed as a means whereby inferior competitors avoid exclusion by species better adapted to resource capture. The distinction between the two viewpoints is important since each proposes a different mechanistic model of how benthic algal assemblages are structured. However, to debate which view is most "correct" seems futile given the paucity of experimental evidence with which to assess the intensity of these different types of interactions within the mat. Indeed, both types of interactions may be found to be important depending on environmental cir-

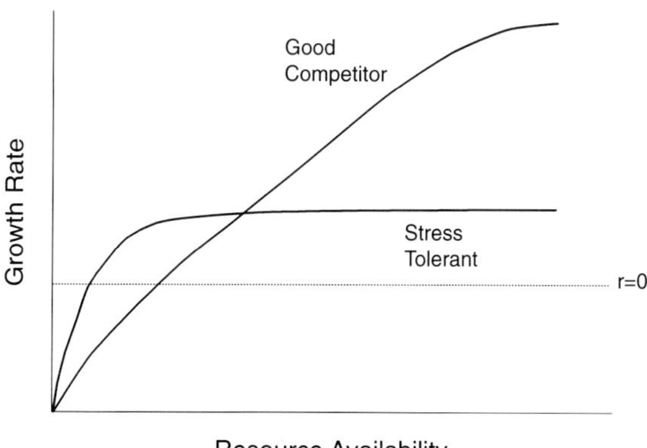

FIGURE 3 Hypothesized differences in nutrient uptake abilities and population growth rates in relation to resource supply for species that are good competitors (*sensu* Grime, 1979) and those that are tolerant of suboptimal resource levels. the dotted line (r=0) represents the zero net growth isocline below which a population declines; here it is assumed that both types of species have similar loss rates. Note the predicted trade-off between maximum resource capture and biomass yield and the minimum resource level required for survival.

cumstances and the degree of vertical structuring or layering in different types of algal mats.

C. Disturbance-Adapted Species

At least two general types of adaptations increase survival in disturbance-prone environments: 1) ability to withstand disturbance and 2) ability to colonize disturbed sites. Given that benthic algal mats are periodically disrupted, either by abiotic disturbance, herbivory, or senescence, both types of adaptations should contribute to a species performance in many benthic habitats.

Given the arguments that have already been presented, morphological adaptations conferring resistance to disturbance are necessarily the opposite of those characteristics that enhance competitive ability: whereas tightly attached, prostrate growth forms are more resistant to disturbance, extended forms should be more competitive in high-biomass environments where density-dependent interactions should be intense. Both abiotic and biotic disturbances disproportionately remove cells in upper layers of the mat. Tight attachment ability and low stature become the dominant adaptive trait in habitats where disturbance is frequent and severe enough to cause considerable scouring of the benthos. For example, epipsammic habitats in streams are prone to frequent disruption and are characterized by low-biomass algal assemblages dominated by prostrate forms that can exist within fine depressions and thus avoid scour effects (Miller *et al.*, 1987). Prostrate algal taxa exhibit constant or positive changes in abundance with increased herbivory, whereas loosely attached forms are invariably reduced in abundance, even under conditions of high resource availability (McCormick and Stevenson, 1989).

To maintain viable populations on the substrate during periods of mat development between disturbances, disturbance-resistant taxa must be capable of tolerating low resource levels within the mat. Several possible strategies, including uptake efficiency, may facilitate persistence under conditions of resource depletion. Luxury uptake of nutrients during times of high availability has been identified as one mechanism enhancing phytoplankton coexistence in variable resource environments (Sommer, 1985). Complete attenuation of light can occur if mat development is sufficient (Meulemans, 1987). Heterotrophy and dormancy are metabolic strategies that may allow taxa to survive extended periods of darkness (e.g., see Tuchman, Chapter 10).

A distinctly different set of adaptive traits enhances a population's ability to invade newly disturbed sites. Removal of benthic mats from the substrate as a result of scour, senescence, or some other form of disturbance creates patches where biomass is low and diffusion of nutrients and light from the water column is high. Two different strategies enhance a species' ability to invade these patches: (1) high immigration potential, that is, the probability that cells in the water column will successfully attach to the substrate, and (2) high emigration potential, that is, the probability that cells on the substrate will contribute to the water column pool of drifting cells. Both strategies have been found to be important in determining patterns of species dominance on newly exposed substrates in streams. Interspecific differences in immigration and emigration potentials have been found for diatom assemblages in several streams (Brettum, 1974; Stevenson and Peterson, 1989, 1991; Barnese and Lowe, 1992). High immigration potentials have been related to several factors, including cell size, shape, and specific gravity (e.g., Stevenson and Peterson, 1989), whereas emigration potential is likely related to a species' ability to reproduce rapidly under low-biomass conditions (i.e., high-resource conditions), thereby producing large numbers of propagules for dispersal. McCormick and Stevenson (1991) found that species that dominated newly exposed substrates included those with high immigration or emigration potentials as well as those capable of fast reproductive rates in low-biomass environments.

If accumulation was governed only by dispersal ability, species with high immigration and emigration potentials would be predicted to dominate undisturbed algal assemblages. In fact, eventual dominance on the substrate is best predicted on the basis of reproductive potential rather than dispersal potential (Stevenson, 1986). Studies of diatom succession on newly exposed substrates in streams have typically found a shift in dominance from species with high immigration/emigration potentials during early stages to those with fast reproductive rates during later stages (Stevenson, 1984; McCormick and Stevenson, 1991). Later-successional species exhibit the attributes of good competitors described earlier, including the ability to maintain relatively high reproductive rates under high-biomass conditions and the capacity to suppress the reproductive rate of other populations (McCormick, 1989; McCormick and Stevenson, 1991). High dispersal ability appears to provide a means of avoiding competitive exclusion by facilitating colonization of disturbed (i.e., low biomass) sites, but does not generally confer dominance in undisturbed locations.

D. Are Trade-Offs Absolute?

The four ecological strategies are summarized in Fig. 4 to emphasize differences in the adaptive traits predicted to confer success along gradients of productivity (i.e., resource supply) and disturbance, including herbivory, within and among freshwater benthic habitats. As discussed earlier, various physiological and morphological constraints generally prevent a species from exhibiting adaptations that allow it to dominate under all possible conditions (e.g., disturbed and undisturbed sites, sites with both very high

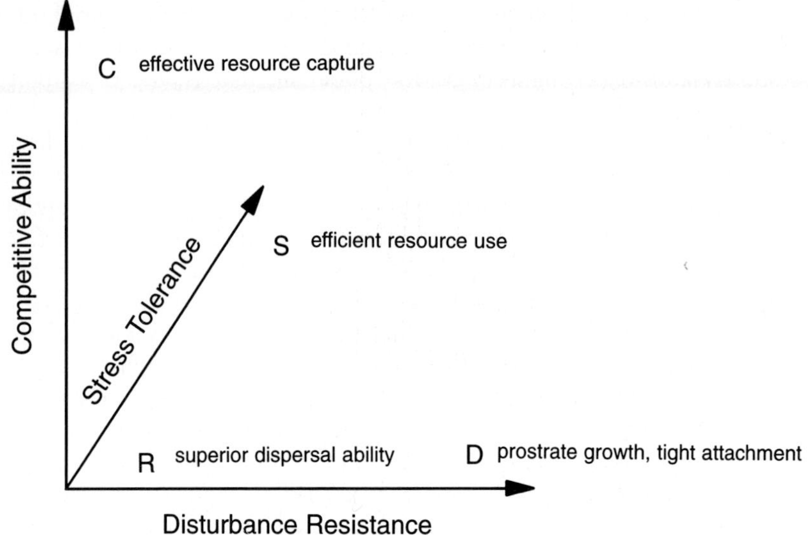

FIGURE 4 Ecological strategies and associated adaptive traits described in text. C, competitors adapted to maximize resource capture and growth rate; S, stress-tolerant species, for example, those adapted to efficiently utilize resources to maintain positive growth rates under conditions of low resource availability; D, disturbance-resistant species possessing morphological adaptations that prevent removal by scouring or herbivory; R, ruderal species adapted for colonizing disturbed sites where resource supply rates are high and density-independent interactions are weak.

and very low resource supply rates). However, it is important to note that traits indicative of different ecological strategies are not always incompatible. For example, algae that are adapted to withstand disturbance based on morphological adaptations (e.g., prostrate growth form) may also be tolerant of low resource levels, an adaptation that is physiological in nature. And, although the dichotomy between dispersal potential and competitive ability holds in some instances described earlier, other studies have found species capable of exhibiting both attributes (e.g., Stevenson *et al.*, 1991). Because of the morphological and physiological plasticity of microbes, an individual species may exhibit a wide range of physiological characteristics among habitats with different environmental conditions (e.g., adaptations to different light regimes). Thus, the categories proposed here should not be construed as being mutually exclusive and individual species may often exhibit adaptations associated with multiple strategies. However, a pivotal assumption of the theory developed in this chapter is that species-rich assemblages will only occur in habitats where spatial and temporal environmental variability at relevant scales (e.g., time frames approximating the life span of individual cells) exceeds the range of conditions under which individual species are predicted to succeed. In the absence of sufficient environmental heterogeneity, the development of a monoculture as observed for laboratory chemostat experiments would be predicted.

V. SUMMARY AND RECOMMENDATIONS FOR FUTURE RESEARCH

There is considerable circumstantial evidence to support the importance of competition as a force structuring benthic algal assemblages. These assemblages have repeatedly been shown to be resource-limited, a condition that is an obvious prerequisite for *resource* competition to occur. Benthic algal succession typically proceeds with the development of a vertically structured mat with high cell densities, conditions that favor the existence of asymmetric competitive interactions and, consequently, competitive exclusion. However, the pronounced spatial and temporal heterogeneity of most benthic habitats favors coexistence of many species that have evolved different ecological strategies for survival in variable environments. Thus, although competition appears to be an important force structuring benthic algal assemblages under certain conditions, it is only one of many processes that influence species richness and composition.

The ecological and evolutionary role of resource competition in benthic algal assemblages has yet to be critically explored. Evidence to support the importance of competition in most assemblages is based on observational data (e.g., collection of experimental substrates after different periods of incubation) that correlate reductions in species performance with increases in biomass. The perceived importance of different adaptive traits and postulated mechanisms of competition have largely been inferred from successional patterns rather than evaluated directly. Unfortunately, the importance of various structuring processes cannot be predicted from ecological patterns alone; in fact, community patterns resulting from competitive interactions are often indistinguishable from those generated at random (see Cale *et al.,* 1989). The literature on terrestrial plant competition illustrates the importance of manipulative experimentation, including variations in the presence and biomass levels of individual species as well as the rates of resource supply in ascertaining the intensity and mechanisms of competition. Experimental trends in other areas of ecological research indicate the need to investigate competition using manipulative experimentation rather than mere induction in order to derive accurate interpretations.

The small size and exceptional dispersal ability of most freshwater benthic algae complicate attempts to conduct the types of manipulative experiments (e.g., additions and removals of selected populations) that are commonly used to study competition in plant communities [see Keddy

(1989) and Goldberg and Scheiner (1993) for review of field methods and experimental designs]. Investigators attempting to study competitive interactions within phytoplankton assemblages have faced similar constraints, but have succeeded in developing a robust predictive theory by relying on extensive laboratory experimentation coupled with a modest amount of observational and manipulative field validation. Laboratory experimentation can provide considerable insight into the adaptive traits and potential mechanisms related to competition in benthic algal assemblages. For example, species from a particular system can be isolated and grown (if culturable) under various environmental conditions in order to determine intrinsic growth rates and the conditions (e.g., temperature, light, and nutrients) under which optimal growth occurs. Pairwise and multispecies addition experiments can be conducted to confirm or refute the importance of different ecological strategies posed here in determining a population's success under different circumstances and to predict species' rankings within competitive hierarchies. Innovative laboratory designs [e.g., the laboratory streams described by Hoffman (1993)] should prove extremely useful for such purposes.

Regardless of the design, laboratory experiments are inevitably constrained in their ability to quantify the importance of various ecological processes because of difficulties in recreating environmental conditions exactly. Ideally, field experimentation should be conducted to (1) quantify the effect of one population on the performance of another, in this case a negative density-dependent effect, and (2) identify the mechanism causing the effect. The most direct measure of the intensity of interspecific effects on species performance is obtained by selectively altering population abundances under reasonably controlled conditions (e.g., similar age and developmental history of different treatments). Although manipulations of individual populations are generally not feasible under field conditions, altering the abundance of groups of species is sometimes possible (McCormick and Stevenson, 1991). Competitive hierarchies can be discerned from such experiments and related to morphological and physiological characteristics ascertained from laboratory experimentation and field observation. Simultaneous manipulations of taxa abundances and resource levels can provide evidence of the relationship between density-dependent interactions and the supply of specific resources. However, resource manipulations in benthic algal assemblages are problematic because increasing nutrient and light levels in the water column may not affect resource availability within the mat.

The ultimate goal of ecological investigation is to achieve a predictive understanding of the natural world. The power of any ecological theory is dependent on the degree to which predictions can be applied to different types of ecosystems and different environmental circumstances. In the absence of such generality, predictive theory becomes idiosyncratic and ad hoc and contributes little to the advancement of understanding in the field. Many of the ideas and predictions in this chapter are based on the results of relatively few manipulative studies that have been performed in a limited number of benthic habitats, principally the epilithon of headwater streams. To develop a robust theory of competition and community structure for benthic algae, it is essential that laboratory and field experimentation be repeated using assemblages in other types of ecosystems. For example, do the adaptive traits that confer competitive dominance differ between epilithic assemblages, where both nutrients and light are supplied from the water column, and epipelic assemblages, where nutrients are also supplied from below? Do the importance and mechanisms of competition differ between lotic systems, where nutrients are continuously replenished, and lentic systems, where water column nutrients are periodically depleted as a result of stratification? Comparative studies should also attempt to assess the importance of competitive interactions and different ecological strategies in determining dominance within benthic algal assemblages across gradients of disturbance, herbivory, and trophic state in ecosystems of the same type within specific regions.

ACKNOWLEDGMENTS

The ideas developed in this manuscript benefitted substantially from conversations with Sue Newman and Chris Peterson. David Hart, Rex Lowe, Al Steinman, and Jan Stevenson provided helpful comments on the draft manuscript.

REFERENCES

Barnese, L. E., and Lowe, R. L. (1992). Effects of substrate, light, and benthic invertebrates on algal drift in small streams. *J. North Am. Benthol. Soc.* **11,** 49–59.

Brettum, P. (1974). The relation between the new colonization and drift of periphytic diatoms in a small stream in Oslo, Norway. *Norw. J. Bot.* **21,** 277–284.

Cale, W. G., Henebry, G. M., and Yeakley, J. A. (1989). Inferring process from pattern in natural communities: Can we understand what we see? *Bioscience* **39,** 600–605.

Carrick, H. J., Lowe, R. L., and Rotenberry, J. T. (1988). Guilds of benthic algae along nutrient gradients: Relationships to algal community diversity. *J. North Am. Benthol. Soc.* **7,** 117–128.

Chapin, F. S., III (1980). The mineral nutrition of wild plants. *Annu. Rev. Ecol. Syst.* **11,** 233–260.

Connell, J. (1978). Diversity in tropical rainforests and coral reefs. *Science* **199,** 1302–1310.

Connell, J. (1983). On the prevalence and relative importance of interspecific competition: Evidence from field experiments. *Am. Nat.* **122,** 661–696.

Darley, W. M. (1982). "Algal Biology: A Physiological Approach." Blackwell, Oxford, UK.

Dayton, P. K. (1975). Experimental evaluation of ecological dominance in a rocky intertidal algal community. *Ecol. Monogr.* **45,** 137–159.

Dayton, P. K. (1984). Processes structuring some marine communities: Are they general? *In* "Ecological Communities: Conceptual Issues and the Evidence" (D. R. Strong, Jr., D.

Simberloff, L. G. Abele, and A. B. Thistle, eds.), pp. 181–200. Princeton Univ. Press, Princeton, NJ.

den Boer, P. J. (1980). Exclusion or coexistence and the taxonomic or ecological relationship between species. *Neth. J. Zoo.* **30,** 278–306.

Fairchild, G. W., Lowe, R. L., and Richardson, W. B. (1985). Algal periphyton growth on nutrient-diffusing substrates: An *in situ* bioassay. *Ecology* **66,** 465–472.

Gause, G. F. (1934). "The Struggle for Existence." Hafner, New York.

Ghilarov, A. M. (1984). The paradox of the plankton reconsidered; or, why do species coexist? *Oikos* **43,** 46–52.

Givnish, T. J. (1982). On the adaptive significance of leaf height in forest herbs. *Am. Nat.* **120,** 353–381.

Goldberg, D. E., and Scheiner, S. M. (1993). ANOVA and ANCOVA: Field competition experiments. *In* "Design and Analysis of Ecological Experiments" (S. M. Scheiner and J. Gurevitch, eds.), pp. 69–93. Chapman & Hall, New York.

Grace, J. B. (1988). The effects of nutrient additions on mixtures of *Typha latifolia* L. and *Typha domingensis* Pers. along a water-depth gradient. *Aqua. Bot.* **31,** 83–92.

Grace, J. B. (1990). On the relationship between plant traits and competitive ability. *In* "Perspectives on Plant Competition" (J. B. Grace and D. Tilman, eds.), pp. 51–66. Academic Press, San Diego, CA.

Grime, J. P. (1979). "Plant Strategies and Vegetation Processes." Wiley, New York.

Grover, J. P. (1991). Dynamics of competition among microalgae in variable environments: Experimental tests of alternative models. *Oikos* **62,** 231–243.

Healey, F. P., and Hendzel, L. L. (1988). Competition for phosphorus between desmids. *J. Phycol.* **24,** 287–292.

Hill, W. R., and Boston, H. L. (1991). Community development alters photosynthesis–irradiance relations in stream periphyton. *Limnol. Oceanogr.* **36,** 1375–1389.

Hoagland, K. D., Roemer, S. C., and Rosowski, J. R. (1982). Colonization and community structure of two periphyton assemblages with emphasis on the diatoms (Bacillariophyceae). *Am. J. Bot.* **69,** 188–213.

Hoffman, J. P. (1993). Investigating hydrodynamic effects on diatoms at small temporal and spatial scales. *J. North Am. Benthol. Soc.* **12,** 313–384.

Holm, N. P., and Armstrong, D. W. (1981). Role of nutrient limitation and competition in controlling the populations of *Asterionella formosa* and *Microcystis aeruginosa* in semicontinuous cultures. *Limnol. Oceanogr.* **26,** 622–634.

Hudon, C., and Bourget, E. (1983). The effect of light on the vertical structure of epibenthic diatom communities. *Bot. Mar.* **26,** 317–330.

Hutchinson, G. E. (1961). The paradox of the plankton. *Am. Nat.* **95,** 137–145.

Hutchinson, G. E. (1967). "A Treatise on Limnology. II. Introduction to Lake Biology and the Limnoplankton." Wiley, New York.

Keddy, P. A. (1989). "Competition." Chapman & Hall, London.

Keddy, P. A. (1990). Competitive hierarchies and centrifugal organization in plant communities. *In* "Perspectives on Plant Competition" (J. B. Grace and D. Tilman, eds.), pp. 266–290, Academic Press, San Diego, CA.

Keddy, P. A., and Shipley, B. (1989). Competitive hierarchies in herbaceous plant communities. *Oikos* **54,** 234–241.

Korte, V. L., and Blinn, D. W. (1983). Diatom colonization on artificial substrata in pool and riffle zones studied by light and scanning electron microscopy. *J. Phycol.* **19,** 332–341.

Lewis, W. M. (1977). Net growth rate through time as an indicator of ecological similarity among phytoplankton species. *Ecology* **58,** 149–157.

Lubchenco, J. (1978). Plant species diversity in a marine intertidal community: Importance of herbivore food preference and algal competitive abilities. *Am. Nat.* **112,** 23–39.

MacArthur, R. H. (1958). Population ecology of some warblers of northeastern coniferous forests. *Ecology* **39,** 599–619.

McCormick, P. V. (1989). Mechanisms of benthic algal succession in streams. Ph.D. Dissertation, University of Louisville, Louisville, KY.

McCormick, P. V. (1990). Direct and indirect effects of consumers on benthic algae in isolated pools of an ephemeral stream. *Can. J. Fish. Aquat. Sci.* **47,** 2057–2065.

McCormick, P. V., and Stevenson, R. J. (1989). Effects of snail grazing on benthic algal community structure in different nutrient environments. *J. North Am. Benthol. Soc.* **8,** 162–172.

McCormick, P. V., and Stevenson, R. J. (1991). Mechanisms of benthic algal succession in lotic environments. *Ecology* **72,** 1835–1848.

Meulemans, J. T. (1987). A method for measuring selective light attenuation within a periphytic community. *Arch. Hydrobiol.* **109,** 139–145.

Miller, A. R., Lowe, R. L., and Rotenberry, J. T. (1987). Succession of diatom communities on sand grains. *J. Ecol.* **75,** 693–709.

Paine, R. T., and Vadas, R. L. (1969). The effects of grazing by sea urchins, *Strongylocentrotus* spp., on benthic algal populations. *Limnol. Oceanogr.* **14,** 710–719.

Pianka, E. R. (1969). Sympatry of desert lizards *(Ctenotus)* in western Australia. *Ecology* **50,** 1012–1030.

Porter, K. G. (1977). The plant–animal interface in freshwater ecosystems. *Am. Sci.* **65,** 159–170.

Price , P. W. (1984). Communities of specialists: Vacant niches in ecological and evolutionary time. *In* "Ecological Communities: Conceptual Issues and the Evidence" (D. R. Strong, Jr., D. Simberloff, L. G. Abele, and A. B. Thistle, eds.), pp. 510–524. Princeton Univ. Press, Princeton, NJ.

Riber, H. H., and Wetzel, R. G. (1987). Boundary-layer and internal diffusion effects on phosphorus fluxcs in lake periphyton. *Limnol. Oceanogr.* **32,** 1181–1194.

Round, F. E. (1981). "The Ecology of Algae." Cambridge University Press, London.

Schoener, T. W. (1983). Field experiments on interspecific competition. *Am. Nat.* **122,** 240–285.

Sommer, U. (1985). Comparison between steady state and non-steady state competition: Experiments with natural phytoplankton. *Limnol. Oceanogr.* **30,** 335–346.

Sommer, U. (1990). Phytoplankton nutrient competition—From laboratory to lake. *In* "Perspectives on Plant Competition" (J. B. Grace and D. Tilman, eds.), pp. 193–213. Academic Press, San Diego, CA.

Sousa, W. P. (1979). Experimental investigations of disturbance and ecological succession in a rocky intertidal algal community. *Ecol. Monogr.* **49,** 227–254.

Sousa, W. P. (1985). Disturbance and patch dynamics on rocky intertidal shores. *In* "The Ecology of Natural Disturbance and Patch Dynamics" (S. T. A. Pickett and P. S. White, eds.), pp. 101–124. Academic Press, New York.

Southwood, T. R. E. (1977). Habitat, the template for ecological strategies? *J. Anim. Ecol.* **46,** 337–365.

Steinman, A. D., and McIntire, C. D. (1986). Effects of current velocity and light energy on the structure of periphyton assemblages in laboratory streams. *J. Phycol.* **22,** 352–361.

Stevenson, R. J. (1984). How current on different sides of substrates in streams affect mechanisms of benthic algal accumulation. *Int. Rev. Gesamten Hydrobiol.* **69,** 241–262.

Stevenson, R. J. (1986). Mathematical model of epilithic diatom accumulation. *In* "Proceedings of the Eighth International Diatom Symposium" (M. Ricard, ed.), pp. 323–336. Koeltz Scientific Books, Koenigstein, Germany.

Stevenson, R. J., and Glover, R. (1993). Effects of algal density and current on ion transport through periphyton communities. *Limnol. Oceanogr.* **38,** 1276–1281.

Stevenson, R. J., and Peterson, C. G. (1989). Variation in benthic diatom (Bacillariophyceae) immigration with habitat characteristics and cell morphology. *J. Phycol.* **25,** 120–129.

Stevenson, R. J., and Peterson, C. G. (1991). Emigration and immigration can be important determinants of benthic diatom assemblages in streams. *Freshwater Biol.* **26,** 279–294.

Stevenson, R. J., Peterson, C. G., Kirschtel, D. B., King, C. C., and Tuchman, N. C. (1991). Succession and ecological strategies of benthic diatoms (Bacillariophyceae): Density-dependent growth and effects of nutrients and shading. *J. Phycol.* **27**, 59–69.

Strong, D. R., Szyska, L. A., and Simberloff, D. (1979). Tests of community-wide character displacement against null hypotheses. *Evolution (Lawrence, Kans.)* **33**, 897–913.

Tilman, D. (1977). Resource competition between planktonic algae: An experimental and theoretical approach. *Ecology* **58**, 338–348.

Tilman, D. (1982). "Resource Competition and Community Structure." Princeton Univ. Press, Princeton, NJ.

Tilman, D. (1989). "Plant Strategies and the Dynamics and Structure of Plant Communities." Princeton Univ. Press, Princeton, NJ.

Tilman, D., Kilham, S. S., and Kilham, P. (1982). Phytoplankton community ecology: The role of limiting nutrients. *Annu. Rev. Ecol. Syst.* **13**, 349–372.

Weiner, J., and Thomas, S. C. (1986). Size variability and competition in plant monocultures. *Oikos* **47**, 221–222.

9

Interactions of Benthic Algae with Their Substrata

JoAnn M. Burkholder

Aquatic Botany and Marine Sciences,
College of Agriculture and Life Sciences,
North Carolina State University,
Raleigh, North Carolina 27695

I. INTRODUCTION: THE INTERFACE NICHE OF BENTHIC ALGAE

Upon initial consideration, benthic algae might be assumed to have relatively unrestricted access to dissolved nutrient substrates and other resources from the water column, with the colonized solid surface or substratum offering limited, more energy-costly resource supplements. Appearance often is not, however, supported by fact—and in many aquatic environments, dissolved nutrients are depauperate in the overlying water relative to their concentrations at or within submersed surfaces (Marshall, 1976; Paerl, 1985). Surfaces provide an interface for concentration of charged and neutral particles, and may also directly provide nutrients (Bitton and Marshall, 1980). They can concentrate various organic and inorganic substances, including phosphate, iron, calcium, copper, other trace metals, amino acids, peptides, lipids, and organic acids (reviewed by Paerl, 1985). Microbial chemotaxis in attraction toward benthic surfaces can—and probably often does—mediate attachment and development of associations among colonizers within a matrix of microbial secretions, and between colonizers and the underlying substratum (Chet and Mitchell, 1976; Round, 1992).

Benthic biofilms of algae, bacteria, fungi, and small animals can more accurately be considered microzones at the liquid–solid interface that function under environmental conditions somewhat distinct from the surrounding aqueous environment (Nikitin, 1973). The overlying "boundary layer" of quiescent water, through which solutes can travel only by slow diffusion, isolates the biofilm community to some degree from the medium (Vogel, 1981). Its thickness is reduced under turbulent conditions, but may be in excess of 1 cm in sheltered lake coves, where the effective isolation of the biofilm community from the water is most extreme (Riber and Wetzel, 1987). Thus, the physical structure of a submersed surface itself creates microhabitat that is chemically distinct and nutrient-enriched relative to the overlying water. The developing biofilm is three dimensionally enmeshed with hydrated glycocalyx and other mucopolysaccharide materials, secreted by bacteria and algae, that act both to sequester ions and to further isolate the microorganisms from the water column (Geesey *et al.,* 1977; Burkholder *et al.,* 1990; Fig. 1). Accumulated sediments and precipitated calcium carbonate, as well as the glycocalyx materials, enhance nutrient enrichment in the habitat by adsorbing phosphorus, ammonium, and various organic substances (Avnimelech *et al.,* 1982; Grobbelaär, 1983; Froelich, 1988; Figs. 2 and 3). Various organics such as glucose and acetate that can be used by benthic microalgae (Droop, 1974) are also released by secreta/excreta from epiphytic micro- and macrofauna, and by decomposition of plant and animal remains that have settled out (Wetzel, 1983).

This chapter addresses known and hypothesized physical/chemical interactions of benthic algae with their substrata. The influences of substrata physical features on benthic algae encompass a long-standing topic of research, and known major effects will be summarized. There exists a wealth of information on use of water-column nutrients by benthic algae (see Borchardt, Chapter 7, this volume) but, because experiments with substance-diffusing natural substrata are methodologically more difficult, identification of the suite of nutrients and other substances available from natural substrata—and the relative importance of those resources to the colonizing algal communities—are poorly understood. In part this discussion is organized by substratum type; it focuses on known or inferred chemical interactions between benthic algae and (1) rock substrata colonized by epilithic and endolithic algae; (2) sand, mud, or other sediments

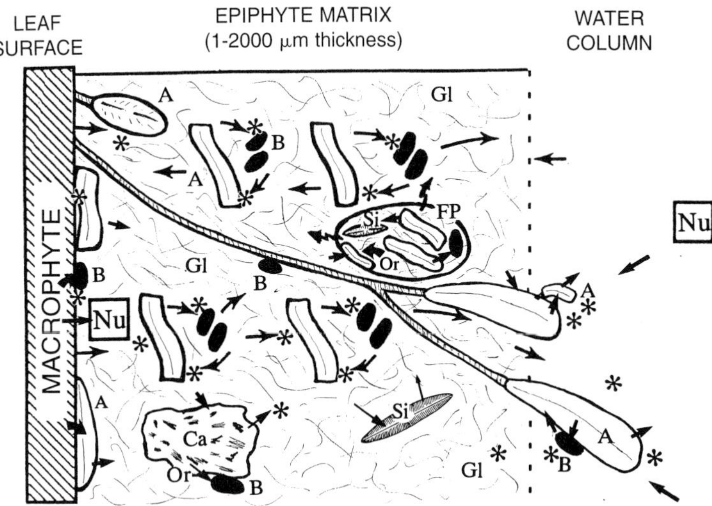

FIGURE 1 Schematic diagram of the supply of nutrients (Nu, as organic and inorganic forms of carbon, phosphorus, and nitrogen) to benthic algae within a biofilm matrix on a biological substratum, in this case a submersed macrophyte leaf. Arrows indicate routes of nutrient cycling from the substratum and the water column, which represent the two major exterior sources of nutrient supply to the epiphyte matrix. Nutrients may be taken up [sometimes with aid of excreted enzymes such as phosphatases (*)], utilized, and released (via leaching, excretions, secretions, or cell lysis) by benthic algae (A) [e.g., showing live diatoms with ornamentation obscured by outer organic covering; Darley (1974)] and decomposers represented by bacteria (B). Macrofauna (e.g., snails, amphipods) and microfauna (e.g., protozoan ciliates) take up nutrients by phagotrophy of bacteria, algae, or other prey, and/or by osmotrophy of dissolved organics. The animals can release phosphatases, and also release nutrients during feeding (e.g., from lysing of algal cells) and excretion [e.g., of fecal pellets (FP) with viable algae and bacteria that have become nutrient-enriched from passage through the animal gut tract (also see Fig. 2B), nutrient waste products, and dead microflora and other organic detritus]. Other nutrient-sequestering/releasing materials include inorganic calcium carbonate (Ca) that is precipitated by photosynthetic processes and silicious frustules of dead diatoms (Si), organic debris [e.g., recently dead algae and animals (Or)], and microbially derived hydrated glycocalyx/mucopolysaccharides (G1). The microflora and fauna can also be colonized and attacked by fungi, especially when moribund, and this process would result in nutrient remineralization and/or release. Note the algae and bacterium with major cell axis in direct contact with the leaf surface in an adnate habit. The other microflora are loosely attached (± stalks or other attachment structures) in the overstory biofilm matrix. (Modified from Burkholder, 1986.)

colonized by edaphic algae; (3) plant substrata colonized by algal epiphytes; and (4) animal substrata colonized by algal epi- or endozooites. Aside from nutrition and other beneficial aspects, known or inferred adverse chemical effects of benthic substrata will also be included, as exemplified by plant substratum release of allelopathic compounds that adversely affect algal colonizers. Associations will be considered accord-

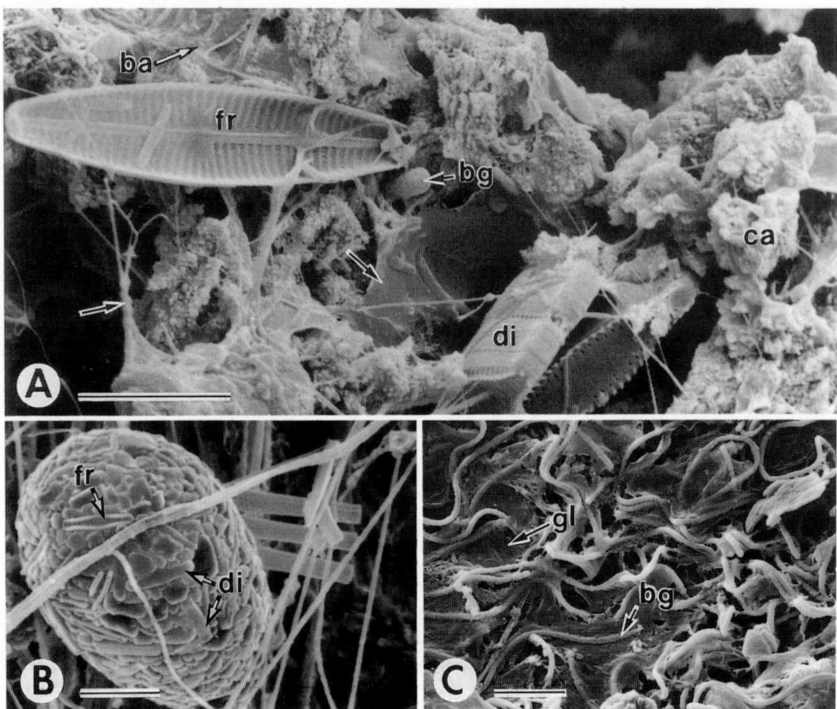

FIGURE 2 Scanning electron micrographs (SEMs) of loosely attached epiphyte communities on substrata, including (**A**) an aging *Potamogeton* leaf, showing the biofilm matrix enmeshed in glycocalyx and other polysaccharide materials (arrows) that have been secreted by the bacteria (ba, uppermost arrow, within glycocalyx) and microalgae (bg = blue-green filament, di = diatoms); also evident are abundant, amorphous precipitated calcium carbonate (ca) and other debris such as the frustule remains of a diatom (fr) (scale bar = 10 μm); (**B**) an artificial substratum (polyethylene strip, colonized ca. 12 weeks during summer) from a control enclosure in a shallow, turbid piedmont impoundment, showing a fecal pellet with empty diatom frustules (fr) and with diatoms (di) that were viable when sampled [with frustule ornamentation obscured by outer membranes; Darley (1974)] (scale bar = 15 μm); and (**C**) a polyethylene strip from a P-enriched enclosure in the same impoundment (mean TP = ca. 70 μg/liter), showing the glycocalyx matrix (gl) with abundant *Anabaena* (bg) that colonized from the water column in this simulated eutrophic condition [see Burkholder and Cuker (1991) for a detailed description of the experiment] (scale bar = 30 μm).

ing to the degree of chemical interaction beginning with loose nonobligate colonizations, and culminating with highly specialized endosymbioses of these algae with plant and animal hosts. In closing, the current state of knowledge and limitations in our understanding of benthic algae/substrata interactions will be summarized, including recommendations for future research directions.

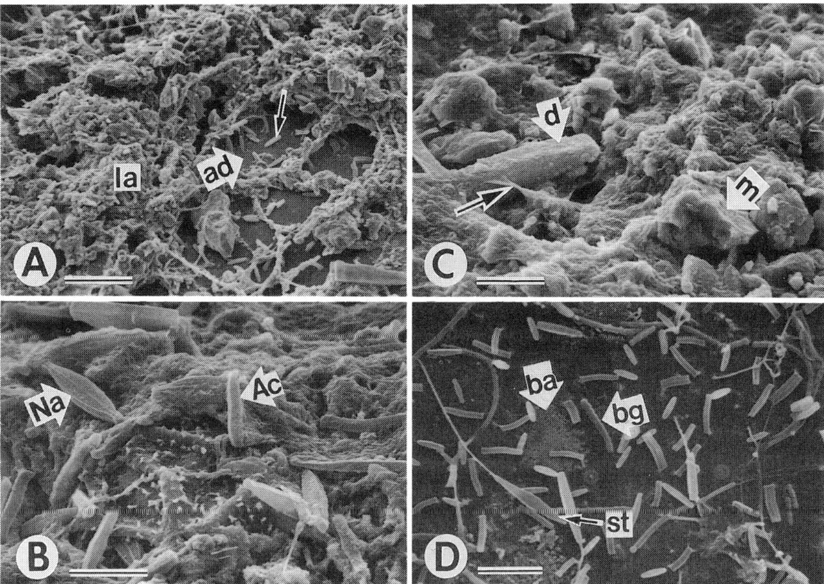

FIGURE 3 The periphyton community from a run-of-river impoundment after 10 weeks of experimentally imposed P enrichment and episodic sediment loading to polyethylene enclosures that each isolated a column of water, open to the surface and the sediments (June–August 1988, North Carolina piedmont region; phosphate added at weekly intervals to effect an initial concentration of ca. 100 μg PO_4^{3-}-P liter^{-1} in the water column; sediment added weekly as a slurry of naturally occurring, hydrated montmorillonite clay, to effect a mean concentration of 25 μg suspended sediments liter^{-1}). The periphyton were sampled from polyethylene strips that were suspended at 0.5 m depth [see Burkholder and Cuker (1991) for details describing similar methods from a 1987 experiment]. SEMs show representative microarchitecture, including (**A**) the loosely attached component (la) with abundant montmorillonite clay debris and the underlying adnate component (ad) with *Achnanthes* (arrow) resting against the substratum surface (scale bar = 50 μm); (**B**) a detailed view of the surface loosely attached or overstory periphyton from an enclosure to which nine hybrid green/bluegill sunfish (*Lepomis cyanellus* Rafinesque/*Lepomis macrochirus* Rafinesque) were added and maintained throughout the 10-week period, showing numerous healthy diatoms [e.g., *Navicula* sp. (Na) and *Achnanthes* sp. (Ac)] among the sediment debris, which is covered with mucopolysaccharide secretions (scale bar = 15 μm); (**C**) a detailed view of the surface loosely attached periphyton from an enclosure where fish were excluded during the experiment, showing a few diatoms (d) among the settled montmorillonite clay (m); note the small clay particles on the diatom surface, and the glycocalyx materials (arrow) (scale bar = 10 μm); and (**D**) the adnate component of the periphyton biofilm from an enclosure with fish after removal of the overlying loosely attached component, showing abundant small diatoms along with bacteria (ba) and filamentous blue-green algae (bg); note also the presence of the stalked diatom (*Gomphonema,* st), which probably was an overstory component in the intact periphyton biofilm (scale bar = 30 μm). In the shallow habitat, suspended sediment additions apparently stimulated periphyton growth by settling out and creating a more nutrient-rich benthic microhabitat. This stimulatory effect on periphyton was enhanced by the sunfish, especially for diatoms; the sunfish added nitrogen to the system, promoted more rapid settling of nutrient-coated clay particles, and foraged on invertebrate predators and grazers in the benthic food web.

II. SUBSTRATUM PHYSICAL INFLUENCES

Some of the oldest literature on benthic algae examines influences of physical properties of the associated substrata. Boundary-layer isolating/insulating effects caused by the physical presence of the substratum surface are more fully discussed by Borchardt (Chapter 7, this volume). A more obvious macrofeature influencing algal settlement and growth is the substratum's physical construction. Substrata with netlike or weblike configuration (e.g., the bryophyte *Fontinalis*) can filter algal colonizers from the water (i.e., slow the water and enable high settling/attachment rates) relative to substrata of simple construction (e.g., twigs; Burkholder and Sheath, 1984). This factor would be expected to be important in algal colonization and growth especially in lotic habitats, where netlike substrata can accumulate and maintain high diversity and abundance of benthic microalgae (Tippett, 1970). Algal abundance and species composition are also controlled by the time available for substratum colonization, and by surface microtopography. For example, a senescing leaf would be expected to have more dense colonization by algae and bacteria than a young leaf on the same plant (Godward, 1934; Burkholder and Wetzel, 1989a; Fig. 4). Moreover, leaves and stems of submersed macrophytes typically support low algal colonization during rapid growth periods; once the structures are fully formed, algal growth can accumulate (Cattaneo and Kalff, 1978; Burkholder and Wetzel, 1989b).

Substrata with changing surface microrelief, such as microcrevices in rock or the "cobblestone" effect of cells comprising a *Potamogeton* leaf, include depressions where spores can settle without being easily dislodged by abrasion or current (Miller *et al.*, 1987; Burkholder and Wetzel, 1989a; Fig. 4). Definitive experimental research to demonstrate the importance of microtopography in algal settlement/germination has focused mostly on the spores of marine macroalgae (e.g., Harlin and Lindbergh, 1977; Norton and Fetter, 1981; Reed *et al.*, 1988), but analogous factors likely occur

FIGURE 4 Adnate microflora (algae and bacteria) epiphytic on *Potamogeton* leaves, showing changes with substratum age progressing from (**A**) a recently mature leaf, colonized ca. 21 days, with numerous bacteria, diatoms, and calcium carbonate crystals (ca); note the blue-green algal filament (lower arrow) and a smaller diatom (*Achnanthes,* upper arrow) epiphytizing the larger diatom epiphytes (*Eunotia*) (scale bar = 15 μm); (**B**) a senescent leaf, colonized for 56 days, with abundant bacteria colonizing the macrophyte surface and the diatom epiphytes; the remnant of a smaller diatom epiphyte also is attached to the larger *Eunotia* cell, and the leaf shows areas of cuticular damage (arrows) where leakage of internal materials would be expected to be higher (scale bar = 15 μm); and (**C**) a recently dead leaf, available for epiphyte colonization for 98 days, with other loss of structural integrity evident as holes and tears in the tissue (arrows) with clusters of bacteria that apparently were attracted to leached nutrients (scale bar = 30 μm) (From Burkholder and Wetzel, 1989a, with permission.)

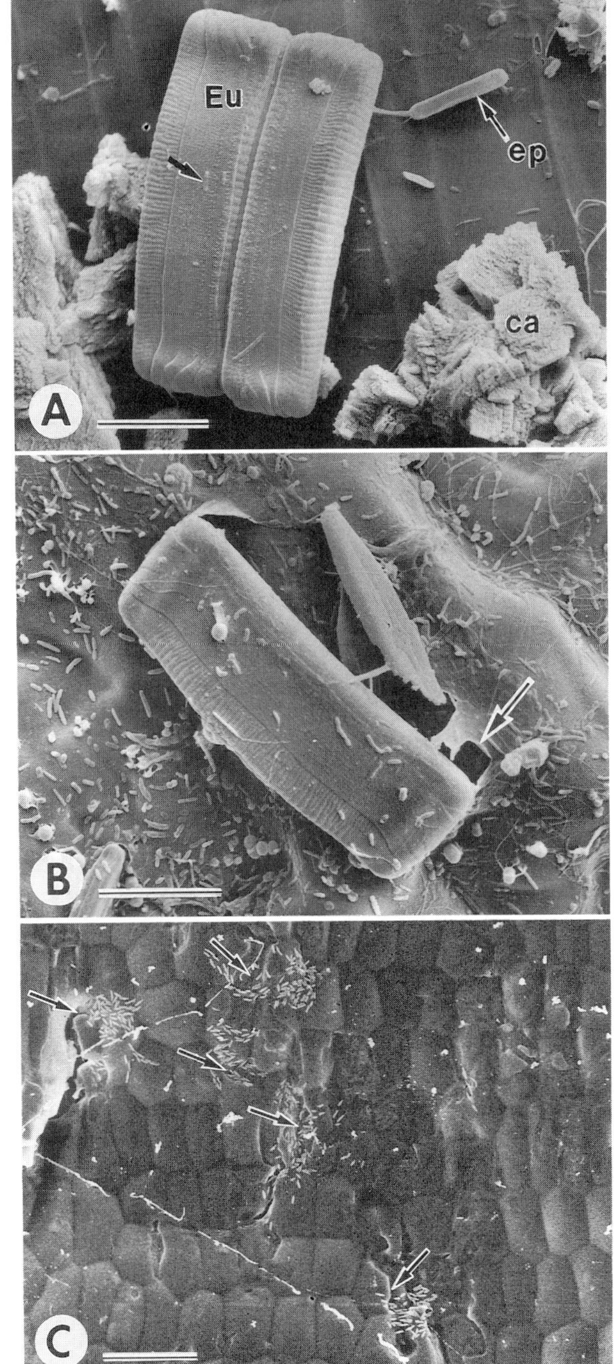
Eu
ep
ca
A
B
C

for algae in freshwater systems (Sheath, 1984). In lotic habitats, for example, R. J. Stevenson (1983) reported that diatom (Bacillariophyceae) immigration or colonizing rates significantly increased when nylon filament was used to reduce shearing stress by interrupting the laminar sublayer of currents near substrata surfaces (unglazed ceramic tiles). The permanence or physical "dependability" of the substratum is also a major factor influencing algal colonization, especially in flowing waters or in wave/splash zones of lakes (Whitton, 1975; Wetzel, 1983). For example, large outcrops of granite flatrock in the piedmont of the southeastern United States open the tree canopy and offer highly dependable substrata that remain stationary through flood scouring events. These substrata are colonized by unusually thick carpetlike growth of the red alga *Paralemanea annularis* (Kütz.) Vis & Sheath (Rhodophyceae), the blue-green *Phormidium subfuscum* Kütz. (Cyanophyceae, or "cyanobacteria"), and other macrophytes (Everitt and Burkholder, 1991). In contrast, small rocks and sediments move with wave or current action, with highly variable accrual rates influenced by wind, rain, or flooding events. Such environments would select for motile microalgae such as diatoms, as opposed to large populations of sessile algae that would be more easily buried or crushed (Hynes, 1970; Whitton, 1975).

Another important factor controlling algal colonization of benthic substrata is physical "preconditioning" by bacteria, adnate diatoms, and other "primary" colonizers found, for example, in macrophyte colonization from both freshwater and marine systems (Sieburth and Thomas, 1973; Burkholder and Wetzel, 1989b). In what is generally regarded as the first of two phases of substratum colonization, weak physical and chemical forces act to either hold or repel microorganisms to or from the surface. This instantaneous, reversible process results from attraction between the substratum and hydrophobic areas of the microbial cells (Marshall and Cruickshank, 1973; Daniels, 1980; Stevenson, 1983). The second phase of colonization involves secretion/excretion of materials that anchor the microorganism to the surface, a process that is considered "irreversible" for bacterial (and, by extension, algal) adsorption pending additional disturbance (Marshall and Cruickshank, 1973; Stevenson, 1983). The bacterial/algal flora provide adhesive glycocalyx and mucopolysaccharides that facilitate settling of secondary colonizers. Stevenson (1983), for example, reported that diatom immigration (colonization) in lotic systems was markedly enhanced when unglazed ceramic tiles were coated with agar, simulating the higher adsorptive potential of surfaces covered by microbial secretions [also see Dillon *et al.* (1989) for marine habitat]. These materials also greatly influence the chemical environment of free-living epibenthic algae which is poorly understood, in part because of technological difficulties in isolating the response of benthic algae to their chemical microhabitat as an intact, undisturbed community.

III. THE CHALLENGE OF EXAMINING SUBSTRATUM CHEMICAL INFLUENCES

Most available assay techniques for benthic algal metabolism have required either consideration of all organisms and debris collectively or separation of the cells from the associated surface (Robinson, 1983). The former approach does not isolate the response of the various components—bacteria, algae, fungi, and animals. Moreover, the metabolic response of the algae in certain assays, such as phosphate uptake, is difficult to interpret because a major portion of the uptake may actually represent adsorption of the ions to debris (Moeller *et al.*, 1988; Burkholder *et al.*, 1990). Physical separation of the community from the surface results in complete disruption of the chemical microhabitat within the biofilm matrix and, often, damage to or death of the cells (Karl, 1980; Delbecque, 1985; Burkholder and Wetzel, 1990).

These difficulties must be confronted when considering the water column as a nutrient source for benthic algae, but they are compounded when attempting to assess the importance of the underlying substrata in supplying nutrients, allelopathic substances, or other compounds because of (1) the often extreme heterogeneity in substratum colonization (Morin and Cattaneo, 1992), (2) the high quantities of "interfering" abiotic/biotic debris (Moeller *et al.*, 1988), and (3) the potential for wounding or otherwise promoting spuriously high release of substratum materials when attempting to remove the algal colonizers (Moeller *et al.*, 1988). The typical experimental approach is to "mark" the substratum source using a radiolabeled tracer for the substance or nutrient of interest (usually phosphorus or carbon, for which such tracers are readily available). Sufficient tracer must be added to account for adsorption by biotic and abiotic debris, as well as the living algal cells. Interpretations become more uncertain; if the periphyton biofilm is left intact, then most of the cells down in the matrix may remain isolated from the radiolabeled nutrient source, whether it is the water column or the underlying substratum (Burkholder and Wetzel, 1990; also see Borchardt, Chapter 7, this volume). If a longer "acclimation" period is used to minimize that problem, the appropriate duration must be estimated while attempting to account for potential interference from other microorganisms such as bacteria and potential recycling of labeled ions. A similar problem is confronted when using either the ^{14}C or oxygen-evolution methods to estimate periphyton productivity, since a critical assumption of both techniques is that the ^{14}C or oxygen should be in diffusional equilibrium between the periphyton layer and the surrounding water (Carlton and Wetzel, 1987). Acclimation periods of hours to days in light or darkness will inevitably involve bacterial as well as algal metabolism. Thus, after a benthic substratum is suitably labeled—more readily accomplished for some substrata than for others—the most desirable technique would provide the means to focus on the living algal cells in the biofilm layer, among all the background debris.

Track light microscope and track scanning electron microscope autoradiography (TLMA and SEMA, respectively) have been used to achieve this focus on viable algal colonizers (Pip and Robinson, 1982a,b; Paul and Duthie, 1988; Burkholder *et al.,* 1990). These powerful techniques offer microscale resolution at the level of individual cells. Equations are available for translating tracks per cell from [^{14}C]-bicarbonate to actual carbon uptake (Pip and Robinson, 1982a). Similar quantitative information can be obtained for phosphorus uptake using ^{33}P, assuming that pools, sources, and sinks can be quantified (Moeller *et al.,* 1988). However, when considering carbon from organic substrates or [^{33}P]-phosphate, it cannot be assumed that one track represents one radiolabeled ion—that is, the relationship between tracks and the number of radiolabeled ions actually taken by the cell vary with radiolabel concentration, incubation period, cell compartmentalization/metabolism of the radiolabeled and nonlabeled substrate pools, and other factors (Rogers, 1979). Large cells become densely labeled before small cells, so that differential exposure periods are needed (along with multiple sets of prepared slides) when data are required for an entire algal community (Burkholder *et al.,* 1990). Other limitations in quantitative interpretation when applying the technique to benthic algal communities are discussed in Moeller *et al.* (1988). Grain-density autoradiography (GDA) can be applied to both algae and bacteria, enabling use of ^{3}H-labeled nitrogenous compounds with weaker energy emissions, as well as other radiolabels (Paerl and Stull, 1979; Paerl, 1982). However, quantitative information from GDA requires enumeration of individual silver grains, rather than tracks, among the debris—a difficult procedure, at best.

In freshwater habitats, most of the definitive experiments to assess chemical interactions between benthic algae and their substrata have been conducted only within the past decade, and much additional work is needed. Benthic algae are strategically positioned to utilize nutrients released from the substratum as well as the water column (Confer, 1972; Bjork-Ramberg, 1985; Fig. 1). Researchers have used nutrient-diffusing artificial substrata ± water-column nutrient enrichments to demonstrate that benthic as well as water-column sources are utilized by benthic algal communities (Fairchild *et al.,* 1985; Carrick and Lowe, 1988; Pringle, 1990). Pringle's (1990) classic experiments also established that the community structure and abundance of overstory versus understory microalgae were highly dependent on whether the primary nutrient source was benthic substrata or the water column. Nutrient and other chemical interactions between benthic algae and most natural substrata have not been rigorously examined, and remain a major challenge in the study of benthic algal ecology.

IV. KNOWN INFLUENCES OF SUBSTRATUM COMPOSITION

A. Rock Substrata and Epilithic/Endolithic Algae

Upon initial consideration, rock substrata would be regarded as "inert" area for colonization by benthic algae, and it might be assumed that periphyton on rock surfaces would derive nutrients either from the overlying water or from microbial regeneration within the periphyton matrix (Stevenson and Glover, 1993). Macroalgae in freshwater ecosystems generally colonize rock substrata by means of rhizoids or filamentous structures that are used for anchoring; these algae are believed to derive their nutrients from the water column (Whitton, 1975; Sheath and Hambrook, 1990). In a model of the nutritional ecology of epilithic microalgae in river systems, Lock *et al.* (1984) invoked these two possible nutrient pools (i.e., the water column and microbial regeneration). Other researchers demonstrated adsorption of nutrients from the overlying water and subsequent release by an epilithic stream microflora community (e.g., amino acids; Armstrong and Barlocher, 1989). The potential for epilithic microalgae to derive nutrients from the rock substratum itself has not been examined, but would be expected to depend on the rock's chemical composition, porosity, crystal size, and other features, as well as the temperature and pH of the medium. For example, silica needed for benthic diatom growth would be more easily dissolved from porous sandstone than from granite (Hiebert and Bennett, 1992).

Occasionally, researchers have noted the presence of pennate diatoms with internal cyclosis and golden pigments, apparently photosynthesizing within large crystals of calcium carbonate (Round, 1981; Burkholder, 1986). Whether these algae would be viable for long periods is not known. Considering an analogy from more extreme habitat with freshwater influence, "crypto-endolithic" algae (reviewed by Bell, 1993) from hot semiarid lands and hot deserts (including intermittent stream areas) were first described in the 1960s. Such communities consist mostly of coccoid and filamentous blue-green algae, although coccoid and filamentous green algae (Chlorophyceae) have been found in cold deserts, as well (Bell, 1993). The harsh microclimate of the exposed rock surfaces inhibits epilithic algal colonization so that light passes relatively unimpeded through the crystal lattices. Porosity enables the organisms to gain entry to the rock interior, and also allows penetration of nutrients and water from the exterior. Although little is known about the nutritional ecology of endolithic algae, it is hypothesized that these organisms rely on atmospheric vapor as a water source. Most crystalline sandstones have calcium carbonate as an adhesive, which probably is solubilized by the endolithic algae; this carbonate removal, together with recycling of respiratory carbon, is thought to provide the carbon source for the slowly growing endoliths. Friedmann and Kiebler (1980) surveyed many hot/cold desert rocks but did not find evidence for algal nitrogen fixation capability. The primary source of nitrogen apparently is nitrate, which may be derived from exterior sources of water and from decomposing materials (Bell, 1993). Phosphorus or other nutrition for these communities has not yet been examined.

B. The Edaphic Habit: Epipsammic and Epipelic Algae among Sands and Other Sediments

On muddy, sandy, or marl sediments in freshwater ecosystems, charophytes (Charophyceae) are the only macroalgae known to have rhizoids that are capable of limited nutrient uptake from the sediments, a feature that allies them closely with terrestrial bryophytes (Raven, 1981; Mishler and Churchill, 1985; Hendricks and White, 1988). Many microalgal periphyton communities, as well as macroalgal mat formers such as the xanthophyte *Vaucheria* (Xanthophyceae) or the blue-green *Phormidium*—along lake shorelines, in streams and near areas of stream entry to lakes, in marshes or shallow run-of-river impoundments—must survive episodic sediment loading/resuspension events. These communities often are inhabited by motile forms such as pennate diatoms, blue-green algae, euglenoids (Euglenophyceae), and dinoflagellates (Dinophyceae) (Round, 1981; Burkholder and Cuker, 1991). Other inhabitants include algae that effectively can extend on a longitudinal axis to avoid permanent burial [e.g., filamentous taxa such as the green alga *Oedogonium inconspicuum* Hirn or the diatom *Eunotia pectinalis* (O.F. Müll.) Rabh., or species that grow on mucilaginous stalks such as the diatom *Gomphonema parvulum* Kütz. (Burkholder *et al.*, 1990); see Figs. 1 and 3)].

Suspended sediments can act as a source or a sink for nutrients and other substances depending on the particle composition, aging, and chemical history (Sozogni *et al.*, 1982; Froelich, 1988). Glycocalyx and other mucilage secretions from the benthic algae and bacteria coat sediment particles (Fig. 3) and adsorb organics, ammonia, phosphate, and other nutritional substances (Grobbelaär, 1983; Avnimelech and McHenry, 1984; Froelich, 1988; Burkholder, 1992). Microbial secretions can also help create anaerobic microenvironments (Paerl, 1990) that enhance solubility and desorption of nutrients such as phosphate for rapid and repeated cycling within the biofilm layer (Lock *et al.*, 1984). Hence, in turbid habitats impacted by sediment loading/resuspension, periphyton actually can thrive in part because of enhanced access to nutrients from close association with the incoming sediments that become an accumulated substratum (Burkholder and Cuker, 1991; Fig. 3).

For some benthic microalgae, correlative evidence for reliance on sediment nutrients has been linked to vertical migration through sediments across steep gradients in oxygen and dissolved nutrients (Revsbech *et al.*, 1983). The movement often follows circadian or diurnal rhythm [e.g., in estuaries, benthic euglenophytes described by Palmer and Round (1965), and pennate diatoms in Round and Palmer (1966); similar phenomena for diatoms have been observed by the author in sand flats along upper Lake Michigan and other freshwater habitats]. The algae vertically migrate down several centimeters through surficial sediments at night and then return to the sediment surface to photosynthesize in daylight [Palmer and Round (1965); in marine intertidal habitats, vertical migration downward at high tide is also believed to enable the algae to avoid being dislodged and removed from the habitat]. A similar phenomenon was documented for filamentous blue-greens (e.g., Richardson and Castenholz, 1987) in the sediments of thermal pools, apparently as an avoidance response to high levels of sulfide (Castenholz, 1976). It is generally believed that in most habitats of occurrence, microalgal vertical migration down into the sediments affords access to higher concentrations of nutrients that are more soluble in hypoxic or anaerobic conditions and, hence, more readily available (Round, 1981).

Mineralization of nutrients in shallow waters occurs mainly at the sediment surface (Wetzel, 1979). The potential for edaphic algae to act as a "sink" for sediment-derived nutrients, thereby reducing or preventing uptake by phytoplankton in the overlying water, is widely recognized (e.g., Confer, 1972; Cahoon *et al.*, 1990). This benthic algal mediation of nutrient cycling between the sediments and the water column is dynamic and fluctuates at least on a diel basis. From elegant experiments using a microscale approach and a novel flow-through system, Carlton and Wetzel (1988) reported that epipelic diatoms in a hard-water lake mediated release of [^{32}P]-phosphate from sediments to the overlying water via daily formation and breakdown of an oxygenated microzone. In the light, surficial sediments were rapidly oxygenated from algal photosynthesis, and release of phosphorus diffusing from deeper sediment layers was inhibited (Fig. 5). In darkness, the microzone became anoxic and phosphorus was released to overlying water at an accelerated rate.

In examining evidence for nutrient resource competition between phytoplankton and epipelic periphyton, Hansson (1988) used laboratory assays to test a hypothesis similar to that of Carlton and Wetzel (1988), namely, that the periphyton mediated outflow of dissolved nutrients from the mineralization zone of the sediment surface. The experiments revealed that substantial [^{33}P]-phosphate from sediments with negligible periphyton (2 μg chlorophyll *a* cm^{-2}) was detected in the overlying water after several days. By contrast, much less of the sediment radiolabeled phosphate reached the water column when periphyton were present (green algae *Ulothrix* sp. and *Scenedesmus* sp.; diatoms *Fragilaria* sp. and *Achnanthes* sp.; 14–25 μg chlorophyll *a* cm^{-2}). The portion of tracer in the water col-

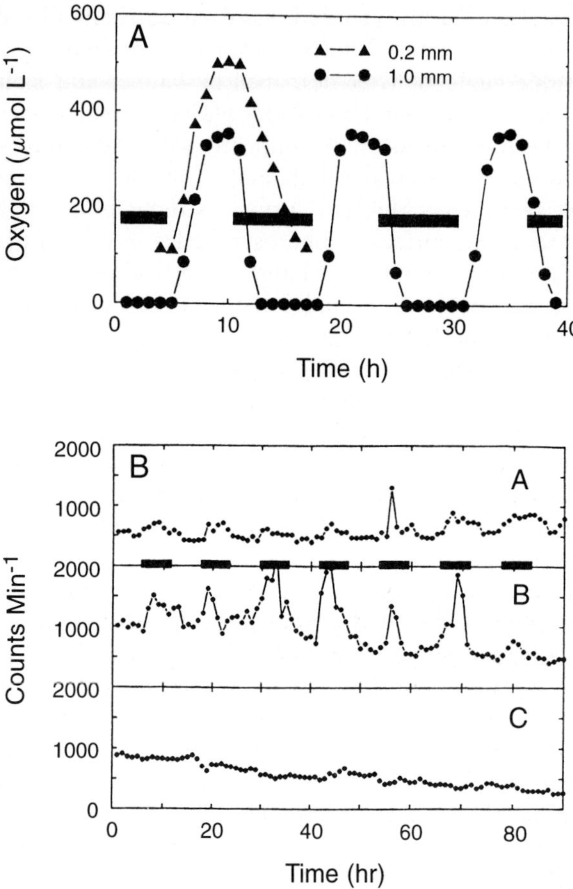

FIGURE 5 Photosynthetic activity (as oxygen release, measured with microelectrodes) and phosphate flux experiments on intact epipelic periphyton communities, conducted in flow-through chambers that permitted manipulation of the phosphate gradient within the sediment, volume and flushing rate of the overlying water, and oxygen concentration of the overlying water. Each experimental unit consisted of an 8- to 12-mm-thick layer of sediment and its intact periphyton isolated on a supported nylon membrane that separated the upper flow-through chamber from the lower artificial pore-water chamber. The data are shown as (**A**) oxygen concentrations through time at depths 0.2 and 1.0 mm below the sediment surface (30 μmol quanta m^{-2} sec^{-1}; 17°C; air saturation, 302 μmol O_2 $liter^{-1}$) and (**B**) phosphorus flux experiment with time course of ^{32}P release from sediments exposed to a 6-hr : 6-hr light/dark cycle (chambers A and B, 30 μmol quanta m^2 sec^{-1}; chamber C, permanently dark; 17°C). In both (A) and (B), bars designate periods of darkness. Similar results were obtained in other repeat trials. Moreover, upon switching to superoxygenated water, P efflux in controls decreased precipitously. (From Carlton and Wetzel, 1988, with permission.)

umn was less than 25% of what was actually released from the sediment, suggesting that the sediment P was rapidly incorporated into periphyton biomass. The focus of the research by Hansson (1988), like that by Carlton and Wetzel (1988), was the function of periphyton in P cycling rather than the importance of the benthic substratum in supplying nutrients to the epipelic periphyton. Although the data indicated that the sediments provided a major P source for the epipelic algae, additional quantitative information is not yet available.

C. Plant Substrata and Epiphytic Algae

1. Positive, Neutral, or Negative Interactions?

Among all types of benthic substrata, the most commonly studied for interactions with algal surface colonizers are plants. An array of plant material is available for colonization, including leaves and stems of submersed vascular plants, stems (and sometimes leaves) of emergent vascular plants; roots, wood, bark, and shed leaves from trees in streams and along lakeshores; and thalli of aquatic bryophytes, ferns, and macroalgae. Several lines of evidence indirectly support the hypothesis of, at least, a loose nonobligate "ectosymbiotic" (Kies, 1992) nutrient interaction between epiphytes and their "host" plants (Wetzel, 1983). The benefits to epiphytes of plant colonization have been reported to include provision of an advantageous location for growth (i.e., elevated in the water column, where there is greater access to light than at the sediment surface) and access to a second source of nutrients (the substratum as well as the water column; Hutchinson, 1975; Sand-Jensen, 1977; Burkholder and Wetzel, 1990). Other evidence indicates a negative interaction, wherein macrophytes compete with algal colonizers for nutrient sources (Sand-Jensen, 1977) and release allelopathic substances that inhibit epiphyte growth (Anthoni *et al.*, 1980). A third hypothesis also has been considered, that macrophytes serve only as a neutral site for attachment or contribute negligibly to epiphyte nutrient supplies (Cattaneo and Kalff, 1979; Stevenson, 1988).

Depending on the season, the plant substratum condition, and the availability of water-column nutrients, the available evidence may support each of these hypothesized interactions (or lack thereof); that is, it is logical that each is valid and may be operative in some situations, but not others. In the course of growth, senescence, death, and decomposition, the chemical influence of plant substrata changes with differences in physiology and structural integrity. The timing of synthesis, storage, and release of allelopathic chemicals varies depending on the host plant species (Ostrofsky, 1993). Young plant tissue is most structurally intact, but leaching increases as tissues age. This process can be enhanced by biological mechanisms, such as: cellulolytic bacteria that dissolve the cuticle of vascular host plants or otherwise damage the substratum leaves (Howard-Williams and Davies, 1978); invertebrates that wound the host plant while grazing microalgal colonizers (Rogers and Breen, 1983); and the colonizing microflora that take up nutrients and other substances lost or secreted/excreted from the underlying plant, sometimes by excreting enzymes that could act to accelerate the leaching process (Roos, 1983; Moeller *et al.,* 1988; Burkholder and Wetzel, 1990).

Most readily usable forms of nutrients (e.g., ammonium, phosphate, and simple sugars) are released from plant tissue during senescence and early decomposition (see discussion in Wetzel, 1983). In microbial succession on leaves and wood in streams, for example, algae may follow bacteria and fungi as early secondary colonizers (Suberkropp and Klug, 1976; Elwood *et al.,* 1981), or may colonize simultaneously with them (e.g., bluegreens and other algae that are known to produce extracellular mucilage in quantity). In the adsorption process, secondary microalgal colonizers would benefit from substratum "conditioning" by bacterial mucilages, although competition for nutrients with the established bacterioflora might act to reduce or limit growth of some algal species (Peterson and Stevenson, 1989). As tree leaves and wood age and become relatively "inert" to further nutrient leaching (Harmon *et al.,* 1986), macroalgae can form thick mats over fungal/bacterial growth on the physical substrata (e.g., Sheath and Burkholder, 1985), which provide chemical substrates for cellulose and lignin decomposers (Suberkropp and Klug, 1976; Harmon *et al.,* 1986), but which probably offer only "neutral" area for attachment to the algal colonizers. Functional tree roots that become exposed for colonization in stream habitats, in contrast, may help to "pull" nutrients from the water that would first become available to associated microflora. There exists little published information about freshwater algal/tree root associations (Hancock, 1985). In soft water streams, some researchers have noted abundant seasonal cover of macroalgae such as the rhodophyte *Audouinella violacea* (Kütz.) Hamel on apparently viable submersed tree roots [Sheath and Burkholder (1985); also see Burkholder and Almodovar (1973) for description of diverse algal communities on mangrove roots in marine systems], but the nature of chemical algal/substratum interactions is unknown.

2. Negative Chemical Interactions: The Case for Allelopathy

Plants may derive some benefit from epiphyte cover acting to protect them from herbivore or pathogen damage (Howard, 1982, in marine systems; Rogers and Breen 1983; Lodge, 1991). But more frequently the underlying host becomes mechanically stressed, light-limited, and carbon-limited with increasing epiphyte colonization [Hutchinson (1975); classic experiments by Sand-Jensen (1977), for a sea grass and its epiphytes]. Hence, the loose "symbiosis" between epiphytic algae and plant substrata is generally considered more beneficial to the colonizers than to the host, and it has been hypothesized that allelopathy and other mechanisms that prevent dense epiphyte growth would increase host productivity and survival [Hutchinson (1975); Hay and Fenical (1988), in marine habitats; experiment by Martin *et al.* (1992), in a freshwater system, indicating that periphyton grazing by herbivorous snails without predaceous fish increased macrophyte production].

Allelopathy may be defined as the release of a chemical, such as an alkaloid, phenolic, or tannin, by a plant that inhibits the growth of nearby flora. Many macroalgae and aquatic vascular plants produce such substances (Ostrofsky, 1993), but the extent to which allelopathy is an operative mechanism that inhibits or reduces epiphytic algal colonization in freshwater habitats is unknown. Because increasing epiphyte biomass is usually deleterious to the underlying plant substratum, it is generally assumed that the plant host would derive benefit from producing allelopathic compounds to inhibit epiphyte growth. The converse has not been examined, probably because known algal toxins (e.g., from blue-green algae) have been reported to affect other microalgae and animals, but have not been found to adversely affect macrophytes. The common approach in early attempts to demonstrate allelopathy from plant substrata toward algal colonizers was to test the effect of the filtrate from a plant culture on the algae (e.g., Fitzgerald, 1969; Szczepańska, 1971). Allelopathic interaction was assessed by change in growth characteristics—for example, length of the lag phase, exponential growth rate, or cell death—of the inhibited or "target" species. A major problem in resolving the question stems from the fact that allelopathic substances are often released in extremely low concentrations, so that it becomes difficult to isolate and chemically characterize the compounds involved, in part because it has not been possible to make these measurements with sufficient microscale resolution. Even if the compound can be identified, it is often difficult to determine whether the inhibitory substance was actively excreted from the plant substratum or was just a by-product of senescence and decomposition (Bonasera *et al.*, 1979; Ostrofsky and Zettler, 1986).

It follows that the evidence for allelopathy by plant substrata toward algal colonizers is mostly correlative, and usually the allelopathic compound that might be involved in the interaction either has not been identified or has not been demonstrated to adversely affect associated species at the "ecologically relevant" concentrations measured in the natural habitat [but see Hay and Fenical (1988) and Hay (1991) for marine macroalgae]. One clear demonstration of allelopathy by a freshwater plant substratum toward algal colonizers is the case of the green macroalga *Chara globularis* Braun, which is commonly associated with a "rank, pungent odor" where it occurs in abundance (Anthoni *et al.*, 1980). This alga is seldom found with epiphytes. Two biologically active sulfur compounds (a dithiolane and a trithiane, simulated with synthetic 4-methylthio-1,2-dithiolane and

5-methylothio-1,2,3-trithiane) were isolated from both freshwater and brackish *C. globularis*. These compounds inhibited photosynthesis of phytoplankton species and of the epiphytic diatom *Nitzschia palea* (Kütz.) W. Smith at concentrations measured within a bed of *C. globularis* in the natural habitat (Wium-Andersen *et al.,* 1982). As improved techniques become available for identifying small concentrations of allelopathic substances from natural water samples and plant tissue, similar experiments will be needed across a broad sample of aquatic macrophyte species before the significance of allelopathy by plant substrata toward algal colonizers can be evaluated.

3. Plant Substrata as Nutrient Sources

Somewhat surprisingly, the potential for plant substrata (microalgae colonized by other microalgae, and macrophytes, that is, macroalgae and vascular plants, colonized by microalgae) to serve as a major nutrient source for epiphytic algae has been a long-standing controversy in aquatic ecology. The available data on benthic algal/plant substrata chemical interactions have been obtained mostly from correlative studies, from "macroscale" experiments that have considered the entire epiphyte biofilm, and from a few "microscale" experiments that have focused on the living algal cells in the biofilm.

Evidence in support of plant substrata providing nutritional benefit for algal colonizers was first derived from observations about epiphyte abundance and community structure among macrophyte species or on natural versus artificial plants (see Cattaneo and Amireault, 1992). Sites of macrophyte tissue damage, where higher leakage of nutrients would be suspected, have been associated with higher epiphytic algal and bacterial abundance than intact substratum tissue (Roos, 1983; Fig. 4C). Apparent substratum preference was documented for epiphytic algae in oligotrophic lakes (e.g., Gough and Woelkering, 1976). In that study, preference of desmids (Chlorophyceae) for the macrophyte *Utricularia* was of additional interest considering that the microalgae often colonized bladders where nitrogen supplies likely were released from captured, dissolving animals. However, microalgal use of this nitrogen source has not been tested. Markedly different epiphyte communities have been observed on natural as opposed to artificial plants in mesotrophic and oligotrophic habitats (Cattaneo and Kalff, 1978; Burkholder and Wetzel, 1989a). But in eutrophic lakes as well as in streams, loosely attached epiphyte communities generally do not exhibit substratum preference (Eminson and Moss, 1980; Fontaine and Nigh, 1983), or algal abundance has been correlated with structural characteristics of the macrophyte (Siver, 1980; Burkholder and Sheath, 1984). The data from these studies may be explained by Eminson and Moss' (1980) hypothesis that macrophytes act as a major nutrient source for epiphytes in nutrient-poor systems, whereas in eutrophic lakes the water column assumes the more important role in supplying nutrients (also see Fig. 1, elements A and C). Habitats with appreciable turbulence would also afford a more nutrient-rich environment, from continuous renewal of nutrients that diffuse across a thin boundary layer (Riber and Wetzel, 1987; Raven, 1992).

To gain insights about the influence of macrophyte substrata on benthic algal nutrition, epiphyte communities on natural versus artificial plants have been compared for their alkaline phosphatase activity as an indicator of phosphorus limitation. Phosphatases are nonspecific monoesterases that catalyze the hydrolysis of many organic P compounds and liberate orthophosphate (Jansson *et al.*, 1988). These enzymes release orthophosphate from particulate and dissolved substrates, so that phosphate becomes available for uptake by algal cells in the preferred inorganic form ($H_2PO_4^-$ or HPO_4^{2-}; Raven, 1980). Alkaline phosphatases are located either on the surface of algal cells or on the plasmalemma (Aaranson and Patni, 1976), and their synthesis is derepressed when inorganic phosphate limits growth (Pettersson, 1980). They may be produced and excreted by associated bacteria and fauna as well as algae (Jansson *et al.*, 1988). Research in oligotrophic and mesotrophic lakes has demonstrated that epiphytes on artificial plants have significantly higher alkaline phosphatase activity than communities on natural plants, suggesting greater P limitation among the artificial plant colonizers (Cattaneo and Kalff, 1979; Burkholder and Wetzel, 1990). The structural integrity of plant substrata generally declines and P release increases with tissue age (Landers, 1982; Fig. 4). Epiphytes on young as well as aging natural leaves apparently can derive some portion of their phosphorus requirement from the substratum throughout the macrophyte growing season (Burkholder and Wetzel, 1990).

4. Insights from Microscale Resolution of Nutrient Interactions

Macrophytes have been found to leach appreciable dissolved organic carbon (e.g., Søndergaard, 1983) and a small percentage of their tissue phosphorus (less than 15%, and often less than 5%; e.g., Carignan and Kalff, 1982; Moeller *et al.*, 1988) to epiphytes, with losses increasing during leaf senescence (Landers, 1982; Morin and Kimball, 1983). In attempts to quantify the importance of this nutrient source to epiphytic algae, a common methodological error has been to estimate algal biomass from fresh weight, dry weight, or chlorophyll content of the total epiphyte biofilm, which typically includes high content of organic and inorganic debris. These methods can overestimate algal biomass by several orders of magnitude (Burkholder, 1986), and likely have led to significant underestimates of the quantitative importance of macrophyte nutrients to microalgal colonizers. In an alternate approach, Moeller *et al.* (1988) used TLMA with radiolabeled [^{33}P]-phosphate to focus on the viable algal cells in the epiphyte matrix on *Najas flexilis* (Willd.) Rostk & Schmidt over a simulated growing season (Table I and Fig. 6). Consideration of algal biomass as ash-free dry weight of the total matrix indicated that the macrophyte was of minor importance as a P source to algal colonizers. The microscale examination of living epiphyte cells made possible by autoradiography indicated, in contrast, that abundant loosely attached algal species (with biovolume quantified by direct microscopic analysis) had derived at least

TABLE I Derivation of the Relative Proportion of Phosphorus Supplied from the Macrophyte *Najas flexilis* versus the Water Column by Epiphytic Microalgae in the Moeller *et al.* (1988) Experiment

Symbols for the experimental system and equations shown in Fig. 9.6 are defined as:

a = Calculated proportion of P supplied from the macrophyte to the viable epiphytic microalgae;

$(1 - a)$ = Proportion (calculated by difference) of P supplied from the water column [i.e., P derived from the sediment via the water (since the sediment was *not* sealed from contact with the water column) and other organic and inorganic dissolved P originally in the water];

A_c = Specific activity (directly measured) of the total loosely attached epiphyte/calcium carbonate-encrusted matrix (or epiphyte "crust") associated with the plants grown in ^{33}P-labeled sediment (A_c^+) or in unlabeled sediment (A_c^-);

A_m = Specific activity (directly measured) of *Najas* plants grown in ^{33}P-labeled sediment (A_m^+) or in unlabeled sediment (A_m^-);

A_s = Specific activity (directly measured) of ^{33}P-labeled (A_s^+) or unlabeled (A_s^- with unlabeled P added to equal the same amount of ^{33}P added to the labeled sediments); and

A_w = Calculated specific activity of the water, that is, the nominal specific activity of the "waterborne" pool of utilizable P.

The specific activities of most interest, namely, of the viable, loosely attached algal cells of each abundant species (sp) within the epiphyte matrix (A_{sp}^+ and A_{sp}^-), could not be measured directly. They were approximated using microautoradiography of tracks from β emission by decaying ^{33}P taken up by living algae. It was assumed that the ratio ($\mathbf{r}$) of specific activities for the living epiphytes was approximately equal to the ratio of tracks from cells on plants grown in labeled versus unlabeled sediment:

$$\mathbf{r}_{sp} = D_{sp}^+ / D_{sp}^- = A_{sp}^+ / A_{sp}^-$$

where

D_s = mean track density (tracks per 10-µm^3 algal biovolume) for three labeled (+) and unlabeled (–) samples of each species (s)

The ratio ($\mathbf{r}_{sp}$) was used to solve for $\mathbf{a}_{sp}$, the proportion of its P that species s had obtained from the host macrophyte, as:

$$\mathbf{a}_{sp} = \frac{(\mathbf{r}_{sp} - 1)\, A_w}{\mathbf{r}_{sp}\,(A_w - A_m^-) - (A_w - A_m^+)}$$

This formula was derived using expressions for A_s^+ and A_s^- analogous to those for A_c^+ and A_c^- (Fig. 9.6A).

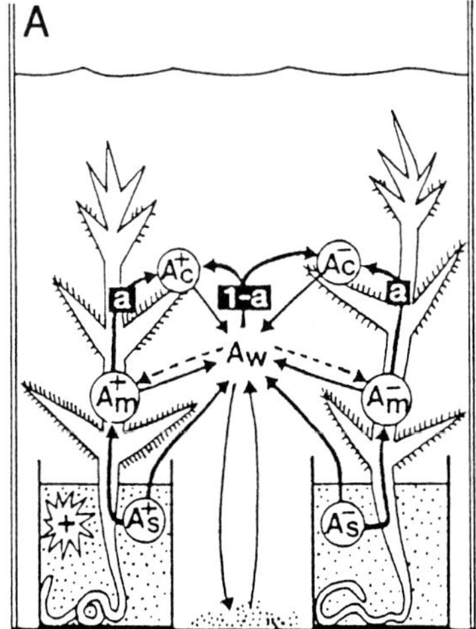

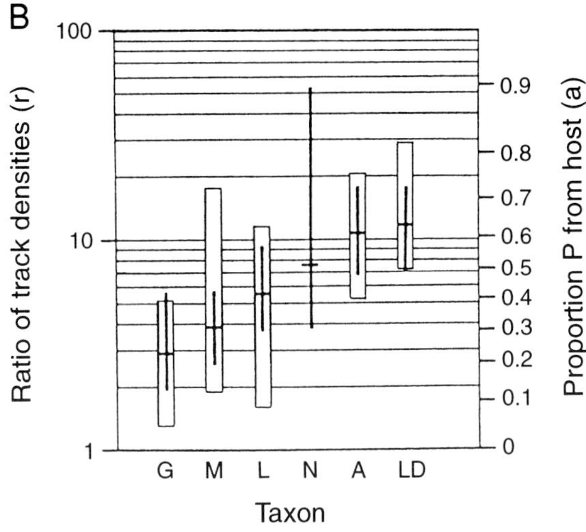

FIGURE 6. Determination of the relative proportion of phosphorus supplies to abundant microalgal epiphytes from the host macrophyte, *Najas flexilis* (*a*), versus from the water column (1-*a*), showing (**A**) the experimental system, with plants in control (–, unlabeled) and labeled (+, with ^{33}P) sediment that was open to the water column, the arrows indicating potential pathways of phosphorus cycling (n = three labeled and three unlabeled plant/sediment systems; see Table II for symbol definitions); and (**B**) results from track microautoradiography, with range of **r** (boxes indicate C.I.'s) and calculated **a** for epiphytic taxa *Gomphonema parvulum* (G), *Mougeotia viridis* (Kütz.) Wittrock (M), *Lyngbya Taylorii* Drouet & Strickland (L), *Navicula* sp. (N), *Achnanthes minutissima* (A), and unidentified large pennate diatoms (LD). Taxa are arranged by increasing **a** (i.e., according to increasing reliance on the macrophyte as a P source). As expected, the highly motile diatom *Navicula* showed the most variability in response, indicated by the undefined C.I. The data indicate that long stalked taxa (e.g., *G. parvulum*) and long filaments (e.g., *M. viridis*) depended more on the external, P-poor lakewater medium (ca. 20% of their P from the macrophyte) than did small adnate forms (e.g., *A. minutissima*, with ca. 60% of their P from the macrophyte). (From Moeller *et al.*, 1988, with permission.)

25–60% of their P supply from the macrophyte. The data also indicated that adnate diatoms such as *Achnanthes minutissima* Kütz. [*Achnanthidium;* Round *et al.* (1990)] had obtained more of their P supply from the macrophyte than had loosely attached overstory algae such as the stalked diatom *Gomphonema parvulum.*

If loosely attached epiphytes, more closely associated with the water column than the plant substratum, can obtain a substantial portion of their nutrient supplies from the benthic source, the substratum's importance in providing nutrients would be expected to increase for adnate microalgae in immediate contact with the macrophyte (Pip and Robinson, 1982b; Fig. 9.1). Burkholder *et al.* (1990) used a combination of TLMA and SEMA with [^{33}P]-phosphate to examine P movement from the water column through a developed epiphyte biofilm with intact chemical microenvironment, and demonstrated that the adnate flora were effectively isolated from the water-column nutrient source (Burkholder *et al.,* 1990). The direct comparison between undisturbed loosely attached and adnate epiphytes demonstrated that there is a physiological assimilation gradient among microalgae coinciding with the position of cells within the biofilm matrix. The adnate microalgae in a mature epiphyte community would be expected to obtain most of their P (and likely other nutrients) from the plant substrata. Comparisons of macrophyte species as nutrient sources for loosely attached epiphytes across lakes of varying trophic status, obtained using microscale techniques that focus on viable algal cells, are not yet available. Moreover, with the exception of the research of Moeller *et al.* (1988), the quantitative reliance of adnate algae on nutrients from plant substrata remains to be determined.

5. From Nonobligate to Specialized Ectosymbiosis

Epiphytism by microalgae on macroalgae is considered to have given rise to closer symbiotic associations often hypothesized, for example, for interactions involving red algal epiphytes and endophytes (= parasites) on red algal hosts (Lee, 1980). Demonstrated preference of an algal species for only one host plant genus or species would be considered an intermediate step between loose epiphyte/plant substratum associations and closely coupled mutualistic symbiosis. In freshwater systems, unlike marine waters, the author knows of no select colonization by one algal species on (only) one particular host plant species. Some microalgal taxa, such as the encapsulated chrysophytes *Epipyxis* and *Halobryon* (Chrysophyceae), may show a degree of preference in epiphytizing filamentous algae rather than other benthic substrata (Round, 1992). Similarly, certain species of the adnate green alga *Coleochaete* are usually found as epiphytes (Prescott, 1973), but other species are known to colonize rock and plastic as well as plants (Prescott, 1973; Cattaneo, 1978). In both Europe and North America, the red algal epiphyte *Audouinella violacea* shows an obvious preference to colonize mature and senescing gametophytes of the red macroalgae *Lemanea* and *Paralemanea* (Sheath and Hambrook, 1990; Everitt and Burkholder, 1991). However, *A. violacea* is also known to colonize other substrata (e.g., tree roots as previously mentioned). Other algal–algal epibiont associations in freshwater systems include the diatoms *Amphora copulata* var. *pediculus* (Kütz.) Schoeman & Archibald and *Synedra parasitica* (W. Smith) Hust. growing epiphytically on *Nitzschia sigmoidea* (Nitzsch) W. Smith, and less commonly on species of *Surirella* and *Campylodiscus* (see review by Round, 1992). These epiphytes have not been found to grow on any other *Nitzschia* species, suggesting that a species-specific chemical attractant may be involved.

A more specialized albeit "one-sided" form of symbiosis is parasitism. Freshwater benthic "algal"/algal associations also include dinoflagellates such as *Cystodinedria inermis* (Geitler) Pascher, *Stylodinium globosum* Klebs, and other Vampyrella-like taxa that parasitize filamentous green algae, especially *Oedogonium* species and members of the Zygnematales (Pfiester and Popovský, 1979; Popovský and Pfiester, 1990). These cryptic "algae-protozoans" have complex life cycles, sometimes with both photosynthetic and nonphotosynthetic stages, and including amoeboid forms that creep along the green algal filaments, bore into the cells, and suck out the cell contents. They are probably widespread among the shallows of freshwater lakes and bogs, but their seasonal dynamics and abundance, and their overall impact on their host plants, are virtually unknown.

D. Animal Substrata and Epizoic Algae

Among the oldest literature in freshwater phycology are anecdotal observations about the apparent preference of some benthic algae for certain animal substrata. The association between the green filamentous alga *Basicladia* and snapping turtles (Edgren *et al.*, 1953) probably made for interesting collection events. At least two turtle genera and two *Basicladia* species can be involved, and the genus *Basicladia* apparently has not been found on any substrata except turtles (Chapman and Waters, 1992). Edgren *et al.* (1953) hypothesized reciprocal evolution between the animals and the algae, or invasion of the niches opened by adaptive radiation of the turtles. Stewart and Schlichting (1966) found 17 species of green algae on 26 aquatic/semiaquatic insect species, and similarly postulated that these epizoic relationships may represent an algal dispersal mechanism (although its importance has not been determined relative to other dispersal mechanisms involving wind and water motion). A nonobligate association between benthic red algae and animals involves the red taxa *Boldia erythrosiphon* Herndon and *Nemalionopsis shawii* Skuja, which frequently are found attached to aquatic snails of the family Pleuroceridae in the southeastern United States (Sheath and Hambrook, 1990). Researchers have suggested that *B. erythrosiphon* may be attracted to the high manganese content in the snail shells (Howard and Parker, 1981). The alga also occurs in an epilithic habit.

Among microalgae, euglenoids and diatoms have been found in abundance on pelagic microcrustacean zooplankton. For example, in one north temperate lake during spring–early summer, the euglenoid *Colacium calvum* Stein and the diatom *Synedra cyclopum* Cunningham preferentially colonize *Daphnia* spp.; and a second euglenoid species, *Colacium vesiculosum* Ehr., occurs most often on cyclopoids (Chiavelli *et al.,* 1993). The association apparently is nonobligate, since the algae also occur in an epiphytic habit. Chiavelli and coworkers found that the prevalence of all three algae in an epizoic habit was directly correlated to *Daphnia* density; in fact, zooplankton species were not colonized by the three epizoic algae unless *Daphnia* was present. The euglenoid *C. vesiculosum* differs in animal substratum preference among lakes, sometimes colonizing mostly nauplii (Green, 1974; Chiavelli *et al.,* 1993). The distribution and abundance of these epizoic algae were hypothesized to be controlled by the balance between algal recruitment and losses on the various host animal species, which periodically molt and shed the carapace with epibionts.

In an alternate approach that encompassed 60 north temperate lakes and marshes, Gaiser and Bachmann (1993) found 49 diatom taxa epizoic on 20 littoral cladoceran zooplankton species, with an average of about 120 diatom cells per animal. These taxa also occurred as epiphytes in littoral habitat. A major epizoic taxon reported in this work as well as in the previously mentioned research was *Synedra cyclopum,* which was most common on large pelagic *Daphnia* species and comprised nearly two-thirds of all epizoic diatom cells. The authors suggested that the seasonality of epizoic diatoms may be controlled by a combination of their temperature and nutrient preferences, and by the cladoceran molt rate which is temperature-dependent.

Epizoic algae probably are common on many animals; for example, epizoic diatoms have also been reported from copepods parasitizing fish, and dinoflagellates apparently are widespread ectoparasites of fish in habitats spanning fresh waters, estuaries, and marine coasts (Fee and Drum, 1965; Cachon and Cachon, 1987). As true for epiphytes, epizoic algae likely derive benefit from the animal substrata by increased access to light and nutrients associated with animal movement that would reduce the boundary zone thickness for nutrient acquisition, which would also potentially be increased by access to animal excreta, leachates, and secretions. The effects of epizoic algae on larger host animals such as snails and turtles are not known; they may help to camouflage the animals from predators or prey. For pelagic zooplankton, however, epizoic algae have been found to impede the animals' movement and reduce their buoyancy, making it more difficult to maintain position in the water column, obtain food, and avoid or escape predators such as planktivorous fish (Willey *et al.*, 1990).

E. Advanced Chemical Interdependence: Algal Endosymbiosis

Although this chapter has described chemical interactions between benthic algae and their substrata as poorly understood in general, an exception to this statement occurs in the case of endosymbiotic associations, which have fascinated freshwater biologists for more than a century. This fascination is reflected in many publications that describe the biology and chemistry of some relationships. Endosymbioses in freshwaters range from nonobligate seasonal associations to tightly coupled interactions in which the animal dies without the alga. The best understood relationships are mutualistic, with both the host and the symbiont deriving benefit. In most algal endosymbioses, the host plant or animal supplies the algal symbiont with both inorganic (carbon dioxide, ammonium) and organic nutrition while also benefiting from carbon, nitrogen, and/or oxygen that has been fixed or produced by the symbiont. The symbiont may derive other benefits as well, such as a protected or optimal environment for growth.

As in the case of epizoic algae on pelagic zooplankton, some endosymbioses provide a means for a "benthic" or attached (as opposed to a free-living planktonic) habitat in a pelagic community. Most freshwater endosymbioses are benthic rather than suspended in the water column, and many occur in mildly acidic softwater, nutrient-depauperate habitats. Alternatively, these associations can be widespread among diverse aquatic habitats that are enriched in some nutrients but limited by other resources that are augmented by the symbiotic interaction. For example, the *Azolla/Anabaena* [*Trichormus;* Komárek and Anagnostidis (1989)] symbiosis (described in the following) is widespread among phosphorus-rich but nitrogen-poor habitats (Peters and Calvert, 1983). Based on endosymbioses that have been examined in detail, these interactions would be expected to be most advantageous in oligotrophic waters, where one study estimated that production of organic matter by the ecosystem can be increased by 100- to 1000-fold relative to a similar system containing only separate autotrophs and heterotrophs [Hallock (1981), considering marine systems]. The relationships likely enhance survival and growth of the interacting benthic algae and all organisms that directly or indirectly depend on them for habitat or nutrition, particularly in food webs of softwater, nutrient-poor freshwater ecosystems.

1. Plant and Fungal Hosts

Diverse algal/plant endosymbioses in freshwaters involve blue-green algae as symbionts within diatoms, cryptomonads, bryophytes, and ferns

(Table II). In these associations the symbiont supplies fixed nitrogen and/or, more rarely, fixed carbon to the host, and the host provides habitat (i.e., optimal oxygen, nutritional, and pH conditions, and optimal oxidation/reduction gradients) for localized growth of the symbiont (Paerl, 1992). Algal photosynthetic activity is usually negligible; instead, the blue-greens rely on organic compounds excreted by the host for energy and growth, thus avoiding the potential for the symbionts to inhibit nitrogen fixation with their own oxygen evolution (Stewart *et al.,* 1980). Although blue-green/fungal associations represent less than 10% of the lichens in the world (Ahmadjian, 1992), many species of coccoid and filamentous blue-greens form these associations with fungi in wetlands, lake shorelines, and other freshwater areas (Table III). They probably serve as important contributors to the nitrogen budgets of the habitat. Lichens are believed to represent symbiotic control of parasitic fungi in which the algae are not inheritable, but develop an association wherein the host and symbionts are almost totally dependent on one another (Ahmadjian, 1992).

The coccoid blue-green *Synechococcus* occurs as endosymbionts in diatoms mainly of the family Epithemiaceae [e.g., *Rhopalodia gibba* (Ehr.) O. Müller and *Epithemia turgida* (Ehr.) Kütz.] in nitrogen-poor lakes, streams, and marshes (Stewart *et al.,* 1980; DeYoe *et al.,* 1992). This association has been shown to fix nitrogen (indicated as acetylene reduction) under light-dependent, microaerobic conditions (Stewart et al., 1980). In one study, the number of endosymbionts increased as ambient nitrogen became limiting, but ambient phosphorus limitation uncoupled the apparent relationship between ambient nitrogen and endosymbiont abundance, likely because the blue-green symbionts became phosphorus-limited (DeYoe *et al.,* 1992).

A well-known, widespread freshwater symbiosis between algae and plants involves planktonic ferns of the genus *Azolla* and the blue-green endosymbiont *Anabaena azollae* Strasburger [*Trichormus anomalus* (Fritsch) Komárek & Anagnostidis, from Komárek and Anagnostidis (1989); note: blue-green algae within the genus *Nostoc* may also be involved in some cases; Tomaselli *et al.* (1988)]. The symbionts colonize mucilage-filled ventral leaf cavities that contain nutrients derived from the fern's photosynthesis. These cavities also contain specific associated bacteria that help promote low-oxygen conditions needed for nitrogen fixation. As much as 25% or more of the symbiont's cells are heterocysts (specialized N_2-fixing cells), and cell-specific rates of N_2 fixation are high relative to those of planktonic *Anabaena* species (Peters and Calvert, 1983). Like other hosts in association with blue-green endosymbionts [e.g., lichens, reviewed in Ahmadjain (1992)], the fern is able to inhibit glutamine synthetase activity of *A. azollae,* forcing the symbionts to release most of the fixed nitrogen as NH_4. The N_2 fixation process can proceed at high rates despite ambient enrichment of NH_4, because most of the nitrogen assimi-

TABLE II Examples of Specialized Freshwater Benthic Algal Associations Ranging from Apparent, Nonobligate Epiphytic and Epizoic Substratum Preference to Endosymbiosis in Plant, Fungal, and Animal Hosts (See Text for Sources)

Algal class	*Symbiosis*	*Host(s)*	*Habitat(s)*
Cyanophyta (blue-greens)	Endosymbiotic *(Synechococcus)*	Diatoms *Epithemia, Rhopalodia, Denticula*	Softwater lakes, marshes
	Endosymbiotic (filamentous species; see Table 9.3)	Fungi (lichen associations)	Marshes and other wetlands, streambanks, upper littoral fringe of lakes
	Cyanelles (approaching chloroplasts)	Glaucocystophytes *Cyanoptyche, Chalarodora, Gloeochaete;* thecate amoeba *Paulinella chromatophora*	Littoral zone epiphytes (glaucocystophytes); muddy sediments of eutrophic lakes, ponds, and ditches (amoeba)
Rhodophyta (red algae)	Nonobligate, epiphytic *(Audouinella)*	Red macroalgae *Lemanea, Paralemanea*	Second- and third-order softwater streams
	Nonobligate, epizoic *(Boldia, Nemalionopsis)*	Snails (family Pleuroceridae)	Lower-order softwater streams
Pyrrhophyta (dinoflagellates)	Parasitic *(Cystodinedria, Stylodinium,* others)	Filamentous greens (*Oedogonium, Spirogyra,* others)	Littoral zone of lakes, ponds, bogs; quiet shallows of streams
	Parasitic (Amylodinium-like)	Finfish	Ponds, bogs
Chrysophyta (diatoms)	Nonobligate, epiphytic *(Amphora sigmoidea, Synedra parasitica)*	Diatoms *Nitzschia sigmoidea, Surirella* spp., *Campylodiscus* spp.	Littoral zone of lakes; lower-order streams
	Nonobligate, epizoic *(Synedra cyclopum)*	Microcrustacean zooplankton (*Daphnia,* others)	Littoral and pelagic zones of *Daphnia*-containing lakes
Euglenophyta (euglenoids)	Nonobligate, epizoic *(Colacium vesiculosum)*	Microcrustacean zooplankton (cyclopoid adults and nauplii)	Littoral and pelagic zones of *Daphnia*-containing lakes
	Obligate?, endozoic (cloacal cavity, *Colacium libellae*)	Damselfly nymphs *(Ischnura verticalis)*	North temperate lakes, moving from littoral zone to deeper waters, winter
Chlorophyta (green algae)	Nonobligate, loosely endozoic *(Chlamydomonas)*	Salamander egg masses *(Ambystoma maculatum, A. gracilis)*	Quiet softwater pools, shallow littoral fringe of soft-water/neutral lakes
	Obligate?, epizoic *(Basicladia)*	Turtle species (e.g., *Sternotherus, Chelydra)*	Littoral of lakes, ponds, streams
	Endosymbiotic (intra- or extracellular; *Chlorella, Symbiococcum, Oocystis*?)	*Hydra* spp., amoebae, ciliates, sponges, coelenterates, molluscs, rotifers, turbellarians (Table 9.4)	Softwater ponds, bogs, lakes (littoral zones); soft-water streams

TABLE III Coccoid and Filamentous Blue-green Algae Involved in Freshwater Lichen Associations[a]

Lichen habitat	*Algal symbionts*
Epilithic, hypolithic, or cryptoendolithic on wet or seepage rock surfaces (wetlands, lakeshores)	*Aphanocapsa* sp., *Entophysalis* sp.
Epilithic, hypolithic, and cryptoendolithic on rocks in hot and cold deserts	*Chroococcidiopsis* sp. (3–4 strains), *Myxosarcina* sp. (1–2 strains; also in caves)
Epilithic on seepage rock surfaces in bog pools	*Chroococcus* sp.
Epilithic, hypolithic, and cryptoendolithic on wet or seepage rock surfaces; submersed in bog pools	*Cyanosarcina* sp., *Gloeocapsa kuetzingiana* Näg. emend. Jaag, *G. muralis* Kütz., *G. sanguinea* Kütz. emend. Jaag
Lake splash zones, hot springs	*Anabaena* sp.
Wet rock surfaces (wetlands, lake splash zones, etc.)	*Fischerella* spp., *Nostoc commune* Vaucher, *N. muscorum* Ag., *N. punctiforme* (Kütz.) Hariot, *N. sphaericum* Vaucher, *Scytonema hoffmannii* Ag., *Tolypothrix* sp., *Calothrix crustacea* Thuret, *C. parietina* (Naeg.) Thuret, *C. pulvinata* Kg., *C. scopulorum* (Weber & Mohr) Ag., *Dichothrix baueriana* (Grun.) Born. & Flah., *D. orsiniana* (Kütz.) Born. & Flah., *Stigonema* sp.

[a]Compiled from references reviewed in Büdel (1992) and Paerl (1992).

lation actually occurs in the host rather than the symbiont (Stewart *et al.*, 1980). The fern supplies to the symbionts, in turn, high levels of reductant and energy, so that the small blue-green population is able to provide the nitrogen requirement for the much larger host. The host carries *A. azollae* inoculum in its overwintering stages, thus ensuring that the symbiosis is perpetuated.

Among the closest physiological and genetic integrations of freshwater benthic algal/"plant" partners are the often overlooked endosymbioses involving the algal class Glaucocystophyceae. This group includes some unicellular or palmelloid heterotrophic protists that contain cyanelles, or obligate blue-green endosymbionts, which are remarkably close in structure and function to true chloroplasts (reviewed in Kies, 1992). Glaucocystophycean "algae" are exclusively freshwater in both plankton and benthic habits, where most species are rare and occur in low density. Palmelloid genera *Cyanoptyche, Chalarodora,* and *Gloeochaete* are epiphytes on filamentous algae, aquatic mosses, and submersed angiosperms. Although the endosymbionts function as chloroplasts, they still retain a peptidoglycan cell wall that is believed to be a remnant structure of the free-living blue-green ancestors.

Cyanelles are regarded as intermediate forms in the evolution of algal chloroplasts; in most features of gene organization/sequence, they are intermediate between blue-green algae and chloroplasts (Bohnert and Loffelhardt, 1992). These symbionts lack α-ketoglutarate dehydrogenase and succinate dehydrogenase; like free-living blue-greens, they have an open tricarboxylic acid cycle that contributes carbon skeletons for amino acid synthesis, but they cannot respire because their respiratory chain lacks cytochrome c oxidase (reviewed in Kies, 1992). Cyanelles cannot fix N_2—assimilation of inorganic nitrogen involves similar cooperation between cyanelle and host organism as between chloroplasts and higher plant cells. The symbionts contain only 10% of the quantity of DNA found in free-living blue-greens; thus, the cyanelle genome is similar in size to genomes of chloroplasts, and the endosymbionts have very limited genetic autonomy. The degree of glaucocystophyte interactions with plant substrata of the external habitat is not known. Close relatives of these epiphytes occur as metaphyton in eutrophic ponds and ditches (Kies, 1992), suggesting that certain nutrient-enriched conditions may be required to sustain the associations.

2. Animal Hosts

Endosymbiotic associations between freshwater algae and animal hosts include many different invertebrate species (Reisser, 1992a; Tables II and IV). The endosymbiotic algae are usually chlorophytes, termed "zoochlorellae," from the genera *Chlorella* or, less commonly, *Chlamydomonas* and *Symbiococcum* (Rahat, 1992), although other algae such as euglenoids also participate in some nonobligate interactions. Among the latter are two associations in which the algae involved are "loosely" endosymbiotic inside the envelope of the animal's eggs, or attached to the animal's outer cloacal cavity.

Eggs of the spotted salamander (*Ambystoma maculatum* Shaw) are usually green because of the presence of symbiotic chlamydomonads that inhabit the envelope of each egg (Bachmann *et al.*, 1986). An oxygen microelectrode was used to measure the effect of algal photosynthesis on oxygen concentrations inside eggs and within the dense gelatinous matrix surrounding them. During darkness, oxygen became severely depleted within the eggs, but oxygen levels rapidly increased upon exposure to light (Fig. 7A). Photosynthetic oxygen production by the chlamydomonads exceeded respiratory consumption by the embryo/algae complex and effected oxygen supersaturation inside eggs, even when water surrounding the egg mass was almost anoxic (Fig. 7B). The advantages to salamander egg and embryo survival afforded from oxygen production by the algal symbionts likely are substantial. Use of oxygen microelectrodes enabled the researchers to demonstrate these effects, as opposed to earlier investigators who lacked this technology and, thus, were not able to measure oxygen evolution at the microscale required to resolve the symbionts' role

TABLE IV Animal Hosts in Freshwater Endosymbiotic Associations Involving Coccoid Chlorophycean Algae (*Chlorella vulgaris* Strains and Other "Zoochlorellae")[a]

PROTOZOA (Amoebae)	(Ciliates)	Ciliates)
Acanthocystis turfacea	*Acaryophyra* sp.	*Paramecium bursaria*
Amphitrema flavum	*Climatocostomum virens*	*Prorodon viridis*
Chaos zoochlorellae	*Coleps hirtus*	*P. ovum*
C. carolinensis	*Disematostoma bütschlii*	*Psilotricha viridis*
Difflugia oblonga	*Euplotes daidaleos*	*Spirostomum viridis*
Heleopara sphagni	*Frontonia leucas*	*Stentor polymorphus*
Hyalosphenia papilio	*F. vernalis*	*S. niger*
Mayorellla viridis	*Halteria bifurcata*	*S. roeseli*
	Holosticha viridis	*Teutophrys trisulca*
	Malacophrys sphagni	*Vorticella* spp.
	Ophrydium versatile	
PORIFERA	COELENTERATA	MOLLUSCA
Ephydatis fluviatilis	*Hydra magnipapillata*	*Anodonta* sp.
Heteromeyenia sp.	*H. viridis*	*Linnaea* sp.
Spongilla lacustris	*H. viridissima*	*Unio pictorum*
ROTATORIA	TURBELLARIA	
Ascomorpha helvetica	*Castrada viridis*	
Cephalodella sp.	*Dalyellia viridis*	
	Phaenocora typhlops	
	Typhloplana viridata	

[a]From Raehat (1992); modified from Reisser (1992a).

(e.g., Hutchinson and Hammen, 1958). In association with egg masses of another salamander (*Ambystoma gracile* Baird), similar algal symbionts absorb ammonia from the perivitelline fluid surrounding the larvae (Goff and Stein, 1978). The ammonia is produced by the larvae as a waste product; its removal and storage in membrane-bound proteinaceous bodies within the algal cells has been correlated with both more vigorous growth of the algae and more rapid larval development.

The euglenoid *Colacium libellae* Rosowski & Willey establishes in the rectum of the damselfly *Ischnura verticalis* (Say) Selys during winter in colder lakes (Rosowski and Willey, 1975). During warm summer months, the damselfly nymphs and the euglenoid live separately, with the euglenoid assuming an epiphytic habit. But with the onset of cold weather, the *C. libellae* population attaches to the cuticle of the rectum of the damselfly larvae; the cells shed their flagella and assume a palmelloid habit, forming a plug that colors the terminal four abdominal segments dark green. As the shoreline areas freeze, the damselfly nymphs move to deeper water to overwinter, forming a protected motile, translucent, and probably nutrient-enriched microhabitat for the euglenoid. In spring the nymphs move to

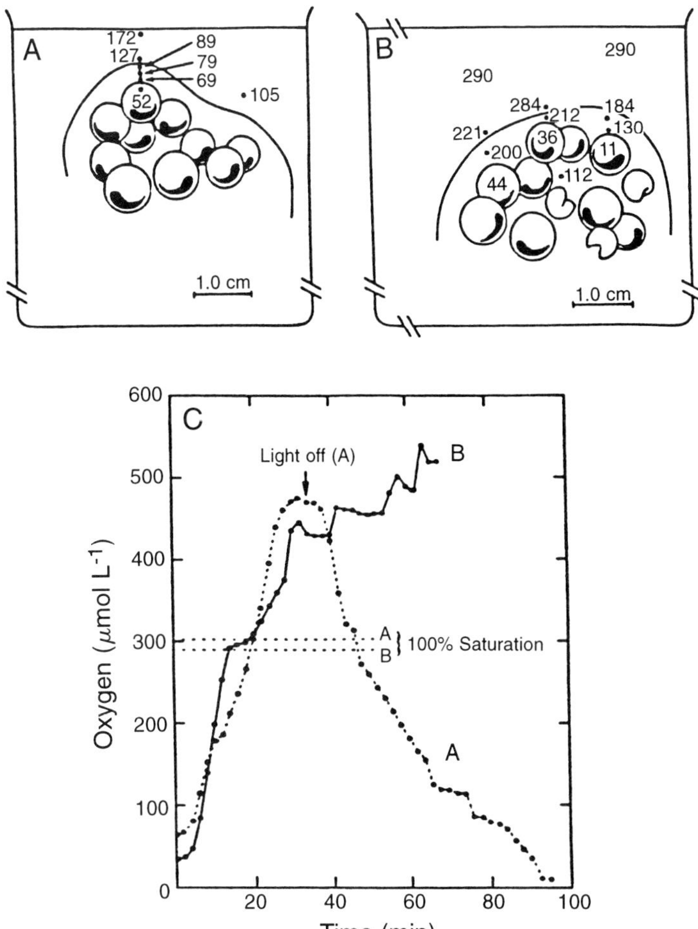

FIGURE 7 Microelectrode measurements of oxygen concentrations (μmol O_2 liter^{-1}) near and within egg masses of the salamander *Ambystoma maculatum,* showing (**A**) oxygen in and around the eggs after 3 hr of darkness in uncirculated water (17°C; 100% air saturation = 302 μmol O_2 liter^{-1}); (**B**) oxygen in and around the eggs after 3 hr of darkness in continuously aerated water (19°C; 100% air saturation = 290 μmol O_2 liter^{-1}); and (**C**) oxygen over time within illuminated eggs after 3 hr of darkness (light on at time zero). Curve A is from the egg mass in hypoxic water (17°C; light off after 34 min elapsed time). Curve B shows the environment of the same egg mass in continuously aerated water (19°C; the steps in the curve were caused by tail movements of the embryo, which mixed and broke down the oxygen gradient in the egg fluid). (From Bachmann *et al.,* 1986, with permission.)

warmer water, and the euglenoids exit from the rectum to become free-living epiphytes throughout the summer. If algal-free nymphs are placed in water with *C. libellae*, the euglenoids establish in the animals' rectal cavities within 36 hours. Whether the host derives benefit from the symbiont is not known, but the symbiosis is quite specific—other euglenoid species, including *Colacium* species, are not able to infect the damselfly nymphs (Willey *et al.*, 1970; Rosowski and Willey, 1975).

Chlorella spp. are involved in nonobligate endosymbioses with animal hosts, in which the symbionts occur intracellularly (in hydra, sponges, protozoa, and some turbellarians and molluscs) or extracellularly (Reisser, 1992a). The symbionts are 2–10 μm in diameter, with neither motile nor sexual forms known; reproduction typically is accomplished by mitosis with multiple fission into 4–64 autospores (Smith and Douglas, 1987). These algae are common and widespread in benthic freshwater habitats. Each algal cell may either be within a host cell food vacuole in various stages of digestion, or within a "perialgal" vacuole in which they are not digested. Zoochlorellae can be isolated from the animals and most can be grown in sterile defined inorganic media supplemented with vitamins B1 and B12. They have no secondary carotenoids and excrete maltose (with traces of alanine and glycolate) to the host by a pH-dependent mechanism.

The proportion of zoochlorellae : host rarely exceeds 1:10 by volume or biomass, and in general neither the host nor the symbiont exhibits a high degree of dependence on the symbiosis (Smith and Douglas, 1987). Symbionts are transferred to the host directly during asexual reproduction (e.g., budding and gemmule production of sponges, or budding of hydra). For animals that reproduce only sexually, the juvenile or adult must acquire free-living algal cells through normal feeding. Among hydras, for example, some species initially are algae-free; others obtain symbionts that were associated with the egg external surface, or they contain algae that were inside the egg prior to fertilization (reviewed in Rahat, 1992). In some cases, such as zoochlorellae of *Hydra viridissima* Pallas, symbionts are transferred by a multistep process that involves a high degree of host–algae specificity. Host cells respond differently depending on the *Chlorella* strain, indicating a definite biochemical "algal recognition" system, yet *H. viridissima* is capable of forming associations with any one of various algal strains. Once a symbiosis is established, however, other algal strains cannot form symbioses; nonnative symbiotic chlorellae are digested if phagocytized by cells that already contain zoochlorellae, whereas similar algae given to animals without symbionts are accepted and retained.

Hydra/zoochlorellae associations are the subject of more than 30 years of intensive research. Whereas symbiotic hydras often lose their symbionts under prolonged periods of darkness, the native zoochlorellae from *H. viridissima* are retained by their host in the dark for years (reviewed in Rahat, 1992). Unlike many strains of symbiotic *Chlorella*, these algae cannot be grown *in vitro* and, thus, are obligate symbionts. Symbiotic hydras use organic carbon fixed by the algae as a supplemental food resource. Experiments have shown no difference in growth of food-replete animals with versus without algae in the light (Douglas and Smith, 1983). With limited food, however, growth of green hydras in the light exceeds that of animals without zoochlorellae (Muscatine and Lenhoff, 1965). Assays with ^{14}C-labeled CO_2 have shown that freshly isolated zoochlorellae from *H. viridissima* liberate 55–85% of photosynthetically fixed carbon into the medium as maltose, glycolic acid, alanine, glucose, and other saccharides (Muscatine, 1965; Cernichiari *et al.*, 1969). Similar nutritional benefit is afforded other animals containing zoochlorellae. For example, the ciliate *Paramecium bursaria* Ehr. usually grows faster with *Chlorella* symbionts, and when the major bacterial food resource is not available, symbiotic cilates display enhanced survival (Reisser and Herbig, 1987). Without zoochlorellae, the freshwater sponge *Spongilla lacustris* L. grew to only 20–40% of the size of animals with zoochlorellae (Frost and Williamson, 1980).

These fascinating associations notwithstanding, it is noteworthy that, as for algal/plant endosymbioses, arguably the most advanced algal/animal association does not involve zoochlorellae; rather, it consists of chloroplast-like cyanelles within the thecate amoeba *Paulinella chromatophora* Lauterborn (Kies, 1992). This amoeba inhabits muddy bottom sediments in the euphotic zone of eutrophic lakes, ponds, and ditches, and often co-occurs with epipelic diatoms such as *Caloneis amphisbaena* (Bory) Cleve. As for the glaucocystophycean "algal" hosts previously discussed, *P. chromatophora* apparently relies entirely on its cyanelles for carbon acquisition.

3. Chemical and Molecular Signals in Symbiont Recognition

Contemporary stable endosymbioses are characterized by a marked specificity of host and symbiont, believed to have resulted from a multistep process evolving toward increasing effectivity of the association (reviewed in Reisser, 1992b). The success of such associations requires a network of chemical signals that allows reaction as a unit. Some progress has been made in identifying signals involved in, or associated with, mechanisms of cell-to-cell recognition. In endosymbioses with hydras, *Paramecium bursaria*, and *Spongilla*, for example, excretion of sugars such as xylose, fructose, glucose, and maltose by the retained algae has adaptive value for establishment of a long-lasting and stable hereditary symbiotic relationship. The carbohydrate patterns in algal-cell wall structures also play a role. For example, the cell-wall surface of *Chlorella* strains differs in surface charge and binding capacity with different lectins and antibodies, and these features are involved in recognition and acceptance versus rejection from the host animal (Reisser, 1992b).

Probably the best understood recognition phenomena are those of the *Paramecium bursaria/Chlorella* association, which involve interactions of the ciliate's membrane-bound lectins with carbohydrate bound to the algal cell-wall surface. In positive recognition, the interaction leads to a stepwise, "zipperlike" enclosure of the alga by the host vacuole membrane (Reisser, 1992b). Accepted algae have enough of certain carbohydrate groups in their cell-wall surfaces to initiate the "zipper" mechanism. If those carbohydrate groups are absent, or if there are insufficient contacts between the algal surface groups and host membrane components, the host does not take up the algal cells. The zipper model could explain why only one species of *Paramecium* is symbiotic with *Chlorella*, since *P. bursaria* has large quantities of membrane-bound agglutination factors relative to other *Paramecium* species.

Signal exchange between multicellular hosts and algal endosymbionts is more complicated, in part because it is more difficult to assign host cell reaction with certainty to an algal stimulus versus stimuli from other host cells that may not have involved the endosymbionts. For example, lectins appear to play a more diverse role in lichen associations. In some cases they are involved in recognition events (Galun and Bubrick, 1984); in other lichen associations, they apparently are not involved. Lectins may also influence fungal morphogenesis and nitrogen partitioning between the lichen host and the algal symbionts (Ahmadjian, 1992). Overall, symbiotic integrations develop through long periods of coevolution that involve genetic changes through adaptive responses of the host and symbiont. Many of these systems are highly coevolved associations wherein the partners are almost totally dependent on each other, communicating by a variety of molecular and chemical signaling mechanisms that we are only beginning to understand and appreciate (Ahmadjain, 1992).

V. CONCLUSIONS AND RECOMMENDATIONS

The cohabitation of unicellular or multicellular organisms of different systematic position is realized as a continuous series of associations, from loose nonobligate ectosymbioses to highly developed obligate endosymbioses (Kies, 1992). This spectrum of interactions occurs among benthic algae and their colonized substrata. Although many of the algal species are habitat generalists or incidental colonizers of various living and nonliving surfaces (Whitton, 1975; Round, 1981), some demonstrate more specialized substratum preferences in nonobligate ectosymbioses; and others become tightly biochemically coupled "partners" in mutualistic endosymbioses with host plants, fungi, or animals (Meeks *et al.,* 1985). Many of these interactions—especially in loose ectosymbioses—appear to involve small concentrations of poorly characterized substances released from the substrata.

Despite considerable research on benthic algal ecology, our knowledge about the range of chemicals, controlling biochemical signals, and overall significance of benthic algal/substratum interactions to freshwater ecosystems remains limited, largely because of limitations in available microscale technology. As application of oxygen microelectrodes, microautoradiography, molecular probes, and other methodologies within the past decade has demonstrated, realistic consideration of chemical interactions involving benthic algae must increasingly focus at the scale of occurrence, namely, the chemical microgradients of the microniche habitat.

Microscale techniques and approaches continually enable resolution of apparent paradoxes. Hutchinson's (1961) famous "paradox of the plankton" is a classic illustration of this point; the paradox of many algal species within "one" apparent habitat was resolved by recognizing the appropriate micro-spatial and temporal scales of the plankton habitat as multidimensional nonequilibrium niches (Richerson *et al.,* 1970; Allen, 1977; Tilman *et al.,* 1982). Microscale measurements of current velocity have made it possible to more accurately describe physical/chemical boundary-layer effects on benthic algal communities and the chemical microenvironment within the periphyton biofilm matrix (Stevenson, 1983; Reiter and Carlson, 1986; Stevenson and Glover, 1993). Microelectrodes enabled realistic evaluation of periphyton productivity and avoided the diffusional ^{14}C or oxygen disequilibrium problems encountered with more traditional methods (Carlton and Wetzel, 1987; see Section III). Microelectrodes used with molecular techniques also allowed discovery of processes such as nitrogen fixation on detrital particle microhabitats, information that has redirected paradigms about the overall or macroscale functioning of nitrogen-limited waters (Paerl, 1990; Currin *et al.,* 1990). Microautoradiography enabled focus on the chemical microhabitat relevant to epiphytic microalgae, and reversed researchers' conceptions about the importance of minor P losses from macrophyte tissue as a significant (Moeller *et al.,* 1988), rather than negligible (Moeller *et al.,* 1985), nutrient supply for epiphytes. Determination of mechanisms for cell-surface recognition among hosts and endosymbionts was beyond our reach until the advent of DNA/RNA probes, confocal microscopy, and related molecular applications that have enabled focus at the appropriate microscale to begin to resolve the biochemical and molecular interactions (White *et al.,* 1987; Reisser, 1992b).

In addition to continued development of and reliance on microscale techniques, fresh conceptual approaches are needed to unravel benthic algal/substratum interactions. There is a certain familiar comfort in invoking well-established paradigms or dogmas to guide the design of hypotheses and experiments about the physiological ecology of benthic algae. Although to some extent such thought processes are unavoidable, the danger in this comfort is that major controlling mechanisms and even major players in these habitats can be completely missed. Previous findings frequently are based on techniques that, with time's passage, become recognized as limited or inadequate for the previous application. It is ironic that, in honoring what is "generally" known to be true, we sometimes fail to see what is actually true for a more specific case or alternate setting.

As an example, a paradigm of low heterotrophic capability of microalgal communities, likely extended beyond the authors' intent, was based on comparative uptake of two dissolved organic substrates by algae and bacteria in surface waters of oligotrophic lakes (Wright and Hobbie, 1966). Later research established phagotrophic phytoplankton as instrumental in the "microbial loop" of oligotrophic–eutrophic surface waters (Gaines and Elbrächter, 1987; Sanders and Porter, 1988; Tranvik *et al.,* 1989; Carrick and Fahnensteil, 1990), and some freshwater phytoplankton also have been found to make efficient use of naturally occurring levels of dissolved organics (Gaines and Elbrächter, 1987; Lewitus and Caron, 1991). If phytoplankton in more light-replete, organic substrate-poor habitats have been found to utilize heterotrophy, it is reasonable to expect that benthic microalgae could also rely on it, at least facultatively (e.g., Pfiester and Popovský, 1979; also see Tuchman, Chapter 10, this volume). The significance of heterotrophy in chemical interactions of benthic algae is a question that merits rigorous evaluation; particulate or dissolved organic resources likely will be found to be important to many benthic microalgae in acquisition of carbon and other nutrients from their associated substrata.

Finally, conceptual approaches to understand the chemical interactions of benthic algae with their substrata will continue to be improved by crossing the freshwater/marine "boundary" that typically is imposed in formal training of aquatic scientists. Benthic algae occur along a continuum from lakes and rivers to estuaries to coastal marine waters and, not surprisingly, these communities share many features in common (Round, 1981; Burkholder and Wetzel, 1989b; Everitt and Burkholder, 1991). Some classic studies on the physiological ecology of benthic algae were reported from marine systems, for example, with direct analogies to freshwater habitats (e.g., Harlin, 1973; Sand-Jensen, 1977; Penhale and Thayer, 1980; Sullivan and Moncreiff, 1990). Freshwater ecology, in turn, has contributed classic findings about the biology of some major groups of benthic algae [e.g., dinoflagellates; Pfiester and Popovský, (1979)], microtechnique applications, and algal/substratum chemical interactions (e.g., Carlton and Wetzel, 1988; Moeller *et al.,* 1988) that have been found to be analogous in marine habitats (e.g., Coleman and Burkholder, 1995; Burkholder *et al.,* 1995). Given that our knowledge of algal/substrata interactions is, in many respects, lacking or limited, our insights will be strengthened and expanded from an appreciation of the ecology of benthic algae across aquatic ecosystems.

ACKNOWLEDGMENTS

I thank J. Stevenson, M. Bothwell, M. Mallin, H. Glasgow, C. Pringle, and two anonymous colleagues for kindly critiquing the manuscript. This chapter is dedicated to Ginny Coleman, in hope of her recovery to continue to enrich this field and the lives of all those around her.

REFERENCES

Aaronson, S., and Patni, J. J. (1976). The role of surface and extracellular phosphatases in the phosphorus requirement of *Ochromonas*. *Limnol. Oceanogr.* **21,** 838–845.

Ahmadjian, V. (1992). Basic mechanisms of signal exchange, recognition, and regulation in lichens. *In* "Algae and Symbioses" (W. Reisser, ed.), pp. 301–324. Biopress Ltd., Bristol, UK.

Allen, T. F. H. (1977). Scale in microscopic algal ecology: The neglected dimension. *Phycologia* **16,** 253–257.

Anthoni, U., Christophersen, C., Madsen, J. O., Wium-Andersen, S., and Jacobson, N. (1980). Biologically active sulphur compounds from the green alga *Chara globularis*. *Phytochemistry* **19,** 1228–1229.

Armstrong, S. M., and Barlocher, F. (1989). Adsorption and release of amino acids from epilithic biofilms in streams. *Freshwater Biol.* **22,** 153–159.

Avnimelech, Y., and McHenry, J. R. (1984). Enrichment of transported sediments with organic carbon, nutrients, and clay. *Soil Sci. Soc. Am. J.* **48,** 259–266.

Avnimelech, Y., Troeger, B. W., and Reed, L. W. (1982). Mutual flocculation of algae and clay: Evidence and implications. *Science* **216,** 63–65.

Bachmann, M. D., Carlton, R. D., Burkholder, J. M., and Wetzel, R. G. (1986). Symbiosis between salamander eggs and green algae: Microelectrode measurements inside eggs demonstrate effects of photosynthesis on oxygen concentrations. *Can. J. Zool.* **64,** 1586–1588.

Bell, R. A. (1993). Cryptoendolithic algae of hot semiarid lands and deserts. *J. Phycol.* **29,** 133–139.

Bitton, G., and Marshall, K. C., eds. (1980). "Adsorption of Microorganisms to Surfaces." Wiley, New York.

Bjork-Ramberg, S. (1985). Uptake of phosphate and inorganic nitrogen by a sediment–algal system in a subarctic lake. *Freshwater Biol.* **15,** 175–183.

Bohnert, H. J., and Loffelhardt, W. (1992). Molecular genetics of cyanelles from *Cyanophora paradoxa*. *In* "Algae and Symbioses" (W. Reisser, ed.), pp. 379-397. Biopress Ltd., Bristol, UK.

Bonasera, J., Lynch, J., and Leck, M. A. (1979). Comparison of the allelopathic potential of four marsh species. *Bull. Torrey Bot. Club* **106,** 217–222.

Büdel, B. (1992). Taxonomy of lichenized procaryotic blue-green algae. *In* "Algae and Symbioses" (W. Reisser, ed.), pp. 301–324. Biopress Ltd., Bristol, UK.

Burkholder, J. M. (1986). Seasonal dynamics, alkaline phosphatase activity and phosphate uptake of adnate and loosely attached epiphytes in an oligotrophic lake. Ph.D. Dissertation, Michigan State University, East Lansing.

Burkholder, J. M. (1992). Phytoplankton and episodic suspended sediment loading: Phosphate partitioning and mechanisms for survival. *Limnol. Oceanogr.* **37,** 974–988.

Burkholder, J. M., and Cuker, B. E. (1991). Response of periphyton communities to clay and phosphate loading in a shallow reservoir. *J. Phycol.* **27,** 373–384.

Burkholder, J. M., and Sheath, R. G. (1984). The seasonal distribution, abundance, and diversity of desmids (Chlorophyta) in a softwater, north temperate stream. *J. Phycol.* **20,** 159–172.

Burkholder, J. M. and Wetzel, R. G. (1989a). Microbial colonization on natural and artificial macrophytes in a phosphorus-limited, hardwater lake. *J. Phycol.* **25,** 55–65.

Burkholder, J. M., and Wetzel, R. G. (1989b). Epiphytic microalgae on a natural substratum in a phosphorus-limited hardwater lake: Seasonal dynamics of community structure, bimoass and ATP content. *Arch. Hydrobiol.* **83**(Suppl.), 1–56.

Burkholder, J. M., and Wetzel, R. G. (1990). Epiphytic alkaline phosphatase activity on natural and artificial plants in a P-limited lake: Re-evaluation of the role of macrophytes as a phosphorus source for epiphytes. *Limnol. Oceanogr.* **35**, 736–746.

Burkholder, J. M., Wetzel, R. G., and Klomparens, K. L. (1990). Direct comparison of phosphate uptake by adnate and loosely attached microalgae within an intact biofilm matrix. *Appl. Environ. Microbiol.* **56**, 2882–2890.

Burkholder, J. M., Glasgow, H. B., Jr., and Steidinger, K. A. (1995). Stage transformations in the complex life cycle of an ichthyotoxic "ambush-predator" dinoflagellate. *In* "Harmful Marine Algal Blooms" (P. Lassus, G. Arzul, E. Erard, P. Gentien, and C. Marcaillou, eds.), pp. 567–572. Elsevier, Amsterdam.

Burkholder, P. R., and Almodovar, L. R. (1973). Studies on mangrove algal communities in Puerto Rico. *Fla. Sci.* **36**, 66–74.

Cachon, J., and Cachon, M. (1987). Parasitic dinoflagellates. *In* "The Biology of Dinoflagellates" (F. J. R. Taylor, ed.), Bot. Monogr. Vol. 21, pp. 571–610. Blackwell, New York.

Cahoon, L. B., Kucklick, J. R., and Stager, J. C. (1990). A natural phosphate source for Lake Waccamaw, North Carolina, USA. *Int. Rev. Gesamten Hydrobiol.* **75**, 339–351.

Carignan, R., and Kalff, J. (1982). Phosphorus release by submerged macrophytes: Significance to epiphyton and phytoplankton. *Limnol. Oceanogr.* **27**, 419–427.

Carlton, R. G., and Wetzel, R. G. (1987). Distributions and fates of oxygen in periphyton communities. *Can. J. Bot.* **65**, 1031–1037.

Carlton, R. G., and Wetzel, R. G. (1988). Phosphorus flux from lake sediments: Effect of epipelic algal oxygen production. *Limnol. Oceanogr.* **33**, 562–571.

Carrick, H. J., and Fahnenstiel, G. L. (1990). Planktonic protozoa in Lakes Huron and Michigan: Seasonal abundance and composition of ciliates and dinoflagellates. *J. Great Lakes Res.* **16**, 319–329.

Carrick, H. J., and Lowe, R. L. (1988). Response of Lake Michigan benthic algae to an *in situ* enrichment with Si, N, and P. *Can. J. Fish. Aquat. Sci.* **45**, 271–279.

Castenholz, R. W. (1976). The effect of sulfide on the blue-green algae of hot springs. I. New Zealand and Iceland. *J. Phycol.* **12**, 54–68.

Cattaneo, A. (1978). The microdistribution of epiphytes on the leaves of natural and artificial macrophytes. *Br. Phycol. J.* **13**, 183–188.

Cattaneo, A., and Amireault, M. C. (1992). How artificial are artificial substrata for periphyton? *J. North Am. Benthol. Soc.* **11**, 244–256.

Cattaneo, A., and Kalff, J. (1978). Seasonal changes in the epiphyte community of natural and artificial macrophytes in Lake Memphremagog (Que. & Vt.). *Hydrobiologia* **60**, 135–144.

Cattaneo, A., and Kalff, J. (1979). Primary production of algae growing on natural and artificial aquatic plants: A study of interactions between epiphytes and their substrate. *Limnol. Oceanogr.* **24**, 1031–1037.

Cernichiari, E., Muscatine, L., and Smith, D. C. (1969). Maltose excretion by the symbiotic algae of *Hydra viridis*. *Proc. R. Soc. London, Ser. B* **173**, 557–576.

Chapman, R. L., and Waters, D. A. (1992). Epi- and endobiotic chlorophytes. *In* "Algae and Symbioses" (W. Reisser, ed.), pp. 619–639. Biopress Ltd., Bristol, UK.

Chet, I., and Mitchell, R. (1976). Ecological aspects of microbial chemotactic behavior. *Annu. Rev. Microbiol.* **30**, 221–239.

Chiavelli, D. A., Mills, E. L., and Threlkeld, S. T. (1993). Host preference, seasonality, and community interactions of zooplankton epibionts. *Limnol. Oceanogr.* **38**, 574–583.

Coleman, V. L., and Burkholder, J. M. (1995). Response of microalgal epiphytes to nitrate enrichment in an eelgrass (*Zostera marina* L.) meadow. *J. Phycol.* **31**, 36–43.

Confer, J. (1972). Interrelations among plankton, attached algae, and the phosphorus cycle in artificial open systems. *Ecol. Monogr.* **42**, 1–23.

Currin, C. A., Paerl, H. W., Suba, G. K., and Alberte, R. S. (1990). Immunofluorescence detection and characterization of N_2-fixing microorganisms from aquatic environments. *Limnol. Oceanogr.* **35**, 59–71.

Daniels, S. L. (1980). Mechanisms involved in sorption of microorganisms to solid surfaces. *In* "Adsorption of Microorganisms to Surfaces" (G. Bitton and K. C. Marshall, eds.), pp. 7–58. Wiley, New York.

Darley, W. M. (1974). Silicification and calcification. *In* "Algal Physiology and Biochemistry" (W. D. P. Stewart, ed.), Bot. Monogr., Vol. 10, pp. 655–675. Blackwell, New York.

Delbecque, E. J. P. (1985). Periphyton on nymphaeids: An evaluation of methods and separation techniques. *Hydrobiologia* **124**, 85–93.

DeYoe, H. R., Lowe, R. L., and Marks, J. C. (1992). Effects of nitrogen and phosphorus on the endosymbiont load of *Rhopalodia gibba* and *Epithemia turgida* (Bacillariophyceae). J. Phycol. **28**, 773–777.

Dillon, P. S., Maki, J. S., and Mitchell, R. (1989). Adhesion of *Enteromorpha* swarmers to microbial films. *Microb. Ecol.* **17**, 39–47.

Douglas, A. E., and Smith, D. C. (1983). The costs of symbionts to their host in green hydra. *In* "Endocytobiology. II. Intracellular Spaces as Oligogenic Ecosystems" (H. E. Schenk and W. Schwemmler, eds.), pp. 631–648. Academic Press, New York.

Droop, M. R. (1974). Heterotrophy of carbon. *In* "Algal Physiology and Biochemistry" (W. D. P. Stewart, ed.), Bot. Monogr., Vol. 10, pp. 530–559. Blackwell, New York.

Edgren, R. A., Edgre, M. K., and Tiffany, L. H. (1953). Some North American turtles and their epizoophytic algae. *Ecology* **34**, 733–740.

Elwood, J. W., Newbold, J. D., Trimble, A. F., and Stark, R. W. (1981). The limiting role of phosphorus in a woodland stream ecosystem: Effects of Penrichment on leaf decomposition and primary producers. *Ecology* **62**, 146–158.

Eminson, D., and Moss, B. (1980). The composition and ecology of periphyton communities in freshwaters. I. The influence of host type and external environment on community composition. *Br. Phycol. J.* **15**, 429–446.

Everitt, D. T., and Burkholder, J. M. (1991). Seasonal dynamics of macrophyte communities from a stream flowing over granite flatrock in North Carolina, U.S.A. *Hydrobiologia* **222**, 159–172.

Fairchild, W., Lowe, R. L., and Richardson, W. B. (1985). Algal periphyton growth on nutrient-diffusing substrates: An *in situ* bioassay. *Ecology* **66**, 465–472.

Fee, E. J., and Drum, R. W. (1965). Diatoms epizoic on copepods parasitizing fishes in the Des Moines River, Iowa. *Am. Midl. Nat.* **81**, 318–373.

Fitzgerald, G. P. (1969). Some factors in the competition or antagonism among bacteria, algae, and aquatic weeds. *J. Phycol.* **5**, 351–359.

Fontaine, T. D., III, and Nigh, D. G. (1983). Characteristics of epiphyte communities on natural and artificial submersed lotic plants: Substrate effects. *Arch. Hydrobiol.* **96**, 293–301.

Friedmann, E. I., and Kiebler, A. P. (1980). Nitrogen economy of endolithic microbial communities in hot and cold desert. *Microb. Ecol.* **6**, 95–108.

Froelich, P. N. (1988). Kinetic control of dissolved phosphate in natural rivers and estuaries: A primer on the phosphate buffer mechanism. *Limnol. Oceanogr.* **33**, 649–668.

Frost, T. M., and Williamson, C. E. (1980). *In situ* determination of the effect of symbiotic algae on the growth of the freshwater sponge *Spongilla lacustris*. *Ecology* **61**, 1361–1370.

Gaines, G., and Elbrächter, M. (1987). Heterotrophic nutrition. *In* "The Biology of Dinoflagellates" (F. J. R. Taylor, ed.), Bot. Monogr., Vol. 21, pp. 224–268. Blackwell, New York.

Gaiser, E. E., and Bachmann, R. W. (1993). The ecology and taxonomy of epizoic diatoms on Cladocera. *Limnol. Oceanogr.* **38**, 628–637.

Galun, M., and Bubrick, P. (1984). Physiological interactions between the partners of the lichen symbiosis. *In* "Encyclopedia of Plant Physiology" (H. F. Linskens and K. Heslop-Harrison, eds.), New Ser. 17, pp. 362–401. Springer-Verlag, New York.

Geesey, G. G., Richardson, W. T., Yeomans, H. G., Irvin, R. T., and Costerton, J. W. (1977). Microscopic examination of natural sessile bacterial populations from an alpine stream. *Can. J. Microbiol.* **23,** 1733–1736.

Godward, M. (1934). An investigation of the causal distribution of algal epiphytes. *Beih. Bot. Zentralbl. Abt. A* **52,** 506–539.

Goff, L. J., and Stein, J. R. (1978). Ammonia: Basis for algal symbiosis in salamander egg masses. *Life Sci.* **22,** 1463–1468.

Gough, S. B., and Woelkerling, W. J. (1976). Wisconsin desmids. II. Aufwuchs and plankton communities of selected soft water lakes, hardwater lakes and calcareous spring ponds. *Hydrobiologia* **49,** 3–25.

Green, J. (1974). Parasites and epibionts of Cladocera. *Trans. Zool. Soc. London* **32,** 417–515.

Grobbelaär, J. U. (1983). Availability to algae of N and P adsorbed on suspended solids in turbid waters of the Amazon River. *Arch. Hydrobiol.* **96,** 302–316.

Hallock, P. (1981). Algal symbiosis: A mathematical analysis. Mar. *Biol. (Berlin)* **62,** 249–255.

Hancock, F. D. (1985). Diatom associations in the aufwuchs of inundated trees and underwater leaves of *Salvinia,* drowned Mwenda River, Lake Kariba, Zimbabwe. *Hydrobiologia* **121,** 65–76.

Hansson, L.-A. (1988). Effects of competitive interactions on the biomass development of planktonic and periphytic algae in lakes. *Limnol. Oceanogr.* **33,** 121–128.

Harlin, M. M. (1973). Transfer of products between epiphytic marine algae and host plants. *J. Phycol.* **9,** 243–248.

Harlin, M. M., and Lindbergh, J. M. (1977). Selection of substrata by seaweeds: Optimal surface relief. *Mar. Biol. (Berlin)* **40,** 33–40.

Harmon, M. E., Franklin, J. F., Swanson, F. J., Sollins, P., Gregory, S. V., Lattin, J. D., Anderson, N. H., Cline, S. P., Aumen, N. G., Sedell, J. R., Lienkaemper, G. W., Cromack, K., Jr., and Cummins, K. W. (1986). Ecology of course woody debris in temperate ecosystems. *Adv. Ecol. Res.* **15,** 133–302.

Hay, M. E. (1991). Marine–terrestrial contrasts in the ecology of plant chemical defenses against herbivores. *Trends Ecol. Evol.* **6,** 362–365.

Hay, M. E., and Fenical, W. (1988). Marine plant–herbivore interactions: The ecology of chemical defense. *Annu. Rev. Ecol. Syst.* **19,** 111–145.

Hendricks, S. P., and White, D. S. (1988). Hummocking by lotic *Chara:* Observations on alterations of hyporheic temperature patterns. *Aquat. Bot.* **31,** 13–22.

Hiebert, F. K., and Bennett, P. C. (1992). Microbial control of silicateweathering in organic-rich ground water. *Science* **258,** 278–281.

Howard, R. K. (1982). Impact of feeding activities on epibenthic amphipods on surface fouling of eelgrass leaves. *Aquat. Bot.* **14,** 91–97.

Howard, R. V., and Parker, B. C. (1981). A non-obligate association between the red alga, *Boldia,* and pleurocerid snails. *Sterkiana* **71,** 18–23.

Howard-Williams, C., and Davies, B. R. (1978). The influence of periphyton on the surface structure of a *Potamogeton pectinatus* L. leaf (an hypothesis). *Aquat. Bot.* **5,** 87–91.

Hutchinson, G. E. (1961). The paradox of the plankton. *Am. Nat.* **95,** 137–146.

Hutchinson, G. E. (1975). "A Treatise on Limnology. III. Limnological Botany." Wiley, New York.

Hutchinson, V. H., and Hammen, C. S. (1958). Oxygen utilization in the symbiosis of embryos of the salamander, *Ambystoma maculatum,* and the alga, *Oophila amblystomatis. Biol. Bull. (Woods Hole, Mass.)* **115,** 483–489.

Hynes, H. B. N. (1970). "The Ecology of Running Waters." Univ. of Toronto Press, Toronto.

Jansson, M., Olsson, H., and Pettersson, K. (1988). Phosphatases: Origin, characteristics and function in lakes. *Hydrobiologia* **170,** 157–175.

Karl, D. M. (1980). Cellular nucleotide measurements and applications in microbialecology. *Microbiol. Rev.* **44**, 739–796.

Kies, L. (1992). Glaucocystophyceae and other protists harbouring procaryotic endocytobionts. *In* "Algae and Symbioses" (W. Reisser, ed.), pp. 353-377. Biopress Ltd., Bristol, UK.

Komárek, J., and Anagnostidis, K. (1989). Modern approach to theclassification system of cyanophytes. 4—Nostocales. *Arch. Hydrobiol. Suppl.* **82**(3), 247–345.

Landers, D. H. (1982). Effects of naturally senescing aquatic macrophytes on nutrient chemistry and chlorophyll *a* of surrounding waters. *Limnol. Oceanogr.* **27**, 428–439.

Lee, R. E. (1980). "Phycology." Cambridge Univ. Press, New York.

Lewitus, A. J., and Caron, D. A. (1991). Physiological responses of phytoflagellates to dissolved organic substrate additions. 1. Dominant role of heterotrophic nutrition in *Poterioochromonas malhamensis* (Chrysophyceae). *Plant Cell Physiol.* **32**, 671–680.

Lock, M. A., Wallace, R. R., Costerton, J. W., Ventullo, R. M., and Charlton, S. E. (1984). River epilithon: Toward a structural-functional model. *Oikos* **42**, 10–22.

Lodge, D. M. (1991). Herbivory on freshwater macrophytes. *Aquat. Bot.* **41**, 195–224.

Marshall, K. C. (1976). "Interfaces in Microbial Ecology." Harvard Univ. Press, Cambridge, MA.

Marshall, K. C., and Cruickshank, R. H. (1973). Cell surface hydrophobicity and the orientation of certain bacteria at interfaces. *Arch. Mikrobiol.* **91**, 29–40.

Martin, T. H., Crowder, L. B., Dumas, C. F., and Burkholder, J. M. (1992). Indirect effects of fish on macrophytes in Bays Mountain Lake: Evidence for a littoral trophic cascade. *Oecologia* **89**, 476–481.

Meeks, J. C., Enderlin, C. S.,Joseph, C. M., Steinberg, N., and Weeden, Y. M. (1985). Use of ^{13}N to study N_2 fixation and assimilation by cyanobacterial–lower plant associations. *In* "Nitrogen Fixation Research Progress" (J. H. Evans, P. J. Bottomley, and W. E. Newton, eds.), pp. 301–308. Martinus Nijhoff, Dordrecht, The Netherlands.

Miller, A. R., Lowe, R. L., and Rotenberry, J. T. (1987). Succession of diatom communities on sand grains. *J. Ecol.* **75**, 693–709.

Mishler, B. D., and Churchill, S. P. (1985). Transition to a land flora: Phylogentic relationships of the green algae and bryophytes. *Cladistics* **1**, 305–338.

Moeller, R. E., Burkholder, J. M., and Wetzel, R. G. (1985). Epiphytic microalgae as mediators of phosphorus transfer from a sediment–macrophyte complex to the water column. *In* "Proceedings of the Bi-annual Conference for Macrophyte Research" (abstr.). Copenhagen, Denmark.

Moeller, R. E., Burkholder, J. M., and Wetzel, R. G. (1988). Significance of sedimentary phosphorus to a rooted submersed macrophyte *(Najas flexilis)* and its algal epiphytes. *Aquat. Bot.* 32, 261–281.

Morin, A., and Cattaneo, A. (1992). Factors affecting sampling variability of freshwater periphyton and the power of periphyton studies. *Can. J. Fish. Aquat. Sci.* **49**, 1695–1703.

Morin, J. O., and Kimball, K. D. (1983). Relationship of macrophyte-mediated changes in the water column to periphyton composition and abundance. *Freshwater Biol.* **13**, 403–414.

Muscatine, L. (1965). Symbiosis of hydra and algae. III. Extracellular products of the algae. *Comp. Biochem. Physiol.* **16**, 77–92.

Muscatine, L., and Lenhoff, H. M. (1965). Symbiosis of hydra and algae. II. Effect of limited food and starvation on growth of symbiotic and aposymbiotic hydra. *Biol. Bull. (Woods Hole, Mass.)* **122**, 316–328.

Nikitin, D. I. (1973). Electron microscope studies of attached microorganisms. *Swed. Nat. Sci. Res. Counc. Ecol. Bull.* **17**, 85–91.

Norton, T.A., and Fetter, R. (1981). The settlement of *Sargassum muticum* propagules in stationary and flowing water. *J. Mar. Biol. Assoc. U. K.* **61**, 929–940.

Ostrofsky, M. L. (1993). Effects of tannins on leaf processing and conditioning rates in aquatic ecosystems: An empirical approach. *Can. J. Fish. Aquat. Sci.* **50,** 1176–1180.

Ostrofsky, M. L., and Zettler, E. R. (1986). Chemical defenses in aquatic plants. *J. Ecol.* **74,** 279–287.

Paerl, H. W. (1982). Feasibility of ^{55}Fe autoradiography as performed on N_2-fixing *Anabaena* spp. populations and associated bacteria. *Appl. Environ. Microbiol.* **43,** 210–217.

Paerl, H. W. (1985). Influence of attachment on microbial metabolism and growth in aquatic ecosystems. *In* "Bacterial Adhesion: Mechanisms and Physiological Significance" (D. C. Savage and M. M. Fletcher, eds.), pp. 363–400. Plenum, New York.

Paerl, H. W. (1990). Physiological ecology and regulation of N_2 fixation in natural waters. *In* "Advances in Microbial Ecology" (K. C. Marshall, ed.), Vol. 2, pp. 261–315. Plenum, New York.

Paerl, H. W. (1992). Epi- and endobiotic interactions of cyanobacteria. *In* "Algae and Symbioses" (W. Reisser, ed.), pp. 537–565. Biopress Ltd., Bristol, UK.

Paerl, H. W., and Stull, E. A. (1979). In defense of grain density autoradiography. *Limnol. Oceanogr.* **24,** 1166–1169.

Palmer, J. D., and Round, F. E. (1965). Persistent vertical-migration rhythms in benthic microflora. I. The effect of light and temperature on the rhythmic behavior of *Euglena obtusa. J. Mar. Biol. Assoc. U.K.* **45,** 567–582.

Paul, B. J., and Duthie, H. C. (1988). Nutrient cycling in the epilithon of running waters. *Can. J. Bot.* **67,** 2302–2309.

Penhale, P., and Thayer, G. W. (1980). Uptake and transfer of carbon and phosphorus by eelgrass (*Zostera marina* L.) and its epiphytes. *J. Exp. Mar. Biol. Ecol.* **42,** 113–123.

Peters, G. A., and Calvert, H. E. (1983). The *Azolla–Anabaena azollae* symbiosis. *In* "Algal Symbiosis" (L. Goff, ed.), pp. 109–145. Cambridge Univ. Press, Cambridge, UK.

Peterson, C. G., and Stevenson, R. J. (1989). Substratum conditioning and diatom colonization in different current regimes. J. Phycol. **25,** 790–793.

Pettersson, K. (1980). Alkaline phosphatase activity and algal surplus phosphorus as phosphorus-deficiency indicators in Lake Erken. *Arch. Hydrobiol.* **89,** 54–87.

Pfiester, L. A., and Popovsk;aay, J. (1979). Parasitic, amoeboid dinoflagellates. *Nature (London)* **379,** 421–424.

Pip, E., and Robinson, G. G. C. (1982a). A study of the seasonal dynamics of three phycoperiphytic communities, using nuclear track autoradiography. I. Inorganic carbon uptake. *Arch. Hydrobol.* **94,** 341–371.

Pip, E., and Robinson, G. G. C. (1982b). A study of the seasonal dynamics of three phycoperiphytic communities, using nuclear track autoradiography. II. Organic carbon uptake. *Arch. Hydrobiol.* **96,** 47–64.

Popovský, J. and Pfiester, L. A. (1990). "Süsswasserflora von Mitteleuropa—Dinophyceae." Fischer, Stuttgart.

Prescott, G. W. (1973). "Algae of the Western Great Lakes Area," rev. ed. Wm. C. Brown, Dubuque, IA.

Pringle, C. M. (1990). Nutrient spatial heterogeneity: Effects on community structure, physiognomy, and diversity of stream algae. *Ecology* **71,** 905–920.

Rahat, M. (1992). Algae/hydra symbioses. *In*: "Algae and Symbioses" (W. Reisser, ed.), pp. 41–62. Biopress Ltd., Bristol, UK.

Raven, J. A. (1980). Nutrient transport in microalgae. *Adv. Microb. Ecol.* **21,** 47–226.

Raven, J. A. (1981). Nutritional strategies of submerged benthic plants: The acquisition of C, N and P by rhizophytes and haptophytes. *New Phytol.* **88,** 1–30.

Raven, J. A. (1992). How benthic macroalgae cope with flowing freshwater: Resource acquisition and retention. *J. Phycol.* **28,** 133–146.

Reed, D. C., Laur, D. R., and Ebeling, A. W. (1988). Variation in algal dispersal and recruitment: The importance of episodic events. *Ecol. Monogr.* **58,** 321–335.

Reisser, W. (1992a). Endosymbiotic associations of algae with freshwater protozoa and invertebrates. *In* "Algae and Symbioses" (W. Reisser, ed.), pp. 1–20. Biopress Ltd., Bristol, UK.

Reisser, W. (1992b). Basic mechanisms of signal exchange, recognition, specificity, and regulation in endosymbiotic systems. *In* "Algae and Symbioses" (W. Reisser, ed.), pp. 657–674. Biopress Ltd., Bristol, UK.

Reisser, W., and Herbig, E. (1987). Studies on the ecophysiology of endocytobiotic associations of ciliates and algae. I. Carbohydrate budgets of green and alga-free *Paramecium bursaria* under laboratory and natural growth conditions. *Endocytbiosis Cell Res.* **4**, 305–316.

Reiter, M. A., and Carlson, R. E. (1986). Current velocity in streams and the composition of benthic algal mats. *Can. J. Fish. Aquat. Sci.* **43**, 1156–1162.

Revsbech, N. P., Bloackburn, T. H., and Cohen, Y. (1983). Microelectrode studies of the photosynthetic and O_2, H_2S, and pH profiles of a microbial mat. *Limnol. Oceanogr.* **28**, 1062–1074.

Riber, H. H., and Wetzel, R. G. (1987). Boundary-layer and internal diffusion effects on phosphorus fluxes in lake periphyton. *Limnol. Oceanogr.* **32**, 1181–1194.

Richardson, L. L., and Castenholz, R. W. (1987). Use of microelectrodes for the analysis of vertical migrations of a cyanobacterium in a sulfide-rich microbial mat. *In* "Proceedings of the Annual Meeting of the American Society of Limnology and Oceanography," abstr. p. 64. *Am. Soc. Limnol. Oceanogr.*, Madison, WI.

Richerson, P., Armstrong, R., and Goldman, C. R. (1970). Contemporaneous disequilibrium, a new hypothesis to explain the "paradox of the plankton." *Proc. Natl. Acad. Sci. U.S.A.* **67**, 1710–1714.

Robinson, G. G. C. (1983). Methodology: The key to understanding periphyton. *In* "Periphyton of Freshwater Ecosystems" (R. G. Wetzel, ed.), pp. 245–252. Dr. W. Junk Publishers, Boston.

Rogers, A. W. (1979). "Techniques of Autoradiography," rev. ed. Elsevier/North-Holland, Amsterdam.

Rogers, K. H. and Breen, C. M. (1983). An investigation of macrophyte, epiphyte and grazing interactions. *In* "Periphyton of Freshwater Ecosystems" (R. G. Wetzel, ed.), pp. 217–226. Dr. W. Junk Publishers, Boston.

Roos, P. J. (1983). Dynamics of periphytic communities. *In* "Periphyton of Freshwater Ecosystems" (R. G. Wetzel, ed.), pp. 5–10. Dr. W. Junk Publishers, Boston.

Rosowski, J. R., and Willey, R. L. (1975). *Colacium libellae* sp. nov. (Euglenophyceae), a photosynthetic inhabitant of the larval damselfly rectum. *J. Phycol.* **11**, 310–315.

Round, F. E. (1981). "The Ecology of Algae." Cambridge Univ. Press, New York.

Round, F. E. (1992). Epibiotic and endobiotic associations between chromophyte algae and their hosts. *In* "Algae and Symbioses" (W. Reisser, ed.), pp. 593–617. Biopress Ltd., Bristol, UK.

Round, F. E., and Palmer, J. D. (1966). Persistent, vertical-migration rhythms in benthic microflora. II. Field and laboratory studies on diatoms from the banks of the river Avon. *J. Mar. Biol. Assoc. U.K.* **46**, 191–214.

Round, F. E., Crawford, R. M., and Mann, D. G. (1990). "The Diatoms—Biology and Morphology of the Genera." Cambridge Univ. Press, New York.

Sand-Jensen, K. (1977). Effect of epiphytes on eelgrass photosynthesis. *Aquat. Bot.* **3**, 55–63.

Sanders, R. W., and Porter, K. G. (1988). Phagotrophic phytoflagellates. *Adv. Microb. Ecol.* **10**, 167–192.

Sheath, R. G. (1984). The biology of freshwater red algae. *Prog. Phycol. Res.* **3**, 89–157.

Sheath, R. G., and Burkholder, J. M. (1985). Characteristics of softwater streams in Rhode Island. II. Composition and seasonal dynamics of macroalgal communities. *Hydrobiologia* **128**, 109–118.

Sheath, R. G., and Hambrook, J. A. (1990). Freshwater ecology. *In* "Biology of the Red

Algae" (K. M. Cole and R. G. Sheath, eds.), pp. 423–454. Cambridge Univ. Press, New York.

Sieburth, J. McN., and Thomas, C. D. (1973). Fouling on eelgrass (*Zostera marina* L.). *J. Phycol.* **9,** 46–50.

Siver, P. A. (1980). Microattachment patterns of diatoms on leaves of *Potamogeton robinsii* Oakes. *Trans. Am. Microsc. Soc.* **99,** 217–220.

Smith, D. C., and Douglas, A. E. (1987). "The Biology of Symbiosis." Arnold, New York.

Søndergaard, M. (1983). Heterotrophic utilization and decomposition of extracellular carbon released by the aquatic angiosperm *Littorella uniflora* (L.) Aschers. *Aquat. Bot.* **16,** 59–73.

Sozogni, W. C., Chapra, S. C., Armstrong, D. E., and Logan, T. J. (1982). Bioavailability of phosphorus inputs to lakes. *J. Environ. Qual.* **11,** 555–563.

Stevenson, J. C. (1988). Comparative ecology of submerged grass beds in freshwater, estuarine and marine environments. *Limnol. Oceanogr.* **33,** 867–893.

Stevenson, R. J.(1983). Effects of current and conditions simulating autogenically changing microhabitats on benthic diatom immigration. *Ecology* **64,** 1514–1524.

Stevenson, R. J., and Glover, R. (1993). Effects of algal density and current on ion transport through periphyton communities. *Limnol. Oceanogr.* **38,** 1276–1281.

Stewart, K. E., and Schlichting, H. E., Jr. (1966). Dispersal of algae and protozoa by selected aquatic insects. *J. Ecol.* **54,** 551–562.

Stewart, W. D. P., Rowell, P., and Rai, A N. (1980). Symbiotic nitrogen-fixing cyanobacteria. *In* "Nitrogen Fixation" (W. D. P. Stewart and J. Gallo, eds.), pp. 239–277. Academic Press, New York.

Suberkropp, K. F., and Klug, M. J. (1976). Fungi and bacteria associated with leaves during processing in a woodland stream. *Ecology* **57,** 707–719.

Sullivan, M. J., and Moncreiff, C. A. (1990). Edaphic algae are an important component of salt marsh food-webs: Evidence from multiple stable isotope analyses. *Mar. Ecol.: Prog. Ser.* **62,** 149–159.

Szczepańska, W. (1971). Allelopathy among the aquatic plants. *Polsk. Arch. Hydrobiol.* **18,** 17–30.

Tilman, D., Kilham, S. S., and Kilham, P. (1982). Phytoplankton community ecology: The role of limiting nutrients. *Annu. Rev. Ecol. Syst.* **13,** 349–372.

Tippett, R. (1970). Artificial surfaces as a method of studying populations of benthic microalgae in fresh water. *Br. Phycol. J.* **5,** 187–199.

Tomaselli, L., Margheri, M. C., Giovannetti, L., Sili, C., and Carlozzi, P. (1988). The taxonomy of *Azolla* spp.—Cyanobionts. *Ann. Microbiol. Enzimol.* **38,** 157–161.

Tranvik, L. J., Porter, K. G., and Sieburth, J. McN. (1989). Occurrence of bacterivory in *Cryptomonas,* a common freshwater phytoplankter. *Oecologia* **78,** 473–476.

Vogel, S. (1981). "Life in Moving Fluids—The Physical Biology of Flow." Princeton Univ. Press, Princeton, NJ.

Wetzel, R. G. (1979). The role of the littoral zone and detritus in lake metabolism. *Arch. Hydrobiol. Beih. Ergeb. Limnol.* **13,** 145–161.

Wetzel, R. G. . (1983). "Limnology," 2nd ed. Saunders College Publisher, Philadelphia.

White, J. G., Amos, W. B., and Fordham, M. (1987). An evaluation of confocal versus conventional imaging of biological structures by fluorescence light microscopy. *J. Cell Biol.* **105,** 41–48.

Whitton, B. A. (1975). Algae. *In* "River Ecology" (B. A. Whitton, ed.), pp. 81–105. Univ. of California Press, Berkeley.

Willey, R. L., Bowen, W. R., and Durban, E. M. (1970). Symbiosis between *Euglena* and damselfly nymphs is seasonal. *Science* **170,** 80–81.

Willey, R. L., Cantrell, P. A., and Threlkeld, S. T. (1990). Epibiotic flagellates increase the susceptibility of some zooplankton to fish predation. *Limnol. Oceanogr.* **35,** 952–959.

Wium-Andersen, S., Anthoni, U., Christophersen, C., and Houen, G. (1982). Allelopathic effects on phytoplankton by substances isolated from aquatic macrophytes (Charales). *Oikos* **39**, 187–190.

Wright, R. T., and Hobbie, J. E. (1966). Use of glucose and acetate by bacteria and algae in aquatic ecosystems. *Ecology* **47**, 447–464.

10 The Role of Heterotrophy in Algae

Nancy C. Tuchman
Department of Biology, Loyola University of Chicago, Chicago, Illinois 60626

I. DEFINING AUTOTROPHY AND HETEROTROPHY

Benthic algae and phytoplankton have largely been considered strictly photoautotrophic in terms of their energy and carbon requirements. In periphyton assemblages, it has been generally recognized that algae comprise the autotrophic component, and the heterotrophic members include bacteria, fungi, protists, and other meiofauna. This categorization is not fully accurate, as some bacteria and protists have autotrophic capabilities, and a large number of algae can metabolize heterotrophically. Autotrophy is generally defined as a mode of nutrition whereby solar radiation serves as the source of energy for generating ATP, and inorganic carbon (e.g., CO_2 or HCO_3^-) is assimilated by the organism to produce organic carbon; the latter is necessary for the synthesis of cellular components and for energy production. In contrast, heterotrophic nutrition does not require light, and utilizes preformed organic compounds from the environment to serve as both an energy source and as a source of carbon-containing building blocks for the organism.

In this discussion, I suggest that most algae are physiologically equipped to metabolize both autotrophically and heterotrophically, and though autotrophic metabolism is employed most frequently and most efficiently by algae, facultative heterotrophy can be a crucial survival mechanism for light-limited algae. This implies that obligate photoautotrophy is probably less common among algae than was conventionally believed. Although heterotrophic metabolism in algae is not a well-studied topic, the experimental evidence that algae do utilize heterotrophy to survive long periods of light deprivation will be elucidated in this chapter.

Many algal physiologists study planktonic species probably because plankton are easier to culture in the laboratory than substratum-associated forms. As a result, most examples of algal heterotrophy in the literature are of planktonic species, and in this chapter they will be referred to as algae that are capable of operating heterotrophically. Since there are several examples of benthic algae utilizing heterotrophy in the literature, it is implied that benthic forms have heterotrophic mechanisms that are physiologically similar to their planktonic counterparts.

To contrast the differences and underscore the overlaps between autotrophic and heterotrophic nutrition, a slightly more detailed definition of strict photoautotrophy and its ecological implications in algae will be given. Strict photoautotrophs acquire their chemical energy from the photosynthetic light-dependent reactions whereby photons of light are absorbed, and through a series of biochemical and biophysical reactions, oxygen is evolved and energy (ATP) and reductant (NADPH) are formed. The products of the light-dependent reactions are subsequently consumed in the light-independent ("dark") reactions where CO_2 fixation occurs. The products of photosynthesis, therefore, provide the organism with both chemical energy (ATP and organic carbon) for metabolic demands and organic carbon for the synthesis of cellular components. Since the light-dependent reactions cannot proceed in the dark, the distribution of obligate photoautotrophic organisms is restricted to well-illuminated habitats for the perpetuation of their metabolic activities, and if strict photoautotrophs are subjected to the dark for extended periods of time, they will either die after using up their storage products or become metabolically inactive (dormant) until environmental conditions become more suitable.

Recent studies have expanded the conventional understanding of photoautotrophy, describing a mechanism whereby the "light" reactions can operate independently of the "dark" reactions under illuminated conditions in algae as well as in higher plants. When photosynthetic reaction centers are subjected to illumination, the light-dependent reactions are automatically activated and will not turn off until the quality/quantity of irradiance decreases below the threshold for photopigment stimulation. However, it has been shown that the light-independent reactions can be deactivated in the light if adequate concentrations of organic carbon substrates are present in the environment for cell uptake (e.g., Sheen, 1990). Under these circumstances, the Calvin cycle is blocked, carbon fixation ceases, and the organism utilizes a more energy-efficient mechanism (active transport) of acquiring exogenously supplied organic carbon sources (Fig. 1). Under this form of nutrition termed photoorganotrophy, the light-dependent reaction products can be (1) funneled into other cellular metabolic pathways,

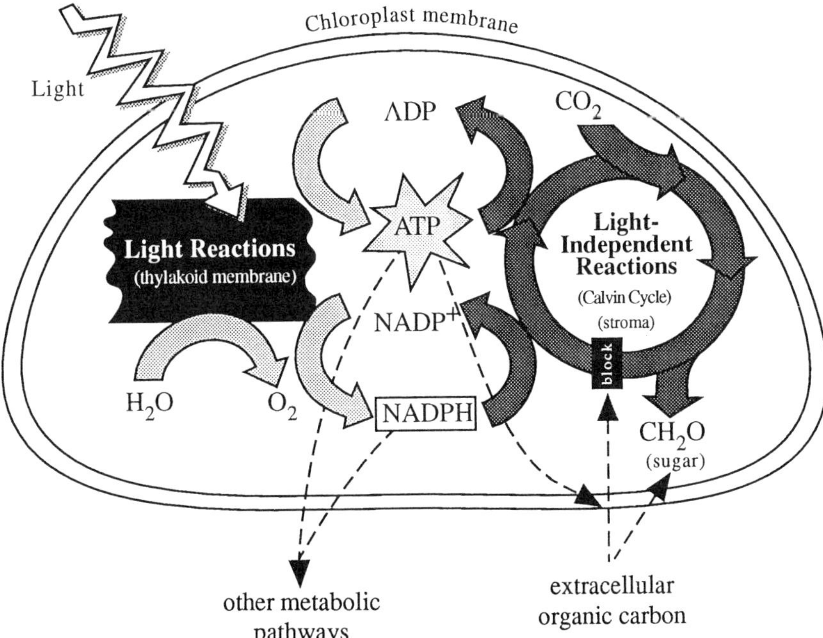

FIGURE 1 An overview of photosynthesis showing integration of the light-dependent and light-independent reactions. In the thylakoid membranes of the chloroplast, the light-dependent reactions utilize solar energy to produce ATP and NADPH, which function as energy and reducing power, respectively, in the Calvin cycle. Within the stroma of the chloroplast, the light-independent reactions assimilate carbon dioxide and produce sugars. Both sets of reactions can occur simultaneously in the light, but the light-independent reactions can be independently regulated by the presence of an exogenous organic carbon source. Dashed lines represent the events that take place in the presence of exogenous organic carbon: the Calvin cycle is blocked and CO_2 fixation ceases while ATP and NADPH are still being produced by the light-dependent reactions. These energy sources can be utilized by other metabolic pathways within the cell, and some ATP may be used to actively transport exogenous DOC sources into the cell. (Modified from "Biology," 3rd ed., by Neil Campbell. Copyright © 1993 by the Benjamin/Cummings Publishing Company. Reprinted by permission.)

(2) used to assemble large, energy-rich storage product compounds, or (3) used to actively transport exogenous organic carbon substrates into the cell. Organic substrates assimilated from the environment can be utilized as a carbon source (building blocks for anabolic pathways) and/or as an energy source that is metabolized through glycolysis, the Krebs cycle, and the electron transport system to generate ATP. To be sure, the photoorganotrophic organism is conserving energy by sequestering exogenously supplied organic carbon rather than fixing its own, as the 18 ATP and 12 NADPH derived from the light-dependent reactions that are required by the Calvin cycle to produce each "glucose" molecule can instead be shifted into other metabolic pathways or used to produce photosynthetic storage products. However, the concentration of organic substrate required to induce photoorganotrophic metabolism is extremely high, exceeding levels that would typically be found in natural open water habitats. Only specific microhabitats, typically those associated with the surfaces of living or dead organisms (e.g., epiphytic or epizooic), or within organic sediments, could provide organic substrate levels high enough to induce photoorganotrophy in nature. The physiological result of algal photoorganotrophic metabolism in light- and organic-rich habitats is an accumulation of cellular storage products (e.g., triglycerides in diatoms and carbohydrates in the Chlorophyceae) and an ultimate increase in photosynthate [dissolved organic carbon (DOC)] exuded by these cells.

The collective modifications of autotrophic and heterotrophic nutrition are several in number (Table I). Photolithotrophy and photoorganotrophy (as previously described) are both considered autotrophic; both utilize light as an energy source for generating ATP, with the former using only CO_2 (obligate photoautotrophy) and the latter utilizing both CO_2 and exogenously supplied simple organic compounds as sources of carbon for the generation of cellular organic materials. Chemoorganotrophy (chemoheterotrophy) is considered to be heterotrophic as it does not require light,

TABLE I Types of Nutritional Procurement Employed by Algae as Defined by the Energy and Carbon Sources Utilized

Nutritional mode	*Energy source(s)*	*Carbon sources(s)*
Photolithotrophy (photoautotrophy)	Solar radiation	CO_2 only
Photoorganotrophy (photoheterotrophy)	Solar radiation	CO_2 and organic C
Chemolithotrophy (chemoautotrophy)	Oxidation of inorganic compounds	CO_2 only
Chemoorganotrophy (chemoheterotrophy)	Organic compounds	Organic C only

but utilizes preformed organic matter as both an energy source and a source of carbon. Facultative heterotrophs possess both photoautotrophic and chemoorganotrophic capabilities, and generally metabolize most efficiently as autotrophs. The ability of algae to gain energy through the oxidation of inorganic compounds and to utilize CO_2 as a carbon source (chemoautotrophy or chemolithotrophy) has rarely been demonstrated, although this mode of nutrition is well known for certain bacteria.

A. Illustrations and Evidence of Facultative Heterotrophy in Algae

Many studies have experimentally demonstrated the ability of numerous planktonic and benthic algal species to grow in culture heterotrophically with organic supplements in complete darkness (Table II). The experimental evidence of algal *growth* in the dark is probably the strongest evidence of algal heterotrophic capacity, since *existence* of viable algae in dark habitats in nature could be a result of resting states and not heterotrophic metabolism. There are, however, numerous examples of algae surviving long-term exposures to darkness in natural environments; light deprivation probably endured by entering dormancy, utilizing heterotrophy, or both. For example, Wasmund (1989) demonstrated the ability of several algal species, including *Scenedesmus quadricauda*, *S. intermedius*, *Monoraphidium contortum*, *Oscillatoria limnetica*, *Lyngbya contorta*, *Merismopedia punctata*, *Fragilaria construens*, and *Nitzschia acicularis*, to resume photosynthetic activity upon reillumination after being buried below the photic zone in shallow marine sediments for an extended period of time. This ability is not unique to one group of algae, as these species represent three taxonomic divisions (Chlorophyta, Cyanophyta, and Bacillariophyta) and both planktonic and benthic growth forms. Several genera of algae have been found to survive extended periods of time in groundwater. Kuehn *et al.* (1992) isolated viable *Chlorella*, *Navicula*, *Ankistrodesmus*, *Oscillatoria*, and *Stichococcus* from an aphotic aquifer. Viable green and blue-green algae were also collected for four successive years from groundwaters below an active streambed (Poulíčková, 1987). In addition, numerous studies have been published on the dark survival of polar algae. Light attenuation by snow-covered ice, sea ice, ice algae, and the water column can decrease irradiance to less than 0.6 μmol quanta m^{-2} s^{-1} at depths of 20 to 30 m in McMurdo Sound, Antarctica. A viable community of shade-adapted algae dominated by the diatom *Trachyneis aspera* was found at this depth (Palmisano *et al.*, 1985). Similar dark-survival has been demonstrated for polar algae that were experimentally kept in complete darkness for 5 months (Palmisano and Sullivan, 1983) and longer (Müller-Haeckel, 1985).

Of all the algae that appear to utilize heterotrophic nutrition, most notable are the chrysophytes and euglenoids, two groups of algae that

TABLE II Examples of Different Algal Taxa That Have Demonstrated the Ability to Grow in the Laboratory in Complete Darkness with Organic Substrates

Taxon	*Organic substrate(s)*	*Source*
Cyanophyceae		
Agemenellum quadruplicatum	Urea, allantoic acid	Oliveira and Huynh (1990)
Anabaena variabilis	Glucose, fructose, sucrose	Bastia *et al.* (1993)
Ankistrodesmus braunii	D-Glucose	Bollman and Robinson (1985)
Aphanocapsa (6 strains)	Glucose	Rippka (1972)
Aulosira prolifica	Glucose, fructose, sucrose	Bastia *et al.* (1993)
Calothrix parietina	Glucose, fructose, sucrose	Bastia *et al.* (1993)
Chlorogloea (1 strain)	Glucose	Rippka (1972)
Microchaete uberrima	Glucose, fructose, sucrose	Bastia *et al.* (1993)
Nostoc sp.	Vanillic acid	Hussein *et al.* (1989)
N. commune	Glucose, fructose, sucrose	Bastia *et al.* (1993)
N. linckia	Glucose, fructose, sucrose	Bastia *et al.* (1993)
Rivularia sp.	Glucose, fructose, sucrose	Bastia *et al.* (1993)
Westiellopsis prolifica	Glucose, fructose, sucrose	Bastia *et al.* (1993)
Chrysophyceae		
Olisthodiscus luteus	Urea	Oliveira and Huynh (1990)
Poterioochromonas malhamensis	Glucose, glycerol, ethanol	Lewitus and Caron (1991)
Dinophyceae		
Amphidinium carterae	Urea, hypoxanthine	Oliveira and Huynh (1990)
Bacillariophyceae		
Achnanthidium rostratum	Acetate, glucose, casamino acids	Panella (1994)
Amphiprora kufferathii	Leucine, glutamic acid, glycine	Rivkin and Putt (1987)
Amphora antarctica	Leucine, glutamic acid, glycine	Rivkin and Putt (1987)
A. coffaeiformis	Glucose, lactate	Lewin and Lewin (1960)
Cocconeis diminuta	Glucose, lactate	Bunt (1969)
C. diminuta	Acetate	Cooksey (1972)
Coscinodiscus sp.	Glucose	White (1974)

Cyclotella cryptica	Glucose	Hellebust (1971)
C. cryptica	Arginine, glutamate, proline	Liu and Hellebust (1974a, b)
C. sp.	Glucose	Lewin and Lewin (1960)
Cylindrotheca fusiformis	Lactate, succinate, fumarate, malate, tryptone, casamino acids, yeast extract	Lewin and Hellebust (1970)
Cymbella pusilla	Glucose	Saks (1983)
Melosira italica	Urea	Cimbleris and Cáceres (1991)
M. nummuloides	Arginine, valine	Hellebust (1970)
Navicula incerta	Glucose	Lewin and Lewin (1960)
N. pavillardi	Glutamate, tryptone, yeast extract	Lewin and Hellebust (1975)
N. pelliculosa	Glucose, glutamate, aspartate, lactate	Jolley and Hellebust (1974)
Nitzschia alba	Glucose, acetate	Linkins (1973)
N. angularis v. *affinis*	Glucose	Lewin and Lewin (1960)
N. closterium	Glucose, lactate	Lewin and Lewin (1960)
N. curvilineata	Lactate	Lewin and Lewin (1960)
N. filiformis	Glucose	Lewin and Lewin (1960)
N. frustulum	Glucose, lactate	Lewin and Lewin (1960)
N. laevis	Glutamate, glucose	Lewin and Hellebust (1978)
N. marginata	Glucose	Lewin and Lewin (1960)
N. obtusa v. *undulata*	Glucose	Saks (1983)
N. ovalis	Arginine	North and Stephens (1972)
N. punctata	Glucose	Lewin and Lewin (1960)
N. tenuissima	Glucose, acetate, lactate	Lewin and Lewin (1960)
Pleurosigma sp.	Leucine, glutamic acid	Rivkin and Putt (1987)
Porosira pseudodenticulata	Leucine, glutamic acid, glycine	Rivkin and Putt (1987)
Thalassiosira nordenskioldii	Urea	Oliveira and Huynh (1990)
T. pseudomonana	Urea	Oliveira and Huynh (1990)
Trachyneis aspera	Glutamic acid, glycine	Rivkin and Putt (1987)
Chlorophyceae		
Chlorella sp. VJ79	Glycine, acetate, alanine	Lalucat *et al.* (1984)
Dunaliella tertiolecta	Urea, allantoic acid, hypoxanthine	Oliveira and Huynh (1990)
Haematococcus pluvialis	Acetate	Kobayashi *et al.* (1992)
Pediastrum duplex	Glycerol, leucine	Berman *et al.* (1977)

commonly utilize heterotrophic metabolism in conjunction with photoautotrophy. Most of these algae utilize several types of organic compounds in the light (photoorganotrophy) or in the dark (chemoorganotrophy), whereas others obtain organic nutrients through bacterivory (mixotrophy). One freshwater chrysophyte, *Poterioochromonas malhamensis,* has an unusually low capacity for photoautotrophic growth, and heterotrophy and bacterivory account for the majority of its nutritional requirements (Lewitus and Caron, 1991). Some genera (e.g., *Euglena* and *Chlorella*) have high photolithotrophic capacities under illuminated conditions in the absence of organic substrates, but when organics are present (such as within the sediments of lakes and streams or on the surface of vascular plants and invertebrates), chloroplast development is notably repressed and heterotrophic metabolism dominates (Ogawa and Aiba, 1981; Monroy and Schwartzbach, 1984). Studies like these suggest that some of the euglenoids and chrysophytes may be as efficient (if not more so) when utilizing heterotrophic nutrition than when they utilize photoautotrophy, and that algae existing in habitats that contain high levels of DOC may utilize DOC as their primary source of carbon.

The challenge of experimentally demonstrating algal facultative heterotrophy in nature is complex, because it is often difficult to distinguish active algal heterotrophic metabolism and dormancy, and to discern algal heterotrophy from bacterial, fungal, or protozoan heterotrophy (see Geider and Osborne, 1992). However, attempts to quantify algal heterotrophy in nature using differential $^{14}CO_2$ and radiolabeled glucose uptake methods have suggested that the utilization of environmental organic carbon by phytoplankton in light-limited conditions plays an important role in the survival of these species (Amblard *et al.,* 1992). Although heterotrophy has not yet been definitively demonstrated for algae in periphyton assemblages, the ratio of glucose : CO_2 molecules assimilated by cells within the canopy is significantly lower (1:25) than that for cells within the light-deprived understory (1:6) in a well-developed stream periphyton community (Fig. 2).

B. Discerning the Difference between Heterotrophic and Dormant Algae

Although there are numerous examples of dark-survival of algae in natural habitats, it remains unclear whether those cells were surviving darkness by operating heterotrophically or if they remained viable by entering a state of dormancy (resting stage). Resting stages are thought to be the mechanisms whereby many plankton overwinter in the deep, cold, dark, and often anoxic waters of lakes and oceans, and whereby benthic algae endure otherwise harsh conditions such as high sediment loading (e.g., Burkholder, 1992). To discern whether algal cells that occur in dark

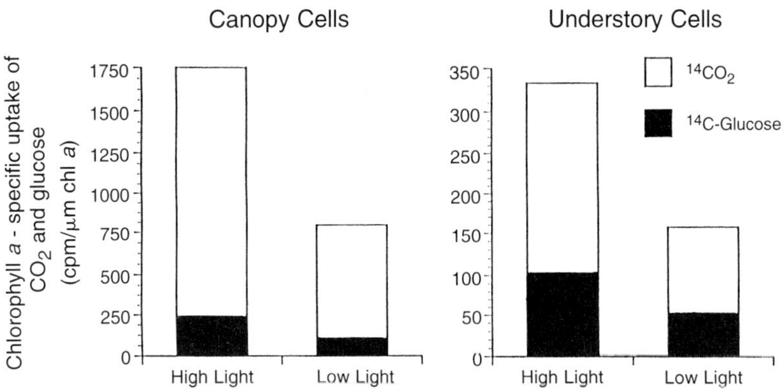

FIGURE 2 The differential assimilation of $^{14}CO_2$ and [^{14}C]-glucose between canopy and basement cells of two periphyton mats grown in an outdoor artificial stream system (one grown in high light, ca. 2000 μmol quanta m^{-2} s^{-1}, and one grown in lower light, ca. 200 μmol quanta m^{-2} s^{-1}). In both light treatments, the quantity of light reaching the understory was 82–91% lower than that in the canopy. Relative to CO_2 uptake, chlorophyll *a*-specific glucose uptake in the understory was 3.0 times greater than that for cells in the canopy. (From N. C. Tuchman and J. Smarrelli, unpublished data.)

habitats are resting or actively metabolizing heterotrophically, the metabolic activity of the algae should be examined. For example, algal cells collected from a dark habitat could be experimentally incubated with radiolabeled organic substrates (e.g., [^{14}C]-glucose, [^{3}H]-leucine) in the dark, and measurable assimilation and incorporation of the labeled substrate would indicate active metabolism. In one such experiment, *Achnanthidium rostratum* cells grown in complete darkness for up to 30 days were able to incorporate [^{3}H]-leucine into proteins at rates similar to those of control cultures grown in full light (Panella, 1994; Fig. 3). If the cells did not incorporate the substrate, they would be either nonviable or in a metabolically inactive dormant state, which could be sorted out by subsequently subjecting the dark-grown cultures to light and nutrient media, which should induce resting cells to rejuvenate and grow autotrophically.

For some algae, physiological resting states (as opposed to resting spores or cysts, which are easily discernible with light microscopy) can be identified using light microscopy. Sicko-Goad *et al.* (1989) identified resting cells in 18 diatom species collected from aphotic sediments in the Laurentian Great Lakes. These cells were characterized as having all of the cytoplasm condensed into one dense mass, which was usually positioned in the center of the frustule. They found that these resting cells could be rejuvenated within 2 hours of reillumination, and that temperature was a major factor in inducing both resting stage formation and subsequent rejuvenation.

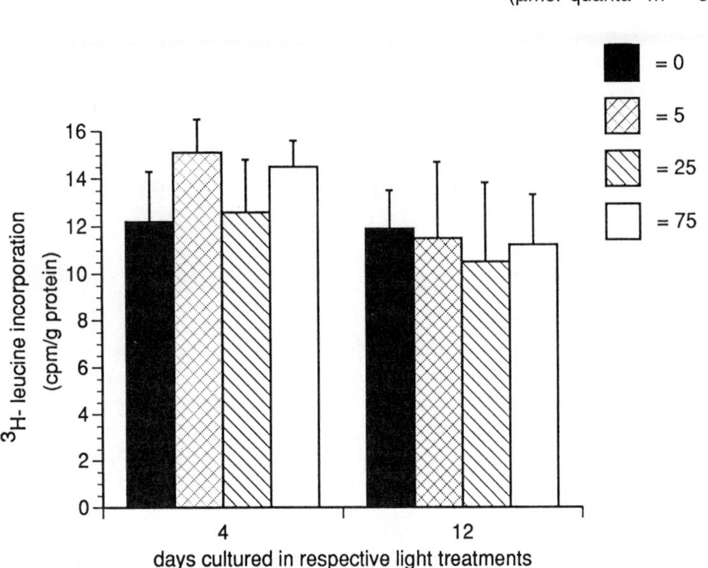

FIGURE 3 A method used to distinguish whether algal cells cultured in complete darkness and low light levels survive by entering dormancy or by metabolizing heterotrophically. The high levels of incorporation of a radiolabeled amino acid into protein by *Achnanthidium rostratum* grown in culture under four light regimes indicate that cells in complete darkness were as metabolically active as those exposed to higher illumination, which suggests that they were employing heterotrophy and were not resting or dormant. (From Panella, 1994.)

C. Mechanisms of Organic Substrate Uptake and Relative Efficiency of Heterotrophic Metabolism in Algae

Heterotrophic nutrition requires that the algae must have a mechanism whereby exogenous organic carbon sources can be effectively taken up by the cell and assimilated. Concentrations of organic substances in the water column of natural aquatic habitats are usually too low for entry into the cells via passive diffusion. However, in benthic sediments, where decomposition of accumulated organic matter provides a greater source of DOC, passive diffusion into certain permeable algal cells may be possible (Stadelmann, 1962).

The existence of specific transport systems that facilitate uptake of certain exogenously supplied organic substances has been documented for many algae. In a review by Hellebust and Lewin (1977), specific transport systems for uptake of organic substrates existed in all the diatom species that were studied. Their review detailed the transport systems in many diatoms and emphasized the difficulty in making a generalized statement concerning the efficiency of the uptake mechanisms. Even though the growth rate of some species was greatly reduced in the dark, certain species were able to assimilate organic compounds very efficiently and grow in the dark at rates equal to their growth in autotrophic conditions. Some species grew faster heterotrophically, as exemplified by Day *et al.* (1991), who developed an industrial-scale process for the heterotrophic production of *Skeletonema, Chaetoceros, Thalassiosira, Isochrysis, Tetraselmis,* and *Nannochloris* for use as a commercial food for cultivating mollusks. Although this process can yield heterotrophic growth rates for these genera that exceed maximal rates of production under autotrophic conditions, the organic substrate concentrations required to achieve such high production are several orders of magnitude higher than would be found in natural waters.

Recent studies (M. Schollett and N. Tuchman, unpublished data) testing the ability of 8 benthic diatom species to metabolize 94 different organic substrates indicate that although each species can actively metabolize all 94 organic substrates to some degree, they zealously metabolize 22 of the organic substrates (oxidizing these substrates from 150 to 600% more than the H_2O control; Table III). These data indicate that a wide array of transport systems are operational in these diatom species, and suggest that diatoms potentially utilize multiple organic compounds.

Transport systems that have been described are diverse among taxa and are generally complex in nature. In some algae, the assimilation of certain organic substrates may require more than one transport system. For example, Cooksey (1972) suggested that complex uptake kinetics of *Cocconeis minuta* indicate the presence of more than one transport system for the assimilation of lactate. Conversely, in *Cyclotella cryptica,* White (1974) demonstrated that glucose and galactose share the same transport system, whereas other hexoses or pentoses were not assimilated via this transport mechanism. In addition to transport systems for organic carbon substrates, many species possess transport systems for the uptake of amino acids (e.g., Bourdier *et al.,* 1989) that can be utilized as a source of nitrogen when exogenous inorganic nitrogen is limiting. There are at least three transport systems for the uptake of amino acids in diatoms, one for each of acidic, neutral, and basic amino acids (see Hellebust and Lewin, 1977).

Since the maintenance of transport systems, like any metabolic process, requires energy and elemental resources, it is advantageous to "turn the system off" when it is not being used. The working capacity of transport systems in facultative heterotrophs appears to be regulated by environmental cues for this purpose. For example, the capacity of the glucose transport system in many algae is closely regulated by illumination. When algae were grown under photosynthetically saturated levels of solar radiation, glucose uptake (regardless of concentration) approached zero; whereas in the dark, glucose uptake capacity was maximized (Lalucat *et al.,* 1984; Hellebust,

TABLE III Relative Rates of Uptake and Oxidation of 21 Organic Substrates Commonly Utilized by Benthic Diatom Species

Organic compound	E.m.p.[a]	N.t.[b]	A.m.[c]	N.p.[d]	A.r.[e]	G.a.[f]	E.m.[g]	N.l.[h]
D-Galactose	++[i]	++	++	++	++	++	++	++
D-Glucose	++	++	++	++	++	++	++	++
D-Mannose	++	++	++	++	++	++	++	++
D-Trehalose	++	++	++	++	++	++	++	++
Acetic acid	+[j]	+	+	++	+	+	++	++
cis-Aconitic acid	++	++	++	++	++	++	++	++
D-Galactonic acid lactone	++	++	++	++	++	++	0[k]	++
Itaconic acid	++	++	++	++	++	++	++	++
Alpha-ketoglutaric acid	++	++	++	++	++	++	++	++
D,L-Lactic acid	++	++	++	++	++	++	++	++
Malonic acid	++	++	++	+	++	+	++	++
Propionic acid	+	+	+	++	+	+	++	+
Quinic acid	++	++	++	++	++	++	++	++
D-Saccharic acid	++	++	++	++	++	++	++	++
Succinic acid	++	++	++	++	++	++	++	++
L-Alanine	++	++	++	++	++	++	++	++
L-Alanyl-glycine	++	++	++	++	++	++	++	++
L-Asparagine	++	++	++	+	++	+	++	++
L-Aspartic acid	++	++	++	++	++	++	++	0
L-Glutamic acid	++	++	++	++	++	++	++	0
L-Serine	++	++	++	++	++	++	++	0

(M. Schollett and N. Tuchman, unpublished data)

[a] *E.m.p.* = *Encyonema minutum* var. *pseudogracilis.* [b] *N.t.* = *Navicula trivialis.* [c] *A.m.* = *Achnanthidium minutissimum.* [d] *N.p.* = *Nitzschia palea.* [e] A.r. = *Achnanthidium rostratum.* [f] *G.a.* = *Gomphonema acuminatum.* [g] *E.m.* = *Encyonema minutum.* [h] *N.l.* = *Nitzschia linearis.* [i] ++ = oxidation rates 150–600% greater than that of the control (water). [j] + = oxidation 15–150% greater than that of the control. [k] 0 = no significant difference from control.

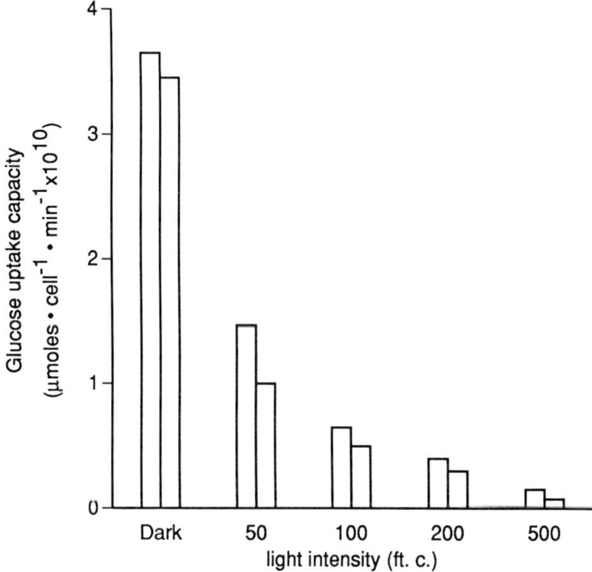

FIGURE 4 The glucose transport system in *Cyclotella cryptica* is down-regulated under high illumination conditions and up-regulated in the dark. These cells were incubated for 3 days in different light conditions in glucose-free media before glucose uptake capacity was measured. Data represent two replicate experiments. (From Hellebust, 1971.)

1971; Fig. 4). Aside from this example, the physiological mechanisms and environmental cues involved in the integration of heterotrophic and autotrophic nutrition within an algal cell have been scarcely explored.

II. THE ECOLOGICAL ROLES OF ALGAL HETEROTROPHY IN BENTHIC ASSEMBLAGES

There are numerous examples of habitats in nature where facultative heterotrophy would provide many advantages over obligatory photoautotrophy for benthic algae. For example, on a large spatial and temporal scale, situations where light limitations are produced by season (e.g., snow-covered ice on lakes, streams, and wetlands, turbid vernal floodwaters, or autogenic shading in eutrophic systems), latitude (e.g., winter in polar and subarctic regions), and depth in a deep lentic or large lotic ecosystem would favor algae that possess heterotrophic capabilities. On a smaller spatial scale, heterotrophy would be beneficial for cells located at the base of a thick periphyton matrix, where access to inorganic nutrients from the water column is impeded and autogenic shading by cells above them is severe. In addition, heterotrophic metabolism would serve benthic algae inhabiting the heavily shaded portions of headwater streams, as well as epipsammon and epipelon temporarily covered with silt or buried in the sediments following a storm event. In the foregoing examples of dark habitats, light limitation is potentially temporary. If algae could survive the temporary darkness by utilizing heterotrophic metabolism, they would not only possess the competitive advantage over obligate photoautotrophs during the extent of dark exposure, but during the initial phase of growth that follows reillumination, the dark-survivors would dominate because their populations would already be established.

In addition to light-limiting habitats, there are many examples of microhabitats rich in dissolved organic compounds that would facilitate photoorganotrophy if light were available, and heterotrophic metabolism if light were limited. For example, microalgae associated with vascular plants, macroalgae, mollusks, crayfish, insects, or other animals acquire high levels of organics from the exudates/feces of their "hosts." Algae associated with detritus (e.g., leaf litter or woody substrata) and organic sediments similarly live in high-DOC environments that are conducive to photoorganotrophic and/or chemoorganotrophic metabolism.

A. Development of a Resource Gradient within a Periphyton Mat and Implications for Cells at the Base of the Mat

In many developing periphyton mats, diatoms dominate the basal cells of the matrix (Korte and Blinn, 1983) whereas stalk-forming diatoms and filamentous Chlorophyceae and blue-green algae comprise a loosely associated upper canopy (e.g., Tuchman and Stevenson, 1991). As cell densities increase in the matrix, internal microhabitats are greatly altered, leading to a series of events that culminate in the sloughing of the mat from the substratum. Often, the early colonizing diatoms stay attached to the substratum as basement cells in the assemblage and remain there beyond the sloughing event (Tuchman, 1988; Johnson, 1996; Fig. 5).

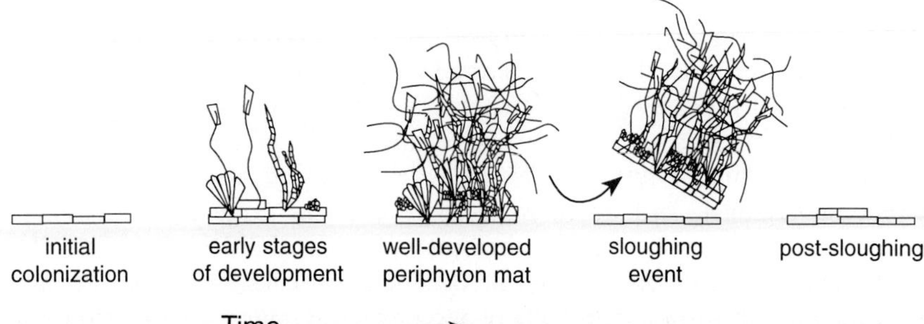

FIGURE 5 The accumulation of benthic algal cells on substrata and the process of periphyton mat development. Adnate, early colonizing cells can remain attached to the substratum at the base of the mat throughout community development and after the sloughing event.

As cell densities within periphytic mats increase during the development process, resources become limiting at the base of the mat, forming a vertical resource gradient. Higher light (Fig. 6) and a more rapid renewal of nutrient supply occur in the canopy, whereas cells at the base of the mat experience low light (Jørgensen *et al.*, 1983; Kühl and Jørgensen, 1992; Dodds, 1992) and lower nutrient levels (Burkholder *et al.*, 1990; Liehr *et al.*, 1990; Blenkinsopp *et al.*, 1991; Stevenson and Glover, 1993). The resource-depleted microhabitat at the base of the mat will have potentially large effects on the basement cells that remain there throughout the process of mat development. One response of the lower-story algae to lower illumination, at least initially, is an increase in phytopigments that theoretically will allow these algae to maximize their photon-sequestering capacities in a dimly lit environment (e.g. Hudon *et al.*, 1987; Lewitus *et al.*, 1991).

The severity and duration of the resource gradient will vary among habitats, with species composition, with the degree of cell packing within a mat, and with disturbance regimes. It is predicted that resource gradients producing low light and low available nutrients in the lower story would induce heterotrophic metabolism in the cells at the base of the mat. Further, in habitats where the duration and/or steepness of the resource gradient is great, algal heterotrophy could be an important survival mechanism for the basement cells.

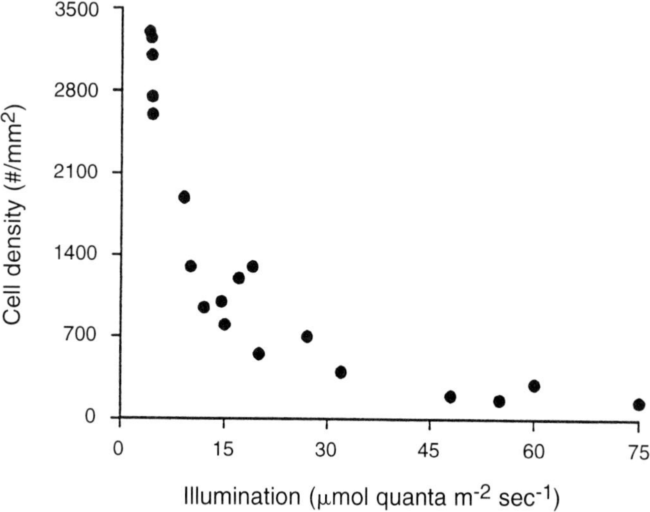

FIGURE 6 The relationship between cell density and light attenuation at the base of a periphyton mat. Stream water was circulated through a series of artificial stream channels lined with clay tiles. Cell density on tiles was assayed weekly for 10 weeks, and light penetration to the base of the mat was measured by placing a photometer sensor against the bottom of a glass window that was built into the bottom of each channel and allowed to be colonized along with the tiles. Each data point represents an average of three measurements of light and cell density. (From N. C. Tuchman, unpublished.)

If an algal cell does not have the means by which to remove itself from a resource-poor microhabitat (e.g., motility or by elevating itself into the canopy via stalk or filament formation), it must endure the poor conditions until it is reintroduced into resource-rich habitats by sediment disturbance or, if at the base of a mat, until the canopy is removed by sloughing, scouring from a flood event, or grazing. These stationary cells could enter a state of dormancy or metabolize heterotrophically until phototrophic conditions improve. The disadvantage of these strategies lies in the inability of dormant cells (and of some species in heterotrophic conditions) to reproduce and perpetuate the population genome during the overgrowth periods. Yet the ecological advantage of surviving the high-density, resource-poor overgrowth period is evident. When cells are reexposed to high light and water column nutrients they will then have the competitive and reproductive advantage in that space until immigration and accumulation of other species occur.

B. The Role of Algal Heterotrophy in Benthic Microbial Carbon Cycling

In addition to being a survival mechanism for algae temporarily subjected to dark, algal heterotrophy is a potentially interesting consideration in the flow of carbon through aquatic food webs. Although it is widely accepted that algal photosynthate can be an important source of DOC to aquatic bacteria, it is unclear to what extent algal facultative heterotrophs (1) utilize the wide array of DOC substrates produced by chemolithotrophic bacteria, (2) can, like bacteria, decompose organic matter in the sediments, and (3) reabsorb their own photosynthates.

As a demonstration of the flow of organic carbon from algae to bacteria, Baines and Pace (1991) analyzed rates of extracellular release of organic carbon from algae in 16 different published studies, and estimated that these algal exudates amounted to nearly half of the carbon required for bacterial growth in most pelagic (marine and freshwater) systems. Relatively few investigations exist concerning the reuse of algal exudates by the algae that produce them, and none of these studies is particularly recent (e.g., Guillard and Hellebust, 1971; Jolley and Jones, 1977; Lancelot and Mathot, 1985; Veldhuis and Admiraal, 1985). Most of these studies investigated a colonial phytoplanktonic alga, *Phaeocystis pouchetii,* which produces a gelatinous matrix about the colony of cells. Photosynthetic exudates produced by the cells are excreted and since diffusion through the viscous mucilage is impeded, many of the exudates remain within the mucilage. Analysis of the chemical components of the exudate reveals that the high-molecular-weight (MW > 1800) compounds were reassimilated by the algae during dark periods, whereas the concentration and composition of low-molecular-weight compounds were unchanged throughout the night. The conclusions were that polymeric compounds were reassimilated by the *P. pouchetii* cells in the dark to cover the carbon and energetic needs of the cells, whereas the monomeric and oligomeric compounds were not used. Sundh (1989) noted that bacteria also prefer to assimilate the high-molecular-weight components of algal exudates, which suggests a potential competition between the algae and associated bacteria for the polymeric substances.

If this information on the flow of DOC between phytoplankton and their associated bacteria is placed in the context of benthic periphyton assemblages, it seems possible that photoautotrophic algae could produce exudates (at a rate dependent on light, nutrient availability, and temperature), which could be utilized by both the bacteria within the assemblage and the light-deprived algal cells that may be metabolizing heterotrophically. Some benthic and epiphytic diatoms (mostly belonging to the genera *Navicula* and *Nitzschia*) have demonstrated the ability to decompose various solid organic compounds such as agar, hydrolyzed casein, and organic phosphate (Tanaka and Ohwada, 1988). The ability of these algae to secrete extracellular enzymes to decompose organic matter suggests that they may be able to utilize the exudates produced by other cells in a different area of a periphyton mat or of the sediments. It is clear that, although some of the questions concerning the flow of carbon in algal/bacterial assemblages have been addressed by growing algae with and without bacteria in culture, we do not understand and potentially underestimate how often algal exudates enter heterotrophic metabolic pathways in periphyton assemblages. Nor do we understand the degree to which algal heterotrophy might influence carbon cycling in benthic food webs.

III. CONCLUSIONS

It is doubtful whether obligate photoautotrophy is as common among algae as was conventionally believed. Since bacteria and protozoa were the evolutionary precursors to the other divisions of algae, heterotrophic nutrition is likely a genetic feature that eukaryotic algae possessed initially and may have retained over the millennia. Evidence from laboratory studies suggests that numerous algal species are facultatively heterotrophic, yet we do not know to what extent heterotrophic metabolism is utilized by algae in nature. Although most facultatively heterotrophic algae exhibit higher growth rates under photoautotrophic conditions, "switching" to heterotrophic metabolism is a highly advantageous survival mechanism for cells that are temporarily subjected to dark conditions.

There are still many important and potentially pivotal questions to be answered concerning the roles of algal heterotrophy in benthic communities, including (1) the potential importance of heterotrophy to algae in dark habitats (including twilight hours), (2) the physiological integration of algal autotrophic metabolism with heterotrophic metabolism and the environmental cues that enable a cell to "switch" from one mode to the other, and (3) the potential role of algal heterotrophy in the processing and cycling of DOC within benthic microbial food webs.

In part, our lack of consideration of the role of algal heterotrophy in benthic microbial assemblages is due to the conventional way in which we study periphyton communities: to scrape a given area of periphyton/epiphyton from substrata or to collect a volume of sediment, homogenize the cells into a slurry, and subsample from the homogenate to make chemical, physical, and taxonomical measurements of the community. Although this sampling technique is quick, and is appropriate for general community-level assays such as aerial-specific biomass estimates (ash-free dry mass) and community taxonomic composition, it may be inappropriate for assays of algal physiology. Since we know that resource conditions vary greatly with vertical position in benthic periphyton assemblages, it follows that physiological conditions of the algae in various positions within assemblages would be different in response to the local microhabitat conditions. Consideration of the position of cells within their natural assemblage, and the physiological mechanisms whereby those cells adapt to the conditions of their microhabitat, is germane to advancing the studies of periphyton community dynamics.

REFERENCES

Amblard, C., Rachiq, S., and Bourdier, G. (1992). Photolithotrophy, photoheterotrophy and chemoheterotrophy during spring phytoplankton development (Lake Pavin). *Microb. Ecol.* **24,** 109–123.

Baines, S. B., and Pace, M. L. (1991). The production of dissolved organic matter by phytoplankton and its importance to bacteria: Patterns across marine and freshwater systems. *Limnol. Oceanogr.* **36,** 1078–1090.

Bastia, A. K., Satapathy, D. P., and Adhikary, S. P. (1993). Heterotrophic growth of several filamentous glue-green algae. *Algol. Stud.* **70,** 65–70.

Berman, T., Hadas, O., and Kaplan, B. (1977). Uptake and respiration of organic compounds and heterotrophic growth in *Pediastrum duplex* (Meyen). *Freshwater Biol.* **7,** 495–502.

Blenkinsopp, S. A., Gabbott, P. A., Freeman, C., and Lock, M. A. (1991). Seasonal trends in river biofilm storage products and electron transport system activity. *Freshwater Biol.* **26,** 21–34.

Bollman, R. C., and Robinson, G. G. C. (1985). Heterotrophic potential of the green alga, *Ankistrodesmus braunii* (Naeg.). *Can. J. Microbiol.* **31,** 549–554.

Bourdier, G., Bohatier, J., Feuillade, M., and Feuillade, J. (1989). Amino acid incorporation by a natural population of *Oscillatoria rubescens.* A microautoradiographic study. *FEMS Microb. Ecol.* **62,** 185–190.

Bunt, J. S. (1969). Observations on photoheterotrophy in a marine diatom. *J. Phycol.* **5,** 37–42.

Burkholder, J. M. (1992). Phytoplankton and episodic suspended sediment loading: Phosphate partitioning and mechanisms for survival. *Limnol. Oceanogr.* **37,** 974–988.

Burkholder, J. M., Wetzel, R. G., and Klomparens, K. L. (1990). Direct comparison of phosphate uptake by adnate and loosely attached microalga within an intact biofilm matrix. *Appl. Environ. Microbiol.* **56,** 2882–2890.

Campbell, N. A. (1993). "Biology," 3rd ed. Benjamin/Cummings, Redwood City, CA.

Cimbleris, A. C. P., and Cáceres, O. (1991). Kinetics of urea uptake by *Melosira italica* (Ehr.) Kütz at different luminosity conditions. *Hydrobiologia* **220,** 211–216.

Cooksey, K. E. (1972). Metabolism of organic acids by a marine pennate diatom. *Plant Physiol.* **50,** 1–6.

Day, J. D., Edwards, A. P., and Rodgers, G. A. (1991). Development of an industrial-scale process for the heterotrophic production of a micro-algal mollusc feed. *Bioresour. Technol.* **38,** 245–249.

Dodds, W. K. (1992). A modified fiber-optic light microprobe to measure spherically integrated photosynthetic photon flux density: Characterization of periphyton photosynthesis–irradiance patterns. *Limnol. Oceanogr.* **37,** 871–878.

Geider, R. J., and Osborne, B. A. (1992). "Algal Photosynthesis; The Measurement of Algal Gas Exchange." Routledge, Chapman & Hall, New York.

Guillard, R. R. L., and Hellebust, J. A. (1971). Growth and the production of extracellular substances by two strains of *Phaeocystis pouchetii. J. Phycol.* **7,** 330–338.

Hellebust, J. A. (1970). The uptake and utilization of organic substances by marine phytoplankters. Occas Publ.—Univ. Alaska, *Inst. Mar. Sci.* **1,** 223–256.

Hellebust, J. A. (1971). Glucose uptake by *Cyclotella cryptica:* Dark induction and light inactivation of transport system. *J. Phycol.* **7,** 345–349.

Hellebust, J. A., and Lewin, J. (1977). Heterotrophic nutrition. *In* "The Biology of Diatoms" (D. Werner, ed.), pp. 169–197. Univ. of California Press, Berkeley.

Hudon, C., Duthie, H. C., and Paul, B. (1987). Physiological modifications related to density increase in periphytic assemblages. *J. Phycol.* **23,** 393–399.

Hussein, Y. A., Shalan, S. N., Abdel-Wahab, S. M., and Hassan, M. E. (1989). Heterotrophic growth from an obligate autotrophic blue-green alga strain. *Phyton (Buenos Aires)* **49,** 9–11.

Johnson, R. E. (1996). The effects of density-dependent resource limitations on the vertical distribution of diatoms in periphyton mats. Master's Thesis, Loyola University of Chicago, Chicago.

Jolley, E. T., and Hellebust, J. A. (1974). Preliminary studies on the nutrition of *Navicula pelliculosa* (Breb.) Hilse., and an associated bacterium, *Flavobacterium* sp. *J. Phycol.* **10,** 7.

Jolley, E. T., and Jones, A. K. (1977). The interaction between *Navicula muralis* Grunow and an associated species of *Flavobacterium. Br. Phycol. J.* **12,** 315–328.

Jørgensen, B. B., Revsbech, N. P., and Cohen, Y. (1983). Photosynthesis and structure of benthic microbial mats: Microelectrode and SEM studies of four cyanobacterial communities. *Limnol. Oceanogr.* **28,** 1075–1093.

Kobayashi, D., Kakizono, M. T., Yamaguchi, K., Nishio, N., and Nagai, S. (1992). Growth and astaxanthin formation of *Haematococcus pluvialis* in heterotrophic and mixotrophic conditions. *J. Ferment. Bioeng.* **1,** 17–20.

Korte, V. L., and Blinn, D. W. (1983). Diatom colonization on artificial substrates in pool and riffle zones studied by light and scanning electron microscopy. *J. Phycol.* **19,** 332–341.

Kuehn, K. A., O'Neil, R. M., and Koehn, R. D. (1992). Viable photosynthetic microalgal isolated from aphotic environments of Edwards Aquifer (central Texas). *Stygologia* **7,** 129–142.

Kühl, M., and Jørgensen, B. B. (1992). Spectral light measurements in microbenthic phototrophic communities with a fiber-optic microprobe coupled to a sensitive diode array detector. *Limnol. Oceanogr.* **37,** 1813–1823.

Lalucat, J., Imperial, J., and Parés, R. (1984). Utilization of light for the assimilation of organic matter in *Chlorella* sp. VJ79. *Biotechnol. Bioeng.* **26,** 677–681.

Lancelot, C., and Mathot, S. (1985). Biochemical fractionation of primary production by phytoplankton in Belgian coastal waters during short and long-term incubations with ^{14}C-bicarbonate. II. *Phaeocystis pouchetii* colonial population. *Mar. Biol. (Berlin)* **86,** 227–232.

Lewin, J. C., and Hellebust, J. A. (1970). Heterotrophic nutrition of the marine pennate diatom, *Cylindrotheca fusiformis. Can. J.Microbiol.* **16,** 1123–1129.

Lewin, J. C., and Hellebust, J. A. (1975). Heterotrophic nutrition of the marine pennate diatom *Navicula pavillardi* Hustedt. *Can. J. Microbiol.* **21,** 1335–1342.

Lewin, J. C., and Hellebust, J. A. (1978). Utilization of glutamate and glucose for heterotrophic growth by the marine pennate diatom *Nitzschia laevis. Mar. Biol. (Berlin)* **47,** 1–7.

Lewin, J. C., and Lewin, R. A. (1960). Auxotrophy and heterotrophy in marine littoral diatoms. *Can. J. Microbiol.* **6,** 127–134.

Lewitus, A. J., and Caron, D. A. (1991). Physiological responses of phytoflagellates to dissolved organic substrate additions. 1. Dominant role of heterotrophic nutrition in *Poterioochromonas malhamensis* (Chrysophyceae). *Plant Cell Physiol.* **32,** 671–680.

Lewitus, A. J., Caron, D. A., and Miller, K. R. (1991). Effects of light and glycerol on the organization of the photosynthetic apparatus in the facultative heterotroph *Pyrenomonas salina* (Cryptophyceae). *J. Phycol.* **27,** 578–587.

Liehr, S. K., Suidan, M. T., and Eheart, J. W. (1990). A modeling study of carbon and light limitation in algal biofilms. *Biotechnol. Bioeng.* **35,** 233–243.

Linkins, A. E. (1973). Uptake and utilization of glucose and acetate by a marine chemoorganotrophic diatom, clone Link 001. Ph.D. Thesis, University of Massachusetts, Amherst.

Liu, M. S., and Hellebust, J. A. (1974a). Uptake of amino acids by the marine centric diatom *Cyclotella cryptica. Can. J. Microbiol.* **20,** 1109–1118.

Liu, M. S., and Hellebust, J. A. (1974b). Utilization of amino acids as nitrogen sources, and their effects on nitrate reductase in the marine diatom *Cyclotella cryptica. Can. J. Microbiol.* **20,** 1119–1125.

Monroy, A. F., and Schwartzbach, S. D. (1984). Catabolite repression of chloroplast development in *Euglena. Proc. Natl. Acad. Sci. U.S.A.* **81,** 2786–2790.

Müller-Haeckel, A. (1985). Shade-adapted algae beneath ice and snow in the Northern Bothnian Sea. *Int. Rev. Gesamten Hydrobiol.* **70,** 325–334.

North, B. B., and Stephens, G. C. (1972). Amino acid transport in *Nitzschia ovalis* Arnott. *J. Phycol.* **8,** 64–68.

Ogawa, T., and Aiba, S. (1981). Bioenergetic analysis of mixotrophic growth in *Chlorella vulgaris* and *Scenedesmus acutus. Biotechnol. Bioeng.* **23,** 1121–1132.

Oliveira, L., and Huynh, H. (1990). Phototrophic growth of microalgae with allantoic acid or hypoxanthine serving as nitrogen source, implications for purine-N utilization. *Can. J. Fish. Aquat. Sci.* **47,** 351–356.

Palmisano, A. C., and Sullivan, C. W. (1983). Physiology of sea ice diatoms. II. Dark survival of three polar diatoms. *Can. J. Microbiol.* **29,** 157–160.

Palmisano, A. C., SooHoo, J. B., White, D. C., Smith, G. A., Stanton, G. R., and Burckle, L. H. (1985). Shade adapted benthic diatoms beneath Antarctica sea ice. *J. Phycol.* **21,** 664–667.

Panella, J. R. (1994). Photoacclimation of *Achnanthidium rostratum* to reduced illumination. Master's Thesis, Loyola University of Chicago, Chicago.

Poulíčková, A. (1987). Algae in ground waters below the active stream of a river (basin of the Morava River, Czechoslovakia). *Arch. Hydrobiol.* **78,** 65–88.

Rippka, R. (1972). Photoheterotrophy and chemoheterotrophy amongunicellular blue-green algae. *Arch. Mikrobiol.* **87,** 93–98.

Rivkin, R. B., and Putt, M. (1987). Heterotrophy and photoheterotrophy by Antarctic microalgae: Light-dependent incorporation of amino acids and glucose. *J. Phycol.* **23,** 442–452.

Saks, N. M. (1983). Primary production and heterotrophy of a pennate and a centric salt marsh diatom. *Mar. Biol. (Berlin)* **76**, 241–246.

Sheen, J. (1990). Metabolic repression of transcription in higher plants. *Plant Cell* **2**, 1027–1038.

Sicko-Goad, L., Stoermer, E. F., and Kociolek, J. P. (1989). Diatom resting cell rejuvenation and formation: Time course, species records and distribution. *J. Plankton Res.* **11**, 375–389.

Stadelmann, E. J. (1962). Permeability. *In* "Physiology and Biochemistry of Algae" (R. A. Lewin, ed.), pp. 493–528. Academic Press, New York.

Stevenson, R. J., and Glover, R. M. (1993). Effects of algal density and current on ion transport through periphyton communities. *Limnol. Oceanogr.* **38**, 1276–1281.

Sundh, I. (1989). Characterization of phytoplankton extracellularproducts (PDOC) and their subsequent uptake by heterotrophic organisms in a mesotrophic forest lake. *J. Plankton Res.* **11**, 463–486.

Tanaka, N., and Ohwada, K. (1988). Decomposition of agar, protein, and organic phosphate by marine epiphytic diatoms. *Nippon Suisan Gakkaishi* **54**, 725–727.

Tuchman, N. C. (1988). Effects of different intensities and frequencies of disturbance by snail herbivory on periphyton succession. Ph.D. Dissertation, University of Louisville, Louisville, KY.

Tuchman, N. C., and Stevenson, R. J. (1991). Effects of selective grazing by snails on benthic algal succession. *J. North Am. Benthol. Soc.* **10**, 430–443.

Veldhuis, M. J. W., and Admiraal, W. (1985). Transfer of photosynthetic products in gelatinous colonies of *Phaeocystis pouchetii* (Haptophyceae) and its effect on the measurement of excretion rate. *Mar. Ecol.: Prog. Ser.* **26**, 301–304.

Wasmund, N. (1989). Live algae in deep sediment layers. *Int. Rev. Gesamten Hydrobiol.* **74**, 589–597.

White, A. W. (1974). Uptake of organic compounds by two facultatively heterotrophic marine centric diatoms. *J. Phycol.* **10**, 433–438.

11 The Stimulation and Drag of Current

R. Jan Stevenson

Department of Biology
University of Louisville
Louisville, Kentucky 40292

I. INTRODUCTION

Currents affect benthic algae in many kinds of aquatic habitats. Currents in streams are obviously significant forces in the ecology of algae in riffles and pools (Ruttner, 1926; Jones, 1951; Zimmerman, 1961). Wave action in lakes can also be important (Stockner and Armstrong, 1971). In addition, close observation in seemingly still water often shows the movement of particles along lake shores or through the plants of wetlands. These currents, whether in streams, lakes, or wetlands, fast or slow, can have important effects on benthic algal ecology.

Currents produce complex spatial and temporal patterns in benthic algae. Currents have stimulatory effects on algal metabolism and make large growths of the filamentous green algae *Cladophora* possible in riffles of streams and along shorelines of lakes (Zimmerman, 1961). However, there is not a simple positive relationship between current velocity and benthic algal metabolism, accumulation processes, and biomass. Algal biomass is greater in fast than in slow currents in some habitats (Ball *et al.*, 1969; Horner *et al.*, 1990), and in other habitats biomass is greater in slow than in fast currents (Antoine and Benson-Evans, 1982; Stevenson, 1984; Lamb and Lowe, 1987; Poff *et al.*, 1990; Uehlinger, 1991). A review of the literature indicates that the highest algal biomasses occur in intermediate current velocities in most habitats (Table I).

TABLE I Reports of Algal Biomass in Different Current Velocities and the Nutrient or Habitat Conditions in Which They Were Found

Rank of algal biomass in different current velocities	*Nutrient/habitat conditions*	*Reference*
78 > 88 > 148		Antoine and Benson-Evans (1982)
30–61 > 61–92 > pool > 92–137	High (urban river)	Ball *et al.* (1969)
10–15 > pool = 18–22 = 37–41	Low P	Ghosh and Gaur (1991)
Fast > slow	Low nutrients	Ghosh and Gaur (1994)
Fast ≥ slow	High nutrients	Ghosh and Gaur (1994)
Fast ≥ slow		Gumtow (1955)
20–50 > 0–20 > 50–80	Moderate to high	Horner and Welch (1981)
0–20 > 20–50 = 50–80	Low	Horner and Welch (1981)
25–70 > 5	Moderate to high	Horner *et al.* (1983)
60 > 20 = 30 > 10 > 80	Moderate to high	Horner *et al.* (1990)
5–7 > 0 = 24–29 > 79–91		von Ivlev (1933)
Pool > riffle	Low (mountain streams)	Keithan and Lowe (1985)
15 > 40	High (Maumee River)	Lamb and Lowe (1987)
38 > 9	High	McIntire (1966a)
35 > 14 > 0	High light	McIntire (1968)
0 > 14 > 35	Low light	McIntire (1968)
Pool > riffle	Heavily grazed	Oemke and Burton (1986)
Slow > fast	Unsheltered from direct current	Peterson (1986)
Fast > slow	Sheltered from direct current	Peterson (1986)
Slow > fast	Sheltered > unsheltered	Peterson (1987)
5–10 = 10–20	Low nitrogen	Peterson and Grimm (1992)
Slow > fast	Low nutrients	Peterson and Stevenson (1992)
1 = 17 > 29 = 41		Poff *et al.* (1990)
24 > 47 > 66		Reisen and Spencer (1970)
~27 > ~33	High	Stevenson (1984)
Fast > slow	*Lemanea* biomass	Thirb and Benson-Evans (1985)
35 – 70 > 70	Moderate to high?	Uehlinger (1991)

The hyperbolic relationship between benthic algal biomass and current implies that two or more forces related to current affect algal biomass: one should be a subsidy at low current speeds and the second should be a stress at high currents (E. P. Odum *et al.,* 1979). Early research in streams indicated the positive effects of currents on algal metabolism (e.g., Ruttner, 1926; H. T. Odum, 1956). Later, McIntire's (1966a) measurements of export increasing with current velocity indicated that the stress was probably drag. Thus, the hyperbolic relationship seems to be due to two direct and opposing forces. First, current stimulates algal metabolism by increasing nutrient transport to cells (e.g., Whitford, 1960; see review by Raven, 1992) and, second, current increases drag on cells, decreases immigration rates, and increases export rates (McIntire, 1966a).

If the effects of current were a relatively simple relationship between one subsidy and one stress, an optimum current should be relatively predictable. However, in some habitats the optimum velocity is low, about 5–7 cm s^{-1} (von Ivlev, 1933), and in other habitats the highest algal biomasses are observed in velocities as high as 60 cm s^{-1} (Horner *et al.,* 1990). The lack of consistency in observed effects of current at specific velocities (Table I) indicates that many factors probably modulate the effects of current on benthic algae.

The objectives of this chapter are to provide a conceptual framework for how and where the diverse effects of current, both direct and indirect, can be important determinants of benthic algal ecology. The direct effects of persistent currents include the positive effect of turbulent mixing on nutrient transport to benthic algal cells and the negative effect of drag on algal attachment to substrata. There are many indirect effects of current, which I group in abiotic and biotic categories. These range from affecting substratum stability and size to affecting herbivore distribution. Finally, a habitat-by-habitat synthesis will be presented.

II. DIRECT EFFECTS

A. Nutrient Transport and Algal Physiology

The positive physiological effects of current were long ago suggested as the reason for great growths of macroscopic benthic algae in streams (Ruttner, 1926). Experiments during the late 1950s and early 1960s showed that current does stimulate photosynthesis and respiration of benthic algae (Odum, 1956; Odum and Hoskin, 1957; Whitford and Schumacher, 1961, 1964; Schumacher and Whitford, 1965). The mechanism for this stimulation was hypothesized to be a decrease in thickness of unmixed layers of water around cells in which nutrients become depleted (Whitford, 1960). Nutrient depletion in these layers was thought to be related to cellular uptake of nutrients exceeding their diffusion into the unmixed layers.

Considerable evidence gathered during the 1970s and 1980s indicates that nutrient depletion does exist in unmixed layers around cells and through the benthic algal matrix. Oxygen gradients in waters overlying periphyton showed that uptake and release of metabolites by cells can exceed diffusion rates of those chemicals through overlying waters (Jørgensen and Revsbach, 1985; Carlton and Wetzel, 1988). Stimulation of carbon and phosphorus uptake by currents (e.g., Whitford and Schumacher, 1961; Sperling and Grunewald, 1969) indicated that similar gradients exist for nutrients. The description of nutrient uptake rates as a power function of current velocity in laminar flows indicated that diffusive boundary layers overlaid periphyton mats and that current reduced the thickness of diffusive boundary layers (Riber and Wetzel, 1987).

Increases in current velocity do stimulate nutrient uptake rates, photosynthesis, respiration, and reproduction rates. A number of studies showed P uptake increasing with current velocity (Whitford and Schumacher, 1961, 1964; Lock and John, 1979; Sperling and Grunewald, 1969; Horner *et al.,* 1990). Many of these same studies and others (McIntire, 1966b; Rodgers and Harvey, 1976; Pfeifer and McDiffett, 1975) showed photosynthesis (^{14}C uptake) and respiration increasing with current velocity. The positive effects of current on metabolism have been shown for many types of algal communities, diatoms (Keithan and Lowe, 1985), filamentous blue-green algae (Sperling and Hale, 1973), *Nostoc* (Dodds, 1989), filamentous green algae (Whitford and Schumacher, 1961; Pfeifer and McDiffett, 1975; Dodds, 1991a), and red algae (Thirb and Benson-Evans, 1985).

No experiments have directly shown that the positive effects of current on algal metabolism result in measured increases in cell division rates. However, considerable evidence indicates that current has a positive effect on cell division, that is, algal reproduction rates. Of course, many studies show that accumulation rates of cells increase with increasing current velocity (e.g., McIntire, 1966a). But others show that accumulation rates decrease with increasing current velocity (e.g., Antoine and Benson-Evans, 1982). Negative effects of current on benthic algal biomass or net accumulation rates do not show that current has a negative effect on reproduction, because export of cells has not been accounted for. Few studies take export into account, thereby rendering accumulation and reproduction rates indistinguishable. McCormick and Stevenson (1991) measured reproduction rates by accounting for immigration and emigration (export) in streamside experimental channels and showed a positive effect of current on benthic algal reproduction. Therefore, three reasons cause me to conclude that current can stimulate benthic algal reproduction rates: (1) modest information that current does stimulate benthic algal reproduction;

(2) compelling evidence that nutrient uptake and metabolism are stimulated by current; and (3) no reasonable alternative hypothesis for greater benthic algal accumulation rates in fast than in slow currents.

There are two important questions for predicting the outcome of currents' opposing effects of enrichment and drag on benthic algae: Do the positive effects of current decrease with increasing current velocity? and Is there a threshold in current velocity above which current does not stimulate algal metabolism? Conceptually the answers to both questions should be "yes." Turbulent mixing of nutrient-rich waters should rinse away the nutrient-depleted water near cells at a specific current velocity. Above that velocity, which will be referred to as the (nutrient) saturating current velocity, further mixing of overlying waters with those near cells should not increase nutrient supply to cells. Whitford (1956) hypothesized that the saturating current velocity was 15 cm s^{-1}. Lock and John (1979) found that it was ≤5.4 cm s^{-1}. Horner *et al.* (1990) showed an increase in areal P uptake in currents ranging from 10 to 50 cm s^{-1}. Dodds (1991a) showed maximum *Cladophora* photosynthetic rates in 8 cm s^{-1} currents. These results show that there is no single saturating current velocity for all algal assemblages in all habitats.

Several factors should, conceptually, affect the saturating current velocity. Algal density negatively affects nutrient transport through the periphyton matrix (Stevenson and Glover, 1993). Many studies show that increases in algal biomass decrease biomass-specific metabolism (e.g., Pfeifer and McDiffett, 1975) and growth rates (Stevenson, 1990) and increase nutrient demand (Bothwell, 1989). Therefore, increasing benthic algal density probably increases the stimulatory effects of current on algal metabolism, because algal metabolism becomes more limited by nutrient transport as density increases. Indeed, measurements of nutrient flux through low- and high-density algal communities show that current can ameliorate the negative effects of periphyton density on nutrient transport through mats (Stevenson and Glover, 1993). Therefore, the saturating current velocity (above which current does not have a more positive effect on algal metabolism) is probably higher in high-density than in low-density algal communities.

The positive effects of current on metabolism increase with nutrient concentrations. Horner and Welch (1981) found that current negatively affected biomass in low-P streams, but enhanced biomass accumulation in high-P stream water. Similarly, Horner *et al.* (1983) and Schulte (1993) found that maximum biomass accrual and growth rates increased with current most in high-nutrient waters. A review of the literature (Table I) indicates that optimum current velocity decreases with nutrient conditions in the habitat. In addition, the positive effects of current may depend on nutrient abundance (see Biggs and Stokseth, 1994).

Positive interactive effects, such as the current–nutrient interaction, are also evident in current–light interactive effects. Steinman and McIntire

(1986) showed that the positive effect of current on benthic algae did not occur in light-limited conditions, but did occur in light-saturated conditions. One of the reasons for the manifestation of the positive current effect in their higher-light treatment was that filamentous green algae were able to grow well only in the high-light treatment.

Some research indicates that current may have negative effects on benthic algal metabolism and growth in high-velocity, low-nutrient habitats. As reported earlier, a 24-h increase in current had a negative effect on subsequent algal growth rates in low-nutrient conditions, when communities recovered in the same current regime (Humphrey and Stevenson, 1992). These negatively affected communities also had lower %N [normalized to ash-free dry mass (AFDM)] than communities that were not exposed to a 24-h increase in current velocity. Current had little effect on the benthic algal communities in high-nutrient conditions. Borchardt *et al.* (Borchardt, 1994; Borchardt *et al.,* 1994; Borchardt, Chapter 7, this volume) reported similar negative effects of high current velocities on P uptake, P storage, and photosynthesis by *Spirogyra.*

Many reasons for negative impacts of current on benthic algal metabolism have been hypothesized but have not been tested. For example, algal cells are leaky (López-Figueroa and Rüdiger, 1991; Marsot *et al.,* 1991), and as nutrients also are actively taken up by cells, nutrients also passively diffuse out of cells. Borchardt *et al.* (1994) suggest that drag on cells may distort cell membrane configuration and increase loss of nutrients from cells. Alternatively, it is hypothesized that current may affect ionic conditions in a way that reduces the abilities of cell membranes to transport nutrients (Borchardt *et al.,* 1994). Another negative effect of current on algal metabolism may be related to exoenzyme activity. Algae, bacteria, and fungi use exoenzymes to digest large organic molecules into nutrients that can be absorbed (see Tuchman, Chapter 10). In nutrient-poor waters, this regeneration and uptake of nutrients within the periphyton may be an important source of nutrition. If regenerated nutrients and exoenzymes are washed away by currents, currents could have negative effects on nutrient availability to benthic algae in low-nutrient conditions.

B. Drag Affects Immigration, Export, and Morphology

Fluids moving past objects attached to a surface produce a shear stress that is a function of current velocity and fluid viscosity. This shear stress can produce form or shear drag on objects, which is a function of shear stress and surface area of an object (Wheeler, 1988). Another force affecting algae on substrata is pressure drag, which is a function of shear stress and cross-sectional area of the alga (Wheeler, 1988). These forces can be predicted when flows are predictable, however, the results of such calculations can be used as guides only when dealing with nonlaminar, turbulent flows (Vogel, 1981), like those in near-bed flows of streams (Young, 1992). Actual flow patterns about objects in streams fluctuate rapidly as eddies swirl past them, even objects as small as algal cells attached to substrata.

Several lines of evidence indicate that drag affects benthic algae close to substrata, even though cells may be considered inside a boundary layer (Silvester and Sleigh, 1985), where frictional forces between water and substratum slow the flow. The growth forms of unicellular algae in fast currents are restricted to small and adnate diatoms attached closely to substrata, whereas larger, apically attached, and stalked cells may be found in slower currents (Stevenson, 1983; Keithan and Lowe, 1985; Lamb and Lowe, 1987). Algal immigration rates are negatively affected and emigration rates were positively affected by current (e.g., McIntire, 1966a). Cells, silt, and fine detritus accumulate in small depressions of substrata and in zones downstream from attached aggregations of cells (Fig. 1).

Some evidence for shear stress affecting benthic algae was obtained by direct observation of cells on substrata in a flume. A number of years ago, I constructed a flume that fit on the stage of my microscope. I saw directly

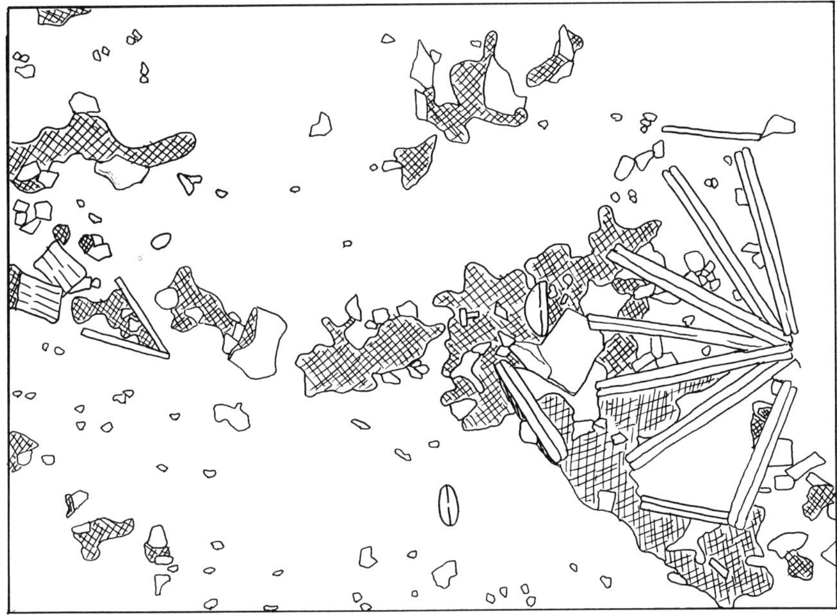

FIGURE 1 Silt and detritus (hatched areas) accumulating downstream from a familial colony of *Synedra* attached to a microscope slide (redrawn from a photomicrograph). Periphyton had developed on the microscope slide for 1 week in Fleming Creek, Michigan, U.S.A. Current flowed from right to left. Periphyton were mounted on microscope slides in HYRAX using vapor substitution. (From Stevenson and Stoermer, 1981.)

the effect of current on cells attached to the surface of the flume. Currents moving past substrata caused stalked organisms to vibrate, thereby sweeping algae from a fan-shaped area downstream from the point of stalk attachment. Effects of current on adnately attached organisms were not evident because they were firmly fixed to the substratum. However, particles suspended in the water showed that currents were significant around these cells that were closely and tightly attached to substrata, as suspended particles (sometimes drifting cells) tumbled around and over attached cells (R. J. Stevenson, unpublished data).

The drag on cells near the substratum surface prevents cells from attaching to substrata and removes cells that have attached. From the perspective of benthic algal accumulation processes, this means that current velocity could decrease benthic algal immigration rates and increase their export rates. Results of recent studies show that current does affect benthic algal immigration and sometimes affects export rates, and that the magnitude of effects is important for understanding patchiness in benthic algal distributions.

Benthic algal immigration rates are negatively and sensitively related to current velocity. Benthic algal immigration rates in a current of 12 cm s^{-1} were 10 times those in 28 cm s^{-1} (Peterson and Stevenson, 1989). In another study comparing community development on different sides of substrata in a current, immigration rates in sheltered conditions (where current was overestimated at 27 cm s^{-1}) were 50 times those in exposed conditions (33 cm s^{-1}) (Stevenson, 1983). The significance of drag as an important force affecting immigration rates was evident in a shift in species composition of immigrating algae from some large and small species in slow currents to only small taxa in fast currents. Negative effects of current on immigration rates are also evident in studies of benthic algal community development. During early stages of development on bare substrata, benthic algal density is lower in fast than in slow current regimes (McIntire, 1966a; Reisen and Spencer, 1970; Stevenson, 1984; Oemke and Burton, 1986; Steinman and McIntire, 1986; Kenney *et al.*, 1991).

In the same way that current velocity and resulting drag decrease immigration rates, they should increase export (emigration) rates. However, experimental evidence provides little evidence of the magnitude of these effects. McIntire measured 4- to 5-fold increases in export between 9 and 38 cm s^{-1} (1966a), little difference between 0 and 14 cm s^{-1} (1968), and 4- to 10-fold increases between 14 and 35 cm s^{-1} (1968). On the other hand, Horner *et al.* (1983, 1990) found little evidence that current stimulated export. Of course, sudden increases in current and sediment load increase export, such as with storm events in streams (Power and Stewart, 1987) and in simulations of such events (Horner *et al.*, 1990; Biggs and Thomsen, 1995).

The relationship between benthic algal export and current is probably not as simple as that between immigration and current. In addition to the effects of drag imposed by current on both immigration and export, export is probably also affected by physiological condition of algae on substrata. The physiological condition of algae is hypothesized to decrease as communities develop and light and nutrient availability decreases within the periphyton matrix (see Borchardt, Chapter 7). Since current maintains nutrient transport to cells, it is reasonable to hypothesize that current maintains the physiological health of cells by transporting nutrients into the periphyton matrix. Therefore, current probably has a negative effect on export rates because it reduces senescence of cells and makes cells more resistant to drag, but it probably positively affects export rate because current increases drag.

Even though the positive effects of current on export have not been measured in some studies, it must be argued that they are probably prevalent. The Horner *et al.* (1983) study, which showed no statistically significant effects of current on export, did show a pattern of increased export with current in low-phosphate conditions. The lack of significance in this study was due to very high variances in measures of export. In addition, there are few alternative hypotheses to explain the common observation of lower algal biomasses in fast-current compared to in slow-current habitats.

Current may also affect the morphology of benthic algae. Increases in current can cause shortening of *Cladophora* cells (Ronnberg and Lax, 1980), the reduction in its branch angle (Whitton, 1975), and an increase in branching (Parodi and Cáceres, 1991). Diatom stalks may also be affected by current. Biggs and Hickey (1994) noted that *Cymbella* and *Gomphoneis* stalks increased in length with increased current velocity. Whether these changes in morphology are functional, caused by nutrient effects, or caused by drag is not known.

III. INDIRECT EFFECTS

Current affects many habitat conditions that themselves directly affect benthic algal metabolism and distribution. These habitat conditions may be abiotic factors like substrate or light. Alternatively, current may affect the distribution of other organisms, such as invertebrate grazers and fish, that eat or disturb benthic algae.

Current does affect substratum size and stability, which greatly affect the kinds, quantity, and thereby the function of benthic algae in habitats. The relation between substratum size, stability, and current velocity was quantitatively established by Nielson (1950), who showed that greater current velocities were necessary to move particles of increasing size. Hynes (1970) describes how the degree to which substrata are embedded among other substrata also affects their stability. In general, larger substrata are found in riffles than in pools.

Large stable substrata, like those occurring in fast-flowing riffles of streams and in wave zones of lakes, seem to be important for establishment and persistence of the benthic macroalgae. Taxa such as *Cladophora, Rhizoclonium, Ulothrix, Draparnaldia,* and *Lemanea* are characteristically found on large stable substrata in fast currents (Power and Stewart, 1987; Dodds, 1991b). Many of these eukaryotic filamentous algae colonize substrata with spores and then grow filaments; so development of filamentous algal assemblages would be greater on stable substrata on which colonists persist compared to on unstable substrata that tumble and have few persisting colonists. The flocculent organic sediments that settle in areas protected from currents can be colonized by motile algae, such as the pennate diatoms *Nitzschia* and *Gyrosigma* and the filamentous blue-green algae *Oscillatoria* and *Microcoleus*. Of course the primary production and food web fates of communities composed of diatoms, filamentous green, or blue-green algae are very different. Thus, substratum size and stability and current affect benthic algal community composition and function in ecosystems.

Indirectly, current also affects light availability to benthic algae. Morphology of stream channels is formed by fast current velocities during floods as substrata are scoured from pools and deposited in riffles. Light varies in riffles and pools because of water depth. In addition, stream width and riparian canopy affect shading and vary with the stream morphology, which is carved by current. Another way that current affects light availability to benthic algae is related to snow and ice cover of streams. Patches of dense benthic diatom communities occurred in open areas of Bear Brook, a small headwater stream in New Hampshire (R. J. Stevenson, personal observation), whereas few algae were observed under the 1-m-thick snow and ice cover in other areas of the stream. Ice and snow cover was low in riffles and therefore seemed to decrease with the increasing frictional force of fast currents in Bear Brook riffles.

Current is known to affect distribution and movement of benthic invertebrates and fish, which may indirectly affect algal abundances and species composition. Many invertebrates and fish eat benthic algae. Many herbivorous caddis flies, mayflies, and beetles live in fast-current areas of streams (Merritt and Cummins, 1984), where these effects seem to be most pronounced (Poff and Ward, 1992). Interaction between herbivores and slow immigration rates in riffles are hypothesized to be the reason for lower algal densities in riffles than in pools of a low-nutrient stream (Harts Run, Kentucky, U.S.A., from R. J. Stevenson, personal observation). Considerable work remains to clarify the interactive effects of current and herbivores on benthic algae and the herbivores themselves.

Benthic algal patterns along depth and current gradients in streams may also be due to disturbance by moving or nesting invertebrates and fish, whose distribution is affected by current. Fish nesting may present relatively small-scale, intensely disturbed patches in the larger stream mosaic. On a larger scale, the greater abundances of algal communities in shallower sections of pools and riffles may be due to algal disturbances by fish movement occurring mostly in the deeper channels of riffles and in the deeper regions of pools (R. J. Stevenson, personal observation).

IV. MANIFESTATIONS OF CURRENT EFFECTS

A. Filamentous Algae and Great Algal Biomasses

The most profound effects of current on benthic algae are probably associated with stimulation of *Cladophora* growth, which is most abundant during warm-water, but not hot-water, seasons (Dodds and Gudder, 1992). The greatest standing crops of benthic algae are usually associated with *Cladophora* growth in moderate to fast currents, either unidirectional currents of streams or multidirectional forces of waves in lakes. *Cladophora* requires currents, as well as sufficient light and nutrients (P), to grow (Zimmerman, 1961; Dodds and Gudder, 1992). Therefore, the greatest standing crops of freshwater benthic algae require current to transport nutrients through thick masses of filaments.

Even though current is required to produce great algal standing crops, benthic algal density is usually higher at intermediate current velocities. Many filamentous green algae, such as *Draparnaldia, Ulothrix,* and *Spirogyra,* are not able to grow long filaments in fast currents, even though their growth is stimulated by current. Only algae like *Cladophora, Lemanea,* and some stalk-forming diatoms (*Gomphoneis,* B. J. F. Biggs, personal communication) are able to withstand the drag of current, modify local current regimes (Reiter and Carlson, 1986; Dodds and Gudder, 1992), and accumulate great biomasses in fast currents. Therefore, the optimum current for manifestation of peak benthic algal biomasses depends on many noncurrent habitat conditions that regulate algal species membership in the habitat.

B. Community Development and Flow Regime

The optimum current velocity for benthic algal accumulation can vary during community development on substrata. During early stages of community development, when algal densities are very low and algal immigration is the dominant mechanism of accumulation, algal biomass decreases with current velocity (Fig. 2) because immigration is negatively related to current velocity. If nutrient concentrations are sufficient to stimulate algal reproduction, accumulation then occurs faster in intermediate-velocity than in low-velocity habitats. Accumulation is not faster in high than in

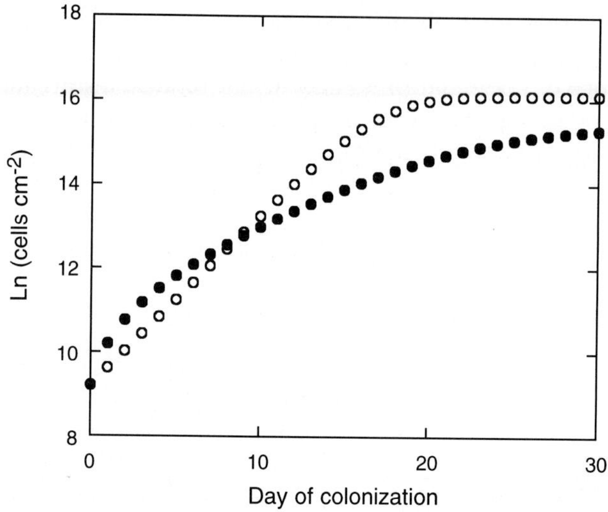

FIGURE 2 Simulated changes in algal density on substrata in slow (solid circles) and fast (open circles) current velocities during community development. Community development was simulated simply with the equation $N_t = N_{t-1}\ (1+d)\ (K - N_{t-1}/K) + I$, where N_{t-1} and N_t are cell densities (cells cm^{-2}) on successive days of community development, D is the reproduction rate as number of divisions per day, K is the maximum areal biomass of the habitat, and I is the immigration rate. The riffle (open circles) had slower immigration rates (I_r=100 cells cm^{-2} d^{-1}) than the pool (I_p=15,000 cells cm^{-2} d^{-1}), but it had faster division rates (d_r=0.3 divisions d^{-1}) than the pool (d_p=0.15 divisions d^{-1}). Maximum areal biomass in the riffle (K_r=10 × 10^6 cells cm^{-2}) was greater than in the pool (K_p=5 × 10^6 cells cm^{-2}). Initial algal densities on substrata were set at 10,000 cells cm^{-2} in both habitats. Simulations were calculated with a Lotus Version 3.4 spreadsheet.

intermediate velocities because cell export increases with current. Therefore, peak biomasses typically occur in intermediate current velocities (Biggs and Gerbeaux, 1993).

Appreciation of the dynamic, disturbance-driven nature of benthic algal ecology is particularly important for understanding why manifested patterns of benthic algal abundance and current velocity are so complex. The effects of current are dependent on the initial density of cells after the disturbance, the time since disturbance, and the nutrient and light supply. Nutrient and light supply affect the resilience of communities to disturbance; but both disturbance intensity and frequency, which affect initial density and time for community development, respectively, may also affect the manifestation of current effects. Therefore, more stable flow regimes favor manifestation of the positive effects of current in intermediate current velocities and development of peak biomasses in communities.

C. Lakes

Benthic algae in lakes are most affected by currents along the shorelines of lakes. Increases in algal density from shallow to deeper littoral waters in lakes are common (Stevenson and Stoermer, 1981) and are probably related to effects of current. In lakes with unstable substrata, benthic algal communities are frequently disturbed by wave action in shallow areas. Frequent disturbance of communities in shallow depths constrains community response to higher light levels, such that highest algal densities are often observed at intermediate depths (Stevenson and Stoermer, 1981).

However, in lakes with stable rock substrata along shorelines, filamentous algae, such as *Cladophora,* may persist through most disturbances and accumulate to great abundances with the stimulatory effects of current. Thus, great biomasses of benthic algae can accumulate in streams and lakes in intense current/wave regimes, however, these great accumulations require stable substrata and other seasonal and nutritionally suitable habitat characteristics for species that can withstand the drag of fast currents.

The more subtle stimulatory effects of nearshore flows and convection currents on benthic algae have not be shown in lakes. However, they have been shown in experimental flumes with lake periphyton and at very low current velocities, like those that would occur in lakes (Riber and Wetzel, 1987). Future research in this area will be valuable.

D. Streams

Peak benthic algal biomasses on natural substrata are usually highest in velocities (10–20 cm s^{-1}) characteristic of runs (Biggs and Gerbeaux, 1993). This is particularly true for benthic algal communities dominated by diatoms, which are relatively susceptible to disturbance by current. Algae in these habitats are less frequently and severely disturbed by storm discharges than are algae in riffles and pools. In addition, algae in runs recolonize rapidly because reproduction is stimulated by modest current velocities, which do not severely limit immigration and excessively promote export. However, if nutrients and light are high enough, filamentous algal growth may be stimulated such that optimum current velocities may be those characteristic of more modest-velocity riffles (30–60 cm s^{-1}).

On a very large spatial scale, differences in algal communities from headwaters to river mouth may be related to current. Benthic algal densities are usually lower in headwaters than in larger streams (Minshall *et al.,* 1983). The river continuum hypothesis predicts that algal densities in headwater streams are low because nutrient concentrations and light availability are low within riparian forest canopies (Vannote *et al.,* 1980).

Although low nutrients and light are surely important reasons for slow accumulation and low algal densities in headwater streams, other factors, such as low immigration rates and perhaps negative effects of current in low-nutrient concentrations, may also be important. Algal immigration rates, which are important for accumulation, are probably lower in headwaters than in larger streams because algae with benthic growth forms have not accumulated in the water column to immigrate. The lack of an immigration pool is probably due to unidirectional flow in streams and to the short upstream distance in which suspended algae can accumulate in the headwaters, and perhaps due to slow reproduction and export from low-nutrient, fast-water habitats. The negative feedback between slow reproduction and export from riffles and lack of immigration to colonize substrata may be ameliorated as nutrient concentrations and light increase downstream and increase algal reproduction and proliferation rates.

Small-scale variation in algal density and species composition, which is caused by current, may be very important for the microdistribution of invertebrates in streams. Current velocity varies on a small spatial scale around substrata in flowing water (Jaag and Ambühl, 1964). Changes in current velocity associated with position around substrata can affect accumulation rates, maximum standing crops, and species composition of most benthic algae (Blum, 1960; Stevenson, 1984). Patchiness in benthic algae may be very important for herbivores and for prey. Many herbivores require or prefer thin algal assemblages (Hart, 1985), perhaps because the food quality of thin assemblages is better than that of thicker, detritus-burdened assemblages or because they can physically handle, move across, or move through thin algal assemblages. Prey may also find refugia from predators in thick algal matrices, as chironomids find refugia from fish (Power, 1990).

Current is probably the most important factor affecting benthic algal community variation among substrata in the same habitat. Precise characterizations of benthic algal communities in flowing waters must consider the effect that current has on algal communities and thereby minimize effects of within-habitat variation. When assessing algal communities with artificial substrata, it is critical that the same current velocities and patterns occur across substrata (see Lowe and Pan, Chapter 22, this volume). When assessing algal communities on natural substrata in flowing water, within-habitat variation is almost always a problem. Composite samples can be collected to improve precision of algal community characterization, and they are made by combining subsamples from many substrata within a habitat and, therefore, including within-habitat variation within the sample. This method of sampling natural waters in streams has been adopted in environmental assessment programs (e.g., USGS, Porter *et al.*, 1993), in which setting and retrieving artificial substrata are impractical.

V. CONCLUSIONS

With the evidence accumulated to date, we can conclude that benthic algal ecology is positively affected by moderate currents, in the 10 to 50 cm s^{-1} range. Slower currents limit algal growth because they do not mix overlying waters through the periphyton matrix and disrupt nutrient-depleted boundary layers caused by the algae themselves (Fig. 3). The drag of faster currents limits immigration rates and enhances export.

The optimum current for a benthic algal assemblage depends on the dominant species, the nutrient concentrations in the habitat, and the density of algae on substrata (Fig. 3). The optimum current velocity is probably higher for filamentous algae that are firmly attached to substrata and that can withstand faster currents than for benthic diatom assemblages. The positive effects of current cannot be manifested unless sufficient light and nutrients are present. The current necessary to thoroughly mix fresh

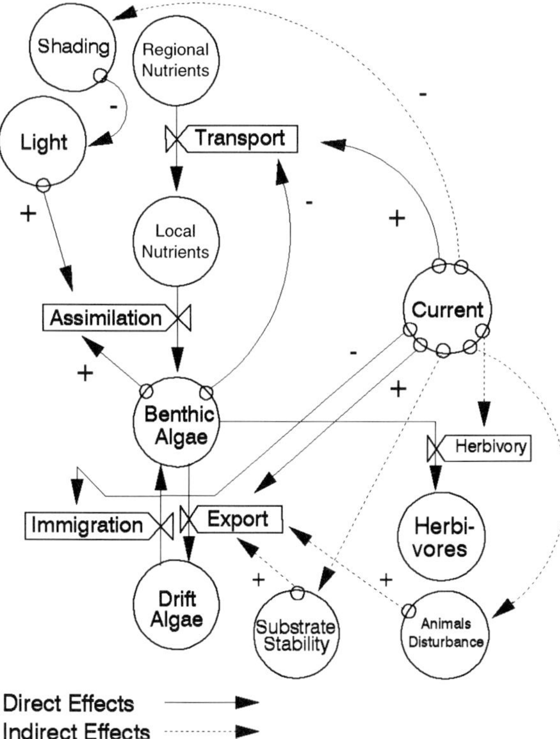

FIGURE 3 Schematic framework for current effects on benthic algae. Direct effects of current on benthic algal processes are shown in solid lines. Indirect effects of current on other ecosystem qualities that affect algal processes are shown in dashed lines.

overlying waters through dense periphyton assemblages is probably higher than that required for thin biofilms, because the positive effects of current on nutrient transport through periphyton are counteracted by negative effects on the same process. Since algal assemblages are probably more dense in high- than in low-nutrient habitats and the stimulatory effect of current probably increases with nutrient concentrations in overlying waters, the optimum current in high-nutrient habitats is probably faster than that in low-nutrient habitats.

The indirect effects of current are also diverse and species specific (Fig. 3). Indirect effects on shading, substrate stability, and herbivore distribution depend on the species of algae and herbivores in a habitat, substrate type, and seasonal occurrence of snow and riparian canopy. However, despite the species specificity of current stimulation or limitation of one species or another, in general, algal biomass and areal-specific production tend to be highest in the moderate current velocities.

The effects of current are evident in many spatial patterns of benthic algae in lakes and streams and the temporal variation of those patterns in aquatic habitats. Patterns of different algae in riffles and pools of streams and along depth gradients in lakes can be attributed to current effects. The relatively large-scale patterns of benthic algae along the river continuum are related to effects of current. Unidirectional flow in streams may limit algal accumulation and production in headwater streams by limiting algal drift. The relatively small-scale patterns of algae in sheltered and unsheltered areas of substrata in streams or along shorelines of lakes are also caused by current effects. Current is an important determinant of benthic algal ecology in freshwater habitats.

REFERENCES

Antoine, S. E., and Benson-Evans, K. (1982). The effect of current velocity on the rate of growth of benthic algal communities. *Int. Rev. Gesamten Hydrobiol.* **67**, 575–583.

Ball, R. C., Kevern, N. R., and Linton, K. J. (1969). The Red Cedar River Report. II. Bioecology. Publications of the Museum, Michigan State University (Lansing, Michigan, USA). *Biol. Ser.* **4**, 106–157.

Biggs, B. J. F., and Gerbeaux, P. (1993). Periphyton development in relation to macro-scale (geology) and micro-scale (velocity) limiters in two gravel-bed rivers, New Zealand. *N. Z. J. Mar. Freshwater Biol.* **27**, 39–53.

Biggs, B. J. F., and Hickey, C. W. (1994). Periphyton response to a hydraulic gradient in a regulated river in New Zealand. *Freshwater Biol.* **32**, 49–49.

Biggs, B. J. F., and Stokseth, S. (1994). Hydraulic habitat of plants in streams. *In:* "Proceedings of the 1st International Symposium on Habitat Hydraulics," pp. 411–429. The Norwegian Institute of Technology, Trondheim, Norway.

Biggs, B. J. F., and Thomsen, H. A. (1995). Disturbance of stream periphyton by perturbations in shear stress: Time to structural failure and differences in community resistance. *J. Phycol.* **31**, 233–241.

Blum, J. L. (1960). The ecology of algae. *In* "The Pymatuning Symposium in Ecology" (C. A. Tryon, Jr. and R. T. Hartman, eds.), Pymatuning Lab. Field Biol., Spec. Publ. No. 2, pp. 11–21. Univ. of Pittsburgh Press, Pittsburgh, PA.

Borchardt, M. A. (1994). Effects of flowing water on nitrogen- and phosphorus-limited photosynthesis and optimum N : P ratios by *Spirogyra fluviatilis* (Charophyceae). *J. Phycol.* **30,** 418–430.

Borchardt, M. A., Hoffman, J. P., and Cook, P. W. (1994). Phosphorus uptake kinetics of *Spirogyra fluviatilis* in flow water. *J. Phycol.* **30,** 403–417.

Bothwell, M. L. (1989). Phosphorus-limited growth dynamics of lotic periphytic diatom communities: Areal biomass and cellular growth rate responses. *Can. J. Fish. Aquat. Sci.* **46,** 1293–1301.

Carlton, R., and Wetzel, R. G. (1988). Distribution and fates of oxygen in periphyton communities. *Can. J. Bot.* **65,** 1031–1037.

Dodds, W. K. (1989). Photosynthesis of different morphologies of *Nostoc parmeloides* (cyanobacteria) as related to current velocities and diffusion patterns. *J. Phycol.* **25,** 258–262.

Dodds, W. K. (1991a). Community interactions between the filamentous *Cladophora glomerata* (L.) Kützing, its epiphytes and epiphyte grazers. *Oecologia* **85,** 572–580.

Dodds, W. K. (1991b). Factors associated with dominance of the filamentous green alga *Cladophora glomerata. Water Res.* **25,** 1325–1332.

Dodds, W. K., and Gudder, D. A. (1992). The ecology of *Cladophora. J. Phycol.* **28,** 415–427.

Ghosh, M., and Gaur, J. P. (1991). Regulatory influence of water current on algal colonization in an unshaded stream at Shillong (Megahalaya, India). *Aquat. Bot.* **40,** 37–46.

Ghosh, M., and Gaur, J. P. (1994). Algal periphyton of an unshaded stream in relation to *in situ* nutrient enrichment and current velocity. *Aquat. Bot.* **47,** 185–189.

Gumtow, R. B. (1955.) An investigation of the periphyton in a riffle of the West Gallatin River, Montana. *Trans. Am. Microsc. Soc.* **74,** 278–292.

Hart, D. D. (1985.) Grazing insects mediate algal interactions in a stream benthic community. *Oikos* **44,** 40–46.

Horner, R. R., and Welch, E. B. (1981.) Stream periphyton development in relation to current velocity and nutrients. *Can. J. Fish. Aquat. Sci.* **38,** 449–457.

Horner, R. R., Welch, E. B., and Veenstra, R. B. (1983). Development of nuisance periphytic algae in laboratory streams in relation to enrichment and velocity. *In* "Periphyton of Freshwater Ecosystems" (R. G. Wetzel, ed.), pp. 121–134. Dr. W. Junk Publishers, The Hague.

Horner, R. R., Welch, E. B., Seeley, M. R., and Jacoby, J. M. (1990). Responses of periphyton to changes in current velocity, suspended sediment and phosphorus concentration. *Freshwater Biol.* **24,** 215–232.

Humphrey, K. P., and Stevenson, R. J. (1992). Responses of benthic algae to pulses of current and nutrients during simulations of subscouring spates. *J. North Am. Benthol. Soc.* **11,** 37–48.

Hynes, H. B. N. (1970). "The Ecology of Running Waters." Liverpool Univ. Press, Liverpool, UK.

Jaag, O., and Ambühl, H. (1964). The effect of the current on the composition of biocoenoses in flowing water streams. *Int. Conf. Water Pollut. Res., London,* pp. 31–49.

Jones, J. R. E. (1951). The ecological study of the River Towy. *J. Anim. Ecol.* **20,** 68–86.

Jørgensen, B. B., and Revsbach, N. P. (1985). Diffusive boundary layers and the oxygen uptake of sediments and detritus. *Limnol. Oceanogr.* **30,** 111–122.

Keithan, E. D., and Lowe, R. L. (1985). Primary productivity and spatial structure of phytolithic growth in streams in the Great Smoky Mountains National Park, Tennessee. *Hydrobiologia* **123,** 59–67.

Kenney, B. C., Kirkland, R., and Mitchell, R. (1991). Effects of supercritical flow and turbulence on the growth of benthic algae in experimental troughs. *Verh.—Int. Ver. Theor. Angew. Limnol.* **24,** 1631–1635.

Lamb, M. A., and Lowe, R. L. (1987). Effects of current velocity on the physical structuring of diatom (Bacillariophyceae) communities. *Ohio J. Sci.* **87,** 72–78.

Lock, M. A., and John, P. H. (1979). The effect of flow patterns on uptake of phosphorus by river periphyton. *Limnol. Oceanogr.* **24,** 376–383.

López-Figueroa, F., and Rüdiger, W. (1991). Stimulation of nitrate uptake and reduction by red and blue light and reversion by far-red light in the green alga *Ulva rigida. J. Phycol.* **27,** 389–394.

Marsot, P., Cembella, A. D., and Colombo, J. C. (1991). Intracellular and extracellular amino acid pools of the marine diatom *Phaeodactylum tricornutum* (Bacillariophyceae) grown on enriched seawater in high-cell density dialysis culture. *J. Phycol.* **27,** 478–491.

McCormick, P. V., and Stevenson, R. J. (1991). Mechanisms of benthic algal succession in different flow environments. *Ecology* **72,** 1835–1848.

McIntire, C. D. (1966a). Some effects of current velocity on periphyton communities in laboratory streams. *Hydrobiologia* **27,** 559–570.

McIntire, C. D. (1966b). Some factors affecting respiration of periphyton communities in lotic environments. *Ecology* **47,** 918–930.

McIntire, C. D. (1968). Physiological-ecological studies of benthic algae in laboratory streams. *J. Water Pollut. Control. Fed.* **40,** 1940–1952.

Merritt, R. W., and Cummins, K. W. (1984). "An Introduction to the Aquatic Insects of North America," 2nd ed. Kendall/Hunt, Dubuque, IA.

Minshall, G. W., Petersen, R. C., Cummins, K. W., Bott, T. L., Sedell, J. R., Cushing, C. E., and Vannote, R. L. (1983). Interbiome comparison of stream ecosystem dynamics. *Ecol. Monogr.* **53,** 1–25.

Nielson, A. (1950). The torrential invertebrate fauna. *Oikos* **2,** 177–196.

Odum, E. P., Finn, J. T., and Franz, E. H. (1979). Perturbation theory and the subsidy–stress gradient. *BioScience* **29,** 349–352.

Odum, H. T. (1956). Primary production in flowing waters. *Limnol. Oceanogr.* **1,** 102–117.

Odum, H. T., and Hoskin, C. M. (1957). Metabolism of a laboratory stream microcosm. *Publ. Inst. Mar. Sci., Univ. Tex.* **4,** 115–133.

Oemke, M. P., and Burton, T. M. (1986). Diatom colonization dynamics in a lotic system. *Hydrobiologia* **139,** 153–166.

Parodi, E. R., and Cáceres, E. J. (1991). Variation in number of apical ramifications and vegetative cell length in freshwater populations of *Cladophora* (Ulvaceae, Chlorophyta). *J. Phycol.* **27,** 628–633.

Peterson, C. G. (1986). Effects of discharge reduction on diatom colonization below a large hydroelectric dam. *J. North Am. Benthol. Soc.* **5,** 278–289.

Peterson, C. G. (1987). Influences of flow regime on development and desiccation response of lotic diatom communities. *Ecology* **68,** 946–954.

Peterson, C. G., and Grimm, N. B. (1992). Temporal variation in enrichment effects during periphyton succession in a nitrogen-limited desert stream ecosystem. *J. North Am. Benthol. Soc.* **11,** 20–36.

Peterson, C. G., and Stevenson, R. J. (1989). Substratum conditioning and diatom colonization in different current regimes. *J. Phycol.* **25,** 790–793.

Peterson, C. G., and Stevenson, R. J. (1992). Resistance and recovery of lotic algal communities: Importance of disturbance timing, disturbance history, and current. *Ecology* **73,** 1445–1461.

Pfeifer, R. F., and McDiffett, W. F. (1975). Some factors affecting primary productivity of stream riffle communities. *Arch. Hydrobiol.* **75,** 306–317.

Poff, N. L., and Ward, J. V. (1992). Heterogeneous currents and algal resources mediate *in situ* foraging activity of a mobile stream grazer. *Oikos* **65,** 465–478.

Poff, N. L., Voelz, N. J., Ward, J. V., and Lee, R. E. (1990). Algal colonization under four experimentally-controlled current regimes in a high mountain stream. *J. North Am. Benthol. Soc.* **9,** 303–318.

Porter, S. G., Cuffney, T. F., Gurtz, M. E., and Meador, M. R. (1993). Methods for collecting algal samples as part of the Water-Quality Assessment Program. *Geol. Surv. Open File Rep. (U.S.)* **93–409,** 1–39.

Power, M. E. (1990). Benthic turfs vs floating mats of algae in river food webs. *Oikos* **58,** 67–79.

Power, M. E., and Stewart, A. J. (1987). Disturbance and recovery of an algal assemblage following flooding in an Oklahoma stream. *Am. Midl. Nat.* **117,** 333–345.

Raven, J. A. (1992). How benthic macroalgae cope with flowing freshwater: Resource acquisition and retention. *J. Phycol.* **28,** 133–146.

Reisen, W. K., and Spencer, D. J. (1970). Succession and current demand relationships of diatoms on artificial substrates in Prater's Creek, South Carolina. *J. Phycol.* **6,** 117–121.

Reiter, M. A., and Carlson, R. E. (1986). Current velocity in streams and the composition of benthic algal mats. *Can. J. Fish. Aquat. Sci.* **43,** 1156–1162.

Riber, H. H., and Wetzel, R. G. (1987). Boundary-layer and internal diffusion effects on phosphorus fluxes in lake periphyton. *Limnol. Oceanogr.* **32,** 1181–1194.

Rodgers, J. H., Jr., and Harvey, R. S. (1976). The effect of current on periphytic productivity as determined using carbon-14. *Water Resour. Bull.* **12,** 1109–1118.

Ronnberg, O., and Lax, P.-E. (1980). Influence of wave action on morphology and epiphytic diatoms on *Cladophora glomerata* (L.) Kütz. *Ophelia Suppl.* **1,** 209–218.

Ruttner, F. (1926). Bemerkungen überden Sauerstoffgehalt der Gewässer und dessen respiratorischen Wert. *Naturwissenschaften* **14,** 1237–1239.

Schulte, R. M. (1993). Differing effect of current on benthic algae in low and high nutrients. M.S. Thesis, University of Louisville, KY.

Schumacher, G. J., and Whitford, L. A. (1965). Respiration and p^{32} uptake in various species of freshwater algae as affected by a current. *J. Phycol.* **1,** 78–80.

Silvester, N. R., and Sleigh, M. A. (1985). The forces on microorganisms at surfaces in flowing water. *Freshwater Biol.* **15,** 433–448.

Sperling, J. A., and Grunewald, R. (1969). Batch culturing of thermophilic benthic algae and phosphorus uptake in a laboratory stream model. *Limnol. Oceanogr.* **14,** 944–949.

Sperling, J. A., and Hale, G. M. (1973). Patterns of radiocarbon uptake by a thermophilic blue-green alga under varying conditions of incubation. *Limnol. Oceanogr.* **18,** 658–662.

Steinman, A. D., and McIntire, C. D. (1986). Effects of current and light energy on the structure of periphyton assemblages in laboratory streams. *J. Phycol.* **22,** 352–361.

Stevenson, R. J. (1983). Effects of current and conditions simulating autogenically changing microhabitats on benthic algal immigration. *Ecology* **64,** 1514–1524.

Stevenson, R. J. (1984). How currents on different sides of substrates in streams affect mechanisms of benthic algal accumulation. *Int. Rev. Gesamten Hydrobiol.* **69,** 241–262.

Stevenson, R. J. (1990). Benthic algal community dynamics in a stream during and after a spate. *J. North Am. Benthol. Soc.* **9,** 277–288.

Stevenson, R. J., and Glover, R. (1993). Effects of algal density and current on ion transport through periphyton communities. *Limnol. Oceanogr.* **38,** 1276–1281.

Stevenson, R. J., and Peterson, C. G. (1989). Variation in benthic diatom (Bacillariophyceae) immigration with habitat characteristics and cell morphology. *J. Phycol.* **25,** 120–129.

Stevenson, R. J., and Stoermer, E. F. (1981). Quantitative differences between benthic algal communities along a depth gradient in Lake Michigan. *J. Phycol.* **17**(1), 29–36.

Stockner, J. G., and Armstrong, F. A. J. (1971). Periphyton of the Experimental Lakes Area, Northwest Ontario. *J. Fish. Res. Board Can.* **28,** 215–229.

Thirb, H. H., and Benson-Evans, K. (1985). The effect of water temperature, current velocity, and suspended solids on the distribution, growth and seasonality of *Lemanea fluviatilis* (C. Ag.), Rhodophyta, in the River Usk and other South Wales rivers. *Hydrobiologia* **127,** 63–78.

Uehlinger, U. (1991). Spatial and temporal variability of the periphyton biomass in a prealpine river (Necker, Switzerland). *Arch. Hydrobiol.* **123**, 219–237.

Vannote, R. L., Minshall, G. W., Cummins, K. W., Sedell, J. R. and Cushing, C. E. (1980). The river continuum concept. *Can. J. Fish. Aquat. Sci.* **37**, 370–377.

Vogel, S. (1981). "Life in Moving Fluids: The Physical Biology of Flow." Princeton Univ. Press, Princeton, NJ.

von Ivlev, V. S. (1933). Ein Versuch zur experimentellen Erforschung der Okologie der Wasserbioconosen. *Arch. Hydrobiol.* **25**, 177–191.

Wheeler, W. H. (1988). Algal productivity and hydrodynamics—A synthesis. *Prog. Phycol. Res.* **6**, 23–59.

Whitford, L. A. (1956). The communities of algae in the springs and spring streams of Florida. *Ecology* **37**, 433–442.

Whitford, L. A. (1960). The current effect and growth of fresh-water algae. *Trans. Am. Microsc. Soc.* **79**, 302–309.

Whitford, L. A., and Schumacher, G. J. (1961). Effect of current on mineral uptake and respiration by a fresh-water alga. *Limnol. Oceanogr.* **6**, 423–425.

Whitford, L. A., and Schumacher, G. J. (1964). Effect of current on respiration and mineral uptake in *Spirogyra* and *Oedogonium*. *Ecology* **45**, 168–170.

Whitton, B. A. (1975). Algae. *In* "River Ecology" (B. A. Whitton, ed.) pp. 81–105. Blackwell, Oxford.

Young, W. J. (1992). Clarification of the criteria used to identify near-bed flow regimes. *Freshwater Biol.* **28**, 383–391.

Zimmerman, P. (1961). Experimentelle Untersuchungen über die ökologische Wirkumg der Strömgeschwindigkeit auf die Lebensgemeinschaften des fliessenden Wassers. *Schweiz. Z. Hydrobiol.* **23**, 1–81.

12

Effects of Grazers on Freshwater Benthic Algae

Alan D. Steinman

Department of Ecosystem Restoration,
South Florida Water Management District,
West Palm Beach, Florida 33416-4680

I. INTRODUCTION

The consumption of living plant material is a critical process in all ecosystems except perhaps for profundal regions of the oceans. Consequently, it is not surprising that a considerable body of literature has been devoted to the process of herbivory. Despite the extensive number of studies conducted, herbivory remains a subject that intrigues researchers and fosters debate. Even fundamental questions, such as whether herbivory results in a positive or negative effect on plants (e.g., McNaughton, 1983; Belsky, 1986), or why does plant matter go unconsumed (Hairston *et al.*, 1960), apparently cannot be answered without controversy. Although less attention has been paid to herbivory in freshwater benthic ecosystems than in terrestrial or marine intertidal systems, there still is a considerable volume of information, including several reviews. Both Gregory (1983) and Lamberti and Moore (1984) have emphasized the reciprocal nature of plant–animal interactions in the freshwater benthos. However, both of these reviews were compiled over a decade ago, and a wealth of new information has since become available. Feminella and Hawkins (1995) in a quantitative review of stream herbivory experiments, concluded that periphyton is strongly regulated by herbivores in streams, and that contrary to popular belief, stream communities apparently are not regulated primarily by abiotic factors. Although this chapter will focus on how grazers influence algae, it should be noted that grazing on freshwater vascular macrophytes also occurs. Reviews by Newman (1991) and Lodge (1991) indicate that the latter process has greater ecological significance than traditionally believed.

The objectives of this chapter are threefold: (1) review the literature covering herbivory on algae in freshwater benthic ecosystems, with special attention paid to studies over the last ten years; (2) provide a synthesis of this information, and search for patterns within a suite of structural and functional algal response variables; and (3) propose future research directions. The focus of this chapter is exclusively on algal responses. For information on benthic grazer responses to herbivory, consult Gregory (1983), Lamberti and Moore (1984), or Lamberti (Chapter 17, this volume).

II. CONCEPTUAL FRAMEWORK

A major goal of this chapter is to quantify the nature and extent of unifying patterns dealing with the way that freshwater benthic algae respond to herbivory. The search for patterns in this field can be an elusive one, as algal responses are influenced by many different factors. Herbivore type (Lamberti *et al.*, 1987a,b; Hill and Knight, 1988; Karouna and Fuller, 1992), density (Colletti *et al.*, 1987; Steinman *et al.*, 1987a,b), and developmental state of the algal community at the time of interaction (DeNicola *et al.*, 1990) can affect the outcome of the algal–herbivore interaction. In addition, abiotic factors such as substratum type (Dudley and D'Antonio, 1991; Karouna and Fuller, 1992; Power, 1992), nutrient concentration (Marks and Lowe, 1989; McCormick and Stevenson, 1991a; Hill *et al.*, 1992), light regime (Steinman *et al.*, 1989; DeNicola and McIntire, 1991), hydraulics (DeNicola and McIntire, 1991), and disturbance history (Steinman *et al.*, 1991; C. G. Peterson *et al.*, 1993) also can influence the interaction. Finally, the response variable examined in the interaction can affect how the interaction is perceived. For example, the removal of a large, slow-growing alga by a herbivore may have a dramatic effect on that species' population dynamics, but if it is replaced by a small, fast-growing alga, not only might community algal biomass not change, but the primary productivity of the system might increase.

Given the number of factors that might influence the outcome of the algal–grazer interaction, the search for patterns quickly becomes a meaningless exercise if its focus is too fine-tuned: separate patterns could be generated from each study. Instead, this chapter focuses on mechanisms that help explain general algal responses observed as a result of herbivory. A mechanism-based analysis provides not only a convenient vehicle for grouping different studies together and assessing patterns, but a foundation for predicting future responses. A literature review was conducted to examine the effects of herbivory on nine different parameters, including both structural and functional components. The journals surveyed included general ecological journals *(Ecology, Ecological Monographs, Oecologia,* and *Oikos),* as well as those specializing in either algae or freshwater ecology *(Journal of the North American Benthological Society, Limnology and Oceanography, Journal of Phycology, Freshwater Biology, Hydrobiologia, Journal of Freshwater Ecology, Canadian Journal of Fisheries and Aquatic Sciences,* and *Archiv für Hydrobiologie).* Algal responses for each parameter were classified as an increase, a decrease, or no change, and were determined for five herbivore types that are commonly used in experimental studies: snails, caddisfly larvae, mayfly larvae, chironomid larvae, and other (usually fish or tadpoles). As will be evident, some parameters have been examined much more frequently (biomass, taxonomic structure) than others (nutrient cycling). The literature on algal–grazer interactions is extensive, so I have focused on those studies that I believe best illustrate the mechanisms in operation.

III. ALGAL RESPONSES TO HERBIVORY

A. Structural Responses

1. Biomass

Algal biomass almost always declines in the presence of herbivores. In 93 studies of algal–grazer interactions, algal biomass was reduced 71 times (Fig. 1). Yet it is useful to recognize that biomass reduction does not always occur (Fig. 1), and that the outcome of the herbivory process varies depending on the alga and herbivore involved. The interaction between algal growth form and grazer morphology helps determine to what degree algal biomass will be removed (Gregory, 1983; Lamberti *et al.,* 1987a; Steinman *et al.,* 1987a). In some systems, mature benthic algal communities have a three-dimensional structure similar to that of terrestrial plant communities (cf. Hoagland *et al.,* 1982), with prostrate and low-profile species at the lowest level, upright or stalked species forming the middle level, and filamentous species emerging from the mat to occupy the upper level. Although these physiognomic levels are rarely that well defined in nature, this simplification is useful in conceptualizing the algal–grazer interaction. The type of mouthpart morphology and

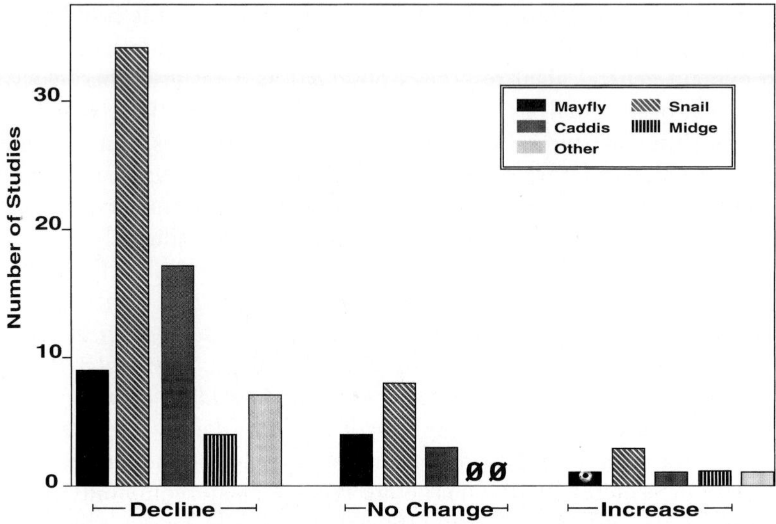

FIGURE 1 Literature survey results dealing with periphyton or benthic algal biomass: the number of studies in which biomass declined, increased, or did not change in value as a result of grazing by different types of herbivores. "Other" category deals mostly with fish or tadpoles.

other structures of the grazer will influence the zone in which it is best adapted to feed (Fig. 2). For example, the larvae of many mayfly species have gathering collector feeding structures (Merritt and Cummins, 1984) and tend to feed at the outer layers, or loosely attached, portions of the periphyton mat (Hill and Knight, 1987, 1988). Caddisfly larvae and snails, with scraping and rasping mouthparts, respectively, are better suited to feed in zones where low-profile, tightly attached algae grow. Lamberti *et al.* (1987a) found that mayfly larvae, caddisfly larvae, and snails all reduced algal biomass compared to ungrazed streams, with caddisflies having the greatest effect despite lower densities per unit area than snails. They attributed the greater impact of caddisflies to their vagility and higher consumption rates.

Reduction of algal biomass due to herbivory has been demonstrated with a variety of grazer types: snails (Hunter, 1980; Mulholland *et al.*, 1983; Jacoby, 1985; Steinman *et al.*, 1987a; Lowe and Hunter, 1988; Osenberg, 1989; Underwood and Thomas, 1990; Bronmark *et al.*, 1991; Tuchman and Stevenson, 1991; Hill *et al.*, 1992; Steinman, 1992; Rosemond *et al.*, 1993), caddisfly larvae (Lamberti and Resh, 1983; McAuliffe, 1984; Hart, 1987; Jacoby, 1987; Lamberti *et al.*, 1987b, 1992; Steinman *et al.*, 1987a; Hill and Knight, 1988), mayfly larvae (Colletti *et al.*, 1987; Hill and Knight, 1987, 1988; Scrimgeour *et al.*, 1991; Karouna and

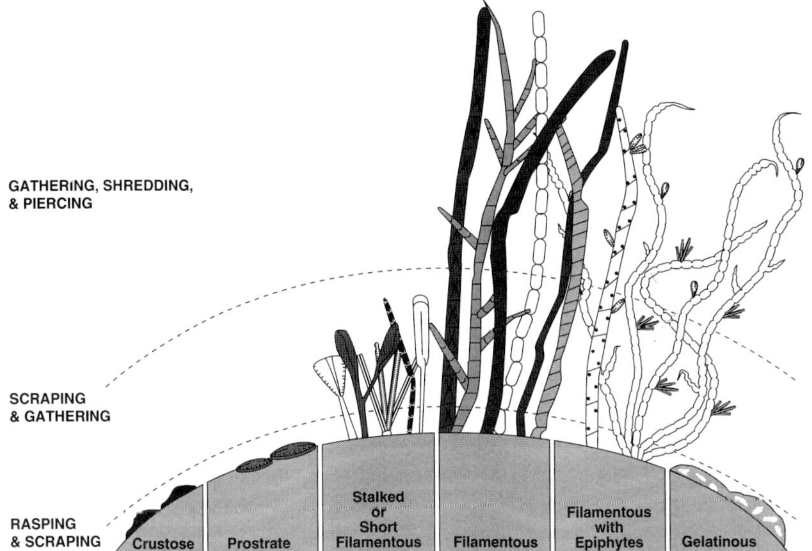

FIGURE 2 Hypothetical schematic representation of key growth forms in benthic algal communities in relation to feeding zones occupied by different types of grazers.

Fuller, 1992), fish (Power and Matthews, 1983; Power *et al.,* 1988; Gelwick and Matthews, 1992; Wootton and Oemke, 1992), shrimp (Pringle *et al.,* 1993), chironomid larvae (Eichenberger and Schlatter, 1978; Power, 1990), and tadpoles (Lamberti *et al.,* 1992). Even in situations where resource limitation has been relieved by additions of nutrients (Elwood *et al.,* 1981; B. J. Peterson *et al.,* 1993) or light (Steinman, 1992), corresponding increases in algal biomass are short-term, as movement of grazers from adjacent areas into the amended reach or increased local recruitment of grazers (i.e., decreased mortality) reduces the increased algal biomass levels.

There are at least three reasons why algal biomass may not decline when herbivores are present: (1) biomass reduction is a density-dependent response, and grazer density and consumption rate are insufficient to result in a measurable decline; (2) the grazer's feeding morphology is not well matched with the dominant algal growth form(s); and (3) resources are so limiting that biomass accrual is constrained irrespective of the presence or absence of grazers. In the first case, Steinman *et al.* (1987a) reported that algal ash-free dry mass (AFDM) declined dramatically once snail densities exceeded 125 m^{-2}. However, algal AFDM was virtually identical in control (i.e., no snail) streams and those with snail densities of 63 m^{-2}. Colletti *et al.* (1987) reported a similar phenomenon with mayfly larvae, with no notable differences in diatom abundance and taxonomic structure between grazed and ungrazed systems when mayfly densities were 800 m^{-2}, although major differences were noted at higher densities. The second case manifests itself when a grazer that feeds optimally in a specific zone (Fig. 2) is situated in an algal assemblage whose physiognomy does not occupy that zone. Karouna and Fuller (1992) demonstrated that mayfly larvae with "brusher" mouthparts had little impact on diatom densities, whereas mayfly larvae with bladelike mandibles were able to reduce diatom densities. Jacoby (1987) reported that although the mayfly *Nixe* was able to ingest small diatoms, herbivory had no detectable impact on algal biomass because small diatoms comprised a small amount of the total standing crop. As an example of the third situation, Feminella *et al.* (1989) attributed the lack of grazer effect on periphyton biomass in Big Canyon Creek, California, to the extremely low inorganic nitrogen levels in the stream; resource limitation presumably slowed algal growth and limited algal biomass to the point that the effect of grazing was not detectable within the time frame of the study.

Algal biomass actually may increase in response to grazing, although these cases are rare (Fig. 1) and restricted primarily to situations with low grazer densities. Several mechanisms may account for an increase: (1) removal of senescent cells results in more external resources reaching viable cells (Lamberti and Resh, 1983; Swamikannu and Hoagland, 1989); (2) herbivory results in nutrient regeneration from within the periphyton matrix, so in cases of nutrient limitation, the positive response of greater nutrient flux may outweigh the negative response of herbivory (McCormick and Stevenson, 1991a); this window where the indirect positive effects of herbivory (i.e., nutrient regeneration) outweigh the direct negative effects (i.e., consumption) appears to be very narrow, however (Stewart, 1987; Mulholland *et al.,* 1991); and (3) analogous to aquatic macrophytes, removal of epiphyton from macroalgae by grazers may benefit the host and result in enhanced biomass (Dudley, 1992).

2. Taxonomic Composition

Generalizations regarding the effect of herbivory on algal taxonomic composition are extremely difficult to make because of the tremendous variety of algal taxa present in freshwater benthic habitats. Although individual studies provide valuable information about how grazing influences taxonomic structure in that system, it is difficult to assess how those changes apply to a system with a totally different algal structure. An alternative approach, that of examining taxonomic changes based on species growth form and community physiognomy (Hoagland *et al.,* 1982; Lowe *et al.,* 1986; Hudon and Legendre, 1987; Steinman *et al.,* 1987a, 1992), appears to have much more heuristic power and will be discussed in more detail later in this chapter. Although studies providing detailed taxonomic information may have limited generality, they still provide valuable insights into the algal–grazer interaction, and the remainder of the taxonomic composition section will highlight these insights.

The topic of selectivity is a popular one in the herbivory literature and, as discussed by Gregory (1983), several stream ecologists have noted that the diets of herbivores differ from the relative availability of food in the stream, suggesting that selectivity may occur in that habitat. However, this index can be a misleading one, as accelerated digestion of certain food items in grazer guts, difficulty in identifying food items in the gut, and misidentification of the true resources of the herbivore all can confound this analysis. Indeed, it is debatable whether selection of periphytic algae physically can take place for many herbivores, as periphytic mats often consist of a large number of taxa in nonordered space. As discussed by Lowe and Hunter (1988), the process of selection as exhibited in terrestrial and marine ecosystems is predicated on the relatively large prey:grazer size ratio in those systems, thereby making prey selectivity more reasonable. But the exceedingly small prey: grazer size ratio for periphyton may render selectivity a moot point except in situations with large prey (macroalgae) or small grazers, such as the apparent rejection of filamentous algae by grazers with mouthparts ill-suited to harvest this growth form (Cummins, 1964; Moore, 1975) and the apparent selectivity of protistan herbivores on diatoms (McCormick, 1991). Thus it is not surprising that Gregory (1983) lists as many studies that found no evidence as those that found evidence for selectivity.

Part of the confusion surrounding selectivity stems from what is meant by the term. Selectivity, in the pure sense, refers to a directed behavior on the part of the herbivore, and there appears to be little evidence that freshwater benthic grazers possess the sensory equipment necessary for discriminating algal taxa (but see Resh and Houp, 1986). Differential efficiency, on the other hand, involves no specialized sensory recognition; grazers exhibit a standard type of behavior that involves harvesting items with different efficiencies. Thus, the removal of overstory growth forms (see the following) most likely reflects differential efficiency and not true selection, as the removal of the algae is related more to morphological constraints of the grazer's feeding apparatus than to sensory recognition per se.

One factor that may influence the degree of "differential efficiency" is herbivore satiation. Calow (1973) noted that differential consumption by herbivores was more likely to be expressed when they were satiated, although even then selectivity was based on gross categories such as detritus, bacteria, and periphyton. Findings by Steinman (1991) were consistent with Calow; Steinman reported that starved snails removed significantly more cells of the diatom *Cocconeis placentula* than did satiated snails. This alga usually is not grazed because of its prostrate growth form, but it was suggested that after more susceptible growth forms were removed, snails had little choice but to graze on these cells.

Several studies have reported how herbivory strongly influences the local taxonomic composition of algal communities in natural streams. Hart (1985) described a system where the caddisfly larva *Leucotrichia pictipes* apparently removed, but did not ingest, filaments of the cyanobacterium *Microcoleus vaginatus* in order to facilitate the growth of an understory assemblage of diatoms and the cyanobacterium *Schizothrix calcicola*. This "weeding" behavior resulted in preferred prey items for the caddisfly and altered algal taxonomic structure in the stream. Kohler and Wiley (1992) reported that populations of the caddisfly larva *Glossosoma nigrior* have collapsed in trout streams in Michigan owing to infection by a microsporidian parasite. In streams where *Glossosoma* abundances declined, abundances of the rhodophyte *Batrachospermum* increased dramatically, despite the fact that this genus had never been observed in these streams before (S. L. Kohler, personal communication). However, *Batrachospermum* abundance quickly declined as other grazers moved into the territory left vacant by *Glossosoma*. Although it is sometimes believed that the mucilaginous growth form of *Batrachospermum* may serve as an effective deterrent to grazers (cf. Steinman *et al.*, 1992), the observations of Wiley and Kohler (1993) and the work of Hambrook and Sheath (1987) show that at least some species of this genus are susceptible to herbivory. Feminella and Resh (1991) demonstrated that two caddisfly larvae coexisted in a northern California stream by grazing on different algal resources; each caddisfly had strong impacts on the algal assemblage it fed on, but the resources were partitioned. *Helicopsyche* grazed on the preferred resource of microalgae, whereas *Gumaga* was restricted to the macroalgal chlorophyte *Cladophora glomerata*.

3. Physiognomy

Physiognomy refers to the study of form and structure in natural communities (Whittaker, 1975). Among benthic algal communities, physiognomy exhibits strikingly consistent patterns in response to herbivory. Results from the literature review indicate that out of 43 studies examined, all but 6 reported a decline in percentage overstory in response to grazing (Fig. 3A). Similarly, 36 out of 41 studies showed an increase in percentage understory in response to grazing (Fig. 3B). The most dramatic effects were observed when the grazers used were either caddisfly larvae (Lamberti and Resh, 1983; Jacoby, 1987; Steinman *et al.*, 1987a; Hill and Knight, 1988; Lamberti *et al.*, 1989; DeNicola *et al.*, 1990; Feminella and Resh, 1991) or snails (Hunter, 1980; Kesler, 1981; Sumner and McIntire, 1982; Cuker, 1983; Lamberti *et al.*, 1987a; Steinman *et al.*, 1987a, 1992; Lowe and Hunter, 1988; Swamikannu and Hoagland, 1989; Tuchman and Stevenson, 1991; Hill *et al.*, 1992; Rosemond *et al.*, 1993). However, this response has also been observed in mayfly larvae (Colletti *et al.*, 1987; Hill and Knight, 1987, 1988), minnows (Power and Matthews, 1983; Power *et al.*, 1985; Gelwick and Matthews, 1992), and crayfish (Vaughn *et al.*, 1993).

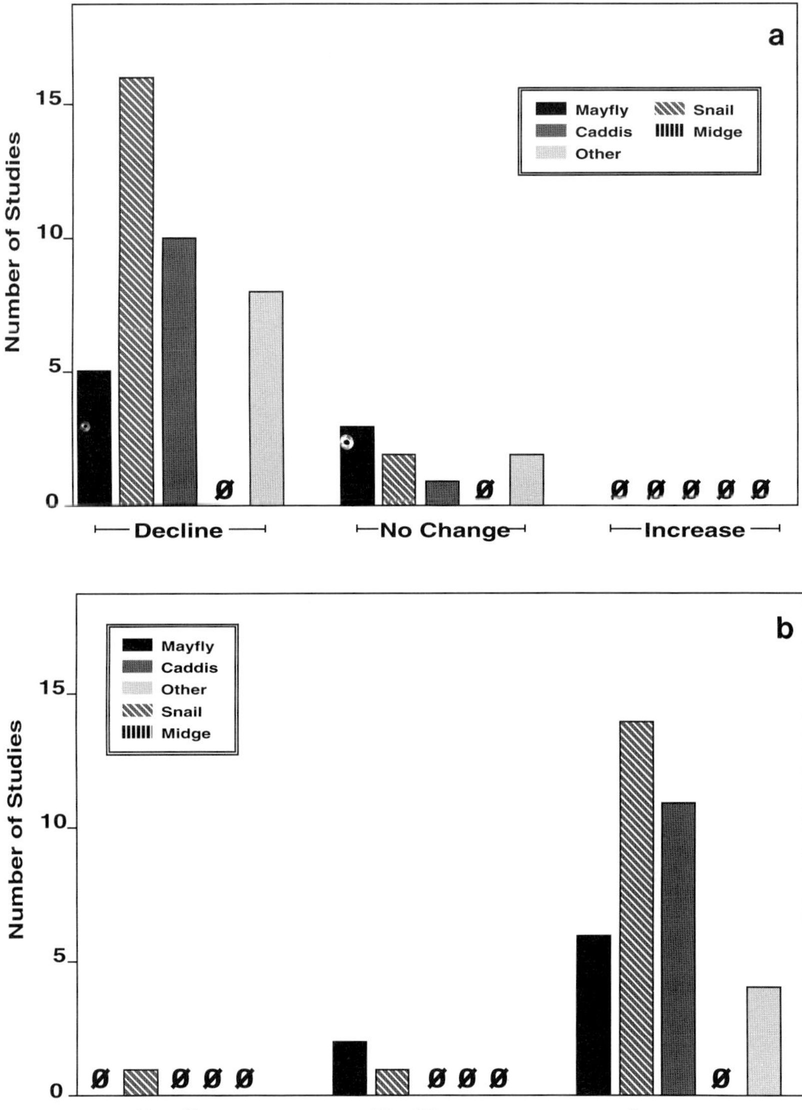

FIGURE 3 Literature survey results dealing with physiognomy. (A) The number of studies in which the percentage of overstory algal growth forms declined, increased, or did not change in value as a result of grazing by different types of herbivores. (B) The number of studies in which the percentage of understory algal growth forms declined, increased, or did not change in value as a result of grazing by different types of herbivores.

The declines in overstory forms are a direct consequence of their vulnerable position in the assemblage. Even if they are not directly grazed, their peripheral location and often loose attachment make them susceptible to dislodgement as grazers maneuver through an assemblage (Hill and Knight, 1987, 1988; Lamberti *et al.*, 1989). Although declines in overstory forms have been quantified in terms of relative abundance, this pattern holds for absolute numbers as well (Lamberti *et al.*, 1987a; Rosemond *et al.*, 1993). Conversely, increases in the percentage of understory forms may result as either an indirect or a direct consequence of grazing activity. The indirect response results from the removal of the more vulnerable overstory form, which may result in either increased percentage abundance of the less vulnerable understory forms or in more resources reaching the understory forms [empirical evidence for this is still equivocal, but see Feminella and Resh (1991) and Mulholland *et al.* (1991)]. Whether this purported increase in resources can stimulate growth rates to the point that they exceed grazer-mediated losses is uncertain, and is in need of rigorous experimentation. The direct response results from nutrient regeneration during the grazing process, which may provide a direct means of stimulating an increase in absolute abundance of understory forms. McCormick and Stevenson (1991a) forwarded this hypothesis to explain increased amounts of *Stigeoclonium* basal cells in response to grazing by the snail *Elimia* [although Hill *et al.* (1992) reported declines in absolute amounts of *Stigeoclonium* basal cells in the presence of snails of the same genus].

Grazers do not always cause a reduction in overstory forms, however. Several researchers working with the filamentous chlorophyte *Cladophora* have reported that grazing may increase overstory biomass (Dudley, 1992; Sarnelle *et al.*, 1993). Apparently, the combination of firm basal attachment, coarse thallus texture, and the grazers' ability to remove epiphytes resulted in a net positive effect of grazing for this alga.

Although grazing may lead to an increased relative abundance of understory forms, the absolute numbers of understory cells usually decline [but because they decline to a smaller degree than overstory cells, their relative abundance increases; Steinman *et al.* (1987a); Mulholland *et al.* (1991); Hill *et al.* (1992)]. Sarnelle *et al.* (1993) reported that mean understory biomass was reduced at moderate snail densities, as expected, but understory biomass was greatest at the highest snail densities. They suggested that the tufted nature of the overstory filaments of *Cladophora* may have physically prevented the snails from reaching the understory.

Mayfly larvae failed to reduce the relative abundance of overstory forms 38% of the time or increase the relative abundance of understory forms 25% of the time (Fig. 3); these percentages are considerably greater than those for either caddisfly larvae or snails. The failure of mayflies to alter physiognomy was due either to a mismatch between mouthpart morphology and community physiognomy or to insufficient densities (Colletti *et al.*, 1987; Karouna and Fuller, 1992).

In one case, grazing reduced the relative abundances of understory forms. Blinn *et al.* (1989) reported that the understory diatom *Epithemia* comprised a much greater proportion of the diet of the limpet *Ferrissia* than what was found in nature. They attributed this apparent "selection" (probably a case of differential efficiency) of understory *Epithemia* to the limpet's shell morphology (ability to push away larger, erect forms) and radular architecture (small, stout teeth of radulae able to scrape adnately attached diatoms).

Our understanding of how grazing influences physiognomy has increased as a result of scanning electron microscopy (SEM). Hudon and Bourget (1981), Hoagland *et al.* (1982), and Korte and Blinn (1983) were among the first to document with SEM the complexity of periphyton assemblages and their three-dimensional nature. Steinman *et al.* (1987a) examined the effects of different grazer densities and types on algal assemblages and, with the aid of SEM, documented the ability of a snail (*Juga*) and a caddisfly larva (*Dicosmoecus*) to change physiognomy. Again, the importance of density was apparent, as *Dicosmoecus* was capable of changing the algal structure from a complex matrix to one of prostrate diatoms even at the caddisfly's lowest density, whereas only at the highest density could *Juga* produce that type of physiognomic structure.

Although physiognomic responses permit researchers to make generalizations about benthic algal responses to herbivory, it is obvious that these generalizations are not without exceptions. Based on the foregoing studies, the ultimate algal response will be dependent on grazer type, algal species and/or growth form, and grazer density.

4. Species Richness and Diversity

Approximately half of the studies examined as part of the literature review indicated that grazing results in a decline in species richness and diversity (Fig. 4). Reductions in diversity usually were achieved primarily through reductions in richness (Lowe and Hunter, 1988; Underwood and Thomas, 1990) as opposed to evenness, but not always (Swamikannu and Hoagland, 1989).

Connell's (1978) intermediate disturbance hypothesis (IDH), as applied to terrestrial and marine ecosystems, predicts that local species diversity is greatest at intermediate levels of disturbance (e.g., herbivory). The reasoning is that grazing of the competitive dominants prevents competitive exclusion, allowing coexistence of competitively dominant and inferior species. As pointed out by Lubchenco and Gaines (1981), however, this scenario should apply only when the competitively dominant alga is preferred by the grazer and a strict competitive hierarchy occurs among algal species.

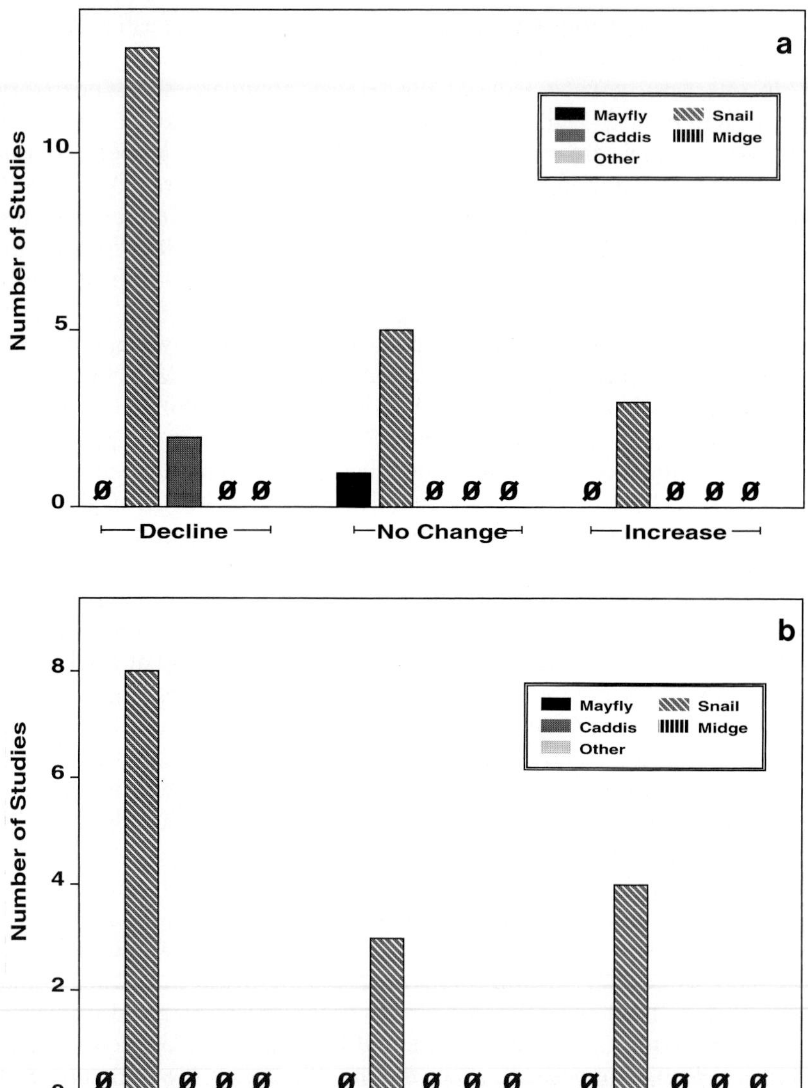

FIGURE 4 Literature survey results dealing with algal species diversity and richness. (A) The number of studies in which diversity declined, increased, or did not change in value as a result of grazing by different types of herbivores. (B) The number of studies in which richness declined, increased, or did not change in value as a result of grazing by different types of herbivores.

Relatively few studies have addressed the applicability of the IDH to freshwater benthic algal communities. In one study, diversity (Shannon–Weiner) declined by 50% in both low and high grazer density treatments relative to ungrazed controls (Lowe and Hunter, 1988). Swamikannu and Hoagland (1989) also failed to observe the largest diversity at intermediate grazing levels, although in their study, diversity was similar at low and intermediate grazer levels, and then dropped at the highest densities. In contrast, McCormick and Stevenson (1989) reported that algal diversity peaked at intermediate grazer densities in unenriched and nitrate-enriched reaches of a stream, but did not change significantly in a phosphate-enriched environment. In their system, the creation of patches by snail grazing apparently increased habitat heterogeneity, thereby increasing diversity; addition of a limiting nutrient (phosphorus), however, increased the population growth rates of some species (*Cocconeis* and *Stigeoclonium*) to the point that the effect of habitat heterogeneity on diversity was obscured.

Clearly more studies are needed to determine the generality of the IDH in freshwater periphyton communities. However, in those cases where diversity patterns fail to follow the patterns outlined in the IDH, several characteristics unique to freshwater periphyton may explain their failure to conform. First, the hypothesis is valid only when the dominant alga is preferred by the key grazer (Lubchenco, 1978). As the dominant species is removed, additional resources become available for other species, thereby resulting in greater richness and diversity. As pointed out earlier, however, it is debatable whether selectivity and preference are useful terms when discussing herbivory in periphyton assemblages because of their small prey: grazer size ratio and the putative absence of the required sensory equipment in grazers. Hence, without preferential consumption of the dominant alga, there is no reason to expect increased diversity. Second, the ability of periphyton assemblages to pack many species into a small space, thereby creating a potentially more diverse community per unit area than macroalgal-dominated marine intertidal systems, allows for grazing on periphyton to effect dramatic declines in diversity (Swamikannu and Hoagland, 1989). Finally, one of the key resources that marine intertidal and terrestrial species compete for is space. In freshwater periphyton communities, competition for primary space does not appear to be as critical a factor as in the other ecosystems because species often are capable of attaching to one another by the use of polysaccharide excretions (Hoagland *et al.*, 1993) or exist as tychoplankton (Steinman and McIntire, 1987). Consequently, the creation of more open space by removal of dominant species, whether by preference or by serendipity, may not have the same ecological implications in periphyton assemblages as in other plant communities.

Huston (1979) suggested that species diversity could be explained mechanistically according to a dynamic equilibrium model, whereby population growth rates (and competitive displacement) are a major determinant of system diversity. Thus, environments that promote high population growth rates (e.g., high resource availability) should have lower diversity than situations with low population growth rates, unless the populations of some species are being eliminated (e.g., through herbivory). Steinman *et al.* (1989) used their data to test Huston's hypothesis and found no correspondence between the two. As light level increased (from severely limiting to photosynthetically saturating), diversity was consistently greater in streams with high light compared to those with low light. In addition, regardless of light level, diversity was consistently greater in streams without grazers than those with grazers. This experiment was conducted in artificial streams with a finite species pool, which prevented immigration by new species and may have confounded the test of Huston's hypotheses.

Although low to moderate grazing pressure may not result in changes in species diversity (e.g., Kehde and Wilhm, 1972; Swamikannu and Hoagland, 1989), intense grazing pressure does reduce diversity (Lowe and Hunter, 1988; Mulholland *et al.,* 1991). However, the degree to which intense grazing pressure reduces diversity may be less severe in lotic ecosystems than in other ecosystems for several reasons. First, the unidirectional flow of water can act as a constant vector for transporting new colonists, which can help replenish species pools at denuded sites. In addition, many algal species reproduce very quickly and can reestablish themselves in relatively short time periods. Finally, several algal species common to lotic ecosystems can exist as small, prostrate filaments or basal cells through much of the year, from which new growth arises when environmental conditions are favorable (Wehr, 1981; Sheath *et al.,* 1986). This growth form has very low susceptibility to grazing (Hill *et al.,* 1992; Steinman, 1992) and persists throughout the year, thereby potentially increasing species diversity. Consequently, the degree to which intense grazing pressure reduces diversity may be less severe in lotic ecosystems compared to other systems.

B. Functional Responses

1. Primary Production

The majority of studies examined in the literature review found that area-specific primary production (ASPP) declined in response to herbivory, whereas biomass-specific primary production (BSPP) exhibited no obvious overall response to grazing (Fig. 5). Because responses were different for each of these parameters, they will be discussed separately.

The process of herbivory, by definition, involves the removal of plant biomass. As a consequence, there is less photosynthetic tissue present to fix carbon over a given area, and under most circumstances there is a con-

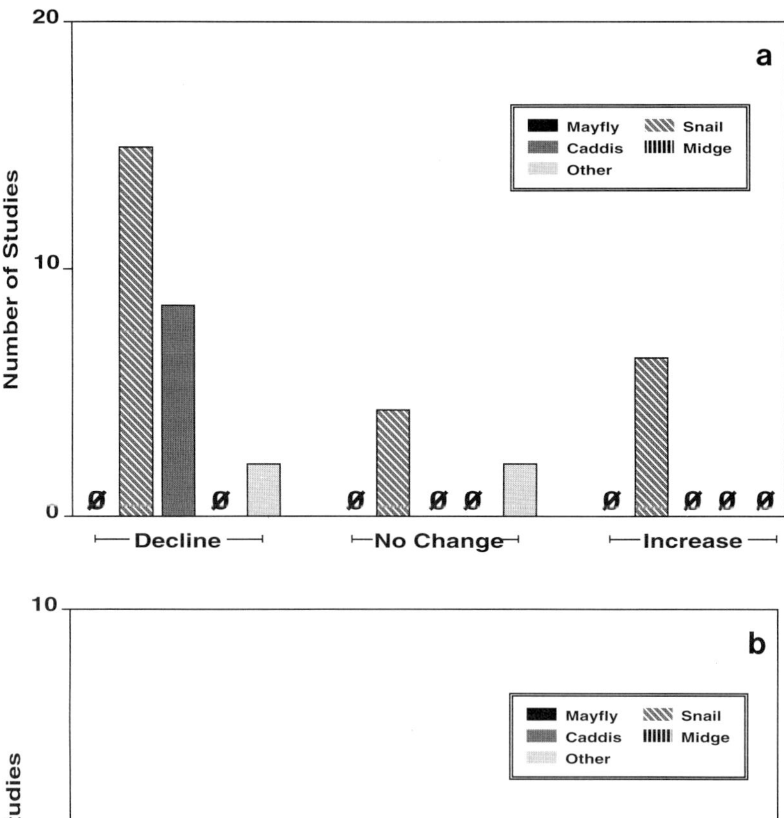

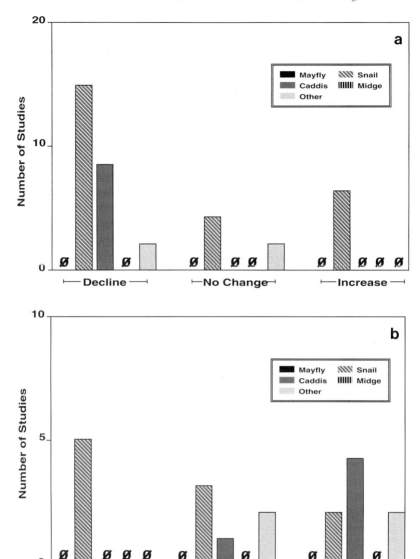

FIGURE 5 Literature survey results dealing with primary production. (A) The number of studies in which area-specific primary production declined, increased, or did not change in value as a result of grazing by different types of herbivores. (B) The number of studies in which biomass-specific primary production declined, increased, or did not change in value as a result of grazing by different types of herbivores.

comitant decline in ASPP (Dickman, 1968; Hunter, 1980; Cuker, 1983; Lamberti and Resh, 1983; Mulholland *et al.*, 1983, 1991; Jacoby, 1987; Stewart, 1987; Hart *et al.*, 1991; Gelwick and Matthews, 1992; Hill *et al.*, 1992; Steinman, 1992; Rosemond *et al.*, 1993). However, some studies have demonstrated either no change (Sumner and McIntire, 1982; Hill and Harvey, 1990) or an increase in ASPP under grazed conditions (Hargrave, 1970; Flint and Goldman, 1975; Cuker, 1983; Gregory, 1983; Lamberti *et al.*, 1987a, 1989). Hill and Harvey (1990) attributed the lack of change in their system to differential effects of grazing within the periphyton matrix. Snails (*Elimia*) reduced productivity in the loosely attached layer but increased productivity in the understory, tightly attached layer, resulting in no net change in ASPP. In studies where ASPP was reported to increase, the stimulation usually was of a very small magnitude and occurred only under conditions of low grazing pressure. Indeed, in the widely cited study of Cuker (1983), the reported stimulation of ASPP by grazing (1) occurred only under low grazing pressure, (2) was unreplicated, and (3) quickly became a *reduction* in ASPP as grazing pressure increased even slightly. Cuker (1983) attributed the lack of consensus in the literature regarding stimulation of ASPP by grazing to the apparent narrowness of conditions under which stimulation can occur.

Lamberti *et al.* (1989) observed that stimulation of ASPP depended not only on grazing pressure, but also on the productive capacity and age of the periphyton assemblage. They reported that stimulation of net primary production (NPP) by grazing occurred at low and intermediate irradiances but not at high irradiances, and after 43 and 75 d of algal development but not after 19 or 27 d. The authors proposed several reasons why grazing may have stimulated NPP in this system: (1) shifts in taxonomic composition to more productive species; (2) increased availability of resources (light, nutrients) to cells in the lower strata of the periphyton matrix; and (3) nutrient inputs via grazer excretion. Quantifying the relative importance of these mechanisms should be a priority of future investigators examining this topic (e.g., see McCormick, 1994).

Results from the literature survey indicated that whenever ASPP either increased or did not change, snails were the herbivore involved; experiments using caddisfly larvae always resulted in declines in ASPP (Fig. 12.5A). Although these results are based on small sample sizes, they are provocative. The lower vagility and consumption rates of snails compared to those of some caddisfly larvae may account for this disparity. As explained earlier, one condition for observing herbivore-mediated increases in ASPP is low grazing pressure; the slower movement rates and lower or less efficient consumption rates of snails result in lower grazing pressure than that of some caddisflies at equivalent grazer densities or biomasses (Steinman *et al.*, 1987a).

Unlike ASPP, biomass-specific primary production (BSPP), or turnover rate, is not positively related to biomass level present in the assemblage. In fact, an inverse relationship between BSPP and biomass level is expected (Gregory, 1983), as increased biomass results in greater barriers to light penetration and nutrient diffusion, leading to a greater proportion of senescent cells. Thus, one mechanism by which herbivory might increase BSPP (in addition to the ones listed for ASPP) is by removing dead or senescent cells, leaving behind a greater percentage of vigorous cells. When productivity is normalized per unit biomass, the elimination of cells that contribute to biomass, but not productivity, results in greater BSPP. Several studies have reported increases in BSPP in response to grazing (Gregory, 1983; Lamberti and Resh, 1983; Lamberti *et al.*, 1987a; Stewart, 1987; Hill and Harvey, 1990; Gelwick and Matthews, 1992), but others have reported either no change (Jacoby, 1987; Mulholland *et al.*, 1991) or a decline (Mulholland *et al.*, 1983; Hill *et al.*, 1992; Rosemond *et al.*, 1993) (Fig. 5B).

All studies that reported a herbivory-induced decline in BSPP involved *Elimia clavaeformis*, a prosobranch snail found at extremely high densities in streams of eastern Tennessee. Hill *et al.* (1992) suggested three possibilities for why grazing by *Elimia* may decrease BSPP: (1) a shift in taxonomic composition to less productive species. This is in direct contrast to the mechanism noted earlier, whereby the removal of senescent cells might result in a more productive assemblage and increased ASPP; studies in Tennessee and Kentucky have revealed that intense grazing by *Elimia* results in dominance by basal cells of *Stigeoclonium* (McCormick and Stevenson, 1991a; Hill *et al.*, 1992; Steinman, 1992; Rosemond *et al.*, 1993), which may have inherently low photosynthetic rates. Hill et al. (1992) speculated that there might be a trade-off between resistance to grazing and growth rate, as has been reported in marine algae (Lubchenco and Cubit, 1980), and was corroborated for *Stigeoclonium* in a later study by Rosemond *et al.* (1993). (2) Mucus deposited by the snail may contain an inhibitor of algal metabolism, although previous studies have reported just the opposite (Calow, 1979; Connor and Quinn, 1984). And (3) the only cells that can avoid predation are those located in microscopic recesses in the substrata, thereby reducing their exposure to light and nutrients and resulting in reduced productivity.

It is important to recognize the potential limitations of normalizing photosynthetic data per unit biomass (i.e., as BSPP); the biomass parameter used to normalize production can influence the data in potentially unexpected and undesired ways (Steinman and Lamberti, 1996). For example, using AFDM as a biomass parameter can introduce bias because not all the organic matter present in the system is photosynthetic (e.g., heterotrophic bacteria, fungi, detritus, meiofauna). Consequently, BSPP of periphyton may increase after the introduction of grazers simply because much of the flocculent nonphotosynthetic material associated with the periphyton mat is ingested or sloughed in the presence of grazers, not because of any change in the absolute photosynthetic rate of the periphyton. In addition, inefficient removal of algae because of their firm attachment to substrata or endolithic nature may bias BSPP if a significant proportion remains unmeasured. This could be especially problematic at low algal densities, such as in the presence of high grazer densities. Alternatively, the use of chlorophyll *a* may introduce problems because chlorophyll levels decline as ambient irradiance levels increase, largely because carbon is allocated away from proteins and toward carbohydrates and lipids. This results in a reduction in the cell quota of pigments and light-harvesting chlorophyll protein complexes (Sukenik *et al.*, 1990; Falkowski and LaRoche, 1991). Consequently, as removal of overstory cells by grazing allows more light to reach the understory layers of the periphyton mat (Jorgensen *et al.*, 1987), there might be a reduction in cellular chlorophyll levels. Although increased light should result in greater primary production (Lamberti *et al.*, 1989), increases in primary production normalized to chlorophyll *a* will be influenced by how the chlorophyll levels respond to the changing environmental conditions.

2. Nutrient Content

The nutrient content in algal tissue can both influence and be influenced by herbivory. Plant nutrient content is often viewed as a key factor influencing herbivory in terrestrial ecosystems (Crawley, 1983; Coley *et al.*, 1985), but virtually no work addressing this question has been conducted at the species level in freshwater periphyton. Although several studies have examined the influence of grazing on enriched versus nonenriched periphyton communities (Cuker, 1983; Stewart, 1987; Marks and Lowe, 1989; Mazumder *et al.*, 1989; Winterbourn, 1990; Hill *et al.*, 1992; Rosemond *et al.*, 1993), the nutrient cell quotas of different algal species were not analyzed.

Even though the influence that algal nutrient content has on herbivory remains largely undefined in freshwater benthos, the role of herbivory on nutrient content has been addressed in select studies. Results from these studies can be summarized into two general findings: (1) absolute nutrient amount per unit area declines in the presence of grazers (Jacoby, 1985; Mazumder *et al.*, 1989). This finding is consistent with the effect of grazing on algal biomass: as biomass declines, it is expected that nutrient content also will decline. (2) Percentage nutrient content increases in response to grazing (Hunter, 1980; Mulholland *et al.*, 1991; Rosemond, 1993; Rosemond *et al.*, 1993); this response has been observed for both nitrogen and phosphorus. The mechanisms behind the latter result may be twofold. First, if grazing leaves behind a layer of metabolically active cells, it is likely that in addition to having greater biomass-specific primary production rates, they also will have greater biomass-specific nutrient contents. Second, as grazers move through periphyton mats, they not only ingest algal cells, but also physically dislodge material, especially loosely attached matter, from the matrix. If most of this material is senescent with low nutrient content, its removal will result in the remaining material having a greater nutrient percentage (cf. Mulholland *et al.*, 1991). In contrast to the foregoing generalizations, which were generated from studies using snails, Gelwick and Matthews (1992) observed that percentage nitrogen of algal mats grazed by the minnow *Campostoma* was significantly lower than that of ungrazed mats. The authors do not discuss this finding, but it is possible that *Campostoma* removed algal species with high nitrogen cell quotas, which were replaced by species with low nitrogen cell quotas.

Carbon:nitrogen ratios have been used as an index of food quality in algae (McMahon *et al.*, 1974), as low ratios (high N) suggest relatively high levels of protein. Most studies that reported an increase in percentage nitrogen with grazing observed, not surprisingly, a corresponding decline in C:N ratio (Hunter, 1980; Gelwick and Matthews, 1992; Rosemond, 1993; Rosemond *et al.*, 1993). However, several studies have questioned whether C:N ratios are reliable indicators of food quality (Harrison and Mann, 1975; Cummins and Klug, 1979; Horn and Neighbors, 1984; Steinman *et al.*, 1987b), as total nitrogen and protein may not be directly related if the nitrogen is stored in refractory or nonproteinaceous material.

Steinman *et al.* (1987b) examined the effect of grazing by snails and caddisfly larvae on algal chemical composition. They found that both grazer types changed the fatty acid and amino acid contents of periphyton assemblages compared to ungrazed assemblages. However, their research analyzed the chemistry of whole communities, not individual species. Steinman *et al.* suggested that more research on this topic may elucidate further relationships between food quality and grazer growth. However, there has been little new work in this area.

3. Nutrient Cycling

Research in terrestrial ecosystems has revealed that herbivory by large grazers substantially increases the rate of nutrient cycling (McNaughton *et al.*, 1988; Pastor *et al.*, 1988). The mechanisms behind this phenomenon include nutrient regeneration by herbivore excretion and physical movement of herbivores preventing the accumulation of litter. Mathematical modeling studies of freshwater benthic systems also have indicated that grazing can result in increased cycling rates (DeAngelis *et al.*, 1989, 1990). However, direct comparisons of grazer-mediated effects on nutrient cycling between terrestrial and lotic ecosystems are confounded; only in stream ecosystems must cycling studies account for downstream transport of elements. Thus, discussions of nutrient cycling in flowing waters, as opposed to most other ecosystems, must discriminate between system-level (i.e., spiralling) cycling and in-place cycling. Mulholland *et al.* (1983) reported that grazing on periphyton increased nutrient regeneration rates, consistent with studies from terrestrial systems. However, grazing also resulted in an overwhelming reduction in algal biomass in the streams, thereby decreasing the total amount of phosphorus taken up in the system. In fact, when measured over the entire experimental length of stream, a net decline in total utilization of phosphorus (i.e., an increase in spiralling length) was measured in the system because reduced demand (due to less biomass) had a greater effect on P dynamics than the increased regeneration rates. Stewart *et al.* (1993) speculated that because grazing on periphyton removes a large surface area for uptake and adsorption in streams, one result of herbivory could be increased downstream movement of dissolved or colloidal toxicants.

Studies that address the effect of grazing on in-place nutrient cycling (i.e., at the periphyton matrix scale) have indicated that grazing reduces nutrient cycling rates, in contrast to terrestrial systems. Alkaline phosphatase enzyme activity has been used as an indicator of phosphorus limitation in aquatic systems (Fitzgerald and Nelson, 1966; Perry, 1972; but see Pick, 1987). The phosphatase enzyme cleaves the inorganic phosphorus moiety from organic phosphorus compounds; elevated concentrations are generally considered indicative of phosphorus limitation. Both Mulholland *et al.* (1991) and Rosemond *et al.* (1993) found that activities were significantly lower in grazed than in ungrazed periphyton assemblages. The authors speculate that heavy grazing pressure reduced algal biomass to the point where nutrient limitation was relieved, so that algae had less need to induce the enzyme. Grazers also can influence nutrient cycling in periphyton matrices simply by altering mat hydraulic characteristics. High-biomass periphyton mats (i.e., ungrazed) have different dispersion coefficients and volumes of transient water storage than low-biomass communities (i.e., ungrazed). Mulholland *et al.* (1994) reported that high biomass resulted in increased size of transient storage zones, which in turn allowed for increased in-place nutrient cycling. Consequently, recycling met a greater fraction of phosphorus demand in high-biomass than in low-biomass communities. Mazumder *et al.* (1989) also reported that phosphorus was retained more tightly in high-biomass than in low-biomass communities. Steinman *et al.* (1995) looked at the interactive effects of biomass, light, and grazing on phosphorus cycling in stream periphyton communities and concluded that: (1) cycling was greater in high biomass than low biomass communities, consistent with the findings of Mulholland *et al.* (1994); (2) cycling was greater in ungrazed than grazed communities, as grazing increased phosphorus turnover because of physical and consumptive losses; and (3) light had no significant effect on cycling.

The physical disruption of plant tissue and excretion of nutrients by herbivores have the potential to increase cycling rates in freshwater benthos, just as in terrestrial ecosystems. However, downstream transport of regenerated nutrients and the importance of periphyton mat integrity for in-place cycling combine to make benthic algal communities unique. Evaluation of nutrient cycling in lotic ecosystems is scale-dependent, and conclusions will differ depending on whether the observations are made at the community or ecosystem level.

4. Export

Studies addressing the effect of grazing on organic matter transport are limited, with most occurring in artificial systems. In the absence of grazers, there is a positive relationship between export and algal biomass (Lamberti *et al.*, 1989; Scrimgeour *et al.*, 1991). In the presence of grazers, however, export can exhibit several different responses. Immediately upon introduction of grazers, there is a pulse of increased export, which quickly tapers off (Lamberti *et al.*, 1987a). Eichenberger and Schlatter (1978) and Lamberti *et al.* (1987a) measured greater export in grazed than in ungrazed artificial streams, even several weeks after grazers had been introduced. However, both Sumner and McIntire (1982) and Mulholland *et al.* (1991) reported that grazer presence did not always elevate export; export was actually reduced in the presence of snails when light levels were elevated, although under both shaded and high-light/high-nitrate conditions snails again enhanced export (Sumner and McIntire, 1982). Mulholland *et al.* (1991) reported lower export in grazed than in ungrazed laboratory streams. However, their streams were exposed to extremely intense grazing pressure, resulting in very low biomass levels to be exported. When export was normalized to the amount of biomass present in the stream (instead of being measured per unit time), export was greater in grazed than in ungrazed streams (A. D. Steinman and P. J. Mulholland, unpublished data). Similar results were reported by Mulholland *et al.* (1983) in a separate study.

Other factors that influence grazer-mediated export include algal taxonomic structure, algal seral stage, and grazer type. Eichenberger and Schlatter (1978) noted that filamentous green algae were more susceptible to export because of their "frail" nature than were cyanobacteria. In addition, algal assemblages composed of firmly attached species, such as some diatoms, will be less susceptible to export because they are able to adhere sessilely to substrata with the aid of extracellular polymeric substances (see Hoagland *et al.*, 1993, for examples). The older the assemblage, the greater the opportunity for biomass to accrue and for cells to senesce. Both of these factors facilitate export; Lamberti *et al.* (1989) reported a positive relationship between export and periphyton age for both grazed and ungrazed assemblages at intermediate and high light levels, but not at low light levels. Finally, both Lamberti *et al.* (1987a) and Dudley (1992) reported that export was greater in the presence of limnephilid and glossosomatid caddisfly larvae, respectively, than with baetid mayfly larvae. It seems likely that differences in both mouthpart morphology and grazer mass account for this difference. The bladelike mandibles of some caddisfly larvae may cover more area per unit time and be less efficient at harvesting resources than the brushlike hairs associated with these mayfly larvae mouthparts; also, large-bodied snails and caddisflies may bulldoze through an algal mat, thereby dislodging more material than a small mayfly.

Although their findings were not directly related to export, Gelwick and Matthews (1992) observed that the grazing minnow *Campostoma* altered the size fractions of particulate organic matter from larger to smaller fractions. This change, owing to either feces production or mechanical fragmentation, resulted in size fractions of organic matter that were more appropriate for gathering collectors than for grazers (e.g., scrapers). Thus grazing can influence not only the flux of organic matter transported from a system, but also how it can be processed.

5. Succession

Given that freshwater benthic algal assemblages follow different successional trajectories depending on local environmental conditions (Hoagland *et al.,* 1982; Roemer *et al.,* 1984; Hamilton and Duthie, 1984; Steinman and McIntire, 1986, 1987), it is not surprising that some studies have suggested that succession in these assemblages cannot be explained by any one mechanistic model (Steinman *et al.,* 1987a; DeNicola *et al.,* 1990; McCormick and Stevenson, 1991b; Tuchman and Stevenson, 1991). Several studies have shown that initial algal colonists are large, elongate diatoms or colonial growth forms, which are followed by smaller, low-lying diatoms, with filamentous forms dominating late seral stages (Oemke and Burton, 1986; Steinman and McIntire, 1986; Peterson and Stevenson, 1989). Other studies have revealed that benthic algal succession is more analogous to higher plant succession, with a directional change in physiognomy from low to high physical stature (Hoagland *et al.,* 1982). The influence of grazing on succession will depend on the prevailing sere, the grazer type and density involved, resource availability, and the timing of the algal–herbivore interaction.

Steinman *et al.* (1987a) found that grazing by caddisfly larvae at all densities and snails at moderate to high densities removed overstory forms, leaving behind a layer of prostrate understory forms. In this situation, grazing can be interpreted as inhibiting succession, because it is preventing the development of the final seral stage. However, because some overstory forms were the initial colonists and were followed by understory forms, an alternative interpretation could be that of grazing facilitating succession. Note that if low-lying, understory cells were the initial colonists in this sere (cf. Hoagland *et al.,* 1982), grazing would have kept the sere in that initial stage, resulting only in inhibition. Complicating this scenario is the influence of grazer density, resource availability, and timing of the plant–grazer interaction. In streams where snail densities were low, grazing had virtually no influence on succession or slowed down the rate and degree to which filamentous algae dominated the final seral stage (Steinman *et al.,* 1987a). When snail density was kept constant, but the light level was changed, grazing had a greater effect on succession at high and intermediate light than at low light (Steinman *et al.,* 1989). This reflects the relative absence of filamentous forms at low light, thereby restricting the ability of grazing to change successional trajectories.

The timing of the algal–grazer interaction can influence the outcome in freshwater benthos because the size and structure of the prey assemblage change over successional time. DeNicola *et al.* (1990) noted that the influence of timing on grazing was more a question of whether or not succession is changed, compared to whether it is retarded or accelerated. They found that the caddisfly *Dicosmoecus* changed the trajectory (to a community of prostrate cells) regardless of what day the grazers were introduced. The snail *Juga* had a patchy effect regardless of what day they were introduced, and baetid mayfly larvae changed succession when introduced at the start of the experiment but had no effect when introduced 16 d after algal growth. Studies dealing with the filamentous chlorophyte *Cladophora* have shown that if grazers are present before *Cladophora* is established, intense herbivory can prevent establishment of this species (Dudley and D'Antonio, 1991). However, if *Cladophora* becomes established before grazers are abundant, herbivory may increase *Cladophora* abundance because grazers remove its epiphytes but not the host, as the *Cladophora* filaments apparently are too large to graze (Dudley, 1992; Sarnelle *et al.,* 1993).

Tuchman and Stevenson (1991) found that the effects of grazing on succession were habitat and assemblage specific. Grazing by snails arrested succession of nondiatom algae by removing the overstory of filamentous cyanobacteria, but accelerated succession among diatoms by ingesting early-succession diatoms and leaving late-succession motile diatoms. Ultimately, these authors concluded that the influence of grazing on rate or direction of succession was dependent on the characteristics of early- and late-successional species, which varied among habitats and assemblages.

These studies suggest strongly that the influence of grazing on freshwater benthic algal succession is highly specific to the habitat under study. It is presumptuous to generalize that grazers facilitate or inhibit succession. As in other ecosystems, consumers may slow, accelerate, or have no effect on the rate of succession (Farrell, 1991); to determine their effect it is necessary to know the successional status of the species being consumed (i.e., early-, middle-, or late-successional species) and the type of successional model operating in the system [i.e., facilitation, tolerance, or inhibition: Connell and Slatyer (1977)]. As with virtually all parameters explored in this chapter, the magnitude of the effect will increase with increasing grazer density.

IV. FUTURE RESEARCH DIRECTIONS AND CONCERNS

Interecosystem Comparisons Our understanding of ecological dynamics can be enhanced by comparing processes across a range of different ecosystems (cf. Cole *et al.,* 1991). This is certainly true for periphyton–herbivore interactions as well, and many of the proposed directions for future research described in the following are based on results from other systems. However, given the potential uniqueness of the periphyton–herbivore interaction in freshwater benthic habitats (i.e., small prey:grazer size ratio, complex community physiognomy and high species diversity, apparent lack of selectivity, apparent lack of chemical defenses in most species), it is plausible that herbivory models developed for other systems may not be appropriate, and attempts to force periphyton–herbivore results into those frameworks will constrain and limit our understanding. This does not mean that all attempts necessarily will fail or that all models are so system specific that they have no applicability elsewhere; rather, we simply have to evaluate carefully their applicability. For example, in terrestrial ecosystems, Coley *et al.* (1985) proposed that environments with low resource availability favor plants with slow growth rates and large investments in antiherbivore defenses, whereas environments with high resource availability favor plants with high growth rates and lower defense levels. Given the relatively fast growth rates of benthic algae, it is unclear how applicable the Coley *et al.* hypothesis is in the freshwater benthos. Rosemond *et al.* (1993) did report a trade-off between algal resistance to grazing and growth rates. They noted that some species that were negatively affected by snail grazing (e.g., *Achnanthes* spp., *Peronia intermedium*) were positively affected by nutrient addition, whereas select species that were positively affected by grazing (e.g., *Chamaesiphon investiens*) were negatively affected by nutrient addition. However, the Rosemond *et al.* study did not address explicitly the question of species nutrient content and herbivory, although they did examine nutrient content of the community as a whole.

Stoichiometry Work in planktonic systems has revealed that different types of zooplankton have different nitrogen and phosphorus requirements (Sterner, 1990; Andersen and Hessen, 1991; Sterner *et al.,* 1992). When zooplankton with high phosphorus needs (e.g., *Daphnia*) ingest phytoplankton with high N:P ratios, there is a disproportionate release of unneeded nitrogen to the system. This, in turn, can affect algal community structure because of changing nutrient ratios in the water column (Urabe, 1993). It seems plausible that if different nutrient stoichiometries exist between different types of zooplankton (Sterner *et al.,* 1992), then different stoichiometries also may exist among snails, caddisflies, and mayflies. It has been shown that nutrient regeneration is mediated by benthic grazers (e.g., Grimm and Fisher, 1986; Mulholland *et al.*, 1991); now studies are needed to determine if there are differential stoichiometries and, if so, whether or not this influences algal community structure.

Algal Food Quality Although the importance of food quality has been recognized for invertebrates in freshwater benthic systems (Anderson and Cummins, 1979; Fuller and Mackay, 1981), surprisingly few studies have addressed its role with respect to the algal–herbivore interaction. Given the importance of food quality in other ecosystems (Butler *et al.*, 1989; Cowles *et al.*, 1988; Freeland and Choquenot, 1990), it is even more surprising that this subject has not been addressed in more detail. Several studies have examined food choice in streams (Vaughn, 1986; Rosillon, 1988) or invertebrate growth under different food regimes (Fuller *et al.*, 1988; Fuller and Fry, 1991), but detailed nutritional analyses of the food items are lacking (but see Lamberti, Chapter 17, this volume). Future work addressing the role of lipids may prove especially rewarding, given the results of Cargill *et al.* (1985) and Steinman *et al.* (1987b).

Selectivity Closely related to questions regarding food quality is the question of whether or not selectivity by herbivores is common in the freshwater benthos. There is a need to distinguish whether the apparent selection noted in studies is a function of true selection or is simply an example of differential efficiency. Ideally, future studies should use whole communities and not single-species choice experiments, as the complex physiognomy of the periphyton community may influence the algal–grazer interaction in nature. However, this may prove to be intractable logistically. Identification of selected prey items may be problematic, but examination of grazer gut contents immediately following experimentation may reduce the uncertainty in identifying ingested food items. Alternatively, radiotracer labeling of food items may allow increased discrimination of what is being eaten, although mere labeling will not resolve whether grazers are feeding in a selective or differentially efficient mode. Specific areas dealing with selectivity that could be addressed include the roles of algal food quality, algal growth form, chemical defense, and herbivore satiation.

Chemical Defense Secondary metabolites in plants are generally recognized as effective deterrents to grazing in terrestrial (Coley *et al.*, 1985) and marine ecosystems (Hay *et al.*, 1987). They have received little attention in the freshwater benthos, except with respect to vascular macrophytes (cf. Newman, 1991). The relatively short life histories exhibited by freshwater algae may reduce the evolutionary pressure to synthesize secondary compounds, but their presence or absence cannot be established unequivocally until studies are conducted.

Pathogens Pathogens, whether in the form of viruses, fungi, or bacteria, have the potential to severely affect algal communities (C. G. Peterson *et al.*, 1993; Reynolds, 1973; Suttle *et al.*, 1990). Studies by Stewart (1988) and C. G. Peterson *et al.* (1993) described pathogenic invasions of periphyton that disrupted community health and integrity. It is unclear how common these invasions are, or if they are detected only when they reach catastrophic stages. Low-level pathogenic invasions may have subtle effects on the algae, which may in turn influence the algal–herbivore interaction. Alternatively, pathogenic invasions of herbivores (cf. Cummins and Wilzbach, 1988; Wiley and Kohler, 1993) may influence their grazing patterns on algae as well.

Herbivore Bias Prosobranch snails and limnephilid and glossosomatid caddisfly larvae are the overwhelming herbivores of choice in herbivory studies conducted in freshwater benthic ecosystems. Indeed, most of the conclusions presented in this chapter are based on studies using these herbivores. Additional studies are needed using different herbivore types to determine whether algal responses are consistent with those effected by these types of snails and caddisflies; otherwise, our present understanding of herbivory in the freshwater benthos will be biased.

ACKNOWLEDGMENTS

I am deeply indebted to the following individuals, each of whom has influenced my thinking on herbivory and its ecological implications: Gary Lamberti, Stan Gregory, Pat Mulholland, Dave McIntire, Jane Lubchenco, Walter Hill, Art Stewart, Amy Rosemond, and Mike Huston. This chapter benefited greatly from comments by Gary Lamberti, Jack Feminella, Ken Cummins, Jan Stevenson, Rex Lowe, Karl Havens, and Susan Gray.

REFERENCES

Andersen, T., and Hessen, D. O. (1991). Carbon, nitrogen, and phosphorus content of freshwater zooplankton. *Limnol. Oceanogr.* **36**, 807–814.

Anderson, N. H., and Cummins, K. W. (1979). Influences of diet on the life histories of insects. *J. Fish. Res. Board Can.* **36**, 335–342.

Belsky, A. J. (1986). Does herbivory benefit plants? A review of the evidence. *Am. Nat.* **100**, 64–82.

Blinn, D. W., Truitt, R. E., and Pickart, A. (1989). Feeding ecology and radular morphology of the freshwater limpet *Ferrissia fragilis*. *J. North Am. Benthol. Soc.* **8**, 237–242.

Bronmark, C., Rundle, S. D., and Erlandson, A. (1991). Interactions between freshwater snails and tadpoles: Competition and facilitation. *Oecologia* **87**, 8–18.

Butler, N. M., Suttle, C. A., and Neill, W. E. (1989). Discrimination by freshwater zooplankton between single algal cells differing in nutritional status. *Oecologia* **78**, 368–372.

Calow, P. (1973). The food of *Ancylus fluviatilis* (Mull.), a littoral stone-dwelling, herbivore. *Oecologia* **13**, 113–133.

Calow, P. (1979). Why some metazoan mucus secretions are more susceptible to microbial attack than others. *Am. Nat.* **114,** 149–152.

Cargill, A. S., Cummins, K. W., Hanson, B. J., and Lowry, R. R. (1985). The roles of lipids, fungi, and temperature on the nutrition of a shredder caddisfly, *Clistoronia magnifica* (Trichoptera: Limnephilidae). *Freshwater Invertebr. Biol.* **4,** 64–78.

Cole, J., Lovett, G., and Findlay, S. (1991). "Comparative Analysis of Ecosystems: Patterns, Mechanisms, and Theories." Springer-Verlag, New York.

Coley, P. D., Bryant, J. P., and Chapin, F. S., III (1985). Resource availability and plant antiherbivore defense. *Science* **230,** 895–899.

Colletti, P. J., Blinn, D. W., Pickart, A., and Wagner, V. T. (1987). Influence of different densities of the mayfly grazer *Heptagenia criddlei* on lotic diatom communities. *J. North Am. Benthol. Soc.* **6,** 270–280.

Connell, J. H. (1978). Diversity in tropical rain forests and coral reefs. *Science* **199,** 1302–1310.

Connell, J. H., and Slatyer, R. O. (1977). Mechanisms of succession in natural communities and their role in community stability and organization. *Am. Nat.* **111,** 1119–1144.

Connor, V. M., and Quinn, J. F. (1984). Stimulation of food species growth by limpet mucus. *Science* **225,** 843–844.

Cowles, T. J., Olson, R. J., and Chisholm, S. W. (1988). Food selection by copepods: Discrimination on the basis of food quality. *Mar. Biol. (Berlin)* **100,** 41–49.

Crawley, M. J. (1983). "Herbivory." Univ. of California Press, Berkeley.

Cuker, B. E. (1983). Grazing and nutrient interactions in controlling the activity and composition of the epilithic algal community of an arctic lake. *Limnol. Oceanogr.* **28,** 133–141.

Cummins, K. W. (1964). Factors limiting the microdistribution of the caddisflies *Pycnopsyche lepida* (Hagen) and *Pycnopsyche guttifer* (Walker) in a Michigan stream (Trichoptera: Limnephilidae). *Ecol. Monogr.* **34,** 271–295.

Cummins, K. W., and Klug, M. J. (1979). Feeding ecology of stream invertebrates. *Ann. Rev. Ecol. Syst.* **10,** 147–172.

Cummins, K. W., and Wilzbach, M. A. (1988). Do pathogens regulate stream invertebrate populations? *Verh.—Int. Ver. Theor. Angew. Limnol.* **23,** 1232–1243.

DeAngelis, D. L., Bartell, S. M., and Brenkert, A. L. (1989). Effects of nutrient recycling and food chain length on resilience. *Am. Nat.* **134,** 778–805.

DeAngelis, D. L., Mulholland, P. J., Elwood, J. W., Palumbo, A. V., and Steinman, A. D. (1990). Biogeochemical cycling constraints on stream ecosystem recovery. *Environ. Manage.* **14,** 685–697.

DeNicola, D. M., and McIntire, C. D. (1991). Effects of substrate relief on the distribution of periphyton in laboratory streams. II. Interactions with irradiance. *J. Phycol.* **26,** 634–641.

DeNicola, D. M., McIntire, C. D., Lamberti, G. A., Gregory, S. V., and Ashkenas, L. A. (1990). Temporal patterns of grazer–periphyton interactions in laboratory streams. *Freshwater Biol.* **23,** 475–489.

Dickman, M. (1968). The effect of grazing by tadpoles on the structure of a periphyton community. *Ecology* **49,** 1188–1190.

Dudley, T. L. (1992). Beneficial effects of herbivores on stream macroalgae via epiphyte removal. *Oikos* **65,** 121–127.

Dudley, T. L., and D'Antonio, C. M. (1991). The effects of substrate texture, grazing, and disturbance on macroalgal establishment in streams. *Ecology* **72,** 297–309.

Eichenberger, E., and Schlatter, F. (1978). The effect of herbivorous insects on the production of benthic algal vegetation in outdoor channels. *Ver.—Int. Ver. Theor. Angew. Limnol.* **20,** 1806–1810.

Elwood, J. W., Newbold, J. D., Trimble, A. F., and Stark, R. W. (1981). The limiting role of phosphorus in a woodland stream ecosystem: Effects of P enrichment on leaf decomposition and primary producers. *Ecology* **62,** 146–158.

Falkowski, P. G., and LaRoche, J. (1991). Acclimation to spectral irradiance in algae. *J. Phycol.* **27,** 8–14.

Farrell, T. M. (1991). Models and mechanisms of succession: An example from a rocky intertidal community. *Ecol. Monogr.* **61,** 95–113.

Feminella, J. W., and Hawkins, C. P. (1995) Interactions between stream herbivores and periphyton: A quantitative analysis of past experiments. *J. North Am. Benthol. Soc.* **14,** 465–509.

Feminella, J. W., and Resh, V. H. (1991). Herbivorous caddisflies, macroalgae, and epilithic microalgae: Dynamic interactions in a stream grazing system. *Oecologia* **87,** 247–256.

Feminella, J. W., Power, M. E., and Resh, V. H. (1989). Periphyton responses to invertebrate grazing and riparian canopy in three northern California coastal streams. *Freshwater Biol.* **22,** 445–457.

Fitzgerald, G. P., and Nelson, T. C. (1966). Extractive and enzymatic analyses for limiting or surplus phosphorus in algae. *J. Phycol.* **2,** 32–37.

Flint, R.W., and Goldman, C. R. (1975). The effects of a benthic grazer on the primary productivity of the littoral zone of Lake Tahoe. *Limnol. Oceanogr.* **20,** 935–944.

Freeland, W. J., and Choquenot, D. (1990). Determinants of herbivore carrying capacity: Plants, nutrients, and *Equus asinus* in northern Australia. *Ecology* **71,** 589–597.

Fuller, R. L., and Fry, T. J. (1991). The influence of temperature and food quality on the growth of *Hydropsyche betteni* (Trichoptera) and *Simulium vittatum* (Diptera). *J. Freshwater Ecol.* **6,** 75–86.

Fuller, R. L., and Mackay, R. J. (1981). Effects of food quality on the growth of three *Hydropsyche* species (Trichoptera: Limnephilidae). *Freshwater Invertebr. Biol.* **59,** 1133–1140.

Fuller, R.L., Fry, T. J., and Roelofs, J. A. (1988). Influence of different food types on the growth of *Simulium vittatum* (Diptera) and *Hydropsyche betteni* (Trichoptera). *J. North Am. Benthol. Soc.* **7,** 197–204.

Gelwick, F. P., and Matthews, W. J. (1992). Effects of an algivorous minnow on temperate stream ecosystem properties. *Ecology* **73,** 1630–1645.

Gregory, S. V. (1983). Plant–herbivore interactions in stream systems. *In* "Stream Ecology" (J. R. Barnes, and G. W. Minshall, eds.), pp. 157–189. Plenum, New York.

Grimm, N. B., and Fisher, S. G. (1986). Nitrogen limitation in a Sonoran desert stream. *J. North Am. Benthol. Soc.* **5,** 2–15.

Hairston, N. G., Smith, F. E., and Slobodkin, L. B. (1960). Community structure, population control and competition. *Am. Nat.* **94,** 421–425.

Hambrook, J. A., and Sheath, R. G. (1987). Grazing of freshwater Rhodophyta. *J. Phycol.* **23,** 656–662.

Hamilton, P. B., and Duthie, H. C. (1984). Periphyton colonization of rock surfaces in a boreal forest stream studied by scanning electron microscopy and track autoradiography. *J. Phycol.* **20,** 525–532.

Hargrave, B. T. (1970). The effect of a deposit-feeding amphipod on the metabolism of benthic microflora. *Limnol. Oceanogr.* **15,** 21–30.

Harrison, P. G., and Mann, K. H. (1975). Chemical changes during the seasonal cycle of growth and decay in eelgrass on the Atlantic coast of Canada. *J. Fish. Res. Board Can.* **32,** 615–621.

Hart, D. D. (1985). Grazing insects mediate algal interactions in a stream benthic community. *Oikos* **44,** 40–46.

Hart, D. D. (1987). Experimental studies of exploitative competition in a grazing stream insect. *Oecologia* **73,** 41–47.

Hart, D. D., Kohler, S. L., and Carlton, R. G. (1991). Harvesting of benthic algae by territorial grazers: The potential for prudent predation. *Oikos* **60,** 329–335.

Hay, M. E., Fenical, W., and Gustafson, K. (1987). Chemical defense against diverse coral-reef herbivores. *Ecology* **68,** 1581–1591.

Hill, W. R., and Harvey, B. C. (1990). Periphyton responses to higher trophic levels and light in a shaded stream. *Can. J. Fish. Aquat. Sci.* **47,** 2307–2314.

Hill, W. R., and Knight, A. W. (1987). Experimental analysis of the grazing interaction between a mayfly and stream algae. *Ecology* **68,** 1955–1965.

Hill, W. R., and Knight, A. W. (1988). Concurrent grazing effects of two stream insects on periphyton. *Limnol. Oceanogr.* **33,** 15–26.

Hill, W. R., Boston, H. L., and Steinman, A. D. (1992). Grazers and nutrients simultaneously limit lotic primary productivity. *Can. J. Fish. Aquat. Sci.* **49,** 504–512.

Hoagland, K. D., Roemer, S. C., and Rosowski, J. R. (1982). Colonization and community structure of two periphyton assemblages, with emphasis on the diatoms (Bacillariophyceae). *Am. J. Bot.* **69,** 188–213.

Hoagland, K. D., Rosowski, J. R., Gretz, M. R., and Roemer, S. C. (1993). Diatom extracellular polymeric substances: Function, fine structure, chemistry, and physiology. *J. Phycol.* **29,** 537–566.

Horn, M., and Neighbors, M. A. (1984). Protein and nitrogen assimilation as a factor in predicting the seasonal macroalgal diet of the monkeyface prickleback. *Trans. Am. Fish. Soc.* **113,** 388–396.

Hudon, C., and Bourget, E. (1981). Initial colonization of artificial substrate: Community development and structure studied by scanning electron microscopy. *Can. J. Fish. Aquat. Sci.* **38,** 1371–1384.

Hudon, C., and Legendre, P. (1987). The ecological implications of growth forms in epibenthic diatoms. *J. Phycology* 23, 434–441.

Hunter, R. D. (1980). Effects of grazing on the quantity and quality of freshwater Aufwuchs. *Hydrobiologia* **69,** 251–259.

Huston, M. A. (1979). A general hypothesis of species diversity. *Am. Nat.* **113,** 81–101.

Jacoby, J. M. (1985). Grazing effects in periphyton by *Theodoxus fluviatilis* (Gastropoda) in a lowland stream. *J. Freshwater Ecol.* **1,** 51–59.

Jacoby, J. M. (1987). Alterations in periphyton characteristics due to grazing in a Cascade foothill stream. *Freshwater Biol.* **18,** 495–508.

Jorgensen, B. B., Cohen, Y., and Des Marais, D. J. (1987). Photosynthetic action spectra and adaptation to spectral light distribution in a benthic cyanobacterial mat. *Appl. Environ. Microbiol.* **53,** 879–886.

Karouna, N. K., and Fuller, R. L. (1992). Influence of four grazers on periphyton communities associated with clay tiles and leaves. *Hydrobiologia* **245,** 53–64.

Kehde, P. M., and Wilhm, J. L. (1972). The effects of grazing by snails on community structure of periphyton in laboratory streams. *Am. Midl. Nat.* **87,** 8–24.

Kesler, D. H. (1981). Periphyton grazing by *Amnicola limosa*: An enclosure-exclosure experiment. *J. Freshwater Ecol.* **1,** 51–59.

Kohler, S. L., and Wiley, M. J. (1992). Parasite-induced collapse of populations of a dominant grazer in Michigan streams. *Oikos* **65,** 443–449.

Korte, V. L., and Blinn, D. L. (1983). Diatom colonization on artificial substrata in pool and riffle zones studied by light and scanning electron microscopy. *J. Phycol.* **19,** 332–341.

Lamberti, G. A., and Moore, J. W. (1984). Aquatic insects as primary consumers. *In* "The Ecology of Aquatic Insects" (V. H. Resh, and D. M. Rosenberg, eds.), pp. 164–195. Praeger, New York.

Lamberti, G. A., and Resh, V. H. (1983). Steam periphyton and insect herbivores: An experimental study of grazing by a caddisfly population. *Ecology* **64,** 1124–1135.

Lamberti, G. A., Ashkenas, L. R., Gregory, S. V., and Steinman, A. D. (1987a). Effects of three herbivores on periphyton communities in laboratory streams. *J. North Am. Benthol. Soc.* **6,** 92–104.

Lamberti, G. A., Feminella, J. W., and Resh, V. H. (1987b). Herbivory and intraspecific competition in a stream caddisfly population. *Oecologia* **73,** 75–81.

Lamberti, G. A., Ashkenas, L. R., Gregory, S. V., Steinman, A. D., and McIntire, C. D. (1989). Productive capacity of periphyton as a determinant of plant–herbivore interactions in streams. *Ecology* **70,** 1840–1856.

Lamberti, G. A., Gregory, S. V., Hawkins, C. P., Wildman, R. C., Ashkenas, L. R., and DeNicola, D. M. (1992). Plant–herbivore interactions in streams near Mount St. Helens. *Freshwater Biol.* **27,** 237–247.

Lodge, D. M. (1991). Herbivory on freshwater macrophytes. *Aquat. Bot.* **41,** 195–224.

Lowe, R. L., and Hunter, R. D. (1988). Effects of grazing by *Physa integra* on periphyton community structure. *J. North Am. Benthol. Soc.* **7,** 29–36.

Lowe, R. L., Golladay, S. W., and Webster, J. R. (1986). Periphyton response to nutrient manipulation in streams draining clearcut and forested watersheds. *J. North Am. Benthol. Soc.* **5,** 221–229.

Lubchenco, J. (1978). Plant species diversity in a marine intertidal community: Importance of herbivore food preference and algal competitive abilities. *Am. Nat.* **112,** 23–39.

Lubchenco, J., and Cubit, J. (1980). Heteromorphic life histories of certain marine algae as adaptations to variations in herbivory. *Ecology* **61,** 676–687.

Lubchenco, J., and Gaines, S. D. (1981). A unified approach to marine plant–herbivore interactions. I. Populations and communities. *Ann. Rev. Ecol. Syst.* **12,** 405–437.

Marks, J. C., and Lowe, R. L. (1989). The independent and interactive effects of snail grazing and nutrient enrichment on structuring periphyton communities. *Hydrobiologia* **185,** 9–17.

Mazumder, A., Taylor, W. D., McQueen, D. J., and Lean, D. R. S. (1989). Effects of nutrients and grazers on periphyton phosphorus in lake enclosures. *Freshwater Biol.* **22,** 405–415.

McAuliffe, J. R. (1984). Resource depression by a stream herbivore: Effects on distributions and abundances of other grazers. *Oikos* **42,** 327–333.

McCormick, P. V. (1991). Lotic protistan herbivore selectivity and its potential impact on benthic algal assemblages. *J. North Am. Benthol. Soc.* **10,** 238–250.

McCormick, P. V. (1994). Evaluating the multiple mechanisms underlying herbivore–algal interactions in streams. *Hydrobiologia* **291,** 47–59.

McCormick, P. V., and Stevenson, R. J. (1989). Effects of snail grazing on benthic community structure in different nutrient environments. *J. North Am. Benthol. Soc.* **8,** 162–172.

McCormick, P. V., and Stevenson, R. J. (1991a). Grazer control of nutrient availability in the periphyton. *Oecologia* **86,** 287–291.

McCormick, P. V., and Stevenson, R. J. (1991b). Mechanisms of benthic algal succession in lotic environments. *Ecology* **72,** 1835–1848.

McMahon, R. F., Hunter, R. D. and Russell-Hunter, W. D. (1974). Variations in aufwuchs at six freshwater habitats in terms of carbon biomass and carbon: nitrogen ratio. *Hydrobiologia* **45,** 391–404.

McNaughton, S. J. (1983). Serengeti grassland ecology: The role of composite environmental factors and contingency in community organization. *Ecol. Monogr.* **53,** 291–320.

McNaughton, S. J., Ruess, R. W., and Seagle, S. W. (1988). Large mammals and process dynamics in African ecosystems. *BioScience* **38,** 794–800.

Merritt, R. W., and Cummins, K. W. (1984). "An Introduction to the Aquatic Insects of North America," 2nd ed. Kendall/Hunt, Dubuque, IA.

Moore, J. W. (1975). The role of algae in the diet of *Asellus aquaticus* L. and *Gammarus pulex* L. *Ecology* **44,** 719–730.

Mulholland, P. J., Newbold, J. D., Elwood, J. W., and Hom, C. L. (1983). The effect of grazing intensity on phosphorus spiralling in autotrophic streams. *Oecologia* **58,** 358–366.

Mulholland, P. J., Steinman, A. D., Palumbo, A. V., Elwood, J. W., and Kirschtel, D. B. (1991). Role of nutrient cycling and herbivory in regulating periphyton communities in laboratory streams. *Ecology* **72,** 966–982.

Mulholland, P. J., Steinman, A. D., Marzolf, E. R., Hart, D. R., and DeAngelis, D. L. (1994). Effect of periphyton biomass on hydraulic characteristics and nutrient cycling in streams. *Oecologia* **98,** 40–47.

Newman, R. M. (1991). Herbivory and detritivory on freshwater macrophytes by invertebrates: A review. *J. North Am. Benthol. Soc.* **10,** 89–114.

Oemke, M. P., and Burton, T. M. (1986). Diatom colonization dynamics in a lotic system. *Hydrobiologia* **139,** 153–166.

Osenberg, C. W. (1989). Resource limitation, competition and the influence of life history in a freshwater snail community. *Oecologia* **79,** 512–519.

Pastor, J., Naiman, R. J., Dewey, B., and McInnes, P. (1988). Moose, microbes, and the boreal forest. *BioScience* **38,** 770–777.

Perry, M. J. (1972). Alkaline phosphatase activity in subtropical Central North Pacific waters using a sensitive fluorometric method. *Mar. Biol. (Berlin)* **15,** 113–119.

Peterson, B. J., Deegan, L., Helfrich, J., Hobbie, J. E., Hullar, M., Moller, B., Ford, T. E., Hershey, A., Hiltner, A., Kipphut, G., Lock, M. A., Fiebig, D. M., McKinley, V., Miller, M. C., Vestal, J. R., Ventullo, R., and Volk, G. (1993). Biological responses of a tundra river to fertilization. *Ecology* **74,** 653–672.

Peterson, C. G., and Stevenson, R. J. (1989). Substratum conditioning and diatom colonization in different current regimes. *J. Phycol.* **25,** 790–793.

Peterson, C. G., Dudley, T. L., Hoagland, K. D., and Johnson, L. M. (1993). Infection, growth, and community-level consequences of a diatom pathogen in a Sonoran Desert stream. *J. Phycol.* **29,** 442–452.

Pick, F. R. (1987). Interpretations of alkaline phosphatase activity in Lake Ontario. *Can. J. Fish. Aquat. Sci.* **44,** 2087–2094.

Power, M. E. (1990). Effects of fish in river food webs. *Science* **250,** 811–814.

Power, M. E. (1992). Habitat heterogeneity and the functional significance of fish in river food webs. *Ecology* **73,** 1675–1688.

Power, M. E., and Matthews, W. J. (1983). Algae-grazing minnows (*Campostoma anomalum*), piscivorous bass (*Micropterus* spp.), and the distribution of attached algae in a small prairie-margin stream. *Oecologia* **60,** 328–332.

Power, M. E., Matthews, W. J., and Stewart, A. J. (1985). Grazing minnows, piscivorous bass, and stream algae: Dynamics of a strong interaction. *Ecology* **66,** 1448–1456.

Power, M. E., Stewart, A. J., and Matthews, W. J. (1988). Grazer control of algae in an Ozark Mountain stream: Effects of short-term exclusion. *Ecology* **69,** 1894–1898.

Pringle, C. M., Blake, G. A., Covich, A. P., Buzbu, K. M., and Finley, A. (1993). Effects of omnivorous shrimp in a montane tropical stream: Sediment removal, disturbance of sessile invertebrates and enhancement of understory algal biomass. *Oecologia* **93,** 1–11.

Resh, V. H., and Houp, R. E. (1986). Life history of the caddisfly *Dibusa angata* and its association with the red alga *Lemanea australis*. *J. North Am. Benthol. Soc.* **5,** 28–40.

Reynolds, C. S. (1973). The seasonal periodicity of planktonic diatoms in a shallow eutrophic lake. *Freshwater Biol.* **3,** 89–110.

Roemer, S. C., Hoagland, K. D., and Rosowski, J. R. (1984). Development of a freshwater periphyton community as influenced by diatom mucilages. *Can. J. Bot.* **62,** 1799–1813.

Rosemond, A. D. (1993). Interactions among irradiance, nutrients, and herbivores constrain a stream algal community. *Oecologia* **94,** 585–594.

Rosemond, A. D., Mulholland, P. J., and Elwood, J. W. (1993). Top-down and bottom-up control of stream periphyton: Effects of nutrients and herbivores. *Ecology* **74,** 1264–1280.

Rosillon, D. (1988). Food preference and relative influence of temperature and food quality on life history characteristics of a grazing mayfly, *Ephemerella ignita* (Poda). *Can. J. Zool.* **66,** 1474–1481.

Sarnelle, O., Kratz, K. W., and Cooper, S. D. (1993). Effects of an invertebrate grazer on the spatial arrangement of a benthic microhabitat. *Oecologia* **96,** 208–218.

Scrimgeour, G. J., Culp, J. M., Bothwell, M. L., Wrona, F. J., and McKee, M. H. (1991). Mechanisms of algal patch depletion: Importance of consumptive and non-consumptive losses in mayfly–diatom systems. *Oecologia* **85**, 343–348.

Sheath, R. G., Burkholder, J. M., Morison, M. O., Steinman, A. D., and Van Alstyne, K. L. (1986). Effect of tree canopy removal by gypsy moth larvae on the macroalgal community of a Rhode Island headwater stream. *J. Phycol.* **22**, 567–570.

Steinman, A. D. (1991). Effects of herbivore size and hunger level on periphyton communities. *J. Phycol.* **27**, 54–59.

Steinman, A. D. (1992). Does an increase in irradiance influence periphyton in a heavily-grazed woodland stream? *Oecologia* **91**, 163–170.

Steinman, A. D., and Lamberti. G. A. (1996). Biomass and pigments of benthic algae. *In* "Methods in Stream Ecology" (F. R. Hauer, and G. A. Lamberti, eds). Academic Press, San Diego, CA.

Steinman, A. D., and McIntire, C. D. (1986). Effects of current velocity and light energy on the structure of periphyton assemblages in laboratory streams. *J. Phycol.* **22**, 352–361.

Steinman, A. D., and McIntire, C. D. (1987). Effects of irradiance on algal community structure in laboratory streams. *Can. J. Fish. Aquat. Sci.* **44**, 1640–1648.

Steinman, A. D., McIntire, C. D., Gregory, S. V., Lamberti, G. A., and Ashkenas, L. (1987a). Effect of herbivore type and density on taxonomic structure and physiognomy of algal assemblages in laboratory streams. *J. North Am. Benthol. Soc.* **6**, 175–188.

Steinman, A. D., McIntire, C. D., and Lowry, R. R. (1987b). Effect of herbivore type and density on chemical composition of algal assemblages in laboratory streams. *J. North Am. Benthol. Soc.* **6**, 189–197.

Steinman, A. D., McIntire, C. D., Gregory, S. V., and Lamberti, G. A. (1989). Effects of irradiance and grazing on lotic algal assemblages. *J. Phycol.* **25**, 478–485.

Steinman, A. D., Mulholland, P. J., Palumbo, A. V., Flum, T. F., and DeAngelis, D. L. (1991). Resilience of lotic ecosystems to a light-elimination disturbance. *Ecology* **72**, 1299–1313.

Steinman, A. D., Mulholland, P. J., and Beauchamp, J. J. (1995) Effects of biomass, light, and grazing on phosphorus cycling in stream periphyton communities. *J. North Am. Benthol. Soc.* **14**, 371–381.

Steinman, A. D., Mulholland, P. J., and Hill, W. R. (1992). Functional responses associated with growth form in stream algae. *J. North Am. Benthol. Soc.* **11**, 229–243.

Sterner, R. W. (1990). N:P resupply by herbivores: Zooplankton and the algal competitive arena. *Am. Nat.* **136**, 209–229.

Sterner, R. W., Elser, J. J., and Hessen, D. O. (1992). Stoichiometric relationships among producers and consumers in food webs. *Biogeochemistry* **17**, 49–67.

Stewart, A. J. (1987). Responses of stream algae to grazing minnows and nutrients: A field test of interactions. *Oecologia* **72**, 1–7.

Stewart, A. J. (1988). A note on the occurrence of unusual patches of senescent periphyton in an Oklahoma stream. *J. Freshwater Ecol.* **4**, 395–399.

Stewart, A. J., Hill, W. R., and Boston, H. L. (1993). Grazers, periphyton and toxicant movement in streams. *Environ. Toxicol. Chem.* **12**, 955–957.

Sukenik, A., Bennett, J., Mortain-Bertrand, A., and Falkowski, P. G. (1990). Adaptation of the photosynthetic apparatus to irradiance in *Dunaliella tertiolecta. Plant Physiol.* **92**, 891–898.

Sumner, W. T., and McIntire, C. D. (1982). Grazer–periphyton interactions in laboratory streams. *Arch. Hydrobiol.* **93**, 135–157.

Suttle, C. A., Chan, A. M., and Cottrell, M. T. (1990). Infection of phytoplankton by viruses and reduction of primary productivity. *Nature (London)* **347**, 467–469.

Swamikannu, X., and Hoagland, K. D. (1989). Effects of snail grazing on the diversity and structure of a periphyton community in a eutrophic pond. *Can. J. Fish. Aquat. Sci.* **46**, 1698–1704.

Tuchman, N. C., and Stevenson, R. J. (1991). Effects of selective grazing by snails on benthic algal communities. *J. North Am. Benthol. Soc.* **10**, 430–443.

Underwood, G. J. C., and Thomas, J. D. (1990). Grazing interactions between pulmonate snails and epiphytic algae and bacteria. *Freshwater Biol.* **23**, 505–522.

Urabe, J. (1993). N and P cycling coupled by grazers' activities: Food quality and nutrient release by zooplankton. *Ecology* **74**, 2337–2350.

Vaughn, C. C. (1986). The role of periphyton abundance and quality in the microdistribution of a stream grazer, *Helicopsyche borealis* (Trichoptera: Helicopsychidae). *Freshwater Biol.* **16**, 485–493.

Vaughn, C. C., Gelwick, F. P., and Matthews, W. J. (1993). Effects of algivorous minnows on production of grazing stream invertebrates. *Oikos* **66**, 119–128.

Wehr, J. P. (1981). Analysis of seasonal succession of attached algae in a mountain stream, the North Alouette River, British Columbia. *Can. J. Bot.* **59**, 1465–1474.

Whittaker, R. H. (1975). "Communities and Ecosystems." Macmillan, New York.

Wiley, M. J., and Kohler, S. L. (1993). Community responses to microsporidian induced collapse of *Glossosoma nigrior* populations in Michigan trout streams. *Bull. North Am. Benthol. Soc.* **10**, 171 (abstr.)

Winterbourn, M. J. (1990). Interactions among nutrients, algae, and invertebrates in a New Zealand mountain stream. *Freshwater Biol.* **23**, 463–474.

Wootton, J. T., and Oemke, M. P. (1992). Latitudinal differences in fish community trophic structure, and the role of fish herbivory in a Costa Rican stream. *Environ. Biol. Fishes* **35**, 311–319.

13 Response of Benthic Algal Communities to Natural Physical Disturbance

Christopher G. Peterson
*Department of Natural Science,
Loyola University, Chicago
Chicago, Illinois 60626*

I. INTRODUCTION

As other chapters within this volume demonstrate, benthic algal communities in freshwater ecosystems are affected by a myriad of physical, chemical, and biological factors that characterize a given habitat at a given time. These factors interact to varying degrees to yield spatial and temporal heterogeneity in benthic algal biomass, species composition, physiology, and physiognomy. Physical disturbance occurs, to some degree, in all aquatic ecosystems, and can alter the form, intensity, and ultimate outcome of such interactions. Thus, the frequency, magnitude, season of occurrence, and type of disturbance to which a system is exposed can greatly influence patterns of benthic algal distribution within and among streams, lakes, and, to a lesser extent, wetlands.

Disturbance has become an integral component of ecological theory aimed at understanding pattern and process in ecological systems (e.g., Huston, 1979; Odum *et al.,* 1979; Fahrig, 1990; Fisher and Grimm, 1991). Temporal and spatial variation in disturbance frequency and intensity is instrumental in shaping community structure and modifying community dynamics in a wide diversity of systems, including terrestrial plant communities (e.g., Wilson and Tilman, 1991), marine intertidal zones (Sousa, 1984), and marine phytoplankton assemblages (Estrada *et al.,* 1987). The role of disturbance in generating large-scale patterns in freshwater benthic algal communities is addressed in this volume by Biggs (Chapter 2) for lotic systems and by Lowe and Pan (Chapter 22) for lentic systems. In this chapter, I will focus on the underlying mechanisms that dictate the response of benthic algae to physical disturbance. I do not address disturbance via exposure to organic or inorganic chemical toxins (see Hoagland *et al.,* Chapter 15, and Genter, Chapter 14, respectively) or thermal extremes (see DeNicola, Chapter 6), as these are also addressed elsewhere.

Disturbance is defined by Huston (1994) as "any process or condition external to the natural physiology of living organisms that results in sudden mortality of biomass in a community on a time scale significantly shorter (e.g. several orders of magnitude faster) than that of accumulation of biomass." Thus, even though disturbances can be very destructive, a disturbance event need not have catastrophic effects; in fact, algal community dynamics can be altered considerably by disturbance events for which cell mortality and/or export is minimal. If an event does not result in a direct physical effect on biological components in a system but, instead, induces a physiological change, it is considered a *stress* rather than a disturbance (Odum *et al.,* 1979; Pickett *et al.,* 1989). I will concentrate here on the effects of three types of physical disturbance to which benthic algae in fresh waters are commonly exposed: scour, emersion and subsequent desiccation, and light deprivation. Although desiccation and light deprivation can induce stress, without producing physical effects, these events often generate physical disruption and, thus, warrant discussion in this chapter.

Scouring of benthic algae from substrata by abrupt increases in current velocity and turbulence commonly occurs during spates in streams (Rounick and Gregory, 1981; Biggs and Close, 1989; Grimm and Fisher, 1989) and via wave action in lentic systems (Young, 1945; Luttenton and Rada, 1986; Peterson and Hoagland, 1990). Though the effects of spate-induced scour and wave disturbance are generally widespread, more localized, small-scale disruption of algal communities can be generated in both lotic and lentic systems by the activity of macrofauna, either aquatic (e.g., macroinvertebrates, fish, beaver) or terrestrial (e.g., moose, cattle, humans).

Desiccation, which is not as well studied as disturbance by scour, affects benthic algae in a wide range of systems. Seasonal drying can expose large expanses of substrata in intermittent streams (Morison and Sheath, 1985; Stanley and Fisher, 1992) or along margins of ponds and reservoirs (Evans, 1958; Pieczyńska and Banas, 1984; Hawes *et al.*, 1992). Periodic algal emersion is also common in tailwaters below hydroelectric dams as discharge varies with diel or seasonal change in electrical power needs (Blinn and Cole, 1991; Blinn *et al.*, 1995), in temperate streams during discharge reduction induced by daytime transpiration of riparian vegetation (Kobayashi *et al.*, 1990) or during periods between storm events (Poff and Ward, 1990), or in Antarctic streams, where water levels can change rapidly in response to variation in cloud cover or diel change in solar radiation (Vincent and Howard-Williams, 1986).

Light deprivation of algal cells can occur autogenically, as algal biomass that accumulates during community development progressively blocks penetration of light to cells in lower strata of attached algal mats (see Hill, Chapter 5). Benthic algae can also be exposed to extended periods of darkness or greatly reduced light availability as a consequence of physical disturbance. In lotic systems, substratum-mobilizing spates result in burial of attached algae as small gravel, sand, and/or silt is redeposited when floodwaters recede (Power and Stewart, 1987; Peterson *et al.*, 1994), or as cobbles are overturned (cf. Englund, 1991). Similarly, suspension and subsequent redeposition of sediments occur in lentic systems as a result of wave action (Luettich *et al.*, 1990). In addition, disturbance-induced increases in turbidity in both lotic (Davis-Colley *et al.*, 1992) and lentic habitats (Hellström, 1991) can cut light penetration to the benthos, limiting algal productivity; such conditions can persist for hours, days, or weeks.

Benthic algal response to any disturbance, regardless of form, is dependent not only on the magnitude and/or duration of the event, but also on physiognomic, taxonomic, and physiological properties of the community. These characteristics are forged by individual and interactive influences of factors such as current and nutrient regimes, the type and extent of grazing pressure, community age, and recent disturbance history. It is in this context that I couch my discussion of disturbance effects on benthic algae in fresh waters. In the text that follows, I separate assessment of disturbance response into two discrete components: factors influencing community *resistance* (the ability to withstand displacement by disturbance) and those affecting community *resilience* (the ability to return to the predisturbance state).

II. FACTORS INFLUENCING RESISTANCE TO SCOUR

A. Influence of Substratum Size and Surface Irregularity

Algal resistance to scour is strongly dependent on the size and stability of benthic substrata (Fig. 1). In stream and lake beds dominated by bedrock or boulders, removal of attached algae by storm flows can be patchy, with stands of periphyton recessed within depressions (Dudley and D'Antonio, 1991) or on the leeward side of boulders (Uehlinger, 1991; Peterson *et al.,* 1994) shielded from the full scouring force of the event. The extent of algal removal from these immobile substrata is increased if communities are barraged by entrained sand and small gravel (Horner *et al.,* 1990; Blinn and Cole, 1991), which can dislodge or damage attached algal cells upon impact (see micrographs in Blenkinsopp and Lock, 1994).

The loss of algal biomass incurred by a spate depends, in part, on the propensity for mobilization of benthic substrata. In an extensive 4-yr study of the benthic diatom community of Belle Grange Beck (Lancashire, U.K.), Douglas (1958) showed that flood events reduced diatom populations on cobble-sized stones to a much greater degree than those attached to bedrock and that epiphytic communities on bryophytes were most resistant, likely because of the more flexible nature of plant substrata relative to mineral surfaces. Power and Stewart (1987) observed persistent macroscopic algal residues on the edges of cobbles after these substrata had been

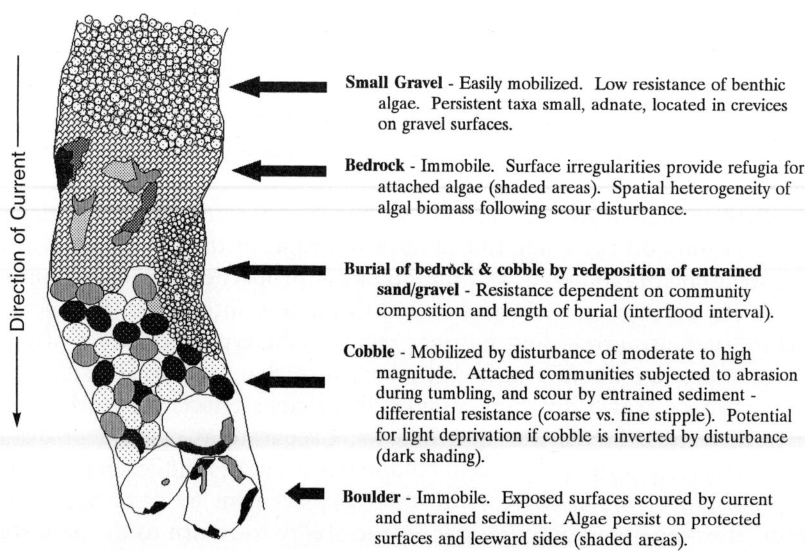

FIGURE 1 Mobilization potential, algal retention patterns, and postspate condition expected for benthic substrate of different size.

translocated by a large spate and following their simulations of spate-induced tumbling in a cement mixer. In contrast, macroscopic algal growths on sand and gravel are decimated by disturbance events that mobilize these smaller substrata (Grimm and Fisher, 1989). Significant shifts in the taxonomic structure of epipsammic and epipelic assemblages occur as a result of spate-induced changes in grain-size distribution of stream sediments (Cox, 1988; 1990). Raised surfaces of sand grains are scoured clean of algal cells during transport, with persistent individuals of prostrate species recessed within crevices on particle surfaces (Krejci and Lowe, 1986; Miller *et al.*, 1987).

In lentic systems, scour imposed by wind-induced turbulence is often less sustained and certainly less unidirectional than that associated with spates in lotic systems. Thus, scour effects on stable substrata in lakes are generally less severe than in streams. Cobble-sized substrata in lakes, if mobilized, are more likely to be simply overturned than to be tumbled repeatedly, subjecting attached algae to light deprivation. Sandy substrata in lentic systems can be stabilized by epipsammic algal/bacterial films, allowing them to withstand displacement by small increases in flow that would mobilize sediments in the absence of such films (Grant *et al.*, 1986; Dade *et al.*, 1990; Delgado *et al.*, 1991). The potential for such autogenic stabilization to reduce transport of sand and gravel in lotic systems is minimal given the sustained, directional shearing forces associated with high-discharge events in streams.

B. Community-Related Factors

1. Overview

Extremely large scour events remove the vast majority of algal biomass from the benthos, regardless of substratum size or algal community characteristics. As disturbance intensity decreases away from a high extreme, however, community characteristics become increasingly influential in dictating the immediate effects of scour. Most disturbance events, therefore, do not result in complete removal of benthic algae but, instead, leave a heterogeneous array of algal patches that differed in resistance based on their location and/or biological properties. This phenomenon is well illustrated by data from Biggs and Close (1989) (Fig. 2) on resistance of benthic algae in 9 New Zealand rivers to 46 distinct spates over a 13-month period. These investigators found that, except for conditions in which spate magnitude was large or a spate was of moderate intensity but prespate algal biomass was high, algal resistance appeared to be unrelated to disturbance intensity (delimited data points in Fig. 2).

The ability to assess benthic algal resistance can vary greatly depending on the amount of time since the last disturbance. As this time increases,

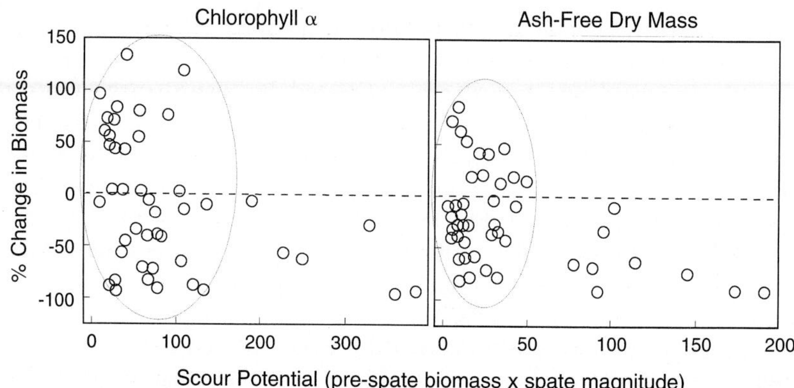

FIGURE 2 Percentage change in AFDM and chlorophyll *a* on stream cobbles plotted against scour potential for 46 spates occurring in 9 New Zealand rivers. Dotted ellipsoids demark events for which percentage change is not predictable based on spate magnitude alone. (Modified from Biggs and Close, 1989 with permission from Blackwell Scientific Publications Ltd.).

community properties reflect components of both resistance and resilience that become increasingly difficult to separate. Algal recolonization of substrata scoured clean of periphyton after a severe spate or sustained wave action is often rapid. Resistance data from the spates studied by Biggs and Close (1989) show increases in chlorophyll after 18 events and increases in ash-free dry mass (AFDM) after 12 (Fig. 2), results attributed to high algal resistance with subsequent postspate increases in biomass. Grimm and Fisher (1989) reported that large increases in diatom biomass on gravel substrata across 14 of 27 spates in a desert stream were attributable to rapid diatom recolonization within 10 d postspate, and that diatom resistance to those events was, in fact, very low.

Variation in resistance of benthic algae to scour disturbance can often be explained by examining characteristics of the community itself. In the sections that follow, I will explore the properties of benthic algal communities that confer resistance or increase susceptibility to scour disturbance.

2. Influences of Community Physiognomy and Taxonomic Representation

Benthic algal communities that exhibit highest resistance to disruption by scour do so by virtue of a low vertical profile, strong adhesion, and/or an interwoven, cohesive physiognomy. Algal taxa differ considerably in attachment strength (Harper and Harper, 1967; Pyne *et al.*, 1984; Cooksey and Cooksey, 1986), and such differences can dictate disturbance-induced change in community structure. For example, Power and Stewart (1987) reported that stronger attachment of *Rhizoclonium* relative to *Spirogyra* was responsible for the differential resistance of these two filamentous chlorophytes to a large flood in an Oklahoma stream.

Hoagland and Peterson (1990) performed transfer experiments in a large reservoir and demonstrated that an epilithic algal community dominated by stalk-forming *Cymbella* and rosette-forming species of *Fragilaria* and *Synedra* was maintained, in part, by periodic removal of motile, nonattaching diatoms (e.g., *Nitzschia* and *Navicula*) by wind-induced turbulence.

During scour events, attached algae are dislodged by shear stress imposed by elevated current, or via impact of entrained sediment particles. Given this, it is not surprising that tightly adherent, adnate taxa (e.g., *Achnanthes, Cocconeis,* small *Navicula* spp., basal cells of heterotrichous green algae) typically dominate persistent algal residues after severe scour events (Rounick and Gregory, 1981; Peterson and Hoagland, 1990; Blenkinsopp and Lock, 1994). Adnate taxa are prevalent on surfaces exposed to chronic wave disturbance (Luttenton and Rada, 1986), in epipsammic assemblages in habitats where sand grains shift frequently (Stevenson and Stoermer, 1981; Miller *et al.,* 1987), and in communities subjected to heavy grazing pressure (see Steinman, Chapter 12, this volume). Such assemblages exhibit high resistance to repeated scour disturbance (Mulholland *et al.,* 1991, Peterson and Stevenson, 1992) (Fig. 3, label 1).

The current regime in which an algal community develops strongly influences community physiognomy (see Stevenson, Chapter 11) and these attributes, in turn, affect resistance (Fig. 3, label 2). In lentic systems, or in lotic habitats with reduced flow, algal community development is not constrained by current shear, and high-biomass, flocculent assemblages are often established in which a large portion of resident cells are unattached or attached to secondary substrata (cf. Lamb and Lowe, 1987). This unanchored algal biomass is easily displaced by turbulence (Fig. 3, label 1). In areas of fast flow, current shear removes most unattached cells, leaving a community with a tightly appressed physiognomy and resident algal cells that are attached strongly to solid substrata, or protected within an attached matrix (Santos *et al.,* 1991; Biggs and Hickey, 1994; Blenkinsopp and Lock, 1994). Because of these properties, as well as some physiological differences considered in the next section, fast-current assemblages are more resistant to abrupt increases in current velocity and turbulence than are communities from more placid, slow-current regimes (Lindström and Traaen, 1984; Peterson and Stevenson, 1992; Biggs and Thomsen, 1995).

Although spatial variation in algal resistance can result from differences in environmental characteristics among habitats, temporal variation in resistance is induced by changes in community physiognomy that occur during benthic algal succession. Patterns of successional change in attached algal communities depend on a complex suite of interactions between physical habitat characteristics, allogenic biological constraints such as grazing, autogenic changes in community condition, and species representation. On horizontal surfaces in lentic systems, early-successional

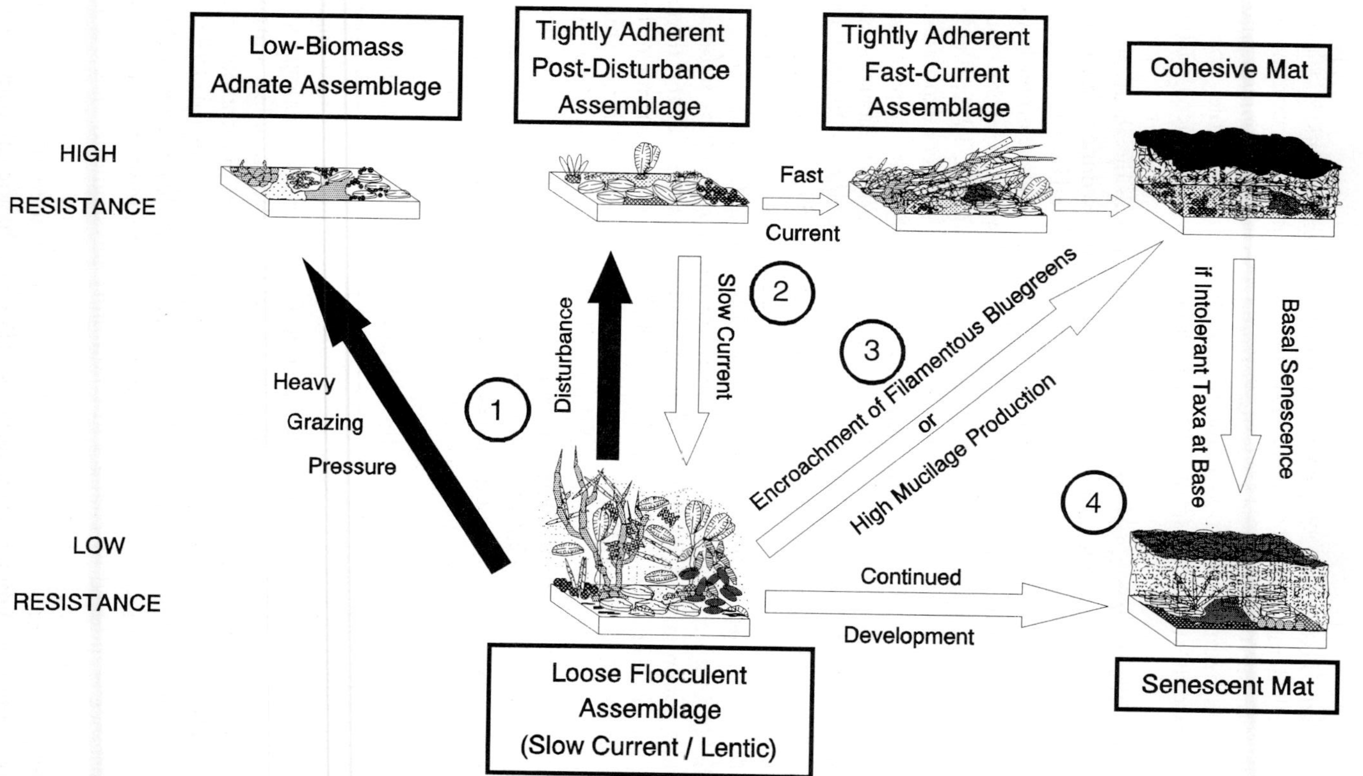

FIGURE 3 Summary of algal community physiognomies that impart high and low resistance to scour disturbance, and the factors that cause transitions between these two states. Numbers demark four transitionary stages referred to in the text.

communities are loose, flocculent, and easily disrupted by turbulence (Peterson *et al.,* 1990); because algal colonists accrue via sedimentation, attachment is not a prerequisite for colonization. This is not the case in flowing waters (Peterson and Stevenson, 1992) or on vertically oriented surfaces in reservoirs (Hoagland, 1983), where early-successional communities exhibit relatively high resistance to scour. The influence of substratum orientation can diminish with successional time, depending on disturbance regime. Cattaneo (1990) reported that in areas of lake bed exposed to frequent wave action, the amount of algal biomass on vertical and horizontal surfaces did not differ. In areas sheltered from prevailing winds, however, horizonal surfaces supported significantly higher biomass.

In placid flow regimes, changes in susceptibility with successional time depend strongly on species composition. Loose assemblages of filamentous zygnematalean green algae in these habitats, for example, are highly susceptible to dislodgement (Power and Stewart, 1987; Peterson and Stevenson, 1992), as are diatom assemblages in which mucilage-producing species are scarce (Peterson *et al.,* 1994). If stalk-forming diatoms dominate midsuccessional communities, however, the copious mucilage they produce contributes to a more cohesive community matrix, thereby increasing resistance (Roemer *et al.,* 1984; Peterson *et al.,* 1990; Biggs and Hickey, 1994). Incorporation of large amounts of detritus into mucilage-rich algal biofilms reduces this binding capacity and lowers resistance (Delgado *et al.,* 1991). Benthic mats that are most resistant to disruption by scour appear to be those dominated by an interwoven matrix of filamentous cyanobacteria, those covered by thin, cohesive, blue-green surface films (Neumann *et al.,* 1970, Peterson *et al.,* 1990, 1994), or those with relatively uniform surfaces (Blenkinsopp and Lock, 1994) (Fig. 3, label 3).

3. Influences of Senescence

In the absence of disturbance or heavy grazing activity, attached algal biomass can accumulate to produce thick, well-developed mats. As mats thicken, access of cells embedded within to light (see Hill, Chapter 5, this volume) and water-column nutrients (see Borchardt, Chapter 7, this volume) is reduced, thereby increasing the intensity of density-dependent interactions (see McCormick, Chapter 8, this volume). If such conditions persist, cells at the community base can senesce, increasing the probability that overlying algal biomass may slough (Meulemans and Roos, 1985; Power, 1992) (Fig. 3, label 4).

Temporal changes in resistance depend, in part, on the extent to which physiological condition of cells within algal mats degenerates during succession, and this varies with taxonomic representation within the basal layer and with ambient current regime. Algal mats dominated by cyanobacteria, for example, can attain very high biomass and persist for extended periods without sloughing (Peterson and Grimm, 1992), presumably because many species within this group can grow heterotrophically (see Tuchman, Chapter 10), fix atmospheric nitrogen, and engage in anoxygenic photosynthesis (Lee, 1980). In contrast, basal strata of diatom-dominated communities from placid flow regimes often show signs of late-successional senescence (e.g., Meulemans and Roos, 1985; Peterson *et al.*, 1990) (Fig. 3, label 4). Peterson and Stevenson (1992) found that resistance (measured as change in diatom taxonomic structure) to a simulated spate of algal communities developed in slow current (12 $cm \cdot s^{-1}$) inversely correlated with predisturbance community phaeophytin (a chlorophyll degradation product) content. Resistance of taxonomically similar, fast-current (29 $cm \cdot s^{-1}$) communities correlated weakly with predisturbance biomass, but showed no relationship with phaeophytin. These results suggest that physiological deterioration is more pronounced in slow-current habitats than in fast. In the latter current regime, aging, senescent cells are likely lost through export, nutrient-supply rates are higher, and fresh biomass is maintained by rapid regrowth under these physiologically richer conditions (see Stevenson, Chapter 11, this volume).

Exposure to periodic disturbance can delay senescence within algal communities and potentially increase resistance to subsequent disturbance. This phenomenon is well illustrated by data from Peterson *et al.* (1990) (Fig. 4). In that study, we exposed benthic algae on artificial substrata in a large reservoir to simulated wave disturbance at different stages of community development (disturbed after 6, 12, 18, or 24 d, or undisturbed). After 24 d, the chlorophyll biomass of these communities did not differ, but phaeophytin was significantly higher in communities that had not been recently disturbed (Fig. 4). Presumably, nutrients infused into algal mats during disturbance replenished an internal nutrient pool that had become depleted below maintenance levels. Infusion of nutrient-rich stream water into algal mats during periods of elevated discharge in artificial streams has been documented by Humphrey and Stevenson (1992) and was inferred by Stevenson (1990) following a natural spate, based on observations of large postspate reductions in diatom sexual activity (an activity induced by nutrient depletion). Similarly, McCormick (1994) reported that physical disruption of epilithic biofilms increased availability to water-column nutrients and stimulated algal growth. Release from light limitation can occur by a similar mechanism; Meulemans (1987) showed that movement of the overstory of attached algal mats by wind-induced turbulence caused transient increases in light penetration to the community base.

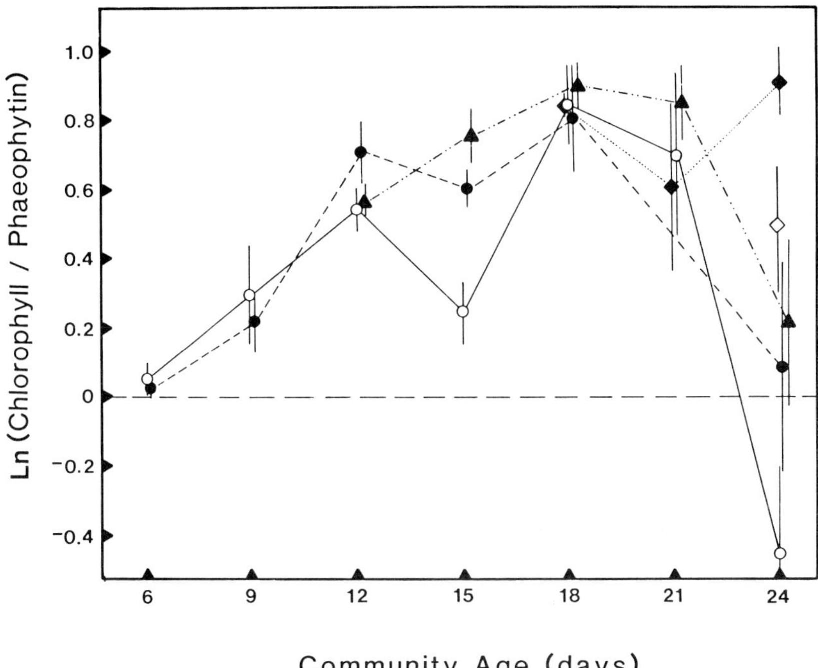

FIGURE 4 Chlorophyll: phaeophytin ratios (±1 SE) [calculated as ln(chl *a* + 1) - ln(phae +1]) from algal communities developed on artificial substrata in Lake McConaughy and subjected to simulated wave disturbance at different stages of community development. Open circles, communities not disturbed; solid circles, disturbed after 6 d of development; solid triangles, disturbed after 12 d; solid diamonds, disturbed after 18 d; open diamond, disturbed after 24 d. (From Peterson *et al.*, 1990 with permission from Blackwell Publications Ltd.)

III. FACTORS INFLUENCING RESILIENCE FOLLOWING SCOUR

Only the most severe disturbance events denude all benthic substrata of attached algae. As detailed previously, algal resistance to scour is influenced by many factors, including current regime, community age, grazing history, degree of substratum exposure, and the propensity for substratum mobilization. Thus, in most systems, postscour recovery of benthic algae initiates from multiple starting points, spatially distributed across the stream or lake bed, and dictated by the resistance of algae in residence prior to disturbance. These initial conditions from which recovery ensues range from substrata that have been scoured clean of algal biomass to surfaces that vary in the amount, condition, and taxonomic content of persistent algal residues. These patches would be expected to differ in recovery rate and pattern based on differences in activity of persistent cells and the rates of immigration and subsequent reproduction of new colonists. Though a large and growing body of information exists on algal immigration and species compositional changes in communities developing on clean substrata, relatively few studies have examined the role of persistent cells in driving postscour recovery or immigration dynamics into persistent biofilms.

A. Contribution of Persistent Cells to Recovery

1. *Change in Biomass*

Growth rates of benthic algal populations following scour events are inversely related to the amount of biomass retained on a substratum through the event (Stevenson, 1990; Uehlinger, 1991; Peterson *et al.*, 1994). The mechanism behind this pattern likely involves renewed access of cells within persistent residues to light and/or water-column nutrients. Attached algae often exhibit accelerated growth or metabolic activity following removal of overlying biomass by simulated or natural spates (Power and Stewart, 1987; Blenkinsopp and Lock, 1992; Peterson and Stevenson, 1992), or by grazing invertebrates (see Steinman, Chapter 12, this volume), suggesting release from growth-limiting conditions. Uptake of water-column nutrients by algal mats, though enhanced at higher current velocities and higher nutrient concentrations, is limited primarily to cells at the mat surface (see Borchardt, Chapter 7, this volume). Thus, communities that have lost a large percentage of their biomass to scour contain a larger percentage of residual cells at or near the biofilm surface that have access to allogenic resources; these lower-biomass communities would be expected to exhibit more rapid recovery. Exceptions to this pattern might occur in nutrient-poor systems in which maintenance of high algal standing crops relies on existence of tight nutrient-recycling pathways within algal mats (Mulholland *et al.*, 1991). Once disrupted by scour, such pathways are not quickly reestablished and, without adequate external nutrient supplies, recovery to predisturbance biomass levels would be slow.

Recovery rate and the ultimate benthic algal biomass attained (barring further disturbance) vary with availability of allogenic nutrients. Short-term water-column enrichment, as often occurs during spates (e.g., Grimm, 1992), apparently has little long-term effect on algal resilience (Humphrey and Stevenson, 1992; Peterson *et al.*, 1994). Prolonged enrichment, however, does influence recovery, although this relationship can be mitigated by taxonomic constraints, as will be discussed later. Lohman *et al.* (1992) tracked algal biomass accrual for 42 d in three streams of differing trophic status after concurrent spates in all systems reduced benthic chlorophyll *a* below 0.1 $mg{\cdot}m^{-2}$. Algal biomass in highly enriched and moderately enriched systems increased steadily during the 6-wk recovery period, whereas biomass accrual in a nutrient-poor stream plateaued between 14 and 28 d. Similar relationships occur *within* systems owing to spatial variability in hydrologic exchange patterns between subsurface and surface waters. Valett *et al.* (1994) reported that postspate accrual of algal biomass (as chlorophyll *a*) in Sycamore Creek, a desert stream in central Arizona, was significantly higher in upwelling areas, where nutrient-rich hyporheic water percolated through the gravel streambed into the surface stream, than in zones of downwelling or no exchange. In lentic systems, growth of attached algae can be greatly enhanced by nutrients supplied via wind-induced turbulence, accounting for luxuriant algal growths on the windward side of islands (Cattaneo, 1990) and wave-exposed littoral zones of large lakes (Reuter *et al.,* 1986).

2. Importance of Species-Specific Patterns

Studies of benthic algal resilience typically employ biomass measures to assess recovery. Taxonomic representation and performance of algal populations within persistent residues are less frequently considered, yet variation at this level can affect recovery patterns and such data can, at times, be essential for accurate interpretation of patterns observed at coarser levels of resolution. For example, Peterson *et al.* (1994) noted no increase in epilithic algal biomass following a large scouring spate in Sycamore Creek, Arizona, despite prolonged (2 wk) spate-induced nitrate enrichment of the water column. When taxa within the three major algal divisions were analyzed separately, however, we found that densities of diatom and green algal species did, in fact, increase significantly but that density of filamentous blue-green algae, which comprised the bulk of algal biomass, remained unchanged, thus accounting for the nonresponse of total biomass. If resistant taxa have autecological characteristics that dictate reliance on recycled nutrients and that allow slow, steady growth in the low-resource environment of late-successional mats (cf. Stevenson *et al.,* 1991), then these taxa would not likely respond rapidly when reexposed to resource-replete conditions. Many cyanobacteria seem to fit this pattern. However, if the strategy employed by basal taxa is one of tolerance of low-resource availability, rather than growth, exposure by scour should stimulate growth.

Recovery patterns of lotic algal communities may vary depending on autecological differences among persistent taxa and may be further influenced by local variation in current velocity. Peterson and Stevenson (1990) estimated *in situ* reproduction of cells that had remained attached through a simulated spate in outdoor artificial streams and compared postspate cell accrual from this source to that attributable to immigration and reproduction of new colonists in both fast- (29 $cm \cdot s^{-1}$) and slow-current (12 $cm \cdot s^{-1}$) channels. As expected, assemblages of persistent cells were dominated by taxa that adhered tightly to the substratum with mucilaginous attachment structures, but interspecific variation in cell-accumulation rates between flow regimes suggested that mode of attachment affected retention of newly produced sibling cells. *Achnanthes minutissima,* though present in the highest densities on these newly scoured substrata, accrued much more rapidly in slow current than in fast. Accrual rates of *Meridion circulare* and *Gomphonema angustatum,* in contrast, did not differ between flow regimes (Fig. 5). *Meridion* and *Gomphonema* extrude mucilage from apical pore fields that, upon division, cement both parent and sibling cell to the substratum. *Achnanthes* spp. adhere strongly to surfaces along their raphe valve (Rosowski *et al.,* 1986) or by means of a subapical mucilaginous stalk (Roemer *et al.,* 1984). Newly produced sibling cells of *Achnanthes,* therefore, are not directly attached to the substratum (see Fig. 5) and could be more prone to removal in areas of high current shear (Rosowski *et al.,* 1986). Although *A. minutissima* can reach high densities in areas of fast flow (Duncan and Blinn, 1989), these apparent interspecific differences in propensity for emigration may dictate successional patterns early in community development, before microeddies produced by accrual of new colonists begin to shield dividing cells from shear forces (see Stevenson, Chapter 11, this volume).

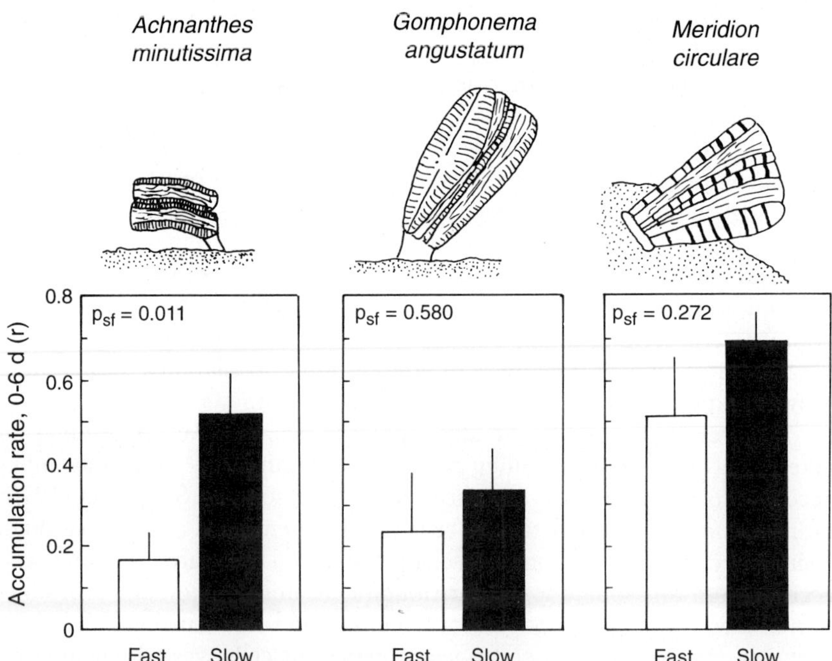

FIGURE 5 Immigration-corrected accumulation rates (+1 SE) of three diatom taxa, which differ in mode of attachment, in fast- (29 cm•s^{-1}) and slow-current (12 cm•s^{-1}) artificial stream channels over 6 d immediately following a simulated spate that left a low-biomass persistent assemblage. (Data from Peterson and Stevenson, 1990.)

B. Contribution of New Immigrants to Recovery

The importance of immigration in dictating the rate and pattern of recovery varies with the amount of persistent biomass retained on substrata after scour, the characteristics (i.e., density, species composition, physiological condition) of the immigration pool, and distance of disturbed substrata from a source of immigrants. Obviously, patches of substrata that are scoured clean of algal biomass will recolonize via immigration alone, since no persistent cells would remain to contribute to recovery. In contrast, communities that have retained much of their biomass through the scour event are less affected by immigration. An influx of new cells, even at relatively high immigration rates (e.g., >5000 cells$\cdot$cm$^{-2}\cdot$d^{-1}), would contribute little to biomass accrual in communities where cell densities exceed 10^6 cells$\cdot$cm^{-2} (cf. Stevenson, 1986).

The surfaces of algal mats are not inert. Thus, interactions between biofilm surfaces and algal immigrants can affect which taxa have greatest success following immigration. For example, Peterson and Stevenson (1989) found that in slow-current habitats, *Nitzschia acicularis,* a lightly silicified diatom, proliferated rapidly on clean substrata, but grew much more slowly on substrata, in the same flow regime, that supported a thin organic biofilm. Presumably, bacteria within the organic film outcompeted *N. acicularis* for nutrients. These effects were not evident on substrata in fast current, where water flow likely increased supply of nutrients or, perhaps, removed newly produced cells via export. Because of interspecific differences in immigration and competitive abilities, a single algal drift pool will likely produce multiple immigration assemblages based on both physical and biological substratum characteristics. Insightful research has been conducted on the influences of physical substratum characteristics on algal immigration success and subsequent reproduction (e.g., Stevenson, 1983). The potential significance of interactions between new immigrants and the biological component of benthic substrata is an area of research that deserves more attention.

With the exception of filamentous blue-green algae, which are highly motile (Donker *et al.,* 1993) and can rapidly encroach onto cleared areas from adjacent substrata, the water column provides the proximate source of algal colonists of newly denuded surfaces (cf. Hoagland *et al.,* 1986). Some disturbances—dislodgement of algal patches by wandering cattle, for example—have little or no effect on the algal immigration pool. Recolonization of these areas is strongly influenced by the taxonomic content (cf. McCormick and Stevenson, 1991) and proximity (Kaufman, 1980) of neighboring algal assemblages.

Disturbances that act on a larger scale, such as wind-induced wave activity in lentic systems and spates in streams and rivers, can significantly alter the density and taxonomic content of the immigration pool. It is also likely that suspension of benthic algae into the water column enhances cell productivity, perhaps on a species-specific basis, although few studies have directly examined this question (but see Shaffer and Sullivan, 1988). In lentic systems, sustained winds beyond a critical velocity (which varies among systems based on basin morphometry and fetch) greatly increase densities of benthic algae in the water column (Demers *et al.*, 1987; de Jonge and van Beusekom, 1992), much of which is rapidly redeposited on the benthos (Gabrielson and Lukatelich, 1985; Hoagland *et al.*, 1986; Peterson and Hoagland, 1990). Unattached diatoms, such as *Nitzschia, Navicula,* and chain-forming *Fragilaria,* are most susceptible to displacement (Cattaneo and Kalff, 1978; Peterson *et al.*, 1990). Sediment-trap data show, however, that suspended *Navicula* and *Fragilaria* resettle rapidly, replacing cells removed by turbulence, whereas lightly silicified *Nitzschia* remain in suspension much longer (Peterson *et al.*, 1990). Such differential resettlement can alter recovery patterns and affect community structure.

The effects of spates on the immigration pool in lotic systems are less well known, and likely less predictable. Unlike wave disturbance in lakes, in which benthic algae are suspended into a volume of water that does not change, stream volume can increase greatly during spates, potentially diluting concentrations of suspended algae and flushing them downstream. Hamilton and Duthie (1987) reported storm-induced shifts in species composition of the drift pool in three eastern Canadian rivers, but found drift densities to be positively correlated with stream discharge only in the two smaller systems. Cazaubon (1988) recorded large increases in algal drift densities during a spate in southern France. In contrast, Uehlinger (1991) reported that chlorophyll *a* and ash-free dry mass in the seston of a Swiss mountain stream were minimal immediately following a flood. Stevenson (1990) speculated that the slow algal immigration rates he measured after a large spate in a Kentucky stream may have resulted from depletion of the algal drift pool via washout of potential immigrants from the system.

In some systems, recolonization of severely scoured substrata is extremely rapid. In Sycamore Creek, for example, a visible film of diatoms is apparent within 1–2 d of recession of spates that scour and rework the large expanses of pea-gravel substrata in that system (Fisher *et al.*, 1982; C. G. Peterson, personal observation). Such rapid recolonization is certainly not due to reproduction of persistent cells, nor does it likely stem, at least directly, from immigration of cells suspended by the spate itself. More likely, these patterns of postspate cell accrual are the product of a cascade effect whereby emigrants from the rapidly reproducing initial immigration assemblage augment the drift pool and accelerate recovery via reimmigration. Stevenson (1990) reported a large increase in immigration activity 9 d after a spate in a Kentucky stream, relative to that measured just after the event. I have measured a similar lag following a large summer spate in Sycamore Creek, concurrent with an increase in diatom drift density (C. G.

Peterson, unpublished data). Such cascade effects are plausible under conditions favorable for diatom growth, following disturbance events that thoroughly scour a large percentage of the streambed. This effect should be less pronounced after spates that leave a large amount of attached algae in place, since reproduction and subsequent emigration from these pockets of persistent biomass would not be as great as that observed on clean substrata.

C. Model for Temporal Change in Resilience Based on Change in Community Condition

Just as benthic algal resistance changes with successional time, it is also likely that autogenic changes that occur during community development would affect recovery. The model presented in Figure 6 illustrates one possible mechanism for such change: specifically, that postscour recovery potential of attached algal communities will change through successional time because of differential tolerance of species at the community base to reductions in availability of allogenic resources.

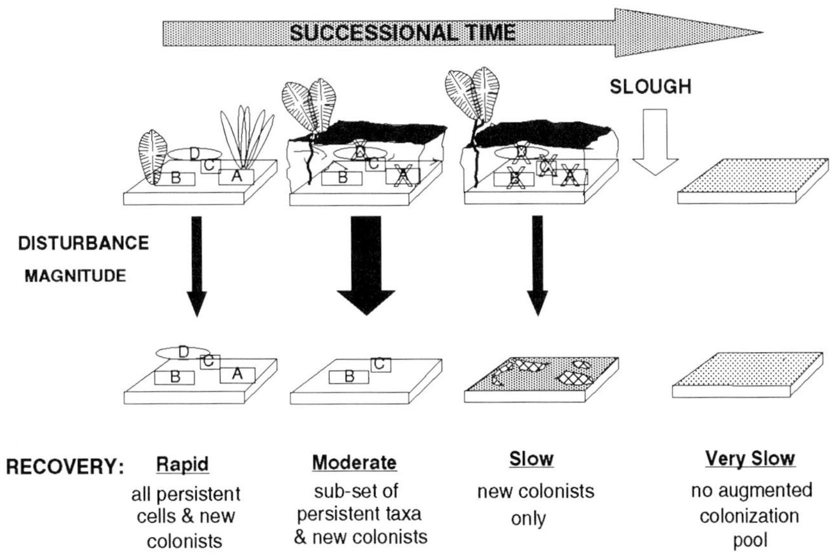

FIGURE 6 Proposed model for successional changes in recovery potential of cells within basal strata of attached algal communities following scour disturbance competent to remove all overlying biomass. The disturbance magnitude required to expose basal cells (indicated by arrow width) varies over successional time because resistance of algal communities to disruption by disturbance changes as with stage of community development (Peterson *et al.*, 1990; Peterson and Stevenson, 1992). Four taxa are depicted in the basal layer (species A–D). An "X" over a species signifies an inability to survive resource limitation at a given stage of development. "Slough" refers to an autogenically induced senescence and detachment of all basal cells.

There is good reason to believe that autogenic changes in microenvironmental conditions in basal strata would generate temporal variation in the recovery potential of cells residing therein. We know that access of basal cells to light and water-column nutrients decreases as algal mats thicken. If disturbance removes overstory biomass before such limitation becomes too severe, all persistent taxa within the community's basal layer (species A–D, left panel, Fig. 6) should be viable and contribute to recovery; immigration of drifting cells scoured from substrata upstream would also contribute. In the absence of disturbance or heavy grazing activity, continued biomass accrual eventually leads to senescence and detachment of basal cells, resulting in sloughing of algal biomass. Recovery from sloughing events can be very slow (Moore, 1976; Biggs and Close, 1989), presumably because no persistent cells survive to drive recovery (right panel, Fig. 6). Additionally, detachment induced by senescence would not be accompanied by an increase in algal drift densities, since sloughing typically occurs in discrete, localized patches across the benthos, rather than simultaneously from all stream substrata, as would occur with large-scale scour disturbance.

It is unlikely that the onset of senescence is simultaneous for all taxa, but rather that different species lose viability (or, if motile, leave the basal stratum) at different points in community development because of autecological differences. For example, we know that algal taxa differ in their ability to proliferate under different spectral regimes (see Hill, Chapter 5, this volume), and in their ability to survive prolonged light deprivation (see Tuchman, Chapter 10, this volume). Variation in patterns of succession under different nutrient regimes suggests that similar interspecific differences in tolerance thresholds exist for autogenically induced nutrient limitation as well. Given this, it follows that the subset of taxa surviving at the community base should diminish as time between disturbance events increases (middle panels, Fig. 6), culminating in sloughing when the threshold for the most tolerant taxon is surpassed.

The hypothesis detailed here has yet to be tested, but its consideration opens a number of avenues for further experimentation that would yield insight into the mechanisms behind variation in postdisturbance recovery patterns of benthic algae. For example, if species-specific tolerance thresholds exist, do resultant changes in species representation within persistent algal residues influence community properties later in recovery? Do species-specific tolerance thresholds differ among current regimes and, if so, can they be altered by changes in water-column nutrient concentrations? An understanding of these mechanisms would greatly enhance our ability to interpret temporal and spatial variation in algal response to disturbance in natural systems, and has important implications for predicting potential consequences of changes in disturbance regimes, whether induced by direct human intervention (e.g., as with regulation of lotic systems) or by changes in global climate patterns.

IV. FACTORS INFLUENCING RESPONSE TO EMERSION

The ability to resist long-term desiccation has been reported for a wide variety of freshwater microalgae, from all major classes. Davis (1972) summarized published accounts of desiccation resistance for >130 species of Chlorophyta, 8 Charophyta, over 60 species of Cyanophyta, 36 diatom species, and 23 taxa from other groups. Viability of zygotes, akinetes, cysts, and vegetative cells upon rewetting was noted after months, years, and decades of desiccation; one specimen of *Nostoc commune* reportedly withstood more than a century of dry storage on an herbarium sheet. Despite this impressive compilation, cell survival during prolonged periods of desiccation is by no means certain. Hostetter and Hoshaw (1970) recovered culturable diatoms from approximately 50% of 91 sediment samples that had been dry for 2 yr, whereas only 5 of 42 samples (12%) from the same site but dried for 6 yr yielded viable diatoms. Attempts by Hawes *et al.* (1992) to culture algal cells from 10-cm^2 sections of cyanobacterial mats that had desiccated on an Antarctic ice shelf for 3 yr yielded only 2–5 live colonies per section; these authors surmised, however, that even a small pool of surviving cells could serve as a propagule source to reestablish mats following pond refilling. Desiccation-tolerant cells, blown by wind or buried in dry sediments, likely contribute substantially to reestablishment of benthic algae in newly wetted basins following prolonged dry periods.

Benthic microalgae employ a variety of mechanisms to withstand short-term (e.g., seasonal or diurnal) emersion; the degree of success varies among taxa and with community structure and condition, and depends on the rate at which drying proceeds. The onset of drying can induce physiological changes in algal cells that enhance resistance. Desiccation-induced changes in the filamentous green alga *Klebsormidium rivulare,* reported by Morison and Sheath (1985), included large increases in cell-wall thickness and in cellular concentrations of low-molecular-weight solutes (and consequent changes in osmotic potential) and a shift from a protein-synthesis metabolism to one of carbohydrate and lipid accumulation. Hostetter and Hoshaw (1970) noted that cells within diatom cultures that were dried gradually exhibited much higher desiccation resistance than those dried rapidly, suggesting that physiological acclimation is time-dependent. Interestingly, slow-dried cells exhibited much greater tolerance to high temperatures than did nondesiccated cells.

Some algal taxa appear less able to withstand desiccation via physiological means, but instead escape the physiological stress of emersion in refugia within complex algal mats. Hawes *et al.* (1992) reported that Antarctic cyanobacterial mats dominated by *Nostoc* dried completely within 5 h of initial atmospheric exposure, but could return to predesiccation rates of photosynthesis within just 10 min of rewetting, regardless of the length of emersion. Clearly, rapid recovery of these communities arose from high physiological resistance. In contrast, mats dominated by the filamentous cyanobacterium *Phormidium* recovered much more slowly (>10 d), with recovery apparently driven by migration of trichomes from the mat interior to the surface, rather than by renewed growth of surface trichomes.

Resistance to periodic emersion of *Cladophora*-dominated communities in tailwaters below hydroelectric dams is similarly influenced by community physiognomy, as water trapped within the mass of highly branched filaments of this green alga protects cells at the interior from desiccation (Usher and Blinn, 1990). Angradi and Kubly (1993), in a series of manipulative experiments on natural *Cladophora* communities below Glen Canyon Dam, Arizona, determined that the extent of chlorophyll *a* degradation and associated reductions in gross primary productivity were directly related to the number of daylight hours these communities were exposed to the atmosphere. Primary productivity per unit chlorophyll, however, was unaffected by exposure length, again indicating that a portion of the community was shielded from the destructive effects of emersion.

Extracellular polysaccharides (i.e., mucilage) produced by many benthic microalgae are thought to serve multiple functions, among them enhanced resistance to desiccation (Hoagland *et al.*, 1993). Mucilage produced by *Gloeocystis*, a colonial blue-green alga, can serve as a reservoir to keep cells hydrated during short periods of atmospheric exposure (Shephard, 1987), making modification of internal physiology unnecessary. Peterson (1987) presented data indicating that mucilage production within epilithic, riverine algal communities was stimulated by short-term (4 h) emersion, and suggested that this higher mucilage content should confer increased resistance to repeated desiccation. On substrata exposed to fast, direct current, this higher mucilage content enhanced cell accrual, presumably by increasing retention of immigrants and decreasing rates of cell export. Communities developed in placid flow were much less resistant to short-term desiccation than those developed in fast current. In fast-current communities, the same properties that conferred resistance to scour (i.e., a tightly adherent, cohesive physiognomy) also increased resistance to desiccation, presumably due to a greater capacity for water retention compared to that of the more loosely associated communities developed in sheltered habitats.

Algal communities with physiognomies that slow water loss and shield resident algal cells from exposure to direct solar radiation can confer resistance to short-term, infrequent emersion. In habitats subjected to chronic exposure, however, persistent communities must be composed of species with physiological tolerance. Thus, *Cladophora*, a taxon tolerant of periodic emersion, is replaced by the filamentous cyanobacterium *Oscillatoria* in shallow areas below hydroelectric dams that are exposed daily by dam-induced discharge fluctuations (Angradi and Kubly, 1993).

V. FACTORS INFLUENCING RESPONSE TO LIGHT DEPRIVATION AND BURIAL

Disturbance-induced light deprivation of attached algae can occur in two ways. Suspension of fine sediments by wind-generated turbulence in lentic systems, or flood events in streams and rivers, can greatly increase turbidity in the water column and block penetration of light to the benthos and affect algal productivity (Burkholder and Cuker, 1991; Hellström, 1991). Such events are typically short-lived and should not result in major taxonomic shifts. The effects of algal burial, and subsequent light deprivation, by redeposition of suspended material can be more long-lasting and can induce significant changes in both productivity and species composition.

Deposition of fine sediments over benthic algae following recession of flood waters can have significant deleterious short-term effects on algal productivity, with the degree to which productivity is compromised directly related to the amount of sediment deposited (Power, 1990). Power made these measurements after she had uncovered buried algae, suggesting that the observed reduction in productivity was physiologically based and not just a function of lowered light availability. In most systems, the probability of extended burial by fine sediments is relatively low since such particles are easily resuspended by the activity of fish (Power, 1990) or invertebrates (Pringle *et al.,* 1993), or by minor increases in discharge. Moreover, the probability of disturbance-induced burial by fine sediments can vary temporally, depending on disturbance regime. Successive spates can flush fine particulates out of a system, lowering the probability of postspate siltation of the benthos (C. G. Peterson, personal observation).

Benthic algae attached to overturned cobbles, or buried under coarse sediments that had been mobilized by turbulence and redeposited, are more likely to be subjected to extended periods of light deprivation. If algae remain buried indefinitely, they eventually decompose and enter the detrital pool. Alternatively, additional disturbance events can reexpose these cells after a length of time no shorter than the return interval of another substratum-mobilizing disturbance. To withstand extended burial, algal cells must maintain viability under conditions of light deprivation and, in many cases, anoxia; these are conditions that can develop, autogenically, at the base of undisturbed periphyton mats (cf. Carlton and Wetzel, 1987). Algae exhibit considerable interspecific variation in their ability to persist under prolonged darkness, and in the mechanisms they employ toward this end (see Tuchman, Chapter 10, this volume). The same taxa capable of tolerating conditions of low or no light at the base of late-successional algal mats are also likely to exhibit high resistance to disturbance-induced light deprivation.

Although the ability of benthic algae to resist and recover from such disturbance has not been extensively studied, a varied body of evidence suggests a potential for species-specific resistance that would, in turn, affect recovery patterns upon reexposure. Wasmund (1989) used autoradiography to demonstrate that algal cells extracted from sediment cores could resume photosynthesis upon reexposure to light; variation in autoradiographic grain density among species from different algal classes suggested differential resistance based on taxonomic identity. Wasmund also reported that gross primary productivity and chlorophyll/phaeophytin ratios decreased with the sediment depth from which samples were extracted, suggesting limits to the amount of time that buried algae can maintain viability. Steinman *et al.* (1990, 1991) covered laboratory streams in black plastic for 92 d to examine response of benthic algae to extended light deprivation. Chlorophyll *a* decreased in these channels by only 37–60% over the 92 d of darkness, but reductions in chlorophyll-specific carbon fixation rates, relative to predisturbance measurements, indicated that persistent algae were physiologically stressed by the disturbance. These authors noted significant reductions in filamentous algae (e.g., *Spirogyra, Mougeotia,* chain-forming *Fragilaria*) with prolonged light deprivation and high resistance of prostrate taxa, primarily basal cells of the heterotrichous green alga *Stigeoclonium.* Entwisle (1989) also reported high survivability of *Stigeoclonium* and *Cladophora* to extended darkness (≤3 and 6 months, respectively). In *Stigeoclonium,* however, high resistance to light deprivation was noted only in filaments containing basal cells—individuals without basal attachment structures died within a month.

Resistance of benthic algae to burial by sediments may be less dependent on duration of light deprivation than on the onset of anoxia. The long-term persistence of chlorophyll *a* noted by Steinman *et al.* (1990) occurred in flow-through stream channels carrying oxygenated water to light-deprived channel sections. Oxygen limitation greatly increases the probability of extensive sloughing in attached bacterial biofilms (Applegate and Bryers, 1991). The same would be expected of algal biofilms. Many algal taxa tolerate transient anoxic conditions that periodically develop within thick algal mats—conditions that are abated when photosynthetic activity in the upper strata resumes during the day (Carlton and Wetzel, 1987). Apparently, few can withstand extended periods of anoxia. Moss (1977) reported that motile diatoms, typical of epipelic habitats, were extremely intolerant to anoxia. Epipsammic diatoms, though more tolerant, could not survive beyond 22 d under these conditions. Similarly, Poulícková (1987) reported that, although she collected many viable algae from hyporheic sediments beneath the Morava River and was able to culture them *in situ,* few taxa survived after her buried culture vessels became anoxic after 36 d of incubation. These observations suggest that the length of time that buried algae can remain viable should vary with the extent and direction of interstitial water movement. In areas of downwelling, in which oxygen-rich surface water infuses into sediment, buried algae may exhibit higher resistance than in stagnant areas or in zones of upwelling of low-oxygen groundwater. Similarly, algae attached to overturned cobbles might persist for longer periods if some water flow across the overturned surface is maintained.

VI. CONCLUDING REMARKS

Disturbance is a key factor determining pattern and process in freshwater benthic algal communities, but the precise nature of this influence is complex and not easily compartmentalized into discrete categories of effects. A single disturbance event can have many consequences depending on the biological characteristics of the communities it touches, and these characteristics are shaped by both internal and external forces. Studies conducted in artificial streams, or those employing simulations of natural disturbance events *in situ,* have been and will continue to be extremely useful in understanding the mechanisms behind algal community response to disturbance. However, given the dramatic effects that many natural disturbance events have on the algal immigration pool, nutrient regime, and bed morphometry, it is imperative that manipulative studies of response mechanisms be tied to field studies in natural systems of natural disturbance events.

Relatively little attention has been paid to the influence of disturbance history in dictating the response of algal communities to future disturbance, or in shaping their structure and dictating their functional attributes. Despite the rapid generation times of the organisms involved, algal communities in natural systems can carry a significant legacy of past disturbance; the persistent algal residues that withstand one disturbance can contribute to recovery and influence community response to the next. Synergistic effects between disturbance types are, most likely, also common. An attached algal community exposed to short-term emersion or elevated concentrations of a chemical pollutant, for example, will likely respond quite differently to an increase in stream discharge than a community of the same biomass, developed under the same current regime, but not so exposed. Finally, seasonal changes in the timing of disturbance can influence algal communities at a number of levels. Power (1992) suggested that the effects of spring floods on establishment of populations of invertebrate grazers in northern California rivers greatly influenced benthic algal dynamics throughout the summer that followed. Such disturbance-related interactions between invertebrate and algal population dynamics are, no doubt, quite common and transcend seasonal and annual boundaries. Seasonal timing of disturbance should also affect algal communities directly, based on differences in the available algal colonization pool at the time of the disturbance. I contend that an understanding of these historical effects could contribute much to our understanding of spatial and temporal heterogeneity in the structure and function of benthic algal communities.

REFERENCES

Angradi, T. R., and Kubly, D. M. (1993). Effects of atmospheric exposure on chlorophyll *a,* biomass, and productivity of the epilithon of a tailwater river. *Regul. Rivers: Res. Manage.* **8,** 345–358.

Applegate, D. H., and Bryers, J. D. (1991). Effects of carbon and oxygen limitations and calcium concentrations on biofilm removal processes. *Biotechnol. Bioeng.* **37,** 17–25.

Biggs, B. J. F., and Close, M. E. (1989). Periphyton biomass dynamics in gravel bed rivers: The relative effects of flows and nutrients. *Freshwater Biol.* **22,** 209–231.

Biggs, B. J. F., and Hickey, C. W. (1994). Periphyton responses to a hydraulic gradient in a regulated river in New Zealand. *Freshwater Biol.* **32,** 49–59.

Biggs, B. J. F., and Thomsen, H. A. (1995). Disturbance of stream periphyton by perturbations in shear stress: Time to structural failure and differences in community resistance. *J. Phycol.* **31,** 233–241.

Blenkinsopp, S. A., and Lock, M. A. (1992). Impact of storm-flow on electron transport system activity in river biofilms. *Freshwater Biol.* **27,** 397–404.

Blenkinsopp, S. A., and Lock, M. A. (1994). The impact of storm-flow on river biofilm architecture. *J. Phycol.* **30,** 807–818

Blinn, D. W., and Cole, G. A. (1991). Algal and invertebrate biota in the Colorado River: Comparison of pre- and post-dam conditions. *In* "Colorado River Ecology and Dam Management," pp. 102–123. National Academy Press, Washington, DC.

Blinn, D. W., Shannon, J. P., Stevens, L. E., and Carder, J. P. (1995). Consequences of fluctuating discharge for lotic communities. *J. N. Am. Benthol. Soc.* **14,** 233–248.

Burkholder, J. M., and Cuker, B. E. (1991). Response of periphyton communities to clay and phosphate loading in a shallow reservoir. *J. Phycol.* **27,** 373–384.

Carlton, R. G., and Wetzel, R. G. (1987). Distributions and fates of oxygen in periphyton communities. *Can. J. Bot.* **65,** 1031–1037.

Cattaneo, A. (1990). The effect of fetch on periphyton spatial variation. *Hydrobiologia* **206,** 1–10.

Cattaneo, A., and Kalff, J. (1978). Seasonal changes in the epiphyte community of natural and artificial macrophytes in Lake Memphremagog (Que. and Vt.). *Hydrobiologia* **60,** 135–144.

Cazaubon, A. (1988). Crue et communautés diatomiques (Moyenne-Durance, Provence). *Cah. Biol. Mar.* **29,** 233–246.

Cooksey, K. E., and Cooksey, B. (1986). Adhesion of fouling diatoms to surfaces: Some biochemistry. *In* "Algal Biofouling" (L. V. Evans and K. D. Hoagland, eds.), pp. 41–53. Elsevier Press, Amsterdam.

Cox, E. J. (1988). Has the role of the substratum been underestimated for algal distribution patterns in freshwater ecosystems? *Biofouling* **1,** 49–63.

Cox, E. J. (1990). Studies on the algae of a small softwater stream. III. Interaction between discharge, sediment composition and diatom flora. *Arch. Hydrobiol. Suppl. Monogr. Beitr.* **83,** 567–584.

Dade, W. B., Davis, J. D., Nichols, P. D., Nowell, A. R. M., Thistle, D., Trexler, M. B., and White, D. C. (1990). Effects of bacterial exopolymer adhesion on the entrainment of sand. *Geomicrobiol. J.* **8,** 1–16.

Davis, J. S. (1972). Survival records in the algae, and the survival role of certain algal pigments, fat, and mucilaginous substances. *Biologist* **54,** 52–93.

Davis-Colley, R. J., Hickey, C. W., Quinn, J. M., and Ryan, P. A. (1992). Effects of clay discharges on streams. I. Optical properties and epilithon. *Hydrobiologia* **248,** 215–234.

de Jonge, V. N., and van Beusekom, J. E. E. (1992). Contribution of resuspended microphytobenthos to total phytoplankton in the Ems Estuary and its possible role for grazers. *Neth. J. Sea Res.* 30, 91–105.

Delgado, M., de Jonge, V. N., and Peletier, H. (1991). Experiments on resuspension of natural microphytobenthos populations. *Mar. Biol. (Berlin)* **108,** 321–328.

Demers, S., Therriault, J.-C., Bourget, E., and Bah, A. (1987). Resuspension in the shallow sublittoral zone of a macrotidal estuarine environment: Wind influence. *Limnol. Oceanogr.* **32,** 327–339.

Donker, V. A., Amewowor, D. H. A. K., and Häder, D.-P. (1993). Effects of tropical solar radiation on the motility of filamentous cyanobacteria. *FEMS Microbiol. Ecol.* **12,** 143–148.

Douglas, B. (1958). The ecology of the attached diatoms and other algae in a small stony stream. *J. Ecol.* **46,** 295–322.

Dudley, T. L., and D'Antonio, C. M. (1991). The effects of substrate texture, grazing, and disturbance on macroalgal establishment in streams. *Ecology* **72,** 297–309.

Duncan, S. W., and Blinn, D. W. (1989). Importance of physical variables on the seasonal dynamics of epilithic algae in a highly shaded canyon stream. *J. Phycol.* **25,** 455–461.

Englund, G. (1991). Effects of disturbance on stream moss and invertebrate community structure. *J. North Am. Benthol. Soc.* **10,** 143–153.

Entwisle, T. J. (1989). Phenology of the *Cladophora-Stigeoclonium* community in two urban creeks of Melbourne. *Aust. J. Mar. Freshwater Res.* **40,** 471–489.

Estrada, M., Alcaraz, M., and Marrasé, C. (1987). Effects of turbulence on the composition of phytoplankton assemblages in marine microcosms. *Mar. Ecol.: Prog. Ser.* **38,** 267–281.

Evans, J. H. (1958). The survival of freshwater algae during dry periods. Part I. An investigation of the algae of five small ponds. *J. Ecol.* **46,** 149–167.

Fahrig, L. (1990). Interacting effects of disturbance and dispersal on individual selection and population stability. *Comments Theor. Biol.* **1,** 275–297.

Fisher, S. G., and Grimm, N. B. (1991). Streams and disturbance: Are cross-ecosystem comparisons useful? *In* "Comparative Analyses of Ecosystems: Patterns, Mechanisms, and Theories" (J. Cole, G. Lovett, and S. Findlay, eds.) pp. 196–221. Springer-Verlag New York.

Fisher, S. G., Gray, L. J., Grimm, N. B., and Busch, D. E. (1982). Temporal succession in a desert stream ecosystem following flash flooding. *Ecol. Monogr.* **52,** 93–110.

Gabrielson, J. O., and Lukatelich, R. J. (1985). Wind-related resuspension of sediments in the Peel–Harvey estuarine system. *Estuarine, Coastal Shelf Sci.* **20,** 135–145.

Grant, J., Bathmann, U. V., and Mills, E. L. (1986). The interaction between benthic diatom films and sediment transport. *Estuarine, Coastal Shelf Sci.* **23,** 225–238.

Grimm, N. B. (1992). Biogeochemistry of nitrogen in Sonoran Desert streams. *J. Ariz.-Nev. Acad. Sci.* **26,** 139–155.

Grimm, N. B., and Fisher, S. G. (1989). Stability of periphyton and macroinvertebrates to disturbance by flash floods in a desert stream. *J. North Am. Benthol. Soc.* **8,** 293–307.

Hamilton, P. B., and Duthie, H. C. (1987). Relationship between algal drift, discharge and stream order in a boreal forest watershed. *Arch. Hydrobiol.* **110,** 275–289.

Harper, M. A., and Harper, J. F. (1967). Measurement of diatom adhesion and their relationship with movement. *Br. Phycol. Bull.* **3,** 195–207.

Hawes, I., Howard-Williams, C., and Vincent, W. F. (1992). Desiccation and recovery of Antarctic cyanobacterial mats. *Polar Biol.* **12,** 587–594.

Hellström, T. (1991). The effect of resuspension on algal production in a shallow lake. *Hydrobiologia* **213,** 183–190.

Hoagland, K. D. (1983). Short-term standing crop and diversity of periphytic diatoms in a eutrophic reservoir. *J. Phycol.* **19,** 30–38.

Hoagland, K. D., and Peterson, C. G. (1990). Effects of light and wave disturbance on vertical zonation of attached microalgae in a large reservoir. *J. Phycol.* **26,** 450–457.

Hoagland, K. D., Zlotsky, A., and Peterson, C. G. (1986). The source of algal colonizers on rock substrates in a freshwater impoundment. *In* "Algal Biofouling" (L. V. Evans and K. D. Hoagland, eds.), pp. 21–39. Elsevier, Amsterdam.

Hoagland, K. D., Rosowski, J. R., Gretz, M. R., and Roemer, S. C. (1993). Diatom extracellular polymeric substances: Function, fine structure, chemistry, and physiology. *J. Phycol.* **29,** 537–566.

Horner, R. R., Welch, E. B., Seeley, M. R., and Jacoby, J. M. (1990). Responses of periphyton to changes in current velocity, suspended sediment and phosphorus concentration. *Freshwater Biol.* **24,** 215–232.

Hostetter, H. P., and Hoshaw, R. W. (1970). Environmental factors affecting resistance to desiccation in the diatom *Stauroneis anceps. Am. J. Bot.* **57,** 512–518.

Humphrey, K. P., and Stevenson, R. J. (1992). Responses of benthic algae to pulses in current and nutrients during simulations of subscouring spates. *J. North Am. Benthol. Soc.* **11,** 37–48.

Huston, M. (1979). A general hypothesis of species diversity. *Am. Nat.* **113,** 81–101.

Huston, M. (1994). "Biological Diversity: The Coexistence of Species on Changing Landscapes." Cambridge Univ. Press, Cambridge, UK.

Kaufman, L. H. (1980). Stream aufwuchs accumulation processes: Effects of ecosystem depopulation. *Hydrobiologia* **70,** 75–81.

Kobayashi, D., Suzuki, K., and Nomura, M. (1990). Diurnal fluctuation in stream flow and in specific electric conductance during drought periods. *J. Hydrol.* **115,** 105–114.

Krejci, M. E., and Lowe, R. L. (1986). Importance of sand grain mineralogy and topography in determining micro-spatial distribution of epipsammic diatoms. *J. North Am. Benthol. Soc.* **5,** 211–220.

Lamb, M. A., and Lowe, R. L. (1987). Effects of current velocity on the physical structuring of diatom (Bacillariophyceae) communities. *Ohio J. Sci.* **87,** 72–78.

Lee, R. E. (1980). "Phycology." Cambridge Univ. Press, Cambridge, UK.

Lindström, E.-A., and Traaen, T. S. (1984). Influence of current velocity on periphyton distribution and succession in a Norwegian soft water river. *Verh.—Int. Ver. Theor. Angew. Limnol.* **22,** 1965–1972.

Lohman, K., Jones, J. R., and Perkins, B. D. (1992). Effects of nutrient enrichment and flood frequency on periphyton biomass in northern Ozark streams. *Can. J. Fish. Aquat. Sci.* **49,** 1198–1205.

Luettich, R. A., Jr., Harleman, D. R. F., and Somlyódy, L. (1990). Dynamic behavior of suspended sediment concentrations in a shallow lake perturbed by episodic wind events. *Limnol. Oceanogr.* **35,** 1050–1067.

Luttenton, M. R., and Rada, R. G. (1986). Effects of disturbance on epiphytic community architecture. *J. Phycol.* **22,** 320–326.

McCormick, P. V. (1994). Evaluating the multiple mechanisms underlying herbivore–algal interactions in streams. *Hydrobiologia* **291,** 47–59.

McCormick, P. V., and Stevenson, R. J. (1991). Mechanisms of benthic algal succession in lotic environments. *Ecology* **72,** 1835–1848.

Meulemans, J. T. (1987). A method for measuring selective light attenuation within a periphytic community. *Arch. Hydrobiol.* **109,** 139–145.

Meulemans, J. T., and Roos, P. J. (1985). Structure and architecture of the periphytic community on dead reed stems in Lake Maarsseveen. *Arch. Hydrobiol.* **102,** 487–502.

Miller, A. R., Lowe, R. L., and Rotenberry, J. T. (1987). Succession of diatom communities on sand grains. *J. Ecol.* **75,** 693–709.

Moore, J. W. (1976). Seasonal succession of algae in rivers. I. Examples from the Avon, a large slow-flowing river. *J. Phycol.* **12,** 342–349.

Morison, M. O., and Sheath, R. G. (1985). Responses to desiccation stress by *Klebsormidium rivulare* (Ulotrichales, Chlorophyta) from a Rhode Island stream. *Phycologia* **24,** 129–145.

Moss, B. (1977). Adaptations of epipelic and epipsammic freshwater algae. *Oecologia* **28,** 103–108.

Mulholland, P. J., Steinman, A. D., Palumbo, A. V., DeAngelis, D. L., and Flum, T. E. (1991). Influence of nutrients and grazing on the response of stream periphyton communities to a scour disturbance. *J. North Am. Benthol. Soc.* **10,** 127–142.

Neumann, A. C., Gebelein, C. D., and Scoffin, T. P. (1970). The composition, structure and erodability of subtidal mats, Abaco, Bahamas. *J. Sediment. Petrol.* **40,** 274–297.

Odum, E. P., Finn, J. T., and Franz, E. H. (1979). Perturbation theory and the subsidy–stress gradient. *BioScience* **29,** 349–352.

Peterson, C. G. (1987). Influences of flow regime on development and desiccation response of lotic diatom communities. *Ecology* **68,** 946–954.

Peterson, C. G., and Grimm, N. B. (1992). Temporal variation in enrichment effects during periphyton succession in a nitrogen-limited desert stream ecosystem. *J. North Am. Benthol. Soc.* **11,** 20–36.

Peterson, C. G., and Hoagland, K. D. (1990). Effects of wind-induced turbulence and algal mat development on epilithic diatom succession in a large reservoir. *Arch. Hydrobiol.* **118,** 47–68.

Peterson, C. G., and Stevenson, R. J. (1989). Substratum conditioning and diatom colonization in different current environments. *J. Phycol.* **25,** 790–793.

Peterson, C. G., and Stevenson, R. J. (1990). Post-spate development of epilithic algal communities in different current regimes. *Can. J. Bot.* **73,** 2092–2102.

Peterson, C. G., and Stevenson, R. J. (1992). Resistance and resilience of lotic algal communities: Importance of disturbance timing and current. *Ecology* **73,** 1445–1461.

Peterson, C. G., Hoagland, K. D., and Stevenson, R. J. (1990). Timing of wave disturbance and the resistance and recovery of a freshwater epilithic microalgal community. *J. North Am. Benthol. Soc.* **9,** 54–67.

Peterson, C. G., Weibel, A. C., Grimm, N. B., and Fisher, S. G. (1994). Mechanisms of benthic algal recovery following spates: Comparison of simulated and natural events. *Oecologia* **98,** 280–290.

Pickett, S. T. A., Kolasa, J., Armesto, J. J., and Collins, S. L. (1989). The ecological concept of disturbance and its expression at various hierarchical levels. *Oikos* **54,** 129–136.

Pieczyńska, E., and Banas, D. (1984). Zonation in periphyton colonizing concrete blocks on the shore of Lake Zegrzyńskie. *Ekol. Pol.* **32,** 533–552.

Poff, N. L., and Ward, J. V. (1990). Physical habitat template of lotic systems: Recovery in the context of historical pattern of spatiotemporal heterogeneity. *Environ. Manage.* **14,** 629–645.

Poulícková, A. (1987). Algae in ground waters below the active stream of a river (basin of the Morava River, Czechoslovakia). *Arch. Hydrobiol., Suppl.* **78** (*Algol. Stud.* 46) 65–88.

Power, M. E. (1990). Resource enhancement by indirect effects of grazers: Armored catfish, algae, and sediment. *Ecology* **71,** 897–904.

Power, M. E. (1992). Hydrologic and trophic controls of seasonal algal blooms in northern California rivers. *Arch. Hydrobiol.* **125,** 385–410.

Power, M. E., and Stewart, A. J. (1987). Disturbance and recovery of an algal assemblage following flooding in an Oklahoma stream. *Am. Midl. Nat.* **117,** 333–345.

Pringle, C. M., Blake, G. A., Covich, A. P., Buzby, K. M., and Finley, A. (1993). Effects of omnivorous shrimp in a montane tropical stream: Sediment removal, disturbance of sessile invertebrates and enhancement of understory algal biomass. *Oecologia* **93,** 1–11.

Pyne, S., Fletcher, R. L., and Jones, E. B. G. (1984). Attachment studies of three common fouling diatoms. *Proc. Int. Congr. Mar. Corros. Foul., Mar. Biol. Vol.,* Athens, Greece, *1984,* pp. 99–112.

Reuter, J. E., Loeb, S. L., and Goldman, C. R. (1986). The physiological ecology of nuisance algae in an oligotrophic lake. *In* "Algal Biofouling" (L. V. Evans and K. D. Hoagland, eds.), pp. 115–127. Elsevier, Amsterdam.

Roemer, S. C., Hoagland, K. D., and Rosowski, J. R. (1984). Development of a freshwater periphyton community as influenced by diatom mucilages. *Can. J. Bot.* **62,** 1799–1813.

Rosowski, J. R., Roemer, S. C., Palmer, J., and Hoagland, K. D. (1986). Extracellular association and adaptive significance of the bas-relief mucilage pad of *Achnanthes lanceolata* (Bacillariophyceae). *Diatom Res.* **1,** 113–129.

Rounick, J. S., and Gregory, S. V. (1981). Temporal changes in periphyton standing crop during an unusually dry winter in streams of the Western Cascades, Oregon. *Hydrobiologia* **83,** 197–205.

Santos, R., Callow, M. E., and Bott, T. R. (1991). The structure of *Pseudomonas fluorescens* biofilms in contact with flowing systems. *Biofouling* **4,** 319–336.

Shaffer, G. P., and Sullivan, M. J. (1988). Water column productivity attributable to displaced benthic diatoms in well-mixed shallow estuaries. *J. Phycol.* **24,** 132–140.

Shephard, K. L. (1987). Evaporation of water from the mucilage of a gelatinous algal community. *Br. Phycol. J.* **22,** 181–185.

Sousa, W. P. (1984). Intertidal mosaics: Patch size, propagule availability, and spatially variable patterns of succession. *Ecology* **65,** 1918–1935.

Stanley, E. H., and Fisher, S. G. (1992). Intermittency, disturbance, and stability in stream ecosystems. *In* "Aquatic Ecosystems in Semi-arid Regions: Implications for Resource Management" (R. D. Robarts and M. L. Bothwell, eds.), N.H.R.I. Symp. Ser. 7, pp. 271–280. Environment Canada, Saskatoon.

Steinman, A. D., Mulholland, P. J., Palumbo, A. V., Flum, T. E., Elwood, J. W., and DeAngelis, D. L. (1990). Resistance of lotic ecosystems to a light elimination disturbance: A laboratory stream study. *Oikos* **58,** 80–90.

Steinman, A. D., Mulholland, P. J., Palumbo, A. V., Flum, T. E., and DeAngelis, D. L. (1991). Resilience of lotic ecosystems to a light-elimination disturbance. *Ecology* **72,** 1299–1313.

Stevenson, R. J. (1983). Effects of current and conditions simulating autogenically changing microhabitats on benthic diatom immigration. *Ecology* **64,** 1514–1524.

Stevenson, R. J. (1986). Mathematical model of epilithic diatom accumulation. *In* "Proceedings of the Eighth International Diatom Symposium" (M. Ricard, ed.), pp. 323–335. Koeltz Scientific Books, Koenigstein, Germany.

Stevenson, R. J. (1990). Benthic algal community dynamics in a stream during and after a spate. *J. North Am. Benthol. Soc.* **9,** 277–288.

Stevenson, R. J., and Stoermer, E. F. (1981). Quantitative differences between benthic algal communities along a depth gradient in Lake Michigan. *J. Phycol.* **17,** 29–36.

Stevenson, R. J., Peterson, C. G., Kirschtel, D. B., King, C. C., and Tuchman, N. C. (1991). Density-dependent growth, ecological strategies, and effects of nutrients and shading on benthic diatom succession in streams. *J. Phycol.* **27,** 59–69.

Uehlinger, U. (1991). Spatial and temporal variability of the periphyton biomass in a prealpine river (Necker, Switzerland). *Arch. Hydrobiol.* **123,** 219–237.

Usher, H. D., and Blinn, D. W. (1990). Influence of various exposure periods on the biomass and chlorophyll *a* of *Cladophora glomerata* (Chlorophyta). *J. Phycol.* **26,** 244–249.

Valett, H. M., Fisher, S. G., Grimm, N. B., and Camill, P. (1994). Vertical hydraulic exchange and ecological stability of a desert stream ecosystem. *Ecology* **75,** 548–560.

Vincent, W. F., and Howard-Williams, C. (1986). Antarctic stream ecosystems: Physiological ecology of a blue-green algal epilithon. *Freshwater Biol.* **16,** 219–233.

Wasmund, N. (1989). Live algae in deep sediment layers. *Int. Rev. Gesamten Hydrobiol.* **74,** 589–597.

Wilson, S. D., and Tilman, D. (1991). Interactive effects of fertilization and disturbance on community structure and resource availability in an old-field plant community. *Oecologia* **88,** 61–71.

Young, O. W. (1945). A limnological investigation of periphyton in Douglas Lake, Michigan. *Trans. Am. Microsc. Soc.* **64,** 1–20.

14

Ecotoxicology of Inorganic Chemical Stress to Algae

Robert Brian Genter

Johnson State College,
Johnson, Vermont 05656

I. INTRODUCTION

Inorganic chemical stress affects algae at biochemical, cellular, population, and community levels of biological organization. Adsorption and uptake of inorganic chemicals by the cell involves metabolism-independent and -dependent phases that precede the stress response by the alga and that have practical application for biological monitoring and wastewater metal remediation of anthropogenic wastes. Effects of inorganic chemicals on the cell influence growth rate, development, and abundance of algal populations that may be reflected in changes in relative abundance in communities. Algae differ in tolerance to stress owing to the activity of inorganic chemicals on intracellular and cell-surface binding sites. Chemical, physical, and biological factors influence the relative toxicity of inorganic chemicals to algae, so effects at the community level of biological organization may be quite complex.

Most elements of the periodic table are classified as metals that are characterized by their opacity, ductility, conductivity, and luster. Metalloids (B, Si, Ge, As, Sb, Te) and nonmetals (the upper right portion of the periodic table) comprise smaller groups (Silberberg, 1996). The "heavy metals" are a group of approximately 40 elements with a density greater than 5. Whitton (1984) advocates continued use of the term "heavy metal" for pollution studies, despite its lack of precision, to better communicate with administrators and the general public. Most elements occur as compounds in which two or more elements are chemically combined. The compounds toxic to algae include different chemical species of metals and also nonmetallic compounds like ozone and cyanide. All elements and compounds are potentially toxic, but an interesting characteristic of some metals is that they are nutrients at low concentrations (e.g., Cu, Mo, Fe, and Zn).

Very thorough reviews of inorganic chemical stress to algae were performed by Rai *et al.* (1981) and by Whitton (1984). More specific reviews of individual inorganic stressors to algae have been written for Cd (Wong, 1987), Cr (Wong and Trevors, 1988), Al (Havas and Jaworski, 1986), chlorine (Steinman *et al.,* 1992), As (Blanck *et al.,* 1989), and ozone (Heath, 1994). Algae also have been used as model systems for plant cells containing solid walls (Heath, 1994; Puiseux-Dao, 1989). Comparative reviews of the environmental chemistry, accumulation, and toxicity of different inorganic stressors to algae and other organisms appear in a number of publications (Eisler, 1985a,b, 1986, 1987; Moore and Ramamoorthy, 1984; Moore, 1991).

II. ADSORPTION AND UPTAKE

Algae concentrate inorganic ions to amounts several thousandfold greater than in external dilute solutions by a variety of biological, chemical, and physical mechanisms involving adsorption, precipitation, and metabolism-dependent processes (Gadd, 1988). Inorganic chemicals can be concentrated by living cells, dead cells, and biochemical products like excreted metabolites, polysaccharides, and constituents of the cell surfaces (Wong *et al.*, 1984). A significant proportion of algal biomass may contain metals if the chemical conditions are optimal. *Anabaena cylindrica* accumulated 2.4 and 3.3% of its dry weight after 6 and 24 h, respectively, when exposed to 190 μ*M* Al. Accumulated Sn was 10% of the dry cell mass for 3 mg liter^{-1} (dry weight) *Ankistrodesmus falcatus* in 6 mg liter^{-1} Sn (Wong *et al.*, 1984). The final concentrations of Zn and Mn in a "cyanobacteria mat" exposed to 14 mg liter^{-1} of each element were 10.13 and 10.30 mg g^{-1}, respectively (Bender *et al.*, 1994).

A. Two Phases of Uptake

Microbial metal uptake often follows two phases (Rai *et al.*, 1981; Gadd, 1988; Cho *et al.*, 1994; Collard and Matagne, 1994): first is a rapid metabolism-independent phase with binding or adsorption to cell walls and external surfaces; second is a slower metabolism-dependent phase with transport across the cell membrane. Either phase may be obscured with actively growing cells owing to metabolism-effected changes in the external growth medium, excretion of substances that complex or precipitate metals, or by internal detoxification mechanisms. Hence, in microbial systems, several mechanisms may operate simultaneously or in sequence (Gadd, 1988). The relative contribution of these two uptake mechanisms, and the specificity of uptake sites to particular metals, influences why periphyton removed Ni, Cr, Fe, and Mn continuously and Cu, Pb, Cd, and Co rapidly during the first two hours of exposure and then only slightly thereafter (Vymazal, 1984).

1. Metabolism-Independent Uptake

The metabolism-independent phase (also called biosorption) involves the accumulation of inorganic chemicals to cell-wall components. Biosorption of metals is often rapid, reversible, and usually complete in 5–10 min in algae (Gadd, 1988; Zhang and Majidi, 1994). Biosorption is generally not influenced by light, temperature, or the presence of metabolic inhibitors (Garnham *et al.*, 1992). Killed or metabolically inactive algae have been used to estimate the magnitude of biosorption, but disruption of membrane permeability must be considered. Most metal accumulated this way is easily removed by washing algae with distilled water alone or with a chelator (EDTA).

Biosorption of metals to algae may involve two steps (Roy *et al.*, 1993): (1) diffusion of chemical through a liquid film to the cell surface and interskeletal spaces, and (2) binding to specific binding sites with adsorption capacity leveling off as these sites are occupied. The exact relationship can be affected by other cations, especially protons, which indicates that metals adsorb to algal surfaces by an ion-exchange phenomenon based on electrostatic attractions to negatively charged sites (Gadd, 1988; Crist *et al.*, 1988, 1990, 1992).

Biosorption is generally less important than metabolism-dependent accumulation. Biosorption accounted for 33, 25, and 50% of the total amounts of Co, Mn, and Zn accumulated by *Chlorella salina*, respectively (Garnham *et al.*, 1992). Algal species differ in the relative importance of biosorption to total uptake primarily because of characteristics of binding sites on cell walls and extracellular matrices (Gadd, 1988).

Accumulation of some inorganic ions consists solely of the single rapid metabolism-independent phase. This may be a characteristic of elements that are not micronutrients, like the accumulation of pertechnetate ions by blue-green algae (Garnham *et al.*, 1993b), zirconium by green algae and blue-green algae (Garnham *et al.*, 1993c), and tributyltin by blue-green algae (Avery *et al.*, 1993). The uptake process may be more complicated than this simple two-phase model, for the green alga *Chlorella emersonii* continued to accumulate tributyltin into a second phase that was not considered to be metabolism-dependent intracellular accumulation (Avery *et al.*, 1993).

2. *Metabolism-Dependent Uptake*

The metabolism-dependent phase is often slower (lasting hours or days) and is inhibited by low temperatures, absence of energy sources (light), metabolic inhibitors, and uncouplers, and may be influenced by the health of the cells and the characteristics of the growth medium (Gadd, 1988; Garnham *et al.*, 1992). An important route for intracellular uptake, which deserves further study, is by passive diffusion owing to increased permeability of cell membranes from stress (Gadd, 1988).

Some nonessential metals are taken up by transport systems intended for essential metals or nutrients with similar chemical structure, but there are still many questions regarding the transport mechanisms involved (Gadd, 1988). For example, arsenate competes with phosphate for membrane uptake sites and interferes with phosphate metabolism, but Kuwabara (1992) found that creek periphyton showed preferential cell sorption of orthophosphate over arsenate despite results of abiotic sorption experiments that demonstrated the similar absorptive characteristics of these two chemicals.

B. Sites of Accumulation

Accumulation may be "adsorption" (or biosorption) at the cell surface (wall, membrane, or external mucilage) or internal "absorption" to organelles, cytoplasmic ligands, and cytoplasmic structures.

1. Cell Surfaces

Metals bind to specific functional groups including hydroxyl (—OH), phosphoryl (—PO_3O_2), amino (—NH_2), carboxyl (—COOH), sulfhydryl (—SH), and thiol groups (Rai *et al.*, 1981). For example, *Chlorella vulgaris* appears to have more adsorption sites for Zn(II) than for Cd(II), with carboxyl groups being the major sites of metal adsorption to the cell wall and with amine groups playing some role (Cho *et al.*, 1994). Rijstenbil *et al.* (1994) noted that stress may increase surface area and roughness of algal cell walls so that metal adsorption capacity increases.

2. Organelles and Subcellular Components

Once inorganic stressors are inside cells, the ions may bind to intracellular components or precipitate (Gadd, 1988). Biological macromolecules and enzymes with appropriate functional groups or metal cofactors are impacted by metal activity. For example, the distribution of Sn in *Ankistrodesmus falcatus* was about 85% in the cellular polysaccharide fraction, 15% in the protein fraction, and 0.2% in the lipid and low-molecular-weight metabolite fractions (Wong *et al.*, 1984). Metals may be detoxified by accumulating in polyphosphate bodies and to intracellular metal-binding proteins of blue-green algae and eukaryotic algae (Zhang and Majidi, 1994) and within vacuoles of some eukaryotic algae (Gadd, 1988; Garnham *et al.*, 1992). Accumulation of metals may damage cellular ultrastructure (Prevot and Soyer-Gobillard, 1986; Puiseux-Dao, 1989).

The presence of one inorganic chemical can change the distribution of others among cellular components. For example, mixtures of Cd and Cr influenced each other's concentration and distribution among membrane, cell wall, soluble, and miscellaneous fractions in cells of *Chlorella ellipsoidea* (Okamura and Aoyama, 1994).

C. General Factors Affecting Uptake

The magnitude of chemical adsorption and uptake is influenced by (1) exposure duration (Nyholm and Källqvist, 1989; Aziz and Ng, 1993; Pascucci and Sneddon, 1993); (2) growth rate, for the concentration of Zn transported into cells of *Chlamydomonas variabilis* increased linearly during the exponential growth phase, but during the stationary phase Zn transport into the cell either reached a plateau, increased linearly, or declined as Zn concentrations increased, respectively (Bates *et al.*, 1985); (3) membrane permeability, for Heath (1994) notes that ozone stress damages the plasmalemma, leading to decreased permeability, disrupted ion distributions, and consequent impairment of photosynthesis; and (4) molecular mass, for biosorption of organotin compounds increased with molecular mass (Avery *et al.*, 1993). The environmental factors affecting uptake will be described later in this chapter.

D. Applied Phycology

1. Biomonitoring

Because of their ability to concentrate metals from dilute aqueous solutions, marine and freshwater algae have great potential as biological monitors for anthropogenic wastes (Rai *et al.*, 1981). Algae tend to accumulate metals at levels representative of the water and comparable to other organisms (Flegal *et al.*, 1993; Stronkhorst *et al.*, 1994). The rate of metal accumulation depends on the selectivity and abundance of binding sites, other metals in solution, and the proximity to sediments (Ramelow *et al.*, 1987). Repeated sampling may be necessary for environmental studies, since changes in water current, nutrient content, and species compositions may alter the metal composition of algae (Flegal *et al.*, 1993). Stephenson and Turner (1993) treated the epilimnion of oligotrophic Lake 382 with 90 ng liter^{-1} Cd and found that equilibrium concentrations of Cd were reached in 2 weeks in water and periphyton.

Bioconcentration factors (BCFs) have been a useful tool for regulatory purposes (e.g., Eisler, 1985a, 1987; Moore, 1991). The BCF is the ratio of the metal concentration in algal biomass (μg g^{-1} dry weight) to the concentration of the metal in the surrounding water (μg ml^{-1}). Skowroński (1986) cautions that this ratio is sensitive to changes in pH, relative biomass of algae in relation to water, and other factors influencing equilibrium concentrations.

Chemical analysis of water by itself may not be sufficient for assessing environmental stress because periphyton can decrease dissolved metal concentrations to background levels (Genter and Amyot, 1994), so measuring inorganic chemical composition of organisms like algae is necessary for environmental assessment.

2. Wastewater Metal Remediation

Algal species tolerant of metal stress are potentially useful for decreasing toxic concentrations of inorganic stressors in polluted waters and for recovering metals for reuse (Gadd, 1988; Wilde and Benemann, 1993; Brady *et al.*, 1994). Living filamentous algae (green algae and blue-green algae) and immobilized cell systems (living or dead) have potential for commercial operation (Gadd, 1988).

Characteristics of specific binding sites determine metal accumulation. For instance, Brady *et al.* (1994) explained that the higher (83–99%) accumulation of Cr^{3+} (as well as Cu^{2+} and Pb^{2+}) over the anionic $Cr2O_7^{2-}$ (8.3–22%) indicated that there are more binding sites for cations than for anions. Bender *et al.* (1994) found that "cyanobacteria mats" developed for metal tolerance and exposed to Zn and Mn each at 14 mg liter^{-1} in mixed solution could remove 96% of the Zn and 85% of the Mn in 3 h. Roy *et al.* (1993) found that dried and pulverized *Chlorella minutissima* quickly removed cationic Cd, Pb, and Zn, and anionic As and Cr, but that cationic Co and Ni were not removed as effectively.

The metabolism-dependent and -independent phases of uptake are important to consider for commercial operations. Garnham *et al.* (1993a) found *Scenedesmus obliquus* to accumulate more Co and Cs than the cyanobacterium *Synechocystis* PCC 6803 in the presence of three clay minerals; initial metal uptake was unaffected by darkness and a metabolic inhibitor, but a slower energy-dependent phase of uptake followed that was inhibited by darkness and a metabolic inhibitor.

Algae are sometimes as effective as commercial resins for removing metals from wastewater. Fehrmann and Pohl (1993) compared 15 algal species (when dry) from six classes in four divisions (in accordance with Sze, 1986) to three standard absorbing materials and found that algae adsorbed more Cd than the standard industrial materials. Holan *et al.* (1993) compared Cd uptake by biomass of nonliving, dried samples of *Sargassum natans, Fucus vesiculosus,* and *Ascophyllum nodosum* and found *A. nodosum* to accumulate the highest amount of Cd, to be reusable, and to outperform a commercial ion-exchange resin. Holan and Volesky (1994) found these same seaweeds to have excellent sorption properties for Pb (Ni was an order of magnitude less), but that a commercial ion-exchange resin had higher Pb uptake than the dried algae.

Chemical characteristics of some metals and solutions decrease the ability of algae to remove metals. Robinson and Wilkinson (1994) found that volatilization of Hg may have decreased bioaccumulation by immobilized algal cell particles. Environmental factors that affect metal uptake will be discussed in the next section.

III. EFFECTS OF INORGANIC CHEMICAL STRESS ON ALGAE

All metals, even those that are nutrients at low concentrations, are toxic to algae at high concentrations (Rai *et al.,* 1981). Toxicologists use acronyms to indicate effects of stressors to aquatic organisms (Rand and Petrocelli, 1985), and these are used in Table I. The LOEC (lowest observed effect concentration) is the lowest chemical concentration to have a significant effect on the organism. The NOEC (no observed effect concentration) is the highest chemical concentration which does not have a significant effect on the organism. The MATC (maximum acceptable toxicant concentration) is an estimate of the toxic threshold and is written as the range between the NOEC and the LOEC. Finally, the "effective concentration" (EC) and the "lethal concentration" (LC) are estimates of the chemical concentrations to have the effect (a decrease in the measured response for EC, or lethality for LC) on a percentage of the population or biological response being measured for tests that have been conducted

TABLE I Reported Concentrations of Inorganic Stressors That Had Adverse Effects (Including the LOEC, EC, and LC) or No Adverse Effects (MATC, NOEC) on Marine and Freshwater Algae[a]

Metal	*Concentration*	*Effects*	*Organism*	*Reference*
Aluminum	15 μ*M*	Growth rate reduced at pH 6	*Asterionella ralfsii* var. *americana*	Gensemer (1991)
	3.7 μ*M*	LOEC for inhibition of growth and nitrogenase activity at pH 6.0 (nitrogenase activity recovered after 100 h as the freguency of heterocysts increased)	*Anabaena cylindrica*	Pettersson *et al.* (1985)
	180 μ*M*	120 h EC_{50} for growth at pH 6.0	*Anabaena cylindrica*	Pettersson *et al.* (1985)
	50 μg $liter^{-1}$	LOEC for decreased abundance at pH 5.7	Freshwater phytoplankton	Pillsbury and Kingston (1990)
Arsenic	80 μg $liter^{-1}$	LOEC for germ tube growth and nuclear migration in gametophytes; inhibition of nuclear migration was not reversible	*Macrocystis pyrifera*	Garman *et al.* (1994)
	0.4 μ*M*	1 h EC_{20} for photosynthesis inhibition	Natural marine periphyton	Blanck and Wängberg (1988a)
	0.6 μ*M*	1 h EC_{20} for photosynthesis inhibition	Marine periphyton established indoors in flow-through aquaria	Blanck and Wängberg (1988a)
		3 wk EC_{20}'s for:		
	0.2 μ*M*	Species composition	Marine periphyton established indoors in flow-through aquaria	Blanck and Wängberg (1988a)
	0.3 μ*M*	Chlorophyll *a*		
	0.5 μ*M*	Total carbon		
	0.8 μ*M*	Total nitrogen		
	0.3 μ*M*	Photosynthesis		

Cadmium	8.9 μ*M*	ATP/ADP ratio increased	*Synechocystis aquatilis*	Pawlik and Skowroński (1994)
	42 μ*M*	Predominantly competitive inhibition of Ca-ATPase activity in cell extract; noncompetitive inhibition of ATPase activity in plasma membrane fractions	*Dunaliella bioculata*	Jeanne *et al.* (1993)
	0.076 mg $liter^{-1}$	NOEC, 24 h, growth inhibition	*Chlamydomonas reinhardtii*	Schäfer *et al.* (1994)
	0.038 mg $liter^{-1}$	NOEC, 72 h, growth inhibition	*Chlamydomonas reinhardtii*	Schäfer *et al.* (1994)
	0.789 mg $liter^{-1}$	EC_{50}, 72 h, growth inhibition	*Chlamydomonas reinhardtii*	Schäfer *et al.* (1994)
	0.069 mg $liter^{-1}$	NOEC, 24 h, effective photosynthesis rate	*Chlamydomonas reinhardtii*	Schäfer *et al.* (1994)
	0.006 mg $liter^{-1}$	NOEC, 24 h, growth inhibition	*Scenedesmus subspicatus*	Schäfer *et al.* (1994)
	0.011 mg $liter^{-1}$	NOEC, 72 h, growth inhibition	*Scenedesmus subspicatus*	Schäfer *et al.* (1994)
	0.032 mg $liter^{-1}$	EC_{50}, 72 h, growth inhibition	*Scenedesmus subspicatus*	Schäfer *et al.* (1994)
	0.009 mg $liter^{-1}$	NOEC, 24 h, effective photosynthesis rate	*Scenedesmus subspicatus*	Schäfer *et al.* (1994)
	4.55 μ*M*	EC_{50}, 96 h, growth	*Dunaliella salina*	Visviki and Rachlin (1994)
	0.025 μ*M*	EC_{50}, 96 h, growth	*Chlamydomonas bullosa*	
	30 and 100 μg $liter^{-1}$	48-h exposure, decreased growth and photosynthesis, higher percentage of carbon synthesis to polysaccharides than to lipid and protein	*Selenastrum capricornutum*	Thompson and Couture (1993)
	15 ppm	Considerable increase in protein and carbohydrate, slight increase in lipid	*Isochrysis galbana*	Wikfors *et al.* (1994)
	60 ppm	Considerable increase in carbohydrate and lipid, decrease in protein	*Dunaliella teriolecta*	Wikfors *et al.* (1994)

(continues)

TABLE I (continued)

Metal	*Concentration*	*Effects*	*Organism*	*Reference*
	60 ppm	Increase in protein, but considerable decrease in carbohydrate and lipid	*Phaeodactylum tricornutum*	Wikfors *et al.* (1994)
	0–22.8 μg $liter^{-1}$	"Safe" concentration (MATC) for female growth and presence of cystocarps (sexual reproduction)	*Champia parvula*	Steele and Thursby (1983)
Chromium	1 mg $liter^{-1}$	Approximate 72 and 168 h EC_{50} for growth; decreased chlorophyll, carotenoid, and protein content, oxygen evolution, and nitrate reductase activity	*Glaucocystis nostochinearum*	Rai *et al.* (1992)
	Many concentrations ($K_2Cr_2O_7$)	72 h EC_{10}'s and EC_{50}'s for photosynthesis inhibition and growth inhibition; diatoms generally most sensitive; sensitivities varied among species	*Phaeodactylum tricornutum, Skeletonema costatum, Thalassiosira pseudonana, Rhodomonas baltica, Dunaliella bioculata,* and natural phytoplankton	Kusk and Nyholm (1992)
Copper	120 μg $liter^{-1}$	Increased proportion of the fatty acids oleate and linoleate	*Fucus vesiculosus*	Jones and Harwood (1993)
	30 μg $liter^{-1}$	Increased proportion of the fatty acids oleate and decreased stearate and myristate	*Ascophyllum nodosum*	

20 and 50 μ*M*	Cu-resistant strain also more resistant to Cd, Zn, and Ni; Cu-resistant strain bound less metal at low concentration and more metal at high concentration than the sensitive strain	*Anabaena variabilis*	Hashemi *et al.* (1994)
200 ppb	Growth stimulated	*Oocystis pusilla*	Chang and Sibley (1993)
>400 ppb	Growth depressed		
20.2–42.8 μg liter^{-1}	LOEC for community structure	Periphyton microcosm	Balczon and Pratt (1994)
24.0–98.5 μg liter^{-1}	LOEC for community structure	"Littoral" periphyton microcosm	Balczon and Pratt (1994)
42.8–310.3 μg liter^{-1}	LOEC for community process	Periphyton microcosm	Balczon and Pratt (1994)
304.7 μg liter^{-1}	Toxicity to community process	"Littoral" periphyton microcosm	Balczon and Pratt (1994)
0.023 mg liter^{-1}	NOEC, 24 h, growth inhibition	*Chlamydomonas reinhardtii*	Schäfer *et al.* (1994)
0.005 mg liter^{-1}	NOEC, 72 h, growth inhibition	*Chlamydomonas reinhardtii*	Schäfer *et al.* (1994)
0.079 mg liter^{-1}	EC_{50}, 72 h, growth inhibition	*Chlamydomonas reinhardtii*	Schäfer *et al.* (1994)
0.636 mg liter^{-1}	NOEC, 24 h, effective photosynthesis rate	*Chlamydomonas reinhardtii*	Schäfer *et al.* (1994)
0.048 mg liter^{-1}	NOEC, 24 h, growth inhibition	*Scenedesmus subspicatus*	Schäfer *et al.* (1994)
0.056 mg liter^{-1}	NOEC, 72 h, growth inhibition	*Scenedesmus subspicatus*	Schäfer *et al.* (1994)
0.120 mg liter^{-1}	EC_{50}, 72 h, growth inhibition	*Scenedesmus subspicatus*	Schäfer *et al.* (1994)
0.041 mg liter^{-1}	NOEC, 24 h, effective photosynthesis rate	*Scenedesmus subspicatus*	Schäfer *et al.* (1994)
5.94 μ*M*	EC_{50}, 96 h, growth	*Dunaliella salina*	Visviki and Rachlin (1994)
0.78 μ*M*	EC_{50}, 96 h, growth	*Chlamydomonas bullosa*	
20 μg liter^{-1}	LOEC for nuclear migration and germ tube growth in gametophytes	*Macrocystis pyrifera*	Garman *et al.* (1994)
16 μg liter^{-1}	72 h EC_{50} (growth) in enriched soft water	*Chlorella pyrenoidosa*	Stauber and Florence (1989)

(continues)

TABLE I (continued)

Metal	*Concentration*	*Effects*	*Organism*	*Reference*
	24 μg $liter^{-1}$	72 h EC_{50} (growth) in U.S. E.P.A. medium	*Chlorella pyrenoidosa*	Stauber and Florence (1989)
	>200 μg $liter^{-1}$	72 h EC_{50} (growth) in MBL medium (minus EDTA)	*Chlorella pyrenoidosa*	Stauber and Florence (1989)
	10 μg $liter^{-1}$	72 h EC_{50} (growth) in seawater	*Nitzschia closterium*	Stauber and Florence (1989)
	3.9–4.7μg $liter^{-1}$	"Safe" concentration (MATC) for tetrasporophyte growth	*Champia parvula*	Steele and Thursby (1983)
Cyanide	0–11 μg $liter^{-1}$	"Safe" concentration (MATC) for female growth and presence of cystocarps (sexual reproduction)	*Champia parvula*	Steele and Thursby (1983)
	500 μ*M*	Complete inhibition of the phosphorylating cytochrome pathway	*Chlamydomonas reinhardtii*	Weger and Dasgupta (1993)
Inorganic lead	9.1–23.3 μg $liter^{-1}$	"Safe" concentration (MATC) for tetrasporophyte growth and weight	*Champia parvula*	Steele and Thursby (1983)
Inorganic mercury	50 and 150 μg $liter^{-1}$ ($HgCl_2$)	Resists toxicity	*Cystoseira barbata*	De Jong *et al.* (1994)
Methyl-mercury	5 and 15 μg $liter^{-1}$ (CH_3HgCl)	More toxic than $HgCl_2$	*Cystoseira barbata*	De Jong *et al.* (1994)
Nickel	100 μ*M*	During germination of resting cells, the germination rate and the protein and carbohydrate content were reduced; during the flagellate stage cell division, phototaxis, pigment concentration, photosynthesis,	*Haematococcus lacustris*	Xyländer and Braune (1994)

		and number of motile cells were reduced; and the formation of resting cells was completely inhibited		
Ozone	3.7 μ*M*	Damage to plasmalemma caused decreased permeability, disrupted ion distributions, and impaired photosynthesis	*Chlorella sorokiniana*	Heath (1994)
Selenium	3 mg $liter^{-1}$	Induced activity of a useful enzyme	*Chlamydomonas reinhardtii*	Takeda *et al.* (1993)
Silver	1.2–1.9 μg $liter^{-1}$	"Safe" concentration (MATC) for cystocarps (sexual reproduction)	*Champia parvula*	Steele and Thursby (1983)
Organic tin (tributyltin)	18.5 n*M*	Cell density and growth rate not affected	*Pavlova lutheri*	Saint-Louis *et al.* (1994)
	74 n*M*	Cell density reduced 40% then growth rate recovered		
	185 n*M*	Chlorophyll *a* decreased in population but not in individual cells, cell density continually decreased		
	5 μg $liter^{-1}$	48 h LC_{100} for growth	*Pavlova lutheri, Dunaliella tertiolecta*	Beaumont and Newman (1986)
	1 μg $liter^{-1}$	Growth inhibition	*Skeletonema costatum*	Beaumont and Newman (1986)
	0.1 μg $liter^{-1}$	Growth inhibition	*Pavlova lutheri, Dunaliella tertiolecta, Skeletonema costatum*	Beaumont and Newman (1986)
	0.33–0.36 μg $liter^{-1}$	72 h EC_{50} growth inhibition for tributyltin compounds	*Skeletonema costatum*	Walsh *et al.* (1985)
	0.59–0.92 μg $liter^{-1}$	72 h EC_{50} growth inhibition for triphenyltin compounds	*Skeletonema costatum*	Walsh *et al.* (1985)

(continues)

TABLE I (continued)

Metal	*Concentration*	*Effects*	*Organism*	*Reference*
	17 >500 μg $liter^{-1}$	72 h EC_{50} growth inhibition for other organotin compounds	*Skeletonema costatum*	Walsh *et al.* (1985)
	1.03–1.28 μg $liter^{-1}$	72 h EC_{50} growth inhibition for tributyltin compounds	*Thalassiosira pseudonana*	Walsh *et al.* (1985)
	1.07–1.34 μg $liter^{-1}$	72 h EC_{50} growth inhibition for triphenyltin compounds	*Thalassiosira pseudonana*	Walsh *et al.* (1985)
	34 >500 μg $liter^{-1}$	72 h EC_{50} growth inhibition for other organotin compounds	*Thalassiosira pseudonana*	Walsh *et al.* (1985)
Zinc	75μg $liter^{-1}$	72 h EC_{50} in seawater, cell division	*Nitzschia closterium*	Stauber and Florence (1989)

[a]The LOEC is the "lowest observed effect concentration," the NOEC is the "no observed effect concentration," the MATC is the "maximum allowable toxicant concentration" to be considered safe for aquatic life, the EC is the "effect concentration" affecting a given percentage of the population (e.g., EC_{50}), and the LC is the "lethal concentration" killing a given percentage of the population.

for a specified duration. For example, Blanck and Wängberg (1988a) estimated that 0.4 μ*M* As was the 1 h EC_{20}, which inhibited photosynthesis by 20% for a 1 h exposure to natural marine periphyton.

A. Effects to the Cell

Effects of inorganic chemical stressors to cellular structure and function have been reviewed for algae by Rai *et al.* (1981) and Puiseux-Dao (1989), and for Al stress to higher plants by Taylor (1988a).

1. Enzyme Activity and Metabolic Pathways

High metal concentrations have a pivotal effect on enzyme systems that control biochemical and physiological functions like photosynthesis, respiration, and the synthesis of biological molecules (Rai *et al.*, 1981; Table I). Nitrogen metabolism is affected by metal stress in heterocystis blue-green algae. Low levels of Al stress inhibited nitrogenase activity in *Anabaena cylindrica* until the plants produced more heterocysts (Pettersson *et al.*, 1985). Nitrate reductase activity was inhibited by Cr in *Glaucocystis* (Rai *et al.*, 1992).

Phosphate metabolism is affected by certain inorganic chemical stressors. Cadmium stress increased the ATP/ADP ratio in *Synechocystis* (Pawlik and Skowroński, 1994). The mode of action may differ within the cell, for Jeanne *et al.* (1993) found that inhibition of Ca-ATPase activity by Cd was competitive with other cations in cell extracts but noncompetitive in cell-membrane fractions. Arsenate acts on photosynthetic phosphorylation because of the similar structures of arsenate and phosphate (Blanck and Wängberg, 1991). Cyanide is used as a standard method to inhibit the phosphorylating cytochrome pathway in order to study alternate respiratory pathways (Weger and Dasgupta, 1993).

One alga's stressor may be another alga's nutrient, for a concentration of Se generally considered to be toxic (3 mg $liter^{-1}$ as sodium selenite; Table II) induced the activity of a useful enzyme in *Chlamydomonas,* since Se is needed as a trace element for the enzyme's active site (Takeda *et al.*, 1993).

Thompson and Couture (1993) suggest that a chain of metabolic events is responsible for the recovery of algae growth to Cd exposure. This chain begins with the recovery of photosynthesis, with enhanced protein and lipid synthesis followed by restoration of oxidative phosphorylation and the use of lipid substrates for respiration.

2. Photosynthesis

Photosynthetic rates are frequently measured to assess short-term inorganic chemical stress in algae, and many inorganic stressors have been found to impair CO_2 uptake or O_2 evolution (Rai *et al.*, 1981; Table I).

TABLE II U.S. E.P.A. 4-Day Average Concentrations for the Protection of Aquatic Life (Unless Otherwise Noted)[a]

Inorganic chemical	*Concentration (μg liter^{-1})*	*Comments*
Mercury	0.025	
Cadmium	0.66 to 2	Range for soft to hard water
Copper	0.66 to 2	Range for soft to hard water
Silver	1.2 to 13	Range for soft to hard water
Lead	1.3 to 1000	Range for soft to hard water
Cyanide	5	
Selenium	35	U.S. E.P.A. 24-h average
Chromium(6+)	50	
Nickel	56 to 160	Range for soft to hard water
Zinc	59 to 190	Range for soft to hard water
Aluminum	87	pH 6.8–9.0
Beryllium	110 to 1100	Range for soft to hard water
Arsenic	190	
Iron	300 to 1000	This is the range for many nations
Barium	5000	

[a]Inorganic chemicals are ranked from most to least toxic based on the lowest concentration given. If the concentration depends on water hardness, then a range is given and mentioned in the comments. Source: Moore (1991).

Impaired photosynthesis may be due to direct chemical damage to photosynthetic pathways, or it may be a general stress response and indirectly due to chemical action on nonphotosystem membrane permeability, disrupted ion distributions, and impaired enzyme activity (Heath, 1994).

3. Pigments

The concentration of photosynthetic pigments is easily measured and frequently used to measure stress for regulatory purposes (Table I). Chlorophyll *a* is the pigment most frequently measured, but carotenoid content may also respond to inorganic chemical stress (Rai *et al.*, 1992). Total biomass estimates, like chlorophyll *a,* by themselves may not respond to stress at the community level if the abundance of sensitive populations is replaced by tolerant populations (Genter *et al.*, 1987, 1988). A decrease in chlorophyll in algal populations may be due to changes in density without any damage to pigments, for Saint-Louis *et al.* (1994) found that decreases in cell density led to decreased chlorophyll concentration for a population but not for individual cells of *Pavlova lutheri.*

4. Biological Macromolecules

Inorganic chemical stress may lead to changes in the relative abundance of organic macromolecules in cells. High levels of Ni decreased protein and carbohydrate content of *Haematococcus* (Xyländer and Braune, 1994). Low levels of Cd stress to *Selenastrum capricornutum* redirected carbon synthesis to a higher percentage of carbohydrate than to lipids or protein (Thompson and Couture, 1993). The response differs among species, for Wikfors *et al.* (1994) found carbon to be allocated to either protein and carbohydrate, to carbohydrate and lipid, or to just protein with decreases in the other biochemical groups for three different microalgae. The relative abundance of fatty acids in marine macroalgae changed because of Cu stress (Jones and Harwood, 1993). These changes are also reflected in different carbon and nitrogen abundances in periphyton (Blanck and Wängberg, 1988a). Alterations in the nutritional value of algae may impact productivity of higher-level consumers as discussed later in this chapter.

5. Life Stage and Development

Algae undergo alternation of generations during their lifetime (Sze, 1986). Stages differ in sensitivity to inorganic chemical stress, and responses are measured in different ways (Xyländer and Braune, 1994). Garman *et al.* (1994) assessed As stress by observing changes in germ tube growth and nuclear migration in gametophytes of *Macrocystis pyrifera*. Steele and Thursby (1983) found the sexual reproductive phase of *Champia parvula* to be a sensitive indicator of metal stress and observed decreases in presence of cystocarps, tetrasporophyte growth and weight, and female growth. Little is known about life-stage sensitivity for the majority of algal species.

6. Growth Rate and Abundance

Toxicant effects at the molecular level often lead to decreases in growth rate or to a delay in the lag phase of growth (Table I). Changed growth rate is generally accepted as an end point for standard toxicity tests because it is easily measured and indicates stress from all chemicals, and because decreased algal growth may impact productivity of higher-level consumers (Rai *et al.*, 1981; Nyholm and Källqvist, 1989). Differential sensitivity among species leads to different growth rates and is expected to alter species composition in communities as discussed later in this chapter.

7. Hormesis

Hormesis is a tendency for low levels of stress to stimulate rather than reduce responses like growth rate. Beaumont and Neuman (1986) observed enhanced growth of *Pavlova lutheri* cultures exposed to 0.1 μg liter^{-1} tributyltin, which suggested that hormesis may have occurred, but errors estimating growth rate may have made this result not repeatable. Heath (1994) found that chlorophyll fluorescence was initially (after 2 min) stimulated by ozone exposure but then was later greatly inhibited (by

8 min). Growth of three green algae was stimulated by low concentrations of Cd, Pb, and Ni (Devi Prasad and Devi Prasad, 1982); this may be due to an increase in enzyme activity or to an increase in free trace metal ion concentrations from ion exchange between Cd, Pb, or Ni and the EDTA chelator used in the growth medium (e.g., Riseng *et al.*, 1991). Hence, stress may cause hormesis, but growth may be stimulated by indirect effects from displaced nutrients and other factors.

B. Relative Toxicity of Inorganic Stressors

Inorganic chemicals differ in toxicity (Table II). It is important to realize that factors like pH and water hardness strongly affect the toxicity of metals, so ranking metals by toxicity only works for specific chemical environments.

Some very toxic metals are nutrients at low concentration, but these metallic micronutrients are not necessarily less toxic than other metals (Table 14.2). Calcium ATPase activity in *Dunaliella bioculata* was most sensitive to Zn stress, with other metals decreasing in toxicity in the order Zn>Cd>Cu>La>Co. Growth in *Chlorella* was inhibited by metals in decreasing toxicity as Cd>Cu>Co (Rachlin and Grosso, 1993). Note the differences between these rankings from algal tests and the U.S. Environmental Protection Agency (E.P.A.) water quality criteria based on tests from species from many trophic levels (Table II).

Relative toxicity of metals may depend on specific uptake sites. For example, Bræk *et al.* (1980) found Cd to be more toxic than Zn to *Skeletonema costatum* (clone Skel-0) and *Thalassiosira pseudonana,* whereas Zn was more toxic than Cd to a different clone of *S. costatum* (clone Skel-5) and to *Phaeodactylum tricornutum;* the authors hypothesized that different uptake sites (even among clones of the same species) lead to differences in metal accumulation and toxicity.

IV. MECHANISMS OF TOLERANCE

Algae may tolerate inorganic chemical stress at the cellular level by a decreased number of binding sites at the cell surface, inhibition of temperature-dependent (phase II) uptake, physiological development of exclusion mechanisms, genetic adaptation, morphological changes, and internal detoxifying mechanisms or safe storage sites (Rai *et al.*, 1981). Tolerance mechanisms may be specific to one inorganic stressor, or tolerance to one chemical may correspond with other chemicals. Rai *et al.* (1981) define "multiple tolerance" as the case where an alga is tolerant of many inorganic chemicals although the mechanisms may differ for each. "Cotolerance" is when a mechanism developed to tolerate one inorganic chemical confers to tolerance to others. Some metal-resistant organisms have actually been more susceptible to other stressors (Klerks and Weis, 1987).

Literature reviewing the tolerance of autotrophs to inorganic chemical stress has focused mainly on heavy metal stress to higher plants, yet this work applies in many ways to algal toxicology. Ernst *et al.* (1992) elucidated the mechanisms of resistance to the heavy metals Cd, Cu, Pb, and Zn and the metalloid As by angiosperms, with an emphasis on accumulation and tolerance by the cell wall, plasma membrane, vacuoles, metallopeptides, organic acids, and enzymes. Taylor's (1988b) review of the physiology of Al tolerance to plants includes exclusion mechanisms (immobilization at the cell wall, selective permeability of the plasma membrane, plant-induced pH barriers, and exudation of chelate ligands) and internal tolerance mechanisms (chelation in the cytoplasm, compartmentation in the vacuole, Al-binding proteins, and evolution of Al-tolerant enzymes).

A. Cell Surfaces

1. Cell Wall and Membrane

Cell surfaces can provide some protection to chemical stress. A laboratory-derived Cu-tolerant population of *Scenedesmus* excluded Cu from the cell interior by an increased Cu-adsorptive capacity of the cell surface (Twiss *et al.*, 1993). Macfie *et al.* (1994) demonstrated that the cell wall affords some protection against metal stress by comparing the toxicity of Cd, Co, Cu, and Ni to two strains of *Chlamydomonas reinhardtii* with and without cell walls. Damage to the cell membrane may be sufficient to cause stress, for failure of energy-dispersive X-ray spectrophotometric analysis to demonstrate any intracellular incorporation of Cd, Cu, or Co suggested that toxicity at EC_{50} concentrations is exerted on the plasma membrane (Rachlin and Grosso, 1993).

2. Algal Extracellular Products

Algal extracellular ligands decrease the toxicity of several metals to the cell (Starodub *et al.*, 1987). Algal extracellular products released during phytoplankton blooms have been found to decrease metal toxicity and uptake (Sanders and Riedel, 1993; Paulson *et al.*, 1994). Some algae excrete secondary metabolites in response to micronutrient limitation, and these metabolites may act as chelators to inorganic chemicals and reduce toxicity (Tables II and III). Iron-chelating siderophores produced by blue-green algae and eukaryotic algae may play an important role in reducing toxicity (Gadd, 1988). For example, *Anabaena* produces the siderophore schizokinen in response to Fe starvation, and the alga actively transports ferric schizokinen to meet its nutritional needs. Schizokinen also binds to

TABLE III Environmental Factors Affecting Metal Toxicity to Marine and Freshwater Algae

Environmental factor	*Chemical*	*Algae*	*Comments*	*Reference*
pH	Aluminum	*Chlorella pyrenoidosa*	Maximum toxicity at pH 5.8 to 6.2 for 5 μg $liter^{-1}$ Al; $Al(OH)_2^+$ implicated as most toxic Al species	Helliwell *et al.* (1983)
	Aluminum	Phytoplankton	Phosphate uptake and photosynthesis decreased more at pH 5.2–6.9 than at pH 4.5	Nalewajko and Paul (1985)
	Aluminum	*Anabaena cylindrica*	Toxicity increased from pH 7.0 to 6.0	Pettersson *et al.* (1985)
	Aluminum	Freshwater phytoplankton	300 μg $liter^{-1}$ Al added to acidified (pH 4.7) water decreased chlorophyll but had little effect on presence of dominant species; hypothesize that H^+ and Al tolerance are linked	Havens and DeCosta (1987)
	Aluminum	Freshwater phytoplankton	Three species-specific responses: (1) acid-tolerant and harmed by Al-acid mixture, (2) not harmed by Al in acidic water, and (3) not harmed by any treatment	Pillsbury and Kingston (1990)
	Aluminum	Freshwater phytoplankton	200 μg $liter^{-1}$ Al added to acidified (pH 4.5) water changed species composition from filamentous cyanobacteria, diatoms, and chrysophytes to *Closterium, Mougeotia,* and *Peridinum inconspicuum;* mean cell size increased	Havens and Heath (1990)

Aluminum	*Asterionella ralfsii* cf. var. *americana*	Toxicity greater per unit of dissolved Al concentration at pH 6 than at pH 5; growth rate reduction independent of pH except near 15 μ*M* Al, where growth decreased more at pH 5 than at pH 6	Gensemer (1991)
Aluminum	Freshwater periphyton	Four species-specific responses: (1) harmed by acid and Al-acid mixtures, (2) harmed by Al-acid mixtures only, (3) stimulated by acid but harmed by Al-acid mixtures, and (4) not affected by any treatment	Genter and Amyot (1994)
Aluminum	*Chlorella pyrenoidosa*	Al effects on cell membrane are highly pH-dependent; 120 h EC_{30} increased markedly from 3 μg $liter^{-1}$ Al at pH 6 to 50 μg $liter^{-1}$ Al at pH 5	Parent and Campbell (1994)
Cadmium	*Scenedesmus quadricauda*	Toxicity to P uptake increases almost 200-fold with increase from pH 5.5 to 8.5; linear relationship between 45 min EC_{50} for P uptake and –Log Cd (m*M*) over this pH range	Peterson *et al.* (1984)
Cadmium	*Synechocystis aquatilis*	Toxicity at pH 7 inhibited CO_2 fixation, carbonic anhydrase activity, and photosynthetic O_2 evolution; no effects at pH 5.5	Pawlik *et al.* (1993)
Copper	*Scenedesmus quadricauda*	Toxicity to P uptake increases almost 76-fold with increase from pH 5.0 to 6.5 and then remains constant at higher pH; linear relationship between 45 min EC_{50} for P uptake and –Log Cu^{2+} (m*M*) over this pH range	Peterson *et al.* (1984)

(continues)

TABLE III (continued)

Environmental factor	*Chemical*	*Algae*	*Comments*	*Reference*
	Zinc	Freshwater periphyton	Four species-specific responses: (1) harmed by acid (pH 6) stress but not Zn stress, (2) harmed by Zn stress but not acid stress, (3) individual Zn and acid treatments were so stressful that the mixture caused no further harm, and (4) not harmed by individual or combined treatments	Genter *et al.* (1988)
	Arsenic, chromium, copper, nickel, zinc	*Selenastrum capricornutum*	Toxicity was least at the optimal pH range for growth	Michnowicz and Weaks (1984)
	Aluminum, copper, iron, manganese, zinc	*Asterionella ralfsii*	Algal growth may be inhibited at higher metal concentrations at certain, usually low, pH levels; increased metal concentration may not affect tolerance to acid stess; result depends on nature and concentration of DOC	Gensemer *et al.* (1993)
	Cadmium, cobalt, copper, nickel	*Chlamydomonas reinhardtii*	Higher toxicity at pH 5 than at pH 7	Macfie *et al.* (1994)
Carbon dioxide	Selenium	*Chlamydomonas reinhardtii*	High CO_2 concentrations inhibit enzyme induction	Takeda *et al.* (1993)
Hardness (general)	Mercury	*Nostoc calcicola*	Ca, Mg, Cu, and Ni decreased toxicity of Hg and methyl-Hg	Singh and Singh (1992)
Calcium	Mercury	*Cystoseira barbata*	Decreased toxicity	De Jong *et al.* (1994)

Manganese	Copper	*Nitzschia closterium*	Pretreatment with Mn prevents toxicity; adding Mn after Cu exposure does not reverse toxicity	Stauber and Florence (1985b)
Salinity (general)	Bromine	*Fucus vesiculosus* zygotes and embryos	Higher toxicity at lower salinity and less effect at higher salinity	Andersson *et al.* (1992)
	Tributyltin	*Synechocystis* PCC 6803, *Plectonema boryanum*, *Chlorella emersonii*	No inhibition of biosorption between 0.05 and 50 m*M* NaCl; 55–65% decrease in biosorption as external NaCl increased from 50 to 500 m*M*	Avery *et al.* (1993)
Chloride	Copper	*Anacystis nidulans*	No difference in toxicity between $CuCl_2$ and $CuSO_4$	Lee *et al.* (1993)
Nutrients				
Iron	Copper	*Nitzschia closterium*	Toxicity increases in Fe-deficient growth medium; colloidal ferric hydroxide may bind to cell membrane, adsorb Cu, and prevent Cu penetration into cell	Stauber and Florence (1985a)
Nitrogen	Copper	*Microcystis* sp.	Toxicity decreases with increasing nitrate; toxicity increases with increasing nitrite or ammonium	Gupta (1989)
Phosphate	Arsenic	Marine periphyton	72 h EC_{20} for inhibition of photosynthesis increased from 0.3 μ*M* to >1 m*M* with PO_4 increase from 0.1 to 0.8 μ*M*	Wängberg and Blanck (1990)
	Aluminum	Phytoplankton	Precipitation of Al-PO_4 particles (>0.45 μm) after Al addition contributed to detrimental effects of Al	Nalewajko and Paul (1985)

(continues)

Table III (continued)

Environmental factor	*Chemical*	*Algae*	*Comments*	*Reference*
	Aluminum	*Scenedesmus obtusiusculus*	Small effects of Al with high P concentration due to precipitation of Al with P in the artificial medium; for low P concentrations, ≥450 μ*M* Al increased cellular Al, and decreased photosynthesis, chlorophyll, and cell division	Greger *et al.* (1992)
	Cadmium	*Macrocystis pyrifera*	Cd uptake was greatly enhanced, yet toxicity was low, when polyphosphate bodies were induced to form	Walsh and Hunter (1992)
	Copper	*Chlorella vulgaris*	Cu toxicity greater under P limitation than under N limitation	Hall *et al.* (1989)
	Copper	*Scenedesmus acutus*	Cellular polyphosphate content decreased toxic effect of Cu on photosynthesis for Cu-tolerant and Cu-sensitive strains of the alga	Twiss and Nalewajko (1992)
	Copper	*Nostoc calcicola*	Toxicity due to Cu-induced phosphate starvation rather than competition between Cu and enzyme; exogenous addition of phosphate antagonizes Cu toxicity; alkaline phosphatase activity increased at low Cu concentration in P-sufficient cells but decreased in P-starved cells	Verma *et al.* (1993)

	Copper	*Anabaena variabilis*	Cu-resistant strain loaded with P could grow in higher Cu concentrations than P-starved cells and had more polyphosphate bodies than the sensitive strain, and Cu did not appear to be localized on polyphosphate bodies	Hashemi *et al.* (1994)
	Mercury	*Selenastrum capricornutum*	Dissolved phosphate reduced Hg toxicity; correlation coefficient of 0.87	Chen (1994)
	Zinc	*Selenastrum capricornutum*	10 μg $liter^{-1}$ Zn may lead to apparent P limitation even though sufficient P is available	Kuwabara (1985)
	Zinc	*Selenastrum capricornutum*	P decreased Zn inhibition of stationary-phase cell density but had minimal effect on growth rate and duration of lag phase	Kuwabara *et al.* (1986)
Silica	Aluminum	*Asterionella ralfsii* cf. var. *americana*	Si-limited growth conditions; Al decreased chlorophyll and *in vivo* fluorescence at pH 6; combined effects of Al and pH 5 more severe than individual effects; Al at pH 6 increased maximum uptake rate eightfold and half-saturation constant twofold for Si	Gensemer *et al.* (1993)
Sulfate	Copper	*Anacystis nidulans*	No difference in toxicity between $CuSO_4$ and $CuCl_2$	Lee *et al.* (1993)
	Selenate	*Selenastrum capricornutum*	Increasing sulfate reduced selenate uptake and toxicity	Williams *et al.* (1994)

(continues)

Table III *(continued)*

Environmental factor	*Chemical*	*Algae*	*Comments*	*Reference*
Chelators and organic substances	Cadmium	*Selenastrum capricornutum*	Toxicity decreased when humus present; 72 h EC_{50} increased from 6 μg $liter^{-1}$ without humus to 150 μg $liter^{-1}$ with humus (8.1 mg $liter^{-1}$ C)	Sedlacek *et al.* (1983)
	Copper (aluminum)	*Scenedesmus quadricauda*	Cu toxicity inhibits growth and alkaline phosphatase activity due to indirect competitive effects of Al displacing Cu from a chelator	Rueter *et al.* (1987)
	Copper	*Selenastrum capricornutum*	Complexed copper is "biologically available" to a significant extent	Tubbing *et al.* (1994)
	Potassium dichromate	*Skeletonema costatum*	72 h EC_{10} decreased from 2.8 to 0.17 mg $liter^{-1}$ when less Na_2EDTA and lower cell density were used	Kusk and Nyholm (1992)
Metal complexation	Aluminum	*Chlorella pyrenoidosa*	Polynuclear Al contributes to monomeric Al toxicity	Parent and Campbell (1994)
Temperature	Cadmium	*Synechocystis aquatilis*	CO_2 fixation inhibited 50% at 30°C and only 5% at 5°C	Pawlik and Skowroński (1994)
Illumination	Cadmium	*Anabaena variabilis*	Photosynthetic rate decreased and photosynthetic membrane energization enhanced in both illuminated and dark-adapted cells; content of photosystem II reaction centers and rate of electron transport decreased more quickly in illuminated than in dark-adapted cells	Gorbunov and Gorbunova (1993)

	Cadmium	*Synechocystis aquatilis*	ATP/ADP ratio increased with duration of exposure in light much more than in dark	Pawlik and Skowroński (1994)
	Tetraethyllead (TEL)	*Poterioochromonas malhamensis*	No toxicity in darkness, even at very high concentrations; illumination converts TEL to a very toxic derivative; photolytic toxification inhibits growth, mitosis, and cytokinesis; the chrysophyte is not able to metabolize TEL to the toxic agent	Röderer (1980)
Season (general)	Boron	Nanoplankton (20-μm fraction)	30 mg liter^{-1} B stimulated primary production and carbon assimilation rates during winter (high nutrients, low temperature), but photosynthesis was inhibited in summer (low nutrients, high temperature); may be a species-specific response based on growth phase	Subba Rao (1981)
Algal extracellular products	Arsenic, copper	Phytoplankton	Phytoplankton growth rate and abundance influenced metal speciation	Sanders and Riedel (1993)
	Copper	*Anabaena* sp. 7120	Toxicity abolished by production of the siderophore schizokinen during iron starvation	Clarke *et al.* (1987)
	Copper, lead, zinc	*Scenedesmus quadricauda*	Extracellular ligands produced by the alga decrease single and combined toxicity	Starodub *et al.* (1987)
Population density	Potassium dichromate	*Skeletonema costatum*	72 h EC_{10} decreased from 2.8 to 0.17 mg liter^{-1} when less Na_2EDTA and lower cell density were used	Kusk and Nyholm (1992)

Cu, so low concentrations of Cu may not inhibit growth because the cupric schizokinen is not taken up by the alga (Clarke *et al.*, 1987). Clarke *et al.* (1987) found that *Anabaena* grown in high concentrations of Fe that are generally inhibitory to siderophore production had an increased sensitivity to Cu stress. These results were in marked contrast to the finding of Arceneaux *et al.* (1984: cited in Clarke *et al.*, 1987) that the diatom *Bacillus megaterium* actively took up the Cu siderophore so that Cu stress dramatically increased.

B. Organelles and Subcellular Components

1. Polyphosphate

It has been hypothesized that when sufficient PO_4 is available for growth, cellular polyphosphate binds to Zn (Bates *et al.*, 1985) and Cd (Walsh and Hunter, 1992), but when low concentrations of PO_4 limit growth and polyphosphate is metabolized, these metals are released into the cell, causing a stress response. Wong *et al.* (1994), using X-ray microanalysis, found that polyphosphate bodies in *Chlorella* contained elevated levels of Al, Fe, Cu, and Zn when exposed to a municipal effluent containing elevated levels of Al, Mn, Fe, Ni, Cu, and Zn. However, Cu-resistant *Anabaena* had increased the number of internal polyphosphate bodies, but X-ray microanalysis did not show a direct localization of Cu on polyphosphate bodies (Hashemi *et al.*, 1994). Twiss and Nalewajko (1992) concluded that polyphosphate plays a passive role in protecting cells from Cu stress and that polyphosphate bodies are relatively unimportant mechanisms for tolerance since a Cu-tolerant strain of *Scenedesmus acutus* was less dependent on cellular polyphosphate than a Cu-sensitive strain.

2. Cytoplasm

Microorganisms may tolerate inorganic chemical stress by maintaining low intracellular concentrations, and mechanisms to do this include energy-driven efflux pumps, enzymatic detoxification, synthesis of intracellular metal-binding polymers, binding to cell surfaces, and precipitation of insoluble metal complexes at the cell surface (Wood and Wang, 1985). Asthana *et al.* (1993) found that decreased cellular Ni uptake in a Ni-resistant strain was due to altered membrane transport properties, and Cd-resistant mutants of *Chlamydomonas* accumulated less metal (Collard and Matagne, 1994).

Other tolerant algae still grow with high intracellular concentrations of stressors, which may be possible when metals are sequestered intracellularly (Gadd, 1988).

Cytoplasmic chelators, like proteinaceous metallothioneins, play an important role in detoxifying metal stress in blue-green algae (Gadd,

1988). Collard and Matagne (1994) found a Cd-binding protein in *Chlamydomonas* that may have been induced by Cd since it was not found when external Cd concentrations were low.

3. Genetic Tolerance

Klerks and Weis (1987) critically reviewed the literature to determine whether evidence exists that populations have genetically adapted to heavy metal stress. They concluded that, in general, some bacteria, algae, and fungi were shown to have a genetic basis for resistance, whereas no distinction could be made between acclimation and adaptation in metazoa. Asthana *et al.* (1993) found a wild-type Ni-sensitive strain of *Nostoc muscorum* to spontaneously yield mutants resistant to Ni stress at a frequency of about 10^{-7}. Twiss *et al.* (1993) demonstrated selection in the laboratory by developing a Cu-resistant strain of *Scenedesmus* from a Cu-sensitive strain by repeated subculturing under sublethal concentrations of Cu. This Cu resistance did not confer Co or Ni cotolerance. Genetic resistance was not developed by two marine diatoms exposed to sequentially higher concentrations of organotins for 12 weeks in laboratory culture (Walsh *et al.*, 1985). Klerks and Weis (1987) caution that the paucity of reports failing to find adaptation may be due to a general reduced interest in negative results and to the difficulty in getting such results published.

If natural selection acts on resistance mechanisms, then evolutionary patterns may emerge in which organisms more closely related to each other respond more similarly to stressors than organisms that are more distantly related. This "phylogenetic hypothesis" did not hold true at the family level, but it did hold true at higher taxonomic levels of algae (Visviki and Rachlin, 1994). Wängberg and Blanck (1988) concluded that there is a relation between algal phylogeny and sensitivity to chemical stress. They examined toxicity data for 19 chemical compounds and 16 microalgal strains (from Chlorophyceae, Cyanophyceae, and Xanthophyceae) and found that the Chlorococcales and Cyanophyceae differed significantly in sensitivity to chemical stress. The remaining Chlorophyceae and Xanthophyceae species were not similar in sensitivity to either the Chlorococcales or Cyanophyceae. Toxicity is often related to uptake of the stressor, and Corder and Reeves (1994) suggest that blue-green algae have a greater ability than green algae to bind Ni. Florence *et al.* (1994) compared temperate (22°C) and tropical (27°C) strains of *Nitzschia closterium* and found the temperate strain to be more sensitive to stress from Ni and an ore leachate, but both strains to have similar sensitivity to stress from Cr and a different ore leachate. Genter *et al.* (1987, 1988) consistently found diatoms to be replaced by green algae as Zn stress increased. The diversity of responses, even among species within larger taxonomic groups, is potentially useful for making algal community composition a useful measure of anthropogenic stress.

C. Acclimation and Cotolerance of Communities

Populations may occur in polluted habitats because of (1) "acclimation," in which a physiological change occurs during exposure to increase tolerance, (2) "adaptation," in which natural selection acts on genetically based individual variation so that populations evolve increased resistance, or (3) a change in species composition to less sensitive species (Klerks and Weis, 1987). Wang (1986) transferred periphyton between low-Zn and high-Zn environments and found that communities adjusted to have similar tolerance to Zn stress as the community into which they were transferred. It was concluded that adaptation, acclimation, and changes in species composition may have acted singly or in concert to give this result.

Algal communities may develop tolerance to a particular chemical, but this does not necessarily confer cotolerance to other chemical stressors. Blanck and Wängberg (1991) established reference and As-treated periphyton communities and then exposed samples of each to a suite of photosynthetic inhibitors. They found that although arsenate tolerance was 16,000-fold higher in the As-treated community, significant cotolerance was conveyed to only one of six photosynthetic inhibitors. They concluded that cotolerance is not ubiquitous nor common in periphyton because different chemicals have different specific modes of action.

V. ENVIRONMENTAL FACTORS AFFECTING METAL TOXICITY AND UPTAKE

A survey of environmental factors affecting metal toxicity and uptake are given in Tables III and IV.

A. Water Chemistry

1. Carbon Dioxide and pH

In batch-culture algal toxicity tests, an insufficient supply of CO_2 leads to high pH levels and potential CO_2-limited growth. Flasks must be sufficiently shaken or continuously aerated (preferably with 0.1% CO_2 in air), but excessive CO_2 concentrations may inhibit metal-dependent enzyme systems (Nyholm and Källqvist, 1989; Takeda *et al.,* 1993).

Hydrogen ions directly influence metal speciation (Campbell and Stokes, 1985; Table III), and pH is strongly influenced by dissolved CO_2 concentrations. As pH decreases, Cd, Cu, and Zn become less toxic, and Pb becomes more toxic to algae (Campbell and Stokes, 1985). Aluminum appears to be most toxic near the range of pH 5.8–6.2 (Helliwell *et al.,* 1983; Nalewajko and Paul, 1985; Pettersson *et al.,* 1985; Gensemer, 1991; Parent and Campbell, 1994). Acidification has little effect on the speciation of Ag and Mn, and even though Hg stress is related to pH, insufficient data exists to make conclusions regarding Co, Hg, and Ni (Campbell and Stokes, 1985).

Hydrogen ions may decrease the toxicity of inorganic ions (and induce acid stress) by competitively excluding them from binding to cell-surface ligands (Peterson *et al.*, 1984; Campbell and Stokes, 1985; Gensemer, 1991; Parent and Campbell, 1994). For example, Cd stress is greater at higher pH's which correspond to increased uptake of the metal (Skowroński, 1986; King *et al.*, 1992; Pawlik *et al.*, 1993; Cho *et al.*, 1994).

2. Hardness

High concentrations of Ca, Mg, and Mn reduce toxicity of a variety of heavy metals by preventing them from passing through the membrane (Table III). This may be due to competition by ion exchange for cellular binding sites; or it may be by precipitation or complexation by carbonate, bicarbonate, or hydroxides of Ca or Mg (Rai *et al.*, 1981). Hence, less metal accumulates in the alga (Table IV). Prior exposure to increased concentrations of cations may reduce toxicity (Garnham *et al.*, 1993c), but increasing concentrations after exposure may not (Stauber and Florence, 1985b). Pertechnetate accumulation appears to be an exception to this rule, for Garnham *et al.* (1993b) found that accumulation of pertechnetate increased when concentrations of Ca, Mg, Mn, Ce, Sr, bicarbonate, and carbonate increased.

3. Salinity

Metals may be more toxic outside the normal salinity ranges for freshwater and marine algae (Andersson *et al.*, 1992; Avery *et al.*, 1993). Salt concentration may influence rates of metal adsorption and uptake and indifferent electrolytes can decrease adsorption if the electrolyte concentration is high enough (Cho *et al.*, 1994). Sodium ions decreased bioaccumulation of cationic metals (Garnham *et al.*, 1993a; Corder and Reeves, 1994), but the concentration of K did not inhibit Cd transport in *Synechocystis* (Pawlik and Skowroński, 1994). Pertechnetate accumulation increased with external osmotic potential but decreased with increasing sulfate concentration (Garnham *et al.*, 1993b). Sulfate and selenate compete for specific uptake sites on cell membranes, so increased sulfate can decrease selenate toxicity (Williams *et al.*, 1994).

Salinity itself can be a form of inorganic chemical stress altering osmotic balance, but this is generally not due to anthropogenic reasons. The effect of osmotic stress on algae is a large area of research and probably deserves a chapter to itself (Robarts *et al.*, 1992; Blinn, 1993; Close and Lammers, 1993; Cumming and Smol, 1993; Fernandes *et al.*, 1993; Jeanjean *et al.* 1993; Rai and Abraham, 1993; Schubert *et al.*, 1993;

TABLE IV Examples of Chemical, Physical, and Biological Environmental Factors Affecting Metal Adsorption and Uptake by Marine and Freshwater Algae

Environmental factor	*Chemical*	*Algae*	*Comments*	*Reference*
pH	Cadmium	*Stichococcus bacillaris*	Rapid increase in adsorption from pH 4 to 8	Skowroński (1986)
	Cadmium	Periphyton	Bioaccumulation of Cd decreased in lake acidified to pH 5.1 from pH 6.1; Al and Fe bioaccumulation unaffected	King *et al.* (1992)
	Cadmium	*Synechocystis aquatilis*	Maximum transport into cell at pH 7.5; no transport at pH 5.5	Pawlik *et al.* (1993)
	Technetium	*Synechocystis* PCC 6803, *S.* PCC 6301, *Plectonema boryanum, Anabaena variabilis, Oscillatoria* sp.	Decreasing pH increased accumulation	Garnham *et al.* (1993b)
	Tributyltin	*Synechocystis* PCC 6803,	Maximal uptake at pH 5.5	Avery *et al.* (1993)
		Plectonema boryanum,	Maximal uptake at pH 6.5	Avery *et al.* (1993)
		Chlorella emersonii	Little effect on biosorption	Avery *et al.* (1993)
	Zirconium	*Synechococcus* PCC 6301, *S.* PCC 6803, *Plectonema boryanum, Chlorella emersonii, C. reinhardtii, Scenedesmus obliquus*	Decreased accumulation with decreased pH	Garnham *et al.* (1993c)
	Cadmium, zinc	*Chlorella vulgaris*	Fraction of bound metal increased 87% from pH 3.3 to 7.2	Cho *et al.* (1994)
	Copper, lead, zinc	*Chlorella vulgaris*	Negligible adsorption at pH 3; maximum adsorption was pH 4.5 for Pb, Ph 4.1 for Cu, pH 5.0 for Zn	Pascucci and Sneddon (1993)

Hardness (general)	Zirconium	*Synechococcus* PCC 6301, *S.* PCC 6803, *Plectonema boryanum*, *Chlorella emersonii*, *C. reinhardtii*, *Scenedesmus obliquus*	Prior exposure to Na, K, Cs, Ca, Mg, and Sr decreased accumulation	Garnham *et al.* (1993c)
Calcium	Cadmium	*Synechocystis aquatilis*	Inhibited Cd transport	Pawlik and Skowroński (1994)
	Technetium	*Synechocystis* PCC 6803, *Oscillatoria* sp.	Increased accumulation	Garnham *et al.* (1993b)
	Cadmium, copper, zinc	*Cystoseira barbata*	Decreased toxicity and accumulation for certain mixtures of metals	Pellegrini *et al.* (1993)
Bicarbonate and carbonate	Technetium	*Anabaena variabilis*, *Oscillatoria* sp.	Decreased accumulation	Garnham *et al.* (1993b)
Magnesium	Cadmium, zinc	*Phaeodactylum tricornutum*	Increased Mg reduced Cd and Zn sorption	Braek *et al.* (1980)
	Cadmium	*Synechocystis aquatilis*	Inhibited Cd transport	Pawlik and Skowroński (1994)
	Technetium	*Synechocystis* PCC 6803	Increased accumulation	Garnham *et al.* (1993b)
Manganese	Cadmium	*Synechocystis aquatilis*	Did not inhibit Cd transport	Pawlik and Skowroński (1994)
	Technetium	*Synechocystis* PCC 6803, *S.* PCC 6301, *Plectonema boryanum*, *Anabaena variabilis*, *Oscillatoria* sp.	Increased accumulation	Garnham *et al.* (1993b)
Salinity				
Ionic strength	Cadmium	*Chlorella valgaris*	An indifferent electrolyte ($NaClO_4$) did not decrease adsorption of Cd until the electrolyte concentration was higher than the metal concentration	Cho *et al.* (1994)

(continues)

TABLE IV (*continued*)

Environmental factor	*Chemical*	*Algae*	*Comments*	*Reference*
	Technetium	*Synechocystis* PCC 6803, *S.* PCC 6301, *Plectonema boryanum, Anabaena variabilis,* and a red *Oscillatoria* sp.	Accumulation increased with increasing external osmotic potential	Garnham *et al.* (1993b)
Potassium	Cadmium	*Synechocystis aquatilis*	Did not inhibit Cd transport	Pawlik and Skowroński (1994)
Sulfate	Technetium	*Synechocystis* PCC 6803	Decreased accumulation	Garnham *et al.* (1993b)
Nutrients				
Nitrate	Copper, cadmium, zinc	*Aphanocapsa pulchra*	Nitrate enhanced uptake of each metal	Subramanian *et al.* (1994)
Phosphate	Aluminum	*Scenedesmus obtusiusculus*	Low P increases Al uptake; high P causes precipitation of Al with P	Greger *et al.* (1992)
	Cadmium	*Macrocystis pyrifera*	Cd uptake was greatly enhanced when polyphosphate bodies were induced to form	Walsh and Hunter (1992)
	Copper	*Chlorella vulgaris*	Cu uptake increased by P limitation	Hall *et al.* (1989)
Chelators and organic substances	Cadmium	*Selenastrum capricornutum*	Log(Cd bioaccumulated by algae) is proportional to Log(algal biomass); humus changes chemical speciation of Cd to prevent toxicity and bioaccumulation	Sedlacek *et al.* (1983)
	Calcium, lithium, magnesium, potassium, sodium, strontium	Alginates from *Laminaria hyperborea* and *Macrocystis pyrifera*	Algal alginates show greater uptake and selectivity than bacterial alginates; order of selectivity is Sr>Ca>Mg>K>Na>Li (bivalent selected over univalent cations)	Geddie and Sutherland (1994)

	Copper, chromium, cobalt, cadmium, iron, manganese, nickel, lead, zinc	*Cladophora glomerata*	Humic substances significantly decrease uptake by alga; highest decrease of uptake was for Cu, Cr, Co, and Cd	Vymazal (1984)
Other pollutants	Cadmium, nickel, lead, zinc	*Chlorella minutissima*	For dried and pulverized cells, the adsorption process was noncompetitive	Roy *et al.* (1993)
	Copper, lead, zinc	*Chlorella vulgaris*	As metal concentration increased, the proportion of bound Cu and Zn decreased, and the proportion of bound Pb remained fairly constant	Pascucci and Sneddon (1993)
	Cadmium	*Synechocystis aquatilis*	Transport inhibited by Zn	Pawlik and Skowroński (1994)
	Chromium	*Selenastrum* sp.	Decreased adsorption of Cr possibly due to oxidation or to competition with organic chelators in tannery effluent	Brady *et al.* (1994)
	Technetium	*Synechocystis* PCC 6803	Accumulation incrased by Ce and Sr	Garnham *et al.* (1993b)
	Tin ($SnCl_4$)	*Ankistrodesmus falcatus*	Uptake and release were not affected by increased concentration of other metals (As, Cd, Cu, Hg, Pb, Sb, Zn)	Wong *et al.* (1984)
Inorganic chemical concentration	Cadmium	*Chlorella minutissima*	Increased Cd concentration resulted in greater binding per unit cell mass	Roy *et al.* (1993)
Temperature	Cadmium	*Stichococcus bacillaris*	Relatively small increase in accumulation with temperature	Skowroński (1986)
	Cadmium	*Chlamydomonas reinhardtii*	Decreased uptake at 4°C suggests that the "slow" phase is temperature dependent	Collard and Matagne (1994)

(continues)

TABLE IV (continued)

Environmental factor	*Chemical*	*Algae*	*Comments*	*Reference*
	Cadmium	*Synechocystis aquatilis*	Cd transport increased with increase in temperature from 10° to 30°C; uptake followed Michaelis–Menten kinetics	Pawlik and Skowroński (1994)
	Cobalt, manganese, zinc	*Chlorella salina*	The second, "metabolic" phase of uptake was inhibited in the light at low temperature	Garnham *et al.* (1992)
Illumination	Cadmium	*Synechocystis aquatilis*	Cells in darkness took up considerably less Cd than cells in illumination	Pawlik and Skowroński (1994)
	Cadmium, copper, zinc	*Aphanocapsa pulchra*	Darkness inhibited uptake 25% for Cu and 21% for Cd, and stimulated uptake 22% for Zn	Subramanian *et al.* (1994)
	Cobalt, manganese, zinc	*Chlorella salina*	The second, "metabolic" phase of uptake was inhibited by darkness	Garnham *et al.* (1992)
Algal extracellular products	Copper	Phytoplankton	Uptake decreases when organisms release dissolved organic ligands that bind to Cu	Paulson *et al.* (1994)
	Manganese	Microbial mat dominated by *Oscillatoria* sp.; developed in the laboratory	Galacturonic acid and glucuronic acid may account for metal-binding properties of the mat	Bender *et al.* (1994)
	Copper, cadmium, lead, manganese, nickel, zinc	*Gracilaria tikvahiae, Gelidium pusillum, Agardhiella subulata, Chondrus crispus*	Living and lyophilized thalli show different capacities to adsorb metals; metal accumulation did not differ between carageenan- or agar-producing algae	Burdin and Bird (1994)

Location in habitat	Mixture of metals	*Fucus ceranoides*, *Fucus vesiculosus*	Vertical position on shore did not affect apical metal concentrations but showed correlations with Mn, Co, Ni, Zn, and Cu in mature tissues	Barreiro *et al.* (1993)
Population density	Cadmium	*Stichococcus bacillaris*	Cd removal from solution is proportional to algal abundance	Skowroński and Przytocka-Jusiak (1986)
	Cadmium	*Chlorella minutissima*	For dried and pulverized cells, increased biomass resulted in decreased binding per unit cell mass	Roy *et al.* (1993)
	Copper	*Oocystis pusilla*	Cu uptake increases with Cu concentration; cellular Cu declines as population density increases while the total Cu bound to algae remains much more constant	Chang and Sibley (1993)
	Copper, lead, zinc	*Chlorella vulgaris*	Fraction of metal adsorbed steadily increased until a plateau was reached at an algal mass of 1.4 g liter^{-1}	Pascucci and Sneddon (1993)

Tominaga *et al.* 1993; Waber and Wylie, 1993; Fugii and Hellebust, 1994; Giordano *et al.* 1994; Leon and Galvan, 1994; Pearson and Davison, 1994; Rothschild *et al.*, 1994; Venkataraman and Kaushik, 1994).

4. Nutrients

The concentration of PO_4 directly influences metal toxicity to algae (Rai *et al.*, 1981; Chen, 1994). High external PO_4 concentrations decrease metal stress either by (1) precipitating with metals like Fe and Al in the external solution (Rai *et al.*, 1981; Greger *et al.*, 1992) or by (2) associating with increased intracellular polyphosphate (Walsh and Hunter, 1992; Hashemi *et al.*, 1994) and PO_4 concentrations (Verma *et al.*, 1993). Low external PO_4 concentrations increase Al stress indirectly when Al precipitates with PO_4 to cause nutrient limitation (Nalewajko and Paul, 1985). Metals may cause apparent nutrient limitation by interfering with uptake of other nutrients, as Zn did for PO_4 uptake (Kuwabara, 1985; Verma *et al.*, 1993) and as Al did for Si uptake (Gensemer *et al.*, 1993b). On the other hand, limiting PO_4 concentrations were found to increase uptake and toxicity of Cu to *Chlorella* (Hall *et al.*, 1989). Equilibrium between Zn and P and their particulate and dissolved fractions influences growth rate and metal toxicity (Kuwabara *et al.*, 1986).

Inorganic chemical stressors with structure similar to that of nutrients may interfere with nutrient uptake sites or enzymes. For example, similar atomic structure allows arsenate to interfere with phosphorus metabolism which is particularly important during nutrient limitation. Wängberg *et al.* (1991) found that periphyton in a phosphorus-limited lake were about one order of magnitude more sensitive to arsenate stress than marine communities that were limited by nitrogen (Blanck and Wängberg, 1988a). Wängberg and Blanck (1990) found periphyton sensitivity to arsenate stress to be correlated to phosphorus status of the community but not to their nitrogen or chlorophyll contents. Copper stress to *Chlorella* was greater under P limitation than under N limitation (Hall *et al.*, 1989).

Rai *et al.* (1981) indicated that very little work had been done on the influence of NO_3 on metal stress, but that it was found to decrease Cd stress in a marine diatom and have no effect on Zn stress to two algae. The chemical form of N compounds may be very important, for although nitrate decreased toxicity, Cu stress increased with higher concentrations of nitrite or ammonium for *Microcystis* (Gupta, 1989). Uptake of Cu, Cd, and Zn by *Aphanocapsa* increased as NO_3 concentration increased (Subramanian *et al.*, 1994). Little is known regarding the difference in toxicity between nitrate and nonnutrient (Cl, SO_4) salts of metals (e.g., $CuNO_3$ versus $CuCl_2$ or $CuSO_4$).

Many metals are nutrients at low concentrations, and little is known regarding the biological response to metal stress when another metal is limiting. Nakano *et al.* (1978) found that Zn-starved *Euglena gracilis* did not replace Zn with Cd as an essential element for metabolism and growth, and that Cd bioaccumulation increased in Zn-starved cells. Iron-deficient growth medium led to increased Cu stress in *Nitzschia closterium,* but stress decreased with high Fe concentrations owing to adsorption of Cu to colloidal ferric hydroxide that accumulated on the cell membrane (Stauber and Florence, 1985a).

Landis (1986) modified Tilman's (1982) graphical model to examine how toxicants may alter nutrient uptake, uptake efficiency, and growth rate and ultimately change competitive relationships. He points out that little is known regarding the relationship between resource availability and toxicity.

5. Chelators and Humic Substances

Amino acids, organic matter, humic acids, fulvic acid, EDTA, and other organic substances can bind to metals and reduce their toxicity (Rai *et al.,* 1981; Sedlacek *et al.,* 1983; Vymazal, 1984). Commonly-used chelators in laboratory growth media will drastically decrease metal toxicity (Nyholm and Källqvist, 1989; Kusk and Nyholm, 1992). However, complexed metals may still be biologically available (Kuwabara *et al.,* 1986; Tubbing *et al.,* 1994), and Rueter *et al.* (1987) showed that apparent Al stress was actually due to Cu being displaced by Al from a chelator. Most ligands that form water-soluble complexes with Cu decrease its toxicity toward algae, but lipid-soluble Cu complexes are usually more toxic than the free ion because they promote diffusion into the cell, where Cu may exert a direct toxic effect (Florence and Stauber, 1986). Humic substances can decrease the toxicity of lipid-soluble Cu complexes in solution and prevent accumulation by the alga (Florence *et al.,* 1992).

6. Metal Polymerization

Aluminum is an unusual inorganic stressor for many reasons, one of which is its tendency to aggregate to a polynuclear form (Parker and Bertsch, 1992). Parent and Campbell (1994) have shown that this polynuclear Al significantly augments monomeric Al toxicity to *Chlorella.*

B. Physical Factors

1. Temperature

It is not clear whether suboptimal temperatures increase vulnerability to inorganic chemical stress (Nyholm and Källqvist, 1989). Research suggests that cooler temperatures decrease metal stress, which may be a consequence of inhibited metabolism-dependent uptake mechanisms (Skowroński, 1986; Garnham *et al.,* 1992; Collard and Matagne, 1994; Pawlik and Skowroński, 1994).

2. Illumination

Algal growth increases with light intensity to a saturation level, but little is known about how toxicity to algae is influenced by light intensity. Nyholm and Källqvist (1989) indicate that "self-shading" by high algal biomass reduces effective light intensity, particularly in cultures that are not well agitated, and is an important problem to be considered for toxicity tests involving algae. Nyholm and Källqvist (1989) recommend determining whether continuous light or light–dark cycles should be used for algal toxicity tests. The second "metabolic" phase of uptake (for Cd, Cu, Co, Mn, Zn) may be inhibited by darkness (Garnham *et al.*, 1992; Pawlik and Skowroński, 1994), but Zn uptake was higher in darkness even though Cd and Cu uptakes decreased (Subramanian *et al.*, 1994).

Photosynthetic rates are measured to assess inorganic chemical stress, and the molecular mechanisms are being investigated. Cadmium inactivates ATPase and Calvin cycle enzymes and leads to decreased protein synthesis and photoinhibition of photosystem II reaction centers in *Anabaena* (Gorbunov and Gorbunova, 1993). Decreased ATP consumption by Cd-stressed enzymes can cause ATP/ADP ratios to increase (Pawlik and Skowroński, 1994).

Illumination can directly alter speciation and toxicity of some metals, for light converts tetraethyllead to a very toxic derivative (Röderer, 1980).

C. Biological Effects

Algae influence the toxicity of inorganic stressors to themselves by their influence on the surrounding water.

1. Biomass

Larger algal populations lead to more membrane binding sites being available, which leads to less metal being bound per alga and, hence, less toxicity (Rai *et al.*, 1981; Nyholm and Källqvist, 1989; Kusk and Nyholm, 1992; Chang and Sibley, 1993; Roy *et al.*, 1993). Pascucci and Sneddon (1993) found that optimizing population density led to a greater percentage of metal being adsorbed than optimizing pH on the adsorption of a mixture of Cu, Pb, and Zn to *Chlorella vulgaris*. The amount of metal accumulated by algae increases with biomass, but at high biomass concentrations the percentage of metal accumulated may decrease or remain the same (Skowroński and Przytocka-Jusiak, 1986; Chang and Sibley, 1993; Pascucci and Sneddon, 1993).

As stated earlier in this chapter, algal extracellular products increase the tolerance of algae to several metals.

2. Biotransformation

Biotransformation is the enzyme-catalyzed conversion of one chemical into another that may be more or less toxic (Lech and Vodicnik, 1985). Mercury is a well-known example of an inorganic chemical that undergoes biotransformation by bacteria to methyl-mercury (Rai *et al.*, 1981; Moore and Ramamoorthy, 1984; Moore, 1991). Algae also play an important role in the cycling of organic metals in aquatic systems.

Algae biotransform some metals to less toxic organic compounds (Moore, 1991). Marine green and brown algae biologically methylate arsenate to produce less toxic dimethylarsenic derivatives (Edmonds and Francesconi, 1981; Wrench and Addison, 1981; Cullen *et al.*, 1994). Algae, such as *Ankistrodesmus,* biotransform very toxic organic metals like tetraethyllead to less toxic compounds (Wong, *et al.*, 1987). Maguire *et al.* (1984) have shown that *Ankistrodesmus falcatus* begins the detoxification process for tributyltin, which depends on temperature, the abundance and species composition of tributyltin-degrading algae, and their nutritional state.

Algae may biotransform some metals to more toxic organic compounds. Besser *et al.* (1994) found that *Chlamydomonas reinhardtii* converts inorganic Se to more toxic Se-amino acid and nonamino-organoselenium compounds, which are released during periods following rapid decrease in algal abundance.

D. Season

The combined effect of the different chemical, physical, and biological factors mentioned in the foregoing can be expected to alter inorganic chemical stress as seasons change (Winner and Owen, 1991a,b). Subba Rao (1981) found B to inhibit primary production in winter yet stimulate it in summer, possibly due to different nutrient, temperature, and species abundance relationships in each season. Genter *et al.* (1987) found Zn concentrations near the U.S. E.P.A. water quality criterion to alter the abundance of benthic algae in three seasons.

E. A Schematic Model of Metal Uptake by the Algal Cell

Metals (including other inorganic chemicals) may follow many pathways before exerting any effect on growth rate (Fig. 1). The process begins with dissolved metal in labile (chemically active) and nonlabile forms. Labile metal directly and indirectly affects algal growth.

Biosorption of metals occurs rapidly to organic metal ligands on the cell wall and cell membrane. Algae that produce a mucilage may have a certain amount of metal adsorb to it. The metabolism-dependent phase of

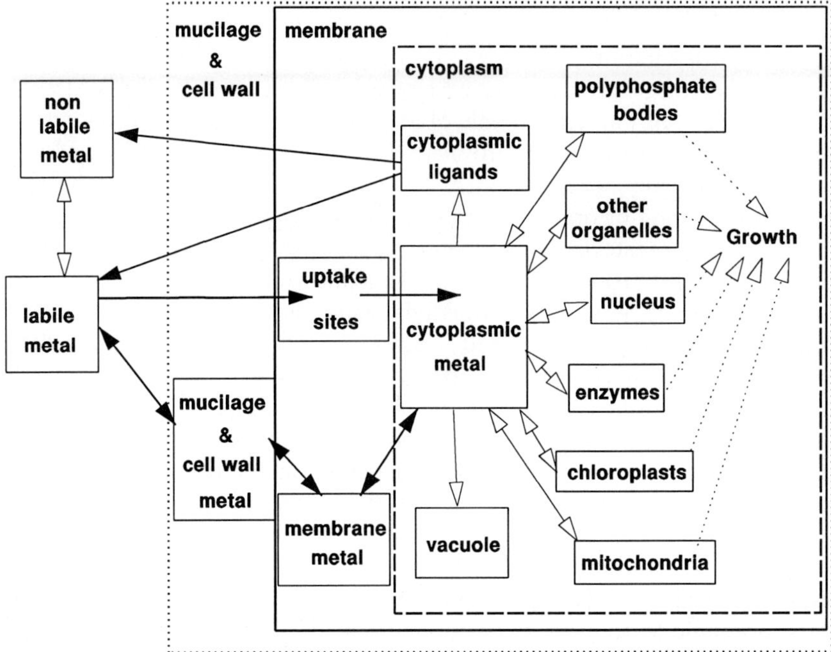

FIGURE 1 Schematic model of pathways of metal taken up by an algal cell. Solid arrows represent direct pathways between dissolved metal and algal surfaces and cytoplasm. Open arrows with solid lines are pathways between forms of dissolved metal and between cytoplasm and intracellular structures. Dotted arrows are factors that affect algal growth. The sizes of the boxes and arrows are not scaled to represent the importance of different components.

metal uptake involves specific uptake sites, and it may involve diffusion through cell surfaces into the cytoplasm (Fig. 1).

Cytoplasmic metal has many fates. Some metal is sequestered to the vacuole or is bound to cytoplasmic ligands, where it no longer alters metabolism. If metals are stored in polyphosphate bodies then metals may be released to the cytoplasm when polyphosphate reserves are being depleted. Biologically active metal may alter enzyme systems and organelles and ultimately affect metabolism (photosynthesis and respiration) and factors like growth, reproduction, and development (Fig. 1). Genetic changes by metal stress on nuclear and extranuclear nucleic acids may result in long-term modifications on algal metabolism. The model in Fig. 1 is tentative, for little is known about chemical equilibria of metals between cytoplasm and intracellular structures.

The algal cell also influences metal toxicity to itself (Rai *et al.*, 1981). Chemical equilibrium between labile and nonlabile metal (Fig. 1) is controlled by photosynthesis, respiration, and algal extracellular products that chelate metals. Photosynthesis consumes CO_2 (decreasing pH) and produces O_2 (increasing REDOX potential), and respiration has the opposite effects. Hence, metal activity can change diurnally and be very different near the cell than in the overlying water (also see Allen, 1977).

Bioaccumulation relationships tend to be curvilinear and approach an asymptote. Theoretical maximum biosorption levels at complete cell-surface saturation may be estimated using Freundlich isotherms or Langmuir isotherms (Gadd, 1988; Avery *et al.*, 1993; Garnham *et al.*, 1993b, 1993c; Corder and Reeves, 1994). Uptake has been observed to follow Michaelis–Menten kinetics for Cd (Pawlik and Skowroński, 1994), Co, Mn, and Zn (Garnham *et al.*, 1992). Thompson and Couture (1993) noted that increasing Cd cell quotas in *Selenastrum capricornutum* cultures were associated with decreased growth and photosynthesis and disrupted patterns of carbon allocation to macromolecules. Lineweaver–Burk plots describe uptake in two phases for Co, Mn, and Zn (Garnham *et al.*, 1992).

VI. INTERACTIONS AMONG INORGANIC STRESSORS

Inorganic chemical stressors almost always occur as mixtures in nature. Three outcomes are possible for the toxicity of a mixture of two metals. As shown in Fig. 2, if metal 1 decreases growth by a certain amount (M_1) and metal 2 decreases growth by a certain amount (M_2) then one possible outcome is that growth is decreased in a simple additive manner ($M_1 + M_2$). This additive decrease in growth occurs for two metals with similar effects. Synergism is when the decrease in growth is greater than the sum of the individual metal effects (Fig. 2). Antagonism is when the metals interfere with each other's toxic effects; growth will be greater than the additive response and may even be similar to or greater than the control (Fig. 2). The biological reasons for these responses are not well known. Rai *et al.* (1981) found that the same alga may behave synergistically to one metal combination and antagonistically to another metal combination, and that a particular metal combination may act antagonistically on one metabolic process and synergistically on another metabolic process for the same alga. Hence, the uniqueness of various physiological and biochemical processes is important in determining the stress response.

Inorganic chemical stressors may also influence each other's uptake and toxicity.

A. Uptake

Inorganic chemicals are taken up by algae by (1) a competitive, ion-exchange process, (2) a noncompetitive process, or (3) a combination of

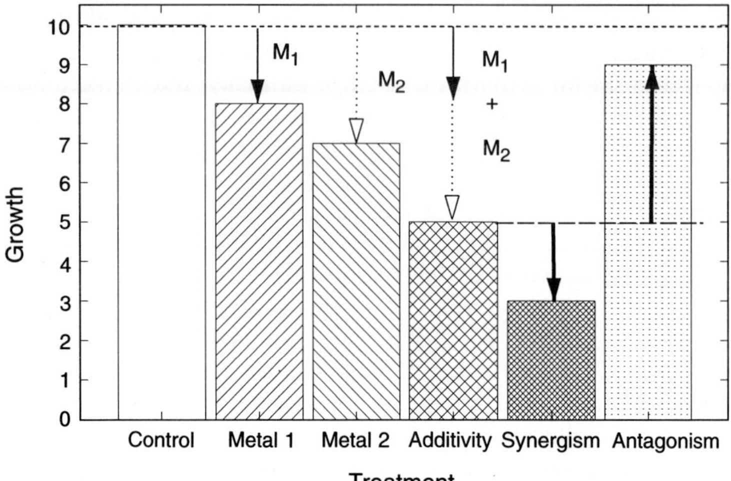

FIGURE 2 Examples of three possible outcomes for growth inhibition of an alga in response to stress from mixtures of two metals (M_1 and M_2) relative to a control. Decreased growth owing to the metal mixture may be additive (decreased as the sum of the effects of each metal separately), synergistic (decrease more than additive), or antagonistic (decrease less than additive and may be greater than the control).

competitive and noncompetitive uptake that is not reciprocal between chemicals.

Competitive uptake occurs when the presence of one chemical influences the uptake of another chemical. For example, transport of Cd was inhibited by Zn in *Synechocystis* (Pawlik and Skowroński, 1994), and the bivalent cations of Mg, Mn, Co, and Zn competed with Cd for cellular binding sites (Skowroński, 1986).

Noncompetitive metal adsorption and uptake occur when uptake of a metal is unaffected by increased concentrations of other metals. This was observed by Roy *et al.* (1993) for previously dried and pulverized *Chlorella;* four metals (Cd, Ni, Pb, Zn) in mixture were taken up in similar manner to the way they were alone in solution. Wong *et al.* (1984) found that Sn uptake by *Ankistrodesmus* was unaffected by increased concentrations of seven other metals.

Both competitive and noncompetitive uptake systems may be found simultaneously. *Chlorella salina* may have a specific uptake system for the micronutrient Co (uptake is unaffected by equal concentrations of Mn, Mg, and Zn) and a shared uptake system with different affinities for Mn, Mg, Zn, and Cd (these elements inhibited each other's uptake, but were not inhibited by Co) (Garnham *et al.*, 1992). Pascucci and Sneddon (1993) found that Cu and Zn inhibited each other's adsorption, but adsorption of Pb was unaffected for *Chlorella.*

Some uptake systems are not reciprocal. Subramanian *et al.* (1994) found that Zn interfered with Cd uptake but that Cd did not interfere with Zn uptake for the blue-green alga *Aphanocapsa pulchra.* The other metal combinations (mixtures of Cu, Cd, Zn) did not show inhibitory effects, and Cu actually enhanced Zn uptake.

Metal concentrations and environmental conditions influence adsorption and uptake of mixtures of inorganic chemicals. Uptake of Cd and Cr by *Chlorella ellipsoidea* was influenced by the concentration of each metal and by the conditions under which the experiment was performed. In batch culture, uptake of each metal increased with its concentration in the medium and with the concentration of the other metal in the medium (Okamura and Aoyama, 1994). In flow-through continuous culture (associated with nutrient limitation), the uptake of Cd increased with its concentration in the medium, but the uptake of Cr by the cell was relatively small; when the metals were in mixture, Cr did not influence uptake of Cd, but Cd increased uptake of Cr (Aoyama and Okamura, 1993).

Inorganic stressors can influence each other's allocation among cellular components. Okamura and Aoyama (1994) found that Cd tended to accumulate in the soluble fraction of algal cells and that Cr influenced the distribution of Cd among cellular components; Cr accumulated equally among cellular components, but in the presence of Cd most of the Cr accumulated in the cell-wall fraction.

Despite the complicated relationships involving metal adsorption and uptake, algae are comparable to other aquatic organisms as biomonitors of metals in water (Moore, 1991).

B. Toxicity

Water quality criteria are based on toxicity data with single inorganic stressors. It is important to distinguish between toxicity of individual chemicals and the effect that these chemicals have in mixtures. Wong and coworkers (1978, 1982) demonstrated that 10 metals (As, Cd, Cr, Cu, Fe, Pb, Hg, Ni, Se, and Zn) at concentrations that were not toxic to algae individually (and satisfied the Great Lakes Water Quality Objectives levels) strongly inhibited primary production of *Scenedesmus quadricauda, Chlorella pyrenoidosa, Anabaena flos-aquae, Navicula pelliculosa,* and natural phytoplankton when the metals were present together.

Toxicity tests with algae have shown that whether mixtures of inorganic chemicals behave in an additive, synergistic, or antagonistic manner depends on many factors. First, interaction may depend on metal concentration (Okamura and Aoyama, 1994). Pellegrini *et al.* (1993) found that Cd and Zn act in synergy or in antagony, depending on their exogenous concentrations for chlorophyll *a* and on carotenoid synthesis. Prevot and Soyer-Gobillard (1986) found interaction between Cd and Se to be antagonistic when they were both at low concentrations and to be synergistic if either one was at high concentration. High concentrations of metal mixtures may be most toxic at acidic pH (Gensemer *et al.*, 1993a; Macfie *et al.* 1994).

Second, interaction may depend on the combination of metals in the mixture. Devi Prasad and Devi Prasad (1982) found antagonism between Ni–Cd and Pb–Cd mixtures and additivity for the Ni–Pb mixture for three freshwater green algae. Gotsis (1982) demonstrated that Se–Hg and Se–Cu interactions were antagonistic to the marine planktonic green alga *Dunaliella minuta*.

Third, toxicity to chemical mixtures can differ between algal species and between clones of the same species. Bræk *et al.* (1980) proposed that competition between Zn and Cd for uptake sites led to less-than-additive toxicity for *Phaeodactylum tricornutum* and *Skeletonema costatum* (clone Skel-0) and synergistic toxicity to the growth of *Thalassiosira pseudonana* and *Skeletonema costatum* (clone Skel-5).

Fourth, interaction may depend on the biological response being measured. Two cultures of the diatom *Ditylum brightwellii* were grown at different Zn concentrations and then exposed to Cu stress. Elevated Zn levels decreased cell division rates and chlorophyll *c* content; but decreased chlorophyll *a*, photosynthetic O_2 evolution, and cell division rates and the increased number of deformed and broken cells were attributed to Cu independent of, and in addition to Zn stress (Rijstenbil et al., 1994). Pellegrini *et al.* (1993) found that Cd and Zn act in synergy or in antagony for chlorophyll *a* and on carotenoid synthesis, and Zn was antagonistic to Cd and Cu in the Cd–Cu–Zn mixture for weight change.

Fifth, interaction may depend on the duration of exposure (Prevot and Soyer-Gobillard, 1986).

Sixth, interaction may depend on culture and water conditions (Okamura and Aoyama, 1994). Starodub *et al.* (1987) demonstrated that the type of toxic interaction between Cu, Zn, and Pb toward algal growth is influenced by the concentration of each metal, abundance of complexing ligands, and pH. Wang (1985) found that the type of interaction response differed for three periphyton communities exposed to mixtures of Fe and Zn. Toxicity of a mixture of As, Cr, Co, Ni, and Zn was least at the optimal pH for growth of *Selenastrum capricornutum* (Michnowicz and Weaks, 1984). These results support a need for site-specific testing for environmental effects.

Seventh, interaction may depend on the mathematical method to determine interactive effects (Okamura and Aoyama, 1994).

C. A Schematic Model of Algal Response to Metal Interaction

The combinations of concentrations of two metals that will allow growth can be represented using a modification of Tilman's (1982) model with resource-dependent growth isoclines. We will begin by considering each metal separately, and then we will consider the effect of the mixture of two metals with the additive model shown in Fig. 3. It is assumed that each metal is a nutrient at low concentrations and prevents growth at high concentrations. Low concentrations of each metal are necessary for growth, and net growth will be zero at some minimum amount for each essential metal (X_1 for metal 1 and Y_1 for metal 2 in Fig. 3). Net growth will be positive as the concentration of each metal rises until the concentration is high enough to completely inhibit growth (X_2 for metal 1 and Y_2 for metal 2 in Fig. 3). Hence, growth is possible between the minimum concentration needed for growth and the maximum concentration that ceases growth for each metal. Note that growth is inhibited by both low and high concentrations of each metal.

The solid lines in Fig. 3 are the zero-net-growth-isoclines (ZNGIs), which represent the range of concentrations where growth is zero for one metal when the other metal is at the minimum concentration needed for maintenance (zero growth). For example, when the concentration of metal

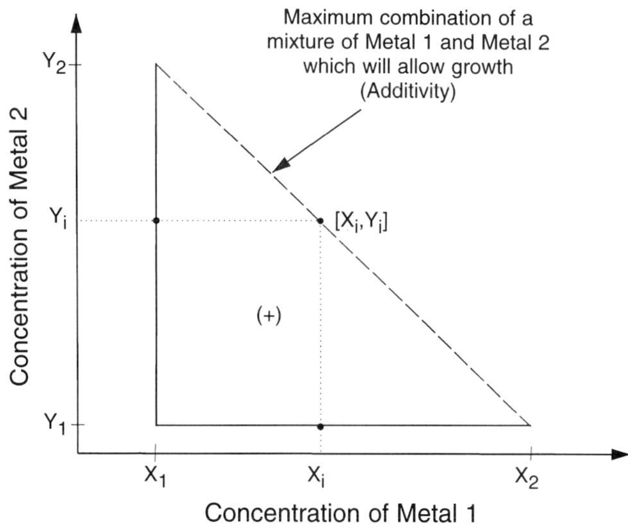

FIGURE 3 Modification of Tilman's (1982) model showing the region of growth possible for two metals that are nutrients at low concentrations and stressors at high concentrations. The solid lines represent the minimum concentration of each essential metal (X_1 for metal 1, Y_1 for metal 2) necessary to maintain the population at a growth rate of zero. The length of each solid line is limited by the maximum concentration of each metal that inhibits growth owing to stress (X_2 for metal 1, Y_2 for metal 2). The dashed line represents the maximum combination for mixtures of concentrations of two metals (X_i, Y_i) that limits growth owing to stress (zero growth). Net positive growth (+) occurs in the region bounded by these lines. This is an example of an additive stress response for mixtures of two essential metals.

1 is at the minimum needed as a nutrient for zero growth (X_1) then zero growth will occur for concentrations of metal 2 between Y_1 and Y_2.

Even metals whose combined effect as stressors is in an additive manner lead to a quite complex array of potential results. At moderate concentrations of metal 1 (X_i), growth will occur between a particular range of concentrations of metal 2 with zero growth at Y_1 and Y_i (Fig. 3). The simple additive model leads to a linear ZNGI for toxicity represented by the dashed line on Fig. 3. The additive condition is quite idealistic, and most mixtures of metals will probably result in interaction. Antagonistic interactions occur when the mixture of metals is less stressful than the additive condition and results in an expanded region for growth (Fig. 4). Synergistic interactions occur when the mixture of metals is more stressful than the additive condition and results in a constricted region for growth (Fig. 4).

The mixture of metals will give an optimum concentration for growth as represented by the (+) in Fig. 3. The thoughtful examiner of Fig. 3 will conclude that there are four possible regions of inhibited growth for mixtures of metals around the optimum: (1) where both metals act as limiting nutrients, (2) where both metals act as stressors, and (3, 4) where one metal inhibits growth as a limiting nutrient and the other limits growth as a stressor.

This model assumes that the living system will behave in a simple mathematical way, but biological systems tend to respond in more complex

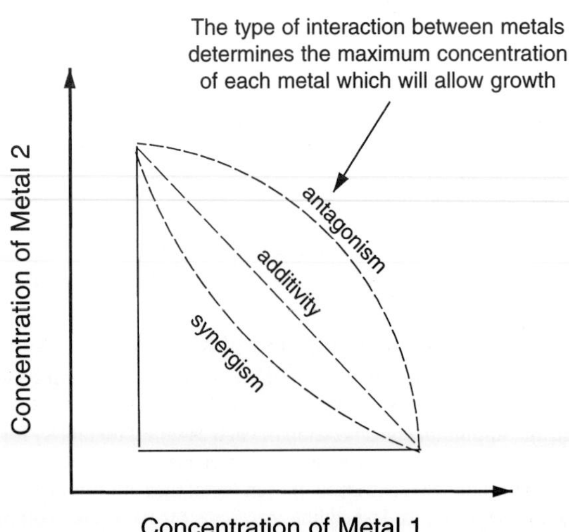

FIGURE 4 Growth regions possible because of different kinds of interaction between mixtures of two metals. Refer to Figs. 2 and 3 for additional information.

ways, so the model may be modified as more tests are performed. Maguire (1973) proposed a version of this model in which the region of growth was represented by closed curves rather than triangles.

Regions of growth may be compared for different species. Figure 5 shows a hypothetical case for a species A that responds in a synergistic manner and for a species B that responds in an additive manner to mixtures of two metals. At least seven regions can be identified in Fig. 5. Only species B grows in region 1 because of the toxic effect of the metal mixture on species A, and only species B grows in region 2 because the concentration of metal 1 is lower than necessary for nutritional requirements of species A. Similarly, only species A grows in region 3 because there is not enough of metal 2 to meet nutritional needs of species B. Both species can grow in region 4. Only species A can grow in region 5. Neither species can grow in region 6 because metal concentrations are too low for nutritional needs, and neither species grows in region 7 because of toxicity of the metal mixture. The reader is invited to ponder the variety of conditions that influence the growth rates of algae in a community in response to different combinations of metals, all of which could result in assignment to any particular region in Fig. 5.

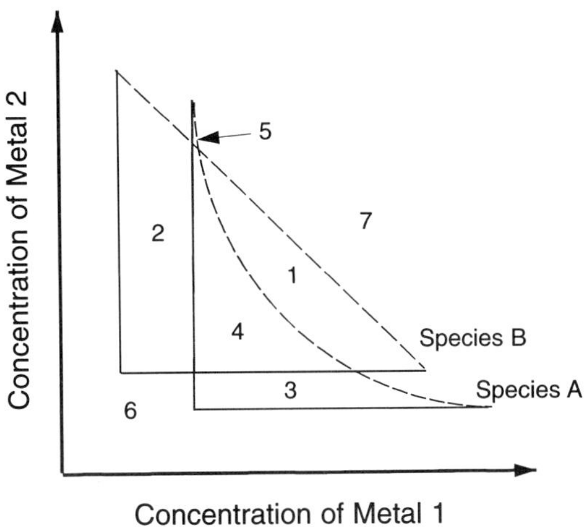

FIGURE 5 Growth regions for two species of algae for mixtures of two metals. Species A responds to the two metals in a synergistic manner, and species B responds to the two metals in an additive manner. Only species B grows in region 1 because of the toxic effect of the metal mixture on species A, and only species B grows in region 2 because the concentration of metal 1 is lower than necessary for nutritional requirements of species A. Similarly, only species A grows in region 3 because there is not enough of metal 2 to meet nutritional needs of species B. Both species can grow in region 4. Only species A can grow in region 5. Neither species can grow in region 6 because metal concentrations are too low for nutritional needs, and neither species grows in region 7 because of toxicity of the metal mixture. Refer to Fig. 4 for additional information.

VII. RESPONSE OF ALGAL COMMUNITIES TO INORGANIC STRESSORS

A. Pattern and Process

Effects at the community level generally involve chronic (long term) exposures to stress that are more realistic of what algae will experience in nature. Exposure to inorganic chemical stress often places a selection pressure on the community that decreases abundance of pollution-sensitive species and either increases or does not change abundance of pollution-tolerant species (Patrick, 1978; Rai *et al.,* 1981; Genter *et al.,* 1987; Blanck and Wängberg, 1988b). A dose–response relationship can occur at the community level so that a gradient in metal stress is correlated with a shift to more tolerant species (Leland and Carter, 1984; Genter *et al.,* 1987; Blanck and Wängberg, 1988a; Wängberg *et al.,* 1991). Functional factors of the community, like photosynthesis and nitrogen metabolism, may also be inhibited (Rai *et al.,* 1981; Leland and Carter, 1985). Multivariate models have also shown diatom species abundance to be a useful variable to identify high or low metal concentrations in streams (Rushforth *et al.,* 1981).

The shift in community composition is important, for no single species has been found to be an indicator of stress by a specific inorganic chemical. It is important to understand that changes in species composition may or may not be associated with decreases or increases in total community biomass (Genter *et al.* 1987, 1988), but may be associated with changes in nutritional value of periphyton to higher-level consumers.

Vertical position on the shore is an important factor influencing the amount of metal accumulated by macroalgae (Barreiro *et al.,* 1993); little is known as to whether vertical position in the periphyton mat influences metal accumulation and toxicity.

B. Periphyton Communities

Leland and Carter (1984, 1985) found that slight increases in Cu stress caused significant decreases in sensitive algal taxa, significant increases in tolerant taxa, and changes in community functional variables during a one-year exposure in a natural stream. Genter *et al.* (1987) found community composition to shift from diatoms, to filamentous green algae, to unicellular green algae as Zn concentration increased for experiments conducted in three seasons. Individual and combined treatments of Zn and acid stress led to similar shifts (Genter *et al.,* 1988). Community-level toxicity tests using periphyton have shown that inorganic stressors at concentrations near the Water Quality Criteria of the U.S. E.P.A. alter species composition (Genter *et al.,* 1987, 1988). Inorganic chemical stress may lead to decreased species diversity (Rai *et al.,* 1981), but diversity indices are sensitive to the number of algal cells counted (Weitzel and Bates, 1981).

When a factorial experimental design is used, community composition can be used to distinguish between the control, inorganic stressor A, inorganic stressor B, and the mixture of stressors A and B. This is demonstrated by a number of examples. Benthic algal species exposed to individual and combined treatments of Zn and acid (pH 6) stress could be divided into four categories: (1) species harmed by acid stress but not Zn stress, (2) species harmed by Zn stress but not acid stress, (3) individual Zn and acid treatments were so stressful that the combined treatment caused no further harm, and (4) species not harmed by individual or combined treatments (Genter *et al.*, 1988). Periphyton exposed to acidity or acid–Al mixtures showed four species-specific responses: (1) taxa harmed by acid and Al–acid mixtures, (2) taxa harmed by Al–acid mixtures only, (3) taxa stimulated by acid but harmed by Al–acid mixtures, and (4) taxa not harmed by any treatment (Genter and Amyot, 1994). Mathematical models can be used to identify lakes of high, intermediate, and low metal concentrations; and diatom species have been classified as being (1) metal-tolerant acid-water taxa, (2) metal-sensitive acid-water taxa, and (3) metal-tolerant circumneutral taxa (Dixit *et al.*, 1991).

The variety of responses of individual species in natural communities of algae exposed to environmental stress depends on a number of factors. Changes in abundance may be due to direct effects of the inorganic stressors on the cell, and abundance may be altered by indirect effects to higher-level biological interactions (competition, predation, etc.) or to effects of water chemistry on metal speciation. Different locations in the periphyton mat may explain why larger, long, thin, pennate diatoms responded early to stress compared to less distinct responses of smaller, adnate diatoms (Genter and Amyot, 1994). Community-level responses may also depend on previous exposure to chemical stress. Niederlehner and Cairns (1993) found that preexposure to Zn stress was harmful to periphyton respiration but beneficial to periphyton primary productivity and relative algal biomass when periphyton were later exposed to acid stress (pH 4.0–4.5). Other structural and functional parameters were not affected by previous Zn treatment when periphyton were later exposed to acid stress, and severe acid stress (pH 3.0–3.5) was so harmful that no structural or functional parameters were affected by preexposure to Zn.

C. Phytoplankton Communities

Phytoplankton communities also shift species composition by decreasing abundance of sensitive taxa, and either increasing or having no effect on the abundance of tolerant taxa when exposed to sublethal concentrations of metals (Reinke and DeNoyelles, 1985). Havens and coworkers (Havens and DeCosta, 1987; Havens and Heath, 1990) were able to distinguish between stress from acidity alone and from Al–acid mixtures by the dominant phytoplankton species present. Mallin *et al.* (1994) attributed a decline in diatom abundance and an increase in green algal abundance to significant decrease in Zn and Cu over a 7-year period in an acidic blackwater reservoir.

Nalewajko and Paul (1985) studied Al stress to phytoplankton from two lakes of different pH. The different responses of the two biotas to Al treatment were influenced by differences in lake pH (hence Al speciation and toxicity), differential susceptibility by different phytoplankton species composition between the two lakes, and differences in the ability of the algal species to use phosphates complexed with Al.

D. Seaweeds

Marine brown algae accumulate metals so efficiently that they have been found to be convenient biomonitors for heavy metal pollution (Rai *et al.*, 1981; Sears *et al.*, 1985; Ho, 1987; Söderlund *et al.*, 1988). Newly grown tissues attain metal concentrations similar to those of the surrounding water, so increases and decreases of heavy metal concentrations in the water are reflected in the metal content of different age groups of tissue. Eide *et al.* (1980) transferred whole plants between polluted and nonpolluted areas, examined metal accumulation in new tissues, and found that changes in metal content can be followed for a considerable period of time, but that individual metals show different patterns of uptake and release during the year (particularly in winter). Estabrook *et al.* (1985) concluded that metals that have accumulated in sediments must be considered in addition to the distance from known pollution sources for metal accumulation by freshwater macrophytes.

E. Importance to Higher-Level Consumers

Algal community composition may be indirectly altered solely by the effects that chemical stressors have on higher-level consumers. If grazers are important in structuring algal community composition, then stress at the grazer level may increase abundance of previously cropped algal species that are tolerant of the stressor (Rai *et al.*, 1981).

Stewart *et al.* (1993) propose that the movement of toxic chemicals in streams can be increased by the action of grazers on periphyton in a manner similar to the nutrient spiraling concept of Newbold *et al.* (1983). Periphyton retard the net movement of toxicants downstream because of their tremendous adsorptive capacity. Effective grazers may increase the downstream movement of inorganic stressors by dislodging particles while feeding and by digestive activities. Hence, the absence of grazers may increase toxicant accumulation in periphyton, whereas the presence of grazers will increase downstream movement of toxicants.

Algae may be an important source of metal stress to higher-level consumers. Stephenson and Turner (1993) found that the amphipod *Hyalella azteca* derived more of its Cd from periphyton food (58%) than from water and that 80% of the ingested Cd was assimilated. Absil *et al.*

(1994a,b) used high concentrations of EDTA to minimize biological availability of dissolved Cu and demonstrated that uptake through food *(Phaeodactylum tricornutum)* was at least as efficient as uptake from the water. Wikfors *et al.* (1994) found that Cd-contaminated algae were detrimental to the survival and growth of oysters and clams owing to stress induced by direct uptake of Cd from food, and the toxic effects of Cd were more severe when more digestible algal species were fed.

Concentration of metal in algal food may affect metal assimilation by grazers. Reinfelder and Fisher (1994) found that assimilation efficiency of nine elements (Ag, Am, C, Cd, Co, P, S, Se, and Zn) was proportional to the fraction of each element in the cytoplasm of algae ingested by bivalve mollusc larvae. The cytoplasmic fraction of Se, Zn, Cd, and protein increased with slower algal *(Isochrysis galbana)* growth rate, so mollusc larvae may assimilate proportionately more of these elements and protein when feeding on senescent cells than when feeding on rapidly dividing cells. Copper concentration in *Daphnia* fed *Oocystis* contaminated with 50 ppb Cu decreased with exposure time, and Cu in *Daphnia* fed *Oocystis* contaminated with 200 ppb Cu remained relatively constant and was associated with impaired fecundity (Chang and Sibley, 1993). Food-borne Se was bioaccumulated by *Daphnia magna* fed *Chlamydomonas reinhardtii* and bluegill sunfish in turn bioaccumulated Se from their daphnid food (Besser *et al.* 1993). *Daphnia magna* fed *Scenedesmus acutus* cultured in 1 mg liter^{-1} Cr^{6+} had increased growth and fecundity, but *Daphnia* fed algae cultured in 5 mg liter^{-1} showed no negative effects, and those fed algae cultured in 10 mg liter^{-1} Cr^{6+} had drastically reduced growth and fecundity (Gorbi and Corradi, 1993). Absil *et al.* (1994a) found that algae spiked with Cu may release enough metal to the water to cause stress in the bivalve *Macoma balthica*.

Micronutrient metals may be bioaccumulated to different extents than nonnutritional metals, for Watras *et al.* (1985) found that there was no biomagnification for *Daphnia magna* fed Ni-loaded *Scenedesmus obliquus* for short-term (72 h) and long-term (13 d) exposures. Direct uptake of Ni from solution, rather than uptake from ingested algae, was the primary accumulation vector (Watras *et al.*, 1985).

Metal stress may make contaminated algae less nutritional to higher-level consumers. Decreased fecundity and growth in *Daphnia magna* were attributed to reduced nutritional value of the algal food since the algae cultured at different Cr^{6+} concentrations supplied *D. magna* with similar amounts of Cr. Changes in the taxonomic composition of periphyton may lead to significant changes in food quality to grazers. Herbivore assimilation efficiency of green algae may be lower than that of diatoms (Lamberti and Moore, 1984), so production of higher-level consumers may be reduced (Stewart *et al.*, 1993).

Tolerant species of algae may be crucial for recovery from stress by the higher-level consumers. Tolerance of the green alga *Oocystis* to Cu stress was crucial for initiating recovery in a microcosm since *Oocystis* survived the Cu insult, was a suitable food for *Daphnia,* and controlled DOC and pH, which are very important in determining Cu toxicity (Meador *et al.,* 1993).

Metal stress may cause blue-green algae to release neurotoxins and hepatotoxins at concentrations high enough to poison wildlife. Traditional treatment with copper sulfate greatly increased the release of algal toxins, whereas lime treatment did not cause increased toxin release (Kenefick *et al.,* 1993).

F. A Schematic Model of Metal Fate in Periphyton

The direct pathway that available metals take to reach benthic algae is relatively simple, but the web of pathways that ultimately influence algal abundance is quite complex. Periphyton contains benthic algae and grazers (macrobenthos and protozoa) that feed on algae and the polysaccharide matrix (PsM). The PsM includes bacteria, fungi, and mucilage (Fig. 6).

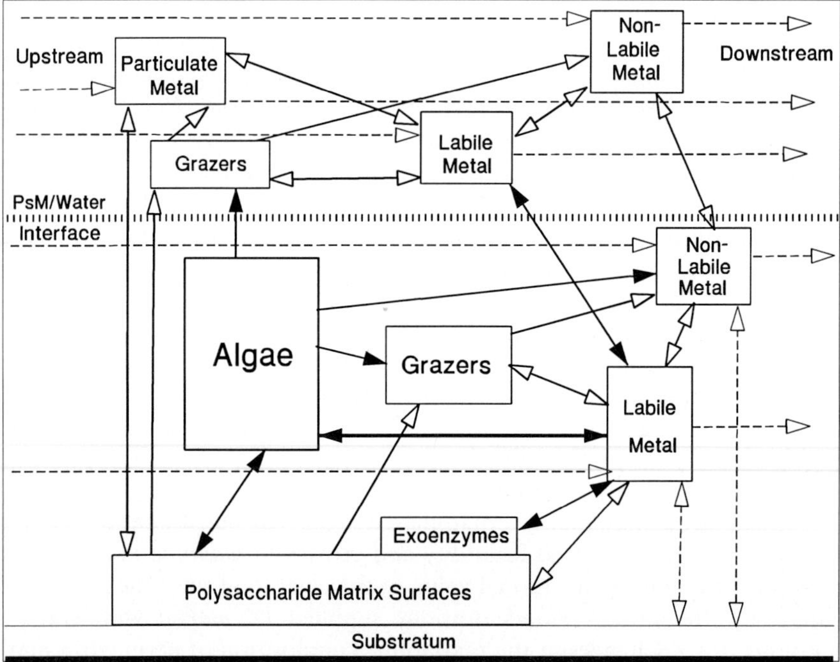

FIGURE 6 Schematic model of pathways of a metal between stream water and biotic and abiotic components of periphyton. The polysaccharide matrix (PsM) includes bacteria, fungi, and mucilage. The periphyton is divided into an overstory and an understory attached to a substratum. Solid arrows represent direct pathways to and from algae, open arrows are indirect pathways through labile metal affecting algae, and dotted arrows are inputs and outputs of metal for the system. The sizes of the boxes and arrows are not scaled to represent the importance of different components.

The overlying water contains labile, nonlabile, and particulate forms of metal. Diffusion of labile and nonlabile metals into the PsM may be enhanced by turbulence, but more work needs to be done to elucidate this (see Stevenson, Chapter 11, this volume; Nowell and Jumars, 1984; Hickey, 1988). Labile metal, after diffusing into PsM, can directly affect algae and exoenzymes. Exoenzymes are potentially important to metabolism in the PsM (Locke, 1981). Metal stress acts on enzyme systems in algae (Rai *et al.* 1981), so metals are likely to alter exoenzyme function and indirectly alter algal growth. Motile algae gliding over surfaces may adsorb metal directly from PsM surfaces. Particulate metal may become adsorbed to the PsM or become dislodged by sloughing. Large grazers will remove PsM and algae with associated metals, and excretion by these grazers releases particulate and dissolved metal.

The reader is invited to ponder the variety of outcomes possible with many algal species responding to direct effects from different combinations of metals and to indirect effects of metals on grazers, exoenzymes, and other components of the PsM with different tolerances to stress.

VIII. ALGAL BIOASSAY OF INORGANIC STRESSORS

A. Static and Flow-Through Methods

Batch-culture toxicity methods for algae have been critically reviewed by Nyholm and Källqvist (1989). Results from different laboratories differ a great deal, in some cases by more than three orders of magnitude, probably due to different physical, chemical, and biological conditions under which experiments are performed. Reproducible results do not require the use of strictly standardized procedures, but certain experimental parameters must be understood and controlled. In particular, Nyholm and Källqvist (1989) recommend that excessively high biomass must be avoided, that sufficient gas exchange be maintained, that tests be of relatively short duration and restricted to the initial period of exponential growth (lasting 2–4 d), that there be no lag phase in control cultures, and that the composition of the test medium be adequately quantified. There is certainly a need to obtain good standard species from groups other than green algae in order to make up suitable test batteries (Wängberg and Blanck, 1988; Nyholm and Källqvist, 1989).

Nyholm and Källqvist (1989) recommend performing batch-culture toxicity tests with algae growing exponentially, hence under conditions with excess nutrients. This may lead to a more conservative test since toxicity is often less under these conditions. Nyholm and Källqvist (1989) point out that total nutrient content must be the same in all experimental units and that test duration should be restricted to the initial period when exponential growth prevails in the controls. They argue that nutrient composition should not mask or obscure toxicity test results as they have in the case of some studies involving complex wastes, where increased final yield was interpreted as a stimulatory growth response.

A goal of toxicity testing is to develop short-term tests that can be used to predict long-term environmental effects. Wängberg *et al.* (1991) and Blanck and Wängberg (1988a) found that a short-term photosynthesis test employing phytoplankton or periphyton was a good predictor of long-term arsenate stress to these communities.

Continuous cultures are more likely than batch cultures to show toxic responses at low chemical concentrations. Hall *et al.* (1989) concluded that the three- to four-order increase in sensitivity of continuous culture experiments may be due to the continuous renewal of stressor (Cu) and to the low concentrations of nutrients necessary to maintain growth, and that increased toxicity was a consequence of being less able to detoxify Cu. Aoyama and Okamura (1993) found that the toxicity of Cd was detected with greater sensitivity when the dilution rate, hence the relative growth rate, was lower in continuous culture. Community-level toxicity tests involving algae often involve flow-through artificial streams. Results of some of these tests were described earlier in this chapter; they improve environmental realism and have great potential for site-specific tests.

B. Assessing Toxicity of Sediments

The science of sediment toxicology has increased dramatically owing to recent concerns about the contamination of sediments from anthropogenic wastes. Phytoplankton exposed to elutriates, interstitial water, overlying water, or whole sediment samples (Santiago *et al.*, 1993) and periphyton in flow-through artificial streams have been used (Burton, 1991). There is clearly a need for more toxicity tests involving algae given their important role as primary producers, as primary food sources to grazing and detrital food webs, and in cycling nutrients and toxicants, and the tendency of metals to accumulate in sediments (Rai *et al.*, 1981).

ACKNOWLEDGMENTS

I thank William Balance and Gwen Balance for their contributions to this chapter.

REFERENCES

Absil, M. C. P., Kroon, J. J., and Wolterbeek, H. T. (1994a). Availability of copper from phytoplankton and water for the bivalve *Macoma balthica*. 1. Separation of uptake pathways using the radiotracer ^{64}Cu. *Mar. Biol. (Berlin)* **118,** 123–127.

Absil, M. C. P., Kroon, J. J., and Wolterbeek, H. T. (1994b). Availability of copper from phytoplankton and water for the bivalve *Macoma balthica*. 2. Uptake and elimination from ^{64}Cu-labelled diatoms and water. *Mar. Biol. (Berlin)* **118,** 129–135.

Allen, T. F. H. (1977). Scale in microscopic algal ecology: A neglected dimension. *Phycologia* **16,** 253–257.

Andersson S., Kautsky, L., and Kautsky, N. (1992). Effects of salinity and bromine on zygotes and embryos of *Fucus vesiculosus* from the Baltic Sea. *Mar. Biol. (Berlin)* **114,** 661–665.

Aoyama, I., and Okamura, H. (1993). Interactive toxic effect and bioconcentration between cadmium and chromium using continuous algal culture. *Environ. Toxicol. Water Qual.* **8,** 255–269.

Asthana, R. K., Sing, A. L., and Sing, S. P. (1993). Comparison of Ni-sensitive and Ni-resistant strains of *Nostoc muscorum. World J. Microbiol. Biotechnol.* **9,** 323–327.

Avery, S. V., Codd, G. A., and Gadd, G. M. (1993). Biosorption of tributyltin and other organotin compounds by cyanobacteria and microalgae. *Appl. Microbiol. Biotechnol.* **39,** 812–817.

Aziz, M. A., and Ng, W. J. (1993). Industrial wastewater treatment using an activated algae-reactor. *Water Sci. Technol.* **28,** 71–76.

Balczon, J. M., and Pratt, J. R. (1994). A comparison of the responses of two microcosm designs to a toxic input of copper. *Hydrobiologia* **281,** 101–114.

Barreiro, R., Real, C., and Carballeira, A. (1993). Heavy-metal accumulation by *Fucus ceranoides* in a small estuary in north-west Spain. *Mar. Environ. Res.* **36,** 39–61.

Bates, S. S., Tessier A., Campbell P. G. C., and Létourneau, M. (1985). Zinc–phosphorus interactions and variation in zinc accumulation during growth of *Chlamydomonas variabilis* (Chlorophyceae) in batch culture. *Can. J. Fish. Aquat. Sci.* **32,** 86–94.

Beaumont, A. R., and Newman, P. B. (1986). Low levels of tributyl tin reduce growth of marine micro-algae. *Mar. Pollut. Bull.* **17,** 457–461.

Bender, J., Gould, J. P., Vatcharapijarn, Y., Young, J. S., and Phillips, P. (1994). Removal of zinc and manganese from contaminated water with cyanobacteria mats. *Water Environ. Res.* **66,** 679–683.

Besser, J. M., Canfield, T. J., and La Point, T. W. (1993). Bioaccumulation of organic and inorganic selenium in a laboratory food chain. *Environ. Toxicol. Chem.* **12,** 57–72.

Besser, J. M., Huckins, J. N., and Clark, R. C. (1994). Separation of selenium species released from Se-exposed algae. *Chemosphere* **29,** 771–780.

Blanck, H., and Wängberg, S.-Å. (1988a). Validity of an ecotoxicological test system: Short-term and long-term effects of arsenate on marine periphyton communities in laboratory systems. *Can. J. Fish. Aquat. Sci.* **45,** 1807–1815.

Blanck, H., and Wängberg, S.-Å. (1988b). Induced community tolerance in marine periphyton established under arsenate stress. *Can. J. Fish. Aquat. Sci.* **45,** 1816–1819.

Blanck, H., and Wängberg, S.-Å. (1991). Pattern of cotolerance in marine periphyton communities established under arsenate stress. *Aquat. Toxicol.* **21,** 1–14.

Blanck, H., Holmgren, K., Landner, L., Norin, H., Notini, M., Rosemarin, A., and Sundelin, B. (1989). Advanced hazard assessment of arsenic in the Swedish environment. *In* "Chemicals in the Aquatic Environment—Advanced Hazard Assessment" (L. Landner, ed.), pp. 256-328. Springer Verlag, New York.

Blinn, D. W. (1993). Diatom community structure along photochemical gradients in saline lakes. *Ecology* **74,** 1246–1263.

Brady, D., Letebele, B., Duncan, J. R., and Rose, P. D. (1994). Bioaccumulation of metals by *Scenedesmus, Selenastrum* and *Chlorella* algae. *Water SA* **20,** 213–218.

Bræk, G. S., Malnes, D., and Jensen, A. (1980). Heavy metal tolerance of marine phytoplankton. IV. Combined effect of zinc and cadmium on growth and uptake in some marine diatoms. *J. Exp. Mar. Biol. Ecol.* **42,** 39–54.

Burdin, K. S., and Bird, K. T. (1994). Heavy metal accumulation by carrageenan and agar producing algae. *Bot. Mar.* **37,** 467–470.

Burton, G. A., Jr. (1991). Assessing the toxicity of freshwater sediments. *Environ. Toxicol. Chem.* **10,** 1585–1627.

Campbell, P. G. C., and Stokes, P. M. (1985). Acidification and toxicity of metal to aquatic biota. *Can. J. Fish. Aquat. Sci.* **42,** 2034–2049.

Chang C., and Sibley, T. H. (1993). Accumulation and transfer of copper by *Oocystis pusilla. Bull. Environ. Contam. Toxicol* **50,** 689–695.

Chen, C. Y. (1994). Theoretical evaluation of the inhibitory effects of mercury on algal growth at various orthophosphate levels. *Water Res.* **28,** 931–937.

Cho, D. Y., Lee, S. T., Park, S. W., and Chung, A. S. (1994). Studies on the biosorption of heavy metals onto *Chlorella vulgaris. J. Environ. Sci. Health Part A,* **A29,** 389–409.

Clarke, S. E., Stuart, J., and Sanders-Loehr, J. (1987). Induction of siderophore activity in Anabaena spp. and its moderation of copper toxicity. *Appl. Environ. Microbiol.* **53,** 917–922.

Close, T. J., and Lammers, P. J. (1993). An osmotic stress protein of cyanobacteria is immunologically related to plant dehydrins. *Plant Physiol.* **101,** 773–779.

Collard, J. M., and Matagne, R. F. (1994). Cd^{2+} resistance in wild-type and mutant strains of *Chlamydomonas reinhardtii. Environ. Exp. Bot.* **34,** 235–244.

Corder, S. L., and Reeves, M. (1994). Biosorption of nickel in complex aqueous waste streams by cyanobacteria. *Appl. Biochem. Biotechnol.* **45,** 847–859.

Crist, R. H., Oberholser, K., Schwartz, D., Marzoff, J., Ryder, D., and Crist, D. R. (1988). Interactions of metals and protons with algae. *Environ. Sci. Technol.* **22,** 755–760.

Crist, R. H., Martin, J. R., Guptill, P. W., Eslinger, J. M., and Crist, D. R. (1990). Interaction of metals and protons with algae. 2. Ion exchange in adsorption and metal displacement by protons. *Environ. Sci. Technol.* **24,** 337–342.

Crist, R. H., Oberholser, K., McGarrity, J., Crist, D. R., Johnson, J. K., and Brittsan, J. M. (1992). Interaction of metals and protons with algae. 3. Marine algae, with emphasis on lead and aluminum. *Environ. Sci. Technol.* **26,** 496–502.

Cullen, W. R., Li, H., Pergantis, S. A., Eigendorf G. K., and Harrison, L. G. (1994). The methylation of arsenate by a marine alga, *Polyphysa peniculus,* in the presence of L-methionine-methyl-D_3. *Chemosphere* **28,** 1009–1019.

Cumming, B. F., and Smol, J. P. (1993). Development of diatom-based salinity models for paleoclimatic research from lakes in British Columbia. *Hydrobiologia* **269,** 179–196.

De Jong, L., Diana, C., Campos, J. R., Arnoux, A., and Pellegrini, L. (1994). Toxicity of methyl mercury and mercury (II) chloride to a brown alga *Cystoseira barbata* (Fucales) under laboratory culture conditions—Detoxifying role of calcium. *Bot. Mar.* **4,** 367–379.

Devi Prasad, P. V., and Devi Prasad, P. S. (1982). Effect of cadmium, lead and nickel on three freshwater green algae. *Water, Air, Soil Pollut.* **17,** 263–268.

Dixit, S. S., Dixit, A. S., and Smol, J. P. (1991). Multivariate environmental inferences based on diatom assemblages from Sudbury (Canada) lakes. *Freshwater Biol.* **26,** 251–266.

Edmonds, J. S., and Francesconi, K. A. (1981). Arseno-sugars from brown kelp *(Ecklonia radiata)* as intermediates in cycling of arsenic in a marine ecosystem. *Nature (London),* **289,** 602–604.

Eide, I., Myklestad, S., and Melsom, S. (1980). Long-term uptake and release of heavy metals by *Ascophyllum nodosum* (L.) Le Jol. (Phaeophyceae) *in situ. Environ. Pollut., Ser. A* **23,** 19–28.

Eisler, R. (1985a). Cadmium hazards to fish, wildlife, and invertebrates: A synoptic review. *U. S. Fish. Wildl. Serv. Biol. Rep.* **85** (1.2).

Eisler, R. (1985b). Selenium hazards to fish, wildlife, and invertebrates: A synoptic review. *U. S. Fish. Wildl. Serv., Biol. Rep.* **85**(1.5).

Eisler, R. (1986). Chromium hazards to fish, wildlife, and invertebrates: A synoptic review. *U. S. Fish. Wildl. Serv., Biol. Rep.* **85**(1.6).

Eisler, R. (1987). Mercury hazards to fish, wildlife, and invertebrates: A synoptic review. *U. S., Fish. Wildl. Serv., Biol. Rep.* **85**(1.10).

Ernst, W. H. O., Verkleij, J. A. C., and Schat, H. (1992). Metal tolerance in plants. *Acta Bot. Neerl.* **41,** 229–248.

Estabrook, G. F., Burk, D. W., Inman, D. R., Kaufman, P. B., Wells, J. R., Jones, J. D., and Ghosheh, N. (1985). Comparison of heavy metals in aquatic plants on Charity Island, Saginaw Bay, Lake Huron, U.S.A., with plants along the shoreline of Saginaw Bay. *Am. J. Bot.* **72,** 209–216.

Fehrmann, C., and Pohl, P. (1993). Cadmium adsorption by the non-living biomass of microalgae grown in axenic mass culture. *J. Appl. Phycol.* **5,** 555–562.

Fernandes, T. A., Iyer, V., and Apte, K. (1993). Differential responses of nitrogen-fixing cyanobacteria to salinity and osmotic stresses. *Appl. Environ. Microbiol.* **59,** 899–904.

Flegal, A. R., Garrison, D. L., and Niemeyer, S. (1993). Lead isotopic disequilibria between plankton assemblages and surface waters reflect life cycle strategies of coastal populations within a northeast Pacific upwelling regime. *Limnol. Oceanogr.* **38,** 670–678.

Florence, T. M., and Stauber, J. L. (1986). Toxicity of copper complexes to the marine diatom *Nitzschia closterium. Aquat. Toxicol.* **8,** 11–26.

Florence, T. M., Powell, H. K. J., Stauber, J. L., and Town, R. M. (1992). Toxicity of lipid-soluble copper(II) complexes to the marine diatom *Nitzschia closterium:* Amelioration by humic substances. *Water Res.* **9,** 1187–1193.

Florence, T. M., Stauber, J. L., and Ahsanullah, M. (1994). Toxicity of nickel ores to marine organisms. *Sci. Total Environ.* **148,** 139–155

Fugii, S., and Hellebust, J. A. (1994). Growth and osmoregulation of *Boekelovia hooglandii* in relation to salinity. *Can. J. Bot.* **72,** 823-828.

Gadd, G. M. (1988). Accumulation of metals by microorganisms and algae. *In* "Biotechnology" (H. J. Rehm, ed.) Vol. 60, pp. 401–434. VCH, Weinheim.

Garman, G. D., Pillai, M. C., and Cherr, G. N. (1994). Inhibition of cellular events during early algal gametophyte development: Effects of select metals and an aqueous petroleum waste. *Aquat. Toxicol.* **28,** 127–144.

Garnham, G. W., Codd, G. A., and Gadd, G. M. (1992). Kinetics of uptake and intracellular location of cobalt, manganese and zinc in the estuarine green alga *Chlorella salina. Appl. Microbiol. Biotechnol.* **37,** 270–276.

Garnham, G. W., Codd, G. A., and Gadd, G. M. (1993a). Uptake of cobalt and cesium by microalgal-clay and cyanobacterial-clay mixtures. *Microb. Ecol.* **25,** 71–82.

Garnham, G. W., Codd, G. A., and Gadd, G. M. (1993b). Accumulation of technetium by cyanobacteria. *J. Appl. Phycol.* **5,** 307–315.

Garnham, G. W., Codd, G. A., and Gadd, G. M. (1993c). Accumulation of zirconium by microalgae and cyanobacteria. *Appl. Microbiol. Biotechnol.* **39,** 666–672.

Geddie, J. L., and Sutherland, I. W. (1994). The effect of acetylation on cation binding by algal and bacterial alginates. *Biotechnol. Appl. Biochem.* **20,** 117–129.

Gensemer, R. W. (1991). The effects of pH and aluminum on the growth of the acidophilic diatom *Asterionella ralfsii* var. *americana. Limnol. Oceanogr.* **36,** 123–131.

Gensemer, R. W., Smith, R. E. H., Duthie, H. C., and Schiff, S. L. (1993a). pH tolerance and metal toxicity in populations of the planktonic diatom *Asterionella:* Influences of synthetic and natural dissolved organic carbon. *Can. J. Fish. Aquat. Sci.* **50,** 121–132.

Gensemer, R. W., Smith, R. E. H., and Duthie, H. C. (1993b). Comparative effects of pH and aluminum on silica-limited growth and nutrient uptake in *Asterionella ralfsii* var. *americana* (Bacillariophyceae). *J. Phycol.* **29,** 36–44.

Genter, R. B., and Amyot, D. J. (1994). Freshwater benthic algal population and community changes due to acidity and aluminum–acid mixtures in artificial streams. *Environ. Toxicol. Chem.* **13,** 369–380.

Genter, R. B., Cherry, D. S, Smith, E. P., and Cairns, J., Jr. (1987). Algal–periphyton population and community changes from zinc stress in stream mesocosms. *Hydrobiologia* **153,** 261–275.

Genter, R. B., Cherry, D. S., Smith, E. P., and Cairns, J., Jr. (1988). Attached-algal abundance altered by individual and combined treatments of zinc and pH. *Environ. Toxicol. Chem.* **7,** 723-733.

Giordano, M., Davis, J. S., and Bowes, G. (1994). Organic carbon release by *Dunaliella salina* (Chlorophyta) under different growth conditions of CO_2, nitrogen, and salinity. *J. Phycol.* **30**, 249–257.

Gorbi, G., and Corradi, M. G. (1993). Chromium toxicity on two linked trophic levels. 1. Effects of contaminated algae on *Daphnia magna. Ecotoxicol. Environ. Saf.* **25**, 64–71.

Gorbunov, M. Y., and Gorbunova, E. A. (1993). The effect of cadmium ions on photosynthesis and the delayed fluorescence of chlorophyll in cyanobacterium *Anabaena variabilis* Kutz. *Russ. J. Plant Physiol. (Engl. Transl.)* **40**, 656–659.

Gotsis, O. (1982). Combined effects of selenium/mercury and selenium/copper on the cell population of the alga *Dunaliella minuta. Mar. Biol. (Berlin)* **71**, 217–222.

Greger, M., Tillberg, J.-E., and Johansson, M. (1992). Aluminum effects on *Scenedesmus obtusiusculus* with different phosphorus status. II. Growth, photosynthesis and pH. *Physiol. Plant.* **84**, 202–208.

Gupta, S. L. (1989). Interactive effects of nitrogen and copper on growth of cyanobacterium *Microcystis. Bull. Environ. Contam. Toxicol.* **42**, 270–275.

Hall, J., Healey F. P., and Robinson, G. G. C. (1989). The interaction of chronic copper toxicity with nutrient limitation in two chlorophytes in batch culture. *Aquat. Toxicol.* **14**, 1–14.

Hashemi, F., Leppard, G. G., and Kushner, D. J. (1994). Copper resistance in *Anabaena variabilis:* Effects of phosphate nutrition and polyphosphate bodies. *Microb. Ecol.* **27**, 159–176.

Havas, M., and Jaworski, J. F., eds. (1986). "Aluminum in the Canadian Environment," NRCC/CNRC No. 24759. National Research Council of Canada, NRCC Associate Committee on Scientific Criteria for Environmental Quality.

Havens, K. E., and DeCosta, J. (1987). The role of aluminium contamination in determining phytoplankton and zooplankton responses to acidification. *Water, Air, Soil Pollut.* **33**, 277–293.

Havens, K. E., and Heath, R. T. (1990). Phytoplankton succession during acidification with and without increasing aluminum levels. *Environ. Pollut.* **68**, 129-145.

Heath, R. L. (1994). Possible mechanisms for the inhibition of photosynthesis by ozone. *Photosynth. Res.* **39**, 439–451.

Helliwell, S., Batley, G. E., Florence, T. M., and Lumsden, B. G. (1983). Speciation and toxicity of aluminium in a model fresh water. *Environ. Technol. Lett.* **4**, 141–144.

Hickey, C. W. (1988). Oxygen uptake kinetics and microbial biomass of river sewage fungus biofilms. *Wat. Res.* **11**, 1365–1373.

Ho, Y. B. (1987). Metals in 19 intertidal macroalgae in Hong Kong waters. *Mar. Pollut. Bull.* **18**, 564–566.

Holan, Z. R., and Volesky, B. (1994). Biosorption of lead and nickel by biomass of marine algae. *Biotechnol. Bioeng.* **43**, 1001–1009.

Holan, Z. R., Volesky, B., and Prasetyo, I. (1993). Biosorption of cadmium by biomass of marine algae. *Biotechnol. Bioeng.* **41**, 819-825.

Jeanjean, R., Matthijs, H. C. P., Onana, B., Havaux, M., and Joset, F. (1993). Exposure of the cyanobacterium *Synechocystis* PCC 6803 to salt stress induces concerted changes in respiration and photosynthesis. *Plant Cell Physiol.* **34**, 1073–1079.

Jeanne, N., Dazy, A. C., and Moreau, A. (1993). Cadmium interactions with ATPase activity in the euryhaline alga *Dunaliella bioculata. Hydrobiologia* **252**, 245–256.

Jones, A. L., and Harwood, J. L. (1993). Lipid metabolism in the brown marine algae *Fucus vesiculosus* and *Ascophyllum nodosum. J. Exp. Bot.* **44**, 1203–1210.

Kenefick, S. L., Hrudey, S. E., Peterson, H. G., and Prepas, E. E. (1993). Toxin release from *Microcystis aeruginosa* after chemical treatment. *Water Sci. Technol.* **27**, 433–440.

King, S. O., Mach, C. E., and Brezonik, P. L. (1992). Changes in trace metal concentrations in lake water and biota during experimental acidification of Little Rock Lake, Wisconsin, USA. *Environ. Pollut.* **78**, 9-18.

Klerks, P. L., and Weis, J. S. (1987). Genetic adaptation to heavy metals in aquatic organisms: A review. *Environ. Pollut.* **45**, 173–205.

Kusk, K. O., and Nyholm, N. (1992). Toxic effects of chlorinated organic compounds and potassium dichromate on growth rate and photosynthesis of marine phytoplankton. *Chemosphere* **25**, 875–886.

Kuwabara, J. S. (1985). Phosphorus–zinc interactive effects on growth by *Selenastrum capricornutum* (Chlorophyta). *Environ. Sci. Technol.* **19**, 417–421.

Kuwabara, J. S. (1992). Associations between benthic flora diel changes in dissolved arsenic, phosphorus, and related physico-chemical parameters. *J. North Am. Benthol. Soc.* **11**, 218–228.

Kuwabara, J. S., Davis J. A., and Chang C. C. Y. (1986). Algal growth response to particle-bound orthophosphate and zinc. *Limnol. Oceanogr.* **31**, 503–511.

Lamberti, G. A., and Moore, J. W. (1984). Aquatic insects as primary consumers. *In* "The Ecology of Aquatic Insects" (V. H. Resh and D. M. Rosenberg, eds.), pp. 164–195. Praeger.

Landis, W. G. (1986). Resource competition modeling of the impacts of xenobiotics on biological communities. *In* "Aquatic Toxicology and Environmental Fate" (T. M. Poston and R. Purdy, eds), ASTM STP 921, Vol. 9, pp. 55-72. Am. Soc. Test. Mater., Philadelphia.

Lech, J. J., and Vodicnik, M. J. (1985). Biotransformation. *In* "Fundamentals of Aquatic Toxicology: Methods and Applications" (G. M. Rand and S. R. Petrocelli, eds.), pp. 526–557. Hemisphere Publishing, New York.

Lee, L. H., Lustigman, B., and Maccari, J. (1993). Effect of copper on the growth of *Anacystis nidulans. Bull. Environ. Contam. Toxicol.* **50**, 600–607.

Leland, H. V., and Carter, J. L. (1984). Effects of copper on species composition of periphyton in a Sierra Nevada, California, stream. *Freshwater Biol.* **14**, 281–296.

Leland, H. V., and Carter, J. L. (1985). Effects of copper on production of periphyton, nitrogen fixation, and processing of leaf litter in a Sierra Nevada, California, stream. *Freshwater Biol.* **15**, 155–176.

Leon, R., and Galvan, F. (1994). Halotolerance studies on *Chlamydomonas reinhardtii—Glycerol excretion by free and immobilized cells. J. Appl. Phycol.* **6**, 13–20.

Locke, M. A. (1981). River epilithon—A light and organic energy transducer. *In* "Perspectives in Running Water Ecology" (M. A. Locke and D. D. Williams, eds), pp. 3–40. Plenum, New York.

Macfie, S. M., Tarmohamed, Y., and Pamperl, M. A. (1994). Phytoplankton community assessments of seven southeast cooling reservoirs. *Water Res.* **28**, 665–673.

Maguire, B., Jr. (1973). Niche response structure and the analytical potentials of its relationship to the habitat. *Am. Nat.* **107**, 213-246.

Maguire, R. J., Wong, P. T. S., and Rhamey, J. S. (1984). Accumulation and metabolism of tri-*n*-butyltin cation by a green alga, *Ankistrodesmus falcatus. Can. J. Fish. Aquat. Sci.* **41**, 537–540.

Mallin, M. A., Stone, K. L., and Pamperl, M. A. (1994). Phytoplankton community assessments of seven U.S. cooling reservoirs. *Water Res.* **28**, 665–673.

Meador, J. P., Taub, F. B., and Sibley, T. H. (1993). Copper dynamics and the mechanism of ecosystem level recovery in a standardized aquatic microcosm. *Ecol. Appl.* **3**, 139–155.

Michnowicz, C. J., and Weaks, T. E. (1984). Effects of pH on toxicity of As, Cr, Cu, Ni and Zn to *Selenastrum capricornutum* Printz. *Hydrobiologia* 118, 299-305.

Michnowicz, C. J., and Weaks, T. E. (1984). Effects of pH on toxicity of As, Cr, Cu, Ni and Zn to *Selenastrum capricornutum* Printz. *Hydrobiologia* **118**, 299–305.

Moore, J. W. (1991). "Inorganic Contaminants of Surface Water: Research and Monitoring Priorities," Springer Ser. on Environ. Manage. Springer-Verlag, New York.

Moore, J. W., and Ramamoorthy, S. (1984). "Heavy Metals in Natural Waters: Applied Monitoring and Impact Assessment," Springer Ser. Environ. Manage. Springer-Verlag, New York.

Nakano, Y., Okamoto, K., Toda, S., and Fuma K. (1978). Toxic effects of cadmium on *Euglena gracilis* grown in zinc deficient and zinc sufficient media. *Agric. Biol Chem.* **42,** 901–907.

Nalewajko, C., and Paul, B. (1985). Effects of manipulations of aluminum concentrations and pH on phosphate uptake and photosynthesis of planktonic communities in two Precambrian shield lakes. *Can. J. Fish. Aquat. Sci.* **42,** 1946–1953.

Newbold, J. D., Elwood, J. W., O'Neill, R. V., and Sheldon, A. L. (1983). Phosphorus dynamics in a woodland stream ecosystem: A study of nutrient spiralling. *Ecology* **64,** 1249–1265.

Niederlehner, B. R., and Cairns, J. (1993). Effects of previous zinc exposure on pH tolerance of periphyton communities. *Environ. Toxicol. Chem.* **12,** 743–753.

Nowell, A. R. M., and Jumars, P. A. (1984). Flow environments of aquatic benthos. *Ann. Rev. Ecol. Syst.* **15,** 303–328.

Nyholm, N., and Källqvist, T. (1989). Methods for growth inhibition toxicity tests with freshwater algae. *Environ. Toxicol. Chem.* **8,** 689-703.

Okamura, H., and Aoyama, I. (1994). Interactive toxic effect and distribution of heavy metals in phytoplankton. *Environ. Toxicol. Water Qual.* **9,** 7–15.

Parent, L., and Campbell, P. G. C. (1994). Aluminum bioavailability to the green alga *Chlorella pyrenoidosa* in acidified synthetic soft water. *Environ. Toxicol. Chem.* **13,** 587–598.

Parker, D. R., and Bertsch, P. M. (1992). Formation of the "Al_{13}" tridecameric polycation under diverse synthesis conditions. *Environ. Sci. Technol.* **26,** 914–921.

Pascucci, P. R., and Sneddon, J. (1993). A simultaneous multielement flame atomic absorption study for the removal of lead, zinc, and copper by an algal biomass. *J. Environ. Sci. Health, Part A* **A28,** 1483–1493.

Patrick, R. (1978). Effects of trace metals in the aquatic ecosystem. *Am. Sci.* **66:** 185–191.

Paulson, A. J., Curl, H. C., and Gendron, J. F. (1994). Partitioning of Cu in estuarine waters. 2. Control of partitioning by the biota. *Mar. Chem.* **45,** 81–93.

Pawlik, B., and Skowroński, T. (1994). Transport and toxicity of cadmium: Its regulation in the cyanobacterium *Synechocystis aquatilis. Environ. Exp. Bot.* **34,** 225-233.

Pawlik, B., Skowroński, T., Ramazanow, Z., Gardestrom, P., and Samuelsson, G. (1993). pH-dependent cadmium transport inhibits photosynthesis in the cyanobacterium *Synechocystis aquatilis. Environ. Exp. Bot.* **33,** 331–337.

Pearson, G. A., and Davison, I. R. (1994). Freezing stress and osmotic dehydration in *Fucus distichus* (Phaeophyta): Evidence for physiological similarity. *J. Phycol.* **30,** 257–267.

Pellegrini, M., Laugier, A., Sergent, M., Phantanlu, R., Valls, R., and Pellegrini, L. (1993). Interactions between the toxicity of the heavy metals cadmium, copper, zinc in combinations and the detoxifying role of calcium in the brown alga *Cystoseira barbata. J. Appl. Phycol.* **5,** 351–51.

Peterson, H. G., Healey, F. P., and Wagemann, R. (1984). Metal toxicity to algae: A highly pH dependent phenomenon. *Can. J. Fish. Aquat. Sci.* **41,** 974–979.

Pettersson, A., Hällbom, L., and Bergman, B. (1985). Physiological and structural responses of the cyanobacterium *Anabaena cylindrica* to aluminium. *Physiol. Plant.* **63,** 153–158.

Pillsbury, R., and Kingston, J. C. (1990). The pH-independent effect of aluminum on cultures of phytoplankton from an acidic Wisconsin lake. *Hydrobiologia* **194,** 225–233.

Prevot, P., and Soyer-Gobillard, M.-O. (1986). Combined action of cadmium and selenium on two marine dinoflagellates in culture, *Prorocentrum micans* Ehrbg. and *Crypthecodinium cohnii* Biecheler. *J. Protozool.* **33,** 42–47.

Puiseux-Dao, S. (1989). 'Phytoplankton model' in ecotoxicology. *In* "Aquatic Ecotoxicology: Fundamental Concepts and Methodologies" (A. Boudou and F. Ribeyre, eds.), Vol. 2, pp. 163-185. CRC Press, Boca Raton, FL.

Rachlin, J. W., and Grosso, A. (1993). The growth response of the green alga *Chlorella vulgaris* to combined divalent cation exposure. *Arch. Environ. Contam. Toxicol.* **24,** 16–20.

Rai, A. K., and Abraham, G. (1993). Salinity tolerance and growth analysis of the cyanobacterium *Anabaena doliolum*. *Bull. Environ. Contam. Toxicol.* **51**, 724–731.

Rai, L. C., Gaur, J. P., and Kumar, H. D. (1981). Phycology and heavy-metal pollution. *Biol. Rev. Cambridge Philos. Soc.* **56**, 99–151.

Rai, U. N., Tripathi, R. D., and Kumar, N. (1992). Bioaccumulation of chromium and toxicity on growth, photosynthetic pigments, photosynthesis, *in vivo* nitrate reductase activity and protein content in a chlorococcalean green alga *Glaucocystis*. *Chemosphere* **25**, 1721–1732.

Ramelow, G. J., Maples, R. S., Thompson, R. L., Mueller, C. S., Webre, C., and Beck, J. N. (1987). Periphyton as monitors for heavy metal pollution in the Calcasieu River estuary. *Environ. Pollut.* **43**, 247–261.

Rand, G. M., and Petrocelli, S. R. (1985). Introduction. *In* "Fundamentals of Aquatic Toxicology: Methods and Applications" (G. M. Rand and S. R. Petrocelli, eds), pp. 1–28. Hemisphere Publishing, New York.

Reinfelder, J. R., and Fisher, N. S. (1994). The assimilation of elements ingested by marine planktonic bivalve larvae. *Limnol. Oceanogr.* **39**, 12–20.

Reinke, D. C., and DeNoyelles, F., Jr. (1985). The species-specific effects of sublethal concentrations of cadmium on freshwater phytoplankton communities in a Canadian shield lake. *Can. J. Bot.* **63**, 1997–2003.

Rijstenbil, J. W., Derksen, J. W. M., Gerringa, L. J. A., Poortvliet, T. C. W., Sandee, A., Vandenberg, M., Vandrie, J., and Wijnholds, J. A. (1994). Oxidative stress induced by copper: Defense and damage in the marine planktonic diatom *Ditylum brightwellii* grown in continuous cultures with high and low zinc levels. *Mar. Biol. (Berlin)* **119**, 583–590.

Riseng, C. M., Gensemer, R. W., and Kilham, S. S. (1991). The effect of pH, aluminum, and chelator manipulations on the growth of acidic and circumneutral species of *Asterionella*. *Water, Air, Soil Pollut.* **60**, 249–261.

Robarts, R. D., Evans, M. S., and Arts, M. T. (1992). Determinants of phytoplankton production in two saline prairie lakes with high sulphate concentrations. *Can. J. Fish. Aquat. Sci.* **49**, 2281–2290.

Robinson, P. K., and Wilkinson, S. C. (1994). Removal of aqueous mercury and phosphate by gel entrapped *Chlorella* in packed-bed reactors. *Enzyme Microb. Technol.* **16**, 802–807.

Röderer, G. (1980). On the toxic effects of tetraethyl lead and its derivatives on the chrysophyte *Poterioochromonas malhamensis*. *Environ. Res.* **23**, 371–384.

Rothschild, L. J., Giver, L. J., White, M. R., and Mancinelli, R. L. (1994). Metabolic activity of microorganisms in evaporites. *J. Phycol.* **30**, 431–438.

Roy, D., Greenlaw, P. N., and Shane, B. S. (1993). Adsorption of heavy metals by green algae and ground rice hulls. *J. Environ. Sci. Health, Part A,* **A28**, 37-50.

Rueter, J. G., Jr., O'Reilly, K. T., and Petersen, R. R. (1987). Indirect aluminum toxicity to the green alga *Scenedesmus* through increased cupric ion activity. *Environ. Sci. Technol.* **21**, 435–438.

Rushforth, S. R., Brotherson, J. D., Fungladda, N., and Evenson, W. E. (1981). The effects of dissolved heavy metals on attached diatoms in the Uintah Basin of Utah, U.S.A. *Hydrobiologia* **83**, 313–323.

Saint-Louis, R., Pelletier, É., Marsot, P., and Fournier, R. (1994). Distribution et effets du chlorure de tributylétain et de ses produits de dégradation sur la croissance de l'algue marine *Pavlova lutheri* en culture continue. *Water Res.* **28**, 2533–2544.

Sanders, J. G., and Riedel, G. F. (1993). Trace element transformation during the development of an estuarine algal bloom. *Estuaries* **16**, 521–532.

Santiago, S., Thomas, R. L., Larbaigt, G., Rossel, D., Echeverria, M. A., Tarradellas, J., Loizeau, J. L., Mccarthy, L., Mayfield, C. I., and Corvi, C. (1993). Comparative ecotoxicity of suspended sediment in the lower Rhone river using algal fractionation, Microtox® and *Daphnia magna* bioassays. *Hydrobiologia* **252**, 231-244.

Schäfer, H., Hettler, H, Fritsche, U., Pitzen, G., Röderer, G., and Wenzel, A. (1994). Biotests using unicellular algae and ciliates for predicting long-term effects of toxicants. *Ecotoxicol. Environ. Saf.* **27,** 64–81.

Schubert, H., Fulda, S., and Hagemann, M. (1993). Effects of adaptation to different salt concentrations on photosynthesis and pigmentation of the cyanobacterium *Synechocystis* sp. PCC-6803. *J. Plant Physiol.* **142,** 291–295.

Sears, J. R., Pecci, K. J., and Cooper, R. A. (1985). Trace metal concentrations in offshore, deep-water seaweeds in the western north Atlantic Ocean. *Mar. Pollut. Bull.* **16,** 325–328.

Sedlacek, J., Källqvist, T., and Gjessing, E. (1983). Effect of aquatic humus on uptake and toxicity of cadmium to *Selenastrum capricornutum* Printz. *In* "Aquatic and Terrestrial Humic Materials" (R. F. Christman and E. T. Gjessing eds.), pp. 495-516. Ann Arbor Sci. Publ., Ann Arbor, MI.

Silberberg, M. (1996). Chemistry: "The Molecular Nature of Matter and Change." Mosby-Year Book, St. Louis, MO.

Singh, C. B., and Singh, S. P. (1992). Protective effects of Ca^{2+}, Mg^{2+}, Cu^{2+}, and Ni^{2+} on mercury and methylmercury toxicity to a cyanobacterium. *Exotoxicol. Environ. Saf.* **23,** 1–10.

Skowroński, T. (1986). Adsorption of cadmium on green microalga *Stichococcus bacillaris. Chemosphere* **15,** 69–76.

Skowroński, T., and Przytocka-Jusiak, M. (1986). Cadmium removal by green alga Stichococcus bacillaris. *Chemosphere* **15,** 77–79.

Söderlund, S., Forsberg, Å., and Pedersén, M. (1988). Concentrations of cadmium and other metals in *Fucus vesiculosus* L. and *Fontinalis dalecarlica* Br. Eur. from the northern Baltic Sea and the southern Bothnian Sea. *Environ. Pollut.* **51,** 197–212.

Starodub, M. E., Wong, P. T. S., Mayfield, C. I., and Chau, Y. K. (1987). Influence of complexation and pH on individual and combined heavy metal toxicity to a freshwater green alga. *Can. J. Fish Aquat Sci.* **44,** 1173–1180.

Stauber, J. L., and Florence, T. M. (1985a). The influence of iron on copper toxicity to the marine diatom *Nitzschia closterium* (Ehrenberg) W. Smith. *Aquat. Toxicol.* **6,** 297-305.

Stauber, J. L., and Florence, T. M. (1985b). Interactions of copper and manganese: A mechanism by which manganese alleviates copper toxicity to the marine diatom *Nitzschia closterium* (Ehrenberg) W. Smith. *Aquat. Toxicol.* **7,** 241–254.

Stauber, J. L., and Florence, T. M. (1989). The effect of culture medium on metal toxicity to the marine diatom *Nitzschia closterium* and the freshwater green alga *Chlorella pyrenoidosa. Water Res.* **23,** 907–911.

Steele, R. L., and Thursby, G. B. (1983). A toxicity test using life stages of *Champia parvula* (Rhodophyta). *In* "Aquatic Toxicology and Hazard Assessment: Sixth Symposium (W. E. Bishop, R. D. Cardwell, and B. B. Heidolph, eds.), ASTM STP 802, pp. 73–89. Am. Soc. Mater., Philadelphia.

Steinman, A. D., Mulholland, P. J., Palumbo, A. V., Deangelis, D. L., and Flum, T. E. (1992). Lotic ecosystem response to a chlorine disturbance. *Ecol. Appl.* **2,** 341–355.

Stephenson, M., and Turner M. A. (1993). A field study of cadmium dynamics in periphyton and in *Hyalella azteca* (Crustacea, Amphipoda). *Water, Air, Soil Pollut.* **68,** 341–361.

Stewart, A. J., Hill, W. R., and Boston, H. L. (1993). Grazers, periphyton and toxicant movement in streams. *Environ. Toxicol. Chem.* **12,** 955–957.

Stronkhorst, J., Vos, P. C., and Misdorp, R. (1994). Trace metals, PCBs and PAHs in benthic (epipelic) diatoms from intertidal sediments—A pilot study. *Bull. Environ. Contam. Toxicol.* **52,** 818–824.

Subba Rao, D. V. (1981). Effect of boron on primary production of nanoplankton. *Can. J. Fish. Aquat. Sci.* **38,** 52–58.

Subramanian, V. V., Sivasubramanian, V., and Gowrinathan, K. P. (1994). Uptake and recovery of heavy metals by immobilized cells of *Aphanocapsa pulchra* (Kutz) Rabenh. *J. Environ. Sci. Health Part A* **A29,** 1723–1733.

Sze, P. (1986). "A Biology of the Algae." Wm. C. Brown, Dubuque, Iowa.

Takeda, T., Nakano, Y., and Shigeoka, S. (1993). Effects of selenite, CO_2 and illumination on the induction of selenium-dependent glutathione peroxidase in *Chlamydomonas reinhardtii*. *Plant Sci.* **94,** 81–88.

Taylor, G. J. (1988a). The physiology of aluminum phytotoxicity. Met. Ions Biol Syst. **24,** 123–163.

Taylor, G. J. (1988b). The physiology of aluminum tolerance. Met. Ions Biolo. Syst. **24,** 165–198.

Thompson P. A., and Couture, P. (1993). Physiology of carbon assimilation in a green alga during exposure to and recovery from cadmium. *Ecotoxicol. Environ. Saf.* **26,** 205–215.

Tilman, D. (1982). "Resource Competition and Community Structure." Princeton Univ. Press, Princeton, N J.

Tominaga, N., Takahata, M., and Tominaga, H. (1993). Effects of NaCl and KNO_3 concentrations on the abscisic acid content of *Dunaliella* sp. (Chlorophyta). *Hydrobiologia* **267,** 163–168.

Tubbing, D. M. J., Admiraal, W., Cleven, R. F. M. J., and Iqbal, M. (1994). The contribution of complexed copper to the metabolic inhibition of algae and bacteria in synthetic media and river water. *Water Res.* **28,** 37–44.

Twiss, M. R., and Nalewajko, C. (1992). Influence of phosphorus nutrition on copper toxicity to three strains of *Scenedesmus acutus* (Chlorophyceae). *J. Phycol.* **28,** 291–298.

Twiss, M. R., Welbourn, P. M., and Schwartzel, E. (1993). Laboratory selection for copper tolerance in *Scenedesmus acutus* (Chlorophyceae). *Can. J. Bot.* **71,** 333–338.

Venkataraman, S., and Kaushik, B. D. (1994). Fatty acid profile of a marine *Nostoc calcicola* under saline and non-saline conditions. *Curr. Sci.* **67,** 120–122.

Verma, S. K., Singh, R. K., and Singh, S. P. (1993). Copper toxicity and phosphate utilization in the cyanobacterium *Nostoc calcicola. Bull. Environ. Contam. Toxicol.* **50,** 192–198.

Visviki, I., and Rachlin, J. W. (1994). Acute and chronic exposure of *Dunaliella salina* and *Chlamydomonas bullosa* to copper and cadmium—Effects on growth. *Arch. Environ. Contam. Toxicol.* **26,** 149–153.

Vymazal, J. (1984). Short-term uptake of heavy metals by periphyton algae. *Hydrobiologia* **119,** 171–179.

Waber, J., and Wylie, J. C. (1993). Changes in RNA and protein synthesis associated with the adaptation of *Synechococcus* sp. to 0.4 *M* NaCl. *Environ. Exp. Bot.* **33,** 539–543.

Walsh, G. E., McLaughlan, L. L., Lores, E. M., Louie, M. K., and Deans, C. H. (1985). Effects of organotins on growth and survival of two marine diatoms, *Skeletonema costatum* and *Thalassiosira pseudonana. Chemosphere* **14,** 383–392.

Walsh, R. S., and Hunter, K. A. (1992). Influence of phosphorus storage on the uptake of cadmium by the marine alga *Macrocystis pyrifera. Limnol. Oceanogr.* **37,** 1361–1369.

Wang, W. (1985). Effect of iron and zinc interaction on algal communities. *In* "Aquatic Toxicology and Hazard Assessment: Seventh Symposium" (R. D. Cardwell, R. Purdy, and R. C. Bahner, eds.), ASTM STP 854, pp. 187–201. Am. So. Test. Mater., Philadelphia.

Wang, W. (1986). Acclimation and response of algal communities from different sources to zinc toxicity. *Water, Air, Soil Pollut.* **28,** 335–349.

Wängberg, S.-Å., and Blanck, H. (1988). Multivariate patterns of algal sensitivity to chemicals in relation to phylogeny. *Ecotoxicol. Environ. Saf.* **16,** 72–82.

Wängberg, S.-Å., and Blanck, H. (1990). Arsenate sensitivity in marine periphyton communities established under various nutrient regimes. *J. Exp. Mar. Biol. Ecol.* **139,** 119–134.

Wängberg, S.-Å., Heyman, U., and Blanck, H. (1991). Long-term and short-term arsenate toxicity to freshwater phytoplankton and periphyton in limnocorrals. *Can. J. Fish. Aquat. Sci.* **48,** 173–182.

Watras, C. J., MacFarlane, J., and Morel, F. M. M. (1985). Nickel accumulation by *Scenedesmus* and *Daphnia:* Food-chain transport and geochemical implications. *Can. J. Fish. Aquat. Sci.* **42,** 724–730.

Weger, H. G., and Dasgupta, R. (1993). Regulation of alternative pathway respiration in *Chlamydomonas reinhardtii* (Chlorophyceae). *J. Phycol.* **29,** 300–308.

Weitzel, R. L., and Bates, J. M. (1981). Assessment of effluent impacts through evaluation of periphyton diatom community structure. *In* "Ecological Assessments of Effluent Impacts on Communities of Indigenous Aquatic Organisms" (J. M. Bates and C. I. Weber eds.), ASTM STP 730, pp. 142–165. Am. Soc. Test. Mater., Philadelphia.

Whitton, B. A. (1984). Algae as monitors of heavy metals in freshwaters. *In* "Algae as Ecological Indicators" (L. E. Shubert, ed.) pp. 257-280. Academic Press, London.

Wikfors, G. H., Twarog, J. W., Ferris, G. E., Smith, B. C., and Ukeles, R. (1994). Survival and growth of post-set oysters and clams on diets of cadmium-contaminated microalgal cultures. *Mar. Environ. Res.* **37,** 257–281.

Wilde, E. W., and Benemann, J. R. (1993). Bioremoval of heavy metals by the use of microalgae. *Biotechnol. Adv.* **11,** 781–812.

Williams, M. J., Ogle, R. S., Knight, A. W., and Burau, R. G. (1994). Effects of sulfate on selenate uptake and toxicity in the green alga *Selenastrum capricornutum. Arch. Environ. Contam. Toxicol.* **27,** 449–453.

Winner, R. W., and Owen, H. A. (1991a). Seasonal variability in the sensitivity of freshwater phytoplankton communities to a chronic copper stress. *Aquat. Toxicol.* **19,** 73–88.

Winner, R. W., and Owen, H. A. (1991b). Toxicity of copper to *Chlamydomonas reinhardtii* (Chlorophyceae) and *Ceriodaphnia dubia* (Crustacea) in relation to changes in water chemistry of a freshwater pond. *Aquat. Toxicol.* **21,** 157–170.

Wong, P. T. S. (1987). Toxicity of cadmium to freshwater microorganisms, phytoplankton, and invertebrates. *In* "Cadmium in the Aquatic Environment" (J. O. Nriagu and J. B. Sprague, eds.), Vol. 19, pp. 117–138. Wiley, New York.

Wong, P. T. S., and Trevors, J. T. (1988). Chromium toxicity to algae and bacteria. *In* "Chromium in the Natural and Human Environments" (J. O. Nriagu and E. Nieboer, eds.), Vol. 20, pp. 305-316. Wiley, New York.

Wong, P. T. S., Chau, Y. K., and Luxon, P. L. (1978). Toxicity of a mixture of metals on freshwater algae. *J. Fish. Res. Board Can.* **35,** 479–481.

Wong, P. T. S., Chau, Y. K., and Patel, D. (1982). Physiological and biochemical responses of several freshwater algae to a mixture of metals. *Chemosphere* **11,** 367–376.

Wong, P. T. S., Chau, Y. K., Yaromich, J. L., and Kramar, O. (1987). Bioaccumulation and metabolism of tri- and dialkyllead compounds by a freshwater alga. *Can. J. Fish. Aquat. Sci.* **44,** 1257–1260.

Wong, P. T. S., Maguire, R. J., Chau, Y. K., and Kramar, O. (1984). Uptake and accumulation of inorganic tin by a freshwater alga, *Ankistrodesmus falcatus. Can. J. Fish. Aquat. Sci.* **41,** 1570-1574.

Wong, S. L., Nakamoto, L., and Wainwright, J. F. (1994). Identification of toxic metals in affected algal cells in assays of wastewaters. *J. Appl. Phycol.* **6,** 405–414.

Wood, J. M., and Wang, H. F. (1985). Strategies for microbial resistance to heavy metals. *In* "Chemical Processes in Lakes" (V. Stumm, ed.), pp. 81-98. Wiley, New York.

Wrench, J. J., and Addison, R. F. (1981). Reduction, methylation, and incorporation of arsenic into lipids by the marine phytoplankton *Dunaliella tertiolecta. Can. J. Fish. Aquat. Sci.* **38,** 518–523.

Xyländer, M., and Braune, W. (1994). Influence of nickel on the green alga *Haematococcus lacustris* Rostafinski in phases of its life cycle. *J. Plant Physiol.* **144,** 86–93.

Zhang, W. X., and Majidi, V. (1994). Monitoring the cellular response of *Stichococcus bacillaris* to exposure of several different metals using *in vivo* P-31 NMR and other spectroscopic techniques. *Environ. Sci. Technol.* **28,** 1577–1581.

15 Effects of Organic Toxic Substances

Kyle D. Hoagland, Justin P. Carder,
and Rebecca L. Spawn

Department of Forestry, Fisheries and Wildlife
University of Nebraska
Lincoln, Nebraska 68583

I. INTRODUCTION

A wide variety of organic toxicants have been detected in freshwater systems, including herbicides, insecticides, surfactants, polycyclic aromatic hydrocarbons (PAHs), polyhalogenated biphenyls, and other organic compounds related to industry (e.g., detergents, dyes, oils, solvents, and resins). Because this array of chemicals occurs in surface waters as a result of both point-source inputs from industrial and municipal sources and nonpoint-source (NPS) inputs from agricultural lands and the atmosphere, its impacts can be localized and site specific or widespread throughout an entire region. Though the source and fate of these toxicants differ widely among localities based on land use, reports of their occurrence in aquatic ecosystems have increased dramatically over the past decade. This is in part due to increased monitoring efforts by state and federal agencies as a result of increased public concern for water quality.

The occurrence of some organic toxicants is widespread in both lotic and lentic systems. A recent survey by the U.S. Geological Survey of surface waters in the midwestern United States, comprising 149 sites in 122 river basins across a 10-state area, revealed that 55% of the basins exceeded the U.S. Environmental Protection Agency (E.P.A.)-promulgated maximum contaminant level (PMCL) for the herbicide atrazine (Thurman *et al.*, 1992). Atrazine and other pesticides have also been detected in rainwater in North America and Europe (Arthur *et al.*, 1976; Richards *et al.*, 1987; Buser, 1990) and in groundwater in the United States and Canada (Frank *et al.*, 1987; Ritter, 1990). Consequently, fresh waters throughout the Northern Hemisphere are subject to periodic low-level pesticide inputs. PAHs may also be widespread in aquatic environments, because they occur as a result of large-scale processes such as forest fires, volcanic activity, and fossil-fuel combustion (Bruno *et al.*, 1982). Unfortunately, state and federal agencies often do not include PAHs as part of their routine monitoring programs. Other organic toxicants in streams and lakes are typically monitored only on a case-by-case basis.

Levels of nonpoint-source organic toxicants (e.g., pesticides) during stream baseflows are generally low or undetectable. For example, the mean atrazine concentration in the Thurman *et al.* (1992) survey appears to have been between 0.2 and 0.5 μg liter^{-1}. The rapid dilution of other point-source toxicant inputs in lotic systems undoubtedly also results in low-level exposures in benthic algal communities at increasing distances downstream, although the levels of some toxicants in sediments may increase with stream order. However, pesticide levels in streams draining agricultural lands can reach relatively high levels during spring storm runoff events. Spalding and Snow (1989) reported atrazine concentrations as high as 87 μg liter^{-1} in Shell Creek, Nebraska, during a spring storm event, and Thurman *et al.* (1991) found >100 μg liter^{-1} during spring flushes as part of their 1989–1990 midwestern United States survey. The median total herbicide concentration in 1989 exceeded 10 μg liter^{-1} in postapplication stream samples. Other organic toxicants can also reach high levels in nature, for example, anionic surfactants as high as 12.6 mg liter^{-1} have been reported (Rapaport and Eckhoff, 1990). Thus, benthic algal communities may be exposed to acute levels of toxicants over periods commensurate with one generation time or less.

An extensive literature on the effects of organic toxicants on algae has developed over the past 10–15 years, though a significant percentage of the research has been directed toward planktonic species such as the green algae *Chlorella* sp. or *Selenastrum capricornutum,* common subjects of standard single-species bioassays [see O'Brien and Dixon (1976), Butler (1977), Lal (1984), Padhy (1985), and Munawar *et al.* (1988) for previous reviews on the effects of various groups of organic toxicants on algae, including planktonic and marine forms]. Nevertheless, some organic toxicants have been relatively well studied for their effects on benthic algae, particularly atrazine and surfactants. The goal of this chapter is not to present an exhaustive summary of all organic toxicants known to occur in fresh waters and their effects on attached algae, rather the focus is on the most prevalent toxicants, with generalizations concerning their ecological impacts on benthic algae at the population and community levels.

II. CONCEPTUAL FRAMEWORK

Both individual algal populations and entire benthic algal communities can be affected by exposure to organic toxicants. Population-level impacts often are referred to as direct effects, whereas community-level effects include a significant indirect component. From an ecological perspective, grazing and physical disturbance (see Steinman, Chapter 12 and Peterson, Chapter 13, respectively, this volume), can produce similar outcomes, depending on the timing (both frequency and occurrence), duration, and intensity of "exposure" to the disturbance or grazing. Similarly, differential species tolerance to pesticides or grazing could result in a common outcome in terms of community composition and biomass. Nevertheless, there are some important physiological- and cellular-level differences in the effects of organic toxicants, based in part on a particular toxicant's mode of action. Interestingly, the mode of action for many of these toxicants is essentially unknown for algae.

In this chapter, the impacts of organic toxicants on the ecology of benthic algae will be explored at several scales. In addition to the obvious potential for the direct effects of individual chemicals at the subcellular and whole-organism levels, the impacts of chemical mixtures and indirect effects via food web interactions will be considered. Physiological- and cellular-level effects will also be discussed. New areas of inquiry are identified in an effort to stimulate further research and a better understanding of this environmentally important class of toxicants.

III. DIRECT EFFECTS

A. Pesticides

1. Atrazine

Atrazine is one of the two most heavily used pesticides in the United States today, with over 35 million kilograms of active ingredient applied annually, primarily for the control of broadleaf weeds in corn and sorghum (Gianessi, 1987). It is also the most intensively studied organic toxicant with respect to freshwater algae; nevertheless, generalizations are difficult because algal responses to atrazine vary widely depending on the concentrations used, duration of exposure, conditions of exposure (e.g., field versus laboratory), and algal species tested (Table I). Atrazine is highly toxic to algae in culture at concentrations ≥100 μg $liter^{-1}$ (Butler *et al.*, 1975; Hersh and Crumpton, 1987; Stratton, 1984) and levels as low as 1–10 μg $liter^{-1}$ are inhibitory (Butler *et al.*, 1975; deNoyelles *et al.*, 1982), however, fewer single-species assays have been conducted on benthic algae (see Solomon *et al.* 1996). O'Kelley and Deason (1976) assayed 36 river algal isolates for 12 pesticides and found that 1 μg $liter^{-1}$ atrazine inhibited growth by 10–50% or more after 2 weeks in 26 isolates, whereas Torres and O'Flaherty (1976) found that the xanthophytes *Tribonema* sp. and *Vaucheria geminata* and the filamentous green alga *Stigeoclonium tenue* were inhibited by 41–67% after 1 week at 1 μg $liter^{-1}$. Similarly, Hughes *et al.* (1988) reported a 5-d EC_{50} (the "effective concentration" that results in a sublethal response such as immotility or altered physiology in 50% of the population or community) for the diatom *Navicula pelliculosa* of 60 μg $liter^{-1}$. On the other hand, EC_{50} values for atrazine for clones of the centric diatom *Cyclotella meneghiniana* were much higher (99–243 μg $liter^{-1}$), based on inhibition of O_2 evolution over a 5-min period. For *Ulothrix fimbriata* (green alga) and *Plectonema boryanum* (blue-green alga), minimum inhibitory concentrations (10 d) on diffusion gradient agar plates were 1.2 and 17.1 mg $liter^{-1}$, respectively. Although differences in methodologies make conclusions drawn from these studies untenable, it does seem clear that atrazine introduced into single-species benthic algal cultures under laboratory conditions can negatively affect photosynthesis and growth in some algal species at low, environmentally realistic concentrations that are nontoxic toward other algal species.

Species-specific responses to atrazine have also been shown in whole-community studies. Several researchers have indicated that diatoms are generally more tolerant of atrazine (Herman *et al.*, 1986; Hoagland *et al.*, 1993), as well as other triazine herbicides (Gurney and Robinson, 1989; Kasai *et al.*, 1993), yet Herman *et al.* (1986) focused only on attached algae in this regard. Some species of periphytic blue-green algae (Herman *et al.*, 1986) or green algae (Hamala and Kollig, 1985; Hamilton *et al.*, 1987) have been found to be more sensitive than other groups of algae to atrazine, however, Hamilton *et al.* (1987) also stated that the green alga *Scenedesmus acutiformis* was the least affected species examined. Based on the limited data available thus far, it appears that benthic algal responses to atrazine, and likely other toxicants as well, are species specific. Indeed, significant clonal differences were found in *Cyclotella meneghiniana*, with EC_{50} values of 99, 105, and 243 μg $liter^{-1}$ for the three clones assayed (Millie and Hersh, 1987). Paterson and Wright (1988) also found differen-

TABLE I Comparisons of the Effects of Atrazine on Periphytic Algae in Cultures, Laboratory Microcosms, and Artificial Enclosures[a]

Atrazine concentration (*μg liter*$^{-1}$)	*Exposure duration*	*Organism(s)/system*	*Effects*	*Reference*
Single-Species Culture Assays				
1–25,000	14 d	37 isolates	All isolates inhibited at 1000 μg liter^{-1}; 72% at 1 μg liter^{-1}	O'Kelley and Deason (1976)
1–1000	7 d	4 filamentous species	Chlorophyll production inhibited at ≥ 1000 μg liter^{-1} in all spp.	Torres and O'Flaherty (1976)
1–338	5 min	*Cyclotella* (3 races)	EC_{50} = 74–243 μg liter^{-1} depending on race and EC_{50} model	Millie and Hersh (1987)
50–2000	7–8 d	4 blue-green species	Growth inhibited at ≥50 μg liter^{-1}	Shabana (1987a,b)
100–3200	5 d	*Navicula pelliculosa*	EC_{50} = 60 μg liter^{-1}; 1710 μg liter^{-1} phytostatic	Hughes *et al.* (1988)
2000–18,000	10 d	16 isolates	Mean minimum inhibitory conc. = 1.2 – 17.1 mg liter^{-1}	Paterson and Wright (1988)
Community Assays				
10–10,000	21 d	Artificial streams	Inhibition at 1000 μg liter^{-1} little evidence for induced resistance	Kosinski and Merkle (1984)
100	15–37 d	Microcosms	Productivity, biomass, diversity declined	Hamala and Kollig (1985)
25	30 d	Artificial streams	Periphyton biomass, productivity, and respiration unaffected	Lynch *et al.* (1985)
100 (80–170)[b]	1–329 d	Limnocorrals	Periphyton production inhibited and community composition changed	Herman *et al.* (1986)
20–5000	1–136 d	Multiple[c]	EC_{50} = 37 – 308 μg liter^{-1} in culture; significant inhibition at 1000 μg liter^{-1} in microcosms; little effect at 20 μg liter^{-1} in ponds, inhibition at 100 μg liter^{-1}	Larsen *et al.* (1986)
100–10,000	7 d	Artificial streams	Significant declines in net community productivity at conc. ≥100 μ liter^{-1} at 3 days and 7 days	Moorhead and Kosinski (1986)
100–2000 (80–1560)[b]	4–223 d	PVC tubes in lake	Shift in periphyton composition; productivity declined 21–82% then returned to control levels; evidence for induced resistance	Hamilton *et al.* (1987)
24 or 134	4–12 d	Artificial streams	Significant inhibition of biomass at 10° and 25°C at 134 μg liter^{-1}	Krieger *et al.* (1988)
2, 30, 100	24 h (×2)	Microcosms	No significant effect on cell density at 100 μg liter^{-1}	Jurgensen and Hoagland (1990)

[a]Modified from Jurgensen and Hoagland (1990).
[b]Nominal concentration listed, with actual concentration in parentheses.
[c]Cultures, laboratory enclosures, field enclosures.

tial atrazine tolerance among varieties and strains of green and blue-green algae.

Consequently, from a toxicological standpoint, the response of a particular benthic algal assemblage is dependent on the composition of the community when the atrazine exposure occurs (Herman *et al.,* 1986). Differential species responses could result in a shift in periphyton species composition, with no significant change in algal biovolume. Thus, many stream and lake systems in the midwestern United States and Europe may already contain benthic algal assemblages that reflect artificial selection via herbicide inputs. Hamilton *et al.* (1987) found that periphyton diversity declined in littoral enclosures exposed to atrazine twice over a 35–d period. Periphyton physiognomy, which is a function of species composition as well, also may influence the overall community response to toxicants by creating diffusion gradients within the biofilm, as has been demonstrated for nutrients (Wetzel, 1993). Thus, additional longer-term experiments at the community level are needed to determine the effects of atrazine on previously unexposed periphyton assemblages at ecologically relevant herbicide concentrations and exposure durations.

There is increasing evidence that benthic algae "recover" from atrazine treatment, that is, they return to preexposure levels of growth, biomass, or photosynthetic rate at some time after exposure, despite the presence of previously inhibitory levels of the parent herbicide. Kosinski and Merkle (1984) reported that photosynthetic inhibition of artificial stream periphyton exposed to 1000 μg liter^{-1} recovered completely after 1 week during a 3-week experiment, and Hamilton *et al.* (1987) observed periphyton productivity recovery from 1560 μg liter^{-1} atrazine 21 d after treatment in a 55-d study [it is important to note that reports of atrazine dissipation rates vary, with half-lives ranging from 7 d (Wauchope, 1978) to 231 d (Kruger *et al.,* 1993), depending on soil and water conditions]. Lozano and Pratt (1994) showed that periphyton exposed to diquat under low-nutrient conditions had limited ability to recover, whereas under medium or high nutrients the community recovered to control levels after 23 d. These algistatic responses (i.e., inhibitory, not lethal, effects) are known to occur after up to 76 d (Hamala and Kollig, 1985) and at concentrations as high as 1710 μg liter^{-1} (Hughes *et al.,* 1988). Freshwater phytoplankton assemblages exhibit very similar recovery responses following atrazine inputs (Brockway *et al.,* 1984; Hoagland *et al.,* 1993), thus the phenomenon is characteristic of the algae not the benthic habitat.

The mechanism of recovery is unknown, but may involve changes in the cell wall and/or membranes (see Torres and O'Flaherty, 1976), adsorption outside the cell (Vérber *et al.,* 1981), and modification of key photosynthetic enzymes (Hamilton *et al.,* 1987), as well as induction of detoxification enzyme systems, increased production of the enzyme(s) inhibited by the herbicide, and modification of the herbicide binding site. There is little evidence for atrazine metabolism by algae (O'Kelley and Deason, 1976) and detoxification systems have not been characterized. It is reasonable to assume that some reports of recovery from atrazine treatment may be "apparent recovery" due to photodegradation and microbial degradation and/or adsorption of the parent compound to sediments over time (i.e., not all studies include atrazine monitoring throughout the duration of the experiment). This may also explain the observation that in some field bioassays, atrazine had no effect even at 24–100 μg liter^{-1} (Krieger *et al.*, 1988; Jurgensen and Hoagland, 1990). Some taxa may also avoid the direct algicidal effects of herbicides by their ability to grow heterotrophically (Goldsborough and Robinson, 1986), although this too would reduce overall growth rates.

Despite their apparent ability to recover from atrazine exposure, there is little evidence that prior exposure confers resistance in benthic algae (Kosinski and Merkle, 1984; Hamala and Kollig, 1985) or phytoplankton (Brockway *et al.*, 1984), although prior studies in other systems have shown indications of resistance to pesticides and other toxicants (e.g., deNoyelles *et al.*, 1982) and DCMU-resistant strains of blue-green algae have been produced in culture (Bisen and Shanthy, 1993). In some instances, low levels of atrazine may actually enhance growth of some benthic algae (Hamala and Kollig, 1985; Hamilton *et al.*, 1987), perhaps by supplying nitrogen via atrazine degradation (in both studies, smaller cells were enhanced), or by indirect effects. Shabana (1987b) found that atrazine also increased heterocyst formation in the blue-green genera *Aulosira*, *Anabaena*, and *Nostoc*. Clearly, induced resistance is an area that warrants further research considering its potential ecological importance to many benthic algal assemblages.

2. Other Herbicides

Butler (1977) provided a thorough review of earlier literature on the effects of other herbicides (as well as insecticides and fungicides) on algae, including compounds no longer available in the United States (e.g., DDT). Again, a majority of previous research has focused on common planktonic algae and not benthic species. Although relatively little information is available for most herbicides assayed for attached algae, simazine and glyphosate have been examined in several studies (Table II). Another member of the *s*-triazine family of herbicides, simazine (Pricep), appears to be slightly less toxic than atrazine to attached algae, with EC_{50} values of approximately 1.0 μg liter^{-1} (Torres and O'Flaherty, 1976; O'Neal and Lembi, 1983; Goldsborough and Robinson, 1986). O'Neal and Lembi (1983) found that the toxicity of simazine was dependent on light intensity, with higher light required for algicidal effects within green algal mats. Millie *et al.* (1992) reported a similar increase in photosynthetic inhibition by simazine with increasing light acclimation levels.

TABLE II Comparison of the Effects of Other Herbicides on Periphytic Algae[a]

Herbicide	*Exposure duration*	*Organism(s)/system*	*Effects*	*Reference*
Single-Species Culture Assays				
2,4-D, simazine	7 d	4 filamentous species	2, 4-D had no effect on chlorophyll production at ≤100 mg liter^{-1}; simazine effective in all but one species at 1 mg liter^{-1}	Torres and O'Flaherty (1976)
Terbuthylazine, prometryne, alachlor, dinoseb, fluometuron, profluralin, 2,4-D	? min	2 blue-green fil., 1 green unicell	2, 4-D and profluralin had no effect on photosynthetic rate at ≤10 μ*M*; alachlor, dinoseb, and fluometuron effective only at 1–10 μ*M*; terbuthylazine and prometryne inhibited at 0.1–1 μ*M*	Hawxby *et al.* (1977)
Picloram, prometryn, fluometuron, dinoseb	48 or 72h	*Lyngbya birgei*	Nutrients, esp. nitrogen, increased toxic effects by increasing growth	Tubea *et al.* (1981)
Simazine	5–6 min/15–45 d	3 filamentous greens	*Spirogyra* most susceptible, photosynthetic inhibition at 1.1 μ*M*; 5 μ*M* strongly inhibited growth of all taxa	O'Neal and Lembi (1983)
Diuron, glyphosate, paraquat, norflurazon	14 d	13 isolates	EC_{100} = 0.13 mg liter^{-1} (diuron), 11 mg liter^{-1} (glyphosate)	Blanck *et al.* (1984)
Terbutryn, diuron, monuron	10 d	16 isolates	Terbutryn and diuron most toxic	Paterson and Wright (1988)
Simetryne	4 d, 7d	56 isolates	EC_{50} = 6.5–1500 μg liter^{-1} depending on species and strain	Kasai *et al.* (1993)
Community Assays				
Ametryne, desmetryne, prometryne, terbutryne, cyanatryn, diquat, HZ 5914, metflurazone, WSCP, asulam	1 d–27 weeks	Multiple	Triazines effective in lab and field assays, esp. terbutryne and cyanatryn (0.1 mg liter^{-1} in lab, higher in field)	Robson *et al.* (1976)

Herbicide	Duration	System	Effect	Reference
Trifluralin, MSMA, paraquat	21 d	Artificial streams	Trifluralin had no effect on productivity at 10 mg liter^{-1}; MSMA inhibited production at 10 mg liter^{-1} paraquat variable	Kosinski and Merkle (1984)
Simazine, terbutryn	9 d–6 weeks	Limnocorrals	Simazine reduced algal biovolume significantly at 1 and 5 mg liter^{-1}; terbutryn more toxic, with $LC_{50} < 0.01$ mg liter^{-1}	Goldsborough and Robinson (1986)
Mixture of alachlor, atrazine, metolachlor, metribuzin	4 d	Artificial streams	Biomass significantly reduced at 10°C, but not 25°C, at 259 μg liter^{-1} total conc.	Krieger *et al.* (1988)
Simazine, terbutryn	87 d	Limnocorrals	Haptobenthos photosynthesis inhibited during first 2 weeks at 2 mg liter^{-1} simazine or 0.01 mg liter^{-1} terbutryn; herpobenthos inhibited throughout experiment	Gurney and Robinson (1989)
Hexazinone, simazine, sethoxydim, difenzoquat, glyphosate	3–4 h lab, 6–10 week enclosures, long term	Multiple[b]	EC_{50} low for hexazinone and simazine (<1 mg liter^{-1}), sethozydim intermediate, difenzoquat and glyphosate high (35–70 mg liter^{-1}); hexazinone inhibited biomass and productivity in field at 0.2 mg liter^{-1}; simazine had no negative effect	Goldsborough (1993)
Diquat	3–21 d	Microcosms	Inhibition highly species specific at 0.3–30 mg liter^{-1}	Melendez *et al.* (1993)
Gardoprim, gesapax	10 d	Grab sample	Gardoprim inhibited chl. *a* and growth at 50–100 μg liter^{-1}; gesapax had no effect	Shehata *et al.* (1993)

[a]Some taxa included may occur in other habitats such as soil, however, planktonic taxa have been excluded.
[b]Laboratory, linmocorrals, whole pond.

Benthic algae also exhibit a range of sensitivities to simazine; EC_{50} values for three species of filamentous green algae varied by a factor of more than three (O'Neal and Lembi, 1983) and diatoms were more tolerant than green algae in periphyton assemblages from limnocorrals (Gurney and Robinson, 1989). Goldsborough and Robinson (1986) found greater relative abundances of the diatom *Cocconeis placentula* in enclosures treated with 1.0 and 5.0 mg liter^{-1} simazine. They concluded that periphyton succession to a three-dimensional mat may be averted by short simazine exposures. Recovery to control levels (or above) after 2 weeks of growth has also been reported for simazine (Gurney and Robinson, 1989). It is interesting to note that epipelon photosynthetic rates did not recover in the same treatments and were more strongly inhibited, presumably due to simazine adsorption to sediment particles and resultant higher levels of exposure throughout the experiment. Induced resistance to simazine has also been reported, and although it can occur within 7 d at relatively high concentrations, the resistance is short-lived after removal of the herbicide (Goldsborough and Robinson, 1988).

Glyphosate (Roundup), however, is generally much less toxic to periphytic algae than atrazine and other herbicides, with EC_{50} values of 8.9–89 mg liter^{-1} [Goldsborough and Brown (1988); Goldsborough (1993) reported a range of 35–70 mg liter^{-1}]. Austin *et al.* (1991) suggested that glyphosate may actually serve as a source of phosphorus to benthic algae. In the longest experiment to date, Goldsborough (1993) indicated in preliminary results that glyphosate dissipates rapidly in the environment to ineffective levels.

Approximately 25 other herbicides have been examined for their effects on benthic algae, however, relatively few data exist for any single compound. This includes alachlor (Lasso), the most widely used pesticide in the United States in 1987 (Gianessi, 1987) and a herbicide often detected in agricultural streams (Thurman *et al.*, 1992). The range of toxicities of these herbicides is quite broad, ranging from no effect at relatively high levels of 2,4-D (Torres and O'Flaherty, 1976; Hawxby *et al.*, 1977) or profluralin (Hawxby *et al.*, 1977), to terbutryn which is generally more toxic than simazine (Goldsborough and Robinson, 1986) or atrazine, (Paterson and Wright, 1988). Hawxby *et al.* (1977) demonstrated that the relative inhibition of seven herbicides among four algal species varied by a factor of up to 45, and that the most toxic herbicides are those that specifically inhibit photosynthesis rather than other metabolic processes. Diquat (Aquacide), a herbicide commonly used for aquatic macrophyte control, was significantly more toxic to diatoms than to green algae in laboratory microcosms (Melendez *et al.*, 1993). In a study on the toxicity of 19 different chemicals (including three herbicides) on 13 freshwater algae, Blanck *et al.* (1984) found that species-dependent algal sensitivity varied by 3.3 orders of magnitude. Thus without specific data for a given herbicide, generalizations are not possible owing to the high degree of variability in toxic response as a result of differences in mode of action, carrier solvent, cell permeability, rate of degradation or metabolism, and other factors.

3. *Insecticides*

Relatively few insecticides have been studied with respect to their impacts on benthic algal communities. O'Kelley and Deason (1976) assayed 37 algal strains (including six or more potentially benthic taxa) for nine insecticides and found 10% or more growth inhibition in two or more strains at concentrations as low as 10 μg liter^{-1}. As is the case with herbicides, the sensitivity of benthic algae to the nine insecticides varied significantly depending on the taxon tested. Diazinon, carbaryl, and toxaphene were generally the most toxic to the algae assayed. Sicko-Goad and Andresen (1993a,b) demonstrated that the effects of trichlorobenzene (used to control termites and aquatic weeds) on diatoms were dependent on the time of exposure, as a function of diel lipid composition. Torres and O'Flaherty (1976) found that malathion inhibited chlorophyll production in *Stigeoclonium, Tribonema,* and *Vaucheria* by 100% at 1 μg liter^{-1} and presented direct evidence that algae can degrade malathion in the presence of light. Benthic algae may also bioconcentrate insecticides in the environment, as has been demonstrated for dieldrin [by a factor of 100–10,000×; Rose and McIntire (1970)], permethrin (Sundaram and Curry, 1991), and a wide variety of other common insecticides [see Butler (1977) for a review of the early literature and Heckman (1994) for a more recent summary, including herbicides].

B. Surfactants

1. *Introduction*

Surface-active agents (surfactants) are important ingredients in many domestic and industrial products ranging from detergents, pharmaceuticals and cosmetics to drilling muds, pesticides, and oil-spill dispersants, with approximately 2,150,000 metric tons of surfactants produced annually in the United States [Stanford Research Institute (SRI), 1984, as reported by Lewis and Hamm, 1986). Rosen (1988) defines a surfactant as a substance that, when present at low concentration in a system, adsorbs onto the surfaces or interfaces of the system. Surfactants are organic-based compounds consisting of both a hydrophobic and hydrophilic portion and are classified into four major groups depending on the nature of the hydrophilic group: anionic, cationic, zwitterionic, and nonionic (Rosen, 1988). Moreover, detergent builders, which enhance surfactant performance by reducing water hardness and so preventing redeposition of ions, and emulsifying oils (Woltering and Bishop, 1989) represent up to 15% of detergent products (Anderson *et al.,* 1985). The occurrence of detergent builders in the aquatic environment also warrants their inclusion in this discussion.

As a result of domestic and industrial usage, surfactants typically end up in municipal sewage systems. Because of their high-volume usage, concern over surfactant toxicity and biodegradability within publicly owned sewage treatment works (POTWs) has received much attention (LaPoint, 1993). Although many surfactants are either readily degraded or adsorbed within POTWs, most concern over surfactant contamination comes as a result of their occurrence in treatment plant effluents. Investigations into the effects of surfactants on nontarget aquatic organisms span several decades; however, the goal here is to provide an overview of the current literature.

Many standardized toxicity assays include planktonic green and blue-green algal species; however, most POTW effluent is discharged into stream systems where attached species likely predominate. Although toxicity data for attached algae will take precedence in the following discussion, toxicity data for planktonic species are included where pertinent. Woltering and Bishop (1989) found that toxicity data for detergent chemical effects on laboratory cultures of planktonic algae often overestimate the effects on periphytic species exposed to treated effluent. Because of the brief nature of this discussion, toxicity data for the most heavily produced and most frequently detected surfactants within each class will be discussed. A series of prior review papers summarize chronic toxicities of surfactants to aquatic organisms (Lewis, 1990, 1991, 1992; Nalewajko, 1994). In addition, Belanger (1994) provided a comprehensive review of community-level surfactant studies.

2. Anionic Surfactants

Anionic surfactants (the surface-active portion of the molecule bears a negative charge) comprise approximately 73% of the total U.S. surfactant consumption (Rosen, 1988) and are common constituents of toothpastes, cosmetics, shampoos, and wool-washing agents (Sivak *et al.,* 1982). Linear alkylbenzenesulfonate (LAS) accounts for 26% of the total U.S. surfactant consumption and is the most extensively used anionic surfactant in both granular and liquid detergents (Lewis *et al.,* 1993). LAS is the most rapidly biodegrading of all surfactants used in household detergent products (Sivak *et al.,* 1982) and several studies have shown that partially degraded LAS is not as toxic to aquatic organisms as the initial product (Painter and Zabel, 1988, as reported in Kimerle, 1989). Environmental degradation studies indicate that LAS homologs have a half-life of <2 d in river water (Fairchild *et al.,* 1993), and extensive environmental monitoring performed by Rapaport and Eckhoff (1990) suggests that mean LAS concentrations in river water below effluent discharges in the United States, Canada, and

Germany are approximately 0.1 ± 0.09 mg liter^{-1}. An extensive data base addressing environmental concentrations of anionic surfactants suggests that levels are generally well below 0.5 mg liter^{-1} in waters receiving POTW effluents (Sivak *et al.*, 1982).

Effective concentrations in algal cultures range from 1.4 to 70 mg liter^{-1}, indicating substantial variability in the sensitivities of the various periphytic species (Table III). This level of variability is to be expected, as Kimerle (1989) reports that approximately 90% of all algal LC_{50} values for LAS are between 0.1 and 100 mg liter^{-1}, a 1000-fold range. Algal stimulation has also been observed at low concentrations of 0.24–5.0 mg liter^{-1} (Hicks and Neuhold, 1966). Chawla *et al.* (1988) reported decreased protein levels in *Nostoc muscorum* grown in concentrations of LAS ranging from 1.0 to 50.0 mg liter^{-1}. In addition, at 10 mg liter^{-1} heterocyst formation was initially inhibited (but recovered), and was completely inhibited at 50 mg liter^{-1}. Fragmentation of *Nostoc* filaments was observed at all concentrations, but scanning electron microscopy revealed no morphological damage to cells.

Other studies have addressed the effects of anionic surfactants (LAS) on entire periphyton communities. A study by Fairchild *et al.* (1993) did not reach the First Observable Effect Concentration (FOEC) using 0.36 mg liter^{-1}, but Lewis *et al.* (1993) reported a FOEC of 3.3 mg liter^{-1}, for communities above an effluent outfall and 16.6 mg liter^{-1} below. Lewis *et al.* (1993) indicated that changes in photosynthetic rates and species richness were the most sensitive of the parameters measured above the outfall, and changes in photosynthesis and chlorophyll concentration were the most sensitive parameters measured below the outfall. Therefore, the importance of monitoring several different parameters is evident. Although Lewis *et al.* (1993) reported that their FOEC values are comparable to those reported for previous community-level studies, all previous studies addressed the effects of LAS on planktonic communities. Because of the inherent differences in chemical, physical, and biological dynamics within lotic and lentic systems, direct comparisons between the two types of studies are thus somewhat limited. Because an extensive data base on community-level effects of anionic surfactants in lotic systems remains to be established, and because most POTW effluents are discharged into rivers, more of these types of studies are needed.

3. Cationic Surfactants

Cationic surfactants (the surface-active portion of the molecule bears a positive charge) account for approximately 6% (45 million kg) of the total U.S. surfactant consumption (Rosen, 1988). QASs (quaternary ammonium surfactants; TMAC-trimethyl ammonium chloride and CTAB-cetyl trimethyl ammonium bromide) are important ingredients in fabric softeners, oil-based drilling muds, biocides, disinfectants, and emulsifying agents

TABLE III Anionic Surfactant Effects on Attached Algae

Surfactant	*Species tested*	*Effect measured*	*Effective conc. (mg liter^{-1})*	*Reference*
ABS[a]	*Nitzschia linearis*	120 h growth	10 (EC_{50})	Patrick and Cairns (1968)
ABS	*Cladophora* sp. *Vaucheria* sp. (communities)	96 h (^{14}C assimilation)	5 (stimulatory) 25–100 (inhibitory)	Hicks and Neuhold (1966)
ABS	*Cladophora glomerata*	3 d growth	2.5 (no inhibition) 5–10 (inhibitory) 12.5 (algicidal)	Whitton (1967)
SLS[b]	*Cladophora glomerata*	3 d growth	60 (no inhibition) 70–100 (inhibitory) 125 (algicidal)	Whitton (1967)
SLS	*Tribonema aequale*	14 d growth	500 (median EC_{100}) 31–4000 (range EC_{100}) reported range is for numerous species (mostly planktonic)	Blanck *et al.* (1984)
SDS[c]	*Nitzschia actinastroides*	5 d growth	<5	Nyberg (1985), as reported by Lewis (1990)
LAS[d]	*Nostoc muscorum*	15 d biomass, heterocyst number, protein content, and growth	>10 (inhibitory)	Chawla *et al.* (1988)
$C_{11.6}$ LAS[e]	*Nitzschia fonticola*	3 d growth	20–50 (EC_{50})	Yamane *et al.* (1984)
$C_{13.3}$ LAS	*Navicula pelliculosa*	4 d growth	1.4 (EC_{50})	Lewis and Hamm (1986)
$C_{11.9}$ LAS	Periphyton community in experimental streams	45 d chlorophyll, AFDW	0.4 (no inhibitory effects)	Fairchild *et al.* (1993)
$C_{11.2}$ LAS	*Plectonema boryanum*	Growth	30 (FOEC)[f]	Dhaliwal *et al.* (1977)
$C_{11.9}$ LAS	Natural periphyton community (*in situ*)	21 d photosynthesis and species diversity	3.3 (FOEC above the effluent outfall) 16.6 (first effect in combination with 20–30% effluent—below the effluent outfall)	Lewis *et al.* (1993)

[a]Alkylbenzene sulfonate.
[b]Sodium lauryl sulfate.
[c]Sodium dodecyl sulfate.
[d]Linear alkylbenzene sulfonate.
[e]11.6 indicates mean alkyl chain length.
[f]First observable effect concentration.

(Woltering and Bishop, 1989), whereas dialkyl cationics (DTDMAC-ditallow dimethyl ammonium chloride) are the most frequently used surfactants in antistatic agents and fabric softeners.

Quaternary ammonium surfactants in the soluble phase are typically reduced by 90% under normal POTW processing (Boethling, 1984). The majority of the QASs are relatively insoluble in water, can complex with dissolved organics, and adsorb strongly to solids, particularly those with negatively charged surfaces (Lewis and Wee, 1983). Although some monoalkyl QASs are readily biodegradable, the adsorptive characteristics of these compounds lead to their ability to be removed by POTWs as complexes on solid surfaces of sludge. In addition, the high incidence of QAS sorption in aquatic systems limits their biodegradation and potential to cause toxicity to aquatic organisms.

Studies investigating the effects of cationic surfactants on periphytic algae are summarized in Table IV. Effective concentrations for attached algal cultures ranged from 0.07 to >100 mg liter^{-1}, with EC_{50} values of 0.07–0.20 mg liter^{-1}. Low levels of cationic surfactants have also exhibited stimulatory effects. Lewis *et al.* (1986) reported that exposure of a natural periphyton assemblage to 0.25 mg liter^{-1} TMAC increased cell density and diversity, and Bélanger *et al.* (1994) observed that species richness of a periphyton community within a model stream increased with increasing concentrations of TMAC up to 1.25 mg liter^{-1}. At the community level, no significant effects were observed for exposure of laboratory-derived periphyton up to 0.513 mg liter^{-1} monoalkyl quaternary surfactant (MAQ). Generally, cationic surfactants are more toxic to algae than either anionic or nonionic surfactants, although they typically occur at lower concentrations (Fig. 1).

4. Nonionic Surfactants

Nonionic surfactants (the surface-active portion of the molecule bears no ionic charge) account for approximately 21% of the total U.S. surfactant consumption (Rosen, 1988), with about 845 million kg produced annually [U.S. International Trade Commission (USITC), 1979 as reported by Sivak *et al.*, 1982)]. Because of their superior cleaning of synthetic fibers, tolerance of water hardness, and low foaming properties, nonionic surfactants are increasingly used in detergent formulations (Sivak *et al.*, 1982).

Alcohol ethoxylate (AE), linear alcohol ethoxylate (LAE), and alkylphenol ethoxylate (APE) have all been shown to undergo complete, rapid degradation in POTWs and in the aquatic environment (Sivak *et al.*, 1982; Dorn *et al.*, 1993). Monitoring data reported by Dorn *et al.* (1993) indicate that levels of nonionic surfactants in U.S. effluent-receiving rivers range from 0.0042 to 0.8 mg liter^{-1}. Results from studies investigating the effects of nonionic surfactants on attached algae are summarized in Table 5. Insufficient data are available to draw sound conclusions regarding nonionic surfactant toxicity to attached algae.

TABLE IV Cationic Surfactant Effects on Attached Algae

Surfactant	*Species tested*	*Effect measured*	*Effective conc. (mg liter^{-1})*	*Reference*
C_{12} MAQ[a]	Community in flow-through artificial streams	35 d chlorophyll, cell densities, and species diversity	0.513 (no significant effects)	Woltering and Bishop (1989)
FNC[b]	*Anabaena cylindrica* *Oscillatoria tenuis* *Stigeoclonium* sp.	7 d growth	0.2–100, 0.2 –75, 0.2–75 (algistatic, respectively)	Hueck *et al.* (1966)
3 QASs[c]	*Gomphonema parvulum* *Nitzschia palea*	21 d growth	2.0 (only slightly toxic, most cultures recovered at 21 d)	Palmer and Maloney (1955)
C_{122} TMAC[d]	*Navicula pelliculosa*	96 h growth	0.2 (EC_{50}) 0.47 (algistatic)	Lewis *et al.* (1986)
C_{12} TMAC	Natural periphyton community *(in situ)*, 122 algal species	21 d biomass, chlorophyll, and community composition	6.9 (FOEC below sewage effuent outfall) 0.96 (FOEC above sewage effluent outfall)	Lewis *et al.* (1986)
$8C_{12}$ TMAC	Periphyton community in flow-through lab streams, >200 algal species	8 week standing crop biomass and species composition	>0.25 (significant community effects)	Belanger *et al.* (1994)
C_{12} TMAC	Preexposed periphyton community in flow-through lab streams, 88 algal species	Cell densities (live:dead ratio), species richness, species composition, community similarity	0.05 and 0.25 (significant effects only in community composition)	McCormick *et al.* (1991)
C_{12} TMAC	Unperturbed periphyton community in flow-through lab streams, 74 algal species	Cell densities (live:dead ratio), species richness, species composition, community similarity	0.05 and 0.25 (significant growth stimulation and shifts in species composition)	McCormick *et al.* (1991)
DTDMAC	*Navicula pelliculosa*	96 h growth	0.07 (EC_{50})	Lewis and Hamm (1986)
DTDMAC[e]	*Navicula seminulum*	12 d growth	0.5–10 (algistatic and algicidal)	Lewis and Wee (1983)

[a]Monoalkyl quaternary surfactants.
[b]Fatty nitrogen compounds (mono-, di-, tri-, and polyquaternary surfactants).
[c]Quaternary ammonium surfactants: methyldodecylbenzyl trimethyl ammonium chloride, cetyldimethyl ammonium bromide, and dodecylacetamido dimethyl benzyl ammonium chloride.
[d]Lauryl trimethyl ammonium chloride.
[e]Ditallow dimethyl ammonium chloride.

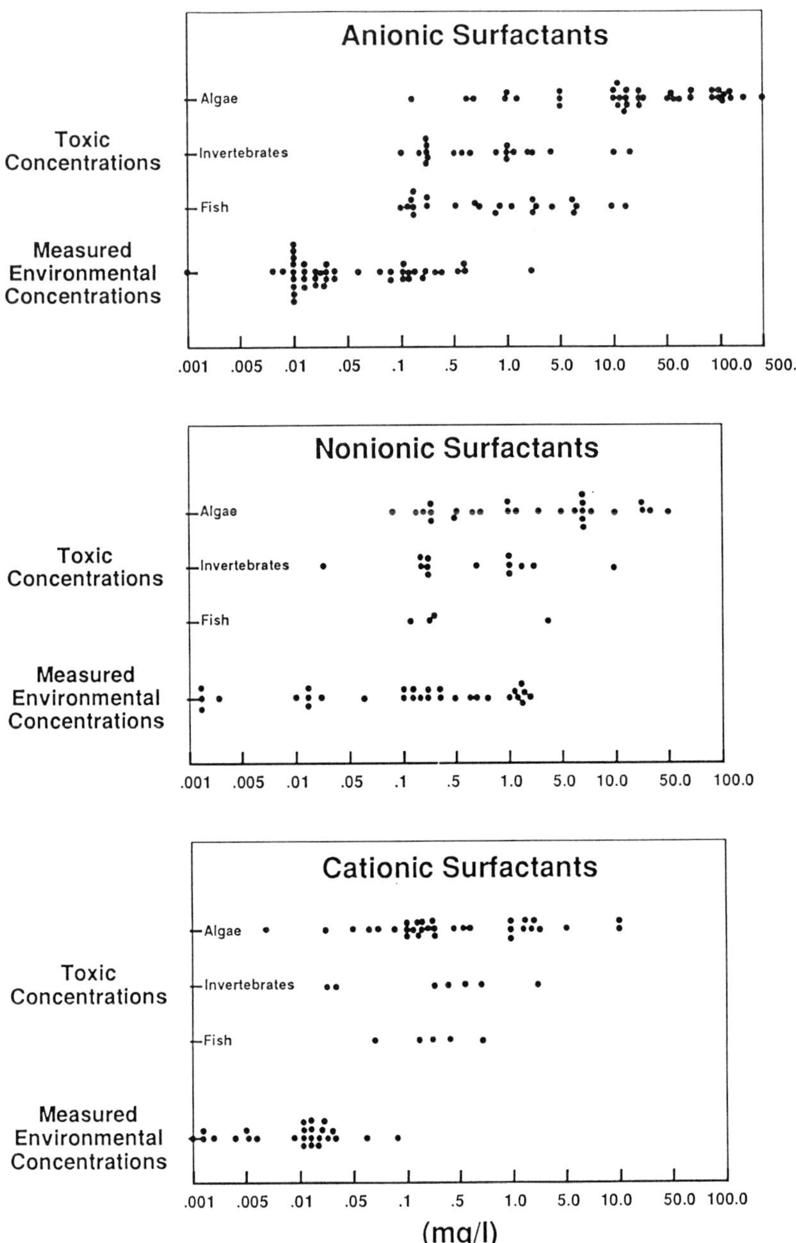

Figure 1 Comparison of measured environmental concentrations and reported chronic toxicity values. (Unpublished table provided by M. A. Lewis.)

TABLE V Nonionic Surfactant Effects on Attached Algae

Surfactant	*Species tested*	*Effect measured*	*Effective conc. (mg liter^{-1})*	*Reference*
AE[a]	*Nitzschia fonticola*	3 d growth	5–10 (EC_{50})	Yamane *et al.* (1984)
C_{14-15} AE_6	*Navicula pelliculosa*	96 h growth	0.28 (EC_{50})	Lewis and Hamm (1986)
C_{14-15} AE_6	*Navicula seminulum*	5 d growth	5–10 (algistatic)	Payne and Hall (1979)

[a]Polyoxyethylene alkyl ether (EO:9).

5. Detergent Builders

Detergent builders (or chelating agents) are important components of soaps and detergents in that they complex with polyvalent cations in aqueous solution and eliminate the harmful effects of metallic impurities (Erickson *et al.*, 1970). A number of field studies have indicated that nitrilotriacetic acid (NTA) is readily biodegradable in POTWs and within aquatic systems (Anderson *et al.*, 1985). Maki and Macek (1978) estimated concentrations of zeolite within effluent-receiving waters to remain below 0.2 mg liter^{-1}. Monitoring studies in both the United States and Europe indicate limited potential for NTA to accumulate in the environment (Anderson *et al.*, 1985).

Toxicity data with respect to the effects of detergent builders on attached algae are summarized in Table VI. In addition, stimulation of algal productivity has been observed at concentrations of 0.5 mg liter^{-1}

TABLE VI Effects of Detergent Builders on Attached Algae

Builder	*Species tested*	*Effect measured*	*Effective conc. (mg liter^{-1})*	*Reference*
NTA[a]	*Navicula seminulum*	96 h growth	185 (EC_{50}) soft water 477 (EC_{50}) hard water	Anderson *et al.* (1985)
NTA	Periphyton community in experimental streams	Species composition and productivity	2.0 (no observable effects)	Bott *et al.* (1978), as reported in Anderson *et al.* (1985)
Zeolite	*Navicula seminulum*	5 d growth	50–100 (algistatic)	Maki and Macek (1978)

[a]Nitrilotriacetic acid.

NTA (Burgi, 1974, as reported in Anderson *et al.*, 1985). However, when compared to the potential levels of productivity expected with phosphate-based pollutants, Burgi concluded that NTA was a much more suitable substitute. Anderson *et al.* (1985) also reported that NTA's ability to sequester polyvalent metal ions may actually protect aquatic organisms from high levels of toxic metals.

6. Mode-of-action

The extensive variability in algal sensitivity to sufactants and detergent builders is most likely attributed to considerable variation in cell-wall structure among algae. Ukeles (1965) reported that, in some cases, cell-wall thickness helped to protect algal cells from damage, whereas other studies indicate that physical thickness is secondary in importance to cell-wall structure. For example, increased algal toxicity to cationic surfactants has been attributed to the interaction of these chemicals with negatively charged structures on the cell surface (Armstrong, 1957, as reported by Nyberg, 1976). This may also explain, in part, differential tolerances to surfactants among algal divisions based on differences in cell composition. In support of this hypothesis, Ukeles (1965) demonstrated that several species of algae maintained their overall negative charge in varying pH environments. Internalized surfactants can impact photosynthetic processes by making the chlorophyll–protein matrix more water soluble (Hicks and Neuhold, 1966).

Lipid concentration in algal cell walls has also been shown to influence toxicity, as more hydrophobic surfactants are readily solubilized in algal lipids with respect to the surrounding aqueous environment. A positive correlation between increasing surfactant hydrophobicity and increased algal toxicity has been demonstrated (Ukeles, 1965). Bock (1965) reported that the toxic concentration of surfactants is inversely proportional to their ability to reduce surface tension (as reported by Lewis, 1990).

The majority of information available on detergent builder mode-of-action in algal toxicity suggests that the primary effects of toxicity are indirect. Erickson *et al.* (1970) suggested that the chelating capacity of detergent builders may sequester trace metals required for algal growth. However, chelators are also known to increase cellular absorption of metals and may allow trace metals to enter algal cells at levels that are toxic to the organism (Erickson *et al.*, 1970). Maki and Macek (1978) demonstrated that concentrations of zeolite in excess of 50 mg liter^{-1} can limit algal growth by the formation of zeolite complexes with essential nutrients.

An important consideration when comparing laboratory-derived toxicity results with field-derived data is the exposure history of the field population, as algal adaptation to several surfactants has been observed. The results of the Chawla *et al.* (1988) study indicated that *Nostoc muscorum* required approximately 7 d to adapt to 5 mg liter^{-1} LAS. Schwab *et al.*

(1992) reported that turnover rates of up to 1.25 mg liter^{-1} C12TMAC were comparable to those of naturally occurring amino acids within 4 weeks of the first exposure. Lewis *et al.* (1986) detected significant differences between FOECs calculated for indigenous periphyton communities exposed to TMAC both above and below a sewage effluent outfall. Additionally, Woltering and Bishop (1989) report that indigenous periphyton communities below a sewage outfall were much less sensitive to cationic MAQs than the relatively unperturbed community above the outfall.

C. Other Organic Toxicants

The toxicity of a wide variety of other organic toxicants has been assayed for benthic algae, but few toxicants have been examined in detail. A majority of studies have employed planktonic species, despite the fact that aquatic environments that frequently receive inputs of these contaminants are lotic rather than lentic systems, that is, habitats often dominated by benthic algae. Further, the toxicants that have been assayed represent a small fraction of the hundreds of compounds released into surface waters annually. For example, the large class of polychlorinated biphenyls (PCBs) has received little attention. Lynch *et al.* (1985) conducted a laboratory stream experiment to assess hexachlorobiphenyl (a component of the complex mixture of chloro-compounds that constitute PCBs) inputs of 100 ng liter^{-1} on periphyton and found no effect on periphyton production or biomass. In some treated stream channels, attached algal biomass actually increased significantly, perhaps due to increases in invertebrate drift. Hall *et al.* (1991) assayed the effects of pulp mill effluent, which contains chlorinated phenolic compounds as well as fatty acids, resin acids, and chlorinated resin acids, and found that low levels increased periphyton production via nutrient enrichment. Clearly, additional studies are needed to evaluate the potential impacts of a wider variety of these chemicals on benthic algal communities under controlled laboratory and field conditions.

Blanck *et al.* (1984) included several organic toxicants in their study of the effects of 19 chemicals on freshwater algae, including fungicides and nematicides. They reported broad ranges in EC_{100} values as a result of species-specific responses. Blanck (1985) reported similar EC_{50} values for some of the same toxicants, using a community-level assay for periphyton. A major class of organic toxicants, polycyclic aromatic hydrocarbons, was examined by Bruno *et al.* (1982). They found convincing evidence for bioaccumulation of PAHs, particularly in the mucilage sheaths of algae, but not for biotransformation. Giddings *et al.* (1983) described an algal assay procedure, including an assessment of the water-soluble fraction of coal-derived oil (resulting from a coal liquefaction process), and reported that the diatom *Nitzschia palea* was among the most sensitive algae assayed. The responses of natural epiphyte assemblages were similar to those of single-species cultures. Other hydrocarbons such as gasoline can actually increase periphyton biomass, primarily as a result of increases in heterotrophs (Pontasch and Brusven, 1987). Putnam *et al.* (1981) found that trinitrotoluene (TNT) manufacturing waste had minimal effects on the periphyton at total munitions concentrations of 50–100 μg liter^{-1}.

Only Shukla *et al.* (1994) examined the impacts of textile dyes on a benthic alga, despite the prevalence of these chemicals in many aquatic systems (Jinqi and Houtian, 1992). They reported an IC_{50} (Inhibitory Concentration) of 82.5 mg liter^{-1} for Metomega Chrome Orange GL, based on photosynthetic inhibition in *Nostoc muscorum*. Concomitant dye concentrations in rivers running through textile industrial regions in India were 120–160 mg liter^{-1}.

D. Mixtures of Toxicants

Pesticide runoff in agricultural regions (e.g., Thurman *et al.,* 1992), pulp mill effluents (Hall *et al.,* 1991), and other organic toxicant inputs typically comprise two or more chemicals simultaneously. For example, K. D. Hoagland and M. Langan *et al.* (unpublished) found 11 pesticides in a single water sample during a storm runoff event in central Nebraska. Consequently, benthic algal communities are often exposed to mixtures rather than individual toxicants. Nevertheless, few studies have assessed whether such combinations of chemicals act synergistically, additively, or antagonistically. Torres and O'Flaherty (1976), for example, found significant three-way interactions for simazine, atrazine, and malathion. In one alga, combinations of simazine and atrazine resulted in synergistic inhibition, whereas atrazine alone stimulated chlorophyll production; however, in other species, mixtures of these herbicides acted antagonistically. Evidence from plankton communities also indicates that combinations of pesticides can produce effects that could not be predicted based an single-pesticide assays alone (Stratton, 1984; Hoagland *et al.,* 1993). Given that mixtures of organic toxicants represent an environmentally realistic phenomenon, significantly more single- and multiple-compound bioassays at the single-species and community levels are warranted.

IV. INDIRECT EFFECTS

In light of current interest in food web interactions in pelagic communities, it is somewhat surprising that the indirect effects of organic toxicants on benthic algal communities have received little attention. As Giddings *et al.* (1983) pointed out, "the interdependence of populations in an ecosystem makes it possible for even temporary inhibition of algal photosynthesis to have a measurable impact on other organisms, particularly if the other organisms are also experiencing toxic effects." Other examples from periphyton studies serve to illustrate this point. Insecticides such as chlorpyrifos and lindane may not have direct toxic effects on benthic algal communities at lower concentrations (35 and 20.3 μg liter^{-1}, respectively), however such communities may be subject to positive indirect effects owing to the impact on grazers (Brock *et al.*, 1992; Mitchell *et al.*, 1993). Chemical control of the macrophyte *Hydrilla* using diquat and cutrine stimulated epiphyton chlorophyll levels, presumably due to nitrogen release from the deteriorating host (Hodgson and Linda, 1984). Indirect effects of surfactants on attached algae due to modifications of algal/animal interactions have also been noted (Parshikova and Negrutskiy, 1988). Generally, algae tend to be more sensitive to surfactants, particularly cationic forms, than are aquatic animals (Lewis, 1990), but widespread variation in sensitivities of aquatic organisms to various surfactant chemicals lends itself to more situation-specific investigations [a review of chronic and sublethal effects of surfactants on aquatic animals by Lewis (1991) is a good starting point for these types of inquiries]. Microcosm or mesocosm studies including whole communities, such as those described by Huggins *et al.* (1994), are clearly needed to develop models that incorporate indirect effects and thus allow more realistic and robust predictions of toxicant impacts on natural aquatic ecosystems.

V. CONCLUSIONS AND FUTURE DIRECTIONS

Substantial variability in the sensitivity of attached algae to organic toxicants is evident from this survey. For example, Blanck *et al.* (1984) showed that variation among algae in their sensitivity to the same chemical can reach more than three orders of magnitude. Nevertheless, several generalizations are possible: (1) a variety of organic toxicants, particularly herbicides, can have dramatic impacts on benthic algal communities at concentrations in the μg liter^{-1} range, (2) although benthic algal biomass and/or productivity may not be affected at the whole-community level, shifts in species composition are common in response to toxicant exposure; (3) effects on periphyton communities can be both direct and indirect, thus single-species *and* community-level bioassays are important; and (4) freshwater benthic algal species and communities have received far less research attention with regard to the impacts of organic toxicants than is needed, particularly in light of their importance in lotic systems and littoral habitats and the prevalence of these toxicants in the environment.

Future research is especially critical in the following areas: (1) the importance of environmental modifying factors such as temperature, light, pH, nutrients, and so on; (2) interaction effects resulting from combinations of two or more toxicants; (3) indirect community-level effects; (4) acute toxicity of numerous commonly occurring compounds, such as alachlor, and their metabolites; (5) specific modes-of-action; (6) mechanisms of community and species recovery; and (7) mechanisms of tolerance by some taxa to some toxicants. The latter three areas are important not only from a physiological and biochemical standpoint, but also from an ecological perspective because they will undoubtedly elucidate species-specific responses to organic toxicants and resultant shifts in community composition and structure. A final critical gap in our knowledge is the long-term chronic effects of many of these chemicals. Nearly all current toxicity data are based on short-term acute bioassays, despite the fact that baseflow or background contaminant levels are below the EC_{50}. As primary producers, the benthic algae should receive more intensive research efforts to more fully understand their role in aquatic communities subjected to organic toxicants.

REFERENCES

Anderson, R. L., Bishop, W. E., Campbell, R. L., and Becking, G. C. (1985). A review of the environmental and mammalian toxicology of nitrilotriacetic acid. *CRC Crit. Rev. Toxicol.* **15**, 1–102.

Armstrong, W. McD. (1957). Surface-active agents and cellular metabolism. 1. The effect of cationic detergents on the production of acid and CO_2 by baker's yeast. *Arch. Biochem. Biophys.* **71**, 137–147.

Arthur, R. D., Cain, J. D., and Barrentine, B. R. (1976). Atmospheric levels of pesticides in the Mississippi Delta. *Bull. Environ. Contam. Toxicol.* **15**, 129–134.

Austin, A. P., Harris, G. E., and Lucey, W. P. (1991). Impact of an organophosphate herbicide (Glyphosate) on periphyton communities developed in experimental streams. *Bull. Environ. Contam. Toxicol.* **47**, 29–35.

Belanger, S. E. (1994). Review of experimental microcosm, mesocosm, and field tests used to evaluate the potential hazard of surfactants to aquatic life and the relation to single species data. *In* "Freshwater Field Tests for Hazard Assessment of Chemicals" (I. R. Hill, F. Heimbach, P. Leeuwangh, and P. Matthiessen, eds.), pp. 287–314. Lewis Publishers, Boca Raton. FL.

Belanger, S. E., Barnum, J. B., Woltering, D. M., Bowling, J. W., Ventullo, R. M., Schermerhorn, S. D,. and Lowe, R. L. (1994). Algal periphyton structure and function in response to consumer chemicals in stream mesocosms. *In* "Aquatic Mesocosm Studies in Ecological Risk Assessment" (R. L. Graney, J. H. Kennedy, and J. H. Rodgers, eds.), pp. 535–568. Lewis Publishers, Boca Raton. FL.

Bisen, P. S., and Shanthy, S. (1993). Characterization of a DCMU-resistant mutant of the filamentous, diazotrophic cyanobacterium *Anabaena doliolum*. *J. Plant Physiol.* **142**, 557–563.

Blanck, H. (1985). A simple, community level, ecotoxicological test system using samples of periphyton. *Hydrobiologia* **124**, 251–261.

Blanck, H., Wallin, G., and Wängberg, S. (1984). Species-dependent variation in algal sensitivity to chemical compounds. *Ecotoxicol. Environ. Saf.* **8**, 339–351.

Bock, K. (1965). Uber die Wirkung von Waschrostoffen aud Fische. *Arch. Fischereiwiss.* **17**, 68–77.

Boethling, R. S. (1984). Environmental fate and toxicity in wastewater treatment of quaternary ammonium surfactants. *Water Res.* **18**, 1061–1076.

Bott, T. L., Patrick, R., Larson, R., and Rhyne, C. (1978). "The Effect of Nitrilotriacetate (NTA) on the Structure and Functioning of Aquatic Communities in Streams," Rep. to U.S. EPA-ERL, Duluth, Minn., Contract No. R-801951. Environmental Protection Agency, Washington, DC.

Brock, T. C. M., van den Bogaert, M., Bos, A. R., van Breukelen, S. W. F., Reiche, R., Terwoert, J., Suykerbuyk, R. E. M., and Roijackers, R. M. M. (1992). Fate and effects of the insecticide Dursban 4E in indoor *Elodea*-dominated and macrophyte-free freshwater model ecosystems. II. Secondary effects on community structure. *Arch. Environ. Contam. Toxicol.* **23**, 391–409.

Brockway, D. L., Smith, P. D., and Stancil, F. E. (1984). Fate and effects of atrazine in small aquatic microcosms. *Bull. Environ. Contam. Toxicol.* **32**, 345–353.

Bruno, M. G., Fannin, T. E., and Leversee, G. J. (1982). The disposition of benzo(*a*)pyrene in the periphyton communities of two South Carolina streams: Uptake and biotransformation. *Can. J. Bot.* **60**, 2084–2091.

Burgi, H. R. (1974). Die Wirkuing von NTA auf das Wachstum des Phytoplanktons uner besonderer Berucksichtigung des Eisens als Microelement. *Schweiz. Z. Hydrol.* **36.**

Buser, H.-R. (1990). Atrazine and other *s*-triazine herbicides in lakes and in rain in Switzerland. *Environ. Sci. Technol.* **24**, 1049–1058.

Butler, G. L. (1977). Algae and pesticides. *Residue Rev.* **66**, 19–61.

Butler, G. L., Deason, T. R., and O'Kelley, J. C. (1975). The effect of atrazine, 2,4–D, methoxychlor, carbaryl and diazinon on the growth of planktonic algae. *Br. Phycol. J.* **10**, 371–376.

Chawla, G., Viswanathan, P. N., and Devi, S. (1988). Phytotoxicity of linear alkylbenzene sulfonate. *Ecotoxicol. Environ. Saf.* **15**, 119–124.

deNoyelles, F., Kettle, W. D., and Sinn, D. E. (1982). The response of plankton communities in experimental ponds to atrazine, the most heavily used pesticide in the United States. *Ecology* **63**, 1285–1293.

Dhaliwal, A. S., Campione, A., and Smaga, S. (1977). Effect of linear alkylbenzene sulfonate (11.2 LAS) on the morphology and physiology of *Plectonema boryanum* and *Chlamydomonas reinhardii*. *J. Phycol., Suppl.* **13**, 18.

Dorn, P. B., Salanitro, J. P., Evans, S. H., and Kravetz, L. (1993). Assessing the aquatic hazard of some branched and linear nonionic surfactants by biodegradation and toxicity. *Environ. Toxicol. Chem.* **12**, 1751–1762.

Erickson, S. J., Maloney, T. E., and Gentile, J. H. (1970). Effect of nitrilotriacetic acid on the growth and metabolism of estuarine phytoplankton. *J. Water Pollut. Control Fed.* **42**, R329–R335.

Fairchild, J. F., Dwyer, F. J., LaPoint, T. W., Burch, S. A., and Ingersoll, C. G. (1993). Evaluation of a laboratory-generated NOEC for linear alkylbenzene sulfonate in outdoor experimental streams. *Environ. Toxicol. Chem.* **12**, 1763–1775.

Frank, R., Clegg, B. S., Ripley, B. D., and Braun, H. E. (1987). Investigations of pesticide contaminations in rural wells, 1979–1984, Ontario, Canada. *Arch. Environ. Contam. Toxicol.* **16**, 9–22.

Gianessi, L. P. (1987). Lack of data stymies informed decisions on agricultural pesticides. *Resources* **89**, 1–4.

Giddings, J. M., Stewart, A. J., O'Neil, R. V., and Gardner, R. H. (1983). An efficient algal bioassay based on short-term photosynthetic response. *ASTM Spec. Tech. Publ.* **802**, 445–459.

Goldsborough, L. G. (1993). Studies on the impact of agricultural and forestry herbicides on non-target aquatic plant communities. Proc. Third Prairie Conser. Endangered Species Workshop, *Nat. Hist. Occas. Pap.* **19**, 49–57.

Goldsborough, L. G., and Brown, D. J. (1988). Effect of glyphosate (Roundup formulation) on periphytic algal photosynthesis. *Bull. Environ. Contam. Toxicol.* **41**, 253–260.

Goldsborough, L. G., and Robinson, G. G. C. (1986). Changes in periphytic algal community structure as a consequence of short herbicide exposures. *Hydrobiologia* **139**, 177–192.

Goldsborough, L. G., and Robinson, G. C. C. (1988). Functional responses of freshwater periphyton to short simazine exposures. *Verh.—Int. Ver. Theor. Angew. Limnol.* **23**, 1586–1593.

Gurney, S. E., and Robinson, G. G. C. (1989). The influence of two triazine herbicides on the productivity, biomass and community composition of freshwater marsh periphyton. *Aquat. Bot.* **36**, 1–22.

Hall, T. J., Haley, R. K., and LaFleur, L. E. (1991). Effects of biologically treated bleached kraft mill effluent on cold water stream productivity in experimental stream channels. *Environ. Toxicol. Chem.* **10**, 1051–1060.

Hamala, J. A., and Kollig, H. P. (1985). The effects of atrazine on periphyton communities in controlled laboratory ecosystems. *Chemosphere* **14**, 1391–1408.

Hamilton, P. B., Jackson, G. S., Kaushik, N. K., and Solomon, K. R. (1987). The impact of atrazine on lake periphyton communities, including carbon uptake dynamics using track autoradiography. *Environ. Pollut.* **46**, 83–103.

Hawxby, K., Tubea, B., Ownby, J., and Basler, E. (1977). Effects of various classes of herbicides on four species of algae. *Pestic. Biochem. Physiol.* **7**, 203–209.

Heckman, C. W. (1994). Pesticide chemistry and toxicity to algae. *Arch. Hydrobiol. Beih., Ergebn. Limnol.* **42**, 205–234.

Herman, D., Kaushik, N. K., and Solomon, K. R. (1986). Impact of atrazine on periphyton in freshwater enclosures and some ecological consequences. *Can. J. Fish. Aquat. Sci.* **43**, 1917–1925.

Hersh, C. M., and Crumpton, W. G. (1987). Determination of growth rate depression of some green algae by atrazine. *Bull. Environ. Contam. Toxicol.* **39**, 1041–1048.

Hicks, C. E., and Neuhold, J. M. (1966). Alkyl benzene sulfonate effects on stream algae communities. *Bull. Environ. Contam. Toxicol.* **1**, 225–236.

Hoagland, K. D., Drenner, R. W., Smith, J. D., and Cross, D. R. (1993). Freshwater community responses to mixtures of agricultural pesticides: Effects of atrazine and bifenthrin. *Environ. Toxicol. Chem.* **12**, 627–637.

Hodgson, L. M., and Linda, S. B. (1984). Response of periphyton and phytoplankton to chemical control of hydrilla in artificial pools. *J. Aquat. Plant Manage.* **22**, 48–52.

Hueck, H. J., Adema, D. M. M., and Weigmann, J. R. (1966). Bacteriostatic, fungistatic and algistatic activity of fatty nitrogen compounds. *Appl. Microbiol.* **14**, 308–319.

Huggins, D. G., Johnson, M. L., and deNoyelles, F., Jr. (1994). The ecotoxic effects of atrazine on aquatic ecosystems: An assessment of direct and indirect effects using structural equation modeling. *In* "Aquatic Mesocosm Studies in Ecological Risk Assessment" (R. L. Graney, J. H. Kennedy, and J. H. Rodgers, eds.), pp. 653–692. Lewis Publishers, Boca Raton, FL.

Hughes, J. S., Alexander, M. M., and Balu, K. (1988). An evaluation of appropriate expressions of toxicity in aquatic plant bioassays as demonstrated by the effects of atrazine on algae and duckweed. *ASTM Spec. Tech. Publ.* **971**, 531–547.

Jinqi, L., and Houtian, L. (1992). Degradation of azo dyes by algae. *Environ. Pollut.* **75**, 273–278.

Jurgensen, T.A., and Hoagland, K.D. (1990). Effects of short-term pulses of atrazine on attached algal communities in a small stream. *Arch. Environ. Contam. Toxicol.* **19**, 617–623.

Kasai, F., Takamura, N., and Hatakeyama, S. (1993). Effects of simetryne on growth of various freshwater algal taxa. *Environ. Pollut.* **79**, 77–83.

Kimerle, R. A. (1989). Aquatic and terrestrial ecotoxicology of linear alkylbenzene sulfonate. *Tenside, Surfactants, Deterg.* **26**, 169–176.

Kosinski, R. J., and Merkle, M. G. (1984). The effect of four terrestrial herbicides on the productivity of artificial stream algal communities. *J. Environ. Qual.* **13**, 75–82.

Krieger, K. A., Baker, D. B., and Kramer, J. W. (1988). Effects of herbicides on stream Aufwuchs productivity and nutrient uptake. *Arch. Environ. Contam. Toxicol.* **17**, 299–306.

Kruger, E. L., Somasundaram, L., Kanwar, K., and Coats, J. R. (1993). Persistence and degradation of [^{14}C]atrazine and [^{14}C]deisopropylatrazine as affected by soil depth and moisture condition. *Environ. Toxicol. Chem.* **12**, 1959–1967.

Lal, S. (1984). Effects of insecticides on algae. *In* "Insecticide Microbiology" (R. Lal, ed.), pp. 203–236. Springer-Verlag, Berlin.

LaPoint, T. W. (1993). Aquatic hazard assessment of surfactants: How much information is needed to judge environmental safety? *Environ. Toxicol. Chem.* **12**, 1749.

Larsen, D. P., deNoyelles, F., Jr., Stay, F., and Shiroyama, T. (1986). Comparisons of single-species, microcosm and experimental pond responses to atrazine exposure. *Environ. Toxicol. Chem.* **5**, 179–190.

Lewis, M. A. (1990). Chronic toxicities of surfactants and detergent builders to algae: A review and risk assessment. *Ecotoxicol. Environ. Saf.* **20**, 123–140.

Lewis, M. A. (1991). Chronic and sublethal toxicities of surfactants to freshwater and marine animals: A review and risk assessment. *Water Res.* **25**, 101–113.

Lewis, M. A. (1992). The effects of mixtures and other environmental modifying factors on the toxicities of surfactants to freshwater and marine life. *Water Res.* **26**, 1013–1023.

Lewis, M. A., and Hamm, B. G. (1986). Environmental modification of the photosynthetic response of lake plankton to surfactants and significance to a laboratory–field comparison. *Water. Res.* **20**, 1575–1582.

Lewis, M. A., and Wee, V. T. (1983). Aquatic safety assessment for cationic surfactants. *Environ. Toxicol. Chem.* **2**, 105–118.

Lewis, M. A., Taylor, M. J., and Larson, R. J. (1986). Structural and functional response of natural phytoplankton and periphyton communities to a cationic surfactant with considerations on environmental fate. *In* "Community Toxicity Testing" (J. Cairns, Jr., ed.), ASTM STP 920, pp. 241–268. Am. Soc. Test. Mat., Philadelphia.

Lewis, M. A., Pittinger, C. A., Davidson, D. H., and Ritchie, C. J. (1993). *In situ* response of natural periphyton to an anionic surfactant and an environmental risk assessment for phytotoxic effects. *Environ. Toxicol. Chem.* **12**, 1803–1812.

Lozano, R. B., and Pratt, J. R. (1994). Interaction of toxicants and communities: The role of nutrients. *Environ. Toxicol. Chem.* **13**, 361–368.

Lynch, T. R., Johnson, H. E., and Adams, W. J. (1985). Impact of atrazine and hexachlorobiphenyl on the structure and function of model stream ecosystems. *Environ. Toxicol. Chem.* **4**, 399–413.

Maki, A. W., and Macek, K. J. (1978). Aquatic environmental safety assessment for a nonphosphate detergent builder. *Environ. Sci. Technol.* **12**, 573–580.

McCormick, P. V., Cairns, J. Jr., Bélanger, S. E., and Smith, E. P. (1991). Response of protistan assemblages to a model toxicant, the surfactant C12–TMAC (dodecyl trimethyl ammonium chloride), in laboratory streams. *Aquat. Toxicol.* **21**, 41–70.

Melendez, A. L., Kepner, R. L., Jr., Balczon, J. M., and Pratt, J. R. (1993). Effects of diquat on freshwater microbial communities. *Arch. Environ. Contam. Toxicol.* **25**, 95–101.

Millie, D. F., and Hersh, C. M. (1987). Statistical characterizations of the atrazine-induced photosynthetic inhibition of *Cyclotella meneghiniana*. *Aquat. Toxicol.* **10**, 239–249.

Millie, D. F., Hersh, C. M., and Dionigi, C. P. (1992). Simazine-induced inhibition in photoacclimated populations of *Anabaena circinalis* (Cyanophyta). *J. Phycol.* **28**, 19–26.

Mitchell, G. C., Bennett, D., and Pearson, N. (1993). Effects of lindane on macroinvertebrates and periphyton in outdoor artificial streams. *Ecotoxicol. Environ. Saf.* **25**, 90–102.

Moorhead, D. L., and Kosinski, R. J. (1986). Effect of atrazine on the productivity of artificial stream algal communities. *Bull. Environ. Contam. Toxicol.* **37**, 330–336.

Munawar, M., Wong, P. T. S., and Rhee, G.-Y. (1988). The effects of contaminants on algae: An overview. *In* "Toxic Contamination in Large Lakes," Vol. 1, pp.113–160. Lewis Publishers, Chelsea.

Nalewajko, C. (1994). Effects of surfactants on algae. *Arch. Hydrobiol. Beih., Ergebn. Limnol.* **42**, 195–204.

Nyberg, H. (1976). The effects of some detergents on the growth of *Nitzschia holsatica* Hust. (Diatomeae). *Ann. Bot. Fenn.* **13**, 65–68.

O'Brien, P. Y., and Dixon, P. S. (1976). The effects of oils and oil components on algae: A review. *Br. Phycol. J.* **11**, 115–142.

O'Kelley, J. C., and Deason, T. R. (1976). "Degradation of Pesticides by Algae. EPA-600/3-76-022. U.S. Environ. Prot. Agency, Athens, GA.

O'Neal, S. W., and Lembi, C. A. (1983). Effect of simazine on photosynthesis and growth of filamentous algae. *Weed Sci.* **31**, 899–903.

Padhy, R. N. (1985). Cyanobacteria and pesticides. *Residue Rev.* **95**, 1–44.

Painter, H. A., and Zabel, T. F. (1988). "Review of the Environmental Safety of LAS," 1659-M/1/EV 8658. Water Research Center, Henley Road, Medmenham, CO.

Palmer, C. M., and Maloney, T. E. (1955). Preliminary screening for potential algicides. *Ohio J. Sci.* **55**, 1–8.

Parshikova, T. V., and Negrutskiy, S. F. (1988). Effect of surfactants on algae (a review). *Hydrobiol. J.* **6**, 47–58.

Paterson, D. M., and Wright, S. J. L. (1988). Diffusion gradient plates for herbicide toxicity tests on micro-algae and cyanobacteria. *Lett. Microbiol.* **7**, 87–90.

Patrick, R., and Cairns, J., Jr. (1968). The relative sensitivity of diatoms, snails, and fish to twenty common constituents of industrial wastes. *Prog. Fish-Cult.* **30**, 137–140.

Payne, A. G., and Hall, R. H. (1979). A method for measuring algal toxicity and its application to the safety assessment of new chemicals. *In* "Aquatic Toxicology" (L. L. Marking and R. A. Kimerle, eds.), ASTM STP 667, pp. 171–180. Am. Soc. Test. Mat., Philadelphia.

Pontasch, K. W., and Brusven, M. A. (1987). Periphyton response to a gasoline spill in Wolf Lodge Creek, Idaho. *Can. J. Fish. Aquat. Sci.* **44**, 1669–1673.

Putnam, H. D., Sullivan, J. H., Jr., Pruitt, B. C., Nichols, J. C., Keirn, M. A., and Swift, D. R. (1981). Impact of trinitrotoluene wastewaters on aquatic biota in Lake Chickamauga, Tennessee. *In* "Ecological Assessments of Effluent Impacts on Communities of Indigenous Aquatic Organisms" (J. M. Bates and C. I. Weber, eds.), ASTM STP 730, pp. 220–242. Am. Soc. Test. Mat., Philadelphia.

Rapaport, R. A., and Eckhoff, W. S. (1990). Monitoring linear alkyl benzene sulfonate in the environment: 1973–1986. *Environ. Toxicol. Chem.* **9**, 1245–1257.

Richards, R. P., Kramer, J. W., Baker, D. B., and Krieger, K. A. (1987). Pesticides in rainwater in the northeastern United States. *Nature (London)* **327**, 129–131.

Ritter, W. F. (1990). Pesticide contamination of ground water in the United States—A review. *J. Environ. Sci. Health, Part B* **B25**, 1–29.

Robson, T. O., Fowler, M. C., and Barrett, P. R. F. (1976). Effect of some herbicides on freshwater algae. *Pestic. Sci.* **7**, 391–402.

Rose, F. L., and McIntire, C. D. (1970). Accumulation of dieldrin by benthic algae in laboratory streams. *Hydrobiologia* **35**, 481–493.

Rosen, M. J. (1988). Characteristic features of surfactants. *In* "Surfactants and Interfacial Phenomena" (M. J. Rosen, ed.), pp. 1–25. Wiley, New York.

Schwab, B. S., Maruscik, D. A., Ventullo, R. M., and Palmisano, A. C. (1992). Adaptation of periphytic communities in model streams to a quaternary ammonium surfactant. *Environ. Toxicol. Chem.* **11**, 1169–1177.

Shabana, E. F. (1987a). Use of batch assays to assess the toxicity of atrazine to some selected cyanobacteria. I. Influence of atrazine on the growth, pigmentation and carbohydrate contents of *Aulosira fertilissima*, *Anabaena oryzae*, *Nostoc muscorum*, and *Tolypothrix tenuis*. *J. Basic Microbiol.* **27**, 113–119.

Shabana, E. F. (1987b). Use of batch assays to assess the toxicity of atrazine to some selected cyanobacteria. II. Effect of atrazine on heterocyst frequency, nitrogen and phosphorus metabolism of four heterocystous cyanobacteria. *J. Basic Microbiol.* **27**, 215–223.

Shehata, S. A., El-Dib, M. A., and Abou-Waly, H. F. (1993). Effect of triazine compounds on freshwater algae. *Bull. Environ. Contam. Toxicol.* **50**, 369–376.

Shukla, S. P., T., A. K., Tiwari, D. N., Mishra, B. P., and Gupta, G. S. (1994). Assessment of the effect of the toxicity of a textile dye on *Nostoc muscorum* ISU, a diazotrophic cyanobacterium. *Environ. Pollut.* **84**, 23–25.

Sicko-Goad, L., and Andresen, N. A. (1993a). Effect of lipid composition on the toxicity of trichlorobenzene isomers to diatoms. I. Short-term effects of 1,3,5-trichlorobenzene. *Arch. Environ. Contam. Toxicol.* **24**, 236–242.

Sicko-Goad, L., and Andresen, N. A. (1993b). Effect of diatom lipid composition on the toxicity of trichlorobenzene. II. Long-term effects of 1,2,3-trichlorobenzene. *Arch. Environ. Contam. Toxicol.* **24,** 243–248.

Sivak, A., Goyer, M., Perwak, J., and Thayer, P. (1982). Environmental and human health aspects of commercially important surfactants. *In* "Solution Behavior of Surfactants" (K. L. Mittal and E. J. Fendler, eds.), Vol. 1, pp. 161–188. Plenum, New York

Solomon, K. R., Baker, D. B., Richards, R. P., Dixon, K. R., Klaine, S. J., La Point, T. W., Kendal, R. J., Weisskopf, C. P., Giddings, J. M., Giesy, J. P. Hall, L. W., Jr., and Williams, W. M. (1966). Ecological risk assessment of atrazine in North American surface waters. *Environ. Toxicol. Chem.* **15,** 31–76.

Spalding, R. F., and Snow, D. D. (1989). Stream levels of agrichemicals during a spring discharge event. *Chemosphere* **19,** 1129–1140.

Stanford Research Institute (SRI) (1984). "Surface-Active Agents," 583.8000A. Chemical Economics Handbook, Menlo Park, CA.

Stratton, G. W. (1984). Effects of the herbicide atrazine and its degradation products, alone and in combination, on phototrophic microorganisms. *Arch. Environ. Contam. Toxicol.* **13,** 35–42.

Sundaram, K. M. S., and Curry, J. (1991). Partitioning and uptake of permethrin by stream invertebrates and periphyton. *J. Environ. Sci. Health, Part B* **B26,** 219–239.

Thurman, E. M., Goolsby, D. A. Meyer, M. T., and Kolpin, D. W. (1991). Herbicides in surface waters of the midwestern United States: The effect of spring flush. *Environ. Sci. Technol.* **25,** 1794–1796.

Thurman, E. M., Goolsby, D. A., Meyer, M. T., Mills, M. S., Pomes, M. L., and Kolpin, D. W. (1992). A reconnaissance study of herbicides and their metabolites in surface water of the midwestern United States using immunoassay and gas chromatography/mass spectrometry. *Environ. Sci. Technol.* **26,** 2440–2447.

Torres, A. M., and O'Flaherty, L. M. (1976). Influence of pesticides on *Chlorella, Chlorococcum, Stigeoclonium* (Chlorophyceae), *Tribonema, Vaucheria* (Xanthophyceae) and *Oscillatoria* (Cyanophyceae). *Phycologia* **15,** 25–36.

Tubea, B., Hawxby, K., and Mehta, R. (1981). The effects of nutrient, pH and herbicide levels on algal growth. *Hydrobiologia* **79,** 221–227.

Ukeles, R. (1965). Inhibition of unicellular algae by synthetic surface-active agents. *J. Phycol.* **1,** 102–110.

U.S. International Trade Commision (USITC) (1979). "Synthetic Organic Chemicals, United States Production and Sales of Surface-active Agents." USITC, Washington, D.C.

Vérber, K., Zahradník, J., Breyl, I., and Krédl, F. (1981). Toxic effect and accumulation of atrazine in algae. *Bull. Environ. Contam. Toxicol.* **27,** 872–876.

Wauchope, R. D. (1978). The pesticide content of surface water draining from agricultural fields—A review. *J. Environ. Qual.* **7,** 459–472.

Wetzel, R. G. (1993). Microcommunities and microgradients: Linking nutrient regeneration, microbial mutualism, and high sustained aquatic primary production. *Neth. J. Aquat. Ecol.* **27,** 3–9.

Whitton, B. A. (1967). Studies on the growth of riverain *Cladophora* in culture. *Arch. Mikrobiol.* **58,** 21–29.

Woltering, D. M., and Bishop, W. E. (1989). Evaluating the environmental safety of detergent chemicals: A case study of cationic surfactants. *In* "The Risk Assessment of Environmental and Human Health Hazards: A Textbook of Case Studies" (D. J. Paustenbach, ed.), pp. 345–389. Wiley, New York.

Yamane, A. N., Okada, M., and Sudo, R. (1984). The growth inhibition of planktonic algae due to surfactants used in washing agents. *Water Res.* **18,** 1101–1105.

16

Acidification Effects

D. Planas

GEOTOP, Département des Sciences Biologiques,
Université du Québec à Montréal,
Montréal, Québec H3C 3P8, Canada

I. INTRODUCTION

Acid inputs to aquatic ecosystems can occur either through natural processes, such as volcanic emissions (Stothers and Raupino, 1983), thermal effluents (Brock, 1973), peat bog drainage inflow (Cook *et al.*, 1987; Kullberg *et al.*, 1993), and oxidation in geological areas with sulfur-containing rocks (Sheath *et al.*, 1982), or through anthropogenic causes. More recent ecosystem acidification studies have focused on anthropogenic activities such as mining and waste leachates (Tease and Coler, 1984), alteration in land use (Soulsby, 1982; Patrick and Stevenson, 1990), and, most importantly, the increase in atmospheric sulfur and nitrogen oxides originating from smelters and fossil fuel power plants. These oxides, precursors of the strong acids (Husar *et al.*, 1978), affect the pH of dry and wet precipitation and contribute to the acidification of the surface waters of large, poorly buffered regions of Europe and North America (Likens, 1976; Overrein *et al.,* 1981; Harvey, 1989). Since the large amount of research developed over the past two decades in the Northern Hemisphere deals mainly with this anthropogenic acid deposition, it will form the core of this review.

The chemical modifications induced by acidification affect water alkalinity, pH, and metal solubilization (Campbell and Stokes, 1985), but may also directly or indirectly alter other chemical variables such as nutrient cycling and availability (Rudd *et al.,* 1988).

Existing literature on the response of benthic algae to acidification and the mechanisms that underlie this response has focused either on the decrease in herbivores, that is, top-down control (e.g., see Baker and Christensen, 1990), or on changes in resource availability, that is, bottom-up control (e.g., see Fairchild and Sherman, 1990).

Of these studies relating to periphyton, most consider the autotrophic component, the algae, but less frequently the heterotrophic component, the bacteria and fungi, which can be important elements in the response of algae to changes in the physical and chemical variables of the environment. The importance of the heterotrophic component lies not only in its effect on nutrient cycling within the algal mat (Wetzel, 1993), but also in its potential role as a competitor for the nutrients required by the autotrophic organisms. This competition may be for a common resource, such as inorganic nutrients, or for substratum space when the algal component of the community is dominated by adnate forms.

In this review, an ecosystem will be considered acidic when its mean pH is < 5.5 and its alkalinity values are around or below 0 μeq liter^{-1}. The subjects to be developed are: (1) general responses to the loss of the acid-neutralizing capacities of the ecosystem, including (i) a brief overview of physical and chemical alterations; (ii) changes in algal community structure and biomass, and (iii) changes in community metabolism; and (2) hypotheses that might explain the effects of pH decrease on algal community structure and metabolism. Both abiotic and biotic factors will be taken into account in this section. The abiotic factors include (i) the effect of increases in hydrogen ions (H^+) or metal solubility, (ii) alterations in nutrient availability, and (iii) changes in physical variables directly related to acidification, such as light, as well as others not directly related to a pH decrease, such as water movement and substratum. The latter variables could explain contradictory results found in different acidic ecosystem studies. In relation to the biotic factors, the following will be considered: (i) decrease in abundance or shift in species of macroinvertebrate grazers and (ii) reduction in competition between the autotrophic (algae) and heterotrophic (bacteria) components of the periphyton for a limiting resource.

Relatively recent reviews exist on the effects of acidification on benthic communities (e.g., see Elwood and Mulholland, 1989, for streams; Harvey, 1989, and Stokes *et al.*, 1989, for lakes). In this chapter the accent will be put on more recent studies, which have added new data to the importance of pH shifts and nutrient availability in shaping benthic algal structure and/or productivity.

Studies on the benthic algal communities of both lakes and streams will be considered. These studies include (1) spatial surveys, comparing ecosystems with different pH and/or metal concentrations, as is frequently done in lakes, though rarely in rivers (in the latter, spatial and temporal changes in community structure are sometimes reported along a pH gradient), and (2) the experimental acidification of whole ecosystems (lakes) or river reaches, as well as of outdoor mesocosms. Both the direct effects of pH as well as the indirect effects, particularly those involving a disruption of trophic pathways and/or of nutrient availability, will also be considered in this chapter.

II. GENERAL RESPONSES TO THE LOSS OF THE ACID-NEUTRALIZING CAPACITIES OF ECOSYSTEMS

A. Physical and Chemical Changes in Aquatic Ecosystems Associated with Anthropogenic Pollutants

In areas subjected to acid precipitation, significant amounts of hydrogen, sulfate, nitrate, and ammonium are supplied to the surface water (Table I). The effect of these pollutants on freshwater ecosystems will depend on the systems' capacity to neutralize acid deposition; this is referred to as the acid-neutralizing capacity (ANC) or alkalinity. The ANC of a system will depend on (1) bedrock and surficial geology (amount of calcium and magnesium present); (2) physiographic characteristics of the watershed, such as soil, vegetation, land use, and topography; (3) hydrological characteristics, that is, drainage, groundwater flow-through seep-

TABLE I Annual Inputs of Total (Wet and Dry) Deposition of Sulfate (SO_4), Ammonium (NH_4), and Nitrate (NO_3) in the Adirondack Mountains (U.S.A.) and on the west coast of Sweden

Region	*SO_4 (eq ha^{-1} yr^{-1})*	*NH_4 (eq ha^{-1} yr^{-1})*	*NO_3 (eq ha^{-1} yr^{-1})*
Adirondacks[a]	750	170	590
Sweden[b]	492	—	—

[a]From Driscoll *et al.* (1990).
[b]From Almer *et al.* (1978).

age, and groundwater discharge lakes; (4) climatology, for example, the quantity of precipitation; and (5) other chemical characteristics of the aquatic ecosystem, such as dissolved organic carbon concentration, and biological processes, such as sulfate reduction and denitrification by bacteria (Schindler, 1986; Munson and Gherini, 1990).

Measuring changes in the water chemistry of lakes and streams in response to acid deposition involves quantification of carbonate ANC ($HCO_3 - H^+$), base cations (Ca^{2+}, Mg^{2+}), inorganic aluminum (IAl), and possible organic acid anions (Sullivan, 1990). Some modifications in the lake chemistry of acidic waters alter the penetration of incident radiation and hence water transparency and temperature.

1. Chemical Effects of Acidification

In general, the direct effect of aquatic ecosystem acidification, in lakes with ANC ≤ 50 μeq liter^{-1}, is a significant decrease in pH and ANC with increasing lake sulfate (SO_4) concentration. The decrease is more pronounced at SO_4 concentrations > 75 μeq liter^{-1}. The relationship between occurrence of acidic lakes and wet SO_4 deposition is not linear; many factors, as mentioned earlier, control ANC (Baker and Christensen, 1990) (Table II).

Aquatic ecosystem acidification may directly or indirectly alter other chemical variables such as nutrient cycling or availability. Higher hydrogen ion (H^+) concentrations bring about decreases in the availability of dissolved inorganic carbon (DIC) and increases in dissolved inorganic nitrogen (DIN) concentrations. Thus increases in DIN can occur either directly, through nitrogen oxide inputs (Gahnström *et al.*, 1993), or indirectly by decreasing the activity of nitrifying bacteria in the nitrogen cycle (Rudd *et al.*, 1988), sometimes leading to ammonia (NH_4) accumulation. The concentrations of Ca^{2+} and Mg^{2+} can also be modified (Sullivan, 1990).

Other indirect chemical consequences of acidification are metal solubilization at pH < 5, particularly aluminum (Al), manganese (Mn), cadmium (Cd), and mercury (Hg) (Almer *et al.*, 1978; Campbell and Stokes, 1985). A growing body of evidence also suggests a decrease in organic acid anion concentrations associated with a decrease in pH (Almer *et al.*, 1974; Krug and Frink, 1983).

2. Physical Effects of Acidification

One of the physical consequences of lake acidification is the increase in light transparency. This has been reported since early field surveys (Schofield, 1972; Grahn *et al.*, 1974; Almer *et al.*, 1974; Yan, 1983) and has also been observed in whole-lake manipulations, when the pH decreased from above 6 to less than 5 (Schindler *et al.*, 1985; Shearer *et al.*, 1987).

This increase in light transparency may indirectly influence other physical and chemical characteristics or biological responses. Thus changes in

TABLE II Range of Some Chemical Characteristics of Selected Acidic Lakes and Streams[a]

Region[b]	*pH*	*ANC*	SO_4	Ca^{2+}	NO_3	*DOC*	$Al_{(ex)}$	$Al_{(tot)}$
Adirondacks[1] (171 lakes)	4.7–5.1	−24–[−6]	103–130	45–61	—	131–427	2.6–7.7	—
Adirondacks[2] (4 streams)	4.9–5.4	−6–11	94–109	40–55	4–8	4–7[c]	—	200–500[c,d]
Maine[3] (128 lakes)	5.1–7.2	−6–150	28–52	20–126	—	170–441	—	—
Florida[4] (150 lakes)	5.3–7.4	−5–382	59–475	57–486	0.2–7	250–990	1.4–2.1	—
Catskill Mountains[5] (7 streams)	5.1–6.0	10–29	119–126	87–155	22–31	117–250	2–19[e]	—
Norway[6] (lakes)	4.4–4.5	—	—	5.2–64	—	—	—	30–295[c,f]
Wales[7] (1 reservoir)	5.3	—	63	40	14.3	—	—	169[c,f]

[a] All units in μeq liter^{-1}, except dissolved organic carbon (DOC) and extractable aluminum ($Al_{(ex)}$) in μ*M*, and total aluminum ($Al_{(tot)}$) in mg liter^{-1}.
[b] Sources: [1]Driscoll *et al.* (1990); [2]Mulholland *et al.* (1992); [3]Kahl *et al.* (1990); [4]Pollman and Canfield (1990); [5]Stoddard and Murdoch (1990); [6]Henriksen, 1982; [7]Goenaga and Williams (1990).
[c] μg liter^{-1}.
[d] Total monomeric Al.
[e] Mean concentration in six major drainage basins of the Catskill region.
[f] Total dissolved Al.

lake transparency may increase epilimnion depth and hypolimnion heating (Yan, 1983; Effler and Field, 1983), leading to increases in the vertical mixing of dissolved substances to the hypolimnion (e.g., oxygen) or increases in nutrient cycling rates between epilimnetic water and sediments (Shearer *et al.,* 1987). Increases in light transparency and epilimnion depth may favor phytoplankton primary productivity at lower depths (Shearer and DeBruyn, 1986; Shearer *et al.,* 1987).

In the Scandinavian countries, increased light transparency has been attributed to a decrease in phytoplankton numbers (Grahn *et al.,* 1974). However, in North America, where phytoplankton abundance does not seem to decrease as a result of acidification, the change in water transparency seems related more to changes in dissolved organic carbon (DOC) concentration or in the chemical characteristics and light absorption capacity of dissolved organics (Effler *et al.,* 1985; Schindler *et al.,* 1985).

B. Changes in Algal Commnity Structure and Biomass

Benthic algal primary production and biomass do not generally decrease with acidification. Increases in biomass have often been observed in the littoral of lakes (Stokes *et al.,* 1989) and in rivers (Elwood and Mulholland, 1989). These increases occur quite rapidly, at least in experimental acidification studies (Hendrey, 1976; Hall *et al.,* 1980; Schindler *et al.,* 1985; Planas *et al.,* 1989), and are usually accompanied by changes in community composition (Jackson *et al.,* 1990) and decreases in species richness (Turner *et al.,* 1991).

1. Species Composition

Certain algal species can be found at very low pH and high metal concentrations (e.g., see Bennett, 1969; Hargreaves and Whitton, 1976). The Zygnemataceae are known to be very abundant in natural acidic lakes (Prescott, 1962; Yung *et al.,* 1986) and the Cyanophyceae also seem to be affected by acidification. The most information on pH specificity is available for the Bacillariophyceae; they have been used as pH indicators since the late 1930s (Hustedt, 1939).

A. Filamentous Greens One of the more striking facts pertaining to acidification is the increase in benthic filamentous green Zygnemataceae algae, particularly *Zygogonium, Mougeotia, Spirogyra,* and *Zygnema,* although other greens such as *Ulothrix* and *Oedogonium* can also become abundant (Table III). In a study of 32 Ontario (Canada) lakes with a pH range of 4.8 to 8.7, Wei *et al.* (1989) found that of the 21 possible Zygnemataceae species, only 4 species were important in acidic lakes: *Zygogonium tunetanum* (pH 4.8–6.8), *Temmogametum tirupatiensis* (pH

5.0–6.6), *Mougeotia quadragulata* (pH 5.6), and *Spirogyra fennica* (pH 5.6). Dense filamentous green growths have been observed, although unfortunately rarely quantified, in the epilithon, epiphyton, and metaphyton of acidic lakes (Stokes, 1981; Schindler *et al.*, 1985; Jackson *et al.*, 1990; Turner *et al.*, 1987), in rivers (Herrmann *et al.*, 1993), and in experimentally acidic channels (Hendrey, 1976; Planas and Moreau, 1989). In some lakes, filamentous benthic algal growth became so dense that the mat detached from the bottom, forming clouds of algae in the littoral zone (Stokes, 1981; Schindler and Turner, 1982; Howell *et al.*, 1990). In acidic lotic systems, coccoid green algae are also frequently reported (Mulholland *et al.*, 1986; Planas *et al.*, 1989; Winterbourn *et al.*, 1992).

In two lake acidification experiments, Lake 223 in northwestern Ontario, Canada (Schindler *et al.*, 1985; Turner *et al.*, 1987), and Little Rock Lake in Wisconsin, U.S.A. (Watras and Frost, 1989), *Mougeotia* appeared when water pH was lowered to around pH 5.6. In an anthropogenically acidic lake where *Mougeotia* and *Zygogonium* covered 10 and 100% of the shoreline, respectively, neutralization of the lake (pH increased from 4.8–5.5 to 6.3–6.7) essentially eliminated both species. Reacidification to pH 5.7–6.7 in the subsequent year was accompanied by an increase in *Zygogonium* but not *Mougeotia* (Jackson *et al.*, 1990).

B. Diatoms One of the better documented algal groups in regard to pH are the diatoms. The existence of acid-tolerant species has been reported by many authors (Hustedt, 1939; Nygaard, 1956; Patrick and Reimer, 1966; Moss, 1973) and has been widely used in paleolimnology studies to determine past acidification processes in lakes and ponds (Renberg and Hellberg, 1982; Charles, 1985; Smol *et al.*, 1986; Fritz *et al.*, 1990).

Alkaliphilic and alkalibiontic diatom species may disappear during acidification or may simply avoid colonizing the substratum in acidic environments. The latter case was observed in an experimental acidification study using stream channels, where species considered acidophilic, such as *Tabellaria flocculosa* and *Eunotia pectinalis,* were the first to colonize the acidic channels, whereas other species present in the control (e.g., *Achnantes minutissima, Synedra vaucheria*) did not settle. This was also true of certain desmids (Paquet, 1993). Acidophilic diatoms, such as *Eunotia,* seem to decrease when lakes are limed (Lazarek, 1986), which is in agreement with the pH tolerance reported for this genus (Table III). A survey of various streams spanning a pH gradient showed *Eunotia* to be unaffected by pH (Winterbourn *et al.*, 1992). It must be noted that all the rivers, or sites within the rivers, included in this survey were below the neutral mark (pH 5.0 to 6.9).

Although diatom decline has been reported in the littoral of lakes or mesocosms subjected to low pH (Muller, 1980; Stevenson *et al.*, 1985; Roberts and Boylen, 1988), diatoms can be the dominant epiphytic species

TABLE III Class, Genus, or Species (If Available) of Benthic Algae Often Found in Acidic Lakes and Streams as Well as the Community Type (in Relation to the Substratum Occupied) and the pH[a, b]

Taxon	*Substratum*	*pH (natural)*	*pH categories*
Bacillariophyceae			
Achnanthes marginulata[3]	Epipelon	4.6–5.0	ACF
Anomoeoneis sp.[1]	Epipelon		
Anomoeoneis serians		4.5–5.0	ACB
Anomoeoneis serians brachysira[20]		4.5	
Diatoma sp.[6]		≤5.0	
Eunotia sp.[8]		5.4–5.8	
Eunotia bactriana		4.5–5.0	ACB
Eunotia curvata[16,17]	Epipelon, epilithon	4.5–5.2	ACF
Eunotia exigua[14,20,22]	Epilithon, epipelon	4.8–5.5	ACB
Eunotia incisa[4]	Epilithon	4.5–6.0	ACB
Eunotia pectinalis[17]	Epilithon	4.5	ACF
Eunotia pectinalis var. *minor*[18,22]	Epilithon	4.4–5.3	ACF
Eunotia tenella[2,18,22]		5.1–5.9	ACB
Eunotia vanheurkii[2]		4.6–5.3	ACF
Eunotia veneta[20]		4.5–5.0	ACF
Fragilaria acidobiontica[1,16]	Epipelon	4.8–5.2	ACB
Fragilaria virescens[6,18]		≤5.0	
Fragilaria virescens var.[3]	Epipelon	4.6–5.0	ACB
Frustulia sp.[1]	Epipelon		
Frustulia rhomboides[4,22]		4.5–7.0	ACF
Frustulia rhomboides var. *crassinervia*[18]	Epilithon	4.4	ACF
Navicula cumbriensis[5]			ACB
Navicula hoeflen[5]			ACB
Navicula subtilissima[20]			ACB
Navicula tennicephala[16]	Epipelon	4.8–5.2	
Neidium affine[4]		4.5–7.5	ACF
Neidium iridis amphigomphius[22]		4.8	ACF
Neidium ladogense desentiatum	Epilithon	4.5–5.0	

(Continues)

of some lakes (Lazarek, 1985). In rivers, diatoms are generally the most abundant group (Patrick *et al.*, 1968; Arnold *et al.*, 1981; Duthie and Hamilton, 1983; Mulholland *et al.*, 1986; Planas *et al.*, 1989; see also review by Elwood and Mulholland, 1989). Among the diatom species, several taxa are acidobiontic (Table III). The genus *Eunotia* seems to be especially favored in acidic lotic systems (Moss, 1973; Mulholland *et al.*, 1986; Marker and Willoughby, 1988; Planas *et al.*, 1989; Winterbourn *et al.*, 1992; Paquet, 1993).

TABLE III *(continued)*

Taxon	Substratum	pH (natural)	pH categories
Pinnularia abaujensis[1,16]	Epipelon	4.8–5.2	ACF
Stauroneis gracillima[3]	Epipelon	4.8	ACB
Tabellaria binalis[5,20,22]		4.5–5.0	ACB
Tabellaria fenestrata[6,7,20]		≤5.0–5.8	ACF
Tabellaria quadriseptata[2,4]		4.5–6.2	ACF
Chlorophyceae			
Bulbochaete sp.[7]		≥5.0	
Microspora sp.[2]			5.6
Mougeotia sp.[6-12,15,18]	Epiphyton, epilithon	≤5.6	
Mougeotia quadragulata [21]			5.6
Spirogyra sp.[7,11]	Epiphyton, epilithon	<5.0[>6.0]	
Spirogyra fennica[21]			5.6
Oedogonium sp.[7,8,17,18]			4.4–7.5
Ulothrix sp.[2]		5.6	
Temmogametum tirupatiensis[21]		5.0–6.6	
Zygnema sp.[7]		<5.0–6.3	
Zygogonium sp.[7,12]			<5.0
Zygogonium tunetanum[15,21]			5.0–5.6

[a]Natural pH refers to the range or mean pH measured or inferred. for diatom species, pH categories as defined by Hustedt (1939) are included: ACB = acidobiontic and ACF = acidophilic.

[b]Sources: [1]Stevenson *et al.* (1985); [2]Ormerod and Wade (1990); [3]Charles (1985); [4]Cook and Jagger (1990); [5]Fritz *et al.* (1990); [6]Hendrey (1976); [7]Stokes (1981); [8]Stokes (1984); [9]Deleted in proof; [10]Schindler *et al.* (1985); [11]Turner *et al.* (1987); [12]Detenbeck and Brezonik (1991); [13]Lazarek (1982b); [14]Lowe (1974); [15]Howell *et al.* (1990); [16]Roberts and Boylen (1989); [17]Planas *et al.* (1989); [18]Fairchild and Sherman (1990); [19]Hultberg and Andersson (1982); [20]Renberg and Wallin (1985); [21]Wei *et al.* (1989); [22]Fairchild and Sherman (1993).

C. Cyanophyceae The blue-greens are among the benthic algae most sensitive to acidification. From an extensive survey of lakes and rivers of different pH, Brock (1973) reported a tolerance limit of about pH 4.8 for Cyanophyceae. Stevenson *et al.* (1985) did not find a strong correlation between the presence of Oscillatoriaceae and pH in their study of 20 lakes with a pH range between 4.46 and 7.29. On the other hand, the latter group was found to be unimportant in the more acidic sites of a stream pH gradient study (Mulholland *et al.*, 1986). In an experimental lake acidification study, Turner *et al.* (1987) observed nonsignificant decreases in *Anabaena* sp. and *Lyngbya* sp. at pH 5.5, and when the lake was further acidified to pH 5.1, a marked decrease in numbers of both species (Turner *et al.*, 1991). Contrary to these findings, Lazarek (1982a) reported heavy growth of epipelic Oscillatoriaceae in acidic lakes (pH 4.3–4.7) and *Halosiphon pumilus* has been reported to be abundant in acidic environments, particularly in mineral-poor dystrophic bogs (Prescott, 1962; Roberts and Boylen, 1989). These contradictory findings could be related to the higher pH present in mat microenvironments or in the sediments of epipelic habitats. Thus benthic communities dominated by filamentous blue-greens are able to buffer a decrease in pH within 1 mm of the mat surface (Jorgensen and Revsbech, 1983; Sweerts *et al.*, 1986). In fact, Lazarek (1982a) found the pH of the benthic mat (pH 5.5–6.2) to be between 1 and 2 units higher than that of the overlying water. Roberts and Boylen (1989) advanced the hypothesis that the abundance of *H. pumilis* in the epipelon of Woods Lake, a naturally acidic brown water lake, could be explained by the presence of this species in the lake before the onset of anthropogenic acid deposition.

2. Species Richness

Much information is available on the reduction in species richness and diversity of phytoplankton associated with decreases in lake pH (e.g., see Almer *et al.*, 1974; Yan and Stokes, 1978; Schindler *et al.*, 1985; Stokes, 1986). Unfortunately, very few papers on the benthic algal structure of acidic environments report total number of species. In the littoral of acidic lakes where *Mougeotia* becomes dominant, the number of algal species generally decreases (Muller, 1980; Stokes, 1986; Turner *et al.*, 1991). In lotic ecosystems, Mulholland *et al.* (1986) found a lower number of species in the more acidic sites. No significant differences in species numbers were found in experimentally acidified outdoor stream channels fed by a first-order stream (Planas *et al.*, 1989) after 1 or 2 months of acidification, although a low initial algal colonization rate and fewer species were observed in the acidific channels during the first week of acidification (Paquet, 1993).

Despite the limited amount of available data, it appears that the benthic algal richness of acidic systems could be affected by a decrease in available substrata. Such decreases can result from a competition for space brought about by the dense growth of a few species or, in the case of epiphytes, to changes in aquatic plant species. Modifications of macrophyte communities have frequently been mentioned in acidic Scandinavian lakes. In these lakes, decreases in pH were accompanied by a rapid expansion and heavy growth of *Sphagnum* and a consequent reduction in other macrophytes (Grahn, 1977). A result of this change in macrophyte species is the loss of structural support for epiphytes; *Sphagnum* does not seem to be colonized by epiphytic algae. Although *Sphagnum* is present in some acidic lakes of North America (Hendrey and Vertucci, 1980; Stokes, 1986; Catling *et al.*, 1988), invasions such as occur in Scandinavia are rarely reported.

3. Biomass

In general, most studies have indicated that acidification increases the benthic algal biomass of both the littoral of lakes and lotic systems (Parent *et al.*, 1986; Stokes *et al.*, 1989; Elwood and Mulholland, 1989; Turner *et al.*, 1991). In a large study comprising 36 lakes with a pH range of 5.3 to 6.6, a negative relationship was found between pH and metaphytic biomass (France and Welbourn, 1992). Data from Scandinavian countries report a shift in the biomass of primary producers from pelagic to benthic communities (Grahn, 1977, 1985; Hendrey and Vertucci, 1980). A decrease in pelagic algae has generally not been reported in North American lakes (Schindler *et al.*, 1985), although this hypothesis has been advanced for shallow Florida lakes (Crisman *et al.*, 1986).

In lakes with heavy Zygnemataceae growth, little areal density data exist. Howell *et al.* (1990) reported 1.2 g dw m^{-2} at peak biomass for a metaphytic mat of *Zygonium tunetanum*. Between 2 and 3.5 times more epilithic biomass (chlorophyll *a*) was also measured by Mulholland *et al.* (1986) in the acidic sites of a stream following a 6- to 8-week period of low precipitation; these differences were partially masked by scouring of the streambed as a result of large storms. After 1 and 2 months of acidification, 2.8 and 9 times more epilithic algal biomass (chlorophyll *a*) was measured in outdoor acidic stream channels than in the control, respectively (Vallée, 1993). Other studies, however, indicate similar levels of epilithic algal biomass in acidic and circumneutral streams despite differences in species composition (Arnold *et al.*, 1981; Winterbourn *et al.*, 1992).

Seasonal studies on the benthic algal biomass of acidic ecosystems are very rare. Acidification-induced filamentous green algal growth in lakes is probably seasonal, and maximum abundance in a given lake can be irregular from year to year. Few acidic lake studies have simultaneously considered the benthic algal biomass of all substrata (epilithon, epipelon, and epiphyton). Given that epilithic biomass can in fact decline with experimental lake acidification (Turner, 1993), it could be that total benthic algal biomass is actually lower than that of an unacidified lake when calculated on an annual basis.

C. Changes in Community Metabolism

Experimental lake acidification data have shown that community metabolism can be disrupted at low pH. Primary production and respiration ratios of periphyton have been found to be affected when the pH falls below 6.2 (Schindler, 1990). In contrast to the number of studies that deal with species shifts or biomass modifications in acidic systems, primary production data are scarce and the effect of low pH on algal benthic primary productivity is a controversial subject, particularly in lentic systems. The increases in benthic algal biomass observed in acidic ecosystems do not necessarily produce increases in primary production (Elwood and Mulholland, 1989; Stokes *et al.*, 1989).

Seemingly positive effects of acidification on primary production have been reported in different habitats. Thus, chlorophyll-specific primary production rates of 0.9 and 1.3 μg C (μg chl *a*)$^{-1}$ h^{-1} were measured for metaphytic Zygnemataceae mats in a circumneutral and an anthropogenically acidic lake, respectively. For one lake the chlorophyll-specific productivity of epiphytic algae was one order of magnitude lower than that of the metaphyton (Howell *et al.*, 1990). In limed Gårsdjön Lake, epiphytic primary production, measured as ^{14}C fixation, was reported to be one order of magnitude lower after liming than during the acidic stage (Lazarek, 1986). However, in an earlier paper the author claimed that epiphytic productivity was similar in both acidic Gårsdjön Lake and limed Lake Högsjön, (Lazarek, 1982b).

In lotic waters, areal and chlorophyll-specific primary production rates were 10 to 15 times greater in acidic outdoor stream channels than in the control (Parent *et al.*, 1986). In a study of streams across a pH gradient, greatest areal primary production was found in the acidic sites, but no significant difference in chlorophyll-specific primary production was found between sites, although values tended to be lower in the acidic sites (Mulholland *et al.*, 1986). In a comparison of rivers of different pH, the more acidic streams showed average or above average primary production rates (Winterbourn *et al.*, 1992).

In the littoral zones of lakes, acidification can reduce primary production due to inorganic carbon limitation at low pH, as will be discussed later (see Section III, A). The addition of inorganic carbon or increased water movement has been shown to eliminate this effect (Turner *et al.*, 1991). Net primary production in acidic environments can also be reduced by increases in community respiration. This was found in the comparative measurements of primary production and dark respiration in an experimentally acidic lake (pH 5.1) and a circumneutal lake (pH 6.9) (Turner *et al.*, 1991). In this study, the P/R ratio of a given lake varied as a function of the type of substratum (epilithon and epiphyton), depth (0.1–0.2 and 1–2 m), and period of the year. Thus dark respiration was higher in the epiphyton than in the epilithon, and gross primary production was 1.4–2.0 times higher in the epiphyton of the acidic lake versus the circumneutral lake, but lower in the acidic epilithon. In the latter community, P_{max} was lower in the shallow than in the middle littoral depth, where algal cell volume was higher. For the epiphytic community, net primary production in July was lower in the treated lake than in the control, whereas in August no differences were found between lakes.

III. HYPOTHESES EXPLAINING ACIDIFICATION EFFECTS

Changes in algal species composition, biomass, and productivity could be attributed to abiotic and biotic factors. The abiotic factors considered are (1) increases in H^+ and metal concentrations, (2) changes in the limiting nutrient, (3) changes in light penetration, (4) changes in the thickness of the periphytic mat boundary layer, and (5) substratum availability. Among the biotic factors discussed are (1) changes in the littoral macroinvertebrate community and (2) competition for a limiting resource.

A. Abiotic Factors

Ecosystem perturbation by anthropogenic acidification induces changes that cannot be solely attributed to the potential toxicity of increased concentrations of H^+ or metals, particularly aluminum. Other variables, such as the species of inorganic carbon dissolved in the water (free CO_2, bicarbonate), the biogeochemical cycle of nutrients in general, and the quantity and quality of dissolved organic carbon, could also be altered by acidification. Changes in certain chemical or biological components could induce modifications in physical variables such as light penetration, stratification, or substratum availability.

1. pH Effects

The distribution of diatom taxa is known to be strongly correlated with pH. Diatoms have been classified following pH occurrence categories: acidobiontic (optimum pH < 5.5), acidophilic (optimum pH < 7.0), circumneutral or indifferent (pH around 7.0), alkaliphilic (widest distribution at pH > 7.0), and alkalibiontic (only present at pH > 7.0) (Hustedt, 1939).

In contrast to the diatoms, little information can be found in the literature regarding pH categories for other algal taxonomic groups. Decreases, or the disappearance, of the green algae *Mougeotia* and other Zygnemataceae species have been reported with increases in water pH. This has been observed following reductions in industrial emissions (Gunn and Keller, 1990), in the recovery of a former acidic lake when the pH increased to over 5.8 (Schindler *et al.*, 1991), and in limed lakes at pH > 7 (Hultberg and Andersson, 1982; Jackson *et al.*, 1990; Fairchild and Sherman, 1990). In all of these studies, however, it could not be ascertained if the decrease in the dominant species was directly or indirectly attributable to pH.

2. Toxicity

Aluminum (Al) becomes extremely soluble at pH < 6.0 (Likens *et al.*, 1977) and dissolved Al can thus be very high in acidic lakes (Cronan and Schofield, 1979). Severalfold increases in Al were measured in the epilimnion of Lake 223 as the pH was lowered, and mean Al concentrations were negatively correlated to pH (Schindler and Turner, 1982).

The mechanisms by which Al is toxic are unknown (Helliwell *et al.*, 1983; Campbell and Stokes, 1985). Not all Al forms are toxic, thus it is very important to determine the form. Inorganic monomeric Al, for example, is very toxic, but Al–fluoride complexes are not. The latter seems to be the dominant form in some lakes (Driscoll, 1985).

There is little evidence of Al toxicity to algae. Some laboratory experiments have shown toxicity of Al to phytoplankton at 50 μg Al $Liter^{-1}$ (Nalewajko and Paul, 1985; Pillsbury and Kingston, 1990) and decreases in phytoplankton diatom and desmid numbers have been attributed to high Al concentrations in pelagic waters (Hömström *et al.*, 1985). No studies are available regarding the toxicity of aluminum to benthic algae.

3. Nutrient Limitation

It is only since the mid-1980s that the changes in nutrient availability that accompany lake acidification have been taken into consideration. Few studies have actually tested the importance of such changes as a mechanism explaining structural or functional changes in benthic algal communities (Winterbourn *et al.*, 1985, 1992; Mulholland *et al.*, 1986; Keithan *et al.*, 1988; Turner *et al.*, 1987, 1991; Howell *et al.*, 1990; Fairchild and Sherman, 1990, 1993). Turner *et al.* (1991) suggest that adnate diatom forms might be limited by carbon, although limitation by another nutrient such as silica (Si) or phosphorus (P) cannot be excluded, especially for certain species. Limitation by one or another nutrient could affect total biomass as well (Pillsbury, 1993).

A. Carbon Carbon has been identified as potentially limiting to littoral primary production in acidic lakes. Bicarbonate ions (HCO_3) are practically absent in waters below pH 5.0, and diffusion of CO_2 from the atmosphere is low even though the water solubility coefficient is high (Wetzel *et al.*, 1985). In standing water, thickness of the boundary layer could hamper carbon diffusion. We would thus expect the photosynthetic demand of dense stands of algae to soon become carbon-limited, particularly if they are on the bottom, far from the air–water interface. Carbon limitation of benthic algal primary production has been demonstrated in low-pH lakes and rivers using short-term C enrichments in confined samples. In these studies, increasing HCO_3 or CO_2 concentrations stimulated benthic algal photosynthesis (Mulholland *et al.*, 1986; Turner *et al.*, 1987,

1991). Similarly, an enhancement of photosynthesis occurred when water movement was set up in the incubation chamber (Turner *et al.,* 1991).

Primary production measurements in confined environments may not necessarily reflect rates that occur under natural conditions. Even with controlled water movement (Turner *et al.,* 1991), the high algal densities often used in bottles or chambers can further skew the results (Howell *et al.,* 1990). In the case of epipelic and thick mats of epilithic algae, overlying water probably is not the only source of carbon. Wetzel *et al.* (1985) demonstrated the importance of the sediments as a source of CO_2 for macrophytes in acidic lakes. They advanced the hypothesis that benthic algal communities living on or near sites of intense mineralization and increased CO_2 concentrations could benefit from this source.

An experimental enrichment of the littoral of an acidic lake with different combinations of nutrients—carbon, nitrogen, and phosphorus—showed different responses before and after neutralization (Fairchild and Sherman, 1990). Before liming, the order of increasing benthic algal biomass in relation to each of the treatments was C < PC < NC < PNC, as measured by chlorophyll or ash-free or dry weight (AFDW). No response was recorded for NP, N, and P enrichment alone. Carbon seemed to have the most important enhancement effect on benthic algal biomass, with the other nutrients probably becoming limiting as a result of increased carbon availability. After liming, algae responded only to N and P enrichments.

Fairchild and Sherman (1990) found that abundance of *Mougeotia,* the dominant species when the lake was acidic, was significantly increased when the substratum was enriched with N and C but not with C alone. This species seems to be very efficient in the uptake of CO_2 at low pH and alkalinity (Turner *et al.,* 1987), but it may be less efficient in the use of bicarbonate. This fact could explain why this genus, as well as *Zygogonium,* tends to disappear in lakes after liming (Lazarek, 1986; Jackson *et al.,* 1990) and is replaced by other species such as *Spirogyra* and *Oedogonium,* which are probably better adapted to alkaline systems (Fairchild and Sherman, 1990). Simpson and Eaton (1986) have put forth the hypothesis that *Spirogyra* could induce carbonic anhydrase, thus permitting the utilization of HCO_3. Consequently, with the exception of *Spirogyra,* pH or low alkalinity appears to have a greater impact than nutrients (N and P) in controlling abundance of greens. This also seems to be the case for the several species of *Eunotia,* which also become very rare after liming (Fairchild and Sherman, 1990).

B. Nitrogen Nitrogen, in the form of nitrates, is one of the atmospheric acidifying agents that has taken on increased importance as a component of acid deposition. This is particularly true in Europe (Gahnström *et al.,* 1993) (Fig. 1), although concentrations do rise sharply during snowmelt in the cold temperate regions of North America (Munson and

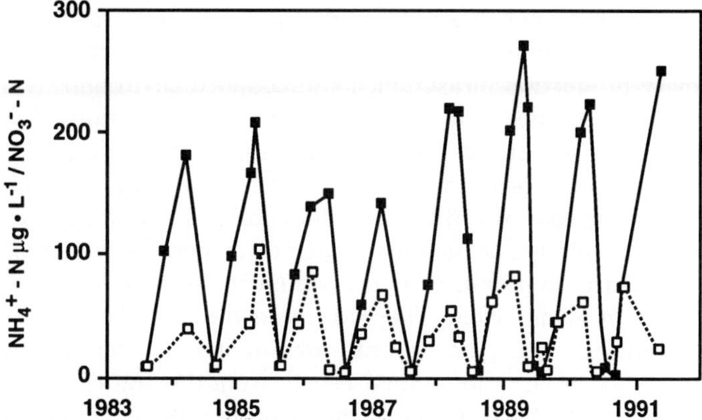

FIGURE 1 Temporal trends (from 1983 to 1991) in nitrate-nitrogen (solid squares) and ammonium-nitrogen (open squares) in a southern Swedish lake. Modified with permission from Gahnström, G., Blomqvist, P., and Fleischer, S. (1993). Are Key Nitrogen Fluxes Changed in the Acidified Aquatic Ecosystem? Ambio 22, 381–424.

Gherini, 1990). The fate of this acidifying nutrient in the study of aquatic ecosystems had been largely ignored in the past. Rudd *et al.* (1990) are among the only researchers to have used nitric acid as an acidifying agent in their experimental acidification project.

Aside from the increased inputs of nitrogen resulting from acid deposition, acidification also appears to alter the in-lake nitrogen cycle. In their experimental acidification of two lakes with sulfuric acid, Rudd *et al.* (1988) reported an accumulation of NH_4 during the winter with no subsequent increase in NO_3, suggesting an inhibition of nitrification. Both NH_4 and NO_3 can be used as nitrogen sources by algae, with NH_4 usually being the source of preference (Wetzel, 1983a). Changes in total inorganic nitrogen concentration (TIN), as well as in the more readily available nitrogen species (NH_4 or NO_3), could lead to a shift in algal species composition.

Results of a Swedish mountain stream study showed that nitrate pulses in early spring are associated with increases in the number and biomass of benthic algae as well as with shifts in species composition, from bluegreens to green algae (mainly Zygnemataceae along with other green species such as *Ulothrix* and *Microspora*). These shifts were not observed in other streams of the same region that were richer in Ca but also experiencing nitrate pulses (Herrmann *et al.,* 1993).

In the multiple-nutrient enrichment experiment mentioned earlier, the lack of response of the benthic algal species usually present in the acidic lake to P and N additions alone (Fairchild and Sherman, 1990) seems to point to an adequate supply of these nutrients in the environment. In fact, nitrate concentrations in the pelagic zone of the studied lake were relatively high, although phosphorus (SRP) was quite low (Table IV).

TABLE IV The pH and Nutrient Concentration of Some Acidic Systems

	Lake Earnest[a]		WCP[b]		Outdoor stream channels[c]	
	1987	1989	Acid	Circumneutral	Acid	Control
pH	4.4 ± 0.0	7.5 ± 0.1	4.5	6.4	4.3	6.7
P-SRP[d] (μg liter^{-1})	1.9 ± 0.0	4.3 ± 0.3	0.3	1.7	2.8 ± 1.6	2.5 ± 1.8
$NO_3 + NO_2$ (μg liter^{-1})	67 ± 39	3.0 ± 0.7			18.3 ± 4.3	19.1 ± 3.0
NH_3 (μg liter^{-1})	83 ± 26	13.0 ± 0.6			9.0 ± 0.7	10.0 ± 1.1
DIC[e] (μmol)	10				25 ± 2.3	28 ± 2.3

[a]From Fairchild and Sherman (1990, 1992).
[b]From Mulholland *et al.* (1986); WCP = Walker Camp Prong River.
[c]From Vallée (1993), 1988 data.
[d]SRP – soluble reactive phosphorus.
[e]DIC = dissolved inorganic carbon.

C. Phosphorus Phosphorus is usually considered to be the most limiting nutrient in temperate oligotrophic ecosystems. The majority of aquatic ecosystems in regions sensitive to acidification are oligotrophic, since their watersheds are on thin soils with granitic or gneiss bedrock. Some early lake acidification studies linked decreases in pH with water oligotrophication due to the coprecipitation of P with aluminum (Grahn *et al.*, 1974; Dickson, 1978; Driscoll, 1985). No clear evidence as to a decrease in P has emerged from other acidic lake survey studies (Dillon *et al.*, 1979; Olsson and Pettersson, 1993) nor from lake or river reach acidification experiments (Hall *et al.*, 1980; Schindler *et al.*, 1991).

These contradictory results regarding the increase or decrease of P in acidic systems could be explained by any of the following: differences in the mobility of aluminum in lake watersheds, thickness of the water path through the soils, or differences in pH between the inflow water and the lake (Driscoll *et al.*, 1990). It is likely that P availability is related not so much to the present acid/base status of aquatic systems but rather to the acidity of incoming waters relative to "*in situ*" pH. Under conditions where the inputs are more acidic than the receiving water body, the aluminum concentration (and perhaps also P) of the incoming water is likely to be above that which can be maintained at the higher pH of the receiving system. It thus precipitates and takes P with it, resulting in a reduction in P concentration. This was observed in a stream drainage system study, where the elevated Al and P levels measured in the highly acidic headwaters were reduced by precipitation and absorption in the downstream sections of the drainage system that were more circumneutral than the headwaters (Mulholland *et al.*, 1986; Elwood and Mulholland, 1989).

For the epipelic and epilithic algae of acidic running waters there is indirect evidence of a greater phosphorus availability. Lower biomass-specific (chlorophyll or ATP) phosphatase activity and lower P turnover rates have been measured in the benthic algae of acidic as compared to more circumneutral environments (Mulholland *et al.*, 1986; Planas and Moreau, 1986; Vallée, 1993). However, differences in the rates or activities of these P-availability indicators could be influenced by other variables, such as increased light (which could decrease the chlorophyll : biomass ratio) or reduced bacterial activity (McKinley and Vestal, 1982; Francis *et al.*, 1984: Burton *et al.*, 1985; Allard and Moreau, 1986a; Marmorek *et al.*, 1986).

The greater P availability suggested by the former studies does not seem to correspond to any measurable difference in water concentration (Table IV). It may be that benthic algae are not solely dependent on pelagic water P, but, as was suggested before for carbon (Wetzel *et al.*, 1985), may receive substantial nutrition from the substratum (Stevenson *et al.*, 1985; Hansson, 1989; see Wetzel, Chapter 20, this volume). Phosphorus could be adsorbed onto sediment surfaces (e.g., clays or fine organic matter) from the surrounding water, thereby augmenting the supply to colonizing benthic algae (Burkholder and Cuker, 1991). Storm events and other turbulence-producing disturbances may greatly modify the steady-state diffusion of pore-water nutrients (Riber and Wetzel, 1987) and simultaneously mix nutrient-rich sediment pore water up into the sediment– water interface (Peters and Cattaneo, 1984). During such nutrient pulses, large quantities of P can be stored for future use by the benthic algae. Decomposition could also result in a steady diffusion of P out of organic sediments (P. J. Mulholland, personal communication). In acid-sensitive regions of North America, snowmelt and storm events are known to produce acid pulses (Marmorek *et al.*, 1986; Galloway *et al.*, 1987; Jeffries, 1990), which could also lead to the release of P from the sediments, however, this hypothesis requires further testing.

To our knowledge, P fluxes have not been measured at the sediment–water interface of acidic aquatic ecosystems. However, results of the P-fractionation analysis of sediments accumulated in outdoor stream channels indicated that the NaOH-extracted P fraction (which is considered to estimate the P associated with Al and Mn) and the total inorganic P fraction had decreased significantly in the sediments of the acidic (pH 4.3) channels as compared to the controls (pH 6.6) (C. Vallée and D. Planas, unpublished). Solubility of variscite ($AlPO_4$) is higher at pH 4.3, thus it is probable that acidification mobilizes the P associated with Al when this mineral is present. In the preceding stream channel experiment, total dissolved Al, as well as total nonexchangeable and exchangeable Al, was significantly higher in the acidic than in the control channels (Vallée, 1993). Increases in algal benthic biomass recorded in these acidic channels were higher or as high as that observed in concurrently P-enriched channels at a mean pH of 6.6. Thus it is possible that acidification mobilizes the P associated with Al, and that it is rapidly taken up by the algae. Increases in P in streams subjected to acid pulses have already been mentioned by Hall *et al.* (1987).

D. Other Nutrients or Micronutrients Because it is less soluble in acidic lakes, silica seems to limit the growth of euplanktonic diatoms at lake pH ≤ 5.8–6.0 (Charles, 1985). In addition to a decrease in Si availability at low pH, the transport of Si across cell membranes may also be reduced (Azam *et al.,* 1974). The fact that benthic diatoms, in contrast to the euplanktonic, can exist at lake pH ≤ 5.8–6.0 can probably be attributed to differences between habitats, and probably to Si availability. Benthic diatoms are able to obtain Si directly from Si-containing substrata (Hutchinson, 1957).

Silica concentrations are often higher in the littoral zone than in the pelagic because it is released from the sediments. Thus the decrease in diatoms in the littoral of lakes cannot be explained simply by Si limitation. Competition by other species, such as filamentous greens, should also be taken into account. The filamentous algae may have the advantage of reaching beyond the sediment boundary layer to compete for light, pelagic nutrients, or DIC compared to adnate or stalked periphytic growth forms (B. W. Pillsbury, personal communication).

It is known that increases in selenium concentrations in the pelagic zone have been responsible for planktonic Chrysophyceae bloom formations in soft-water and acidic lakes (Wehr and Brown, 1985). It could be that this or other micronutrients with high solubility at low pH could benefit algal growth or favor one species over another. On the whole, however, little information exists on the effects of acidification on nutrients other than those mentioned in the foregoing discussion.

4. Physical Variables

Among the physical variables that can influence the abundance and metabolism of benthic algae are light, temperature, water movement, and substratum availability.

A. Light In the shallow depths of lakes, changes in transparency due to acidification could benefit benthic algae by deepening the light compensation point. Many studies have noted sharp increases in filamentous green algae with increases in light in oligotrophic waters (Shortreed and Stockner, 1983; Sheath *et al.,* 1986; see Hill, Chapter 5, this volume). Changes in light penetration in acidic ecosystems could increase the availability of colonization space and could thus enhance areal benthic primary production (Stokes *et al.,* 1989). Alterations in water transparency could also result in a greatly altered zonation pattern of benthic algal species or biomass zonation in lakes (Stevenson *et al.,* 1985).

B. Water Movement A variable that is often forgotten when comparing changes in algal community structure between natural ecosystems of varying pH is wave action in the case of lakes and current in rivers (see Stevenson, Chapter 11 this volume, and Peterson, Chapter 13, this volume).

Water movement may enhance benthic algal development by modifying the boundary layer around the cells and increasing diffusion. Thus in lakes, as mentioned before, water movement appears to increase the efficiency of filamentous Zygnemataceae in absorbing dissolved carbon (Turner *et al.,* 1987). Excessive turbulence, however, can be detrimental to attached algae; lower abundances or absence of filamentous algae are observed in running water when water velocity exceeds a threshold at which mechanical abrasion of turbulence surpasses the resistance limit of the algae. Interactive effects of water movement and shoreline features could also contribute to differences in the structure and abundance of benthic algae among acidic lakes (Howell *et al.,* 1990).

Diatoms dominate in the fast-current areas of lotic systems and filamentous, mostly green algae dominate at low current speed (McIntire, 1966; Horner *et al.,* 1983). In an 8-year experimental acidification study carried out in outdoor stream channels with controlled water quality, light exposure, substratum, and velocity (20 cm s^{-1}), the diatom *Eunotia pectinalis* was always the dominant species in the acidic channel (Planas *et al.,* 1989). The exception to this rule occurred the year an inclosure–exclosure experiment was carried out, at which point *Mougeotia* became dominant (Planas and Moreau, 1989; Paquet, 1993). The more plausible explanation for this shift is the decrease in flow caused by the exclosure device.

C. Substratum Substratum is an important variable to consider in comparative acidification effect studies. Nearly all substrata are highly dynamic in their physical characteristics and in their chemical contribution to the attached microflora (Wetzel, 1993). Not only do the epipelic algae live in the very steep microgradient existing within the first few centimeters of sediment (Wetzel, 1983b), but they may actually influence certain characteristics of the environment. Sharp increases in pH have been reported within the upper centimeters of acidic lake sediments as a result of alkalinity generation within the sediments (Cook *et al.,* 1986, 1987; Carlton and Wetzel, 1988).

Some studies on the epipelic algae of different lakes spanning a wide pH range, or of a given lake before and after liming, have shown a species-specific relationship between algal species biovolume and pH or ANC that may change in relation to depth in the littoral (Stevenson *et al.,* 1985; Roberts and Boylen, 1988, 1989). In the case of periphytic bacteria, Palumbo *et al.* (1987) and Mulholland *et al.* (1992) reported that in contrast to the bacteria found on rocks or decomposing leaves, epipelic bacterial production was not affected by low water pH. This is presumably a result of the sediments having a higher pH or buffering capacity, or because of the higher nutrient concentrations found there, or both.

For epiphytic algae, the proximity of the photosynthesizing tissue of plants could have both advantages and disadvantages. For example, there could be competition for CO_2, or the pH of the algal environment could vary daily as photosynthesis proceeds (Wetzel, 1983b).

5. Interaction of Chemical and Physical Variables and Succession

Many of the abiotic variables that could explain the presence or absence of any given species vary synchronously through succession. These changes interact with acidification effects, which may also vary with time. For example, the pH of streams and of the littoral of small lakes receiving acid precipitation declines in early spring (during and shortly after snowmelt), whereas flow, light quantity, and temperature increase.

Research on pollutant effects on algae seldom accounts for the natural changes in species that occur during the annual succession. Light quantity as well as quality could be important (see Hill, Chapter 5, this volume), as could changes in nutrient availability through time (Fairchild and Sherman, 1990; see Borchardt, Chapter 7, this volume). In streams, changes in nutrient cycling and hence nutrient availability occur as biomass increases during succession (see Mulholland, Chapter 19, this volume). In cold temperate regions, the decrease in pH associated with snowmelt in early spring probably also results in increased nutrient availability. This is liable to be particularly true of nitrogen owing to the anthropogenic nitrogen accumulated in the snowpack during the winter (Jones and Sochanska, 1985). In the acid-sensitive areas of Canadian Shield lotic systems, benthic algal communities are dominated by *Tabellaria* in the spring. If acid conditions persist, *Eunotia pectinalis* becomes dominant later on in the succession in rapid-flowing waters and Zygnematales, particularly *Mougeotia,* in slower-moving waters (S. Paquet and D. Planas, unpublished data). Turner *et al.* (1991) mention that *Mougeotia* growth coincides with periods of water clarity, as well as increased water temperature.

In conclusion, physical or chemical variables influence community structure and processes because algal species are differentially sensitive to pH as well as to physicochemical factors both related and unrelated to pH.

B. Biotic Factors

Modifications in food web interactions could also be important in determining the observed response of benthic algae to acidification. A decrease in grazing pressure is one of the mechanisms that has often been advanced to explain increases in benthic algal biomass. One point that has not been explored much in relation to ecosystem perturbations, and particularly in regard to the harsh conditions caused by acidification, is a change in resource competition (e.g., see Hart and Robinson, 1990).

1. Control by Grazers

The hypothesis often put forward to explain massive benthic algal growth in acidic or acidifying ecosystems is a decrease in macroinvertebrate grazing pressure. This hypothesis has been supported mainly by the widely observed disappearance of acid-sensitive macrobenthos taxa (crustacea, molluscs, leaches, and insects) with declines in median pH to below 5.5 (Harvey, 1989; Schindler *et al.,* 1991). Certain enclosure/exclosure and transplant experiments seem to corroborate this hypothesis (e.g., see Parent *et al.,* 1986; Rosemond *et al.,* 1992). Herbivorous macroinvertebrates, particularly those such as gastropods and crustaceans, which need calcium for their shells or exoskeletons, tend to be highly affected by acidification, becoming rare at pH ≤ 6 (Okland and Okland, 1986; France, 1987). Previous reviews have shown that the density and species richness and diversity of insect larvae, particularly mayflies (Ephemoptera), are negatively affected by acidification (e.g., see reviews by Harvey, 1989; Stokes *et al.,* 1989; Elwood and Mulholland, 1989).

The effects of acidification on total insect biomass or productivity (Allard and Moreau, 1986b; Baker and Christensen, 1990) and the relationship between grazer abundance and algal development are less clear. In a lotic system study of the macrofloral and invertebrate assemblages of 88 sites located in soft-water streams with a pH range of 4 to 7, Ormerod *et al.* (1987) found strong empirical links between benthic communities and stream acidity, but the precise cause was unclear. In a survey of five lakes, which included mainly acidic lakes, no consistent relationship was found between abundances of Ephemeroptera, Trichoptera, and other insect grazers and the presence or absence of metaphytic algae (France *et al.,* 1991).

In many acidic aquatic ecosystems the dominant microinvertebrates are dipterans, and particularly chironomids. Although those species deemed as herbivorous, that is, those belonging to the Orthocladiinae and Tanytarsii tribes, seem to be more sensitive to a lowering of pH (Hall *et al.,* 1980; Allard and Moreau, 1987; Harvey, 1989), chironomid density and biomass as a whole appear to be independent of pH (Allard and Moreau, 1986b; Harvey, 1989; Winterbourn *et al.,* 1992; Junger and Planas, 1993).

The decrease in herbivores seen in acidic systems could explain changes in the algal community microstructure from bidimensional to tridimensional (e.g., see Planas *et al.,* 1989). On the other hand, direct modification of benthic algal food quality or palatability could disrupt invertebrate grazing. This has been observed for the mayfly *Baetis rhodoni* (Sjöström, 1990) and other mayfly species (Collier and Winterbourn, 1990). The latter authors found that the epilithic grazing rates of *Deleatidium* spp. were significantly higher on stones taken from acidic streams, although these higher rates did not translate into higher productivity. The traditional view, which holds that changes in the macroinvertebrate community could structure the composition of benthic algae (Harvey, 1989), implies that herbivores are selective. Grazer selectivity is related to the physiognomy of the periphytic community, with palatability and food quality probably also playing a part (Steinman *et al.*, 1987; see Steinman, Chapter 12, this volume).

Not many studies supporting the top-down hypothesis (control by grazers) as a cause of increased benthic algal growth in acidic water have looked at what the pH-sensitive insect larvae eat. Several of the insect larvae species in the taxa considered to be acid sensitive, the Ephemeroptera and Chironomidae, for example, are in the collector-gatherer functional feeding group, with variable and ill-defined diets (Anderson and Cummins, 1979). When examined, as in the study by Collier and Winterbourn (1990), it has been found that a great proportion (>60%) of stomach content is detritus, independently of stream pH. This amorphous detritus could originate from diverse sources (Wallace *et al.*, 1987). Using carbon isotopes as a marker, Junger and Planas (1993) found that the relative contribution of autochthonous (epilithon) versus allochthonous food sources to total invertebrate biomass was a function of the length of time of acidification. Certain invertebrate taxa exhibited feeding shifts after a month of acidification, thus providing evidence that acidification modifies trophic interactions between benthic algae and primary consumers.

Few experiments have been done to specifically test the hypothesis that the increase in benthic algae in acidic ecosystems is related to changes in trophic interactions. Planas and Moreau (1989), in an inclosure/exclosure experiment carried out in stream channels at natural (6.7) and acidic pH (4.3), found a highly significant increase in algal biomass (chlorophyll) and a decrease in the production/biomass ratio in the acidic channels where macroinvertebrates were excluded, but failed to show any significant increase in algal biomass in the control channels where macroinvertebrates were also excluded. In a study aimed at determining the relative importance of physiological resistance, macrograzers, and competitive exclusion in regulating epilithic algae during the recovery of acid-stressed lakes, Vinebrooke (1995) found that acid-stress resistance was the primary factor regulating species composition and biomass accrual. Macrograzers were found to be of secondary importance, affecting biomass, physiognomy, and species composition only under low acid-stressed (pH<6) conditions.

Furthermore, the grazer hypothesis does not explain the rapid changes in benthic algae observed as a result of liming. Nor does it explain the increases in Zygnemataceae that occur in oligotrophic lakes with increasing light, where grazers are present (Shortreed and Stockner, 1983; Sheath *et al.*, 1986).

Thus the observation of increased periphyton standing stock at low pH cannot definitely be attributed to a release from the consumptive processes of grazers, that is, the top-down effect. More information is needed on grazer food sources, their activity, and the physiognomy of the benthic algal community.

2. Decrease in Competition

A decrease in competition from other algae and/or other members of the periphytic biofilm has also been advanced as a mechanism to explain changes in benthic algae (Stokes, 1986). The increase in benthic algal biomass could be a consequence of a decrease in heterotrophic activity, with the growth/decomposition ratio favoring the buildup of Zygnemataceae. The balance between the various heterotrophic and autotrophic populations that make up the periphyton could result in a modification in nutrient recycling and in decreased competition for a common resource (Reice, 1981; Peckarsky, 1983). The role of changing nutrient availability on the shift of benthic algal species composition and productivity has been considered only very recently in acidified systems (Mulholland *et al.,* 1986; Planas and Moreau, 1986; Turner *et al.,* 1987; Elwood and Mulholland, 1989; Fairchild and Sherman, 1990; Vallée, 1993; Herrmann *et al.,* 1993).

Reduced decomposition rates have been reported in certain acidic lakes (McKinley and Vestal, 1982; Andersson, 1985) as well as in lotic systems (Burton *et al.,* 1985; Kimmel *et al.,* 1985; MacKay and Kersey, 1985; Allard and Moreau, 1986a). Few data are available on bacterial activity and production in acidic environments (Osgood and Boylen, 1992). In streams, epilithic bacteria and bacteria associated with decomposing leaves seem to have a lower biomass and production rate than the bacteria inhabiting circumneutral habitats (Palumbo *et al.,* 1989; Osgood and Boylen, 1992). The bacteria living in fine-grained sediments do not seem to be affected by acidification (Palumbo *et al.,* 1987, 1989; Elwood and Mulholland, 1989; Mulholland *et al.,* 1992).

In circumneutral waters, bacteria may outcompete algae for the limited supply of P, as has been shown in pelagic communities (see review by Jansson, 1988). In a study of a clear-water lake, before and after liming, Bell and Tranvik (1993) found that bacterial response to changes in nutrient concentrations and pelagic food web structure were indirectly mediated by the presence or absence of *Merismopedia tenuissima.* In this study, bacteria were colimited by DOC and inorganic N/P during acidic conditions. In a P-limited outdoor stream channel experiment, epilithic bacteria were relatively more abundant in the control than in the acidic channels. A complementary P-enrichment experiment carried out at circumneutral pH showed that bacterial growth (as measured by density increase) was more favorably affected by the enrichment than was algal growth. Algal growth in the P-enriched channel was lower than that measured in the acid-treated channels (C. Vallée and D. Planas, unpublished). This difference could not be attributed to grazing, since total macrobenthic biomass, as well as potential grazer biomass, was not significantly different between the P-treated and the control channels (D. Planas and M. Junger, unpublished). These results suggest a reduction in numbers of the more competitive bacteria in acid waters, leaving a greater availability of P to the algae.

IV. CONCLUSIONS

A. Summary

The effects of acidification on benthic algal structure and processes can be attributed to "physical and chemical variable changes" and/or "decreases in grazing pressure or competition." However, it is not clear which factor or combination of factors is more significant. Some tendencies do emerge from the literature: (1) In the Northern Hemisphere, the shift in species composition related to acidification usually follows the same pattern. Zygnemataceae, and mainly *Zygogonium* and *Mougeotia,* become predominant in lakes and low-flow running waters, whereas diatoms, particularly *Eunotia,* dominate in more turbulent lotic systems. (2) The toxicity of hydrogen ions and metals, particularly aluminum, could be directly responsible for the change in species. (3) Heightened light transparency and water temperature could increase areal primary productivity in acidic ecosystems. (4) Evidence of carbon as a limiting nutrient is growing, mainly in acidic lentic waters, as is its influence on species composition (due to bicarbonate loss and reduced free CO_2). (5) Nitrogen becomes more available owing to increased HNO_3 inputs from precipitation and/or changes in chemical speciation; NH_4 accumulation as a consequence of the inhibition of nitrification could influence species composition and productivity. (6) Changes in phosphorus availability could be responsible for changes in primary production and biomass. (7) Alteration of microbial–algal interactions could result in changes in nutrient availability. (8) Potential control by herbivores cannot be excluded.

B. Research Needs

Further information is needed on the community structure and processes of benthic algae in acidic ecosystems as well as in their natural environment. Most studies have examined only straight cause–effect relationships between two or a few variables (e.g., pH and presence or absence of a species) without taking into consideration other physical, chemical, and biological variables that might influence their presence or absence. These other variables might include substratum type, water movement, light transmission, temperature, changes in nutrient availability in the water or sediments, frequency of naturally occurring perturbing agents such as spates in rivers, presence or absence of grazers, preferential food sources of the consumer, and/or decreased competition for a common resource at the lower levels of the food web.

Long-term data, including the period before anthropogenic acidification of a given ecosystem, are often lacking. This type of data would provide much-needed temporal controls since spatial replication is often overlooked (e.g., see Underwood, 1994). Alternatively, when experiments are carried out in laboratory micro- or mesocosm, negative effects are more often noticed (Bell and Tranvik, 1993).

Few studies have dealt with the species richness and diversity of benthic algae in acidic systems. More information pertaining to the enhancement of dark respiration in acidic lakes would allow for a better understanding of net primary production changes. Studies must also be done through an annual or seasonal (e.g., for ice-free season) cycle since changes in the benthic algal community structure could be driven by changes in the environment unrelated to the perturbation. One must also consider the particular characteristics of the substratum that can modify the algal microenvironment. Modifications in the balance between the heterotrophic and autotrophic populations of the periphytic mat could result in alterations in nutrient recycling and in decreased competition for a common resource. Finally, the interaction between the pelagic and benthic communities cannot be ignored, particularly in small lakes.

REFERENCES

Allard, M., and Moreau, G. (1986a). Leaf decomposition in an experimentally acidified stream channel. *Hydrobiologia* **139**, 109–117.

Allard, M., and Moreau, G. (1986b). Influence of acidification and aluminium on the density and biomass of lotic benthic invertebrates. *Water, Air, Soil Pollut.* **30**, 673–679.

Allard, M., and Moreau, G. (1987). Effects of experimental acidification on a lotic macroinvertebrate community. *Hydrobiologia* **144**, 37–49.

Almer, B., Dickson, W., Ekström, C., Hörnström, E., and Miller, U. (1974). Effects of acidification on Swedish lakes. *Ambio* **3**, 30–36.

Almer, B., Dickson, W., Ekström, C., and Hörnström, E. (1978). Sulfur pollution and the aquatic ecosystem. *In* "Sulfur in the Environment. Part 2. Ecological Impacts" (J. O. Nriagu, ed.), pp. 271–311. Wiley, New York.

Anderson, N. H., and Cummins, K. W. (1979). Influence of diet on the life history of aquatic insects. *Can. J. Fish. Aquat. Sci.* **36**, 335–342.

Andersson, G. (1985). Decomposition of alder leaves in acid lake waters. *Ecol. Bull.* **37**, 293–299.

Arnold, D. E., Bender, P. M., Hale, A. B., and Light, R. W. (1981). Studies on infertile, acidic Pennsylvania streams and their benthic communities. *In* "Effects of Acid Precipitation on Benthos" (R. Singer, ed.), pp. 15–33. North Am. Benthol. Soc., Springfield, IL.

Azam, F., Hemmingsen, B., and Volcani, B. E. (1974). Role of silicon in diatom metabolism. V. Silica acid transport and metabolism in the heterotrophic diatom *Nitzchia alba*. *Arch. Mikrobiol.* **97,** 103–114.

Baker, J. P., and Christensen, S. W. (1990). Effects of acidification in biological communities in aquatic ecosystems. *In* "Acid Deposition and Aquatic Ecosystems: Regional Case Studies" (D. F. Charles, ed.), pp. 83–106. Springer-Verlag, New York.

Bell, R. T., and Tranvik, L. (1993). Impact of acidification and liming on the microbial ecology of lakes. *Ambio* **22,** 325–330.

Bennett, W. H. (1969). Algae in relation to mine water. *Castanea* **34,** 306–328.

Brock, T. D. (1973). Lower pH limit for the existence of blue-green algae: Evolutionary and ecological implication. *Science* **179,** 480–483.

Burkholder, J. M., and Cuker, B. E. (1991). Response in periphyton communities to clay and phosphate loading in a shallow reservoir. *J. Phycol.* **27,** 373–384.

Burton, T. M., Stanford, R. M., and Allen, J. W. (1985). Acidification effects on stream biota and organic matter processing. *Can. J. Fish. Aquat. Sci.* **42,** 669–675.

Campbell, P. G. C., and Stokes, P. M. (1985). Acidification and toxicity of metals to aquatic biota. *Can. J. Fish. Aquat. Sci.* **42,** 2034–2049.

Carlton, R. G., and Wetzel, R. G. (1988). Phosphorus flux from lake sediments: Effect of epipelic algal oxygen production. *Limnol. Oceanogr.* **33,** 562–570.

Catling, C. P., Freedman, B., Stewart, C., Kerekes, J. J., and Lefkovitch, L. P. (1988). Aquatic plants of acid lakes in Kejimkujik National Park, Nova Scotia; Floristic composition and relation to water chemistry. *Can. J. Bot.* **64,** 724–729.

Charles, D. F. (1985). Relationships between surface sediment diatom assemblages and lakewater characteristics in Adirondack lakes. *Ecology* **66,** 994–1011.

Collier, K. J., and Winterbourn, M. J. (1990). Population dynamics and feeding of mayfly larvae in some acid and alkaline New Zealand streams. *Freshwater Biol.* **23,** 181–189.

Cook, R. B., and Jagger, H. L. (1990). Upper Midwest. *In* "Acid Deposition and Aquatic Ecosystems: Regional Case Studies" (D. F. Charles, ed.), pp. 421–466. Springer-Verlag, New York.

Cook, R. B., Kelly, C. A., Schindler, D. W., and Turner, M. A. (1986). Mechanisms of hydrogen ion neutralization in an experimentally acidified lake. *Limnol. Oceanogr.* **31,** 134–148.

Cook, R. B., Kelly, C. A., Kingston, J. C., and Krei, R. G., Jr. (1987). Chemical limnology of soft water lakes in the Upper Midwest. *Biogeochemistry* **4,** 97–117.

Crisman, T. L., Clarkson, C. L., Kellar, A. E., Garren, R. A., and Riennert, R. W. (1986). A preliminary assessment of the importance of littoral and benthic autotrophic communities in acidic lakes. *In* "Impact of Acid Rain and Deposition on Aquatic Biological Systems" (B. G. Isom, S. M. Dennis, and J. M. Bates, eds.), ASTM STP 928, pp. 17–27. Am. Soc. Test. Mater., Philadelphia.

Cronan, C. S., and Schofield, C. L. (1979). Aluminum leaching response to acid precipitation: Effects on high-elevation watersheds in the northeast. *Science* **204,** 304–306.

Detenbeck, N. E., and Brezonik, P. L. (1991). Phosphorus sorption by sediments from a softwater seepage lake. 2. Effects of pH and sediment composition. *Environ. Sci. Technol.* **25,** 403–409.

Dickson, W. (1978). Some effects of the acidification of Swedish lakes. *Verh.—Int. Ver. Theor. Angew. Limnol.* **20,** 851–856.

Dillon, P. J., Yan, N. D., Scheider, W. A., and Conroy, N. (1979). Acidic lakes in Ontario, Canada: Characterization, extent and responses to base and nutrient additions. *Arch. Hydrobiol. Beih. Ergebn. Limnol.* **13,** 317–336.

Driscoll, C. T. (1985). Aluminum in acidic surface waters: Chemistry, transport, and effects. *Environ. Health Perspect.* **63,** 93–104.

Driscoll, C. T., Newton, R. M., Gubala, C. P., Baker, J. P., and Cristensen, S. W. (1990). Adirondack Mountains. *In* "Acid Deposition and Aquatic Ecosystems: Regional Case Studies" (D. F. Charles, ed.), pp. 130–202. Springer-Verlag, New York.

Duthie, H. C., and Hamilton, P. B. (1983). Studies on periphyton community dynamics of acidic streams using track autoradiography. *In* "Periphyton in Freshwater Ecosystems" (R. G. Wetzel, ed.), pp. 185–190. Dr. W. Junk Publishers, The Hague, The Netherlands.

Effler, S. W., and Field, S. D. (1983). Vertical diffusivity in the stratified layers of the mixolimnion of Green Lake, Jamesville, N.Y. *J. Freshwater Ecol.* **2,** 273–286.

Effler, S. W., Schafran, G. C., and Driscoll, C. T. (1985). Partitioning light attenuation in an acidic lake. *Can. J. Fish. Aquat. Sci.* **42,** 1707–1711.

Elwood, G. W., and Mulholland, P. J. (1989). Effects of acidic precipitation on stream ecosystems. *In* "Acid Precipitation" (D. C. Adrian and A. H. Johnson, eds.), Vol. 2, pp. 85–135. Springer-Verlag, New York.

Fairchild, G. W., and Sherman, J. W. (1990). Effects of liming on nutrient limitation of epilithic algae in an acid lake. *Water, Air, Soil Pollut.* **52,** 133–147.

Fairchild, G. W., and Sherman, J. W. (1992). Linkage between epilithic algal growth and water column nutrients in softwater lakes. *Can. J. Fish. Aquat. Sci.* **49,** 1641–1649.

Fairchild, G. W., and Sherman, J. W. (1993). Algal periphyton response to acidity and nutrients in softwater lakes: Lake comparisons vs nutrient enrichment approach. *J. North Am. Benthol. Soc.* **12,** 157–167.

France, R. L. (1987). Reproductive impairment of the crayfish *Orconectes virilis* in response to acidification of Lake 223. *Can. J. Fish. Aquat. Sci.* **44,** 97–106.

France, R. L., and Welbourn, P. M. (1992). Influence of lake pH and macrograzers on the distribution and abundance of nuisance metaphytic algae in Ontario, Canada. *Can. J. Fish. Aquat. Sci.* **49,** 185–195.

France, R. L., Howell, E. T., Paterson, M. J., and Welbourn, P. M. (1991). Relationship between littoral grazers and metaphytic algae in five softwater lakes. *Hydrobiologia* **220,** 9–27.

Francis, A. J., Quinby, H. L., and Hendrey, G. R. (1984) Effects of lake pH on microbial decomposition of allochthonous litter. *In* "Early Biotic Responses to Advancing Lake Acidification" (G. R. Hendrey, ed.). Acid Precipitation Ser., Vol. 6, pp. 1–21. Butterworth, Boston.

Fritz, S. C., Kreiser, A. M., Appleby, P. G., and Battarbee, R. W. (1990). Recent acidification of upland lakes in North Wales: Paleolimnological evidence. *In* "Acid Water in Wales" (K. W. Edwards, A. S. Gee, and J. H. Stoner, eds.), pp. 27–37. Kluwer Academic Publishers, Dordrecht, The Netherlands.

Gahnström, G., Blomqvist, P., and Fleisher, S. (1993). Are key nitrogen fluxes changed in acidified aquatic systems? *Ambio* **22,** 381–324.

Galloway, J. N., Hendrey, G. R., Schofield, C. L., Peters, N. E., and Johanes, A. H. (1987). Processes and causes of lake acidification during spring snowmelt in the west-central Adirondack Mountains, New York. *Can. J. Fish. Aquat. Sci.* **44,** 1595–1602.

Goenaga, X., and Williams, D. J. A. (1990). Determination of aluminium speciation in acid waters. *In* "Acid Water in Wales" (K. W. Edwards, A. S. Gee, and J. H. Stoner, eds.), pp. 189–201. Kluwer Academic Publishers, Dordrecht, The Netherlands.

Grahn, O. (1977). Macrophyte succession in Swedish lakes caused by deposition of airborne acid substances. *Water, Air, Soil Pollut.* **7,** 295–305.

Grahn, O. (1985). Macrophyte biomass and production in Lake Gårdsjön—An acidified clearwater lake in SW Sweden. *Ecol. Bull.* **37,** 203–212.

Grahn, O., Hultberg, H., and Landner, L. (1974). Oligotrophication—A self-accelerating process in lakes subjected to excessive supply of acid substances. *Ambio* **3,** 93–94.

Gunn, J. M., and Keller, W. (1990). Biological recovery of an acid lake after reduction in industrial emissions of sulphur. *Nature (London)* **345,** 431–433.

Hall, R. J., Likens, G. E., Fiance, S. B., and Hendrey, G. R. (1980). Experimental acidification of a stream in the Hubbard Brook Experimental Forest, New Hampshire. *Ecology* **61,** 976–989.

Hall, R. G., Driscoll, C. T., and Likens, G. E. (1987). Importance of hydrogen ions and aluminum in regulating the structure and function of stream ecosystems: An experimental test. *Freshwater Biol.* **18**, 17–43.

Hansson, L.-A. (1989). The influence of a periphytic biolayer on phosphorus exchange between substrate and water. *Arch. Hydrobiol.* **115**, 21–26.

Hargreaves, J. W., and Whitton, B. A. (1976). Effect of pH on growth of acid stream algae. *Br. Phycol. J.* **11**, 215–223.

Hart, D. D., and Robinson, C. T. (1990). Resource limitation in a stream community: Phosphorus enrichment effect on periphyton and grazers. *Ecology* **71**, 1494–1502.

Harvey, H. H. (1989). Effects of acidic precipitation on lake ecosystems. *In* "Acid Precipitation" (D. C. Adrian and A. H. Johnson, eds.), Vol. 2, pp. 137–164. Spring-Verlag, New York.

Helliwell, S., Batley, G. E., Florence, T. M., and Lumsden, B. C. (1983). Speciation and toxicity of aluminium in a model freshwater. *Environ. Technol. Lett.* **4**, 141–144.

Hendrey, G. R. (1976). "Effects of pH on the Growth of Periphytic Algae in Artificial Stream Channels," Internal Report IR 25/76. SNSF, Oslo, Norway.

Hendrey, G. R., and Vertucci, F. A. (1980). Benthic plant communities in acidic Lake Golden, New York: Sphagnum and the algal mat. *In* "Ecological Impact of Acid Precipitation" (D. Drablos and A. Tollan, eds.), pp. 314–315. SNSF, Oslo, Norway.

Henriksen, A. (1982). Susceptibility of surface waters to acidification. *In* "Acid Rain/Fisheries" (T. A. Haines and R. E. Johnson, eds.), pp. 103–121. Am. Fish. Soc., Bethesda, MD.

Herrmann, J., Degerman, E., Gerhart, A., Johansson, C., Lingdell, P.-E., and Muniz, I. P. (1993). Acid-stress effects on stream biology. *Ambio* **22**, 298–307.

Horner, R. R., Welch, E. B., and Veenstra, R. B. (1983). Development of nuisance periphytic algae in laboratory streams in relation to enrichment and velocity. *In* "Periphyton in Freshwater Ecosystems" (R. G. Wetzel, ed.), pp. 121–134. Dr. W. Junk Publishers, The Hague.

Hörnström, E., Ekström, C., and Duraini, M. O. (1985). Effects of pH and different levels of aluminium on lake plankton in the Swedish West Coast area. *Inst. Freshwater Res. Drottningholm Rep.* **61**, 115–127.

Howell, E. T., Turner, M. A., France, R., and Stokes, P. M. (1990). Ecology features of acidification-induced growth of metaphyton. *Can. J. Fish. Aquat. Sci.* **47**, 1085–1092.

Hultberg, H., and Andersson, I. B. (1982). Liming of acidified lakes: Induced long-term changes. *Water, Air, Soil Pollut.* **18**, 311–331.

Husar, R. B., Lodge, J. P., Jr., and Moore, D. J., eds. (1978). Sulphur in the atmosphere. *Atmos. Environ.* **12**, 1–3.

Hustedt, F. (1939). Systematische und okologische untersuchungen uber die diatomeen-flora von Java, Bali, und Sumatra nach dem Material der Deutschen Limnologischen Sunda-expedition III. Die okologischen factor in und ihr einfluss auf die diatomeen-flora. *Arch. Hydrobiol., Suppl.* **16**, 274–394.

Hutchinson, G. E. (1957). "A Treatise on Limnology, Geography, Physics and Chemistry," Vol. 1. Wiley, New York.

Jackson, M. B., Vandermeer, E. M., Lester, N., Booth, J. A., Molot, L., and Gray, I. M. (1990). Effects of neutralization and early reacidification on filamentous algae and macrophytes in Bowland Lake. *Can. J. Fish. Aquat. Sci.* **47**, 432–439.

Jansson, M. (1988). Phosphate uptake and utilisation by bacteria and algae. *Hydrobiologia* **170**, 177–189.

Jeffries, D. (1990). Southeastern Canada: An overview of the effect of acidic deposition on aquatic resources. *In* "Acid Deposition and Aquatic Ecosystems: Regional Case Studies" (D. F. Charles, ed.), pp. 273–289. Springer-Verlag, New York.

Jones, H. G., and Sochanska, W. (1985). The chemical characteristics of snow cover in a northern boreal forest during the spring run-off period. *Ann. Glaciol.* **7**, 167–174.

Jorgensen, B., and Revsbech, N. P. (1983). Photosynthesis and structure of benthic microbial mats; microelectrode and SEM studies of four cyanobacterial communities. *Limnol. Oceanogr.* **28,** 1075–1093.

Junger, M., and Planas, D. (1993). Alteration of trophic interactions between periphyton and invertebrates in an acidified stream: A stable carbon isotope study. *Hydrobiologia* **262,** 97–107.

Kahl, J. S., Norton, S. A., Cronan, C. S., Fernandez, I. J., Bacon, L. C., and Haines, T. A. (1990). Maine. *In* "Acidic Deposition and Aquatic Ecosystems: Regional Case Studies" (D. F. Charles, ed.), pp. 203–235. Springer-Verlag, New York.

Keithan, E. D., Lowe, R. L., and Deyoe, H. R. (1988). Benthic diatom distribution in a Pennsylvania stream: Role of pH and nutrients. *J. Phycol.* **24,** 581–585.

Kimmel, W. G., Murphey, D. J., Sharpe, W. E., and DeWalle, D. R. (1985). Macroinvertebrate community structure and detritus processing rates in two southeastern Pennsylvania streams acidified by atmospheric deposition. *Hydrobiologia* **124,** 97–102.

Krug, E. C., and Frink, C. R. (1983). Acid rain on acid soil: A new perspective. *Science* **221,** 520–525.

Kullberg, A., Bishop, K., Hargeby, A., Jansson, M., and Petersen, R. C., Jr. (1993). The ecological significance of dissolved organic carbon in acidified waters. *Ambio* **22,** 331–337.

Lazarek, S. (1982a). Structure and function of a cyanophyton mat community in an acidified lake. *Can. J. Bot.* **60,** 2235–2240.

Lazarek, S. (1982b). Structure and productivity of epiphytic algal communities on *Lobelia dortmanna* L. in acidified and limed lakes. *Water, Air, Soil Pollut.* **18,** 333–342.

Lazarek, S. (1985). Epiphytic algal production in the acidified Lake Gårdsjön, SW Sweden. *Ecol. Bull.* **37,** 213–218.

Lazarek, S. (1986). Responses of the *Lobelia* epiphytes complex to liming of an acidified lake. *Aquat. Bot.* **25,** 73–81.

Likens, G. E. (1976). Acid precipitation. *Chem. Eng. News* **54,** 29–44.

Likens, G. E., Bormann, F. H., Pierce, R. S., Eaton, J. S., and Johnson, N. M. (1977). "Biogeochemistry of a Forested Ecosystem." Springer-Verlag, New York.

Lowe, R. L. (1974). "Environmental Requirements and Pollution Tolerance of Freshwater Diatoms," US EPA Environ. Monit. Ser. 640/4-74-005. USEPA, Cincinnati, OH.

MacKay, R. J., and Kersey, K. E. (1985). A preliminary study of aquatic insect communities and leaf decomposition in acid streams near Dorest, Ontario. *Hydrobiologia* **122,** 3–11.

Marker, A. F. H., and Willoughby, L. G. (1988). Epilithic and epiphytic algae in streams of contrasting pH and hardness. *In* "Algae and the Aquatic Environment" (F. E. Round, ed.), pp. 312–325. Biopress Ltd., Bristol, UK.

Marmorek, D. R., Thornton, K. W., Baker, J. P., Bernard, D. P., Jones, M. L., and Reuber, B. S. (1986). "Acidic Episodes in Surface Waters: The State of the Sciences," Final Report. U.S. Environ. Prot. Agency, Corvallis, OR.

McIntire, C. D. (1966). Some effects of current velocity on periphyton communities in laboratory streams. *Hydrobiologia* **27,** 559–570.

McKinley, V. L., and Vestal, R. (1982). Effects of acid on plant litter decomposition in an arctic lake. *Appl. Environ. Microbiol.* **43,** 1188–1195.

Moss, B. (1973). The influence of environmental factors on the distribution of freshwater algae: An experimental study. II. The role of pH and the carbon dioxide–bicarbonate system. *J. Ecol.* **61,** 157–177.

Mulholland, P. J., Elwood, J. W., Palumbo, A. V., and Stevenson, R. J. (1986). Effects of stream acidification on periphyton composition, chlorophyll, and productivity. *Can. J. Fish. Aquat. Sci.* **43,** 1846–1858.

Mulholland, P. J., Driscoll, C. T., Elwood, J. W., Osgood, M. P., Palumbo, A. V., Rosemond, A. D., Smith, M. E., and Schofield, C. (1992). Relationships between stream acidity and bacteria, macroinvertebrates, and fish: A comparison of north temperate and south temperate mountain streams, USA. *Hydrobiologia* **239,** 7–24.

Muller, P. (1980). Effects of artificial acidification on the growth of periphyton. *Can. J. Fish. Aquat. Sci.* **37,** 355–363.

Munson, R. K., and Gherini, S. A. (1990). Processes influencing the acid–base chemistry of surface waters. *In* "Acidic Deposition and Aquatic Ecosystems: Regional Case Studies" (D. F. Charles, ed.), pp. 9–63. Springer-Verlag, New York.

Nalewajko, C., and Paul, B. (1985). Effects of manipulations of aluminum concentrations and pH on phosphate uptake and photosynthesis of planktonic communities in two Precambrian Shield lakes. *Can. J. Fish. Aquat. Sci.* **42,** 1946–1953.

Nygaard, G. (1956). Ancient and recent flora of diatoms and Chrysophyceae in Lake Gribbso. pp. 32–94. *Folia Limnol. Scand.* **8,** 1–273.

Okland, J., and Okland, K. A. (1986). The effects of acid deposition on benthic animals in lakes and streams. *Experientia* **42,** 471–486.

Olsson, H., and Pettersson, A. (1993). Oligotrophication of acidified lakes—A review of hypotheses. *Ambio* **22,** 312–317.

Ormerod, S. J., and Wade, K. R. (1990). The role of acidity in the ecology of Welsh lakes. *In* "Acid Water in Wales" (K. W. Edwards, A. S. Gee, and J. H. Stoner, eds.), pp. 93–119. Kluwer Academic Publishers, Dordrecht, The Netherlands.

Ormerod, S. J., Wade, K. R., and Gee, A. S. (1987). Macro-floral assemblages in upland Welsh streams in relation to acidity, and their importance to invertebrates. *Freshwater Biol.* **18,** 545–557.

Osgood, M. P., and Boylen, C. W. (1992). Microbial leaf decomposition in Adirondack streams exhibiting pH gradients. *Can. J. Fish. Aquat. Sci.* **49,** 1916–1923.

Overrein, L. N., Seip, H. M., and Tollan, A., eds. (1981). "Acid Precipitation Effects on Forest and Fish," Res. Rep. FR 19/80. SNSF, Oslo, Norway.

Palumbo, A. V., Boyle, M. A., Turner, R. R., Elwood, J. W., and Mulholland, P. J. (1987). Bacterial communities in acidic and circumneutral streams. *Appl. Environ. Microbiol.* **53,** 337–344.

Palumbo, A. V., Mulholland, P. J., and Elwood, J. W. (1989). Epilithic microbial populations and leaf decomposition in acid-stressed streams. *In* "Acid Stress and Aquatic Microbial Interactions" (S. S. Rao, ed.), pp. 69–91. CRC Press, Boca Raton, FL.

Paquet, S. (1993). Effects de l'acidification sur le taux de croissance des algues périphytiques et la modification de leur structure, en présence et absence des brouteurs. M.Sc. Thesis, Université du Québec à Montréal, Montréal.

Parent, L., Allard, M., Planas, D., and Moreau, G. (1986). The effects of short-term and continuous experimental acidification on biomass and productivity of running water periphytic algae. *In* "Impact of Acid Rain and Deposition on Aquatic Biological Systems" (B. G. Isom and J. M. Bates, eds.), ASTM STP 928, pp. 28–41. Am. Soc. Test. Mater., Philadelphia.

Patrick, R., and Reimer, C. W. (1966). "The Diatoms of the United States," Vol. 1, No. 13. Academy of Natural Sciences of Philadelphia, Philadelphia.

Patrick, R., Roberts, N. A., and Davis, B. (1968). The effect of changes in pH on the structure of diatom communities. *Not. Nat. Acad. Nat. Sci. Philadelphia,* **416.**

Patrick, S. T., and Stevenson, A. C. (1990). Acidified Welsh lakes: The significance of land use and management. *In* "Acid Water in Wales" (K. W. Edwards, A. S. Gee, and J. H. Stoner, eds.), pp. 189–201. Kluwer Academic Publishers, Dordrecht, The Netherlands.

Peckarsky, B. L. (1983). Biotic interactions or abiotic limitations? A model of lotic community structure. *In* "Dynamics of Lotic Ecosystems" (T. D. Fontaine, III and S. M. Bartell, eds.), pp. 303–323. Ann Arbor Sci. Publ., Ann Arbor, MI.

Peters, R., and Cattaneo, A. (1984). The effects of turbulence on phosphorus supply in a shallow bay of Lake Memphremagog. *Verh.—Int. Ver. Theor. Angew. Limnol.* **22,** 185–189.

Pillsbury, B. W. (1993). Factors influential on periphyton in acidic lakes. Ph.D. Dissertation, Bowling Green State University, Bowling Green, OH.

Pillsbury, B. W., and Kingston, J. C. (1990). The pH-independent effect of aluminum on cultures of phytoplankton from an acidic Wisconsin lake. *Hydrobiologia* **194,** 225–233.

Planas, D., and Moreau, G. (1986). Reaction of lotic periphyton to experimental acidification. *Water, Air, Soil Pollut.* **30,** 681–686.

Planas, D., and Moreau, G. (1989). Interaction périphyton-benthos en milieu acidifié. *Rev. Sci. Eau* **2,** 607–619.

Planas, D., Lapierre, L., Moreau, G., and Allard, M. (1989). Structural organization and species composition of a lotic periphyton community in response to experimental acidification. *Can. J. Fish. Aquat. Sci.* **46,** 827–835.

Pollman, C. D., and Canfield, D. E., Jr. (1990). Florida. *In* "Acid Deposition and Aquatic Ecosystems: Regional Case Studies" (D. F. Charles, ed.), pp. 367–416. Springer-Verlag, New York.

Prescott, G. W. (1962). "Algae of the Western Great Lakes Areas." W. C. Brown, Co. Dubuque, IA.

Reice, S. R. (1981). Interspecific associations in a woodland stream. *Can. J. Fish. Aquat. Sci.* **38,** 1271–1280.

Renberg, I., and Hellberg, T. (1982). The pH history of lakes in southwestern Sweden, as calculated from the subfossil diatom flora of the sediments. *Ambio* **11,** 30–33.

Renberg, I., and Wallin, J.-E. (1985). The history of the acidification of Lake Gårdsjön as deduced from diatoms and sphagnum leaves in the sediments. *Ecol. Bull.* **37,** 47–52.

Riber, H. H., and Wetzel, R. G. (1987). Boundary-layer and internal diffusion effects on phosphorus fluxes in lake periphyton. *Limnol. Oceanogr.* **32,** 1181–1194.

Roberts, D. A., and Boylen, C. W. (1988). Patterns of epipelic algae distribution in an acidic Adirondack Lake. *J. Phycol.* **24,** 146–152.

Roberts, D. A., and Boylen, C. W. (1989). Effects of liming on the epipelic algal community of Woods Lake, New York. *Can. J. Fish. Aquat. Sci.* **46,** 287–294.

Rosemond, A. D., Rice, S. R., Elwood, J. W., and Mulholland, P. J. (1992). The effects of stream acidity on benthic invertebrate communities in the south-eastern United States. *Freshwater Biol.* **27,** 193–209.

Rudd, J. W. M., Kelly, C. A., Schindler, D. W., and Turner, M. A. (1988). Disruption of the nitrogen cycle in acidified lakes. *Science* **240,** 1515–1517.

Rudd, J. W. M., Kelly, C. A., Schindler, D. W., and Turner, M. A. (1990). A comparison of the acidification efficiencies of nitric and sulphuric acids by two whole lake addition experiments. *Limnol. Oceanogr.* **35,** 663–697.

Schindler, D. W. (1986). The significance of in-lake production of alkalinity. *Water, Air, Soil Pollut.* **30,** 931–944.

Schindler, D. W. (1990). Experimental perturbation of whole lakes as test of hypothesis concerning ecosystem structure and function. *Oikos* **57,** 25–41.

Schindler, D. W., and Turner, M. A. (1982). Biological, chemical and physical responses of lakes to experimental acidification. *Water, Air, Soil Pollut.* **18,** 259–271.

Schindler, D. W., Mills, K. H., Malley, D. F., Findlay, D. L., Shearer, J. A., Davies, I. J., Turner, M. A., Linsey, G. A., and Cruikshank, D. R. (1985). Long-term ecosystem stress: The effect of 5 years of experimental acidification. *Science* **228,** 1395–1401.

Schindler, D. W., Frost, T. M., Mills, K. H., Chang P. S. S., Davies, I. J., Findlay, L., Malley, D. F., Shearer, J. A., Turner, M. A., Garrison, P. J., Watras, C. J., Webster, K., Gunn, J. M., Brezonik, P. L., and Swenson, W. A. (1991). Comparisons between experimentally- and atmospherically-acidified lakes during stress and recovery. *Proc.—Roy. Soc. Edinburgh, Sect. B: Biol. Sci.* **97,** 193–226.

Schofield, C. O. (1972). The ecological significance of air pollution induced changes in water quality of dilute lake districts in the northeast. *Trans. North East Fish Wild. Conf.,* pp. 98–112.

Shearer, J. A., and DeBruyn, E. R. (1986). Phytoplankton productivity responses to direct addition of sulfuric and nitric acids to the waters of a double-basin lake. *Water, Air, Soil Pollut.* **30,** 695–702.

Shearer, J. A., Fee, E. J., DeBruyn, E. R., and DeClercq, D. R. (1987). Phytoplankton primary production and light attenuation responses to the experimental acidification of a small Canadian Shield lake. *Can. J. Fish. Aquat. Sci.* **44**(Suppl. 1), 83–90.

Sheath, R. G., Havas, M., Hellebust, J. A., and Hutchinson, T. C. (1982). Effects of long-term natural acidification on algal communities of tundra ponds at the Smoking Hills, N.W.T., Canada. *Can. J. Bot.* **60,** 58–72.

Sheath, R. G., Burkholder, J. M., Morison, M. O., Steinman, A. D., and Van Alystyne, K. L., (1986). Effect of tree canopy removal by gypsy moth larvae on the macroalgae of a Rhode Island headwater stream. *J. Phycol.* **22,** 567–570.

Shortreed, K. S., and Stockner, J. G. (1983). Periphyton biomass and species composition in a coastal rain forest stream in British-Columbia, Canada: Effects of environmental changes caused by logging. *Can. J. Fish. Aquat. Sci.* **40,** 1887–1895.

Simpson, P. S., and Eaton, J. W. (1986). Comparative studies of the photosynthesis of the submerged macrophyte *Elodea canadensis* and the filamentous algae *Cladophora glomerata* and *Spirogyra* sp. *Aquat. Bot.* **24,** 1–12.

Sjöström, P. (1990). Food and feeding of *Baetis rhodani* (Ephemeroptera) in acidic environments. *In* "The Surface Water Acidification Programme" (B. J. Mason, ed.), pp. 427–428. Cambridge Univ. Press, Cambridge, UK.

Smol, J. P., Battarbee, R. W., Davis, R. B., and Meriläinen, J., eds. (1986). "Diatoms and Lake Acidity." Junk Pub., Dordrecht, The Netherlands.

Soulsby, C. (1982). Hydrological controls on acid runoff generation in an afforested headwater catchment at Llyn Brianne, Mid-Wales. *J. Hydrol.* **138,** 431–448.

Steinman, A. D., McIntire, C. D., Gregory, S. V., Lamberti, G. A., and Ashkenas, L. R. (1987). Effects of herbivore type and density on taxonomic structure and physiognomy of algal assemblages in laboratory streams. *J. North Am. Benthol. Soc.* **6,** 175–188.

Stevenson, R. J., Singer, R., Roberts, D. A., and Boylen, C. W. (1985). Patterns of epipelic algal abundance with depth, trophic status, and acidity of poorly buffered New Hampshire lakes. *Can. J. Fish. Aquat. Sci.* **42,** 1501–1512.

Stoddard, J. L., and Murdoch, P. S. (1990). Catskill Mountains. *In* "Acid Deposition and Aquatic Ecosystems: Regional Case Studies" (D. F. Charles, ed.), pp. 237–271. Springer-Verlag, New York.

Stokes, P. M. (1981). Benthic algal communities in acidic lakes. *In* "Effects of Acid Precipitation on Benthos" (R. Singer, ed.), pp. 119–138. North Am. Benthol. Soc., Springfield, IL.

Stokes, P. M. (1984). pH-related changes in attached algal communities of softwater lakes. *In* "Early Biotic Responses to Advancing Lake Acidification" (G. R. Hendrey, ed.). Acid Precipitation Ser., Vol. 6, pp. 43–61. Butterworth, Boston.

Stokes, P. M. (1986). Ecological effect of acidification on primary producers in aquatic ecosystems. *Water, Air, Soil Pollut.* **30,** 421–438.

Stokes, P. M., Howell, E. T., and Kratzberg, G. (1989). Effects of acidic precipitation on the biota of freshwater lakes. *In* "Acid Precipitation" (D. C. Adrian and A. H. Johnson, eds.), Vol. 2, pp. 273–304. Springer-Verlag, New York.

Stothers, R. B., and Raupino, M. R. (1983). Historical volcanism, European dry fogs, and Greenland acid precipitation, 1500 BP to AD 1500. *Science* **222,** 411–413.

Sullivan, T. J. (1990). Long-term temporal trends in surface water chemistry. *In* "Acidic Deposition and Aquatic Ecosystems: Regional Case Studies" (D. F. Charles, ed.), pp. 615–639. Springer-Verlag, New York.

Sweerts, J. P., Rudd, J. W. M., and Kelly, C. A. (1986). Metabolic activities in flocculent surface sediments and underlying sandy littoral sediments of a Canadian Shield lake. *Limnol. Oceanogr.* **31,** 330–338.

Tease, B., and Coler, R. A. (1984). The effects of mineral acids and aluminum from coal leachate on substrate periphyton composition and productivity. *J. Freshwater Ecol.* **2,** 459–467.

Turner, M. A. (1993). The ecological effects of experimental acidification upon littoral algal associations of lakes in the boreal forest. Ph.D. Thesis, University of Manitoba, Winnipeg.

Turner, M. A., Jackson, M. B., Findlay, D. L., Graham, R. W., DeBruyn, E. R., and Vandermeere, E. M. (1987). Early response of periphyton to experimental lake acidification. *Can. J. Fish. Aquat. Sci.* **44**(Suppl. 1), 135–149.

Turner, M. A., Howell, E. T., Summersby, M., Hesslein, R. H., Jackson, M. B., and Findlay, D. L. (1991). Changes in epilithon and epiphyton associated with experimental acidification of a lake to pH 5.0. *Limnol. Oceanogr.* **36,** 1390–1405.

Underwood, A. J. (1994). On beyond BACI: Sampling designs that might reliably detect environmental disturbance. *Ecol. Appl.* **4,** 3–15.

Vallée, C. (1993). Effet de l'acidification et de l'enrichissement au phosphore, en conditions expérimentales, sur le biofilm épipelique (algues et bactéries) des milieux lotiques. M.Sc. Thesis, Université du Québec à Montréal, Montréal.

Vinebrooke, R. D. (1995). Epilithic algae as indicators of natural recovery in acid stressed lakes. M.Sc. thesis. University of Toronto, Dep. of Botany, Toronto.

Wallace, J. B., Benke, A. C., Lingle, A. H., and Parsons, K. (1987). Trophic pathways of macroinvertebrate primary consumers in subtropical blackwater streams. *Arch. Hydrobiol., Suppl.* **74,** 423–451.

Watras, C. J., and Frost, T. M. (1989). Little Rock Lake (Wisconsin): Perspectives on an experimental ecosystem approach to seepage lake acidification. *Arch. Environ. Contam. Toxicol.* **18,** 157–165.

Wehr, J. D., and Brown, L. M. (1985). Selenium requirements of a bloom-forming planktonic alga from softwater and acidified lakes. *Can. J. Fish. Aquat. Sci.* **42,** 1783–1788.

Wei, Y.-X., Yung, Y.-K., Jackson, M. B., and Sawa, T. (1989). Some Zygnemataceae (Chlorophyta) of Ontario, Canada, including descriptions of two new species. *Can. J. Bot.* **67,** 3233–3247.

Wetzel, R. G. (1983a). "Limnology," 2nd ed. Saunders College Publishers, Philadelphia.

Wetzel, R. G. (1983b). Attached algal–substrata interactions: Fact or myth, and when and how. *In* "Periphyton in Freshwater Ecosystems" (R. G. Wetzel, ed.), pp. 207–215. Dr. W. Junk Publishers, The Hague, The Netherlands.

Wetzel, R. G. (1993). Microcommunities and microgradients: Linking nutrient regeneration, microbial mutualism, and high sustained aquatic primary production. *Neth. J. Aquat. Ecol.* **27,** 3–9.

Wetzel, R. G., Brammer, E. S., Lindström, K., and Forsberg, C. (1985). Photosynthesis of submerged macrophytes in acidified lakes. II. Carbon limitation and utilization of benthic CO_2 sources. *Aquat. Bot.* **22,** 107–120.

Winterbourn, M. J., Hildrew, A. G., and Box, A. (1985). Structure and grazing of stone-surface layers in some acid streams of southern England. *Freshwater Biol.* **15,** 363–374.

Winterbourn, M. J., Hildrew, A. G., and Orton, S. (1992). Nutrients, algae and grazers in some British streams of contrasting pH. *Freshwater Biol.* **2,** 173–182.

Yan, N. D. (1983). Effects of changes in pH on transparency and thermal regimes of Lohi Lake, near Sudbury, Ontario. *Can. J. Fish. Aquat. Sci.* **40,** 612–626.

Yan, N. D., and Stokes, P. M. (1978). Phytoplankton of an acidic lake, and its responses to experimental alterations of pH. *Environ. Conserv.* **5,** 93–100.

SECTION THREE

THE NICHE OF BENTHIC ALGAE IN FRESHWATER ECOSYSTEMS

17

The Role of Periphyton in Benthic Food Webs

Gary A. Lamberti

Department of Biological Sciences
University of Notre Dame
Notre Dame, Indiana 46556

I. INTRODUCTION

The role of benthic algae in aquatic food webs has received relatively little attention when compared to studies of the function of phytoplankton in pelagic food webs (Wetzel, 1983a; McQueen *et al.*, 1989). In freshwater ecosystems, considerable attention in the past two decades has been directed to *lotic* periphyton (Minshall, 1988), for the reason that attached

Algal Ecology

algae usually dominate the algal communities of flowing waters (Hynes, 1970). As a consequence, this review will concentrate on the role of periphyton in streams, but, where appropriate, research on periphyton in lake littoral zones will be discussed.

A prevailing paradigm in stream ecology is that detritus derived from terrestrial sources constitutes the major energetic resource for lotic food webs. The energy budget measured for Bear Brook in New Hampshire (Fisher and Likens, 1973), which was 99% dominated by terrestrial inputs, greatly influenced stream ecologists and reinforced the notion that detritus formed the energy base of lotic systems. In his landmark paper "The Stream and Its Valley," Hynes (1975) stated that in streams (riparian) "vegetation rules the supply of organic matter." In-stream production was considered to have little energetic importance until Minshall (1978) demonstrated that the carbon budgets of some streams could be almost exclusively algal-based. Still, the notion persisted that even algal carbon was mostly routed through the detrital pathway (e.g., Anderson and Sedell, 1979) rather than being consumed directly by stream herbivores.

The river continuum concept (Vannote *et al.*, 1980) proposed that autotrophic production would dominate carbon budgets in midorder streams and that herbivore densities would be highest in those areas. However, strong evidence of the energetic role of periphyton in stream food webs still was largely absent. Fortunately, the fundamental papers of Minshall and Vannote *et al.*, combined with refinements in food web theory (reviewed by Power, 1992) and stream theory (reviewed by Minshall, 1988), focused attention on autotrophic processes. This recognition ushered in a wave of observational and experimental studies in the 1980s concerned with the utilization and importance of lotic periphyton.

In this chapter, I attempt to summarize current knowledge of the functional importance of periphyton in aquatic food webs, with emphasis on stream ecosystems. I concentrate on field studies in which periphyton and at least one additional trophic level were included. Investigations of periphyton and consumers in laboratory streams have been reviewed in a symposium proceedings (Lamberti and Steinman, 1993). I do not address specific effects of grazers on periphyton, because they are reviewed elsewhere in this book (see Steinman, Chapter 12, this volume).

The specific objectives of this chapter are to: (1) summarize studies examining the contribution of periphyton to energy flow or consumer production in benthic systems; (2) review conceptual models in food web theory, relate those models to benthic food webs, and propose alternative models for the operation of periphyton-based food webs; (3) evaluate field studies of benthic food webs, especially those involving bottom-up or top-down experiments; and (4) identify major gaps in our knowledge of periphyton-based food webs and suggest directions for future research.

II. STRUCTURE OF "MACROSCOPIC" BENTHIC FOOD WEBS

The trophic structure of ecosystems can be broadly divided into *producers* and *consumers*, the latter category composed of herbivores, predators, detritivores, and decomposers (Begon *et al.*, 1990). I will not deal with detrital pathways (or detritivores) in this chapter because they do not directly involve living periphyton. Primary producers in freshwater ecosystems include algae, bryophytes, and vascular plants. Benthic algae are the dominant primary producers in most streams (Bott, 1983) and will grow on virtually any submerged surface, inorganic or organic and living or dead, including transient surfaces such as leaves. Common benthic *micro*-algae include diatoms, nonfilamentous green algae, and some blue-green algae. In fresh water, the term *macro*-algae normally is applied to filamentous and thalloid chlorophytes, some blue-green algae, red algae, and other large or mat-forming algae (e.g., *Chara, Vaucheria)*.

Populations of *bryophytes* (mosses and liverworts) may account for a substantial proportion of the autotrophic production in small, hydrologically stable streams (Steinman and Boston, 1993) and in some clear, moderate-sized rivers (Naiman *et al.*, 1987). In most streams, moss distribution tends to be patchy, compared to periphyton, and concentrated on larger, stable substrates (e.g., logs, boulders, bedrock outcrops). *Macrophytes* (rooted aquatic vascular plants) are extremely important components of lake littoral zones (Lodge *et al.*, 1988) and may be the dominant primary producers in large rivers (Vannote *et al.*, 1980). In small, woodland streams, macrophytes may be locally important along protected stream margins with groundwater inflow (Newman, 1991). Streams in agricultural landscapes frequently have dense growths of macrophytes, whose biomass may far exceed that of other primary producers. Aquatic bryophytes and macrophytes tend to be less commonly grazed than periphyton [Gregory (1983), but see Lodge (1991) for a different view], but may serve as important habitat or sources of detritus for invertebrates.

Primary consumers (*herbivores*) in freshwater ecosystems span many taxonomic groups, but insects, molluscs, crustaceans, and fish appear to be particularly important. Snails (Gastropoda), caddis flies (Trichoptera), and mayflies (Ephemeroptera) are conspicuous benthic consumers in streams and thus have been the focus of much study [reviewed by Lamberti (1993) for laboratory streams, see Steinman, Chapter 12, this volume for field studies]. However, chironomid midges (Diptera) may be equally important herbivores owing to their ubiquity, high densities, and short generation times (Berg and Hellenthal, 1992a). Crayfish consume a variety of plant types in both lakes and streams (Lodge *et al.*, 1994). Common herbivorous fish include various minnows (Cyprinidae) in temperate regions (Matthews *et al.*, 1987) and armored catfish (Loricariidae) in tropical systems (Power, 1984a).

Secondary and tertiary consumers (i.e., *predators*) can be illustrated with invertebrate and vertebrate predators, respectively. Common invertebrate predators in streams include stone flies (Plecoptera), megalopterans, and some caddis flies (e.g., Rhyacophilidae). In lake littoral zones, damselflies and dragonflies (Odonata), certain beetles (Coleoptera) and true bugs (Hemiptera), and crayfish (Decapoda) are common predators. Vertebrate predators in lakes and streams encompass many fish taxa (e.g., sculpin, sunfish, and salmonids). Animals that ingest predaceous fish include salamanders, turtles, snakes, and various piscivorous mammals and birds. The foregoing groups are intended only as examples and there is considerable trophic overlap among these categories.

Omnivory may be the basic food habit of many aquatic consumers. Omnivory is the process of feeding at more than one trophic level in a food web. Hynes (1970) recognized early that most stream invertebrates consumed a variety of items because benthic food resources are hopelessly intertwined. For example, periphyton consists of intermingled algae, detritus, bacteria, protists, and inorganic matter. To deal with this trophic quandary, Cummins (1973) developed the concept of functional feeding groups (FFGs), which emphasizes the manner in which benthic animals *obtain* their food rather than the specific food type. The four FFGs (scrapers, collectors, shredders, and predators), however, are still mistakenly interpreted as trophic levels. Benthic consumers in lakes probably also can be considered in terms of FFGs, although this concept has not been explicitly extended to lentic benthos.

Is omnivory the prevalent feeding habit of benthic animals? In a sense, most aquatic consumers probably are functional omnivores because microscopic food items are not easily separated. Some consumers appear to be especially flexible in their feeding habits, such as pleurocerid snails in streams (Hawkins and Furnish, 1987; Steinman, 1992), crayfish in both lakes and streams (Lodge et al., 1994; Charlebois, 1994), and various filter feeders in blackwater streams (Wallace *et al.,* 1987). Many consumers also undergo ontogenetic shifts in their diet. For example, caddis flies of the genus *Dicosmoecus* change their food preferences with age, switching among detritus, periphyton, and animal tissue (Gotceitas and Clifford, 1983; Li and Gregory, 1989). This dietary flexibility has major implications for how we view aquatic food webs.

III. PERIPHYTON IN AQUATIC ENERGY BUDGETS

Organic carbon in aquatic ecosystems derives from two principal sources: material contributed externally by the surrounding terrestrial system (*allochthonous* carbon) or material synthesized internally by autotrophic organisms within the system (*autochthonous* carbon). Lake littoral zones are characterized by diverse energy sources. Autochthonous resources include periphyton, macrophytes, and metaphyton (floating algae). Allochthonous resources include pelagic plankton transported to littoral zones by currents, organic matter carried by inflowing streams, and detritus contributed by shoreline vegetation. In lakes, littoral periphyton production generally is not limited by riparian shading, although suspended matter and macrophyte beds can reduce light penetration and thus benthic primary production (Wetzel, 1983a). Streams contain many of the same carbon resources as lakes, but in differing proportions. For example, phytoplankton generally is less abundant in flowing waters than in lakes, although even small streams can contain "potamoplankton" (Burkholder and Sheath, 1984). In small streams, benthic algal production is inversely correlated with the amount of cover by riparian canopy (Hawkins *et al.*, 1982; Lowe *et al.*, 1986; Feminella *et al.*, 1989; Tait *et al.*, 1994). The river continuum concept predicts that benthic primary production should be maximized in midorder streams (Vannote *et al.*, 1980). At the other extreme, large rivers usually are turbid, which limits periphyton production.

Energy budgets for temperate-zone, aquatic ecosystems must account for seasonal changes in energetic resources related to tree species and phenology. Leaf-fall into streams or lake littoral zones is pulsed in the autumn, when detritivore abundance and growth generally are highest (Cummins *et al.*, 1989). Benthic algae can grow throughout the year (Lyford and Gregory 1975), but conditions during the spring (high irradiance and nutrient loading) often favor high rates of primary production. Spring periphyton "blooms" are common in lakes (Wetzel, 1983b; Cattaneo, 1983) and streams (Power, 1990a; Lamberti *et al.*, 1991), although algal seasonality in streams will vary with canopy type (Sheath and Burkholder, 1985). Berg and Hellenthal (1992b) measured the highest production of invertebrates (mostly Chironomidae) in the winter and spring in an Indiana deciduous woodland stream, which they attributed to the coincident periphyton bloom. Algal biomass and invertebrate production were low for the remainder of the year. Thus, resource seasonality may be especially pronounced in deciduous forest streams where irradiance (and thus primary production) is linked to tree phenology.

In stream ecosystems (even of approximately the same size or found in the same biome), autotrophic production accounts for a wide range of total energy inputs (Fig. 1). Most energy budgets have been developed for small (order 1–2) streams, for which autotrophic production accounts for <1% to >60% of the total energy (Fig. 1A). Primary producers in streams up to fourth-order are dominated by periphyton and bryophytes. With increasing stream size, autotrophic production also increases (as a proportion of total energy), but the relative contribution of periphyton declines as plant communities shift to macrophytes in large rivers (e.g., Naiman *et al.*, 1987).

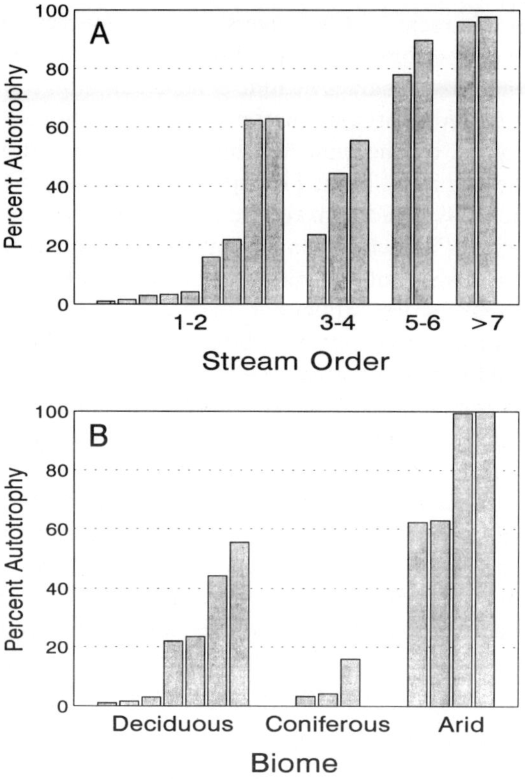

FIGURE 1 Summary of published energy budgets for stream ecosystems, showing autotrophic production as percentage of total annual energy inputs organized by stream order (A) and by biome (B). Each bar represents a *single* stream; sample sizes for (A) and (B) differ because stream order was not always reported and orders > 4 were not included in (B). Budgets were drawn from Minshall (1978), Cummins *et al.* (1983), Naiman *et al.* (1987), and Elwood and Turner (1989).

The degree of autotrophy is also related to biome type. In arid regions, where streams have limited riparian vegetation, carbon budgets predictably are dominated by autotrophic production (Fig. 1B). For streams in deciduous woodlands, energetic dependence on periphyton ranges broadly, which is perhaps related to the high seasonality in energy inputs. There are only a few energy budgets for small streams in coniferous forests, but those budgets indicate a low relative amount of autochthonous production. Year-round shading by conifers probably limits primary production in those systems. Periphyton, however, may be important even in small streams believed to be primarily heterotrophic (see Section IV, A).

For a few streams, there is either a complete seasonal energy budget (e.g., Cummins *et al.,* 1983) or a detailed analysis of consumer diets (e.g., Coffman *et al.,* 1971), but rarely do both elements exist for the same stream. As a result, other indirect methods must be used to determine the energy support system for aquatic food webs. Studies of energy flow using stable carbon isotopes are beginning to elucidate the food base of streams (Rosenfeld and Roff, 1992; see Section IV, A).

In littoral zones of lakes, autotrophic production (from macrophytes, benthic algae, and settled plankton) usually dominates energy inputs to benthic food webs (Wetzel, 1983a). In some cases, however, inputs from terrestrial vegetation, inflowing rivers, or atmospheric deposition can be significant. This allochthonous matter can be transported by currents and wind or moved by gravity to deeper parts of the lake. Benthic algal production can account for a substantial proportion of total lake primary productivity. Wetzel (1983a; his Table 19-7) summarized data from eight lakes and found that littoral, attached algae accounted for anywhere from 1 to 62% ($\bar{x}$ = 21%) of whole-lake primary productivity. Many factors influence this percentage, including basin morphology, shoreline development, depth of the euphotic zone, sediment type, and nutrient loading. In several small lakes in northern Michigan, periphyton on littoral zone sediments and wood (a frequently ignored substratum) accounted for approximately 50% of total epilimnetic chlorophyll *a* and primary production (Y. Vadeboncoeur and D. Lodge, unpublished data).

IV. PERIPHYTON IN BENTHIC FOOD WEBS

A. Fate and Utilization

In benthic systems, the overall availability of periphyton to consumers over space and time is relatively high, compared to other aquatic plants or to allochthonous inputs. Periphyton biomass produced in aquatic ecosystems can be allocated to several possible energetic "compartments" (Fig. 2): (1) *accumulation* as standing crop of algae, (2) *respiration* to CO_2 (i.e., *decomposition*), (3) *consumption* by herbivores (i.e., *grazing*), or (4) *export* to suspended matter. Each of these fates involves different processes and has different implications for energy flow through aquatic food webs.

In laboratory streams, Lamberti *et al.* (1987a) measured the allocation of primary production to these different pathways (except for respiration) under several grazing regimes (Fig. 3). Responses varied depending on grazing intensity, but at higher grazing pressures exerted by *Juga* snails or *Dicosmoecus* caddis flies, virtually all periphyton was either consumed or exported. In the remainder of this chapter, I will deal primarily with consumption and to some extent export, because these are the obvious entry points of periphyton into benthic food webs.

When evaluating energy (carbon) budgets, it is important to keep in mind that carbon budgets do not directly reflect food quality or utilization by consumers. Some carbon is rapidly assimilated (*labile*), whereas other

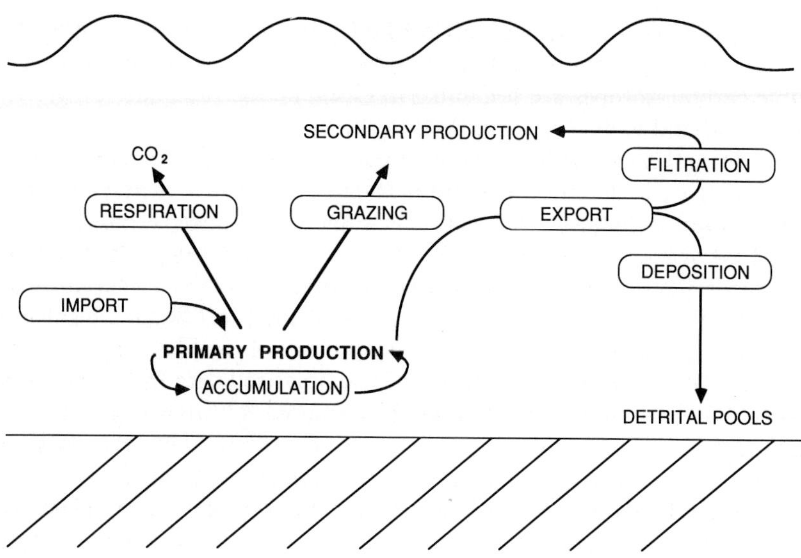

FIGURE 2 Diagram of processes and fates involved in benthic primary production in aquatic ecosystems. The primary avenues of entry of periphyton biomass into food webs are through grazing and export.

carbon forms are slowly assimilated (*refractory*). In small, heavily canopied streams, most carbon input may be allochthonous, yet specific populations or consumer guilds may rely on benthic algal production. For example, more than 99% of the energy input to Bear Brook (New Hamp-

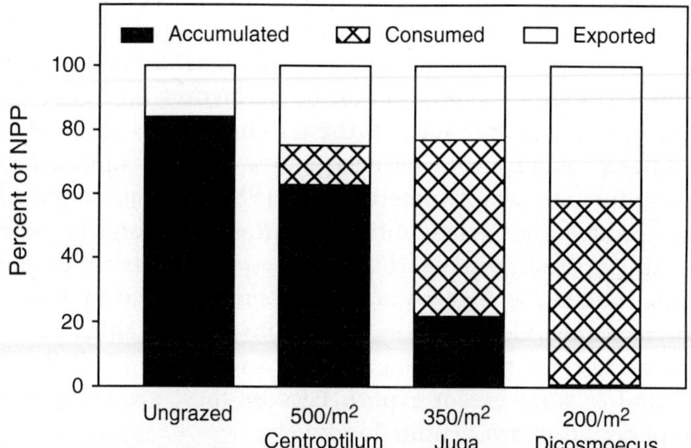

FIGURE 3 Percentage of net primary production (NPP) accumulated, consumed by grazers, or exported in laboratory streams containing *Centroptilum* mayflies, *Juga* snails, *Dicosmoecus* caddis flies, or no grazers. (Data from Lamberti *et al.*, 1987a.)

shire) was allochthonous (Fisher and Likens, 1973), but Mayer and Likens (1987) found that an abundant caddis fly (*Neophylax aniqua*) in that stream relied on periphyton for most of its dietary needs. The digestive tracts of larvae contained >50% algae by volume, which supported an estimated 75% of their growth. In contrast, blackwater streams of the southeastern United States have virtually no benthic grazers, but algae carried in seston (export) accounted for up to 20% of filter-feeder production (Benke and Wallace, 1980; Wallace *et al.*, 1987). Thus, in different systems, consumption of either benthic or resuspended periphyton may support algal-based food webs. However, energy transfer to higher trophic levels is more difficult to assess.

It is difficult to track autotrophically produced resources beyond one trophic level. Stable isotope analysis (especially ^{12}C and its "heavy" isotope ^{13}C) holds promise for revealing the relative importance of autochthonous and allochthonous carbon in aquatic food webs (Rounick *et al.*, 1982; Peterson and Fry, 1987). For this approach to be interpretable, algae must be depleted in ^{13}C compared to terrestrial carbon. For example, Rosenfeld and Roff (1992) saw clear separation in the isotopic ratios of stream algae and allochthonous detritus. They deduced that unforested sites were mostly dependent on the autotrophic carbon of periphyton, and that forested sites were periodically dependent on periphyton, such as in the spring. Enrichment also can induce changes in CO_2 uptake pathways in algae, which can shift the relative abundance of the heavy carbon isotope; this signal can be used to trace autotrophic production through a food web (e.g., Peterson *et al.*, 1993). Stable carbon isotopes have been used to analyze trophic structure in streams (e.g., Winterbourn *et al.*, 1986; Bunn *et al.*, 1989), lakes (e.g., Kling *et al.*, 1992), and salt marshes (e.g., Haines and Montague, 1979).

Most studies of the chemical composition of algae indicate that the food quality of living algae is high compared to other nonanimal benthic food (e.g., detritus). I examined data from nine published studies and reviews on the chemical composition of various algal taxa (Table I). Because of the paucity of information available, I included both freshwater *and* marine taxa, including both benthic *and* planktonic forms, to make generalizations as robust as possible. Keep in mind, however, that particular assemblages of benthic algae are certain to vary in biochemistry, as a function of species composition and habitat. In general, protein and nitrogen content of algae is much higher than that for benthic detritus, and is only slightly lower than that of zooplankton (Table I). Combined protein and lipid content of algae accounts for about 29–65% of cell dry weight, depending on taxonomic group, compared to <5% for fine detritus. Colonial chlorophytes appear to be especially nutritious, with one-third each of protein and lipid by weight, whereas blue-green algae have the highest nitrogen content (possibly related to N fixation by some taxa). Based on biochemical considerations, the food value of blue-

TABLE I Chemical Composition (% Dry Weight) of Several Algal Groups, Benthic Detritus, and Zooplankton. Mean, Minimum, and Maximum Values, and Sample Size (*N*) Are Presented.[a]

Taxon	*Number of species*	*Metric*	*Protein*	*Lipid*	*Carbohydrate*	*Cellulose and lignin*	*Pigment*	*Ash*	*Nitrogen*	*C:N ratio*	*Reference*[b]
Cyanophyceae (blue-green)	17	mean	48.2	4.0	15.8	1.5	0.9	7.2	9.5	5.9	1,2,3,4
		min	18.6	1.6	12.6	0.3	0.9	4.3	8.1	4.3	
		max	68.9	7.5	18.8	5.4	0.9	14.0	12.5	8.3	
		N	17	9	3	8	3	11	11	9	
Chlorophyceae (green) Filamentous	5	mean	14.5	14.5	42.7	12.8	NR	20.3	3.2	NR	1,5
		min	8.0	11.8	35.5	0.6		9.1	2.8		
		max	21.7	16.1	48.5	18.5		26.5	3.5		
		N	5	3	3	5		5	2		
Nonfilamentous	12	mean	35.8	15.6	24.5	10.6	2.1	10.1	7.6	7.0	2,3,4,5,6,7
		min	12.8	3.7	10.3	8.8	1.1	4.5	3.2	4.3	
		max	57.0	28.4	46.0	13.3	3.0	23.8	10.7	13.8	
		N	12	9	7	3	3	7	6	7	
Colonial	3	mean	33.8	31.4	13.3	6.3	NR	8.5	5.4	NR	2,7
		min	20.8	20.8	13.2	6.3		5.6	3.3		
		max	47.6	42.0	13.3	6.3		13.6	7.6		
		N	3	2	2	1		3	3		
Bacillariophyceae (diatoms)	13	mean	30.0	10.0	15.1	NR	1.0	30.4	4.8	5.7	5,6,7,8
		min	13.4	1.8	4.1		0.3	7.6	2.1	4.4	
		max	42.0	25.0	28.0		2.9	57.0	6.7	7.0	
		N	12	8	12		8	12	12	8	

Chrysophyceae (golden-brown)	2	mean	42.6	7.2	17.3	NR	NR	16.8	6.8	NR	5,7
		min	32.6	6.2	9.9			12.0	5.2		
		max	52.5	8.1	24.6			21.5	8.4		
		N	2	2	2			2	2		
Eustigmatophyceae (yellow-green)	5	mean	22.0	13.5	7.5	NR	1.4	13.6	3.5	NR	7,9
		min	18.1	8.2	5.2		0.6	13.6	2.9		
		max	28.6	18.0	11.4		1.7	13.6	4.6		
		N	5	5	5		4	1	5		
Benthic detritus	4 studies	mean	2.5	2.2	16.3	16.9	NR	74.9	0.4	NR	5
		min	1.7	0.8	2.0	8.0		60.0	0.3		
		max	3.9	3.2	30.5	31.5		93.0	0.6		
		N	3	3	2	3		4	3		
Zooplankton (microcrustacea)	11	mean	55.2	NR	NR	6.9	NR	9.7	8.8	NR	2
		min	44.9			4.0		4.1	7.2		
		max	67.1			10.1		18.5	10.7		
		N	10			10		10	10		

[a]NR = not reported. Data are from freshwater and marine, benthic and planktonic studies. *Caution should be exercised when extrapolating to specific algal assemblages.*

[b]Sources: [1]Tiffany (1938); [2]Davis (1955); [3]Ahlgren *et al.* (1992); [4]Ortega-Calvo *et al.* (1993); [5]Lamberti and Moore (1984); [6]Parsons *et al.* (1961); [7]Ben-Amotz *et al.* (1985); [8]Darley (1977); [9]Volkman *et al.* (1993)

green algae approaches that of animals (represented by microcrustaceans), although secondary chemicals may limit the palatability of blue-greens to benthic grazers (Porter, 1977; Gregory, 1983). Filamentous chlorophytes appear to be nutritionally poor, having low protein and lipid content and high ash. Diatoms are generally considered to be good food (Gregory, 1983), but this is not borne out by biochemistry. Diatoms have the highest ash content among the algae surveyed, and combined protein and lipid content is relatively low. However, if ingestion and assimilation rates for diatoms are high, then low nutritive value may not matter (Lamberti *et al.,* 1989).

The mechanical ability of herbivores to harvest algae varies with herbivore type and algal growth form (Lamberti *et al.,* 1987a; see Steinman, Chapter 12, this volume). Some herbivores may have difficulty in harvesting certain algal forms (e.g., large filaments, mucilaginous colonies, tightly appressed cells) regardless of food value. In these instances, ease of harvest (as for erect algal forms) and higher ingestion rates may compensate for lower nutritional value (e.g., Lamberti *et al.,* 1989). Gross chemical composition also cannot reveal the importance of specific dietary requirements of herbivores. For example, certain amino acids and fatty acids are required for herbivore growth and development (Crawley, 1983) and specific algae may contain higher proportions of these molecules (Steinman *et al.,* 1987; Ahlgren *et al.,* 1992).

Virtually all invertebrate feeding groups (and not only herbivores) consume benthic algae in some form (Gregory, 1983; Lamberti and Moore, 1984). For shredders, the microorganisms (which include algae) that grow on decaying leaves may be the most nutritional component of the leaves (Cummins *et al.,* 1989). Nonalgal components of microbial communities may be dependent to some degree on the products of autotrophs, such as extracellular secretions and decomposition products. Benthic algae can enter the drift through sloughing or dislodgement by grazers (Lamberti *et al.,* 1987a, 1989; Stevenson and Peterson, 1989; Barnese and Lowe, 1992). These algae become part of the filterable seston. Filter feeders capture and consume suspended algae, which contribute significantly to animal production. For example, in Appalachian streams, 35% of the annual production of blackflies was attributed to diatom ingestion (Wallace *et al.,* 1987), although only 8% of net-spinning caddis fly production was supported directly by algae (Benke and Wallace, 1980). Lake outflows often support high densities of filter feeders, most likely owing to the abundance of drifting algae and other seston (Richardson and Mackay, 1991). Filter feeding can substantially reduce the concentration of drifting algae and the food quality of seston (Wallace and Merritt, 1980). Drifting algae that are redeposited with fine detritus can be consumed by deposit collectors (Lamberti and Moore, 1984). In two blackwater streams in Georgia, up to 20% of collector diet was algal material (Wallace *et al.,* 1987).

In-depth studies of the trophic basis of invertebrate production indicate a significant reliance on periphyton in some streams. In desert streams of the southwestern United States, invertebrate production is tied to the recovery of periphyton following scouring floods (Grimm and Fisher, 1989). Over an annual cycle in Linesville Creek, Pennsylvania, between 30 and 70% of the benthic fauna ingested primarily periphyton (Cummins *et al.,* 1966; Coffman *et al.,* 1971). In an Indiana woodland stream, over 50% of the invertebrate production was supported by algae (Schwenneker, 1985; Berg, 1989), whereas algae supported <20% of total invertebrate production in blackwater streams (Wallace *et al.,* 1987).

B. Are Benthic Grazers Food-Limited?

Because periphyton is a valuable food resource in aquatic ecosystems, its abundance may influence the physiological fitness of herbivores. Hill *et al.* (1992) demonstrated that *Elimia* snails and *Neophylax* caddis flies were strongly food-limited in a small woodland stream, and grew faster when they were supplied with additional periphyton. In particular, the grazers' lipid reserves were increased by the periphyton-supplemented diet. In a correlative study of 12 streams, Hill (1992) found that periphyton biomass, *Neophylax* mass, and *Neophylax* lipid content were higher in streams lacking *Elimia* than in streams with snails. They postulated that snails created or exacerbated food-limited conditions in these small Tennessee streams, thus having an indirect negative effect on the other herbivores.

Experiments involving direct manipulations of grazer density have shown that benthic herbivores may grow in a density-dependent manner (Hart, 1987; Hill and Knight, 1987; Lamberti *et al.,* 1987b; Feminella and Resh, 1990; Hart and Robinson, 1990), suggesting that growth is limited by algal abundance. Indirect tests of the algal-limitation hypothesis are provided by manipulations of irradiance reaching streams. Artificial shading of a stream riffle reduced both the accumulation of algae and the density of the algivorous mayfly *Baetis tricaudatus,* but did not affect filter feeders or detritivores (Fuller *et al.,* 1986). A similar shading experiment in a New Zealand stream, however, had no effect on benthic invertebrates (Towns, 1981), suggesting that nonautotrophic resources had prevalence in that system. In three northern California streams, heterogeneity of canopy both within and among streams affected periphyton accrual rates and grazer abundances (Feminella *et al.,* 1989). However, standing biomass of periphyton was not related to canopy because grazers controlled algal biomass at all irradiance levels examined. This suggests that algal abundance was limited more by grazers than by light.

In aggregate, these studies suggest that periphyton is an important resource, the supply of which may limit consumer populations in some aquatic systems. Feminella and Resh (1990) warn, however, that physical factors such as disturbance can override periods of algal food limitation because grazer populations can be depressed by the disturbance, which relaxes competition. Of course, periphyton also can be removed by disturbance, but it typically recovers faster than do consumer populations (Steinman and McIntire, 1990; see Peterson, Chapter 13, this volume). A seasonal perspective of biotic and abiotic factors is essential in evaluating the role of periphyton in food webs.

C. Other Associations between Benthic Plants and Animals

Benthic algae, in particular macroalgae, can serve as habitat for benthic invertebrates because they provide complex structures that attract and retain animals (Dudley *et al.*, 1986; Dudley, 1988; Creed, 1994). Periphyton can affect the spatial distribution of herbivores at the scales of substrate particles (Lamberti and Resh, 1983; Lodge, 1986) to stream reaches (Fuller *et al.*, 1986). Herbivores apparently can perceive spatial heterogeneity in their algal food resources and thereby concentrate their feeding activity in rich patches of periphyton (Hart, 1981; Lamberti and Resh, 1983; McAuliffe, 1983; Kohler, 1984; Richards and Minshall, 1988). It is unclear if grazer aggregation in these algal patches is the result of (1) random encounter of patches followed by reduced emigration or (2) nonrandom behavioral attraction to patches by chemosensory or visual means. The converse situation is also possible: herbivores can create patchiness in benthic algal communities at a local scale (e.g., three-dimensional structure of periphyton) or along a longitudinal stream axis (e.g., upstream to downstream gradient) (Sarnelle *et al.*, 1993; see also Steinman, Chapter 12, this volume).

Animals themselves can provide additional habitat for benthic algae, such as on the tubes of chironomid larvae (Pringle, 1985; Hershey *et al.*, 1988), the cases of caddis fly larvae (Hasselrot, 1993; Bergey and Resh, 1994), and the shells of snails (Stock *et al.*, 1987). These epizoic habitats can serve as "gardens" for algal growth, which is supported by the relatively high-nutrient environments surrounding the metabolizing animals. Algae can be grazed by the inhabitants of the domiciles (Hershey *et al.*, 1988; Hasselrot, 1993) or by other benthic herbivores (Bergey and Resh, 1994).

Despite high utilization of periphyton, there are few examples of host-specific associations between herbivores and benthic algae, for reasons that are unclear (Gregory, 1983). The close relationships that have been documented involve an aquatic insect and a macroalga. One example is the association between the micro caddis fly *Dibusa angata* and the freshwater red alga *Lemanea australis* (Resh and Houp, 1986). Larval instars 1–4 of *Dibusa* are found among the basal holdfasts of *Lemanea,* where they consume epiphytic diatoms. Fifth-instar larvae construct cases made of

Lemanea and consume only *Lemanea* tissue. There is no apparent benefit of this association to the alga.

A possible mutualistic relationship exists between the chironomid midge *Cricotopus nostocicola* and the blue-green alga *Nostoc parmelioides*. The endosymbiotic midge larva locates and enters a colony of *Nostoc*, where it feeds only on *Nostoc* cells, in the process changing the colony morphology from a small sphere to a larger earlike shape (Brock, 1960; Ward *et al.*, 1985). The midge pupates within the colony and later emerges as an adult. The association also may benefit the colony, which has a higher photosynthetic rate when the fly is present, possibly due to higher light interception or increased gas exchange by the "ear form" and fertilization by grazer excreta (Ward *et al.*, 1985; Dodds, 1989). The midge also can reattach the colony if it becomes dislodged from the substratum (Brock, 1960; Dodds and Marra, 1989), which probably benefits both the colony and the larva.

D. Interplay of Production and Consumption

The inherent capacity of an ecosystem to produce plant biomass [or *productive capacity* of Warren *et al.* (1964)] is a function of resource availability, temperature, disturbance, and other factors. Productive capacity influences higher trophic levels by providing the template for food web interactions. Oksanen (1988) proposed that herbivore impact on resources (plants) should be related to the level of primary productivity in a unimodal fashion. Herbivores should maximize both resource demand (i.e., use), relative to resource abundance, and their effects on resources at intermediate levels of production. At either extreme (low or high production), demand and impacts on resources should be lower. Nutrient recycling and other indirect effects of grazing can produce unimodal relationships between grazing pressure and plant growth rates in both lakes (Flint and Goldman, 1975; Cuker, 1983; Carpenter and Lodge, 1986) and streams (Lamberti and Resh, 1983; Lamberti *et al.*, 1989; Power, 1990b).

Substantial feedback of herbivores on periphyton can occur, because grazers affect algal biomass, productivity, composition, physiognomy, and nutrient content (see Steinman, Chapter 12, this volume). The potential for benthic consumers to increase rates of primary production is poorly understood but intriguing. The notion that grazing increases primary production has undergone considerable debate in the terrestrial literature (e.g., Belsky, 1986; McNaughton, 1986). There are important differences between terrestrial and aquatic plants that make stimulation of production more plausible in aquatic ecosystems (Gregory, 1983). For example, terrestrial vascular plants grow slowly (compared to algae) and their generation time tends to be longer than for their herbivores. In contrast, algae grow rapidly and their generation time usually is much shorter than that of benthic grazers. Modeling efforts have concluded that a small biomass of microalgae with rapid turnover can support a larger biomass of consumers with slower turnover (McIntire, 1973; see McIntire *et al.*, Chapter 21, this volume). Benthic grazers continue to grow at consumer: algae biomass ratios of 13:1 (McIntire, 1975) and 20:1 (Gregory, 1980) in laboratory streams. Furthermore, for reasons outlined by Lamberti and Resh (1983), it is conceivable that grazers may increase the productivity of periphyton at certain levels of grazing (see also Steinman, Chapter 12, this volume). Specific circumstances appear to be necessary, such as favorable algal growth conditions and low grazer densities, to stimulate periphyton productivity and enhance consumer growth and development.

At a smaller spatial scale, local removal of fine sediments or epiphytes can expose periphyton and increase its productivity. For example, large armored catfish (*Ancistrus spinosus*) in Panamanian streams clear nutrient-poor, fine sediments from rock surfaces, thereby stimulating growth of epilithic algae (Power, 1984a). Small catfish also benefit because they graze algae from the cleared patches. In marshes and streams, grazing by vertebrates exposes underlying blue-green algae and stimulates their growth (Power *et al.*, 1988; Bazely and Jeffries, 1989). Migratory shrimp in Puerto Rican streams have been shown to clear deposits of fine sediments from rock surfaces, which may encourage algal growth (Pringle and Blake, 1994). In an oligotrophic, spring-fed pond in Oregon, colonies of the blue-green alga *Nostoc pruniforme* grow large enough (to 22 cm diameter) to be considered "macrophytes" (Dodds and Castenholz, 1987). Grazing by *Vorticiplex effusa* snails on epiphytes increased the growth of *Nostoc* colonies, although the mechanism was not known. Similar stimulation by snail grazing has been demonstrated repeatedly for vascular macrophyte–epiphyte associations (reviewed by Bronmark, 1989).

V. TOP-DOWN AND BOTTOM-UP REGULATION OF BENTHIC FOOD WEBS

A. Conceptual Framework

The regulation of food web structure has been a topic of long-standing interest to ecologists, who historically have debated the relative importance of abiotic (physical) and biotic (trophic) forces. Among trophic interactions, *bottom-up* control implies that all trophic levels are food-limited, which begins with limitation of the lowest trophic level, normally primary producers (Murdoch, 1966; White, 1978). Increases in resources for primary producers (e.g., light or nutrients) are postulated to boost productivity through the food web. The magnitude of the response, however, should decline with trophic distance from producers owing to inefficiencies in energy transfer (Lindeman, 1942; Hairston and Hairston, 1993). *Top-down* control means that control is exerted from the top level of the food web and alternating trophic levels are either predator- or resource-limited. For example, top predators in the food web exert control on the next lower trophic level (e.g., herbivores). The next lower trophic level (e.g., producers) thus becomes resource-limited because their consumers are controlled by predators. Several models of top-down effects have been advanced, as follows.

1. HSS Hypothesis

Hairston *et al.* (1960), in their famous theory about why "the world is green" (or HSS), envisioned food webs as consisting of three trophic levels (producers, herbivores, and predators). They theorized that abundant green plants dominated the terrestrial landscape because herbivores were limited by predators. Producers thus were limited only by their resources, most notably nutrients.

2. Fretwell–Oksanen Hypothesis

Fretwell, Oksanen, and coworkers (Fretwell, 1977; Oksanen *et al.*, 1981) extended HSS to more than three trophic levels, suggesting that the number of trophic levels in the food web would determine how a specific trophic level was limited. For example, in food webs with an odd number of trophic levels, primary producers would be abundant (*lush* habitats), whereas with an even number of trophic levels, primary producers should be sparse (*barren* habitats). This occurs because of the alternating control of trophic levels by predators and resource competition, which is initiated by the top predators. Fretwell (1987) further postulated that trophic levels would be added sequentially as primary productivity increased, so that over large-scale gradients of productivity, habitats would be either lush or barren depending on productivity and the resultant number of trophic levels.

3. Trophic Cascade

Paine (1980) proposed that some marine food webs display a "cascade of effects," triggered by key predators and involving herbivores and producers. Carpenter *et al.* (1985) applied the concept to lakes, which they described as having "cascading trophic interactions." In their four-level model, piscivorous fish (often the top predators in lakes) initiate a cascade in which zooplanktivorous fish are controlled by predation, zooplankton are limited by phytoplankton abundance, and phytoplankton are limited by zooplankton grazing. Lakes lacking piscivorous fish thus would tend to have a large biomass of phytoplankton, whereas lakes containing piscivorous fish should have a low abundance of phytoplankton.

4. Menge–Sutherland Model

Menge and Sutherland (1976, 1987) extended simple food chain models to include axes that represented disturbance and recruitment. They argued that increasing environmental stress (e.g., disturbance frequency) would tend to shorten food chains and to reduce the effects of predation. They also predicted that a high level of omnivory in food webs would lead to increasing control of lower trophic levels by predation, because of the ability of omnivores to feed at several trophic levels. These attributes are consistent with the keystone species concept (Paine, 1969), which suggests that the removal of a keystone predator (e.g., by disturbance) will diminish top-down control.

B. Top-down Experiments in Benthic Systems

1. General Observations

Because fish normally are the top predators, or *functional* top predators (*sensu* Oksanen, 1988) in aquatic ecosystems, they have been studied in several field experiments. Effects of predaceous fish in streams generally have not been examined beyond the response of prey items (i.e., only two trophic levels). Fish affect the abundance and composition of their benthic prey in some streams (Flecker, 1984; Feltmate and Williams, 1989; Gilliam *et al.,* 1989; Harvey and Hill, 1991; Wiseman *et al.,* 1993) but not in others (Allan, 1982; Culp, 1986; Reice and Edwards, 1986; Holomuzki and Stevenson, 1992). In general, drift-feeding salmonids appear to be less capable of affecting benthic populations than are fish that feed directly on the benthos. In stream pools, however, adult trout can alter the behavior of predaceous stone fly nymphs (Feltmate and Williams, 1989) and lestid damselflies (Wiseman *et al.,* 1993), and force water striders (Gerridae) into stream margins (Cooper, 1984). This harassment can reduce the feeding rates of the fishes' prey, thus potentially affecting at least three trophic levels.

2. Effects on Food Webs

The effects of an aquatic predator on two or more trophic levels, including periphyton, have been examined experimentally in only a few instances (Table II). Power and coworkers (Table III, examples 1–3) documented some degree of top-down control in three different lotic food webs. In prairie streams, predaceous bass (*Micropterus* spp.) determined the local distribution of algivorous minnows (*Campostoma anomolum*), thereby indirectly affecting the standing crop of benthic algae (Fig. 4, example 1). Bass pools tended to have high algal standing crop, whereas minnow pools had low algal standing crop. Piscivorous wading birds initiated a similar three-level "cascade" in stream pools of the Rio Frijoles in Panama (Fig. 4, example 2). Birds such as herons excluded grazing catfish from shallow margins of stream pools, resulting in a bloom of macroalgae in shallow water. In a California stream, juvenile steelhead trout and invertivorous roach triggered a trophic cascade that involved predaceous damselflies, chironomid larvae, and filamentous algae (Fig. 4, example 3).

It is unclear why more cascades have not been detected in studies of benthic food webs. A trophic cascade was not found in a small Tennessee stream, where invertivorous creek chubs were unable to affect grazing snail populations (Hill and Harvey, 1990). In high desert streams of Oregon, steelhead trout had little apparent effect on lower trophic levels, but all trophic levels were highly responsive to abiotic factors (Tait *et al.,* 1994). It is possible that the heterogeneity of the benthic environment makes intermediate trophic levels less susceptible to influence by fish predation than in simpler planktonic systems. Consequently, fish-initiated cascades may be less likely to occur in benthic systems. For example, Holomuzki and Stevenson (1992) found that sunfish in a Kentucky stream had no effect on macrobenthos on cobble (where prey refugia were available), and reduced only one invertebrate species on relatively uniform bedrock. They concluded that sunfish did not initiate a trophic cascade in that stream, at either high- or low-flow conditions.

Bowlby and Roff (1986) found that food web structure in a cluster of Canadian streams generally was consistent with the top-down hypothesis. They sampled multiple trophic levels from 30 sites in 20 southern Ontario streams to test different trophic models emphasizing control by predation, resources, or habitat availability. In their correlative study, Bowlby and Roff recognized four trophic levels: piscivorous brown trout, invertivorous fish, invertebrates, and the "microcommunity." These levels tended to alternate in abundance in a manner consistent with a "cascade." However, effects diminished with "distance" from the top of the food web; periphyton abundance showed no relationship to what fish occupied the top trophic level. McQueen *et al.* (1989) observed a similar outcome in the food webs of lake pelagic systems. Thus, it can be inferred that resources and habitat also constrained lower trophic levels.

C. Bottom-up Experiments in Benthic Systems

1. General Observations

Warren *et al.* (1964) proposed that the inherent primary productivity of a benthic ecosystem should establish the potential levels of productivity for higher trophic levels. Benthic primary production can be limited by both light (Gregory, 1983; Steinman, 1992; see Hill, Chapter 5, this volume) and nutrients (Elwood *et al.,* 1981; Bothwell, 1985; see Borchardt, Chapter 7, this volume). Benthic herbivores respond in abundance and growth to spatial variation in the abundance of periphyton (Lamberti and

TABLE II Features of Selected Experimental and Descriptive Studies Examining Aquatic Food Webs and Inferred to Represent "Top-down," "Bottom-up," or "Intermediate" Regulation.[a]

Reference[a]	*Habitat and location*	*Producer and stimulant (if applicable)*	*Herbivore*	*First-level predator or omnivore (*)*	*Second-level predator*
Top-down Control					
Power and Matthews (1983) (1)	Prairie stream pools; Brier Creek, Oklahoma	Microalgae and macroalgae	*Campostoma* minnows	Bass	NM
Power *et al.* (1989), Power (1984b) (2)	Tropical stream pools; Rio Frijoles, Panama	Microalgae and macroalgae	Armored catfish	Wading birds	NM
Power (1990a) (3)	Coastal mountain stream; Eel River, California	Microalgae and *Cladophora* macroalgae	Chironomid larvae	Lestid damselflies	Streelhead juveniles and roach
Batzer and Resh (1991)	Salt marsh experimental ponds; Solano Bay, California	Microalgae and pickleweed	Chironomids	Hydrophilid beetles	NM
Bowlby and Roff (1986)	20 small streams; southern Ontario	Microcommunity measured as ATP	Total benthos	Invertivorous fish	Brown trout
Bottom-up Control					
Tait *et al.* (1994) (4)	4 high-desert streams; John Day River, Oregon	Microalgae and macroalgae; light	*Dicosmoecus* caddis flies and other taxa	Various minnows and suckers	Rainbow trout and sculpins
Warren *et al.* (1964) (5)	Partitioned woodland stream; Berry Creek, Oregon	Algae and *Sphaerotilus* bacteria; sucrose	Chironomids and other taxa	Crayfish and other taxa	Cutthroat trout
Peterson *et al.* (1985, 1993) (6)	Tundra stream; Kuparuk River, Alaska	Microalgae and macroalgae, bacteria; P	Chironomids, blackflies, and mayflies	*Brachycentrus* caddis flies	Grayling
Perrin *et al.* (1987), Johnston *et al.* (1990) (7)	Coastal forest stream; Keogh River, British Columbia	Microalgae and macroalgae; grain, N, P	Total benthos; taxa not specified	Not specified	Coho salmon and steelhead fry

Intermediate Regulation					
Hawkins and Furnish (1987), Lamberti *et al.* (1989) (8)	Experimental channels; Oak Creek, Oregon	Microalgae and macroalgae; light	Chironomids and other benthos	**Juga silicula* snails	NM (cutthroat trout)
Stewart (1987), Gelwick and Matthews (1992) (9)	Prairie stream pools; Brier Creek, Oklahoma	Microalgae and macroalgae	Unspecified invertebrates	**Campostoma* minnows	NM
Hart (1992) (10)	Woodland stream; Augusta Creek, Michigan	Microalgae and *Cladophora* macroalgae	*Leucotrichia* and *Psychomyia* caddis flies	**Orconectes* crayfish	NM
Hill and Harvey (1990) (11)	Chambers in woodland stream; Ish Creek, Tennessee	Microalgae; light	*Elimia* snails	Creek chubs	NM
Hill (1992) (11)	12 woodland streams; Oak Ridge, Tennessee	Microalgae	*Neophylax* caddis flies	*Elimia* snails	NM
Steinman (1992) (11)	Woodland stream; Walker Branch, Tennessee	Microalgae and macroalgae; light	*Elimia* snails	NM	NM
Rosemond *et al.* (1993) (11)	Woodland stream and channels; Walker Branch, Tennessee	Microalgae and macroalgae; nutrients	*Elimia* snails	NM	NM
Lodge *et al.* (1994) (12)	Littoral zone enclosures; Plum Lake, Wisconsin	Microalgae and macrophytes	Snails and other taxa	**Orconectes* crayfish	NM
Charlebois (1994) (13)	Stream enclosures; Ontonagon River, Michigan	Microalgae	Total benthos	**Orconectes* crayfish	NM
Lamberti *et al.* (1992)	Mountain streams; Mount St. Helens, Washington	Microalgae	Total benthos	**Ascaphus* tadpoles	NM

[a]Parenthetical number under reference refers to diagrammed food web in Figs. 4 and 5. NM = not measured.

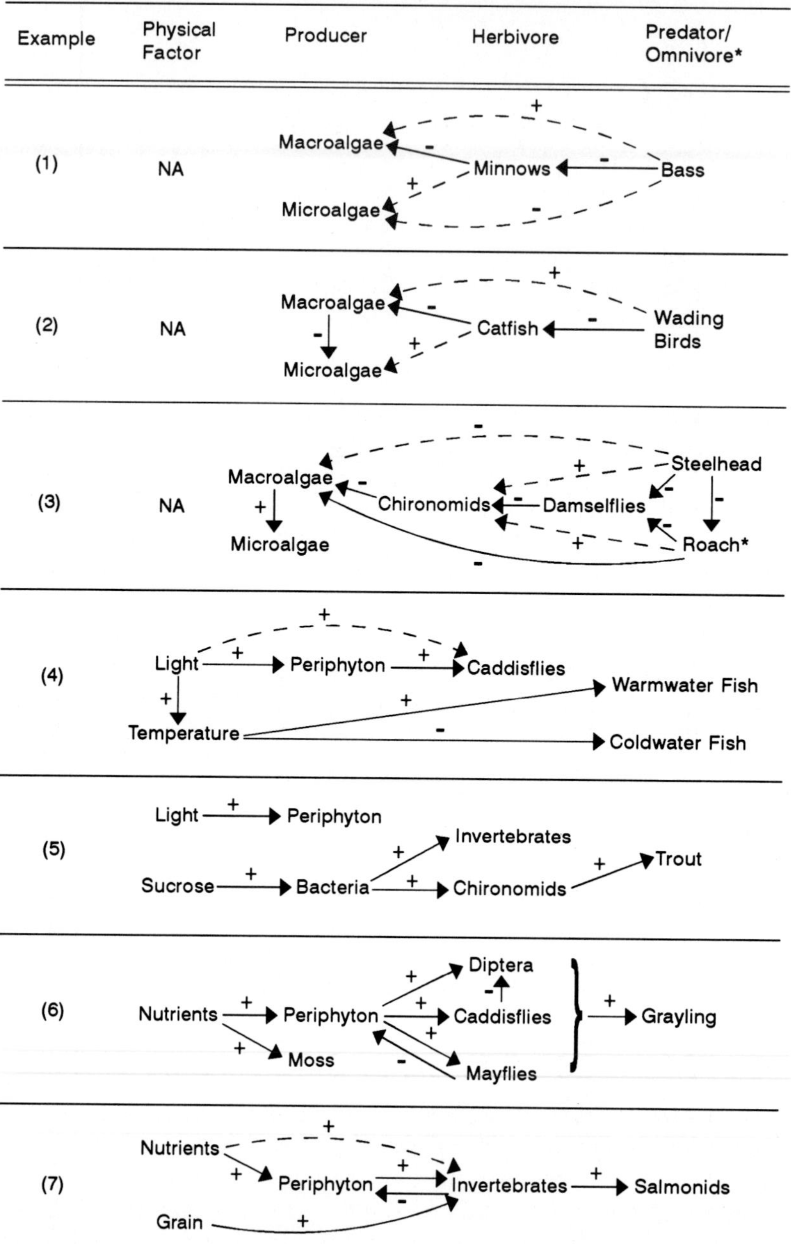

FIGURE 4 Summary food web diagrams (my interpretation) from selected studies of top-down (1–3) and bottom-up (4–7) processes. Arrows depict trophic level effects (solid arrow = direct effect; dashed arrow = indirect effect). Sign indicates direction of effect (+ = positive effect on biomass or density; − = negative effect on biomass or density). See Table II for study details and Table III for references and summary of overall effects. NA = not addressed.

Resh, 1983; Hill, 1992; Hill *et al.*, 1992) and macrophytes (Gregg and Rose, 1985; Dudley, 1988). Correlative studies have identified positive associations between irradiance, benthic primary production, and the biomass of higher trophic levels including predators (Hawkins *et al.*, 1982; Taylor and Roff, 1982). A study in high desert streams (Tait *et al.*, 1994) found a positive association between light and herbivore abundance, but predaceous fish were negatively affected owing to the concurrent increase in water temperature (Fig. 4, example 4).

2. Experiments Examining Two Trophic Levels

Long-term experiments are beginning to establish the relationship between primary productivity and benthic trophic structure in streams. Enrichment of a Tennessee woodland stream with phosphorus increased periphyton standing crop initially, but periphyton then declined slightly (Elwood *et al.*, 1981). This decline was explained by a large increase in grazer abundances (*Elimia clavaeformis* snails) in the enriched section, suggesting that enrichment affected two trophic levels. Hart and Robinson (1990) reported a similar stimulatory effect of localized enrichment on the herbivore trophic level.

Some investigators have directly augmented the food of consumers and have observed increases in fish production [resulting from krill additions by Mason (1976)] and in the density and biomass of benthic invertebrates [resulting from organic matter additions by Williams *et al.* (1977) and Mundie *et al.* (1983)]. These studies concluded that stream animals were limited by food rather than space during summer low-flow conditions. However, Mason (1976) reported that fish losses during the winter nullified the summer gains. Thus, seasonal augmentation of a trophic level may be offset by physical stress exerted during another season, suggesting that population "bottlenecks" should be considered when interpreting experimental results.

3. Experiments Examining More Than Two Trophic Levels

Huntsman (1948) provided the first report from a stream fertilization involving placement of bags of pelletized fertilizer in a Nova Scotia (Canada) stream. Treatment of this "infertile" stream stimulated the downstream growth of filamentous green algae, increased the abundance of "insect larvae," and attracted various fish. However, quantitative data on responses were largely lacking in that study.

The first systematic, controlled stream fertilization apparently was conducted by Warren *et al.* (1964) in Berry Creek, Oregon. Enrichment with sucrose increased the biomass of bacteria (*Sphaerotilus natans*) substantially, thus stimulating the heterotrophically based food web (Fig. 4, example 5). Benthic invertebrates, especially chironomid midges, increased 2- to 10-fold and cutthroat trout production increased by seven times. Autotrophic pro-

TABLE III References and Generalized Effects of Manipulative Experiments Diagrammed in Figures 4 and 5

Example	*Reference*	*Overall effects*
(1)	Power and Matthews (1983), Power *et al.* (1985)	–piscivorous bass exclude minnows from stream pools or subhabitats –minnows suppress macroalgae, thereby encouraging growth of microalgae –bass indirectly determine periphyton composition
(2)	Power *et al.* (1989), Power (1984b)	–piscivorous birds restrict armored catfish to deeper water –catfish suppress macroalgae, except in shallow fished areas –by removing macroalgae and fine sediment, catfish may encourage growth of macroalgae
(3)	Power (1990a)	–steelhead juveniles and roach consume predatory damselflies –damselflies suppress macroalgal tuft-weaving chironomids –in presence of fish, filamentous algae increase, also stimulating microalgae –omnivorous roach also consume macroalgae
(4)	Tait *et al.* (1994), Li *et al.* (1994)	–open canopy (higher light) increases periphyton and indirectly increases grazer biomass –top-down effects by either predators or herbivores are not apparent –higher temperature in open-canopy reaches favors warm-water fish
(5)	Warren *et al.* (1964)	–sucrose addition increases bacterial growth, which stimulates heterotrophic food web up to trout –sevenfold increase in trout production linked to increased chironomid abundance and drift –light enhancement (by canopy removal) increases primary productivity but has no effect on higher trophic levels
(6)	Peterson *et al.* (1985, 1993)	–phosphorus addition initiates shift from heterotrophy to autotrophy –early (years 1–2) effects indicate bottom-up stimulation of food web –later (years 3–4) effects suggest strong top-down feedback of grazers on algae and also competition
(7)	Perrin *et al.* (1987), Johnston *et al.* (1990)	–phosphorus plus nitrogen enrichment increases periphyton growth and indirectly stimulates higher tropic levels –grain has no effect on periphyton but may increase invertebrates –salmonid growth increases by 40–95% in enriched reaches

(continues)

TABLE III (continued)

Example	*Reference*	*Overall effects*
(8)	Hawkins and Furnish (1987), Lamberti *et al.* (1989)	–river snails directly reduce periphyton and indirectly reduce small invertebrates –higher light directly increases periphyton growth and indirectly increases snail growth –high snail abundances may limit trout growth because trout prefer other food items over snails
(9)	Stewart (1987), Gelwick and Matthews (1992)	–minnows negatively affect periphyton, but increase benthic detritus –nutrient effect on periphyton is secondary to strong fish effect –positive invertebrate response is related to higher benthic detritus
(10)	Hart (1992), Creed (1994)	–high current velocity negatively affects crayfish –crayfish directly reduce macroalgae, and indirectly reduce epiphytic invertebrates –macroalgae and sessile grazers exert priority over space; once established, each can suppress the other
(11)	Hill and Harvey (1990), Hill (1992), Steinman (1992), Rosemond *et al.* (1993)	–chubs have no effect on snails, but may be ineffective predators –snails negatively affect periphyton, and indirectly reduce growth rates of grazing caddis flies –increases in light or nutrients positively affect periphyton, but effects can be overridden by grazing
(12)	Lodge *et al.* (1994)	–littoral crayfish directly reduce abundances of macrophytes and grazing snails –crayfish indirectly increase periphyton areal abundance –crayfish reduction of macrophytes reduces total periphyton abundance
(13)	Charlebois (1994)	–stream crayfish directly reduce abundances of benthic invertebrates –crayfish indirectly increase areal periphyton chlorophyll *a* –omnivorous crayfish also may directly reduce periphyton biomass

duction was not stimulated by the sucrose addition, but was increased by an associated light-enhancement experiment (by canopy removal).

A long-term enrichment experiment has been conducted in the Kuparuk River in the Alaskan tundra (summarized by Peterson *et al.*, 1985, 1993). Fertilization of the river with phosphorus has resulted in stimulation of the entire food web (Fig. 4, example 6). Positive effects have been measured for bacteria, algae, bryophytes, invertebrates, and fish in this relatively simple community. However, feedback effects of consumers on food resources, starting in the third year of enrichment, have limited the accumulation of biomass in the stream.

Two elements (N and P) and organic matter (grain) were added to the Keogh River in British Columbia (Perrin *et al.*, 1987; Johnston *et al.*, 1990). Nutrient addition initially increased chlorophyll *a* accrual rate by over 10 times, but periphyton biomass declined after midsummer (Fig. 4, example 7). Grain addition increased alkaline phosphatase activity by about one-third (suggesting P deficiency), but had no effect on periphyton accrual. Juvenile salmonids responded positively to nutrient addition, suggesting a trophic connection from periphyton to fish. Nutrient addition had a greater positive effect on salmonid growth than did grain addition.

Although a number of lake fertilization experiments have been conducted, benthic responses to lake fertilization have been measured only rarely. In the fertilization of Lake N-2 in Alaska, Hershey (1992) concluded that planktonic models did not apply to the benthic response; different species and trophic levels responded differentially and unpredictably. Hershey speculated that lake pelagic and benthic systems responded differently because food can settle and accumulate in the benthic zone but not in the pelagic zone. Furthermore, benthic algal productivity generally is negatively related to water column productivity (Stevenson *et al.*, 1985).

D. Concurrent Top-down and Bottom-up Experiments

McQueen *et al.* (1986, 1989) proposed that freshwater pelagic systems were simultaneously regulated by both bottom-up and top-down processes (the BU:TD model). As demonstrated earlier, both resources and predators can affect food web structure of benthic systems. Perhaps the most promising approach to determining the relative importance of BU:TD effects is to conduct experiments in which both directions are manipulated simultaneously. In experiments involving two benthic trophic levels, various investigators have manipulated light and grazers (e.g., Dodds and Castenholz, 1987; Lamberti *et al.*, 1989; Steinman, 1992) and nutrients and grazers (e.g., Cuker, 1983; Stewart, 1987; Steinman *et al.* 1991; Rosemond *et al.*, 1993). These studies suggest that periphyton productivity is ultimately controlled by abiotic factors, but that grazing exerts proximate control over standing crop. Positive effects on productivity often are manifested in higher growth rates of grazers (Lamberti *et al.*, 1989; Rosemond *et al.*, 1993).

Hill and Harvey (1990) incorporated a third trophic level by simultaneously manipulating predators (creek chubs) and grazers (snails) under different natural light regimes. They found that fish had no effects on lower trophic levels, but that both light regime and grazing simultaneously affected periphyton (Fig. 5, example 11). This dual control of plants by herbivores and resources [or "colimitation" *sensu* Power (1992)] appears to be a recurring deduction from benthic food web experiments.

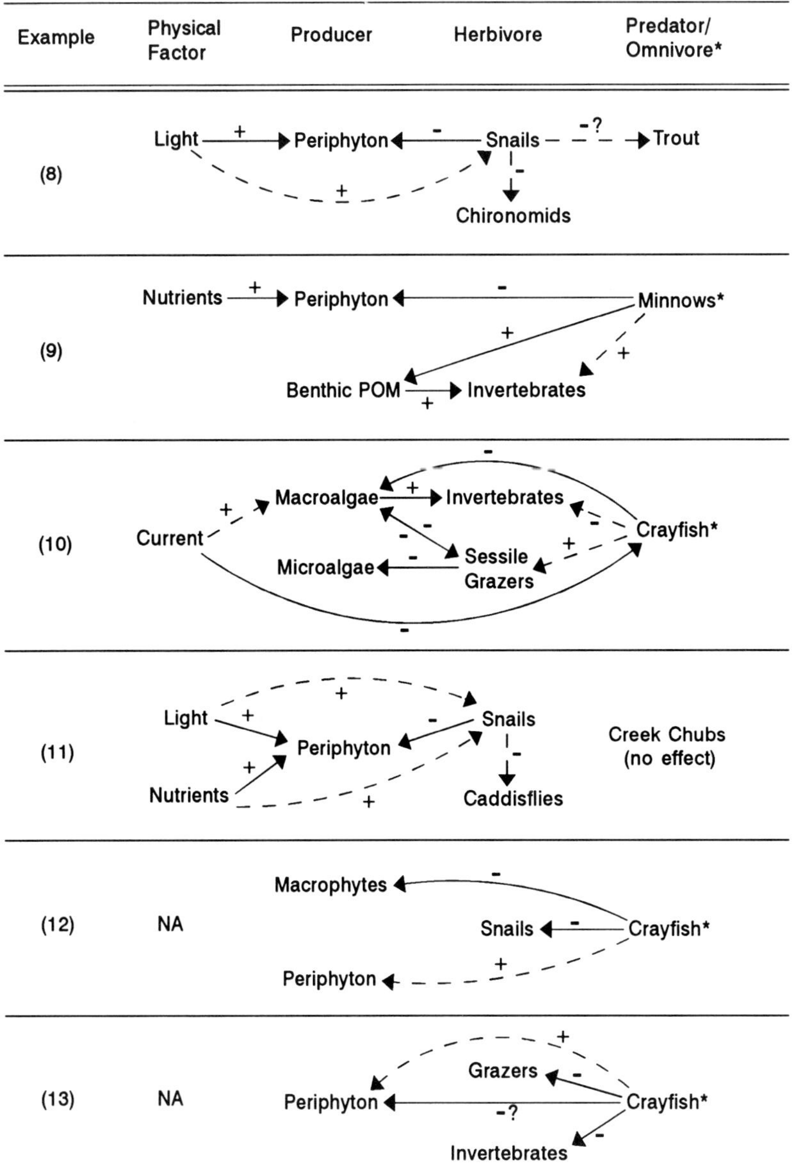

FIGURE 5 Summary food web diagrams (my interpretation) from selected studies of "intermediate" consumers. Arrows depict trophic level effects (solid arrow = direct effect; dashed arrow = indirect effect). Sign indicates direction of effect (+ = positive effect on biomass or density; – = negative effect on biomass or density). See Table II for study details and Table III for references and summary of overall effects. NA = not addressed.

VI. THE CASE FOR "INTERMEDIATE REGULATION"

A central element of the pelagic BU:TD model is the trophic "uncoupling" that occurs at the zooplankton–phytoplankton link (McQueen *et al.*, 1989). That is, predator or resource influence often breaks down at that interface because effects diminish with distance up or down a food web. In benthic systems, by contrast, the herbivore–periphyton linkage appears to be strong. Similar to pelagic systems, however, effects of top predators or abiotic resources tend to weaken in either direction. This suggests a central role for periphyton and grazers in benthic systems.

A. Herbivores and Omnivores

In benthic communities, the combined effects of resources and predators frequently are mediated by primary consumers (herbivores), which may serve as keystone species (*sensu* Paine, 1969). Paine (1966) considered keystone species normally to be predators, such as starfish in the marine intertidal zone or otters in kelp beds, that exert direct control over prey abundance and species composition and thereby indirect control of primary producers. This idea has since been extended to encompass any consumer that exerts an overriding influence on the structure of a food web. Hairston and Hairston (1993) reemphasized the importance of studying species that constitute the "dominant bulk" of the biomass or that exercise "exceptional influence" at each trophic level.

As discussed earlier, there are few examples from stream ecosystems in which top predators regulate the structure of food webs, but many examples that suggest other controls (Table II). Evidence is mounting to suggest that dominant consumers intermediate in the food web can exert control over the food web (Table III). This idea is similar to the "link hypothesis" of Oksanen (1988), in which the outcome of trophic interactions is dependent on the number of links in the food chain (defined as the number of trophic levels for which the highest level strongly influences its resources). Thus, downward control is initiated at the level of a dominating population, not necessarily at the top trophic level. In benthic systems, I propose that the influence of a dominant midlevel consumer also can extend laterally (at the same trophic level) or, indirectly, even upward (to higher trophic levels) in a food web, thus resulting in "intermediate regulation."

Pleurocerid snails appear to act as intermediate regulators in many North American streams. For example, *Elimia clavaeformis* greatly depletes periphyton in eastern Tennessee streams, thereby creating food-limited conditions for its own population and those of other benthic grazers (Fig. 5, example 11). The snail apparently is unaffected by common fish. In many streams of the Pacific Northwest, *Juga silicula* dominates consumer populations and has negative effects on periphyton standing crop and the densities of other invertebrates including competing grazers (Fig. 5, example 8). Furthermore, *Juga* may have negative indirect effects on higher trophic levels (e.g., fish) because it consumes fish prey items whereas it is rarely consumed by fish. In northern California streams, the caddis fly *Helicopsyche borealis* dramatically alters the benthic landscape and reaches densities of 8000 m^{-2}, but has few invertebrate or vertebrate predators (Lamberti and Resh, 1983; Lamberti *et al.,* 1987b). Another example of an "intermediate regulator" may be the well-studied minnow *Campostoma anomolum,* which strongly influences autotrophic production and detritus processing in North American prairie streams (Fig. 5, example 9).

A growing literature suggests that crayfish are important consumers in lake littoral zones and streams. For example, *Orconectes* crayfish in lakes consume fish eggs, invertebrates, macrophytes, and periphyton (Lodge *et al.,* 1985; Lodge and Lorman, 1987; Olsen *et al.,* 1991). This omnivory may lead to intermediate regulation. In a northern Wisconsin lake, *Orconectes rusticus* controlled the abundance of littoral macrophytes, snails, and periphyton (Fig. 5, example 12). *Orconectes propinquus* was a key species in regulating the benthic community structure of a productive, southern Michigan stream (Fig. 5, example 10). Crayfish reduced the abundance of the filamentous green alga *Cladophora glomerata,* thereby clearing space for microalgae and small sessile grazers. In a less productive, northern Michigan stream that did not contain *Cladophora, Orconectes rusticus* reduced the abundances of benthic invertebrates (including grazers), which indirectly increased the productivity of microalgae (Fig. 5, example 13). There was no evidence that crayfish were controlled by fish predators in either Michigan stream. These studies suggest that *Orconectes* can regulate the structure of benthic communities.

B. The Disturbance Template

A continuing debate in aquatic ecology is whether an "equilibrium" condition is ever established in aquatic food webs (Resh *et al.,* 1988; Poff, 1992). Food webs can be simplified by disturbance, thereby reducing connectance and ameliorating top-down effects (Menge and Sutherland 1976, 1987). In aquatic ecosystems, disturbance events frequently "reset" food webs. Resetting disturbances include floods (Power *et al.,* 1985; Grimm and Fisher, 1989), drought (Resh, 1982), and debris flows (Lamberti *et al.,* 1991) in streams, storm-induced wave action and ice scour in lake littoral zones (Wetzel, 1983a), and freezing in shallow wetlands (see Goldsborough and Robinson, Chapter 4). Because of the high frequency of disturbance by spates, streams appear to be at the extreme of the environmental constancy spectrum (Gregory, 1983). Thus, stream food webs would be predicted to have fewer links and lower connectance than other ecosystems (Oksanen, 1988).

The timing, frequency, and intensity of disturbance can affect aquatic trophic levels ranging from periphyton to fish predators (e.g., Lamberti *et al.,* 1991). Gregory (1983) points out that freshwater ecosystems are among the most frequently disturbed habitats. In their review of food webs, Briand and Cohen (1987) found that river and lake benthic food webs on average contained fewer trophic levels ($\bar{x}$ = 2.6 levels) than did pelagic systems ($\bar{x}$ = 3.6 levels). Natural disturbances that negatively affect the abundance and distribution of periphyton may indirectly affect higher trophic levels. Periphyton community recovery following disturbance is a key factor in food web, and hence ecosystem, recovery (Steinman and McIntire, 1990). In recovering food webs, omnivory may be favored because of rapidly changing food resources during succession. This may lead to regulation by omnivores that are "intermediate" in the food web and that can consume periphyton.

VII. GENERAL CONCLUSIONS AND DIRECTIONS FOR FUTURE RESEARCH

(1) Benthic algae clearly are a major component of aquatic food webs. Many, if not most, benthic consumers in aquatic ecosystems ingest periphyton because of a high degree of omnivory. The interplay between the physical template (productive capacity) and utilization by grazers (consumptive efficiency) in time and space may determine the ultimate role of periphyton in aquatic food webs. Pluralistic approaches [developed as factorial experiments, e.g., Rosemond *et al.* (1993)] are now needed to determine the relative importance of different ecological factors in regulating both benthic algae and the food webs that they support.

(2) Plant–herbivore interactions are central to food web structure and energy flow in aquatic ecosystems. Benthic grazers appear to be chronically food-limited, by either the abundance or the quality of their periphyton food. This is most common at the lower end of the productivity gradient for streams, but may also occur under intense grazing at higher productive capacities. Herbivore growth rates frequently are controlled by periphyton availability. Future studies should examine the relative roles of macroalgae, bryophytes, and vascular plants in stream energetics compared to the more commonly studied benthic microalgae. Other studies have suggested important roles for macroscopic lotic plants, ranging from provision of food and habitat to uptake and cycling of nutrients (e.g., see Marzolf *et al.,* 1994).

(3) The relative strengths of top-down and bottom-up effects cannot be generalized among benthic systems. Effects of either force appear to

diminish with distance down or up the food web. Dual control of periphyton by consumers and resources may be the most common condition, and models of interactive effects (e.g., the BU:TD model; McQueen *et al.*, 1986, 1989) appear to have the most promise for predicting the structure of aquatic food webs. In benthic systems, consumers intermediate in the food web ("intermediate regulators") most commonly regulate lower trophic levels and may also indirectly affect predators. In general, trophic proximity to periphyton contributes to dominance by specific consumer populations.

(4) Because of high levels of disturbance by spates, streams may be at the extreme of the environmental constancy spectrum. Lake littoral zones also are frequently disturbed. As a result, benthic food webs would be predicted to have fewer links and lower connectance than in other ecosystems. A high level of omnivory in streams may compensate for reduced connectance, but further study is needed to resolve this issue. Frequent disturbance also may help explain why there are apparently so few host-specific associations between benthic plants and herbivores, compared to in terrestrial ecosystems. An alternative explanation is that aquatic herbivores tend to be larger and live longer than their plant foods (e.g., microalgae), which may preclude selectivity. Further study in this areas seems to be warranted.

(5) There still are relatively few detailed analyses of *benthic* energy budgets or energy flow through food webs, for either streams or lakes. The application of new technologies (e.g., stable isotopes) holds promise for revealing patterns of energy utilization. Existing data suggest that periphyton are an important food resource throughout the river continuum and in lake littoral zones. More detailed information is needed on the relative importance of autotrophic and heterotrophic food resources in different biomes, stream sizes, lake environments, and latitudes. Furthermore, energy flow from aquatic *to* terrestrial habitats is rarely considered, even though a number of terrestrial vertebates consume benthic organisms such as macrophytes and invertebrates.

(6) Global environmental change, as related to atmospheric pollution, threatens to affect aquatic plants and potentially alter entire aquatic food webs. For example, ultraviolet light (UV-B) initially has a negative effect on periphyton in shallow flowing water (Bothwell *et al.*, 1993), but later has a positive *indirect* effect on periphyton because small herbivores (chironomid larvae) are eliminated by UV-B (Bothwell *et al.*, 1994). At the level of food webs, we should wonder whether connections between algae and higher consumers could be severed if small herbivores are removed, thereby inducing the collapse of aquatic food webs. Global changes in greenhouse gases, especially CO_2 and NO_x, could also affect periphyton. Research is needed to predict possible changes in benthic primary production and riparian vegetation, which could combine to set levels of instream autotrophic production that affect entire food webs.

ACKNOWLEDGMENTS

Many individuals have positively influenced my thinking about stream food webs, but I am especially grateful to the following people: Vince Resh, Stan Gregory, Al Steinman, Dave McIntire, Marty Berg, and David Lodge. The ecology graduate students at the University of Notre Dame provided examples and helped to clarify my thinking. Kieu Vu assembled the data presented in Table 17.1. Research in my laboratory contributing to this report was supported by grants from the National Science Foundation (BSR-8907968) and the U.S. Environmental Protection Agency (CR 8200290-01-0).

REFERENCES

Ahlgren, G., Gustafsson, I.-B., and Boberg, M. (1992). Fatty acid content and chemical composition of freshwater microalgae. *J. Phycol.* **28**, 37–50.

Allan, J. D. (1982). The effects of reduction in trout density on the invertebrate community of a mountain stream. *Ecology* **63**, 1444–1455.

Anderson, N. H., and Sedell, J. R. (1979). Detritus processing by macroinvertebrates in stream ecosystems. *Annu. Rev. Entomol.* **24**, 351–377.

Barnese, L. E., and Lowe, R. L. (1992). Effects of substrate, light, and benthic invertebrates on algal drift in small streams. *J. North Am. Benthol. Soc.* **11**, 49–59.

Batzer, D. P., and Resh, V. H. (1991). Trophic interactions among a beetle predator, a chironomid grazer, and periphyton in a seasonal wetland. *Oikos* **60**, 251–257.

Bazely, D. R., and Jefferies, R. L. (1989). Leaf and shoot demography of an Arctic stoloniferous grass, *Puccinellia phryganodes,* in response to grazing. *J. Ecol.* **77**, 811–822.

Begon, M., Harper, J. L., and Townsend, C. R., (1990). "Ecology: Individuals, Populations, and Communities," 2nd ed. Blackwell, Boston.

Belsky, A. J. (1986). Does herbivory benefit plants? A review of the evidence. *Am. Nat.* **127,** 870–892.

Ben-Amotz, A., Tornabene, T. G., and Thomas, W. H. (1985). Chemical profile of selected species of microalgae with emphasis on lipids. *J. Phycol.* **21**, 72–81.

Benke, A. C., and Wallace, J. B. (1980). Trophic basis of production among net-spinning caddisflies in a southern Appalachian stream. *Ecology* **61**, 108–118.

Berg, M. B. (1989). The role of Chironomidae in stream insect secondary production. Ph.D. Thesis, University of Notre Dame, Notre Dame, IN.

Berg, M. B., and Hellenthal, R. A. (1992a). The role of Chironomidae in energy flow of a lotic ecosystem. *Neth. J. Aquat. Ecol.* **26**, 471–476.

Berg, M. B., and Hellenthal, R. A. (1992b). Life histories and growth of lotic chironomids (Diptera: Chironomidae). *Ann. Entomol. Soc. Am.* **85**, 578–589.

Bergey, E. A., and Resh, V. H. (1994). Interactions between a stream caddisfly and the algae on its case: factors affecting algal quantity. *Freshwater Biol.* **31**, 153–163.

Bothwell, M. L. (1985). Phosphorus limitation of lotic periphyton growth rates: An intersite comparison using continuous-flow troughs (Thompson River system, British Columbia). *Limnol. Oceanogr.* **30**, 527–542.

Bothwell, M. L., Sherbot, D., Roberge, A. C., and Daley, R. J. (1993). Influence of natural ultraviolet radiation on lotic periphyton diatom community growth, biomass accrual, and species composition: Short-term versus long-term effects. *J. Phycol.* **29**, 24–35.

Bothwell, M. L., Sherbot, D. M. J., and Pollock, C. M. (1994). Ecosystem response to solar ultraviolet-B radiation: Influence of trophic-level interactions. *Science* **265**, 97–100.

Bott, T. L. (1983). Primary productivity in streams. *In* "Stream Ecology" (J. R. Barnes and G. W. Minshall, eds.), pp. 29–53. Plenum, NY.

Bowlby, J. N., and Roff, J. C. (1986). Trophic structure in southern Ontario streams. *Ecology* 67:1670–1679.

Briand, F., and Cohen, J. E. (1987). Environmental correlates of food chain length. *Science* **238**, 956–960.

Brock, E. M. (1960). Mutualism between the midge *Cricotopus* and the alga *Nostoc*. *Ecology* **41**, 474–483.

Bronmark, C. (1989). Interactions between epiphytes, macrophytes, and freshwater snails: A review. *J. Molluscan Stud.* **55**, 299–311.

Bunn, S. E., Barton, D. R., Hynes, H. B. N., Power, G., and Pope, M. A. (1989). Stable isotope analysis of carbon flow in a tundra river system. *Can. J. Fish. Aquat. Sci.* **46**, 1769–1775.

Burkholder, J. M., and Sheath, R. G. (1984). The seasonal distribution, abundance and diversity of desmids (Chlorophyta) in a softwater, north temperate stream. *J. Phycol.* **20**, 159–172.

Carpenter, S. R., and Lodge, D. M. (1986). Effects of submersed macrophytes on ecosystem processes. *Aquat. Bot.* **26**, 341–370.

Carpenter, S. R., Kitchell, J. F., and Hodgson, J. R. (1985). Cascading trophic interactions and lake productivity. *BioScience* **35**, 634–639.

Cattaneo, A. (1983). Grazing on epiphytes. *Limnol. Oceanogr.* **28**, 124–132.

Charlebois, P. M. (1994). The effects of crayfish (*Orconectes rusticus*) on the macroinvertebrate and algal assemblages in a northern Michigan stream. M.S. Thesis, University of Notre Dame, Notre Dame, IN.

Coffman, W. P., Cummins, K. W., and Wuycheck, J. C. (1971). Energy flow in a woodland stream ecosystem. I. Tissue support trophic structure of the autumnal community. *Arch. Hydrobiol.* **68**, 232–276.

Cooper, S. D. (1984). The effects of trout on water striders in stream pools. *Oecologia* **63**, 376–379.

Crawley, M. J. (1983). "Herbivory." Univ. of California Press, Berkeley.

Creed, R. P., Jr. (1994). Direct and indirect effects of crayfish grazing in a stream community. *Ecology* **75**, 2091–2103.

Cuker, B. E. (1983). Grazing and nutrient interactions in controlling the activity and composition of the epilithic algal community of an arctic lake. *Limnol. Oceanogr.* **28**, 133–141.

Culp, J. M. (1986). Experimental evidence that stream macroinvertebrate community structure is unaffected by different densities of coho salmon fry. *J. North Am. Benthol. Soc.* **5**, 140–149.

Cummins, K. W. (1973). Trophic relations of aquatic insects. *Annu. Rev. Entomol.* **18**, 183–206.

Cummins, K. W., Coffman, W. P., and Roff, P. A. (1966). Trophic relationships in a small woodland stream. *Verh.—Int. Ver. Theor. Angew. Limnol.* **16**, 627–638.

Cummins, K. W., Sedell, J. R., Swanson, F. J., Minshall, G. W., Fisher, S. G., Cushing, C. E., Peterson, R. C., and Vannote, R. L. (1983). Organic matter budgets for stream ecosystems: Problems in their evaluation. *In* "Stream Ecology" (J. R. Barnes and G.W. Minshall, eds.), pp. 299–353. Plenum, New York.

Cummins, K. W., Wilzbach, M. A., Gates, D. M., Perry, J. B., and Taliaferro, W. B. (1989). Shredders and riparian vegetation. *BioScience* **39**, 24–30.

Darley, W. M. (1977). Biochemical composition. *In* "The Biology of Diatoms" (D. Werner ed.), pp. 198–204. Univ. of California Press, Berkeley.

Davis, C. C. (1955). "The Marine and Freshwater Plankton." Michigan State Univ. Press, East Lansing.

Dodds, W. K. (1989). Photosynthesis of two morphologies of *Nostoc parmelioides* (Cyanobacteria) as related to current velocities and diffusion patterns. *J Phycol.* **25**, 258–262.

Dodds, W. K., and Castenholz, W. W. (1987). Effects of grazing and light on the growth of *Nostoc pruniforme* (Cyanobacteria). *Br. Phycol. J.* **23,** 219–227.

Dodds, W. K., and Marra, J. L. (1989). Behaviors of the midge, *Cricotopus* (Diptera: Chironomidae) related to mutualism with *Nostoc parmelioides* (Cyanobacteria). *Aquat. Insects* **11,** 201–208.

Dudley, T. L. (1988). The roles of plant complexity and epiphyton in colonization of macrophytes by stream insects. *Verh.—Int. Ver. Theor. Angew. Limnol.* **23,** 1153–1158.

Dudley, T. L., Cooper, S. D., and Hemphill, N. (1986). Effects of macroalgae on a stream invertebrate community. *J. North Am. Benthol. Soc.* **5,** 93–106.

Elwood, J. W., and Turner, R. R. (1989). Streams: Water chemistry and ecology. *In* "Analysis of Biogeochemical Cycling Processes in Walker Branch Watershed" (D. W. Johnson and R. I. Van Hook, eds.) pp. 301–350. Springer-Verlag, New York.

Elwood, J. W., Newbold, J. D., Trimble, A. F., and Stark, R. W. (1981). The limiting role of phosphorus in a woodland stream ecosystem: Effects of P enrichment on leaf decomposition and primary producers. *Ecology* **62,** 146–158.

Feltmate, B. W., and Williams, D. D. (1989). Influence of rainbow trout (*Oncorhynchus mykiss*) on density and feeding behaviour of a perlid stonefly. *Can. J. Fish. Aquat. Sci.* **46,** 1575–1580.

Feminella, J. W., and Resh, V. H. (1990). Hydrologic influences, disturbance, and intraspecific competition in a stream caddisfly population. *Ecology* **71,** 2083–2094.

Feminella, J. W., Power, M. E., and Resh, V. H. (1989). Periphyton responses to invertebrate grazing and riparian canopy in three northern California coastal streams. *Freshwater Biol.* **22,** 445–457.

Fisher, S. G., and Likens, G. E. (1973). Energy flow in Bear Brook, New Hampshire: An integrative approach to stream ecosystem metabolism. *Ecol. Monogr.* **43,** 421–439.

Flecker, A. S. (1984). The effects of predation and detritus on the structure of stream insect community: A field test. *Oecologia* **64,** 300–305.

Flint, R. W., and Goldman, C. R. (1975). The effects of a benthic grazer on the primary productivity of the littoral zone of Lake Tahoe. *Limnol. Oceanogr.* **20,** 935–944.

Fretwell, S. D. (1977). The regulation of plant communities by the food chains exploiting them. *Perspect. Biol. Med.* **20,** 169–185.

Fretwell, S. D. (1987). Food chain dynamics: The central theory of ecology? *Oikos* **50,** 291–301.

Fuller, R. L., Roelofs, J. L., and Fry, T. J. (1986). The importance of algae to stream invertebrates. *J. North Am. Benthol. Soc.* **5,** 290–296.

Gelwick, F. P., and Matthews, W. J. (1992). Effects of an algivorous minnow on temperate stream ecosystem properties. *Ecology* **73,** 1630–1645.

Gilliam, J. R., Fraser, D. F., and Sabat, A. M. (1989). Strong effects of foraging minnows on a stream benthic invertebrate community. *Ecology* **70,** 445–452.

Gotceitas, V., and Clifford, H. F. (1983). The life history of *Dicosmoecus atripes* (Hagen) (Limnephilidae: Trichoptera) in a Rocky Mountain stream of Alberta, Canada. *Can. J. Zool.* **61,** 586–596.

Gregg, W. W., and Rose. F. L. (1985). Influences of aquatic macrophytes on invertebrate community structure, guild structure, and microdistribution in streams. *Hydrobiologia* **128,** 45–56.

Gregory, S. V. (1980). Effects of light, nutrients, and grazing on periphyton communities in streams. Ph.D. Thesis, Oregon State University, Corvallis.

Gregory, S. V. (1983). Plant–herbivore interactions in stream systems. *In* "Stream Ecology" (J. R. Barnes and G. W. Minshall, eds.), pp. 157–189. Plenum, New York.

Grimm, N. B., and Fisher, S. G. (1989). Stability of periphyton and macroinvertebrates to disturbance by flash floods in a desert stream. *J. North Am. Benthol. Soc.* **8,** 293–307.

Haines, E. B., and Montague, C. L. (1979). Food sources of estuarine invertebrates analyzed using $^{13}C/^{12}C$ ratios. *Ecology* **60**, 48–56.

Hairston, N. G., Jr., and Hairston, N. G., Sr. (1993). Cause–effect relationships in energy flow, trophic structure, and interspecific interactions. *Am. Nat.* **142**, 379–411.

Hairston, N. G., Sr., Smith, F. E., and Slobodkin, L. B. (1960). Community structure, population control, and competition. *Am. Nat.* **94**, 421–425.

Hart, D. D. (1981). Foraging and resource patchiness: Field experiments with a grazing stream insect. *Oikos* **37**, 46–52.

Hart, D. D. (1987). Experimental studies of exploitative competition in a grazing stream insect. *Oecologia* **73**, 41–47.

Hart, D. D. (1992). Community organization in streams: The importance of species interactions, physical factors, and chance. *Oecologia* **91**, 220–228.

Hart, D. D., and Robinson, C. T. (1990). Resource limitation in a stream community: Phosphorus enrichment effects on periphyton and grazers. *Ecology* **71**, 1494–1502.

Harvey, B. C., and Hill, W. R. (1991). Effects of snails and fish on benthic invertebrate assemblages in a headwater stream. *J. North Am. Benthol. Soc.* **10**, 263–270.

Hasselrot, A. T. (1993). Insight into a psychomyiid life. Ph.D. Thesis, Uppsala University, Sweden.

Hawkins, C. P., and Furnish, J. K. (1987). Are snails important competitors in stream ecosystems? *Oikos* **49**, 209–220.

Hawkins, C. P., Murphy, M. L., and Anderson, N. H. (1982). Effects of canopy, substrate composition, and gradient on the structure of macroinvertebrate communities in Cascade Range streams of Oregon. *Ecology* **63**, 1840–1856.

Hershey, A. E. (1992). Effects of experimental fertilization on the benthic macroinvertebrate community of an arctic lake. *J. North Am. Benthol. Soc.* **11**, 204–217.

Hershey, A. E., Hiltner, A. L., Hullar, M. A. J., Miller, M. C., Vestal, J. R., Lock, M. A., Rundle, S., and Peterson, B. J. (1988). Nutrient influence on a stream grazer: *Orthocladius* microcommunities respond to nutrient input. *Ecology* **69**, 1383–1392.

Hill, W. R. (1992). Food limitation and interspecific competition in snail-dominated streams. *Can. J. Fish. Aquat. Sci.* **49**, 1257–1267.

Hill, W. R., and Harvey, B. C. (1990). Periphyton responses to higher trophic levels and light in a shaded stream. *Can. J. Fish. Aquat. Sci.* **47**, 2307–2314.

Hill, W. R., and Knight, A. W. (1987). Experimental analysis of the grazing interaction between a mayfly and stream algae. *Ecology* **68**, 1955–1965.

Hill, W. R., Weber, S. C., and Stewart, A. J. (1992). Food limitation of two lotic grazers: Quantity, quality, and size-specificity. *J. North Am. Benthol. Soc.* **11**, 420–432.

Holomuzki, J. R., and Stevenson, R. J. (1992). Role of predatory fish in community dynamics of an ephemeral stream. *Can. J. Fish. Aquat. Sci.* **49**, 2322–2330.

Huntsman, A. G. (1948). Fertility and fertilization of streams. *J. Fish. Res. Board Can.* **7**, 248–253.

Hynes, H. B. N. (1970). "The Ecology of Running Waters." Univ. of Toronto Press, Ontario.

Hynes, H. B. N. (1975). The stream and its valley. *Verh.—Int. Ver. Theor. Angew. Limnol.* **19**, 1–15.

Johnston, N. T., Perrin, C. J., Slaney, P. A., and Ward, B. R. (1990). Increased juvenile salmonid growth by whole-river fertilization. *Can. J. Fish. Aquat. Sci.* **47**, 862–872.

Kling, G. W., Fry, B., and O'Brien, W. J. (1992). Stable isotopes and planktonic trophic structure in arctic lakes. *Ecology* **73**, 561-566.

Kohler, S. L. (1984). Search mechanism of a stream grazer in patchy environments: The role of food abundance. *Oecologia* **62**, 209–218.

Lamberti, G. A. (1993). Grazing experiments in artificial streams. *J. North Am. Benthol. Soc.* **12**, 337–342.

Lamberti, G. A., and Moore, J. W. (1984). Aquatic insects as primary consumers. *In* "The Ecology of Aquatic Insects" (V. H. Resh and D. M. Rosenberg, eds.) pp. 164–195. Praeger, New York.

Lamberti, G. A., and Resh, V. H. (1983). Stream periphyton and insect herbivores: An experimental study of grazing by a caddisfly population. *Ecology* **64,** 1124–1135.

Lamberti, G. A., and Steinman, A. D., eds. (1993). Research in artificial streams: Applications, uses, and abuses. *J. North Am. Benthol. Soc.* **12,** 313–384.

Lamberti, G. A., Ashkenas, L. R., Gregory, S. V., and Steinman, A. D. (1987a). Effects of three herbivores on periphyton communities in laboratory streams. *J. North Am. Benthol. Soc.* **6,** 92–104.

Lamberti, G. A., Feminella, J. W., and Resh, V. H. (1987b). Herbivory and intraspecific competition in a stream caddisfly population. *Oecologia* **73,** 75-81.

Lamberti, G. A., Gregory, S. V., Ashkenas, L. R., Steinman, A. D., and McIntire, C. D. (1989). Productive capacity of periphyton as a determinant of plant–herbivore interactions in streams. *Ecology* **70,** 1840–1856.

Lamberti, G. A., Gregory, S. V., Ashkenas, L. R., Wildman, R. C., and Moore, K. M. S. (1991). Stream ecosystem recovery following a catastrophic debris flow. *Can. J. Fish. Aquat. Sci.* **48,** 196–208.

Lamberti, G. A., Gregory, S. V., Hawkins, C. P., Wildman, R. C., Ashkenas, L. R., and DeNicola, D. M. (1992). Plant–herbivore interactions in streams near Mount St Helens. *Freshwater Biol.* **27,** 237–247.

Li, H. W., Lamberti, G. A., Pearsons, T. N., Tait, C. K., Li, J. L., and Buckhouse, J. C. (1994). Cumulative effects of riparian disturbances along high desert trout streams of the John Day Basin, Oregon. *Trans. Am. Fish. Soc.* **123,** 627–640.

Li, J. L., and Gregory, S. V. (1989). Behavioral changes in the herbivorous caddisfly *Dicosmoecus gilvipes* (Limnephilidae). *J. North Am. Benthol. Soc.* **8,** 250–259.

Lindeman, R. L. (1942). The trophic-dynamic aspect of ecology. *Ecology* **23,** 399–418.

Lodge, D. M. (1986). Selective grazing on periphyton: A determinant of freshwater gastropod microdistributions. *Freshwater Biol.* **16,** 831–841.

Lodge, D. M. (1991). Herbivory on freshwater macrophytes. *Aquat. Bot.* **41,** 195–224.

Lodge, D. M., and Lorman, J. G. (1987). Reductions in submersed macrophyte biomass and species richness by the crayfish *Orconectes rusticus*. *Can. J. Fish. Aquat. Sci.* **44,** 591–597.

Lodge, D. M., Beckel, A. L., and Magnuson, J. J. (1985). Lake-bottom tyrant. *Nat. Hist.* **94,** 32–37.

Lodge, D. M., Barko, J. W., Strayer, D., Melack, J. M., Mittelbach, G. G., Howarth, R. W., Menge, B., and Titus, J. E. (1988). Spatial heterogeneity and habitat interactions in lake communities. *In* "Complex Interactions in Lake Communities" (S. R. Carpenter, ed.) pp. 181–208. Springer-Verlag, New York.

Lodge, D. M., Kershner, M. W., Aloi, J. E., and Covich, A. P. (1994). Effects of an omnivorous crayfish (*Orconectes rusticus*) on a freshwater littoral food web. *Ecology* **75,** 1265–1281.

Lowe, R. L., Golladay, S. W., and Webster, J. R. (1986). Periphyton response to nutrient manipulation in streams draining clearcut and forested watersheds. *J. North Am. Benthol. Soc.* **5,** 221–229.

Lyford, J. H., Jr., and Gregory, S.V. (1975). The dynamics and structure of periphyton communities in three Cascade Mountain streams. *Verh.—Int. Ver. Theor. Angew. Limnol.* **19,** 1610–1616.

Marzolf, E. R., Mulholland, P. J., and Steinman, A. D. (1994). Improvements to the diurnal upstream–downstream dissolved oxygen change technique for determining whole-stream metabolism in small streams. *Can. J. Fish. Aquat. Sci.* **51,** 1591–1599.

Mason, J. C. (1976). Response of underyearling coho salmon to supplemental feeding in a natural stream. *J. Wildl. Manage.* **40,** 775–788.

Matthews, W. J., Stewart, A. J., and Power, M. E. (1987). Grazing fishes as components of North American stream ecosystems: Effects of *Campostoma anomalum*. *In* "Community and Evolutionary Ecology of North American Stream Fishes" (W. J. Matthews and D. C. Heins, eds.), pp. 128–135. Univ. of Oklahoma Press, Norman.

Mayer, M. S., and Likens, G. E. (1987). The importance of algae in a shaded headwater stream as food for an abundant caddisfly (Trichoptera). *J. North Am. Benthol. Soc.* **6**, 262–269.

McAuliffe, J. R. (1983). Competition, colonization patterns, and disturbance in stream benthic communities. *In* "Stream Ecology" (J. R. Barnes and G. W. Minshall, eds.), pp. 137–156. Plenum, New York.

McIntire, C. D. (1973). Periphyton dnamics in laboratory streams: A simulation model and its implications. *Ecol. Monogr.* **43**, 399–420.

McIntire, C. D. (1975). Periphyton assemblages in laboratory streams. *In* "River Ecology" (B. A. Whitton, ed.), pp. 403–430. Univ. of California Press, Berkeley

McNaughton, S. J. (1986). On plants and herbivores. *Am. Nat.* **128**, 765–770.

McQueen, D. J., Post, J. R., and Mills, E. L. (1986). Trophic relationships in freshwater pelagic ecosystems. *Can. J. Fish. Aquat. Sci.* **43**, 1571–1581.

McQueen, D. J., Johannes, M. R. S., Post, J. R., Stewart, T. J., and Lean, D. R. S. (1989). Bottom-up and top-down impacts on freshwater pelagic community structure. *Ecol. Monogr.* **59**, 289–309.

Menge, B. A., and Sutherland, J. P. (1976). Species diversity gradients: Synthesis of the roles of predation, competition, and temporal heterogeneity. *Am. Nat.* **110**, 351–369.

Menge, B. A., and Sutherland, J. P. (1987). Community regulation: Variation in disturbance, competition, and predation in relation to environmental stress and recruitment. *Am. Nat.* **130**, 730–757.

Minshall, G. W. (1978). Autotrophy in stream ecosystems. *BioScience* **28**, 767–771.

Minshall, G. W. (1988). Stream ecosystem theory: A global perspective. *J. North Am. Benthol. Soc.* **7**, 263–288.

Mundie, J. H., McKinnell, S. M., and Trabe, R. E. (1983). Responses of stream zoobenthos to enrichment of gravel substrates with cereal grain and soybean. *Can. J. Fish. Aquat. Sci.* **40**, 1702–1712.

Murdoch. W. W. (1966). "Community structure, population control, and competition"—A critique. *Am. Nat.* **100**, 219–226.

Naiman, R. J., Melillo, J. M., Lock, M. A., Ford, T. E., and Reice, S. R. (1987). Longitudinal patterns of ecosystem processes and community structure in a subarctic river continuum. *Ecology* **68**, 1139–1156.

Newman, R. M. (1991). Herbivory and detritivory on freshwater macrophytes by invertebrates: A review. *J. North Am. Benthol. Soc.* **10**, 89–114.

Oksanen, L. (1988). Ecosystem organization: Mutualism and cybernetics or plain Darwinian struggle for existence? *Am. Nat.* **131**, 424–444.

Oksanen, L., Fretwell, S. D., Arruda, J., and Niemelä, P. (1981). Exploitation ecosystems in gradients of primary productivity. *Am. Nat.* **118**, 240–261.

Olsen, T. M., Lodge, D. M., Capelli, G. M., and Houlihan, R. J. (1991). Mechanisms of impact of an introduced crayfish (*Orconectes rusticus*) on littoral congeners, snails, and macrophytes. *Can. J. Fish. Aquat. Sci.* **48**, 1853–1861.

Ortega-Calvo, J. J., Mazuelos, C., Hermosin, B., and Saiz-Jimenez, C. (1993). Chemical composition of *Spirulina* and eukaryotic algae food products marketed in Spain. *J. Appl. Phycol.* **5**, 425–435.

Paine, R. T. (1966). Food web complexity and species diversity. *Am. Nat.* **100**, 65–75.

Paine, R. T. (1969). A note on trophic complexity and community stability. *Am. Nat.* **103**, 91–93.

Paine, R. T. (1980). Food webs: Linkage, interaction strength and community infrastructure. *J. Anim. Ecol.* **49**, 667–685.

Parsons, T. R., Stephens, K., and Strickland, J. D. H. (1961). On the chemical composition of eleven species of marine phytoplankters. *J. Fish. Res. Board Can.* **18,** 1001–1016.

Perrin, C. J., Bothwell, M. L., and Slaney, P. A. (1987). Experimental enrichment of a coastal stream in British Columbia: Effects of organic and inorganic additions on autotrophic periphyton production. *Can. J. Fish. Aquat. Sci.* **44,** 1247–1256.

Peterson, B. J., and Fry, B. (1987). Stable isotopes in ecosystem studies. *Annu. Rev. Ecol. Syst.* **18,** 293–320.

Peterson, B. J., Hobbie, J. E., Hershey, A. E., Lock, M. A., Ford, T. E., Vestal, J. R., McKinley, V. L., Hullar, M. A. J., Miller, M. C., Ventullo, R. M., and Volk, G. S. (1985). Transformation of a tundra river from heterotrophy to autotrophy by addition of phosphorus. *Science* **229,** 1383–1386.

Peterson, B. J., Deegan, L., Helfrich, J., Hobbie, J. E., Hullar, M. A. J., Moller, B., Ford, T. E., Hershey, A. E., Hiltner, A., Kipphut, G., Lock, M. A., Fiebig, D. M., McKinley, V. L., Miller, M. C., Vestal, J. R., Ventullo, R. M., and Volk, G. S. (1993). Biological responses of a tundra river to fertilization. *Ecology* **74,** 653–672.

Poff, N. L. (1992). Why disturbances can be predictable: A perspective on the definition of disturbance in streams. *J. North Am. Benthol. Soc.* **11,** 86–92.

Porter, K. G. (1977). The plant–animal interface in freshwater ecosystems. *Am. Sci.* **65,** 159–170.

Power, M. E. (1984a). Depth distributions of armored catfish: Predator-induced resource avoidance? *Ecology* **65,** 523–528.

Power, M. E. (1984b.) The importance of sediment in the grazing ecology and size class interactions of an armored catfish, *Ancistrus spinosus. Environ. Biol. Fishes* **10,** 173–181.

Power, M. E. (1990a). Effects of fish in river food webs. *Science* **250,** 811–814.

Power, M. E. (1990b). Resource enhancement by indirect effects of grazers: Armored catfish, algae, and sediment. *Ecology* 7, 897–904.

Power, M. E. (1992). Top-down and bottom-up forces in food webs: Do plants have primacy? *Ecology* **73,** 733–746.

Power, M. E., and Matthews, W. J. (1983). Algae-grazing minnows (*Campostoma anomalum*), piscivorous bass (*Micropterus* spp.), and the distribution of attached algae in a small prairie-margin stream. *Oecologia* **60,** 328–332.

Power, M. E., Matthews, W. J., and Stewart, A. J. (1985). Grazing minnows, piscivorous bass, and stream algae: Dynamics of a strong interaction. *Ecology* **66,** 1448–1456.

Power, M. E., Stewart, A. J., and Matthews, W. J. (1988). Grazer control of algae in an Ozark Mountain stream: Effects of short-term exclusion. *Ecology* **69,** 1894–1898.

Pringle, C. M. (1985). Effects of chironomid (Insecta: Diptera) tube-building activities on stream diatom communities. *J. Phycol.* **21,** 185–194.

Pringle, C. M., and Blake, G. A. (1994). Quantitative effects of atyid shrimp (Decapoda: Atyidae) on the depositional environment in a tropical stream: Use of electricity for experimental exclusion. *Can. J. Fish. Aquat. Sci.* **51,** 1443–1450.

Reice, S. R., and Edwards, R. L. (1986). The effect of vertebrate predation on lotic macroinvertebrate communities in Québec, Canada. *Can. J. Zool.* **64,** 1930–1936.

Resh, V. H. (1982). Age structure alteration in a caddisfly population after habitat loss and recovery. *Oikos* **38,** 280–284.

Resh, V. H., and Houp, R. E. (1986). Life history of the caddisfly *Dibusa angata* and its association with the red alga *Lemanea australis. J. North Am. Benthol. Soc.* **5,** 28–40.

Resh, V. H., Brown, A. V., Covich, A. P., Gurtz, M. E., Li, H. W., Minshall, G. W., Reice, S. R., Sheldon, A. L., Wallace, J. B., and Wissmar, R. C. (1988). The role of disturbance in stream ecology. *J. North Am. Benthol. Soc.* **7,** 433–455.

Richards, C. and Minshall, G. W. (1988). The influence of periphyton abundance on *Baetis bicaudatus* distribution and colonization in a small stream. *J. North Am. Benthol. Soc.* **7,** 77–86.

Richardson, J. S., and Mackay, R. J. (1991). Lake outlets and the distribution of filter feeders: An assessment of hypotheses. *Oikos* **62,** 370–380.

Rosemond, A. D., Mulholland, P. J., and Elwood, J. W. (1993). Top-down and bottom-up control of stream periphyton: Effects of nutrients and herbivores. *Ecology* **74,** 1264–1280.

Rosenfeld, J. S., and Roff, J. C. (1992). Examination of the carbon base in southern Ontario streams using stable isotopes. *J. North Am. Benthol. Soc.* **11,** 1–10.

Rounick, J. S., Winterbourn, M. J., and Lyon, G. M. (1982). Differential utilization of allochthonous and autochthonous inputs by aquatic invertebrates in some New Zealand streams: A stable carbon isotope study. *Oikos* **39,** 191–198.

Sarnelle, O., Kratz, K. W., and Cooper, S. D. (1993). Effects of an invertebrate grazer on the spatial arrangement of a benthic microhabitat. *Oecologia* **96,** 208–218.

Schwenneker, B. W. (1985). The contribution of allochthonous and autochthonous organic material to aquatic insect secondary production rates in a north temperate stream. Ph.D. Thesis, University of Notre Dame, Notre Dame, IN.

Sheath, R. G., and Burkholder, J. M. (1985). Characteristics of soltwater streams in Rhode Island. II. Composition and seasonal dynamics of macroalgal communities. *Hydrobiologia* **128,** 109–118.

Steinman, A. D. (1992). Does an increase in irradiance influence periphyton in a heavily-grazed woodland stream? *Oecologia* **91,** 162–170.

Steinman, A. D., and Boston, H. L. (1993). The ecological role of aquatic bryophytes in a woodland stream. *J. North Am. Benthol. Soc.* **12,** 17–26.

Steinman, A. D., and McIntire, C. D. (1990). Recovery of lotic periphyton communities after disturbance. *Environ. Manage.* **14,** 589–604.

Steinman, A. D., McIntire, C. D., and Lowry, R. R. (1987). Effects of herbivore type and density on chemical composition of algal assemblages in laboratory streams. *J. North Am. Benthol. Soc.* **6,** 189–197.

Steinman, A. D., Mulholland, P. J., and Kirstel, D. B. (1991). Interactive effects of nutrient reduction and herbivory on biomass, taxonomic structure, and P uptake in lotic periphyton communities. *Can. J. Fish. Aquat. Sci.* **48,** 1951–1959.

Stevenson, R. J., and Peterson, C. G. (1989). Variation in benthic diatom (Bacillariophyceae) immigration with habitat characteristics and cell morphology. *J. Phycol.* **25,** 120–129.

Stevenson, R. J., Singer, R., Roberts, D. A., and Boylen, C. W. (1985). Patterns of benthic algal abundance with depth, trophic status, and acidity in poorly buffered New Hampshire lakes. *Can. J. Fish. Aquat. Sci.* **42,** 1501–1512.

Stewart, A. J. (1987). Responses of stream algae to grazing minnows and nutrients: A field test for interactions. *Oecologia* **72,** 1–7.

Stock, M. S., Richardson, T. D., and Ward, A. K. (1987). Distribution and primary productivity of the epizoic macroalga *Boldia erythrosiphon* (Rhodophyta) in a small Alabama stream. *J. North Am. Benthol. Soc.* **6,** 168–174.

Tait, C. K., Li, J. L., Lamberti, G. A., Pearsons, T. N., and Li, H. W. (1994). Relationships between riparian cover and the community structure of high desert streams. *J. North Am. Benthol. Soc.* **13,** 45–56.

Taylor, B. R., and Roff, J. C. (1982). Evaluation of ecological maturity in three headwater streams. *Arch. Hydrobiol.* **94,** 99–125.

Tiffany, L. H. (1938). "Algae: The Grass of Many Waters." Thomas, Springfield, IL.

Towns, D. R. (1981). Effects of artificial shading on periphyton and invertebrates in a New Zealand stream. *N. Z. J. Mar. Freshwater Res.* **15,** 185–192.

Vannote, R. L., Minshall, G. W., Cummins, K. W., Sedell, J. R., and Cushing, C. E. (1980). The river continuum concept. *Can. J. Fish. Aquat. Sci.* **37**, 130–137.

Volkman, J. K., Brown, M. R., Dunstan, G. A., and Jeffrey, S. W. (1993). The biochemical composition of marine microalgae from the class Eustigmatophyceae. *J. Phycol.* **29**, 69–78.

Wallace, J. B., and Merritt, R. W. (1980). Filter-feeding ecology of aquatic insects. *Annu. Rev. Entomol.* **25**, 103–132.

Wallace, J. B., Benke, A. C., Lingle, A. H., and Parsons, K. (1987). Trophic pathways of macroinvertebrate primary consumers in subtropical blackwater streams. *Arch. Hydrobiol. Suppl.* **74**, 423–451.

Ward, A. K., Dahm, C. N., and Cummins, K. W. (1985). *Nostoc* (Cyanophyta) productivity in Oregon stream ecosystems: Invertebrate influences and differences between morphological types. *J. Phycol.* **21**, 223–227.

Warren, C. E., Wales, J. H., Davis, G. E., and Doudoroff, P. (1964). Trout production in an experimental stream enriched with sucrose. *J. Wildl. Manage.* **28**, 617–660.

Wetzel, R. G. (1983a). "Limnology," 2nd ed. Saunders, Philadelphia.

Wetzel, R. G., ed. (1983b). "Periphyton of Freshwater Ecosystems." Dr. W. Junk Publishers, The Hague, The Netherlands.

White, T. R. C. (1978). The importance of relative shortage of food in animal ecology. *Oecologia* **33**, 71–86.

Williams, D. D., Mundie, J. H., and Mounce, D. E. (1977). Some aspects of benthic production in a salmonid rearing channel. *J. Fish. Res. Board Can.* **34**, 2133–2141.

Winterbourn, M. J., Rounick, J. S., and Hildrew, A. G. (1986). Patterns of carbon resource utilization by benthic invertebrates in two British river systems: A stable carbon isotope study. *Arch. Hydrobiol.* **107**, 349–361.

Wiseman, S. W., Cooper, S. D., and Dudley, T. L. (1993). The effects of trout on epibenthic odonate naiads in stream pools. *Freshwater Biol.* **30**, 133–145.

18

Algae in Microscopic Food Webs

Thomas L. Bott

Stroud Water Research Center
The Academy of Natural Sciences
Avondale, Pennsylvania 19311

I. INTRODUCTION

For the most part, algivory in lotic systems has been analyzed with attention to a restricted range of consumers—macroinvertebrates (particularly insect larvae and snails) and fish. Some imaginative manipulation experiments have been performed to elucidate those biological interactions and their importance relative to physical and chemical factors in controlling algal biomass (e.g., Feminella *et al.*, 1989; Power, 1990). Hildrew (1992) referred to this body of research as "one of the big success stories" of recent stream ecology.

Here, the role of algae as food for microconsumers is considered in the context of a complex microbial food web that includes both algae and bacteria as resources for consumers of microscopic dimensions (protozoa and meiofauna, including temporary meiofauna). Meiofauna are defined operationally as organisms in the size range of 50–500 μm (Fenchel, 1978) or 50–1000 μm (Higgins and Thiel, 1988), and can include large ciliates as well as metazoa (e.g., rotifers, copepods, oligochaetes). Organisms that will eventually reach macrofaunal dimensions (e.g., insect larvae) are considered temporary meiofauna during the portion of their life history that they are of meiofaunal dimensions. Furthermore, specimens of some taxa usually encountered as meiofauna can be >1 mm. The groups of organisms in this microbial web are portrayed in relation to macrofauna in Fig. 1.

Pomeroy (1974) suggested that in marine planktonic systems, energy flow from algae (in the form of excreted photosynthate) → heterotrophic bacteria → protozoa → zooplankton [later called the "microbial loop" by Azam *et al.* (1983)] might be an important supplement to the route from algae → zooplankton → fish. The relative importance of the microbial loop remains controversial, being considered significant by some (e.g., Porter *et al.*, 1979; Fenchel, 1984; Wright and Coffin, 1984; E. B. Sherr and Sherr, 1987) but not by others (Ducklow *et al.*, 1986, Pomeroy and Wiebe, 1988). Some studies have indicated that the microbial loop was more important in oligotrophic than eutrophic or heterotrophic ecosystems (e.g., Carney, 1990; Findlay *et al.*, 1991) but evaluations are limited to plankton. In the benthos, nutrient sources should be more diverse, and Kemp (1990) suspected that food webs analogous to the microbial loop would be especially important there. Strayer (1991) conceptualized an analogous benthic "meiofaunal loop" that introduces additional steps in the flow of energy from microbes to macroorganisms.

My interest in these feeding relationships is connected with our studies of energy flow through the benthic community in White Clay Creek (WCC), a piedmont stream draining a rural watershed in southeastern Pennsylvania. Studies of the function of protozoa and especially meiofauna in streams are rare, although in benthic marine environments, particularly intertidal habitats, ecological studies of meiofaunal groups and individual

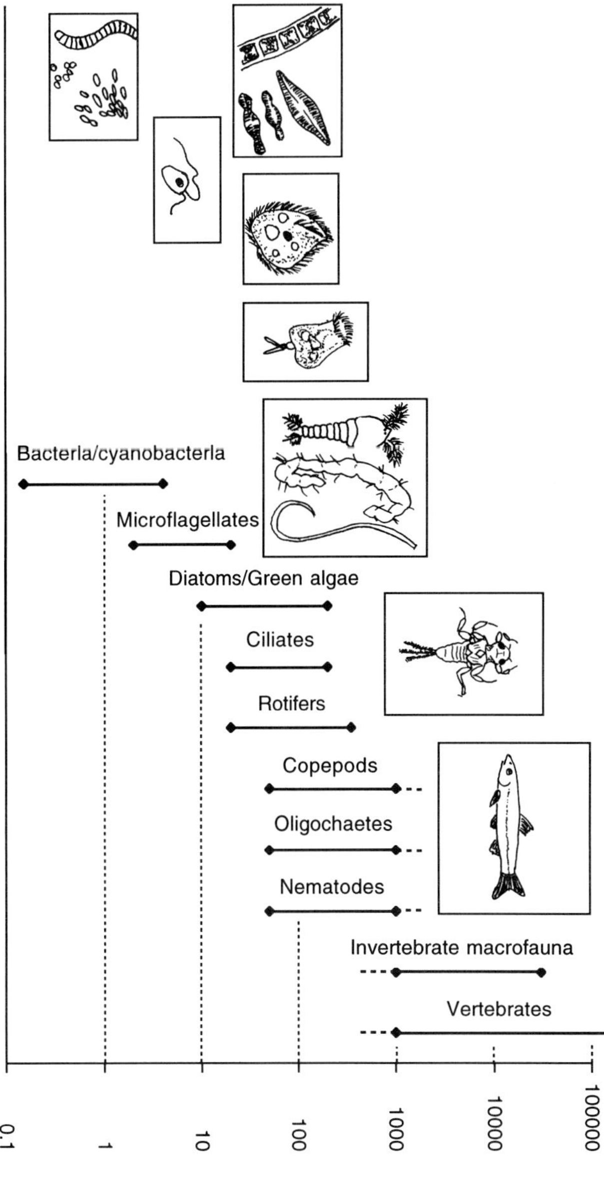

FIGURE 1 Selected components of a benthic aquatic food web arrayed according to size range. Organisms <1000 μm are considered part of the microscopic food web.

taxa have been performed. The literature concerning microconsumer feeding in planktonic communities is more extensive. While recognizing the obvious habitat differences and that the species composition of benthic and planktonic communities will differ, I draw on some of that literature and relevant laboratory studies when they illustrate aspects of interactions that are likely to be operative in the benthos.

II. ALGAL–BACTERIAL INTERACTION

Although most of this chapter will focus on algae as food, algae are also important in microscopic food webs because their excreted photosynthetic products partially support heterotrophic organisms, particularly bacteria, the other food resource for microconsumers considered here. Reported excretion values for phytoplankton range between 5 and 40% of total photosynthate, and data for benthic algae are of the same order of magnitude (Bott and Ritter, 1981; Kaplan and Bott, 1982; Haack and McFeters, 1982). From ≈30–90% of exudates can be used by planktonic (Wolter, 1982; Coveney and Wetzel, 1989) and benthic (Haack and McFeters, 1982; Kaplan and Bott, 1982; J. D. Newbold, T. L. Bott, and L. A. Kaplan, unpublished data) heterotrophic bacteria. However, some researchers have concluded that high levels of excretion are experimental artifacts (Sharp, 1977; Lancelet and Billen, 1985), and Jumars *et al.* (1989) suggest that excretion would not exceed 10% in nature.

Jumars *et al.* (1989) highlighted zooplankton excretion and egestion (see also Riemann and Søndergaard, 1986) in addition to sloppy feeding (Lampert, 1978) as sources of dissolved organic matter (DOM) for planktonic bacteria. Strayer (1988), in a reanalysis of the data of Scavia and Laird (1987), showed that by including recycled organic carbon, total secondary productivity (including bacterial) could be released from restrictive dependence on algal DOM. In fact, total secondary productivity could be greater than primary productivity to a degree related to consumer production efficiency and system retentiveness, with the increase being greatest in more retentive systems. In more open systems, such as streams, inputs of dissolved and particulate organic matter from the watershed are important supplements to internally generated nutrients that support bacterial productivity.

III. DETECTION OF MICROBIAL FEEDING RELATIONSHIPS

A. Immunological Approaches

Immunological techniques have been employed in a few studies of feeding relationships. Feller (1984) used taxon-specific serological tracers to describe trophic interactions in a benthic salt marsh community. Fluorescent antibodies were used to identify cyanobacteria ingested by a planktonic copepod (Goarant *et al.*, 1994) and to trace gut passage of a *Pseudomonas* sp. in a polychaete worm (Plante and Jumars, 1993). Such techniques have the advantages of specificity when cross-reactions have been eliminated, they can be used in both qualitative and quantitative studies, and they allow work with soft-bodied food items, but they have not been used routinely.

B. Correlation Analyses

Temporal or spatial correlations of densities or biomass have been used to establish relationships between microconsumers and food resources. For example, the densities of large (>40 μm long) diatom-ingesting benthic ciliates in the River Itchen (England) peaked during diatom blooms, at which time the percentage of ciliates with diatoms in them was also high (Baldock and Sleigh, 1988). Meiofaunal distributions in some marine sediments were correlated with algae (Montagna *et al.*, 1983; Decho and Fleeger, 1988a; Pinckney and Sandulli, 1990) and bacteria (Gray, 1966; Gray and Johnson, 1970; Decho and Castenholz, 1986). In some of these studies, relationships were established using spatial autocorrelation analyses when simple correlation analyses revealed none.

Given the complexity of the environment, interpretation of correlation analyses must be done cautiously. Meiofauna densities were correlated with diatom distributions, but the absence of a time lag in food and grazer dynamics suggested that both groups were responding to one or more physical factors (Montagna *et al.*, 1983). Blanchard (1990), working in a coastal pond, found that nematodes were associated specifically with the diatoms *Pleurosigma* and *Gyrosigma* spp., and most harpacticoid copepods with *Amphora* spp., but concluded that despite positive spatial autocorrelations between meiofauna and diatoms >40 μm, food resource distributions only partially influenced meiofaunal distribution patterns. Physical-chemical factors such as sediment grain size, pore water content, and dissolved oxygen concentration, as well as biological factors such as macrofaunal grazing of both algae and meiofauna, also affected results. Feeding mode affected results in a study in which one harpacticoid copepod species was correlated with diatom patches in mudflat sediments at low tide, but another was not, presumably because it was a suspension feeder (Decho and Fleeger, 1988a). Interpretation of the results of correlation analyses may also require consideration of algal morphology because filamentous forms can provide habitat and protection from predators as well as a source of food (Hall and Bell, 1993).

C. Feeding Rate Measurements: Methodological Considerations

Various methods have been applied to the measurement of feeding rates, each with advantages and pitfalls. Rates have been measured directly by following the ingestion of added labeled tracer particles, for example, fluorescent microspheres used as surrogates for microorganisms (Borsheim, 1984), cultured bacteria (FLB, B. F. Sherr *et al.* 1987), and algae (FLA, Rublee and Gallegos, 1989; E. B. Sherr *et al.*, 1991) stained with fluorescent dyes, and bacterial minicells (Wikner *et al.*, 1986). Some studies showed that surrogate particles were selected against by many protozoa (Pace and Bailiff, 1987; B. F. Sherr *et al.*, 1987). The FLB technique, developed with planktonic grazers, was extended to sediment communities by Epstein and Shiaris (1992). Because FLB and FLA are heat-killed during preparation, they may be less palatable to some micrograzers (Landry *et al.*, 1991), although another study suggested that labeling did not affect ingestion (E. B. Sherr *et al.*, 1989). Cultured cells are usually larger and may differ qualitatively from cells found *in situ* (Carman and Thistle, 1985; Bott and Kaplan, 1990). Borchardt and Bott (1995) fluorescently labeled algae and bacteria in aliquots of sediment, thereby providing grazers with mixes of organisms as they occurred *in situ.* They also used the fluorochrome Nile Red to stain diatoms (FLD), which retained them in a living state and raphid forms remained motile.

These approaches ensure a direct measurement of ingestion but are limited to prey that are ingested whole and remain recognizable in the animal. Algae in ciliate food vacuoles were observable for 5–24 h and bacteria for 1–2 h (Sawicka *et al.*, 1983, Rassoulzadegan *et al.*, 1988), although perhaps some rapidly digested cells were missed. Additional concerns are that (1) ingested prey may not be assimilated, (2) improperly chosen fixatives may cause expulsion of food particles, as formalin did with protozoa (Sieracki *et al.*, 1987), and (3) procedures, as usually applied, provide estimates of ingestion for numerically important taxa in the community rather than an integrated rate for all meiofauna or protozoa.

In other feeding studies with microconsumers, the food resource was tagged radioisotopically before addition (e.g., Duncan *et al.*, 1974; Nygaard and Hessen, 1990; Burns and Xu, 1990), but again, cultured cells may differ from the native microflora. Another radioisotopic approach allows work with native microbes. Radiolabeled dissolved substrate, for example, bicarbonate or organic compound, is added to sediment cores to preferentially tag either algae or bacteria, respectively, and the subsequent accumulation of radiolabel in grazers of these microbes is measured (Montagna, 1984a; Carman and Thistle, 1985; Montagna and Bauer, 1988). Radioisotopic procedures allow the measurement of both ingestion and assimilation, but must include controls for sorption of radiolabel, uptake by living animals, and uptake by epicuticular microbes (Montagna, 1983,

1984b; Carman, 1990), as well as consider exchange of label between microbial groups. The method of introducing label is critical and injection and pore water replacement techniques were preferred over slurry preparation because disruption of spatial relationships was minimized (Carman *et al.*, 1989). This approach also assumes linear uptake by microbes and hyperbolic kinetics into grazers, assumptions that may not always be the case.

An integrated rate for the entire grazer community can be obtained by comparing temporal changes in food densities when grazer densities are increased (Carrick *et al.* 1992), reduced by sieving (Wright and Coffin, 1984; Borchardt and Bott, 1995) or dilution (Landry and Hassett, 1982), or left unaltered. Shortcomings of such experiments, in addition to the fact that data for individual taxa are not obtained, are that: (1) grazer densities may change significantly in experiments of longer duration; (2) it may be difficult to detect statistically significant changes if prey abundances are low; (3) different models of prey dynamics may give different estimates of grazing rates (Ducklow and Hill, 1985); (4) the technique employed to remove grazers may alter nutrient levels and thereby stimulate growth, yielding a higher estimate of grazing rate (McManus and Fuhrman, 1988); (5) if grazer concentrations are elevated, concentration and confinement may alter grazing rates; (6) grazing rate may change with food densities (P. A. Montagna, personal communication); and (7) food abundances may be altered by the treatment (Borchardt and Bott, 1995). Chemicals should not be used to poison cells in experiments to estimate protozoan grazing because nontarget organisms can be affected (B. F. Sherr *et al.*, 1986; Sanders and Porter, 1986; Tremaine and Mills, 1987).

IV. ALGAL AND BACTERIAL INGESTION BY MICROCONSUMERS

A. Protozoa

The protists, though microscopic in size, exhibit an amazing diversity of energy acquisition mechanisms. Pratt and Cairns (1985) grouped protozoa into six categories: photosynthetic autotrophs, bacterivores-detritivores, DOM utilizers, algivores, omnivores, and predators. Even though Fenchel (1968) over 25 years ago documented that algae, mostly diatoms, made up >75% of the diet of the benthic ciliate *Frontonia arenaria,* algal ingestion rate data remain scarce for protozoa (Table I). Rates for benthic ciliates calculated from the work of McCormick (1991) were at the low end of the range of 3–48 algae·ciliate^{-1}·h^{-1} reported for planktonic forms by E. B. Sherr *et al.* (1991). Ciliates were one of the three most important algivorous meiofauna groups in WCC, and were predominant in experiments in which total meiofaunal grazing pressure on diatoms was great

TABLE I Ingestion Rates for Benthic and Epibenthic Protists and Meiofauna Consuming Algae and Bacteria

Grazer	*Habitat*	*Food resource*	*Method*	*Units*	*Rate*	*Comments*	*Reference*
Ciliates							
Ciliates	Sandy tidal flat, White Sea, Russia	Algae	Ingestion of native taxa	% of estimated daily algal productivity	10	93% of dinoflagellate productivity, 6% of diatom productivity	Epstein *et al.* (1992)
		Bacteria	Ingestion of native taxa	% of estimated daily bacterial productivity	11		
		Algae and bacteria	Ingestion of native taxa	10^{-6} mg·ciliate^{-1}·h^{-1}	2.5–4.5	At field densities	
Chilodonella sp., *Trithigmostoma cucullulus* [and *Pelomyxa* sp. (Sarcodina)]	Wilson Creek, KY, experimental flumes	Diatoms	Ingestion of native taxa	Diatoms·cm^{-2}·h^{-1}	4400	When no macrograzers present; 5 diatoms·ciliate^{-1}·h^{-1}; 0.1% of algal standing crop (SC); 19% of daily algal productivity (P)	McCormick (1991)
		Diatoms	Ingestion of native taxa	Diatoms·cm^{-2}·h^{-1}	4640	At average macrograzer density; 6 diatoms·ciliate^{-1}·h^{-1}; ≈ 0.1% of algal SC	
		Diatoms	Ingestion of native taxa	Diatoms·cm^{-2}·h^{-1}	150	At high macrograzer density; 3 diatoms·ciliate^{-1}·h^{-1}; ≈ 0.01% of algal SC; 1% of daily algal P	

Ciliates	Salt marsh, Sapelo Is., GA	Bacteria	FLB,[a] microspheres	Bacteria·ciliate^{-1}·h^{-1}	37–421		Kemp (1988)
	Saline pond, Bahamas	Bacteria	FLB	Bacteria·ciliate^{-1}·h^{-1}	525		
Ciliate spp.	Arctic tundra pond	Bacteria	Emptying of vacuoles	Bacteria·ciliate^{-1}·h^{-1}	180–600	Ingestion rate varied with growth phase; ingestion ≈ 25% of ciliate SC·d^{-1}, ≈ 0.03% of bacterial SC·d^{-1}	Fenchel (1975)
		Algae	Emptying of vacuoles	Algae·ciliate^{-1}·h^{-1}	2–14	Ingest ≈ 15% of algivorous ciliate SC·d^{-1}; ≈ 0.03% of algal SC·d^{-1}	
Ciliate spp.	River Saale, Germany	Diatoms	Emptying of vacuoles	Diatoms·ciliate^{-1}·h^{-1}	4–100	Data for individual ciliate taxa	Schönborn (1982)
		Bacteria	Emptying of vacuoles	Bacteria·ciliate^{-1}·h^{-1}	300–4100	Data for numerous species, a few outside given range	
Ciliates	Recolonized sediments, White Clay Creek, PA	Bacteria	Change in densities	Bacteria·ciliate^{-1} h^{-1}	414–2113		Bott and Kaplan (1990)
Prorodon sp.	Tidal flat, Boston harbor, MA	Bacteria	FL coliforms[b]	Bacteria·ciliate^{-1} h^{-1}	704		Epstein and Shiaris (1992)
Chlamidodon sp.		Bacteria	FL coliforms		675		
Other ciliates		Bacteria	FL coliforms		108		
Loxodes sp.	Eutrophic pond, England	*Scenedesmus* spp.	From digestion rate	Algae·ciliate^{-1}·h^{-1}	0.4–1.3	Ingested between 0.003% and 0.68% of *Scenedesmus* SC·d^{-1}	Goulder (1972)

(continues)

TABLE I (continued)

Grazer	*Habitat*	*Food resource*	*Method*	*Units*	*Rate*	*Comments*	*Reference*
Microflagellates							
Microflagellates	Recolonized sediments, White Clay Creek, PA	Bacteria	Change in densities	Bacteria·protozoan^{-1}·h^{-1}	9–12		Bott and Kaplan (1990)
Microflagellates	Sediments, Wadden Sea	Bacteria	FLB	Bacteria·protozoan^{-1}·h^{-1}	3–12	Range of min. estimates from 4 dates	Hondeveld *et al.* (1992)
					23–46	Range of max. estimates from 4 dates	
Microflagellates	Tidal flat, Boston harbor, MA	Bacteria	FL coliforms	Bacteria·protozoan^{-1}·h^{-1}	5		Epstein and Shiaris (1992)
Microflagellates (*Bodo* sp., *Monas* sp., *Oikomonas* sp.)	Arctic tundra pond	Bacteria	Emptying of vacuoles			Ingestion ≈ 43% of microflagellate SC·d^{-1}, ≈ 8% of bacterial SC·d^{-1}	Fenchel (1975)
Rotifers							
Rotifers	White Clay Creek, PA	Diatoms	FLD[c]	Diatoms·rotifer^{-1}·h^{-1}	0, 13	Means for 2 morphotypes	Borchardt and Bott (1995)
		Bacteria	FLB	Bacteria·rotifer^{-1}·h^{-1}	237–2125	Range of means of 5 experiments	
Copepods							
Copepods	Salt marsh, SC	Bacteria	[^{14}C]glucose injection	μg C·10 cm^{-2}·h^{-1}	0.34, 0.50	Summer, winter	Montagna (1984a)

		Algae	[^{14}C]bicarbonate injection	μg C·10 cm^{-2}·h^{-1}	1.5, 1.8	Summer, winter	
Harpacticoids	Nanaimo R. estuary, BC	Bacteria	[^{3}H]glucose injection	μg C·g DM^{-1}·h^{-1}	3.3–46		Brown and Sibert (1977)
		Algae	[^{14}C]bicarbonate injection	μg C·g DM^{-1}·h^{-1}	0.37–5.1		
Harpacticoids							
Tisbe holothuriae	Lab experiments	*Skeletonema costatum*	^{14}C-labeled algae	μg C·animal^{-1}·d^{-1}	0.161		Rieper (1985)
				Algae·animal^{-1}·d^{-1}	335		
Paramphiascella vararensis		*S. costatum*		μg C·animal^{-1}·d^{-1}	0.032–0.102		
				Algae·animal^{-1}·h^{-1}	66–213		
Amphiascoides debilis		*S. costatum*		μg C·animal^{-1}·d^{-1}	0.024		
				Algae·animal^{-1}·h^{-1}	50		
T. holothuriae		*Asterionella glacialis*		μg C·animal^{-1}·d^{-1}	0.003		
				Algae·animal^{-1}·h^{-1}	12		
P. vararensis		*A. glacialis*		μg C·animal^{-1}·d^{-1}	0.001–0.007		
				Algae·animal^{-1}·h^{-1}	5–43		
T. holothuriae		Bacteria		μg C·animal^{-1}·d^{-1}	1.03–1.71	Ingestion of ciliates (*Uronema* sp.) = 0.13–0.31 μg C·animal^{-1}·d^{-1}	
		Algae		μg C·animal^{-1}·d^{-1}	0.16		
P. vararensis		Bacteria		μg C·animal^{-1}·d^{-1}	1.53–3.54	Ingestion of ciliates (*Uronema* sp.) = 0.05–0.09 μg C·animal^{-1}·d^{-1}	
		Algae		μg C·animal^{-1}·d^{-1}	0.03–0.10		
Harpacticoids (*T. holothuriae*, *P. vararensis*)	Lab study	Bacteria	Attraction to food	μg C·animal^{-1}·d^{-1}	1–3.5		Rieper (1985)
				10^{6} bacteria·animal^{-1}·h^{-1}	2.08–7.29		Rieper (1982b)

(continues)

TABLE I (continued)

Grazer	Habitat	Food resource	Method	Units	Rate	Comments	Reference
Calanoid (*Boeckella* spp.)	Lab study	*Cryptomonas* sp.	Uptake of ^{14}C-labeled algae	ng C·μg body wt^{-1}·h^{-1}	39–115		Burns and Xu (1990)
		Cyclotella sp.			20–50		
		Anabaena flos-aquae			18–30		
		A. oscillarioides			0–37		
		Nostoc sp. 2			20–50		
		N. calcicola			10–65		
Copepods	Oyster pond, France	Algae	[^{14}C]bicarbonate injection	μg C·10 cm^{-2}·h^{-1}	11.73		Blanchard (1991)
Copepods	White Clay Creek, PA	Diatoms	FLD	Diatoms·$copepod^{-1}$·h^{-1}	4	Cyclopoids 6.5, nauplii 4	Borchardt and Bott (1995)
		Bacteria	FLB	Bacteria·$copepod^{-1}$·h^{-1}	239–907	Range of means from 5 experiments	
Harpacticoids (*Tisbe holothuriae, T. battagliai, T. furcata*)	Lab studies	*Dunaliella tertiolecta*	Uptake of ^{14}C-labeled algae	μg C·$copepod^{-1}$·d^{-1}	0.10–0.78	Based on range of assimilation values (0.05–0.39) and assimilation efficiency of 20–50%	Vanden Berghe and Bergmans (1981)
Harpacticoids	Tidal flat, Boston harbor, MA	Bacteria	FL coliforms	Bacteria·$copepod^{-1}$·h^{-1}	42		Epstein and Shiaris (1992)
Nauplii		Bacteria	FL coliforms	Bacteria·$copepod^{-1}$·h^{-1}	33		
Harpacticoids (*Attheyella* spp.)	Hugh White Creek, NC	Bacteria	Change in cell densities	μg bacterial C·$copepod^{-1}$·d^{-1}	0.03–0.47	Range of 3 experimental averages	Perlmutter and Meyer (1991)

Nematodes							
Nematodes	White Clay Creek, PA	Diatoms	FLD	Diatoms·nematode^{-1}·h^{-1}	0		Borchardt and Bott (1995)
		Bacteria	FLB	Bacteria·nematode^{-1}·h^{-1}	6–320	Range of means from 5 experiments	
Nematodes	Salt marsh, SC	Bacteria	[^{14}C]glucose injection	μg C·10 cm^{-2}·h^{-1}	0.32, 0.43	Summer, winter	Montagna (1984a)
		Algae	[^{14}C]bicarbonate injection	μg C·10 cm^{-2}·h^{-1}	3.6, 7.1	Summer, winter	
Chromadorita tenuis	Lab study	*Nitzschia palea*	Ingestion of diatoms	μg algae·female^{-1} d^{-1}	0.9–3.3		Jensen (1984)
Chromadora macrolaimoides	Lab study	Algae	Ingestion of ^{32}P-labeled foods	No.·animal^{-1}·h^{-1}	0.5–1750		Tietjen and Lee (1973)
		Bacteria			<1–2.2		
Nematodes	Oyster pond, France	Algae	[^{14}C]bicarbonate injection	μg C·10 cm^{-2}·h^{-1}	22.2		Blanchard (1991)
Plectus palustrus	Lab study	Bacteria	Ingestion of ^{14}C-labeled *Acinetobacter* sp.	No.·animal^{-1}·h^{-1}	300,000		Duncan *et al.* (1974)
Diplogasteritus nudicapitatus	Lab study	Mixed bacteria	Pumping rates, bacterial density, lumen volume	μg dw·bacteria·worm^{-1}·d^{-1}	0.07–0.36		Woombs and Laybourn-Parry (1984)
Rhabditis curvicaudata		Mixed bacteria		μg dw·bacteria·worm^{-1}·d^{-1}	0.01–0.41		
Eudiplogaster sp.	Mudflat, Eems-Dollard estuary	*Navicula pygmaea*	Ingestion of ^{14}C-labeled cells	Diatoms·nematode^{-1}·h^{-1}	0.91–1.8		Admiraal *et al.* (1983)
		N. salinarum			4.0–7.5		

(continues)

TABLE I (continued)

Grazer	Habitat	Food resource	Method	Units	Rate	Comments	Reference
Metoncholaimus sp.	Tidal flat, Boston harbor, MA	Bacteria	FL coliforms	Bacteria·nematode^{-1}·h^{-1}	525		Epstein and Shiaris (1992)
Other nematodes					21		
Pelodera chitwoodi	Lab study	Bacteria	Calculation from energetic measurement	Bacteria·nematode^{-1}·h^{-1}	16,146	Average for males	Mercer and Cairns (1973)
					52,083	Average for females	
Oligochaetes							
Amphichaeta sp.	White Clay Creek, PA	Diatoms	FLD	Diatoms·worm^{-1}·h^{-1}	8–11	Range of means from 4 experiments	Borchardt and Bott (1995)
		Bacteria	FLB	Bacteria·worm^{-1}·h^{-1}	0–316	Range of means of 5 experiements	
Chaetogaster sp.		Diatoms	FLD	Diatoms·worm^{-1}·h^{-1}	9	1 experiment	
Other oligochaetes		Bacteria	FLB	Bacteria·worm^{-1}·h^{-1}	83–939	Range of means of 5 experiments	
Limnodrilus spp.		Bacteria	Fluroescent microspheres	Bacteria·worm^{-1}·h^{-1}	11,600		Takada *et al.* (1994)
Branchiura sowerby					116,600		
Chironomid Larvae							
Chironomids	White Clay Creek, PA	Diatoms	FLD	Diatoms·larva^{-1}·h^{-1}	11		Borchardt and Bott (1995)
		Bacteria	FLB	Bacteria·larva^{-1}·h^{-1}	0–468	Range of means of 5 experiments	

Isopod							
Lirceus sp.	Ogeechee River, GA	Bacteria	Uptake of ^{3}H-labeled cells	ng bacterial C·mg wet wt^{-1}·h^{-1}	2.3–17.9		Findlay *et al.* (1984)
Ostracods							
Ostracods	Salt marsh, SC	Bacteria	[^{14}C]glucose injection	μg C·10 cm^{-2}·h^{-1}	0.06, 1.24	Summer, winter	Montagna (1984a)
		Algae	[^{14}C]bicarbonate injection	μg C·10 cm^{-2}·h^{-1}	2.7, 4.4	Summer, winter	
Mixed Meiofauna							
Meiofauna	White Clay Creek, PA	Unicellular diatoms	Change in densities	Diatoms·animal^{-1}·h^{-1}	3–37	Ciliates and rotifers codominants	Borchardt and Bott (1995)
		Melosira varians	Change in densities	Diatoms·animal^{-1}·h^{-1}	128		
		Bacteria	Change in densities	Bacteria·animal^{-1}·h^{-1}	3513–8981	Ciliates and rotifers codominants	
Meiofauna	Salt marsh, SC	Bacteria	[^{14}C]glucose injection	μg bacterial C·animal^{-1}·d^{-1}	0.042, 0.53	Summer and winter values for copepods, nematodes, and ostracods	Montagna (1984a)
		Algae	[^{14}C]bicarbonate injection	μg algal C·animal^{-1}·d^{-1}	0.53, 0.29	Summer and winter values for copepods, nematodes, and ostracods	

[a]FLB = ingestion of fluorescently labeled bacteria.
[b]FL coliforms = ingestion of fluorescently labeled coliforms.
[c]FLD = ingestion of fluorescently labeled diatoms.

(Borchardt and Bott, 1995). Planktonic microflagellates ingested algae at rates in the range of <1–15 cells·h^{-1} (E. B. Sherr *et al.,* 1991), but to date there are no reported algivory rates for benthic microflagellates for comparison. Thirteen taxa representing the Dinophyceae, Chrysophyceae, and Cryptophyceae all supported growth of 5 freshwater ciliates in single-species diets, 15 representatives of Euglenophyceae, Bacillariophyceae, and Chlorophyceae were of variable food value, and cyanobacteria did not support the growth of any of the ciliates (Skogstad *et al.,* 1987). A study of the feeding habits of 18 benthic ciliate species showed that (1) some taxa ingested diatoms and dinoflagellates preferentially, although a few preferred bacteria and heterotrophic flagellates, (2) ciliate abundance and community composition were shaped in part by competition for food, and (3) the effect on diatom community structure was more pronounced than that on total densities, since only 4 of 42 species were extensively grazed (Epstein *et al.,* 1992).

Algae larger than the protozoa can be ingested, sometimes in considerable numbers. Greatly distended ciliates were observed in samples of the WCC epibenthos (Fig. 2), with one containing nearly 50 diatoms. McCormick (1991) reported that individuals of three benthic taxa (two ciliate, one sarcodine) contained from 2 to 139 diatoms and specimens of one of those genera, *Chilodonella,* contained 70–80 diatoms in another study (Sawicka *et al.,* 1983). Even microflagellates are capable of ingesting larger algae, for example, *Paraphysomonas* sp. ingested diatoms (*Synedra* sp.) six times longer than the protozoan mean diameter (Suttle *et al.,* 1986).

There are only a few studies of protozoan bacterivory in sediments, although numerous measurements have been made for planktonic protists found in oceans, lakes, and large rivers. Available ingestion rates for benthic ciliates (Table I) are at the low to upper-middle range of data available for planktonic forms [200–5000 bacteria·ciliate^{-1}·h^{-1}; E. B. Sherr and Sherr (1987), Sanders *et al.* (1989), Carlough and Meyer (1990), Barcina *et al.* (1991)]. The rates reported for benthic ciliates by Kemp (1988) were approximately 10-fold lower than the rates reported by Bott and Kaplan (1990), but they are similar if only ciliates in the same size range are compared between studies. Data for benthic microflagellates (Table I) are limited to the studies of Bott and Kaplan (1990) and Hondeveld *et al.* (1992), who reported average cell-specific ingestion rates of 12 and 15 bacteria·h^{-1}, respectively, at 20°C. These values are at the low end of the ingestion rates reported for planktonic forms [9–266 bacteria·microflagellate^{-1}·h^{-1}; Fenchel (1986), Sanders *et al.* (1989), Carlough and Meyer (1990), Barcina *et al.* (1991), Weisse and Scheffel-Möser (1991)].

Selective feeding has been documented for some protozoa. Although total *Scenedesmus* densities were essentially unaffected by ciliates, *Loxodes magnus* appeared to differentially ingest particular species (Goulder, 1972). Some planktonic ciliates and microflagellates grazed attached bacteria

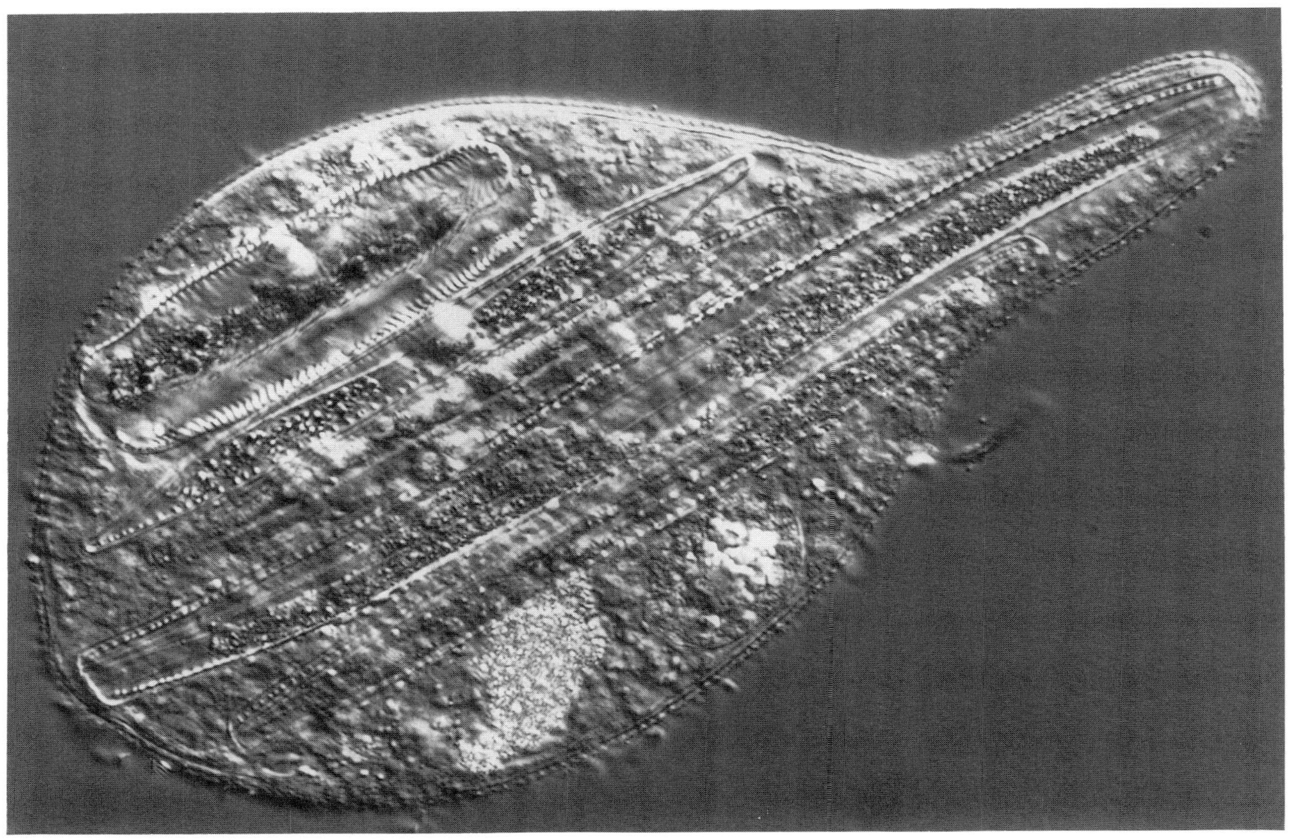

FIGURE 2 An unidentified ciliate (≈60 μm in diameter when undistended) from White Clay Creek, Pennsylvania, with several ingested diatoms viewed under Hoffman interference contrast microscopy. Ingested diatoms were identified as *Pinnularia* sp. (top) and *Nitzschia* spp. (four lower specimens).

whereas others preferred freely suspended cells (Albright *et al.*, 1987; Caron, 1987). Ciliates discriminate between particles in part by mouth size and morphology (Patterson *et al.*, 1989). Small ciliates (≤30 μm long) in natural communities consumed primarily picoplankton (bacteria and cyanobacteria 0.2–2 μm) whereas larger ciliates selected nanoplankton (algae and protozoa 2–20 μm) (Jonsson, 1986; Rassoulzadegan *et al.*, 1988; Dolan, 1991). However, cell size affected diatom selection only for the smallest ciliate taxon studied by McCormick (1991) . The finer the filtering apparatus, the more capable the taxon can be of significant bacterivory (Porter *et al.*, 1979; E. B. Sherr and Sherr, 1987), even at low bacterial densities (E. B. Sherr *et al.*, 1989). Some benthic ciliates discriminated between bacterial cell shapes, consuming rod-shaped fluorescently labeled coliforms at 74–155 times the rate they consumed coccus-shaped cells (Epstein and Shiaris, 1992).

Size-selective feeding has been reported even for microflagellates, although studies are restricted to the plankton. Estuarine planktonic microflagellates (5–10 μm) primarily ingested *Nannochloris* sp. (2 μm, 3.4 μm^3), whereas ciliates (15–60 μm) had lowest clearance rates on *Nannochloris* when compared to *Chlorella* sp. (3.4 μm, 20 μm^3) and *Thalassiosira* sp. (5.4 μm, 80 μm^3) (E. B. Sherr *et al.*, 1991). Microflagellates in four size classes between 21 and 109 μm^3 preferred "large" bacteria (0.8–1.2 μm^3) and selected them even when they were relatively rare in the bacterial community (Chrzanowski and Simek, 1990). The microflagellate *Ochromonas* sp. preferred cells >0.2 μm^3, and as protozoan densities in the northern Baltic Sea increased during the summer months, the median bacterial cell volume decreased (Andersson *et al.*, 1986). Another microflagellate, *Paraphysomonas imperforata,* reduced its size in response to smaller prey items, precluding the ingestion of larger algae, and attempted to maintain a predator : prey radius ratio of 2:1 (Goldman and Dennett, 1990), although other researchers have argued for other ratios. Still others (Monger and Landry, 1991) have claimed that flagellate feeding was only weakly size selective. In addition to size, food selection among protozoa can involve chemosensory cues related to cell-wall composition and the production of extracellular gums and antagonistic compounds, as reviewed for planktonic taxa by Verity (1991). Similar detailed studies are lacking for benthic protozoa, but it is reasonable to expect similar factors to affect interactions there as well.

B. Rotifers

The size and structure of the ciliary apparatus and the mastax (a crushing, grasping, trapping, sucking or pumping organ) determine the size and kind of foods usable by rotifers. Some taxa appear to be exclusively or predominantly algivorous, for example, species of *Epiphanes, Notholca,*

Rhinoglena, and *Euchlanis,* sometimes with specialization on particular groups of algae. Other rotifers use a wider range of resources, including bacteria, detritus, and other animals (Pourriot, 1977). *Epiphanes, Euchlanis,* and *Lecane* (the latter predominantly bacterivorous) were identified in WCC epibenthic samples.

To my knowledge, the only algivory rates for rotifers in benthic communities are our data for those in the WCC epibenthos. Diatoms were ingested by one morphotype but not by another (Table I). In other experiments showing significant total meiofaunal grazing pressure on benthic diatoms, rotifers (along with ciliates) were predominant (Borchardt and Bott, 1995). Grazing experiments conducted *in situ* showed alteration of algal densities, species composition, and size distribution by planktonic rotifers (Scheda and Cowell, 1988). Small (217–223 μm long) and large (276–305 μm long) specimens of *Euchlanis dilatata lucksiana* ingested filamentous cyanobacteria and a morphologically similar prochlorophyte (150–250 μm long), but at slower rates than that for ingestion of lake seston <33 μm (Gulati *et al.,* 1993). Sierszen (1990), also working with planktonic taxa, concluded that rotifers grazed nonselectively, utilizing locally abundant specimens. In contrast, six species of the benthic genus *Notommata* each grazed a different narrow spectrum of algae (and other foods for two predatory species) (Pourriot, 1977).

One study suggested that between 10 and 40% of the rotifer diet could be heterotrophic microorganisms (large bacteria, heterotrophic microflagellates, and small ciliates) (Arndt, 1993). Filter-feeding taxa fed on bacteria, nano- and microflagellates, and small algea and ciliates (0.5–20 μm). Grasping taxa such as synchaetids ingested organisms 1–10 μm long (although they could be much larger) and asplanchnids utilized foods >10–15 μm long, mostly algae and protozoa. Other work showed that *Brachionus quadridentatus* preferentially ingested Chlorophyceae and did not ingest bacteria (*Pseudomonas* sp.) or detritus, whereas *B. plicatilis,* although preferring algae, ingested all food sources (Hlawa and Heerkloss, 1994). Bacterivory rates for two morphologically different WCC benthic taxa were substantially different (Table I) but fell into the wide range (22–5000 bacteria·rotifer^{-1}·h^{-1}) reported by Sanders *et al.,* (1989) and Ooms-Wilms *et al.,* (1993) for planktonic taxa.

C. Copepods

Harpacticoid copepods have been grouped into four feeding groups: deposit feeders, and those that pick items from surfaces, scrape the edges of particles, or sweep food into their mouths (Marcotte, described in Hicks and Coull, 1983). In WCC, benthic copepods ingested ≈4 diatoms·individual^{-1}·h^{-1} and ≈300–900 bacteria·individual^{-1}·h^{-1} (Table I). Selective feeding among planktonic taxa was reported nearly a century ago when

Birge (1898) observed that *Diaptomus* sp. rejected and *Cyclops* sp. selected the dinoflagellate *Ceratium* sp. Some planktonic cyclopoids were described as predominantly "vegetarian" (Fryer, 1957) and the use of diatoms by three benthic harpacticoids, a particularly numerous group in sediment, was documented by Decho (1988). In laboratory studies, calanoids (Burns and Xu, 1990) and harpacticoids (Rieper, 1985) differentiated between algal species, presumably on the basis of food value or cell morphology. Using high-performance liquid chromatography (HPLC) analyses of carotenoid pigments, Swadling and Marcus (1994) reported both selective and nonselective ingestion of algae by different life stages of a calanoid copepod in laboratory experiments and the field. Other laboratory studies with three co-occurring species of *Tisbe* showed that neither radiolabeled bacteria nor algae were selected exclusively, although *T. furcata* preferred bacteria (Vanden Berghe and Bergmans, 1981), as also noted by Brown and Sibert (1977).

Differentiation between algae and bacteria in natural communities was shown experimentally by injecting radiolabeled acetate or bicarbonate into marine sediment cores. *Thompsonula hyaenae* preferred algae, *Halicyclops coulli* preferred bacteria, whereas *Zausodes arenicolus* fed on both (Carman and Thistle, 1985). Follow-up autoradiographic studies with *Z. arenicolus,* however, indicated that although radiolabel from [^{14}C]bicarbonate was concentrated in the copepod guts, indicating ingestion of radiolabeled algae, the label from [^{14}C]acetate was surface associated, indicating uptake by epicuticular microbes (Carman, 1990). Decho and Castenholz (1986) investigated both the ingestion and digestion of algal and bacterial foods and related findings to copepod distributions in a tidal marsh. High densities of *Leptocaris brevicornis* were correlated with dense growths of diatoms. The organism ingested diatoms and associated heterotrophs, but assimilated only heterotrophic carbon and egested intact diatoms. *Oscillatoria* sp. was not ingested. The distribution of *Mesochra lilljeborgi* was related in part to the purple sulfur bacterium *Thiocapsa* sp. *Mesochra lilljeborgi* ingested *Spirulina* sp., *Thiocapsa* sp., and associated heterotrophs, but assimilated only the label from *Thiocapsa* sp. The authors suggested that selective digestion may accomplish the same end as selective ingestion when that becomes energetically costly. Diet shifts with copepod life stage also have been reported. Adult harpacticoid *Nitocra lacustris* ingested and assimilated diatoms, passively ingesting adhering bacteria in the process. Naupliar stages did not ingest diatoms but they used as food the bacteria scraped from the mucus outside diatoms and probably the mucus itself (Decho and Fleeger, 1988b).

Two of the four taxa studied by Rieper (1982a) discriminated between bacterial species, usually selecting gram-negative organisms and avoiding a gram-positive *Micrococcus* sp. Rieper (1978) used *Tisbe holothuriae* and *Paramphiascella vararensis* as representative planktonic and benthic species, respectively. The planktonic copepod grazed bacteria offered in suspension to a greater extent than did the benthic taxon, although both species fed selectively, and grazing rates were directly proportional to bacterial densities. In leaf litter accumulations in a stream, the harpacticoid *Attheyella* spp. selectively grazed larger rod-shaped bacteria (reducing mean bacterial cell size by 17–30%) and bacterial productivity rates increased (Perlmutter and Meyer, 1991). The reported grazing rates were similar to those reported for marine taxa by Vanden Berghe and Bergmans (1981).

Behavioral and physiological mechanisms for selectivity have been explored for some copepods, mostly planktonic species. Some copepods, for example, *Diaptomus* sp., are able to ingest filamentous cyanobacteria and diatoms by biting off pieces from the ends of trichomes. The cladoceran *Daphnia* cannot do this and must clear particles from its food groove after indiscriminant filter feeding (Schaffner *et al.*, 1994). Chemoreceptors for selective feeding have been located on the mouthparts of *Diaptomus pallidus,* a calanoid taxon (Friedman and Strickler, 1975). Discrimination between nutritionally rich and poor cells within a single diatom species has been noted (Cowles *et al.*, 1988). Diversity in the calanoid copepod diet, reviewed by Kleppel (1993), ensures acquisition of essential nutrients.

D. Nematodes

Jensen (1987) assigned free-living nematodes to four feeding groups: nonselective deposit and particulate surface feeders, predators, scavengers, and DOM utilizers. Some nematodes have mouthparts especially adapted for the ingestion of diatoms, such as a tooth to pierce the frustule (Jensen, 1982). Algivory and bacterivory rates are shown in Table I. The nematodes observed in WCC were not significant herbivores and showed significant bacterivory in only one of five experiments (Borchardt and Bott, 1995).

Nematodes consumed numerous algal and bacterial taxa in laboratory studies, but only two diatom species (with associated bacteria) supported growth of *Chromadora macrolaimoides* in single-species diets, implying that dietary diversity sustains natural populations (Tietjen and Lee, 1973). In other experiments, one marine species preferred diatoms over bacteria, another preferred bacteria over diatoms, and a third did not selectively ingest either bacteria or diatoms although it did prefer *Synedra tabulata* over other diatoms (Trotter and Webster, 1984). *Eudiplogaster pararmatus* discriminated between species within a genus, consuming ^{14}C-labeled *Navicula salinarum* at a fourfold faster rate than *N. pygmaea* (Admiraal *et al.*, 1983). *Eudiplogaster pararmatus* required a high threshold density of diatoms for feeding and thus was numerous only in patches where diatom blooms occurred, revealing an ability to differentiate between diatoms and the detritus that was uniformly distributed throughout the mudflat.

Adults of *Plectus palustris* ingested 5000 bacteria·min^{-1} (650% of nematode body weight per day) when fed radiolabeled cultured bacteria (*Acinetobacter* sp.) but had a low assimilation efficiency on that species (Duncan *et al.*, 1974). The distribution of two nematode taxa were interpreted on the basis of physiological traits and *in situ* food abundances (Schiemer, 1983). *Plectus palustris* (with longer reproductive period, slower growth and metabolic rates, greater tolerance of suboptimal conditions) was found in habitats with low, but perhaps more predictable, bacterial biomass. *Caenorhabditis briggsae* (with rapid growth and reproductive rates and ability to withstand harsh conditions) occurred in nutritionally rich but fluctuating habitats. Some nematodes alter behavior to maximize feeding efficiency. For example, *Rhabditis curvcaudata,* found in habitats with a wide range of bacterial densities, responded to a reduction in food supply by reducing its pumping rate and expending effort toward foraging. In contrast, *Diplogasteritus nudicapitatus,* found only in "food-rich" habitats with high bacterial densities, had a higher growth and reproduction rate and pumped at the same rate at all food densities because there was no need to possess a feeding strategy to compensate for unstable food supply (Woombs and Laybourn-Parry, 1984).

Blanchard (1991) reported that nematodes ingested algae two times faster than did copepods (three times faster on the basis of body weight equivalent) and concluded either that copepods had a higher assimilation/ingestion ratio or that other foods (e.g., bacteria) were a more important component of their diet. Austen (1989), however, claimed that copepods were less flexible than nematodes in resource acquisition.

E. Oligochaetes and Chironomid Larvae

Oligochaetes and chironomids were important diatom grazers in the WCC epibenthos (Table I), especially the oligochaete *Amphichaeta* sp. (Borchardt and Bott, 1995). *Amphichaeta* species were considered to be significant algivores in other studies. *Amphichaeta leydigii* fed almost exclusively on diatoms in the benthos of an Italian lake (Mastrantuono, 1988), as did *A. sannio* (along with *Paranais littoralis,* another oligochaete) in a tidal mudflat (Bouwman, in Admiraal *et al.,* 1983). Schönborn (1984) determined that diatoms made up ≈55% of the diet of another oligochaete, *Chaetogaster diastrophus,* in the River Saale (Germany).

Chironomid larvae of meiofaunal dimensions were considered to be significant grazers of epiphytic algae (Botts, 1993). Algal taxa used as food (*Cosmarium* spp. and *Aphanocapsa* spp.) or for case construction (*Lyngbya* sp., *Bulbochaete* spp., and *Oedogonium* spp.) all increased in biovolume when animals were removed in microcosm experiments, although total algal biovolume and the biovolume of large algae did not change. In other exclosure experiments, oligochaetes and chironomids (along with cladocerans) replaced snails as grazers of attached algae and algal cell size was reduced (Cattaneo and Kalff, 1986), results similar to those obtained with plankton communities when large grazers were removed.

In WCC, both of these groups also grazed bacteria, oligochaetes being more important (Table I). Earlier work focusing on three tubificid oligochaetes *(Tubifex tubifex, Limnodrilus hoffmeisteri,* and *Peloscolex multisetosus)* and eight bacterial species provided little evidence for selective ingestion of bacteria, except in one instance involving an *Aeromonas* sp., but suggested that differential digestion of ingested bacteria allowed for coexistence of the oligochaetes in sediments (Brinkhurst and Chua, 1969; Wavre and Brinkhurst, 1971). Ingestion rates for three other oligochaete taxa based on ingestion of fluorescent microspheres were much higher, and ingestion rates varied with size and density of food particles and temperature.

V. USE OF OTHER FOOD RESOURCES, INTRAGUILD PREDATION, AND LINKS TO HIGHER ORGANISMS

Despite demonstrated dietary preferences, few, if any, protozoa or meiofauna taxa are strictly algivorous or bacterivorous and other foods, for example, detritus (Meyer-Reil and Faubel, 1980; Couch, 1989), DOM (Brinkhurst and Chua, 1969; Montagna, 1984b; Jensen, 1987; Patterson *et al.,* 1989), fungi (Spaull, 1973), and other animals (Coull, 1973; Rieper and Flowtow, 1981), may also be used. More research is needed to establish the relative importance of these other resources in microconsumer diets *in situ,* but predatory interactions present a special case. Two consumers that use the same resource belong to the same feeding guild, for example, algivores and bacterivores. However, if one member of a guild can ingest another [intraguild predation *sensu* Polis and Holt (1992)], a competitive feeding interaction is made more complex and the pathway of energy flow is lengthened.

Two of many possible examples of intraguild predation in a microscopic food web are when (1) heterotrophic microflagellates and ciliates graze bacteria and algae, but some ciliates also capture microflagellates (Jonsson, 1986; Weisse, 1990; Epstein *et al.,* 1992) and other ciliates (Fenchel, 1968), and (2) the oligochaete *Chaetogaster* sp. and large ciliates graze algae, but *Chaetogaster* also ingests large ciliates (Schönborn, 1984). Such carnivory can be significant, for example, microflagellates made up 27% of the ration of one benthic ciliate species (Epstein *et al.,* 1992), *Chaetogaster* consumed ≈72% of annual ciliate production (Schönborn,

1984), and marine harpacticoid copepods *(Tisbe holothuriae)* ingested ciliates at a rate of 12–192 ciliates·copepod^{-1}·h^{-1} (Rieper and Flotow, 1981). Again, feeding selectivity has been shown, for example, the rotifers *Brachionus rebens* and *B. angularis* selected the microflagellates *Bodo* and *Spumella* over the alga *Scenedesmus* sp., and distinguished between microflagellates of the same size but with different behavior (Arndt, 1993). Epstein and Gallagher (1992) found that total benthic ciliate densities were usually unaffected by meiofauna, but because some species were grazed more heavily than others, community structure was altered. Cascading effects in benthic communities were noted in experiments in which either rotifers or meiofaunal assemblages reduced ciliate and sarcodine densities (along with dominant algal taxa), through either competition or ingestion, which resulted in increased densities of heterotrophic microflagellates (McCormick and Cairns, 1991).

Links between meiofauna and predatory macroinvertebrates (Sherberger and Wallace, 1971; Benke and Wallace, 1980; Hildrew and Townsend, 1982), juvenile fish, crabs, shrimp, polychaetes and gastropods (Warwick, 1987; Telesh, 1993), and even waterfowl (Gaston, 1992) have been shown. Schönborn (1992) traced three main flows of energy in River Saale: (1) bacteria and microalgae → ciliates → *Chaetogaster diastrophus* → *Erpobdella octoculata* (a leech); (2) detritus and associated bacteria → *Nais* (a detritivorous naidid oligochaete) → tubificids → chironomids → *Erpobdella* and fish; and (3) diatoms, other microbenthos, and detritus → *Hydropsyche angustipennis* (Trichoptera, Insecta) → fish. Other field studies demonstrated that small epibenthic macrofauna could control copepod, although not nematode, densities in a marine subtidal mud (Olafsson and Moore, 1990).

Although there are exceptions, for example, protozoan plasticity, interactions between planktonic taxa often are size-dependent (Capblancq, 1990). Maximum efficiency is gained when microconsumers graze organisms only slightly smaller in size, as when ciliates (Jonsson, 1986) and *Daphnia pulex* (Jürgens *et al.,* 1994) grazed microflagellates at higher rates than bacteria. Although size selectivity is likely to occur in benthic communities as well, Strayer (1991) suspected that it could be more difficult to detect there because numerous refugia and a more diverse fauna are present. Size-dependent feeding maximizes efficiency at the initial step of energy intake, but several size-dependent grazing steps will reduce the overall efficiency of energy transfer to macroconsumers owing to respiration losses at each step. Given their larger size, fewer links are likely to be involved in the transfer of energy from algae to macroorganisms than from bacteria to macroorganisms.

Some grazers, for example, *Daphnia* species, ingest particles ranging in size from bacteria to meiofauna and in turn are consumed by fish. This makes *Daphnia* a potentially important conduit of energy flow (Porter *et al.,* 1988, Jürgens *et al.,* 1994). Some organisms capable of ingesting significant numbers of bacteria in benthic communities might function similarly, for example, *Simulium* spp., *Stenonema* spp., and some meiofauna [the copepod genus *Attheyella* spp., which removed between 1 and 22% of bacterial production in leaf litter per day, Perlmutter and Meyer (1991)].

VI. IMPACT ON PRODUCTION AND STANDING CROPS

There is ample evidence that algae can be an important component of the diets of many microconsumers. It is less clear whether microconsumers can remove significant amounts of algal biomass and/or productivity. In two experiments, WCC meiofauna ingested 0.8 and 2.4% of concurrently measured algal productivity. In two other experiments, grazing-enforced generation times were 1.3 and 5.7 d. Those data equal or approach accepted doubling times for diatoms *in situ* (1.5–3 d) and imply that microconsumers (*Amphichaeta,* large ciliates, and rotifers were predominant) ingested large percentages of diatom productivity at those times. Micrograzers had much less of an impact in two other experiments (Borchardt and Bott, 1995).

In other studies, (1) meiofaunal grazing slightly exceeded benthic algal productivity, suggesting that meiofauna could be food-limited (Blanchard, 1991); (2) meiofauna consumed ≈10% of daily algal productivity in a mudflat in a North Sea estuary (Admiraal *et al.,* 1983); and (3) benthic metazoan meiofauna exerted an estimated food demand amounting to 27% of annual primary production in a sub-Antarctic pond (Bouvy, 1988). Epstein *et al.,* (1992) estimated that ciliates grazed 10% of daily algal productivity in a mud flat during summer, although the percentages of productivity of different algal types ranged from 6% for diatoms to 93% for dinoflagellates.

Others have expressed ingestion as a proportion of algal standing crop. Fenchel (1975) estimated that meiofauna ingested 1.4%, and herbivorous ciliates 0.03%, of the algal standing crop per day at 12°C in an Arctic tundra pond. Meiofauna diatom ingestion in the North Sea estuarine mudflat studied by Admiraal *et al.,* (1983) expressed as a proportion of the algal standing crop was ≈5–10%·d^{-1}. Nematodes and meiofaunal polychaetes ingested an estimated 1% of diatom standing crop·per hour in salt marsh sediments, presumably significantly impacting benthic periphyton, with the caveat that if diatoms survived gut passage the effect of grazing would be reduced (Montagna, 1984a). During the late spring, algivorous protozoa grazed an estimated 0.1% of the benthic diatom standing crop per hour at the mean density of macrograzers in Wilson Creek, Virginia, but only 0.01% at higher macrograzer densities; and from 19% of diatom daily productivity in the absence of macrograzers to 1% at elevated densities (see Table I; McCormick, 1991), although these data were considered overestimates since export was not accounted for in their computation.

These diverse findings were based on measurements made at one or a few time(s) of the year and are not easily generalized. However, it appears that on occasion, microconsumers can ingest a significant proportion of algal productivity, and a smaller proportion of total algal biomass. Montagna (1996), on the other hand, after reviewing several studies, concluded that metazoan meiofauna could potentially ingest 1% of algal standing crop per hour worldwide.

Macrofauna are unlikely to be significant bacterivores when judged by the proportion of bacteria in the diet (Seki, 1969) or calculated amounts of productivity reaching higher trophic levels (Kemp, 1987, 1990). Some insect larvae, for example, *Simulium* sp. (Wotton, 1980; Edwards and Meyer, 1987) and *Stenonema* sp. (Edwards and Meyer, 1990), appear to be exceptions. Some metazoan meiofauna have impressive bacterivory rates. However, their low field densities in some habitats reduced the probability that they could have a significant impact on bacteria *in situ* (Epstein and Shiaris, 1992; Borchardt and Bott, 1995). In contrast, Montagna (1996) calculated that, as for algae, metazoan meiofauna could ingest 1% of bacterial standing crop per hour. Thus, the question of the impact of meiofauna on bacteria remains open.

Ciliates were considered to be an important link from bacteria to metazoa in some planktonic food webs (Porter *et al.,* 1979; E. B. Sherr *et al.,* 1989; Arndt *et al.,* 1990). Although Epstein *et al.* (1992) estimated that ciliates grazed 11% of hourly bacterial production in a tidal flat, Kemp (1988) did not view them as an important link from bacteria to benthic metazoa since predicted ciliate bacterivory amounted to <4% of hourly bacterial production. The primary consumers of bacteria appear to be the microflagellates. Using the annual protozoan bacterivory rate calculated by Bott and Kaplan (1990), protozoa would ingest over half of the current estimate of annual benthic bacterial productivity in WCC (37 g $C{\cdot}m^{-2}{\cdot}yr^{-1}$). Despite a lower cell-specific ingestion rate, microflagellates accounted for 67% of bacterial ingestion because of their greater *in situ* densities. Fenchel (1975) also concluded that microflagellates were the more important bacterivores in an Arctic tundra pond. Even so, because of low densities, Epstein and Shiaris (1992) estimated that microflagellates and ciliates would remove only 0.2 and 0.1% of the bacterial standing stock per day, respectively.

Kemp (1990) concluded that the amount of bacterial productivity reaching macrofauna via meiofaunal bacterivory would amount to no more than 20% of bacterial productivity. This perspective is consistent with other work showing that <2% of bacterial carbon was transferred to zooplankton (Ducklow *et al.,* 1986) and that only 18% of bacterial productivity would be available to larger zooplankton after consumption by flagellates and ciliates (Jonsson, 1986). Pomeroy and Wiebe (1988) discussed the flow of energy to macroorganisms through a planktonic microbial food web and concluded that theoretically 12 or 6% of the photosynthate utilized by bacteria (assumed to be 50%) might be passed on to macroorganisms depending on whether three or four steps were involved, respectively. Similar assessments of the transfer of energy from algae through microconsumers to macroconsumers are unavailable.

VII. NUTRIENT REGENERATION

It is important to note that even though microconsumer grazing represents a loss of resource for herbivorous macroconsumers, the negative effect could possibly be offset by regeneration of nutrients that support primary productivity. For example, although grazing zooplankton reduced phytoplankton densities, nitrogen regeneration increased algal reproduction rates sufficiently to compensate for the loss (Sterner, 1986). The ingestion of bacteria by protozoa releases significant amounts of remineralized nutrient (Johannes, 1965), and Caron and Goldman (1990) document the current perspective concerning the role of protozoa in nutrient regeneration. For example, in laboratory studies, P-limited algae grew only when the bacterivorous microflagellate *Spumella* sp. was present to release bacterial P (Rothhaupt, 1992). Algivorous microflagellates, for example, *Paraphysomonas imperforata,* would regenerate nutrients without bacterial involvement. In one study, 80% of ingested algal C was released as CO_2 and the remainder as particulate organic carbon (POC) and dissolved organic carbon (DOC) (Caron *et al.,* 1985). Meiofaunal grazing can also increase nutrient regeneration through bioturbation effects, for example, nematodes increased decomposition rates of some detrital substrates (Alkemade *et al.,* 1992; Findlay and Tenore, 1982) although not all (Tietjen and Alongi, 1990). The alteration of physical and chemical gradients will be most pronounced where physical disturbance from tides or stream flow is minimal.

VIII. CONCLUSIONS, IMPLICATIONS, AND RESEARCH NEEDS

Kemp (1990) observed that the role of microbes as food in plankton communities was considered mostly from the perspective of the heterotrophic nanoflagellates and in benthic food webs from the perspective of food for macrofauna, and that the study of the microscopic food web was in its infancy. Nevertheless, it is clear that complex food webs exist among microconsumers, just as they do among macroorganisms. Other factors, not considered here, that would introduce even more complexity are, for example, mixotrophic protists (Bird and Kalff, 1986) or viral infestations of natural populations of bacteria and cyanobacteria (Proctor and Fuhrman, 1990; Murray and Eldridge, 1994). Some ingestion rate data and other quantitative information regarding freshwater benthic microbial food webs are available, but most data come from planktonic or marine intertidal systems. This discussion has emphasized ingestion as the link between microconsumers and resources, but it is important to recognize that the food value of each item depends on its assimilation and that factors such as consumer growth efficiency will have effects on the structure of the food web.

Microconsumers appear to be capable of consuming significant amounts of algal primary productivity on occasion and are capable of shaping community structure through selective ingestion of particular taxa. Thus, microconsumers can have some of the same effects on the algal productivity and species composition that macroconsumers have (Hildrew, 1992). The relative importance of the microconsumers will be related to the densities and species composition of both microconsumers and algae, as well as macrograzer densities, nutrients, and physical factors such as light, temperature, sediment properties, and disturbance (e.g., tides and storms).

We have a poor understanding of the flow of energy through microconsumers and of the interface between microconsumers and macrograzers. Even our understanding of algae as a resource for macroconsumers (see Steinman, Chapter 12, and Lamberti Chapter 17, this volume) is based on studies with larger animals that are easily manipulated. Food resources for early instars of insect larvae (temporary meiofauna) as well as the juvenile forms of other taxa are poorly known. No one has explicitly examined the relative importance of benthic microbial food webs over a trophic gradient or a range of sediment organic matter conditions.

As for planktonic microbial food webs, both comparative and experimental approaches are needed in addressing these research questions (Pace, 1991), and many of the suggestions made by Porter *et al.*, (1988) and Power (1992) for the study of planktonic microscopic food webs and herbivory, respectively, should be incorporated into future studies of benthic microscopic food webs. Measurements made *in situ* and field and mesocosm manipulation experiments are needed in which careful attention is paid to habitat heterogeneity and seasonal variation in grazer and resource community structure, and in which sampling is conducted over time frames appropriate for measuring microbial dynamics.

ACKNOWLEDGMENTS

Financial assistance was provided by the Pennswood #2 Trust and the National Science Foundation (Grant No. DEB-9310595). D. Funk photographed the specimen in Fig. 18.2 after M. Borchardt called it to our attention. D. Charles and N. Roberts identified the ingested diatoms. The comments of M. Borchardt, P. Montagna, and R. Lowe on an earlier draft are gratefully acknowledged.

REFERENCES

Admiraal, W., Bouwman, L. A., Hoekstra, L., and Romeyn, K. (1983). Qualitative and quantitative interactions between microphytobenthos and herbivorous meiofauna on a brackish intertidal mudflat. *Int. Rev. Gesamten Hydrobiol.* **68,** 175–191.

Albright, L. J., Sherr, E. B., Sherr, B. F., and Fallon, R. D. (1987). Grazing of ciliated protozoa on free and particle-attached bacteria. *Mar. Ecol.: Prog. Ser.* **38,** 125–129.

Alkemade, R., Wielemaker, A., de Jong, S.A., and Sandee, A. J. J. (1992). Experimental evidence for the role of bioturbation by the marine nematode *Diplolaimella dievengatensis* in stimulating the mineralization of *Spartina anglica* detritus. *Mar. Ecol.: Prog. Ser.* **90,** 149–155.

Andersson, A., Larsson, U., and Hagström, Å. (1986). Size-selective grazing by a microflagellate on pelagic bacteria. *Mar. Ecol.: Prog. Ser.* **33,** 51–57.

Arndt, H. (1993). Rotifers as predators on components of the microbial food web (bacteria, heterotrophic flagellates, ciliates)—A review. *Hydrobiologia* **255/256,** 231–246.

Arndt, H., Jost, G., and Wasmund, N. (1990). Dynamics of pelagic ciliates in eutrophic estuarine waters: Importance of functional groups among ciliates and responses to bacterial and phytoplankton production. *Arch. Hydrobiol. Beih. Ergeb. Limnol.* **34,** 239–245.

Austen, M. C. (1989). Factors affecting estuarine meiobenthic assemblage structure: A multifactorial microcosm experiment. *J. Exp. Mar. Biol. Ecol.* **130,** 167–187.

Azam, F., Fenchel, T., Field, J. G., Gray, J. S., Meyer-Reil, L. A., and Thingstad, F. (1983). The ecological role of water-column microbes in the sea. *Mar. Ecol.: Prog. Ser.* **10,** 257–263.

Baldock, B. M., and Sleigh, M. A. (1988). The ecology of benthic protozoa in rivers: Seasonal variation in numeric abundance in fine sediments. *Arch. Hydrobiol.* **111,** 409–421.

Barcina, I., Ayo, B., Muela, A., Egea, L., and Iriberri, J. (1991). Predation rates of flagellate and ciliated protozoa on bacterioplankton in a river. *FEMS Microbiol. Ecol.* **85,** 141–149.

Benke, A., and Wallace, J. B. (1980). The trophic basis of production among net-spinning caddisflies in a southern Appalachian stream. *Ecology* **61,** 108–118.

Bird, D. F., and Kalff, J. (1986). Bacterial grazing by planktonic algae. *Science* **231,** 493–495.

Birge, E. A. (1898). Plankton studies on Lake Mendota. II. The crustacea of the plankton from July, 1894, to December, 1896. *Trans. Wis. Acad. Sci., Arts Lett.* **11,** 274–448.

Blanchard, G. F. (1990). Overlapping microscale dispersion patterns of meiofauna and microphytobenthos. *Mar. Ecol.: Prog. Ser.* **68,** 101–111.

Blanchard, G. F. (1991). Measurement of meiofauna grazing rates on microphytobenthos: Is primary production a limiting factor? *J. Exp. Mar. Biol. Ecol.* **147,** 37–46.

Borchardt, M. A., and Bott, T. L. (1995). Meiofaunal grazing of bacteria and algae in a Piedmont stream. *J. North Am. Benthol. Soc.* **14,** 278–298.

Borsheim, K. Y. (1984). Clearance rates of bacteria-sized particles by freshwater ciliates measured with monodisperse fluorescent latex beads. *Oecologia* **63,** 286–288.

Bott, T. L., and Kaplan, L. A. (1990). Potential for protozoan grazing of bacteria in streambed sediments. *J. North Am. Benthol. Soc.* **9,** 336–345.

Bott, T. L., and Ritter, F. L. (1981). Benthic algal productivity in a piedmont stream measured by ^{14}C and dissolved oxygen change procedures. *J. Freshwater Ecol.* **1,** 267–278.

Botts, P. S. (1993). The impact of small chironomid grazers on epiphytic algal abundance and dispersion. *Freshwater Biol.* **30,** 25–33.

Bouvy, M. (1988). Contribution of the bacterial and microphytobenthic microflora in the energetic demand of the meiobenthos in an intertidal muddy sediment (Kerguelen Archipelago). *Mar. Ecol.* **9,** 109–122.

Brinkhurst, R. O., and Chua, K. E. (1969). Preliminary investigation of the exploitation of some potential nutritional resources by three sympatric tubificid oligochaetes. *J. Fish. Res. Board Can.* **26,** 2659–2668.

Brown, T. J., and Sibert, J. R. (1977). Food of some benthic harpacticoid copepods. *J. Fish. Res. Board Can.* **34,** 1028–1031.

Burns, C. W., and Xu, Z. K. (1990). Calanoid copepods feeding on algae and filamentous Cyanobacteria: Rates of ingestion, defecation and effects on trichome length. *J. Plankton Res.* **12,** 201–213.

Capblancq, J. (1990). Nutrient dynamics and pelagic food web interactions in oligotrophic and eutrophic environments: An overview. *Hydrobiologia* **207,** 1–14.

Carlough, L. A., and Meyer, J. L. (1990). Rates of protozoan bacterivory in three habitats of a southeastern blackwater river. *J. North Am. Benthol. Soc.* **9,** 45–53.

Carman, K. R. (1990). Mechanisms of uptake of radioactive labels by meiobenthic copepods during grazing experiments. *Mar. Ecol.: Prog. Ser.* **68,** 71–83.

Carman, K. R., and Thistle, D. (1985). Microbial food partitioning by three species of benthic copepods. *Mar. Biol. (Berlin)* **88,** 143–148.

Carman, K. R., Dobbs, F. C., and Guckert, J. B. (1989). Comparison of three techniques for administering radiolabeled substrates to sediments for trophic studies: Uptake of label by harpacticoid copepods. *Mar. Biol. (Berlin)* **102,** 119–125.

Carney, H. J. (1990). A general hypothesis for the strength of food web interactions in relation to trophic state. *Verh.—Int. Ver. Theor. Angew. Limnol.* **24,** 487–492.

Caron, D. A. (1987). Grazing of attached bacteria by heterotrophic microflagellates. *Microb. Ecol.* **13,** 203–218.

Caron, D. A., and Goldman, J. C. (1990). Protozoan nutrient regeneration. *In* "Ecology of Marine Protozoa" (G. M. Capriulo, ed.), pp. 283–306. Oxford Univ. Press, New York.

Caron, D. A., Goldman, J. C., Anderson, O. K., and Dennett, M. R. (1985). Nutrient cycling in a microflagellate food chain. II. Population dynamics and carbon cycling. *Mar. Ecol.: Prog. Ser.* **24,** 243–254.

Carrick, H. J., Fahnenstiel, G. L., and Taylor, W. D. (1992). Growth and production of planktonic protozoa in Lake Michigan: *In situ* versus *in vitro* comparisons and importance to food web dynamics. *Limnol. Oceanogr.* **37,** 1221–1235.

Cattaneo, A., and Kalff, J. (1986). The effect of grazer size manipulation on periphyton communities. *Oecologia* **69,** 612–617.

Chrzanowski, T. H., and Simek, K. (1990). Prey-size selection by freshwater flagellated protozoa. *Limnol. Oceanogr.* **35,** 1429–1436.

Couch, C. A. (1989). Carbon and nitrogen stable isotopes of meiobenthos and their food resources. *Estuarine, Coastal Shelf Sci.* **28,** 433–441.

Coull, B. C. (1973). Estuarine meiofauna: A review, trophic relationships and microbial interactions. *In* "Estuarine Microbial Ecology" (L. H. Stevenson and R. R. Colwell, eds.), pp. 499–511. Univ. of South Carolina Press, Columbia.

Coveney, M. F., and Wetzel, R. G. (1989). Bacterial metabolism of algal extracellular carbon. *Hydrobiologia* **173,** 141–149.

Cowles, T. J., Olson, R. J., and Chisholm, S. W. (1988). Food selection by copepods: Discrimination on the basis of food quality. *Mar. Biol. (Berlin)* **100,** 41–49.

Decho, A. W. (1988). How do harpacticoid grazing rates differ over a tidal cycle? Field verification using chlorophyll-pigment analyses. *Mar. Ecol.: Prog. Ser.* **45,** 263–270.

Decho, A. W., and Castenholz, R. W. (1986). Spatial patterns and feeding of meiobenthic harpacticoid copepods in relation to resident microbial flora. *Hydrobiologia* **131,** 87–96.

Decho, A. W., and Fleeger, J. W. (1988a). Microscale dispersion of meiobenthic copepods in response to food-resource patchiness. *J. Exp. Mar. Biol. Ecol.* **118,** 229–243.

Decho, A. W., and Fleeger, J. W. (1988b). Ontogenetic feeding shifts in the meiobenthic harpacticoid copepod *Nitocra lacustris. Mar. Biol. (Berlin)* **97,** 191–197.

Dolan, J. R. (1991). Guilds of ciliate microzooplankton in the Chesapeake Bay. *Estuarine, Coastal Shelf Sci.* **33,** 137–152.

Ducklow, H. W., and Hill, S. M. (1985). The growth of heterotrophic bacteria in the surface waters of warm core rings. *Limnol. Oceanogr.* **30**, 239–259.

Ducklow, H. W., Purdie, D. A., Williams, P. J. L., and Davies, J. M. (1986). Bacterioplankton: A sink for carbon in a coastal marine plankton community. *Science* **232**, 865–867.

Duncan, A., Schiemer, F., and Klekowski, R. Z. (1974). A preliminary study of feeding rates on bacterial food by adult females of a benthic nematode, *Plectus palustris* de Man 1880. Pol. Arch. Hydrobiol. **21**, 249–253.

Edwards, R. T., and Meyer, J. L. (1987). Bacteria as a food source for black fly larvae in a blackwater river. *J. North Am. Benthol. Soc.* **6**, 241–250.

Edwards, R. T., and Meyer, J. L. (1990). Bacterivory by deposit-feeding mayfly larvae (*Stenonema* spp.). *Freshwater Biol.* **24**, 453–462.

Epstein, S. S., and Gallagher, E. D. (1992). Evidence for facilitation and inhibition of ciliate population growth by meiofauna and macrofauna on a temperate zone sandflat. *J. Exp. Mar. Biol. Ecol.* **155**, 27–39.

Epstein, S. S., and Shiaris, M. P. (1992). Rates of microbenthic and meiobenthic bacterivory in a temperate muddy tidal flat community. *Appl. Environ. Microbiol.* **58**, 2426–2431.

Epstein, S. S., Burkovsky, I. W., and Shiaris, M. P. (1992). Ciliate grazing on bacteria, flagellates, and microalgae in a temperate zone sandy tidal flat: Ingestion rates and food niche partitioning. *J. Exp. Mar. Biol. Ecol.* **165**, 103–123.

Feller, R. J. (1984). Serological tracers of meiofaunal food webs. *Hydrobiologia* **118**, 119–125.

Feminella, J. W., Power, M. E., and Resh, V. H. (1989). Periphyton responses to invertebrate grazing and riparian canopy in three northern California coastal streams. *Freshwater Biol.* **22**, 445–457.

Fenchel, T. (1968). The ecology of marine microbenthos. II. The food of marine benthic ciliates. *Ophelia* **5**, 73–121.

Fenchel, T. (1975). The quantitative importance of the benthic microfauna of an Arctic tundra pond. *Hydrobiologia* **46**, 445–464.

Fenchel, T. (1978). The ecology of micro- and meiobenthos. *Annu. Rev. Ecol. Syst.* **9**, 99–121.

Fenchel, T. (1984). Suspended marine bacteria as a food source. *In* "Flows of Energy and Material in Marine Ecosystems" (M. J. R. Fasham, ed.), pp. 301–315. Plenum, New York.

Fenchel, T. (1986). Protozoan filter feeding. *In* "Progress in Protistology" (J. O. Corliss and D. J. Patterson, eds.), Vol. 1, pp. 65–113. Biopress Ltd., Bristol, UK.

Findlay, S., and Tenore, K. R. (1982). Effect of a free-living marine nematode *(Diplolaimella chitwoodi)* on detrital carbon mineralization. *Mar. Ecol.: Prog. Ser.* **8**, 161–166.

Findlay, S., Meyer, J. L., and Smith, P. J. (1984). Significance of bacterial biomass in the nutrition of a freshwater isopod (*Lirceus* sp.). *Oecologia* **63**, 38–42.

Findlay, S., Pace, M. L., Lints, D., Cole, J. J., Caraco, N. F., and Peierls, B. (1991). Weak coupling of bacterial and algal production in a heterotrophic ecosystem: The Hudson River estuary. *Limnol. Oceanogr.* **36**, 268–278.

Friedman, M. M., and Strickler, J. R. (1975). Chemoreceptors and feeding in calanoid copepods (Arthropoda: Crustacea). *Proc. Natl. Acad. Sci. U. S. A.* **72**, 4185–4188.

Fryer, G. (1957). The food of some freshwater cyclopoid copepods and its ecological significance. *J. Anim. Ecol.* **26**, 263–286.

Gaston, G. R. (1992). Green-winged teal ingest epibenthic meiofauna. *Estuaries* **15**, 227–229.

Goarant, E., Prensier, G., and Lair, N. (1994). Specific immunological probes for the identification and tracing of prey in crustacean gut contents. The example of cyanobacteria. *Arch. Hydrobiol.* **131**, 243–252.

Goldman, J. C., and Dennett, M. R. (1990). Dynamics of prey selection by an omnivorous flagellate. *Mar. Ecol.: Prog. Ser.* **59**, 183–194.

Goulder, R. (1972). Grazing by the ciliated protozoan *Loxodes magnus* on the alga *Scenedesmus* in a eutrophic pond. *Oikos* **23,** 109–115.

Gray, J. S. (1966). The attractive factor of intertidal sands to *Protodrilus symbioticus. J. Mar. Biol. Assoc. U. K.* **46,** 627–645.

Gray, J. S., and Johnson, R. M. (1970). The bacteria of a sandy beach as an ecological factor affecting the interstitial gastrotrich *Turbanella hyalina* Schultze. *J. Exp. Mar. Biol. Ecol.* **4,** 119–133.

Gulati, R. D., Ejsmont-Karabin, J., and Postema, G. (1993). Feeding in *Euchlanis dilatata lucksiana* Hauer on filamentous cyanobacteria and a prochlorophyte. *Hydrobiologia* **255/256,** 269–274.

Haack, T. K., and McFeters, G. A. (1982). Nutritional relationships among microorganisms in an epilithic biofilm community. *Microb. Ecol.* **8,** 115–126.

Hall, M. O., and Bell, S. S. (1993). Meiofauna on the seagrass *Thalassia testudinum:* Population characteristics of harpacticoid copepods and associations with algal epiphytes. *Mar. Biol. (Berlin)* **116,** 137–146.

Hicks, G. R. F., and Coull, B. C. (1983). The ecology of marine meiobenthic harpacticoid copepods. *Oceanogr. Mar. Biol. Annu. Rev.* **21,** 67–175.

Higgins, R. P., and Thiel, H. (1988). "Introduction to the Study of Meiofauna." Smithsonian Inst. Press, Washington, DC.

Hildrew, A. G. (1992). Food webs and species interactions. *In* "The Rivers Handbook" (P. Calow and G. E. Petts, eds.), Vol. 1, pp. 309–330. Blackwell, Oxford.

Hildrew, A. G., and Townsend, C. R. (1982). Predators and prey in a patchy environment: A freshwater study. *J. Anim. Ecol.* **51,** 797–815.

Hlawa, S., and Heerkloss, R. (1994). Experimental studies into the feeding biology of rotifers in brackish water. *J. Plankton Res.* **16,** 1021–1038.

Hondeveld, B. J. M., Bak, R. P. M., and van Duyl, F. C. (1992). Bacterivory by heterotrophic nanoflagellates in marine sediments measured by uptake of fluorescently labeled bacteria. *Mar. Ecol: Prog. Ser.* **89,** 63–71.

Jensen, P. (1982). Diatom-feeding behaviour of the free-living marine nematode *Chromadorita tenuis. Nematologica* **28,** 71–76.

Jensen, P. (1984). Food ingestion and growth of the diatom-feeding nematode *Chromadorita tenuis. Mar. Ecol.: Prog. Ser.* **35,** 187–196.

Jensen, P. (1987). Feeding ecology of free-living aquatic nematodes. *Mar. Ecol.: Prog. Ser.* **35,** 187–196.

Johannes, R. E. (1965). Influence of marine protozoa on nutrient regeneration. *Limnol. Oceanogr.* **10,** 434–442.

Jonsson, P. R. (1986). Particle size selection, feeding rates and growth dynamics of marine planktonic oligotrichous ciliates (Ciliophora: Oligotrichina). *Mar. Ecol.: Prog. Ser.* **33,** 265–277.

Jumars, P. A., Penry, D. L., Baross, J. A., Perry, M. J., and Frost, B. W. (1989). Closing the microbial loop: Dissolved carbon pathway to heterotrophic bacteria from incomplete ingestion, digestion and absorption in animals. *Deep-Sea Res.* **36,** 483–495.

Jürgens, K., Gasol, J. M., Massana, R., and Pedrós-Alió, C. (1994). Control of heterotrophic bacteria and protozoans by *Daphnia pulex* in the epilimnion of Lake Cisó. *Arch. Hydrobiol.* **131,** 55–78.

Kaplan, L. A., and Bott, T. L. (1982). Diel fluctuations of DOC generated by algae in a piedmont stream. *Limnol. Oceanogr.* **27,** 1091–1100.

Kemp, P. F. (1987). Potential impact on bacteria of grazing by a macrofaunal deposit feeder, and the fate of bacterial production. *Mar. Ecol.: Prog. Ser.* **36,** 151–161.

Kemp, P. F. (1988). Bacterivory by benthic ciliates: Significance as a carbon source and impact on sediment bacteria. *Mar. Ecol.: Prog. Ser.* **49,** 163–169.

Kemp, P. F. (1990). The fate of benthic bacterial production. *Rev. Aquat. Sci.* **2,** 109–124.

Kleppel, G. S. (1993). On the diets of calanoid copepods. *Mar. Ecol. Prog. Ser.* **99,** 183–195.

Lampert, W. (1978). Release of dissolved organic carbon by grazing zooplankton. *Limnol. Oceanogr.* **23,** 831–834.
Lancelet, C., and Billen, G. (1985). Carbon–nitrogen relationships in nutrient metabolism of coastal marine ecosystems. *Adv. Aquat. Microbiol.* **3,** 263–321.
Landry, M. R., and Hassett, R. P. (1982). Estimating the grazing impact of micro-zooplankton. *Mar. Biol. (Berlin)* **67,** 283–288.
Landry, M. R., Lehner-Fournier, J. M., Sundstrom, J. A., Fagerness, V. L., and Selph, K. E. (1991). Discrimination between living and heat-killed prey by a marine zooflagellate, *Paraphysomonas vestita* (Stokes). *J. Exp. Mar. Biol. Ecol.* **146,** 139–151.
Mastrantuono, L. (1988). A note on the feeding of *Amphichaeta leydigii* (Oligochaeta, Naididae) in lacustrine sandy shores. *Hydrobiol. Bull.* **22,** 195–198.
McCormick, P. V. (1991). Lotic protistan herbivore selectivity and its potential impact on benthic algal assemblages. *J. North Am. Benthol. Soc.* **10,** 238–250.
McCormick, P. V., and Cairns, J. C., Jr. (1991). Effects of micrometazoa on the protistan assemblage of a littoral food web. *Freshwater Biol.* **26,** 111–119.
McManus, G. B., and Fuhrman, J. A. (1988). Control of marine bacterioplankton populations: Measurement and significance of grazing. *Hydrobiologia* **159,** 51–62.
Mercer, E. K., and Cairns, E. J. (1973). Food consumption of the free-living aquatic nematode *Pelodera chitwoodi. J. Nematol.* **5,** 201–208.
Meyer-Reil, L.-A., and Faubel, A. (1980). Uptake of organic matter by meiofauna organisms and interrelationships with bacteria. *Mar. Ecol.: Prog. Ser.* **3,** 251–256.
Monger, B. C., and Landry, M. R. (1991). Prey-size dependency of grazing by free-living marine flagellates. *Mar. Ecol.: Prog. Ser.* **74,** 239–248.
Montagna, P. A. (1983). Live controls for radioisotope tracer food chain experiments using meiofauna. *Mar. Ecol.: Prog. Ser.* **12,** 43–46.
Montagna, P. A. (1984a). *In situ* measurement of meiobenthic grazing rates on sediment bacteria and edaphic diatoms. *Mar. Ecol.: Prog. Ser.* **18,** 119–130.
Montagna, P. A. (1984b). Competition for dissolved glucose between meiobenthos and sediment microbes. *J. Exp. Mar. Biol. Ecol.* **76,** 177–190.
Montagna, P. A. (1995). Rates of metazoan meiofaunal microbivory: A review. *Vie Milieu* **45,** 1–9.
Montagna, P. A., and Bauer, J. E. (1988). Partitioning radiolabeled thymidine uptake by bacteria and meiofauna using metabolic blocks and poisons in benthic feeding studies. *Mar. Biol. (Berlin)* **98,** 101–110.
Montagna, P.A., Coull, B. C., Herring, T. L., and Dudley, B. W. (1983). The relationship between abundances of meiofauna and their suspected microbial food (diatoms and bacteria). *Estuarine, Coastal Shelf Sci.* **17,** 381–394.
Murray, A. G., and Eldridge, P. M. (1994). Marine viral ecology: Incorporation of bacteriophage into the microbial planktonic food web paradigm. *J. Plankton Res.* **16,** 627–641.
Nygaard, K., and Hessen, D. O. (1990). Use of ^{14}C-protein-labelled bacteria for estimating clearance rates by heterotrophic and mixotrophic flagellates. *Mar. Ecol.: Prog. Ser.* **68,** 7–14.
Olafsson, E., and Moore, C. G. (1990). Control of meiobenthic abundance by macroepifauna in a subtidal muddy habitat. *Mar. Ecol.: Prog. Ser.* **65,** 241–249.
Ooms-Wilms, A. L., Postema, G., and Gulati, R. D. (1993). Clearance rates of bacteria by the rotifer *Filinia longiseta* (Ehrb.) measured using three tracers. *Hydrobiologia* **255/256,** 255–260.
Pace, M. L. (1991). Comparative and experimental approaches to the study of microbial food webs. *J. Protozool.* **38,** 87–92.
Pace, M. L., and Bailiff, M. D. (1987). Evaluation of a fluorescent microsphere technique for measuring grazing rates of phagotrophic microorganisms. *Mar. Ecol.: Prog. Ser.* **40,** 185–193.

Patterson, D. J., Larsen, J., and Corliss, J. O. (1989). The ecology of heterotrophic flagellates and ciliates living in marine sediments. *Prog. Protistol.* **3**, 185–277.

Perlmutter, D. G., and Meyer, J. L. (1991). The impact of a stream-dwelling harpacticoid copepod upon detritally associated bacteria. *Ecology* **72**, 2170– 2180.

Pinckney, J., and Sandulli, R. (1990). Spatial autocorrelation analysis of meiofaunal and microalgal populations on an intertidal sandflat: Scale linkage between consumers and resources. *Estuarine, Coastal Shelf Sci.* **30**, 341–353.

Plante, C., and Jumars, P. (1993). Immunofluorescence assay for effects on field abundance of a naturally occurring pseudomonad during passage through the gut of a marine deposit feeder, *Abarenicola pacifiea. Microb. Ecol.* **26**, 247–266.

Polis, G. A., and Holt, R. D. (1992). Intraguild predation: The dynamics of complex trophic interactions. *Trends Ecol. Evol.* **7**, 151–154.

Pomeroy, L. R. (1974). The ocean's food web, a changing paradigm. *BioScience* **24**, 499–504.

Pomeroy, L. R., and Wiebe, W. J. (1988). Energetics of microbial food webs. *Hydrobiologia* **159**, 7–18.

Porter, K. G., Pace, M. L., and Battey, J. F. (1979). Ciliate protozoans as links in freshwater planktonic food chains. *Nature (London)* **277**, 563–565.

Porter, K. G., Paerl, H., Hodson, R., Pace, M., Priscu, J., Riemann, B., Scavia, D., and Stockner, J. (1988). Microbial interactions in lake food webs. *In* "Complex Interactions in Lake Communities" (S. R. Carpenter, ed.), pp. 209–227. Springer-Verlag, New York.

Pourriot, R. (1977). Food and feeding habits of rotifera. *Arch. Hydrobiol. Beih. Ergeb. Limnol.* **8**, 243–260.

Power, M. E. (1990). Effects of fish in river food webs. *Science* **250**, 811–814.

Power, M. E. (1992). Top-down and bottom-up forces in food webs: Do plants have primacy? *Ecology* **73**, 733–746.

Pratt, J. R., and Cairns, J., Jr. (1985). Functional groups in the protozoa: Roles in the differing ecosystems. *J. Protozool.* **32**, 415–423.

Proctor, L. M., and Furhrman, J. A. (1990). Viral mortality of marine bacteria and cyanobacteria. *Nature (London)* **343**, 60–62.

Rassoulzadegan, F., Laval-Peuto, M., and Sheldon, R. W. (1988). Partitioning of the food ration of marine ciliates between pico- and nanoplankton. *Hydrobiologia* **159**, 75–88.

Riemann, B., and Søndergaard, M. (1986). Regulation of bacterial secondary production in two eutrophic lakes and in experimental enclosures. *J. Plankton Res.* **8**, 519–536.

Rieper, M. (1978). Bacteria as food for marine harpacticoid copepods. *Mar. Biol. (Berlin)* **45**, 337–346.

Rieper, M. (1982a). Feeding preferences of marine harpacticoid copepods for various species of bacteria. *Mar. Ecol.: Prog. Ser.* **7**, 303–307.

Rieper, M. (1982b). Relationships between bacteria and marine copepods. *Bacteriol. Mar., Colloq. Int. C. N. R. S.* **331**, 169–172.

Rieper, M. (1985). Some lower food web organisms in the nutrition of marine harpacticoid copepods: An experimental study. *Helgol. Meeresunters.* **39**, 357–366.

Rieper, M., and Flotow, C. (1981). Feeding experiments with bacteria, ciliates and harpacticoid copepods. *Kiel. Meeresforsch., Sonderh.* **5**, 370–375.

Rothhaupt, K. O. (1992). Stimulation of phosphorus-limited phytoplankton by bacterivorous flagellates in laboratory experiments. *Limnol. Oceanogr.* **37**, 750–759.

Rublee, P. A., and Gallegos, C. L. (1989). Use of fluorescently labeled algae (FLA) to estimate microzooplankton grazing. *Mar. Ecol.: Prog. Ser.* **51**, 221–227.

Sanders, R. W., and Porter, K. G. (1986). Use of metabolic inhibitors to estimate protozooplankton grazing and bacterial production in a monomictic eutrophic lake with an anaerobic hypolimnion. *Appl. Environ. Microbiol.* **52**, 101–107.

Sanders., R. W., Porter, K. G., Bennett, S. J., and DeBiase, A. E. (1989). Seasonal patterns of bacterivory by flagellates, ciliates, rotifers, and cladocerans in a freshwater planktonic community. *Limnol. Oceanogr.* **34**, 673–687.

Sawicka, K., Kaczanowski, A., and Kaczanowska, J. (1983). Kinetics of ingestion and egestion of food vacuoles during cell cycle of *Chilodonella steini. Acta Protozool.* **22**, 157–167.

Scavia, D., and Laird, G. A. (1987). Bacterioplankton in Lake Michigan: Dynamics, controls, and significance to carbon flux. *Limnol. Oceanogr.* **32**, 1017–1033.

Schaffner, W. R., Hairston, N. G., Jr. and Howarth, R. W. (1994). Feeding rates and filament clipping by crustacean zooplankton consuming cyanobacteria. *Verh.—Int. Ver. Theor. Angew. Limnol.* **25**, 2375–2381.

Scheda, S. M., and Cowell, B. C. (1988). Rotifer grazers and phytoplankton: Seasonal experiments on natural communities. *Arch. Hydrobiol.* **114**, 31–44.

Schiemer, F. (1983). Comparative aspects of food dependence and energetics of freeliving nematodes. *Oikos* **41**, 32–42.

Schönborn, W. (1982). Die Ziliatenproduktion in der mittleren Saale. *Limnologica* **14**, 329–346.

Schönborn, W. (1984). The annual energy transfer from the communities of Ciliata to the population of *Chaetogaster diastrophus* (Gruithuisen) in the River Saale. Limnologica **16**, 15–23.

Schönborn, W. (1992). The role of protozoan communities in freshwater and soil ecosystems. *Acta Protozool.* **31**, 11–18.

Seki, H. (1969). Marine microorganisms associated with the food of young salmon. *Appl. Microbiol.* **17**, 252–255.

Sharp, J. H. (1977). Excretion of organic matter by marine phytoplankton: Do healthy cells do it? *Limnol. Oceanogr.* **22**, 381–399.

Sherberger, F. F., and Wallace, J. B. (1971). Larvae of the southeastern species of *Molanna. J. Kans. Entomol. Soc.* **44**, 217–224.

Sherr, B. F., Sherr, E. B., Andrew, T. L., Fallon, R. D., and Newell, S. Y. (1986). Trophic interactions between heterotrophic protozoa and bacterioplankton in estuarine water analyzed with selective metabolic inhibitors. *Mar. Ecol.: Prog. Ser.* **32**, 169–179.

Sherr, B. F., Sherr, E. B., and Fallon, R. D. (1987). Use of monodispersed, fluorescently labeled bacteria to estimate *in situ* protozoan bacterivory. *Appl. Environ. Microbiol.* **53**, 958–965.

Sherr, E. B., and Sherr, B. F. (1987). High rates of consumption of bacteria by pelagic ciliates. *Nature (London)* **325**, 710–711.

Sherr, E. B., Rassoulzadegan, F., and Sherr, B. F. (1989). Bacterivory by pelagic choreotrichous ciliates in coastal waters of the NW Mediterranean Sea. *Mar. Ecol. Prog. Ser.* **55**, 235–240.

Sherr, E. B., Sherr, B. F., and McDaniel, J. (1991). Clearance rates of <6 μm fluorescently labeled algae (FLA) by estuarine protozoa: Potential grazing impact of flagellates and ciliates. *Mar. Ecol.: Prog. Ser.* **69**, 81–92.

Sieracki, M. E., Haas, L. W., Caron, D. A., and Lessard, E. J. (1987). Effect of fixation on particle retention by microflagellates: Underestimation of grazing rates. *Mar. Ecol.: Prog. Ser.* **38**, 295–303.

Sierszen, M. E. (1990). Variable selectivity and the role of nutritional quality in food selection by a planktonic rotifer. *Oikos* **59**, 241–247.

Skogstad, A., Granskog, L., and Klaveness, D. (1987). Growth of freshwater ciliates offered planktonic algae as food. *J. Plankton Res.* **9**, 503–512.

Spaull, V. W. (1973). Distribution of nematode feeding groups at Signy Island, South Orkney Islands, with an estimate of their biomass and oxygen consumption. *Br. Antarct. Surv. Bull.* **37**, 21–32.

Sterner, R. W. (1986). Herbivores' direct and indirect effects on algal populations. *Science* **231**, 605–608.

Strayer, D. (1988). On the limits to secondary production. *Limnol. Oceanogr.* **33**, 1217–1220.

Strayer. D. (1991). Perspectives on the size structure of lacustrine zoobenthos, its causes, and its consequences. *J. North Am. Benthol. Soc.* **10**, 210–221.

Suttle, C. A., Chan, A. M., Taylor, W. D., and Harrison, P. J. (1986). Grazing of planktonic diatoms by microflagellates. *J. Plankton Res.* **8**, 393–398.

Swadling, K. M., and Marcus, N. H. (1994). Selectivity in the normal diets of *Acartia tonsa* Dana (Copepoda: Calanoida): Comparison of juveniles and adults. *J. Exp. Mar. Biol. Ecol.* **181**, 91–103.

Takada, K., Kato, K., and Okino, T. (1994). Image analyses of feeding activity of aquatic oligochaetes using fluorescent latex beads. *Verh.—Int. Ver. Theor. Angew. Limnol.* **25**, 2336–2340.

Telesh, I. V. (1993). The effect of fish on planktonic rotifers. *Hydrobiologia* **255/256**, 289–296.

Tietjen, J. H., and Alongi, D. M. (1990). Population growth and effects of nematodes on nutrient regeneration and bacteria associated with mangrove detritus from northeastern Queensland (Australia). *Mar. Ecol.: Prog. Ser.* **68**, 169–179.

Tietjen, J. H., and Lee, J. J. (1973). Life history and feeding habits of the marine nematode, *Chromadora macrolaimoides* Steiner. *Oecologia* **12**, 303–314.

Tremaine, S. C., and Mills, A. L. (1987). Inadequacy of the eucaryote inhibitor cycloheximide in studies of protozoan grazing on bacteria at the freshwater-sediment interface. *Appl. Environ. Microbiol.* **53**, 1969–1972.

Trotter, D. B., and Webster, J. M. (1984). Feeding preferences and seasonality of free living marine nematodes inhabiting the kelp *Macrocystis integrifolia. Mar. Ecol.: Prog. Ser.* **14**, 151–157.

Vanden Berghe, W., and Bergmans, M. (1981). Differential food preferences in three co-occurring species of *Tisbe* (Copepoda, Harpacticoida). *Mar. Ecol.: Prog. Ser.* **4**, 213–219.

Verity, P. G. (1991). Feeding in planktonic protozoans: Evidence for non-random acquisition of prey. *J. Protozool.* **38**, 69–76.

Warwick, R. M. (1987). Meiofauna: Their role in marine detrital systems. *In* "Detritus and Microbial Ecology in Aquaculture" (D. J. W. Moriarty and R. S. V. Pullin, eds.), pp. 282–295. International Center for Living Aquatic Resources Management, Manila.

Wavre, M., and Brinkhurst, R. O. (1971). Interactions between some tubificid oligochaetes and bacteria found in the sediments of Toronto Harbor, Ontario. *J. Fish. Res. Board Can.* **28**, 335–341.

Weisse, T. (1990). Trophic interactions among heterotrophic microplankton, nanoplankton, and bacteria in Lake Constance. *Hydrobiologia* **191**, 111–122.

Weisse, T., and Scheffel-Möser, U. (1991). Uncoupling the microbial loop: Growth and grazing loss rates of bacteria and heterotrophic nanoflagellates in the North Atlantic. *Mar. Ecol. Prog. Ser.* **71**, 195–205.

Wikner, J., Andersson, A., Normark, S., and Hagström, Å. (1986). Use of genetically marked minicells as a probe in measurement of predation on bacteria in aquatic environments. *Appl. Environ. Microbiol.* **52**, 4–8.

Wolter, K. (1982). Bacterial incorporation of organic substances released by natural phytoplankton populations. *Mar. Ecol.: Prog. Ser.* **7**, 287–295.

Woombs, M., and Laybourn-Parry, J. (1984). Feeding biology of *Diplogasteritus nudicapitatus* and *Rhabditis curvicaudata* (Nematoda) related to food concentration and temperature in sewage treatment plants. *Oecologia* **64**, 163–167.

Wotton, R. S. (1980). Bacteria as food for blackfly larvae (Diptera: Simuliidae) in a lake outlet in Finland. *Ann. Zool. Fenn.* **17**, 127–130.

Wright, R. T., and Coffin, R. B. (1984). Measuring microzooplankton grazing on planktonic marine bacteria by its impact on bacterial production. *Microb. Ecol.* **10**, 137–149.

19

Role in Nutrient Cycling in Streams

Patrick J. Mulholland

Environmental Sciences Division
Oak Ridge National Laboratory
Oak Ridge, Tennessee 37831-6036

I. INTRODUCTION

Nutrient cycling in ecosystems includes a sequence of processes: uptake of inorganic (and in some cases organic) forms of elements by biota, transfer of these elements from one organism to another (through food chains), release back to the environment in available forms (release in soluble forms, remineralization), and element reassimilation by organisms. The rate of nutrient cycling is often quantified as the turnover rate of nutrients (k) within an ecosystem compartment (e.g., water, algae, detritus) per unit time (expressed as $time^{-1}$). Nutrient cycling is also sometimes presented as the flux of organic nutrients remineralized per unit time (expressed as mass $time^{-1}$). However, the latter expression does not provide information on the contribution of remineralization to assimilation, and thus can be misleading as an index of the importance of nutrient cycling in meeting nutrient demands within ecosystems.

If ecosystems were closed to exchanges of material with surrounding systems, primary production would be entirely dependent on nutrient cycling, and the rate of primary production would be a direct function of the rate of nutrient cycling. However, most ecosystems are not closed, receiving additions and losing nutrients to surrounding systems. In open ecosystems, primary production is therefore dependent on the supply of nutrients from outside plus those cycled within. In general, the greater the external supply of nutrients, the less important is nutrient cycling in meeting biological demands. If nutrient inputs are large, nutrients may not be the limiting resource and primary production may become limited by the supply of other resources, such as light.

In this chapter, I discuss the role that benthic algal communities play in nutrient cycling in streams. The following chapter by Wetzel (Chapter 20) presents the role of benthic algae in nutrient cycling in lakes. Here, I take a somewhat broad view of benthic algal communities and include discussion of effects of the microbial and invertebrate organisms closely associated with benthic algae. However, herbivory and the role of benthic algae in food webs are the subjects of other chapters in this volume (see Steinman, Chapter 12; Lamberti, Chapter 17; and Bott, Chapter 18, this volume) and are considered here only in terms of their effects on nutrient cycling in stream benthic algal communities. This chapter considers the direct effects of benthic algae on nutrient cycling, including processes such as nutrient uptake and remineralization, as well as indirect effects of benthic algae on nutrient cycling via their modification of streambed hydraulic properties and algal/herbivore interactions.

A. Importance of Nutrient Cycling to Stream Algae

Streams are open ecosystems that receive large inputs of materials from their catchments (e.g., dissolved load in groundwater discharges, inputs of soil and terrestrial plant materials). This would suggest that nutrient cycling should be of minor importance in meeting nutrient demand in streams. However, a considerable number of studies have demonstrated nutrient limitation of algal biomass accrual or productivity in streams (Stockner and Shortreed, 1978; Elwood *et al.*, 1981; Peterson *et al.*, 1983; Grimm and Fisher, 1986; Pringle, 1987; Stewart, 1987; Perrin *et al.*, 1987; Hill and Knight, 1988; Bothwell, 1989; Tate, 1990; Hill *et al.*, 1992; Rosemond *et al.*, 1993), suggesting that cycling should be important. This apparent contradiction may be at least partially resolved by recognizing the differences in the spatial dimensions of nutrient resources and biological demand in stream ecosystems. Although available nutrients primarily consist of the dissolved pool dispersed within the volume of flowing water, the majority of organisms live attached to the stream bottom and are able to assimilate only those nutrients in close proximity to the bottom. Physical processes, such as turbulence and diffusion, that control nutrient transport into the zone of water adjacent to the stream bottom (and algal surfaces) are therefore important factors controlling nutrient availability to stream algae (discussed further in Section III,A of this chapter; see also Borchardt, Chapter 7, this volume). The importance of nutrient cycling to nutrient availability is amplified because remineralization occurs largely on the stream bottom as well, in close proximity to algal demand.

Groundwater discharge supplies most of the nutrient inputs to streams during much of the year. In the absence of direct human disturbance, catchment geology, atmospheric deposition, and vegetation age/status largely determine the concentrations of nutrients in groundwater discharged to streams (Dillon and Kirchner, 1975; Vitousek, 1977; Mulholland, 1992). Algal growth (Biggs and Gerbeaux, 1993) and species composition (Lay and Ward, 1987) in streams have been related to catchment geology via its effect on external nutrient supplies. Nonetheless, even in streams with relatively large groundwater discharges (e.g., low-order, headwater streams), nutrient cycling may provide an important additional supply of nutrients to stream algae. Groundwater discharge is not spatially uniform and long reaches of streams may receive very little groundwater input. Stream water nutrients can become depleted as water flows downstream, increasing the potential importance of nutrient cycling in meeting algal nutrient demands. This pattern of longitudinal depletion of stream water nutrient concentrations and consequent reliance on nutrient cycling occurs seasonally in Walker Branch, a small, forested stream in eastern Tennessee (Mulholland and Rosemond, 1992). In Walker Branch, declines in stream water nutrient concentrations with distance downstream from springs were consistently accompanied by increased phosphatase activity and reduced phosphorus content of periphyton communities, suggesting both increased phosphorus limitation and cycling.

B. Spatial Context of Nutrient Cycling in Streams

The spatial dimensions of nutrient cycling are particularly important in streams because of the strong advective forces present in these ecosystems. The unidirectional flow of water downhill can displace nutrients some distance downstream as they complete their cycle through the biota. To quantify the process of nutrient cycling in streams in a way that incorporates the spatial dimensions of cycling resulting from advection, the concept of nutrient spiraling was proposed (Webster, 1975; Webster and Patten, 1979) and refined (Newbold *et al.,* 1981, 1982, 1983; Elwood *et al.,* 1983). In this concept, the distance traveled by a nutrient atom while completing one cycle through the ecosystem is called the spiraling length (S), and its inverse is the nutrient cycling rate in the spatial dimension (distance^{-1}). Thus, spiraling length, or more accurately its inverse (S^{-1}), provides a quantitative measure of the efficiency of nutrient cycling in a spatial context. It is analogous to nutrient turnover rate in water, k, presented as a first-order rate coefficient with units of time^{-1}. These two measures of nutrient cycling are related by the average downstream velocity of nutrients, v, as shown by Elwood *et al.* (1983):

$$S^{-1} = k \cdot v^{-1} \tag{1}$$

Although nutrient turnover rate in stream water (k) is dependent only on the size of the stream water nutrient pool and the rate of uptake and remineralization of nutrients by organisms, nutrient cycling rate per unit distance (S^{-1}) is influenced also by the velocity at which nutrients are transported downstream (v).

II. DIRECT EFFECTS: NUTRIENT CYCLING PROCESSES

Direct effects of benthic algae on nutrient cycling in streams include the following: (1) increasing the total supply of nutrients via fixation of atmospheric forms and utilization of substratum nutrients, (2) nutrient uptake from stream water, and (3) nutrient transformation and remineralization (Fig. 1). Although each of these processes also involves the activities of other stream organisms, the role of benthic algae is important in most streams at least during some periods.

A. Nutrient Supply

Stream water may not supply all the nutrients assimilated by benthic algae in streams. Benthic algae can obtain nutrients from the substrata to which they are attached and from the atmosphere, by N fixation, thereby increasing nutrient inputs to streams. Although nutrient inputs increase the mass of nutrients potentially available for cycling in streams, larger inputs are likely to reduce nutrient turnover rates in water because a greater fraction of the nutrient demands of benthic algae are met by "new" nutrients rather than by recycled nutrients.

Benthic algae may acquire nutrients from the inorganic or organic substrata to which they are attached via passive diffusion, ion-exchange processes, or active "mining" of the substratum (see also Burkholder, Chapter 9, this volume). This may be particularly important in streams containing substrata with relatively high nutrient content, such as P-rich volcanic rock or carbonates. Although there are numerous experimental studies using nutrient-diffusing substrata for tests of nutrient limitation in streams (e.g., Pringle, 1987, 1990; Tate, 1990), there is only indirect evidence of nat-

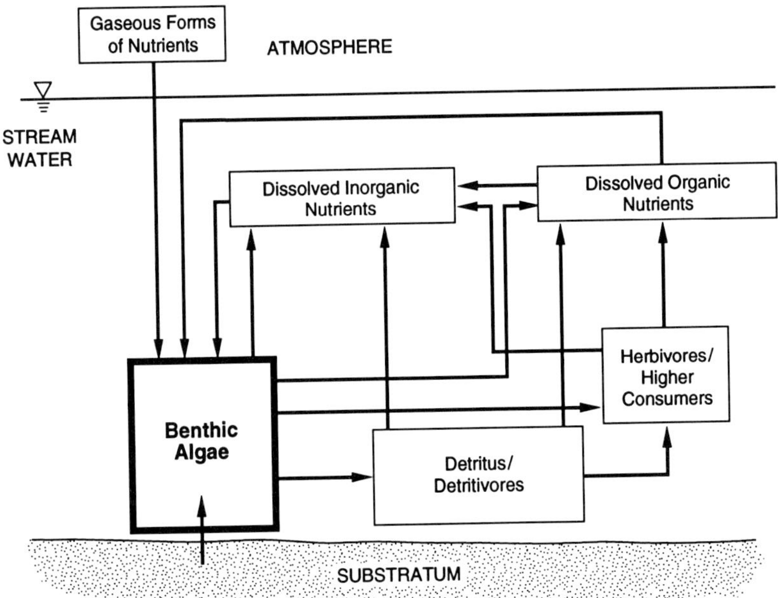

FIGURE 1 Nutrient cycling in streams and the role of benthic algae. Downstream transport of dissolved nutrients in stream water is not shown.

ural substrata providing nutrients directly to benthic algae. In a study of several streams in upstate New York, Klotz (1985) found that stream periphyton was less P-limited when growing on sediments with high P content than on sediments with lower quantities of adsorbed P, regardless of stream water P concentration. Coleman and Dahm (1990) reported greater benthic algal production in a stream with similar water chemistry but greater sediment nutrient content, compared with a nearby stream in New Mexico, although the authors suggest that hyporheic processes were responsible for increased nutrient availability to the algae. Evidence from lakes showing that benthic algae can deplete sediment nutrient content suggests that they can actively mine the substratum for nutrients (Hansson, 1989).

Benthic algae may increase the supply of N to streams via N fixation, although there are few measurements of N fixation by lotic algae and it is difficult to assess the general importance of N fixation in meeting total N demands by benthic algae in streams. In Rocky Creek, California, under relatively high light and high P concentration, the gelatinous cyanophyte *Nostoc* sp. fixed up to 360 mg N m^{-2} yr^{-1} (Horne and Carmiggelt, 1975). Although the fractional contribution of N fixation to total N demand was not reported, annual N fixation in Rocky Creek was about one-half of the low end of the range of total nitrate uptake for another northern California stream, Little Lost Man Creek [computed from data of Triska *et al.*

(1989), assuming a 12-h daily photoperiod]. N fixation rates were measured in five shallow rivers in Alberta, Canada, and ranged from 0.02 to 0.38 mg N m^{-2} h^{-1} (Charlton and Hickman, 1988), values roughly similar to hourly rates in Rocky Creek. Livingstone *et al.* (1984) reported N fixation rates as high as 1×10^{-4} μgN (μg chl *a*)$^{-1}$ min^{-1} by *Rivularia* sp. in a stream in northern England, but estimated that fixation supplied only a small percentage of the alga's N requirements. *Nostoc* is also reported to be common in N-deficient Oregon streams with open canopies (Ward, 1985), and the N-fixing cyanophyte *Calothrix* dominates in N-deficient Sycamore Creek, Arizona, during the late stages of algal succession (Peterson and Grimm, 1992). These reports suggest that N fixation at times may contribute significant quantities of N to the benthic algal communities of streams with low dissolved N concentrations.

Nitrogen fixation rates by benthic algae may also be influenced by consumers. Power *et al.* (1988) showed that grazing fish and macroinvertebrates facilitated the persistence of a benthic algal community dominated by *Calothrix* in an Ozark Mountain (Oklahoma) stream by removing overstory algae (mostly diatoms). Consumers also have been shown to increase the productivity of N-fixing algae via symbiotic relationships. In their studies of *Nostoc parmeloides* in Oregon streams, Ward *et al.* (1985) found that *Nostoc* colonies housing chironomid larvae had a more erect shape than noninfected forms, allowing the alga to be exposed to higher light intensities, thus increasing its photosynthesis and presumably N fixation rates. These examples suggest that N fixation by benthic algae can be controlled by biotic interactions as well as by the physical and chemical characteristics of stream ecosystems.

B. Nutrient Uptake from Stream Water

Nutrient uptake from stream water by benthic algae increases nutrient cycling rates in two ways: (1) slowing the rate of downstream transport of nutrients by immobilizing and retaining streamwater nutrients at one location, thereby increasing the potential number of times a nutrient atom can cycle within a given reach of stream, and (2) depleting the nutrient concentrations in stream water supplied to downstream communities, thus increasing the fractional uptake rate (stream water nutrient turnover rate) and the potential for recycling within those communities. Obviously, nutrient uptake is central to nutrient cycling; if nutrients are not initially taken up, they cannot be cycled. However, the relationship between nutrient uptake by algae and nutrient cycling may not always be positive. For example, nutrient uptake rates expressed in terms of *mass* per unit time may increase with increases in concentration of the limiting nutrient, but higher mass uptake rates may nonetheless remove smaller fractions of the available nutrient pool in stream water under these conditions and thus be equivalent to lower nutrient uptake rates expressed as the *fraction* of total nutrients cycled per unit time or distance (stream water nutrient turnover rate). Further, in lotic ecosystems, the processes of nutrient uptake and remineralization may be decoupled in space. Nutrients taken up at one location may be transported in organic form and remineralized well downstream from where they were taken up. This is referred to as the "nutrient turnover length" in the spiraling concept (Newbold *et al.*, 1981, 1983).

There are methodological problems with the measurement of nutrient uptake rates by benthic algae in streams. Nutrient uptake rates have been determined using closed chambers into which benthic algae are placed (e.g., Whitford and Schumacher, 1964; Steinman *et al.*, 1992) or from measurements of nutrient depletion along stream segments dominated by algae (e.g., Grimm *et al.*, 1981; Tate, 1990). The chamber approach may be inappropriate for algal communities that are difficult to place in a chamber (e.g., those on fine-grained substrata or on bedrock and those with very patchy distributions). The whole-stream measurements suffer from the difficulty of separating algal uptake from uptake by other organisms (e.g., bacteria and fungi). However, perhaps the biggest obstacle for measuring nutrient uptake by benthic algae using either approach is the simultaneous remineralization of nutrients. Algal uptake rates estimated by changes in nutrient concentrations in chambers or over stream segments usually underestimate nutrient uptake because some release of nutrients to water undoubtedly occurs at the same time as uptake. This problem can be overcome by using radiotracer techniques.

The ideal approach for determining nutrient uptake rates by benthic algae involves the use of radiotracers in short-term experiments, because measurements of uptake are not confounded by simultaneous remineralization and because it is not necessary to increase nutrient concentrations during the measurement. The problems of separating algal from nonalgal uptake and computing area-based uptake rates can be overcome, to some extent, by extracting the radiotracer from algae collected from a known area of stream bottom. However, the radiotracer approach is generally not possible for N uptake, because an ecologically useful radiotracer of N does not exist, and P radiotracers cannot be used in many locations. Thus, there are only a few radiotracer measurements of nutrient uptake by algae in streams, in either closed chambers (e.g., Whitford and Schumacher, 1964; Corning *et al.*, 1989; Steinman *et al.*, 1991, 1992) or whole-stream studies (Mulholland *et al.*, 1985, 1990).

An alternative to the radiotracer approach for determining nutrient uptake rates by benthic algae involves nutrient additions to water (in chambers or to whole streams) and measurement of concentration declines over time or distance. Although nutrient additions will often increase mass rates of nutrient uptake from stream water because of stimulation of biological uptake or physicochemical immobilization (particularly for NH_4

and PO_4), it may be possible to estimate ambient mass uptake rates from measurements of nutrient concentration declines during the first hour or so of chamber or whole-stream nutrient addition experiments. This approach involves computing first-order uptake rate coefficients (k) from the measured rates of decline in the concentration of added nutrients (after ambient concentrations are subtracted) over time (chambers) or distance (whole stream). Mass rates of nutrient uptake at ambient nutrient concentrations can then be computed from the uptake rate coefficients (in units of time^{-1} or distance^{-1}), the ambient nutrient concentrations and information on the biomass:water volume ratios in chambers or information on segment length, area, and discharge rate in whole-stream studies, as shown in the footnote to Table I.

The nutrient addition approach assumes that first-order uptake rate coefficients are the same under elevated and ambient nutrient concentrations (i.e., mass uptake rate increases by the same factor as the increase in nutrient concentration). Although this assumption may be reasonable for very small increases in concentration of the limiting nutrient, a comparative study of radiotracer and nutrient addition approaches in Walker Branch indicated problems with the latter (Mulholland *et al.,* 1990). Uptake rate coefficients determined for even relatively small increases in PO_4 concentration (about 5 μg liter^{-1}) in Walker Branch were only about two-thirds as high as uptake rate coefficients measured using radiotracer additions under ambient nutrient concentrations. Thus, ambient mass rates of P uptake were underestimated when calculated from the PO_4 addition experiments. Also, ambient mass rates of P uptake could be overestimated using the PO_4 addition approach if there is significant short-term luxury uptake of P.

There have been surprisingly few stream studies in which ambient rates of nutrient uptake by benthic algae were reported. Mass rates of NO_3 uptake in streams range from near 0 to 51.6 mg N m^{-2} h^{-1}, and rates of PO_4 uptake range from 0.005 to 18 mg P m^{-2} h^{-1} (Table I). Most of the nutrient uptake rates reported in Table I were determined from whole-stream measurements and therefore include uptake by other organisms (e.g., bacteria, fungi, bryophytes) as well as by benthic algae. In many streams in forested catchments, uptake by benthic algae is not the dominant mechanism of nutrient uptake. For example, in Walker Branch, periphyton communities are estimated to account for 5–35% of total PO_4 uptake from stream water, with the higher values generally occurring in spring when light reaching the stream is highest (Newbold *et al.,* 1983; Mulholland *et al.,* 1985, unpublished data). However, the biomass of periphyton is relatively low in Walker Branch [ash-free dry mass (AFDM) = 0.2–0.6 mg cm^{-2}; chlorophyll *a* = 0.8–6 μg cm^{-2}], primarily as a result of intense grazing by snails (Mulholland and Rosemond, 1992; Rosemond, 1994). Further, nutrient uptake by periphyton includes uptake by the heterotrophic components of the peri-

TABLE I Rates of Nutrient Uptake by Benthic Algae in Streams

Stream	*Location*	*Order*	*Canopy*	*Ambient N or P conc. (μg liter^{-1})*	*Uptake rate of N or P (μg m^{-2} h^{-1})*	*Method*	*Reference*
Nitrate							
Sonoran Desert streams	S. Arizona	—	Open	<200 >200	0–7600 5800–51,700	Whole-stream, ambient conc. change (up-down)	Grimm (1992)
Kings Ck. (perennial reach)	Kansas	3	Semiopen	4–8	120–300	Whole-stream, ambient conc. change (up-down)	Tate (1990)
Kings Ck. (intermittent reach)	Kansas	3	Semiopen	12–20	1680–6840	Whole-stream, ambient conc. change (up-down)	Tate (1990)
Little Lost Man Ck.	N. California	3	Semiopen	—	216–1370	Whole-stream, NO_3 incr. 5×	Triska *et al.* (1989)
Hugh White Ck.	N. Carolina	2	Closed	6	102–306[a]	Whole-stream, NO_3 incr. 3×	Munn and Meyer (1990)
WS 2, H. J. Andrews	Oregon	2	Closed	2	72–78[a]	Whole-stream, NO_3 incr. 5×	Munn and Meyer (1990)
Phosphate							
Sonoran Desert streams	S. Arizona	—	Open	—	186–18,000	Whole-stream, ambient conc. change	Grimm *et al.* (1981)
Hugh White Ck.	N. Carolina	2	Closed	1	72–264[a]	Whole-stream, SRP incr. 6×	Munn and Meyer (1990)
WS 2, H. J. Andrews	Oregon	2	Closed	5	36–54[a]	Whole-stream, SRP incr. 3×	Munn and Meyer (1990)
Walker Branch	E. Tennessee	1	Closed	1–5	5–130	Whole-stream, radiotracer, periphyton only	Mulholland *et al.* (1985), Mulholland, unpublished data
					50–300	Chambers, radiotracer, periphyton only	Steinman and Boston (1993)
Boreal forest streams	NE Quebec	1–9	Open-Closed	0.9–2.6	20–180	Chambers, radiotracer, periphyton only	Corning *et al.* (1989)

[a]Computed for ambient concentrations using uptake rate coefficients ($1/S_w$, computed by authors), flow (Q), ambient concentrations (C), and stream width (w) according to the following formula: uptake rate (ambient) = $(Q)(C)/(S_w)(w)$ (see equation 9 in Stream Solute Workshop, 1990). Data from only cobble, riffle, and bedrock reaches were used because these are presumably dominated to a greater extent by periphyton than reaches with debris dams.

phyton as well as algae. In streams with greater algal biomass, such as the Sonoran Desert streams studied by Grimm and coworkers (Grimm *et al.*, 1981; Grimm, 1992), the role of benthic algae in nutrient uptake is likely much greater than in small forested streams such as Walker Branch.

Nutrient uptake by benthic algae can have significant spatial (i.e., longitudinal) effects on nutrient cycling in streams via effects on stream water nutrient concentrations. Grimm *et al.* (1981) reported nutrient concentration declines of up to 90% over stream reaches of varying lengths (80 to 6000 m) below groundwater nutrient sources in Arizona desert streams. These spatial declines in concentration were steepest for nitrate, and presumably intensified N limitation and cycling in the algal communities downstream. There can also be an interaction between spatial and temporal dynamics in nutrient cycling. For example, Grimm, Fisher, and colleagues (Fisher *et al.*, 1982; Grimm, 1987; Valett *et al.*, 1994) have shown that net N retention and rates of longitudinal decline in stream water N concentration increased as benthic algal communities regrew during the period immediately following flood events, thus suggesting progressively more intensive N cycling with time and distance. As mentioned earlier, Mulholland and Rosemond (1992) also reported longitudinal nutrient concentration declines during some seasons in Walker Branch, and found longitudinal changes in nutrient content and phosphatase activity of periphyton communities in response to the longitudinal nutrient gradients (Fig. 19.2). In a laboratory stream study, Mulholland *et al.* (1991) showed that reductions in nutrient input and stream water nutrient concentrations resulted in greater nutrient cycling in streams with a high biomass of benthic algae. Greater nutrient cycling rates compensated for reduced nutrient inputs, thus maintaining algal biomass and primary production rates at levels similar to those in streams with no reductions in nutrient input. In a subsequent study conducted in long laboratory streams, Mulholland *et al.* (1995) found few longitudinal patterns in total periphyton biomass or primary productivity with distance downstream from water input, but distinct longitudinal patterns in phosphorus cycling. Whereas P cycling supplied only 10–25% of P uptake from water in upstream segments, P cycling supplied 60–70% of P uptake in the downstream segments.

An important role of nutrient cycling in ecosystems is support of primary production above levels that can be supported by one-time use of nutrient inputs from external sources alone. In the ocean and in many lakes, most of the primary production is supported by nutrient cycling (Liao and Lean, 1978; Eppley and Peterson, 1979; Caraco *et al.*, 1992). In streams, because of large inputs of nutrients via groundwater, nutrient cycling probably supports a lower fraction of the primary production than in lentic ecosystems. For example, phosphorus inputs via groundwater to the entire first-order portion of Walker Branch (360 m) are approximately 100 μg m^{-2} streambed h^{-1} during baseflow periods [based on flows of 7

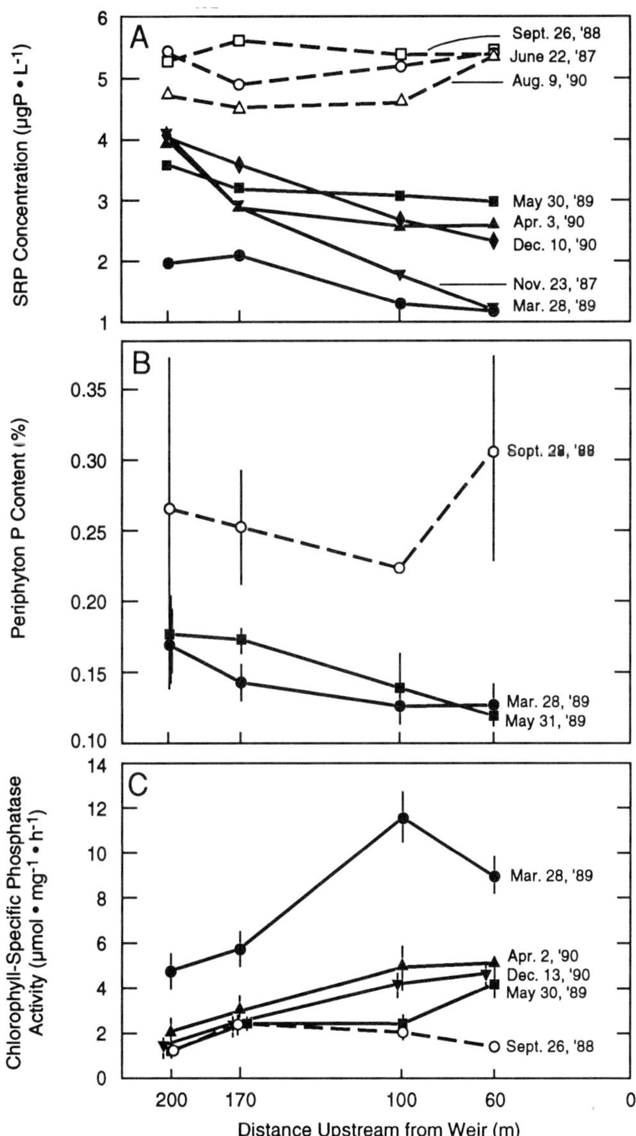

FIGURE 2 Longitudinal gradients in stream water phosphorus concentration and response of parameters related to nutrient cycling in Walker Branch. Phosphatase activity values >3 μmol mg^{-1} h^{-1} indicate slight P deficiency and values >5 μmol mg^{-1} h^{-1} indicate severe P deficiency according to studies of Healey and Hendzel (1979). (Adapted from Mulholland and Rosemond, 1992.)

liters s^{-1} and groundwater soluble reactive phosphorous (SRP) concentrations of 5 μg $liter^{-1}$)], and thus high enough to support most of the P uptake by benthic algae in Walker Branch (Table I). However, total P uptake in Walker Branch ranges from about 150 to 450 μg m^{-2} h^{-1}, considerably higher than uptake by the benthic algae alone. Therefore, most of the total P uptake (including that by benthic algae) must be supported by recycled P in Walker Branch.

C. Nutrient Remineralization

Nutrient remineralization (release to water of nutrients in biologically available forms) is the process that completes the nutrient cycle. Increases in nutrient remineralization rate will tend to increase nutrient cycling, particularly in streams with nutrient-limited algal communities. Nutrient remineralization from benthic algae is accomplished via several mechanisms: (1) excretion or passive leakage of nutrients from algal cells, (2) cell lysis by viruses or physical/chemical agents, (3) hydrolysis of dissolved organic forms of nutrients by exoenzymes produced by algae and bacteria (e.g., phosphatases, proteases), and (4) consumption of algae and release of nutrients by consumers (microconsumers, macroinvertebrates, fish) via excretion or leaching of feces. I consider the first three mechanisms in the following, and the effects of consumers in Section III,B.

Laboratory studies conducted using algal cultures and P radiotracers have indicated that algae can lose up to 10–20% of their cellular P per hour, although sustained rates of P loss are probably <3% per hour (Lean and Nalewajko, 1976; Currie and Kalff, 1984a; Olsen, 1989; Jansson, 1993). It is unclear under what conditions algal loss of P is high. Some studies have indicated the highest rates of P loss when algae become P-starved (e.g., Olsen, 1989), whereas others have shown maximum P loss under P-replete conditions (e.g., Jansson, 1993) as might be expected. In one of the few studies conducted under flowing-water conditions, Borchardt *et al.* (1994) reported P loss rates from *Spirogyra fluviatilis* of generally <1% of cellular P per hour, with maximum loss rates of 3% per hour when the alga was P-replete.

Nutrient loss from benthic algae can be in the form of either inorganic or organic compounds. If nutrients are taken up by benthic algae primarily in inorganic form and released back to stream water in organic form, the availability of these nutrients to downstream communities may be reduced. There have been reports of increases in dissolved organic nutrients and decreases in dissolved inorganic nutrients from upstream to downstream in some streams (Tate, 1990; Mulholland, 1992). Net transformation from inorganic to organic forms of nutrients in stream water may result in reduced rates of uptake and cycling downstream.

Some organic forms of nutrients may become available to benthic algae via production of hydrolytic exoenzymes, thus increasing rates of cycling. For example, stream algae produce phosphatase under conditions of phosphorus deficiency and low stream water phosphate concentrations (Bothwell, 1985; Mulholland *et al.*, 1986; Mulholland and Rosemond, 1992; Rosemond *et al.*, 1993). Organic phosphorus concentrations in stream water (usually determined as the total soluble P unreactive with molybdate prior to persulfate or UV hydrolysis) are often as large or larger than soluble reactive phosphorus concentrations (e.g., Peters, 1978; Mulholland, 1992). Hydrolysis of organic P compounds via phosphatases and subsequent uptake of the cleaved PO_4 by benthic algae enhance the total quantity of P cycled through stream biota. In several upstate New York streams, concentrations of organic phosphorus compounds that could be hydrolyzed by phosphatases (termed phosphatase-hydrolyzable phosphorus) were typically 1–2 μg P $liter^{-1}$, values that in some cases were similar to SRP concentrations (Klotz, 1991). In addition, Klotz's study suggested that phosphatase-hydrolyzable phosphorus adsorbed to streambed sediments may be an important source of phosphorus to benthic algae that have the ability to produce external phosphatases. Organic nitrogen also can be remineralized via hydrolytic exoenzymes within benthic algal communities. Dissolved organic N concentrations often exceed dissolved inorganic N concentrations in undisturbed streams (e.g., Triska *et al.*, 1984; Tate, 1990; Mulholland, 1992) and it is possible that exoenzymes produced by benthic algae increase the cycling of N as well as P in streams. Protease activity associated with periphyton has been detected in laboratory and natural streams at Oak Ridge National Laboratory (A. V. Palumbo, Environmental Sciences Division, Oak Ridge National Laboratory, unpublished data).

III. INDIRECT EFFECTS

A. Formation of Boundary Zones

Benthic algal communities can change the hydraulic characteristics of the stream bottom as they accumulate biomass. Zones of relatively stationary water (boundary zones) develop around and within benthic algal communities, thus partially isolating the algae from the overlying flowing water. The size of these boundary zones is dependent on algal biomass and growth form, as well as the physical characteristics of the streambed (roughness) and the velocity of the flowing water. Wetzel (1993) argued that boundary zones are a general feature of attached communities and are responsible for the high sustained rates of primary productivity in many inland aquatic ecosystems via creation of microhabitat that enhances nutrient conservation and cycling.

The boundary zones established by benthic algal communities in streams are critically important to microbes, meiofauna, and other small-bodied organisms because they provide a relatively quiescent microhabitat in which the need for strong attachment mechanisms is reduced. In this way, boundary zones counteract the fundamental problem for life in lotic ecosystems—downstream displacement by the strong advective forces of flowing water. In their review of elemental dynamics in lotic ecosystems, Meyer *et al.* (1988) suggested that the intimate relationships between algae, heterotrophic microbes, and consumers within benthic boundary zones are critical factors in nutrient cycling and argued for increased study of these zones.

Boundary zones also reduce the exchange of solutes between benthic organisms and the flowing water, thereby allowing a greater level of autogenic control of the immediate chemical environment and enhancing the importance of nutrient cycling. Transport rates within boundary zones are substantially lower than in overlying stream water, resulting in development of steep gradients in nutrient concentrations within them. Transport of nutrients from stream water to organisms residing at lower levels within boundary zones is restricted (Bothwell, 1989; Pringle, 1990; Stevenson and Glover, 1993). Although this partial isolation of benthic algae and their associated heterotrophic community may reduce the external nutrient supply from the flowing water, it can enhance nutrient recycling by reducing the loss of remineralized nutrients to the flowing water (Riber and Wetzel, 1987; Paul and Duthie, 1989; Mulholland *et al.*, 1994; see also review by Raven, 1992).

Nutrient cycling within boundary zones created by benthic algae in streams can be considered in-place cycling–cycling that does not involve downstream displacement of nutrients and is thus an alternative pathway to nutrient spiraling (Fig. 3). If boundary zone cycling is important, then nutrient atoms may complete many more cycles within a given length of stream than would be computed from measurements of spiraling length alone (see Newbold *et al.*, 1981, 1983; Mulholland *et al.*, 1985). In this way, considerably higher rates of productivity can be supported within

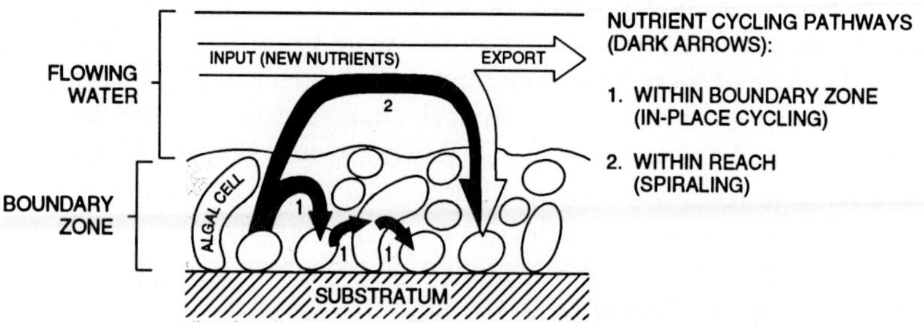

FIGURE 3 Pathways of nutrient cycling involving benthic algal communities within boundary zones in streams. (Adapted with permission from Mulholland *et al.*, 1991.)

streams than estimates based on measurements of nutrient uptake from flowing water.

Other conceptual models of boundary zones formed by benthic algae in streams have been presented. Lock *et al.* (1984) described the structural and functional characteristics of a river epilithon community, emphasizing the importance of the polysaccharide matrix produced by bacteria and algae as a sorption site for solutes as well as a site of intense exoenzyme activity (see also Lock, 1993). Enzyme assays of epilithic biofilms in streams have indicated a rich array of enzymes active within these communities (Sinsabaugh *et al.*, 1991). DeAngelis *et al.* (1995) developed a stream ecosystem model that explicitly includes boundary zones (Fig. 4A). Steady-state simulations of this model have indicated that nutrient cycling is greater in stream reaches with larger boundary zones, as indicated by longer nutrient residence times in these reaches (Fig. 4B).

What is the evidence that boundary zones created by benthic algae do exist in streams? One approach that has been used to quantify boundary zones in streams is that proposed by Bencala and Walters (1983) for the determination of transient storage zones in streams. This approach involves conservative tracer injections to stream water and fitting an advection–dispersion model to the data (see also Stream Solute Workshop, 1990). Transient storage zones defined by Bencala and coworkers include all zones of stationary water that can exchange water and solutes with the flowing water in streams. Although in most streams transient storage zones are dominated by water within streambed sediments (hyporheic zone), boundary zones created by benthic algae can also be important transient storage zones in streams. For example, Kim *et al.* (1990) found that periphyton accumulations on artificial stubstrata placed in flumes set on the bottom of a natural stream produced measurable transient storage zones. Mulholland *et al.* (1994) compared the results of conservative tracer injections made to laboratory streams with thin and thick algal communities and found distinctly greater transient storage zones in the thick community streams, with transient storage zone volumes up to about 18% of the flowing water volume (Fig. 5). Although both of these studies were performed in artificial channels, biomass and growth form of the benthic algal communities were similar to those reported for natural streams, particularly those with high light levels.

The empirical evidence for nutrient cycling within boundary zones created by benthic algae in streams is compelling but largely circumstantial. Bothwell (1989) studied benthic algal growth under different nutrient levels and found that the PO_4 concentrations required to saturate total biomass accumulation rates (about 30 μg P liter^{-1}) were two orders of magnitude greater than those required to saturate cell-specific growth rates of algae in thin-film periphyton communities. Bothwell suggested that physical diffusion controlled nutrient transport from flowing water into the

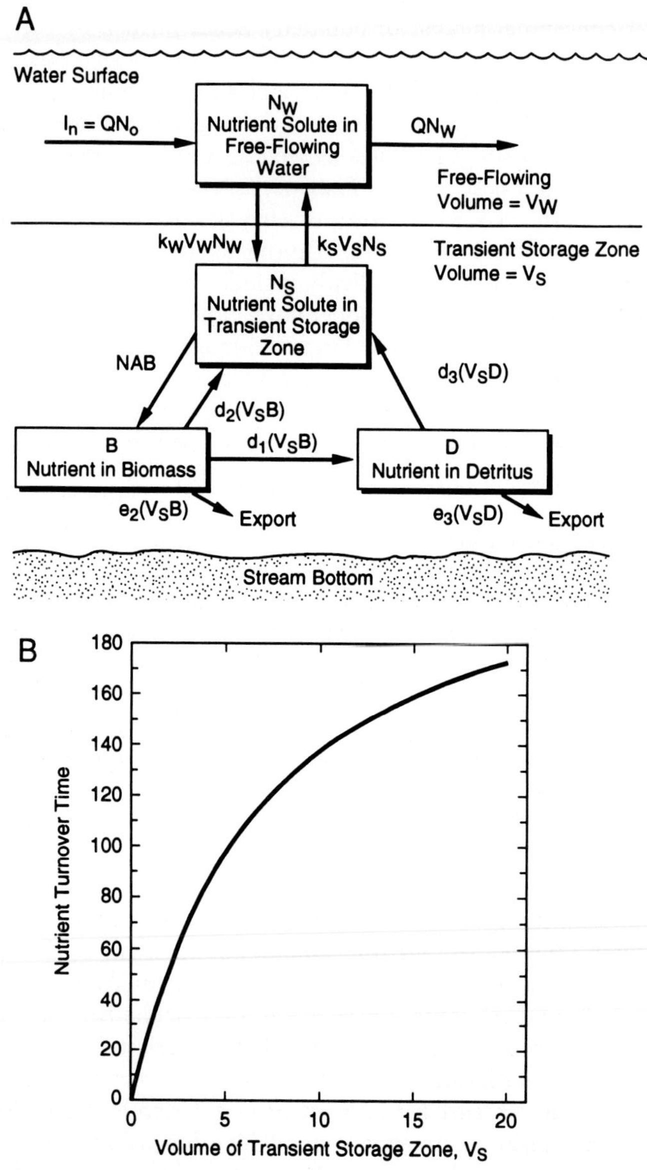

FIGURE 4 (A) Stream model of nutrient cycling in which benthic algae are contained within a boundary zone (transient storage zone). (B) Effect of boundary zone volume on nutrient turnover time within boundary zones. Longer nutrient turnover times within boundary zones are the result of increased recycling within this zone. (Adapted with permission from DeAngelis *et al.*, 1995.)

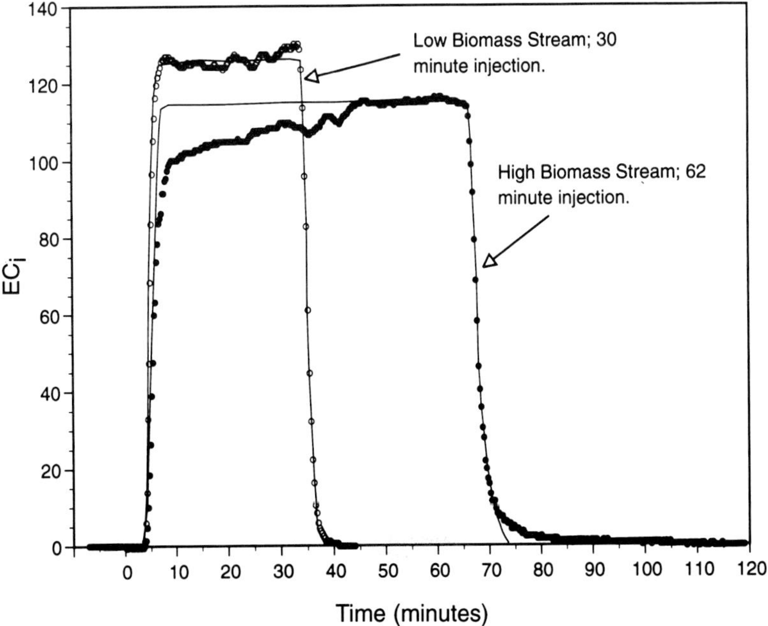

FIGURE 5 Effects of benthic algal biomass on the concentration profile of a conservative tracer (chloride, measured as the change in electrical conductance above background, (EC_i) injected 22 m upstream. The solid lines indicate the EC_i profile simulated by an advection–dispersion model with no transient storage zone (boundary zone). The longer time needed to reach steady-state EC_i during the injection and the more extended tail of the EC_i profile in the high-biomass stream indicate larger transient storage zone volume compared to in the low-biomass stream. Transient storage zone volumes were computed to be 3% of the flowing water volume in the low-biomass stream and 18% of the flowing water volume in the high-biomass stream, based on an advection–dispersion model with a transient storage zone. (Reprinted with permission from Mulholland *et al.,* 1994.)

periphyton matrix, increasingly limiting nutrient availability as algal biomass accumulated. Higher stream water PO_4 concentrations stimulated biomass accumulation by increasing concentration gradients between the flowing water and the algae, and thus the flux of nutrients into the algal matrix. Horner *et al.* (1990) reported that P uptake by benthic algae increased with increases in stream water PO_4 concentration up to about 15 μg P liter^{-1} and with increases in water velocity up to about 50 cm s^{-1}, suggesting that diffusion to interior algal cells limited total P uptake by benthic algal communities. Other studies have also demonstrated a strong linkage between stream water velocity and nutrient uptake by benthic algae (Whitford and Schumacher, 1964; Horner and Welch, 1981; Lock, 1979; Gantzer *et al.,* 1988; although see also Borchardt *et al.,* 1994).

Studies of nutrient uptake by benthic algae in streams have demonstrated the importance of the biomass and physiognomy of the community, attributes that control the size of boundary zones. From his work in laboratory streams, McIntire (1968) postulated that benthic algal communities are composed of an outer layer of actively growing cells that can remove nutrients from the flowing water and an inner layer of older, less metabolically active cells, some of which are decomposing. Subsequent studies of nutrient uptake by benthic algal communities in natural streams have largely confirmed this viewpoint. In thick, well-developed algal communities, uptake rates of overstory cells were considerably greater than those of understory cells (Paul and Duthie, 1989; Pringle, 1990). McCormick and Stevenson (1991) have shown that the response of an understory alga to nutrient additions to stream water increased when an overstory of diatoms was removed by grazers. Similarly, Peterson and Grimm (1992) reported that responses of nutrient-limited algae to nutrient enrichment were greater during early succession when algal biomass was low than later in succession when biomass was high. Peterson and Grimm proposed that the temporal change in response to nutrient enrichment was caused by the development of a thick periphyton mat that reduced the availability of nutrients from the flowing water and from the substratum, and increased reliance on internal nutrient cycling.

Mulholland *et al.* (1994) proposed two indices to quantify nutrient cycling within boundary zones produced by stream algae. The first index involves measurements of primary production and nutrient uptake from flowing water, the latter using radiotracer techniques, and calculation of the nutrient uptake:primary production ratio (N_{upt}:PP). The second index involves determining the apparent turnover rate of nutrients within benthic algal communities, again using radiotracer techniques. Because both of these approaches require the use of radiotracers, they are applicable only to cycling of nutrients for which radiotracers exist (e.g., P). The first index, the N_{upt}:PP ratio, assumes that the total nutrient uptake by autotrophic organisms in the boundary zone is proportional to primary production (i.e., there is a relatively narrow range in the stoichiometric requirements of primary production among algae). If nutrient cycling within the boundary zone is high and nutrient uptake from stream water (N_{upt}) provides only a portion of the nutrient demands associated with primary productivity, the N_{upt}:PP ratio will be lower than if nutrient cycling is low. Testing this index in a laboratory stream study of P dynamics, Mulholland *et al.* (1994) showed that N_{upt}:PP ratios were about an order of magnitude lower in high-biomass benthic algal communities than in low-biomass algal communities exposed to the same stream water, suggesting that much more of the P demand was met by cycling in the high-biomass communities.

The second index of boundary zone nutrient cycling (apparent nutrient turnover rate in algae) was also tested in the laboratory stream study of Mulholland *et al.* (1994). Turnover rates of P in benthic algal communities, determined following radiotracer labeling experiments, were about threefold lower in the high-biomass communities than in the low-biomass communities, again suggesting that P cycling was greater in the high-biomass communities. The lower turnover rates of P in the high-biomass communities were not the result of lower rates of metabolism because gross primary production per unit biomass was actually somewhat higher in the high-biomass communities.

Finally, boundary zones formed by benthic algal communities in streams can enhance the loss of nitrogen via denitrification. Denitrification, a heterotrophic process performed by some bacteria under anaerobic conditions, may be possible within benthic algal communities because of limited water exchange and readily available organic carbon substrates. Denitrification is most likely to occur deep within the algal matrix where light levels (and consequently O_2 production) are low, many of the algal cells are senescent, and anaerobic microzones can be established. In one of the first experimental studies of denitrification in stream epilithic communities, Ventullo and Rowe (1982) demonstrated that denitrification could occur when the bulkwater was well oxygenated, although rates were below the limit of detection unless relatively high concentrations of nitrate or nitrite were added (>50 mg N liter^{-1}). Duff *et al.* (1984) reported denitrification rates of up to 1.3 mmol NO_3 m^{-2} d^{-1} in intact periphyton communities (dominated by *Cladophora* sp.) in an N-rich northern California stream (San Francisquito Creek), but no measurable denitrification in an N-poor stream (Little Lost Man Creek). Denitrification rates of up to about 9 mmol NO_3 m^{-2} h^{-1} have been reported at the base of algal communities growing on hard and soft sediments in N-rich Danish streams (Christensen *et al.*, 1990; Nielsen *et al.*, 1990a,b). Denitrification rates were considerably higher during the night than during the day in all streams, suggesting that anaerobic microsites in the periphyton matrix expand when photosynthetic oxygen production ceases.

B. Effects of Algal–Herbivore Interactions

Interactions between benthic algae and herbivores strongly influence nutrient cycling in lotic ecosystems via: (1) direct consumption of algae and nutrient remineralization by herbivores, (2) differential herbivory on algal species or growth forms with different nutrient uptake kinetics, and (3) stoichiometric imbalances between the herbivore and its algal diet.

Consumption of algae by herbivores can have either positive or negative effects on nutrient cycling in streams, depending on the relative importance of effects on algal biomass (negative) and nutrient remineralization (positive). In a theoretical analysis, Newbold *et al,* (1982) showed that for streams in which nutrient limitation is strong, herbivory should increase nutrient cycling rates (decrease spiraling lengths). However, in an empirical test of this hypothesis conducted in laboratory streams, Mulholland *et al.* (1983) found that herbivorous snails *(Elimia clavaeformis)* decreased nutrient cycling rates (increased spiraling lengths), because their negative effects on algal biomass were more important than their positive effects on nutrient remineralization. Additional studies are needed to determine if other herbivores have a similar effect on nutrient cycling in streams.

Benthic algae are consumed and their nutrients are remineralized by a variety of stream herbivores ranging in size from microscopic protozoa, which consume only the small unicellular forms of algae, to herbivorous fish that can consume even large filamentous forms. A review of nutrient remineralization by aquatic insects is provided by Merritt *et al.* (1984). Herbivores remineralize algal nutrients in several ways: (1) "sloppy feeding," which involves inadvertent breakage of algal cells and release of cell contents during feeding without actual consumption of those cells; (2) excretion of dissolved nutrients to water when nutrient consumption exceeds the metabolic needs of the herbivore; and (3) production of feces and subsequent leaching or microbial release of nutrients from the fecal material. In a laboratory stream study in which herbivorous snails were added at densities typical of local natural streams, Mulholland *et al.* (1991) found that nutrient excretion by snails could supply about 14% of the nutrient demand by benthic algae. In Walker Branch, using radiotracer methods, Jay (1993) found that P remineralization by snails could supply about 25% of the periphyton P demand. In a desert stream in Arizona, from 15 to 70% of the total N taken up by algae was remineralized by macroinvertebrates (Grimm, 1988). There is little information on the role of meio- or microfaunal herbivores in streams or rivers (e.g., microflagellates, ciliates, nematodes), although these organisms are known to be important nutrient regenerators in pelagic and other benthic environments (Johannes, 1965; Fenchel, 1970; Goldman and Caron, 1985).

Differential grazing on stream algal communities has been frequently reported, because not all algal species or growth forms are equally available to stream herbivores (see reviews by Gregory, 1983; Lamberti and Moore, 1984; see Steinman, Chapter 12, this volume). Intense herbivory often results in stream benthic algal communities dominated by species with prostrate growth forms or those that are difficult to consume, such as gelatinous forms (Lamberti and Resh, 1983; Jacoby, 1987; Steinman *et al.*, 1987a, 1992; Rosemond *et al.*, 1993; Rosemond, 1994). This effect of herbivory also can result in changes in nutrient uptake and cycling in the benthic algal community if grazer-resistant species or growth forms have different nutrient uptake kinetics than species preferred by grazers. For example, Steinman *et al.* (1992) showed that some grazer-resistant species of stream algae (e.g., *Oedogonium* sp.) had relatively high P uptake rates, whereas other grazer-resistant species (e.g., *Tetraspora* sp.) had relatively low P uptake rates compared to other species more accessible to grazers. Studies by Rosemond (Rosemond *et al.*, 1993, Rosemond and Brawley, 1996) have suggested that the dominant alga in Walker Branch (*Stigeoclonium* sp.) is a relatively poor nutrient competitor at low (ambient) concentrations, but persists because its basal form is highly resistant to grazers. Consequently, herbivory appears to result in lower rates of nutrient uptake by the algal community in Walker Branch, independent of its effects on total algal biomass. Herbivory also may enhance the importance of N-fixing cyanophytes, as described in Section II,A, thus increasing the input of fixed nitrogen to streams. Finally, grazing has been shown to alter the chemical composition of stream algae, particularly the fatty acid and amino acid composition, which may alter nutrient demand or remineralization rates (Steinman *et al.*, 1987b).

Imbalances between the stoichiometric requirements of herbivores and the stoichiometry of algal biomass can strongly influence the remineralization rates and ratios of nutrients. Although I have found no studies that have addressed effects of differences in algae–herbivore nutrient stoichiometry in lotic ecosystems, there have been a number of such studies in pelagic environments. Of critical importance is the nutrient:C ratio of algae relative to the nutrient requirements for somatic growth and production of herbivores. When herbivores are limited by C the remineralization of other nutrients can be high. This condition will occur if the algal nutrient:C ratio is high relative to the nutrient:C ratio of the herbivore (Olsen *et al.*, 1986; Urabe and Watanabe, 1992). Although nutrient stoichiometry varies relatively little among individuals of the same herbivore species, stoichiometric variability between species can be considerable and changes in the species composition of the herbivore community have been shown to have a strong influence on the rates of N and P remineralized (Andersen and Hessen, 1991; Hessen and Lyche, 1991). Herbivore growth efficiency is also an important determinant of nutrient remineralization rate because only nutrients required for growth are retained by the herbivore. Model studies of nutrient dynamics in algal–herbivore systems have shown that nutrient remineralization is greater at low herbivore growth efficiencies than at high growth efficiencies (Hessen and Andersen, 1992; Landry, 1993).

Stoichiometric relationships among algae and herbivores can influence the ratios as well as the rates of nutrients cycled, and thus the relative availability of the nutrient most strongly limiting algal production. Sterner (1990) showed that imbalances between herbivore and algal N:P ratios will result in lower cycling rates of the nutrient most deficient in the food relative to the needs of the herbivore, thus increasing the potential for that nutrient to become limiting or more limiting to algal production. Because of taxonomic differences in herbivore N:P requirements, shifts in the dominant herbivore species can alter N:P remineralization ratios and thus resource competition among algae (Elser, *et al.,* 1988; Sterner, 1990; Sterner, *et al.,* 1992; Carpenter, *et al.,* 1992). Though these effects are probably greater in ecosystems in which nutrient supply is dominated by recycling (e.g., lakes with relatively long water turnover times), they may also be important in long stream reaches with little groundwater input or in thick periphyton mats that rely heavily on cycling to meet nutrient demands.

Herbivore feces may be a source of remineralized nutrients to benthic algal communities if nutrients are present in leachable forms. However, any leaching of nutrients from herbivore feces is probably short term, because feces have been shown to accumulate nutrients as the bacteria and fungi colonizing this material compete with benthic algae for stream water nutrients (Fisher and Gray, 1983). Evidence from pelagic systems suggests that the role of heterotrophic microbes as nutrient competitors or remineralizers appears to depend on the availability of labile organic carbon (Cotner and Wetzel, 1992; Jansson, 1993).

IV. NUTRIENT TURNOVER RATES AND COMPARISON WITH OTHER ECOSYSTEMS

Nutrient turnover rates in water and in algal biomass are often used to compare the role of algae in nutrient cycling among ecosystems. For streams, nutrient turnover rate in stream water resulting from uptake by benthic algae can be calculated as

$$N_{upt}/(N_w \cdot d) \qquad (2)$$

where N_w is the nutrient concentration in stream water (mg m^{-3}), d is the average stream depth (m), and N_{upt} is the nutrient mass uptake rate by benthic algae (mg m^{-2} h^{-1}). Nutrient turnover rate in benthic algal biomass can be calculated as

$$N_r/(N_a \cdot B) \qquad (3)$$

where N_a is the nutrient concentration in algal biomass (mg mg^{-1}), B is algal biomass per unit area (mg m^{-2}), and N_r is the nutrient mass release rate from benthic algae (mg m^{-2} h^{-1}). At steady state, $N_r = N_{upt}$, and often the latter term is used to compute turnover rate in algae.

Phosphorus turnover rates in water and in periphyton have been computed for Walker Branch, Tennessee, and for Hugh White Creek, North Carolina, using data from radiotracer injection experiments conducted in these streams. They are compared with turnover rates reported in lakes and the ocean (Table II). Phosphorus turnover rates in stream water resulting from P uptake by periphyton were in the low end of the range reported

TABLE II Turnover Rates of Phosphorus in Water (Resulting from Algal P Uptake) and in Algae in Streams, Lakes, and Marine Ecosystems

System	*Water PO_4 turnover rate (h^{-1})*	*Algal P turnover rate (d^{-1})*	*Reference*
Streams			
Walker Branch, TN	0.30–0.62[a]	0.03–0.10[b]	P. J. Mulholland, unpublished data
Hugh White Ck., NC	0.26[a]	0.003[c]	P. J. Mulholland, unpublished data
Lakes			
Jack's Lake	2.0–3.0[d]	6.2[d]	Lean and White (1983)
Lake Mekkojarvi	0.3[e]	0.5[e]	Salonen *et al.* (1994)
Lake Memphremagog	0.3–6.0	—	Peters (1979), Currie and Kaliff (1984b)
Lake Michigan	6.0–18	—	Tarapchak *et al.* (1981)
Canadian and New Zealand lakes	0.01–60	—	White *et al.* (1982)
Shallow Alberta lakes	0.07 30[f]	—	Prepas (1983)
Estuaries/oceans			
Coastal Nova Scotia	0.001–0.02	0.5–1.0	Harrison (1983)
Strait of Georgia	0.04	—	Nalewajiko and Lee (1983)
Sargasso Sea	0.12	—	Nalewajiko and Lee (1983)

[a]Calculated by measuring rate of decline in stream water $^{33}PO_4$ and fraction of total ^{33}P uptake due to periphyton during 2-h radiotracer injections to stream water on three occasions in Walker Branch (April–June) and on one occasion in Hugh White Creek (July).
[b]Calculated by measuring rate of decline in ^{33}P present in periphyton biomass following a 2-h injection of $^{33}PO_4$ to stream water.
[c]Calculated by dividing the periphyton P uptake rate (determined from uptake of $^{33}PO_4$ by periphyton during radiotracer injection) by periphyton total P content.
[d]>1 μm size fraction only.
[e]1–50 μm size fraction only.
[f]Lakes with SRP concentrations < 10 μg liter^{-1}.

for lakes, but were higher than P turnover rates reported in the ocean. This comparison suggests that benthic algae in streams are capable of depleting PO_4 supplies in water as rapidly as the phytoplankton in many lakes, and more than rapidly than phytoplankton in the oceans.

Phosphorus turnover rates in periphyton in Walker Branch are 10–30 times greater than the one measurement made in Hugh White Creek, but are 5 to 100 times lower than P turnover rates in algae in lakes and the ocean (Table II). The more rapid turnover rates of algal P in lakes and the ocean compared to in streams may be a result of more intensive herbivory in the pelagic environments. Alternatively, a greater portion of the nutrient uptake by periphyton communities in streams may go into relatively recalcitrant materials (e.g., mucilage) than in pelagic algal communities.

V. SUMMARY AND CONCLUSIONS

Benthic algae contribute to nutrient cycling in stream ecosystems directly via increasing nutrient supplies, uptake of nutrients from stream water, and release of nutrients back to stream water. In streams in forested catchments, much of the nutrient uptake is by heterotrophic microbes (bacteria, fungi); nonetheless, benthic algae can contribute significantly to nutrient uptake during times of the year when the forest canopy is leafless or in stream reaches with open canopies. Because streams are shallow (high bottom surface:volume ratios), P uptake by benthic algae can result in PO_4 turnover rates in water that are as high as those measured in lakes and higher than those measured in the ocean. Depletion of nutrient concentrations in stream water as a result of nutrient uptake can result in longitudinal increases in nutrient limitation of benthic algae and greater dependence on nutrient cycling to meet algal nutrient demands. In Walker Branch, Tennessee, more of the phosphorus demand of benthic algae is met by recycled P than by groundwater inputs of P.

Benthic algae also contribute to nutrient cycling indirectly through the formation of boundary zones (zones of nearly stationary water) within and immediately surrounding accumulations of algal biomass. By altering the hydraulic characteristics of the stream bottom, boundary zones increase the probability that remineralized nutrients will be recycled before they are transported downstream by flowing water. In this way, boundary zones partially isolate benthic algal communities from the flowing water and enhance the importance of autogenic processes such as nutrient cycling in controlling algal biomass, productivity, and species composition. Boundary zones also provide habitat for herbivores and other consumers, enhancing nutrient remineralization within benthic algal communities. Formation of boundary zones associated with benthic algae in streams may have important implications for more applied issues, such as the effects of stream water toxicants on benthic algae and their fate once assimilated. Additional research is needed to (1) define the physical dimensions of boundary zones in different algal communities, (2) quantify the interactions between heterotrophs and autotrophs within boundary zones, and (3) identify the relative importance of autogenic processes and exchanges with the flowing water in controlling nutrient cycling and solute dynamics within boundary zones.

Considerable progress has been made over the past 25 years in elucidating the role of benthic algae in stream nutrient cycles. In particular, we now have a reasonable understanding of P cycling, although our understanding of the importance of P cycling within benthic algal mats is far from complete. Our knowledge of N cycling is much less advanced than for P, probably because of the lack of a radioactive N isotope suitable for use as a tracer. More creative use of stable N isotopes must be made in studies of N cycling in streams. The rates and controls on the cycling of nutrients other than N and P also deserve further study. In particular, the role of benthic algal communities in the cycling of Si (particularly in streams dominated by diatoms), Ca, Mg, and S, and the trace metals Fe, Mn, and Mo, should be investigated. Though many of the factors controlling the cycling of these nutrients may be similar, there undoubtedly are important differences, and it is the relative rates of cycling of several potentially limiting nutrients that may control ecosystem productivity and species composition.

Finally, interactions between benthic algae and herbivores, such as differential herbivory and imbalances between herbivore and algal stoichiometry, can strongly influence nutrient cycling in benthic algal communities. Recent advances have been made in understanding the influence of herbivore and algal stoichiometry on nutrient dynamics in pelagic systems. Although these findings can be tentatively applied to algal–herbivore interactions in stream nutrient cycles, similar studies should be conducted in streams.

ACKNOWLEDGMENTS

This chapter has benefited from discussions with many of my colleagues and graduate students at Oak Ridge National Laboratory. In particular, I wish to thank Don DeAngelis, Alan Steinman, Amy Rosemond, Erich Marzolf, Deborah Hart, Susan Hendricks, Walter Hill, Art Stewart, Tony Palumbo, Dave Kirschtel, and Terry Flum. Amy Rosemond, Erich Marzolf, Alan Steinman, Jan Stevenson, Rex Lowe, and an anonymous reviewer reviewed the manuscript and provided many suggestions to improve it. Support during manuscript preparation was received from the Ecosystems Studies Program of the National Science Foundation under Interagency Agreement No. DEB-9013883 with the U.S. Department of Energy. Oak Ridge National Laboratory is managed by Lockheed Martin Energy Research, Corp. under Contract DE-AC0584OR21400 with the U.S. Department of Energy. This is Publication Number 4376, Environmental Sciences Division, Oak Ridge National Laboratory.

REFERENCES

Andersen, T., and Hessen, D. O. (1991). Carbon, nitrogen, and phosphorus content of freshwater zooplankton. *Limnol. Oceanogr.* **36**, 807–814.

Bencala, K. E., and Walters, R. A. (1983). Simulation of solute transport in a mountain pool-and-riffle stream: A transient storage model. *Water Resour. Res.* **19**, 718–724.

Biggs, B. J. F., and Gerbeaux, P. (1993). Periphyton development in relation to macro-scale (geology) and micro-scale (velocity) limiters in two gravel-bed rivers, New Zealand. *N. Z. J. Mar. Freshwater Res.* **27**, 39–53.

Borchardt, M. A., Hoffmann, J. P., and Cook, P. W. (1994). Phosphorus uptake kinetics of *Spirogyra fluviatilis* (Charophyceae) in flowing water. *J. Phycol.* **30**, 403–417.

Bothwell, M. L. (1985). Phosphorus limitation of lotic periphyton growth rates: An intersite comparison using continuous-flow troughs (Thompson River system, British Columbia). *Limnol. Oceanogr.* **30**, 527–542.

Bothwell, M. L. (1989). Phosphorus-limited growth dynamics of lotic periphytic diatom communities: Areal biomass and cellular growth rate responses. *Can. J. Fish. Aquat. Sci.* **46,** 1293–1301.

Caraco, N. F., Cole, J. J., and Likens, G. E. (1992). New and recycled primary production in an oligotrophic lake: Insights for summer phosphorus dynamics. *Limnol. Oceanogr.* **37,** 590–602.

Carpenter, S. R., Cottingham, K. L., and Schindler, D. E. (1992). Biotic feedback, in lake phosphorus cycles. *Trends in Ecology and Evolution* **7,** 332–336.

Charlton, S. E. D., and Hickman, M. (1988). Epilithic algal nitrogen fixation, standing crops and productivity in five rivers flowing through the oilsands region of Alberta/Canada. *Arch. Hydrobiol., Suppl.* **79,** 109–143.

Christensen, P. B., Nielsen, L. P., Sorensen, J., and Revsbech, N. P. (1990). Denitrification in nitrate-rich streams: Diurnal and seasonal variation related to benthic oxygen metabolism. *Limnol. Oceanogr.* **35,** 640–651.

Coleman, R. L., and Dahm, C. N. (1990). Stream geomorphology: effects on periphyton standing crop and primary production. *J. North Am. Benthol. Soc.* **9,** 293–302.

Corning, K. E., Duthie, H. C., and Paul, B. J. (1989). Phosphorus and glucose uptake by seston and epilithon in boreal forest streams. *J. North Am. Benthol. Soc.* **8,** 123–133.

Cotner, J. B., Jr., and Wetzel, R. G. (1992). Uptake of dissolved inorganic and organic phosphorus compounds by phytoplankton and bacterioplankton. *Limnol. Oceanogr.* **37,** 232–243.

Currie, D. J., and Kalff, J. (1984a). A comparison of the abilities of freshwater algae and bacteria to acquire and retain phosphorus. *Limnol. Oceanogr.* **29,** 298–310.

Currie, D. J., and Kalff, J. (1984b). The relative importance of bacterioplankton and phytoplankton in phosphate uptake in freshwater. *Limnol. Oceanogr.* **29,** 311–321.

DeAngelis, D. L., Loreau, M., Neergaard, D., Mulholland, P. J., and Marzolf, E. R. (1995). Modeling nutrient–periphyton dynamics in the transient storage zone of streams: the importance of transient storage zones. *Ecol. Model.* **80,** 149–160.

Dillon, P. J., and Kirchner, W. B. (1975). The effects of geology and land use on the export of phosphorus from watersheds. *Water Res.* **9,** 135–148.

Duff, J. H., Triska, F. J., and Oremland, R. S. (1984). Denitrification associated with stream periphyton: Chamber estimates from undisrupted communities. *J. Environ. Qual.* **13,** 514–518.

Elser, J. J., Elser, M. M., MacKay, N. A., and Carpenter, S. R. (1988). Zooplankton-mediated transitions between *N*- and *P*-limited algal growth. *Limnology and Oceanography* **33,** 1–14.

Elwood, J. W., Newbold, J. D., Trimble, A. F., and Stark, R. W. (1981). The limiting role of phosphorus in a woodland stream ecosystem: Effects of P enrichment on leaf decomposition and primary producers. *Ecology* **62,** 146–158.

Elwood, J. W., Newbold, J. D., O'Neill, R. V., and Van Winkle, W. (1983). Resource spiralling: An operational paradigm for analyzing lotic ecosystems. *In* "Dynamics of Lotic Ecosystems" (T. D. Fontaine and S. M. Bartell, eds.), pp. 3–27. Ann Arbor Sci. Publ., Ann Arbor, MI.

Eppley, R. W., and Peterson, B. J. (1979). Particulate organic matter flux and planktonic new production in the deep ocean. *Nature (London)* **282,** 677–680.

Fenchel, T. (1970). Studies on the decomposition of organic detritus derived from the turtle grass, *Thalassia testudinum. Limnol. Oceanogr.* **15,** 14–20.

Fisher, S. G., and Gray, L. J. (1983). Secondary production and organic matter processing by collector macroinvertebrates in a desert stream. *Ecology* **64,** 1217–1224.

Fisher, S. G., Gray, L. J., Grimm, N. B., and Busch, D. E. (1982). Temporal succession in a desert stream ecosystem following flash flooding. *Ecol. Monogr.* **52,** 93–110.

Gantzer, C. J., Rittmann, B. E., and Herricks, E. E. (1988). Mass transport to streambed biofilms. *Water Res.* **22,** 709–722.

Goldman, J. C., and Caron, D. A. (1985). Experimental studies on an omnivorous microflaggelate: Implications for grazing and nutrient regeneration in the marine microbial food chain. *Deep-Sea Res.* **32,** 899–915.

Gregory, S. V. (1983). Plant–herbivore interactions in stream systems. *In* "Stream Ecology" (J. R. Barnes and G. W. Minshall eds.), pp. 157–189. Plenum, New York.

Grimm, N. B. (1987). Nitrogen dynamics during succession in a desert stream. *Ecology* **68,** 1157–1170.

Grimm, N. B. (1988). Role of macroinvertebrates in nitrogen dynamics of a desert stream. *Ecology* **69,** 1884–1893.

Grimm, N. B. (1992). Biogeochemistry of nitrogen in Sonoran Desert streams. *J. Ariz.-Nev. Acad. Sci.* **26,** 139–155.

Grimm, N. B., and Fisher, S. G. (1986). Nitrogen limitation in a Sonoran Desert stream. *J. North Am. Benthol. Soc.* **5,** 2–15.

Grimm, N. B., Fisher, S. G., and Minckley, W. L. (1981). Nitrogen and phosphorus dynamics in hot desert streams of the southwestern U.S.A. *Hydrobiologia* **83,** 303–312.

Hansson, L. A. (1989). The influence of a periphytic biolayer on phosphorus exchange between substrate and water. *Arch. Hydrobiol.* **115,** 21–26.

Harrison, W. G. (1983). Uptake and recycling of soluble reactive phosphorus by marine microplankton. *Mar. Ecol.: Prog Ser.* **10,** 127–135.

Healey, F. P., and Hendzel, L. L. (1979). Fluorometric measurement of alkaline phosphatase activity in algae. *Freshwater Biol.* **9,** 429–439.

Hessen, D. O., and Andersen, T. (1992). The algae–grazer interface: Feedback mechanisms linked to elemental ratios and nutrient cycling. *Arch. Hydrobiol. Beih. Ergebn. Limnol.* **35,** 111–120.

Hessen, D. O., and Lyche, A. (1991). Inter- and intraspecific variations in zooplankton element composition. *Arch. Hydrobiol.* **121,** 343–353.

Hill, W. R., and Knight, A.W. (1988). Nutrient and light limitation of algae in two northern California streams. *J. Phycol.* **24,** 125–132.

Hill, W. R., Boston, H. L., and Steinman, A. D. (1992). Grazers and nutrients simultaneously limit lotic primary productivity. *Can. J. Fish. Aquat. Sci.* **49,** 504–512.

Horne, A. J., and Carmiggelt, C. J. W. (1975). Algal nitrogen fixation in California streams: Seasonal cycles. *Freshwater Biol.* **5,** 461–470.

Horner, R. R., and Welch, E. B. (1981). Stream periphyton development in relation to current velocity and nutrients. *Can. J. Fish. Aquat. Sci.* **38,** 449–457.

Horner, R. R., Welch, E. B., Seeley, M. R., and Jacoby, J. M. (1990). Responses of periphyton to changes in current velocity, suspended sediment and phosphorus concentration. *Freshwater Biol.* **24,** 215–232.

Jacoby, J. M. (1987). Alterations in periphyton characteristics due to grazing in a Cascade foothill stream. *Freshwater Biol.* **18,** 495–508.

Jansson, M. (1993). Uptake, exchange, and excretion of orthophosphate in phosphate-starved *Scenedesmus quadricauda* and *Pseudomonas* K7. *Limnol. Oceanogr.* **38,** 1162–1178.

Jay, E. A. (1993). Effect of snails *(Elimia clavaeformis)* on phosphorus cycling in stream periphyton and leaf detritus communities. M.S. Thesis, University of North Carolina, Chapel Hill.

Johannes, R. E. (1965). Influence of marine protozoa on nutrient regeneration. *Limnol. Oceanogr.* **10,** 434–442.

Kim, B. K., Jackman, A. P., and Triska, F. J. (1990). Modeling transient storage and nitrate uptake kinetics in a flume containing a natural periphyton community. *Water Resour. Res.* **26,** 505–515.

Klotz, R. L. (1985). Factors controlling phosphorus limitation in stream sediments. *Limnol. Oceanogr.* **30,** 543–553.

Klotz, R. L. (1991). Cycling of phosphatase hydrolyzable phosphorus in streams. *Can. J. Fish. Aquat. Sci.* **48,** 1460–1467.

Lamberti, G. A., and Moore, J. W. (1984). Aquatic insects as primary consumers. *In* "The Ecology of Aquatic Insects" (V. H. Resh and D. M. Rosenberg eds.), pp. 164–195. Praeger, New York.

Lamberti, G. A., and Resh, V. H. (1983). Stream periphyton and insect herbivores: an experimental study of grazing by a caddisfly population. *Ecology* **64,** 1124–1135.

Landry, M. R. (1993). Predicting excretion rates of microzooplankton from carbon metabolism and elemental ratios. *Limnol. Oceanogr.* **38,** 468–472.

Lay, J. A., and Ward, A. K. (1987). Algal community dynamics in two streams associated with different geological regions in the southeastern United States. *Arch. Hydrobiol.* **108,** 305–324.

Lean, D. R. S., and Nalewajko, C. (1976). Phosphate exchange and organic phosphorus excretion by freshwater algae. *J. Fish. Res. Board Can.* **33,** 1312–1323.

Lean, D. R. S., and White, E. (1983). Chemical and radiotracer measurements of phosphorus uptake by lake plankton. *Can. J. Fish. Aquat. Sci.* **40,** 147–155.

Liao, C. F. -H., and Lean, D. R. S. (1978). Nitrogen transformations within the trophogenic zone of lakes. *J. Fish. Res. Board Can.* **35,** 1102–1108.

Livingstone, D., Pentecost, A., and Whitton, B. A. (1984). Diel variations in nitrogen and carbon dioxide fixation by the blue-green alga *Rivularia* in an upland stream. *Phycologia* **23,** 125–133.

Lock, M. A. (1979). The effect of flow patterns on uptake of phosphorus by river periphyton. *Limnol. Oceanogr.* **24,** 376–383.

Lock, M. A. (1993). Attached microbial communities in rivers. *In* "Aquatic Microbiology: An Ecological Approach" (T. E. Ford, ed.), pp. 113–138. Blackwell, Boston.

Lock, M. A., Wallace, R. R., Costerton, J. W., Ventullo, R. M., and Charlton, S. E. (1984). River epilithon: Toward a structural-functional model. *Oikos* **42,** 10–22.

McCormick, P. V., and Stevenson, R. J. (1991). Grazer control of nutrient availability in the periphyton. *Oecologia* **86,** 287–291.

McIntire, C. D. (1968). Structural characteristics of benthic algae communities in laboratory streams. *Ecology* **49,** 520–537.

Merritt, R. W., Cummins, K. W., and Burton, T. M. (1984). The role of aquatic insects in the processing and cycling of nutrients. *In* "The Ecology of Aquatic Insects" (V. H. Resh and D. M. Rosenberg, eds.), pp. 134–162. Praeger, New York.

Meyer, J. L., McDowell, W. H., Bott, T. L., Elwood, J. W., Ishizaki, C., Melack, J. M., Peckarsky, B. L., Peterson, B. J., and Rublee, P. A. (1988). Elemental dynamics in streams. *J. North Am. Benthol. Soc.* **7,** 410–432.

Mulholland, P. J. (1992). Regulation of nutrient concentrations in a temperate forest stream: Roles of upland, riparian, and instream processes. *Limnol. Oceanogr.* **37,** 1512–1526.

Mulholland, P. J., and Rosemond, A. D. (1992). Periphyton response to longitudinal nutrient depletion in a woodland stream: Evidence of upstream–downstream linkage. *J. North Am. Benthol. Soc.* **11,** 405–419.

Mulholland, P. J., Newbold, J. D., Elwood, J. W., and Hom, C. L. (1983). The effect of grazing intensity on phosphorus spiralling in autotrophic streams. *Oecologia* **58,** 358–366.

Mulholland, P. J., Newbold, J. D., Elwood, J. W., Ferren, L. A., and Webster, J. R. (1985). Phosphorus spiralling in a woodland stream: Seasonal variations. *Ecology* **66,** 1012–1023.

Mulholland, P. J., Elwood, J. W., Palumbo, A. V, and Stevenson, R. J. (1986). Effect of stream acidification on periphyton composition, chlorophyll, and productivity. *Can. J. Fish. Aquat. Sci.* **43,** 1846–1858.

Mulholland, P. J., Steinman, A. D., and Elwood, J. W. (1990). Measurement of phosphorus uptake length in streams: Comparison of radiotracer and stable PO_4 releases. *Can. J. Fish. Aquat. Sci.* **47,** 2351–2357.

Mulholland, P. J., Steinman, A. D., Palumbo, A. V., Elwood, J. W., and Kirschtel, D. B. (1991). Role of nutrient cycling and herbivory in regulating periphyton communities in laboratory streams. *Ecology* **72,** 966–982.

Mulholland, P. J., Steinman, A. D., Marzolf, E. R., Hart, D. R., and DeAngelis, D. L. (1994). Effect of periphyton biomass on hydraulic characteristics and nutrient cycling in streams. *Oecologia* **98,** 40–47.

Mulholland, P. J., Marzolf, E. R., Hendricks, S. P., Wilkerson, R. V., and Baybayan, A. K. (1995) Longitudinal patterns in Periphyton biomass, productivity, and nutrient cycling: A test of upstream– downstream linkage. *J. North Am. Benthol. Soc.* **14,** 357–370.

Munn, N. L., and Meyer, J. L. (1990). Habitat-specific solute retention in two small streams: an intersite comparison. *Ecology* **71,** 2069–2082.

Nalewajko, C., and Lee, K. (1983). Light stimulation of phosphate uptake in marine phytoplankton. *Mar. Biol. (Berlin)* **74,** 9–15.

Newbold, J. D., Elwood, J. W., O'Neill, R. V., and Van Winkle, W. (1981). Measuring nutrient spiralling in streams. *Can. J. Fish. Aquat. Sci.* **38,** 860–863.

Newbold, J. D., O'Neill, R. V., Elwood, J. W., and Van Winkle, W. (1982). Nutrient spiralling in streams: Implications for nutrient limitation and invertebrate activity. *Am. Nat.* **120,** 628–652.

Newbold, J. D., Elwood, J. W., O'Neill, R. V., and Sheldon, A. L. (1983). Phosphorus dynamics in a woodland stream ecosystem: A study of nutrient spiralling. *Ecol.* **64,** 1249–1265.

Nielsen, L. P., Christensen, P. B., Revsbech, N. P., and Sorensen, J. (1990a). Denitrification and photosynthesis in stream sediment studied with microsensor and whole-core techniques. *Limnol. Oceanogr.* **35,** 1135–1144.

Nielsen, L. P., Christensen, P. B., Revsbech, N. P., and Sorensen, J. (1990b). Denitrification and oxygen respiration in biofilms studied with a microsensor for nitrous oxide and oxygen. *Microb. Ecol.* **19,** 63–72.

Olsen, Y. (1989). Evaluation of competitive ability of *Staurastrum luetkemuellerii* (Chlorophyceae) and *Microcystis aeruginosa* (Cyanophyceae) under P limitation. *J. Phycol.* **25,** 486–499.

Olsen, Y., Jensen, A., Reinertsen, H., Borsheim, K. Y., Heldal, M., and Langeland, A. (1986). Dependence of the rate of release of phosphorus by zooplankton on the P:C ratio in the food supply, as calculated by a recycling model. *Limnol. Oceanogr.* **31,** 34–44.

Paul, B. J., and Duthie, H. C. (1989). Nutrient cycling in the epilithon of running waters. *Can. J. Bot.* **67,** 2302–2309.

Perrin, C. J., Bothwell, M. L., and Slaney, P. L. (1987). Experimental enrichment of a coastal stream in British Columbia: Effects of organic and inorganic additions on autotrophic periphyton production. *Can. J. Fish. Aquat. Sci.* **44,** 1247–1256.

Peters, R. H. (1978). Concentrations and kinetics of phosphorus fractions in water from streams entering Lake Memphremagog. *J. Fish. Res. Board Can.* **35,** 315–328.

Peters, R. H. (1979). Concentrations and kinetics of phosphorus fractions along the trophic gradient of Lake Memphremagog. *J. Fish. Res. Board Can.* **36,** 970–979.

Peterson, B. J., Hobbie, J. E., Corliss, T. L., and Kriet, K. (1983). A continuous-flow periphyton bioassay: Tests of nutrient limitation in a tundra stream. *Limnol. Oceanogr.* **28,** 583–591.

Peterson, C. G., and Grimm, N. B. (1992). Temporal variation in enrichment effects during periphyton succession in a nitrogen-limited desert stream ecosystem. *J. North Am. Benthol. Soc.* **11,** 20–36.

Power, M. E., Stewart, A. J., and Matthews, W. J. (1988). Grazer control of algae in an Ozark Mountain stream: effects of short-term exclusion. *Ecology* **69,** 1894–1898.

Prepas, E. E. (1983). Orthophosphate turnover time in shallow productive lakes. *Can. J. Fish. Aquat. Sci.* **40,** 1412–1418.

Pringle, C. M. (1987). Effects of water and substratum nutrient supplies on lotic periphyton growth: An integrated bioassay. *Can. J. Fish. Aquat. Sci.* **44,** 619–629.

Pringle, C. M. (1990). Nutrient spatial heterogeneity: Effects on community structure, physiognomy, and diversity of stream algae. *Ecology* **71,** 905–920.

Raven, J. A. (1992). How benthic macroalgae cope with flowing freshwater: Resource acquisition and retention. *J. Phycol.* **28,** 133–146.

Riber, H. H., and Wetzel, R. G. (1987). Boundary-layer and internal diffusion effects on phosphorus fluxes in lake periphyton. *Limnol. Oceanogr.* **32,** 1181–1194.

Rosemond, A. D. (1994). Multiple factors limit seasonal changes in periphyton in a forest stream. *J. North Am. Benthol. Soc.* **13,** 333–344.

Rosemond, A. D., Mulholland, P. J., and Elwood, J. W. (1993). Top-down and bottom-up control of stream periphyton: Effects of nutrients and herbivores. *Ecology* **74,** 1264–1280.

Rosemond, A. D., and Brawley, S.H. (1996). Species-specific characteristics explain the persistence of *Stigeoclonium tenue* (Chlorophyta) in a woodland stream. *J. Phycol.* **32,** (in press).

Salonen, K., Jones, R. I., De Haan, H., and James, M. (1994). Radiotracer study of phosphorus uptake by plankton and redistribution in the water column of a small humic lake. *Limnol. Oceanogr.* **39,** 69–83.

Sinsabaugh, R. L., Report, D., Weiland, T., Golladay, S. W., and Linkins, A. E. (1991). Exoenzyme accumulation in epilithic biofilms. *Hydrobiologia* **222,** 29–37.

Steinman, A. D., and Boston, H. L. (1993). The ecological role of aquatic bryophytes in a woodland stream. *J. North Am. Benthol. Soc.* **12,** 17–26.

Steinman, A. D., McIntire, C. D., Gregory, S. V., Lamberti, G. A., and Ashkenas, L. R. (1987a). Effects of herbivore type and density on taxonomic structure and physiognomy of algal assemblages in laboratory streams. *J. North Am. Benthol. Soc.* **6,** 175–188.

Steinman, A. D., McIntire, C. D., and Lowery, R. R. (1987b). Effects of herbivore type and density on chemical composition of algal assemblages in laboratory streams. *J. North Am. Benthol. Soc.* **6,** 189–197.

Steinman, A. D., Mulholland, P. J., and Kirschtel, D. B. (1991). Interactive effects of nutrient reduction and herbivory on biomass, taxonomic structure, and P uptake in lotic periphyton communities. *Can. J. Fish. Aquat. Sci.* **10,** 1951–1959.

Steinman, A. D., Mulholland, P. J., and Hill, W. R. (1992). Functional responses associated with growth form in stream algae. *J. North Am. Benthol. Soc.* **11,** 229–243.

Sterner, R. W. (1990). N:P resupply by herbivores: Zooplankton and the algal competitive arena. *Am. Nat.* **136,** 209–229.

Sterner, R. W., Elser, J. J., and Hessen, D. O. (1992). Stoichiometric relationships among producers, consumers and nutrient cycling in pelagic ecosystems. *Biogeochemistry* **17,** 49–67.

Stevenson, R. J., and Glover, R. (1993). Effects of algal density and current on ion transport through periphyton communities. *Limnol. Oceanogr.* **38,** 1276–1281.

Stewart, A. J. (1987). Responses of stream algae to grazing minnows and nutrients: A field test for interactions. *Oecologia* **72,** 1–7.

Stockner, J. G., and Shortreed, K. R. S. (1978). Enhancement of autotrophic production by nutrient addition in a coastal rainforest stream on Vancouver Island. *J. Fish. Res. Board Can.* **35,** 28–34.

Stream Solute Workshop (1990). Concepts and methods for assessing solute dynamics in stream ecosystems. *J. North Am. Benthol. Soc.* **9,** 95–119.

Tarapchak, S. J., Slavens, D. R., and Maloney, L. M. (1981). Abiotic versus biotic uptake of radiophosphorus in lake water. *Can. J. Fish. Aquat. Sci.* **38,** 889–895.

Tate, C. M. (1990). Patterns and controls of nitrogen in tallgrass prairie streams. *Ecology* **71,** 2007–2018.

Triska, F. J., Sedell, J. R., Cromack, K., Jr., Gregory, S. V., and McCorison, F. M. (1984). Nitrogen budget for a small coniferous forest stream. *Ecol. Monogr.* **54,** 119–140.

Triska, F. J., Kennedy, V. C., Avanzino, R. J., Zellweger, G. W., and Bencala, K. E. (1989). Retention and transport of nutrients in a third-order stream: Channel processes. *Ecology* **70**, 1877–1892.

Urabe, J., and Watanabe, Y. (1992). Possibility of N or P limitation for planktonic cladocerans: An experimental test. *Limnol. Oceanogr.* **37**, 244–251.

Valett, H. M., Fisher, S. G., Grimm, N. B., and Camill, P. (1994). Vertical hydrologic exchange and ecological stability of a desert stream ecosystem. *Ecology* **75**, 548–560.

Ventullo, R. M., and Rowe, J. J. (1982). Denitrification potential of epilithic communities in a lotic environment. *Curr. Microbiol.* **7**, 29–33.

Vitousek, P. M. (1977). The regulation of element concentrations in mountain streams in the northeastern United States. *Ecol. Monogr.* **47**, 65–81.

Ward, A. K. (1985). Factors affecting distribution of *Nostoc* in Cascade mountain streams of western Oregon, U.S.A. *Verh.—Int. Ver. Theor. Angew. Limnol.* **22**, 2799–2804.

Ward, A. K., Dahm, C. N., and Cummins, K. W. (1985). *Nostoc* (Cyanophyta) productivity in Oregon stream ecosystems: Invertebrate influences and differences between morphological types. *J. Phycol.* **21**, 223–227.

Webster, J. R. (1975). Analysis of potassium and calcium dynamics in stream ecosystems on three southern Appalachian watersheds of contrasting vegetation. Ph.D. Dissertation, University of Georgia, Athens.

Webster, J. R., and Patten, B. C. (1979). Effects of watershed perturbation on stream potassium and calcium dynamics. *Ecol. Monogr.* **19**, 51–72.

Wetzel, R. G. (1993). Microcommunities and microgradients: Linking nutrient regeneration, microbial mutualism, and high sustained aquatic primary production. *Neth. J. Aquat. Ecol.* **27**, 3–9.

White, E., Payne, G., Pickmere, S., and Pick, F. R. (1982). Factors influencing orthophosphate turnover times: A comparison of Canadian and New Zealand lakes. *Can. J. Fish. Aquat. Sci.* **39**, 469–474.

Whitford, L. A., and Schumacher, G. J. (1964). Effect of a current on respiration and mineral uptake in *Spirogyra* and *Oedogonium*. *Ecology* **45**, 168–170.

20

Benthic Algae and Nutrient Cycling in Lentic Freshwater Ecosystems

Robert G. Wetzel

Department of Biological Sciences
University of Alabama
Tuscaloosa, Alabama 35487

I. INTRODUCTION

Nutrient cycling implies by definition that nutrients pass among different components of a cell, community, or ecosystem and can be cycled and reutilized by some of these components. It is essential to quantify the rates at which nutrients are assimilated, transferred among biota, and released for subsequent reassimilation. This cycling rate per unit time is an informative parameter of community or ecosystem operations, particularly in how it changes in response to alteration of physical and biotic parameters. Nutrients, however, move within an aquatic ecosystem along an elevational gradient from land to water. Nutrients are lost from sites of photosynthetic and secondary production, and such losses must be replenished by nutrient inputs for sustenance, growth, and net production to continue.

Nutrient cycling occurs at many spatial and temporal scales. These processes must be viewed as relatively self-contained in communities in which much recycling of nutrients is occurring. Many microcommunities with intense rates of recycling are interactively nested along and within fewer larger-scale communities with progressively longer and slower rates of recycling. Because growth and reproduction are preeminent among organisms, rates of nutrient cycling and recycling are paramount to meet metabolic demands. These rates of nutrient cycling are regulated by a complex interplay of physical, chemical, and biotic factors. Environmental and community factors regulating nutrient processes in lentic waters can be similar to some of those discussed for flowing waters (see Mulholland, Chapter 19, this volume), but differ markedly in greater rates and interactive intensities because of the much higher photosynthetic productivity, loading of organic matter, and habitat diversity among lentic waters than is the case in lotic waters (Wetzel, 1983b,c; Wetzel and Ward, 1992).

Evaluation of the importance of nutrient cycling and recycling requires foremost an appreciation that the habitat conditions in lentic (nonriverine) fresh waters mandate nutrient conservation and retention in order to achieve and sustain the very high levels of community productivity. Losses and inactivation of nutrients from sites of photosynthesis are dynamic and occur on both micro- and macroscales. Among planktonic communities, available nutrient concentrations are much lower in an extremely dilute pelagic environment and loss rates of nutrients, particularly by sedimentation of nutrients with producers out of the photic zone, are very high. Although an analogous relationship of a "phycosphere," in which bacteria interact closely with algae, has been suggested for planktonic algae, evidence is generally weak. A clear exception is the essential microzone development of bacterially reduced oxygen within consortia of nitrogen-fixing cyanobacteria and heterotrophic bacteria (Paerl, 1985, 1990; Paerl and Carlton, 1988). The bacteria are reducing oxygen near heterocysts of cyanobacteria, which is necessary physiologically for N_2 fixation, and the photosynthetic cells provide organic substrates as well as electron acceptors. Such direct mutualistic couplings are not common among most actively growing phytoplanktonic algae. Similarly, bacterioplankton tend to be chronically starved for organic substrates and nutrients (e.g., Kuznetsov, 1968, 1970). Thus, planktonic recycling of nutrients through heterotrophic microorganisms, algae, and protista is not very efficient because of large distances among biota. Efficiencies of nutrient utilization, retention, and recycling are much greater among closely aggregated attached algal–microbial communities than is the case among the plankton. As a result, a primary characteristic of benthic algal community structure and assimilation mechanisms is high retention of nutrients after acquisition. This conservation and intensive recycling of nutrients lead to maximal resource utilization, efficiency, and productivity per areal unit among the communities within the lentic water ecosystem.

Although lentic waters are usually open, flow-through ecosystems, fluxes are much greater from sites of production to sediments rather than downstream as in flowing ecosystems. Water movements are generally very slow with concordant reductions of physical nutrient movements and transfers. Hence, lentic ecosystems are much more closed in regard to nutrient cycling than is the case in flowing waters.

II. NUTRIENT CHARACTERISTICS IN LENTIC HABITATS

Benthic algae are generally associated with substrata. These substrata are living or dead, organic or inorganic, and in various combinations. Since photosynthesis is the only significant process for synthesis of organic matter among algae, light is mandatory within both the macrohabitat and the microhabitat of the communities in quantities adequate to allow synthesis of ATP with no net losses. The land–water interface zone is a gradient that extends from detrital masses in hydrosoils and pools of wetlands through littoral zone regions of emergent, floating-leaved, and submersed macrophyte communities, to sediments extending into deeper waters well beyond the macrophytes (littoriprofundal zone, cf. Wetzel, 1983b). Available light shifts from abundant in wetland and shallow littoral areas to low and highly variable seasonally in deeper areas as variable densities of seston occur in the overlying waters. Within this zone of adequate light, substratum diversity is highly variable, but can be organized into several general zones in regard to potential substratum nutrient availability (Fig. 1).

1. Upland wetland saturated hydrosoils and pools among emergent macrophytes largely consist of standing and collapsed dead macrophytes and dense aggregations of particulate detritus among the understory. Although many critical plant nutrients are retained within living plant tissues and translocated to rooting tissues at the end of growth periods of individual cohorts, soluble organic matter and organic nutrients, particu-

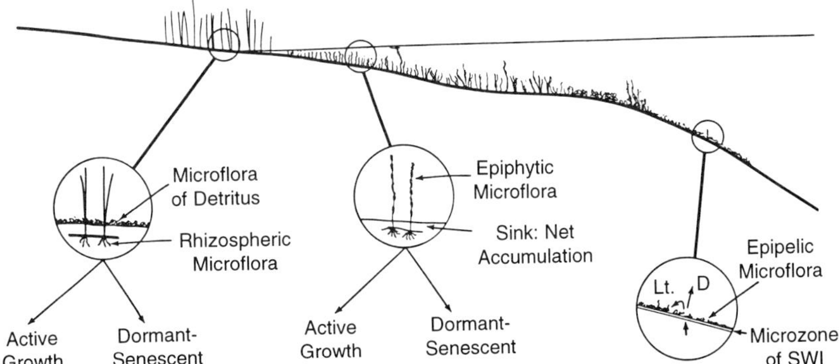

FIGURE 1 Primary habitat and attached algal community components along a wetland–littoral gradient of standing waters. SWI = sediment–water interface; D = dark; Lt = light.

larly organic phosphorus and nitrogen, are released from the highly productive emergent macrophytes during and following active growth (e.g., Godshalk and Wetzel, 1978a,b). Many of these herbaceous perennials have more or less continuous leaf or culm turnover of many cohorts per year in both northern and southern regions [e.g., Dickerman and Wetzel (1985) for *Typha;* Wetzel (1996) for *Juncus*]. Thus, relatively constant partial senescence of leaves occurs within a population with release of cellular organic matter and nutrients.

The very high productivity of the emergent macrophytes results in high detrital organic matter accumulation at the sediment–water interface. This loading, which contains relatively high proportions of moderately recalcitrant structural tissues from emergent plants, contributes to high substrata surface area for microbial colonization in relatively shallow water of high light and temperatures. The high loading of organic matter results in predominantly anaerobic conditions with increased solubility of critical nutrients, particularly P and N, in the interstitial waters near and at the sediment–water interface. Much combined nitrogen as nitrate is commonly denitrified by facultative anaerobic bacteria in such detrital-sediment masses and released to the atmosphere as N_2 (Triska and Oremland, 1981; Bowden, 1987).

2. Submersed macrophytes within standing waters of lakes, ponds, and floodplain pools generally possess a morphology of thin, finely divided, and reticulated leaves. The resulting increased surface area of leaves markedly enhances exchange of gases with those of the water and interceptions of light (Sculthorpe, 1967). This greatly increased ratio of surface area to volume results in enormously increased substrata for colonization by epiphytic algae. For example, the leaf surface area available for colonization by epiphytic algae on the submersed linear-leaved macrophyte *Scirpus subterminalis* Torr. averaged 24 m^2 m^{-2} of bottom in the moderately developed littoral zone of a lake in southwestern Michigan (Burkholder and Wetzel, 1989). The living macrophytic substrata are of course competing for similar resources, particularly light, with the attached algae but have several mechanisms available, as discussed extensively in the following, that allow noncompeting mutualism for various periods of time (Phillips *et al.,* 1978; Wetzel, 1983a). The extensive area of these macrophytic surfaces project myriad diverse microhabitats with attendant attached algal communities upward into littoral environments of relatively abundant light, dissolved gases from photosynthesis (O_2) and decomposition (CO_2), and nutrients diffusing from high decomposition of interstitial waters of detritus and sediments. As a result, the productivity of epiphytic algae and benthic algae on detritus frequently greatly exceeds that of the supporting macrophytes (Wetzel, 1983a,b, 1990). Additionally, the diversity of surfaces and morphological configurations within and among the submersed parts of littoral macrophyte communities provides very large diversity in microhabitats. Most

(>80%) freshwater species of algae are attached, sessile forms in this diverse, dynamic, and highly productive habitat (Round, 1981).

3. The benthic habitat in lentic fresh waters can be a physically hostile yet nutritionally advantageous environment. Epilithic and epipsammic algae, though frequently having the advantages of more active water movements and nutrient exchanges than is the case among epiphytic communities, are exposed to molar action by particles such as sand and potential burial by sediments with reduction or removal of light (Meadows and Anderson, 1966; Moss and Round, 1967). Most lakes and ponds are small with modest fetch and wind-induced water movements (Wetzel, 1983b, 1990). Water turbulence also decreases rapidly with increasing depth (Imberger and Patterson, 1990), and submersed macrophyte communities greatly attenuate water flows. For example, vertical and horizontal flows external to plant beds were dissipated markedly within 10–15 cm of the outer plant-bed boundary even under severe external flow-rate conditions (external flows of >30 cm s^{-1} were reduced to an average of 0.07 cm s^{-1} within the macrophyte bed and exhibited strictly laminar flow movements) (Losee and Wetzel, 1993). As a result, even loosely aggregated organic-rich sediments can serve as substrata for epipelic algae when receiving adequate light. The unstable nature of such sediments and often high rates of deposition from settling seston increase the risk of algal burial and light reduction. The high nutrient availability from interstitial waters of sediments, however, is a distinct advantage. Many epipelic algal species migrate, often in rhythmic fashion, to compensate for light attenuation by sediment (cf. Round, 1981).

4. Loosely aggregated algae in the littoral zone of many lakes, ponds, and wetlands, the metaphyton, are neither strictly attached to substrata nor truly suspended. *Metaphyton* (Behre, 1956) is basically synonymous with the terms *tychoplankton* and *pseudoplankton* (Naumann, 1931) and *pseudoperiphyton* (Sládečková, 1960). Many colloquial and provincial terms of similar ilk occur in the literature (e.g., *flab*) (Hillebrand, 1983). Metaphyton communities originate from fragmentation of dense epiphytic or epipelic algal communities or occasionally from dense phytoplanktonic communities that aggregate and become clumped and loosely attached in littoral areas by wind-induced surface water movements, and they can form dense microbial accumulations with intense internal nutrient recycling. The collective metabolism of metaphyton can radically alter the nutrient cycling of littoral zones.

III. NUTRIENT GRADIENTS AND LIMITATIONS

A. Macrogradients

Nutrient gradients (concentrations; mass) and cycling (rates) have been examined at large scales within lentic waters. Lakes and reservoirs have been treated almost uniformly in the context of planktonic reactor systems, with nutrient inputs, outputs, and various regulators internally within the pelagic region (cf. detailed reviews of Reckhow and Chapra, 1983; DeAngelis, 1992). Losses of nutrients via sedimentation and outflows can often be coupled to external and internal loadings of nutrients in dynamic nutrient budgets that allow reasonable evaluation of controlling physical and biotic parameters. Nearly always, and certainly historically, the quantitative roles of nonplanktonic producers were ignored as insignificant in these budgets. Regeneration and recycling of nutrients from sedimentary sources (=internal loadings) generally also ignored the attached algae, or included effects of these sessile organisms in the composite budgets.

Early evaluations of carbon budgets of lakes (e.g., Wetzel, 1964; Wetzel *et al.,* 1972) demonstrated the overwhelming, often dominating contributions of attached algal communities to the overall productivity of many lakes (cf. review of Wetzel, 1983b). Because most lakes are relatively small and shallow, the wetland–littoral complex of macro- and microphytes produces the major source of organic matter of most freshwater ecosystems (reviews of Wetzel, 1983b, and particularly 1990). Most of the particulate organic matter is decomposed within the wetland and littoral interface regions, with intensive recycling and conservation of nutrients within these regions. Large quantities of dissolved organic matter, however, are exported from the littoral areas to the pelagic, and these carbon and energy sources supplement or dominate the pelagic bacterial metabolism and nutrient recycling mechanisms (Wetzel *et al.,* 1972; Coveney and Wetzel, 1995).

Phosphorus fluxes at the ecosystem level have been examined in a few cases in which the littoral macrophytes and attached algae have been collectively included. In a small acidic lake with a well-developed littoral zone of rooted aquatic plants, the littoral region was the most important contributor to the turnover of phosphorus in the epilimnion (Rigler, 1956, 1964, 1973). Phosphorus was lost to this compartment 10 times more rapidly than via plankton to the hypolimnion and sediments, and 50 times more rapidly than its loss through the lake outflow. As found in more recent studies discussed in the following, most of this uptake presumably occurred by the attached algae and bacteria rather than the macrophytes. Return of phosphorus from the littoral zone during the summer was about 20% greater than losses. Other studies in lakes to which tracer phosphorus was added indicated the importance of littoral macrophytes in assimilating and retaining phosphorus (e.g., Coffin *et al.,* 1949; Hayes and Coffin, 1951). Such whole-lake studies reveal little about the controlling organisms, but clearly indicate that the turnover time of phosphorus in the epilimnion, which ranged from 20 to 45 days, was inversely correlated with the areas of the lakes and estimates of the development of littoral vegetation (Table I). Recent information clearly indicates that the attached algae and other microflora are the primary mediators in these exchanges.

TABLE I Estimated Rates of Phosphorus Transport in Three Lakes of Differing Littoral Development[a]

Lake	*Area (ha)*	*Littoral vegetation (rank)*	*P turnover time (d)*	*Rate constants* k *(out of epilimnion*	k *(sedimentation from epilimnion)*
Toussaint	4.7	1	20	0.05	0.01
Upper Bass	5.8	2	27	0.04	—
Linsley Pond	9.4	3	45	0.02	0.02

[a]Lakes were ranked in order of decreasing amounts, subjectively estimated, of littoral vegetation. Modified from Rigler (1973).

Numerous studies have demonstrated the impacts of nutrient loadings, particularly P and N, on the development of benthic algae on many different substrata. For example, nitrified enrichment led not only to an increase in epilithic algal abundance but to a shift from a diatom/N_2-fixing cyanobacteria-dominated community to one dominated by diatoms (Hawes and Smith, 1993). Where nitrogen insufficiency is maintained, N_2 fixation by cyanobacteria appears to be a successful strategy for survival in N-deficient environments (Reuter *et al.,* 1983). Species composition of algal periphyton is often relatively unresponsive to small changes in nutrient concentrations and dynamics in the overlying water (e.g., Pringle, 1987; Riber and Wetzel, 1987; Fairchild and Sherman, 1993; Niederhauser and Schanz, 1993). Chemical conditions within the periphyton communities differ considerably from those of the immediate surroundings, and boundary-layer development is a key factor affecting the metabolism of periphyton communities (see Section III, B).

Often studies have addressed quantitative capabilities of the attached algae to sequester nutrient loadings from external sources along wetland–littoral gradients and to retain such nutrients within the periphyton–detritus–sediment communities. Indeed, many types of constructed wetland–periphyton systems are being developed to cope with nutrient loadings from wastewaters of various human activities (e.g., Moshiri, 1993). Among natural lentic waters, for example, detailed nutrient budgets demonstrated that epiphytic algae of plants and particulate plant detritus assimilate significant fractions of available carbon, nitrogen, and phosphorus during their collective growth (Pieczyńska, 1972; Wetzel *et al.,* 1972; Howard-Williams, 1981).

The algal–microbial communities have variable but distinct capacities for nutrient assimilation and retention. For example, growth responses of a natural epiphytic algal community of the submersed macrophyte *Scirpus subterminalis* of an oligotrophic, hard-water lake of southwestern Michigan were examined after exposure to a continuous and localized addition of phosphorus to the littoral water (Fig. 2, Site A) (Moeller *et al.*, 1996). Epiphytic algal growth was compared in another area of the littoral (Site B) in which nutrients were administered continuously in the sediment to stimulate phosphorus release from growing macrophyte tissue, which can be a direct though small source of phosphorus for overlying epiphytic algae (Carignan and Kalff, 1982; Burkholder and Wetzel, 1990). Epiphytic growth increased immediately where phosphate was released above the sediment, and after 10 weeks a 40-fold increase of epiphytic biomass occurred at the center of the plot as compared to biomass immediately outside of the plot (Fig. 2). Changes in epiphytic biomass were very minor as a result of the below-sediment enrichment at Site B, although additional

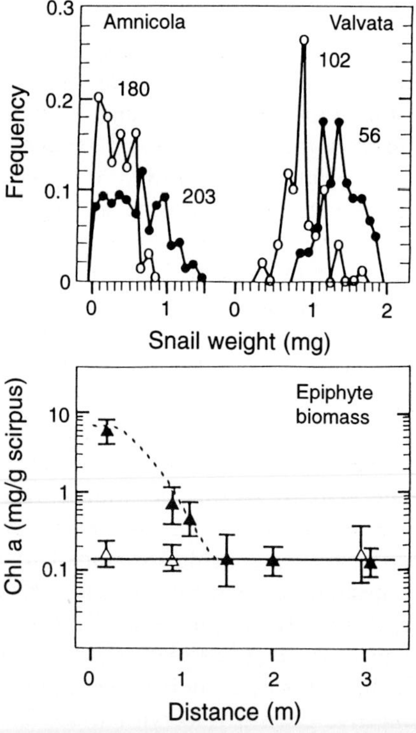

FIGURE 2 Response of epiphytic algal biomass and snails to 10 weeks of continuous P enrichment. (Bottom) Mean epiphytic biomass (±95% C.I., *N*=7) increased within Site A (▲) but not at Site B (△). (Top) Weight–frequency distributions of the two most common snails (*Amnicola limosa* and *Valvata tricarinata*) shifted to the right within Site A (●, enriched zone 1 cm from center of plot; ○, control zone >2m from center; *N*=56–203 snails). (From Moeller *et al.*, 1996.)

phosphorus was incorporated by the macrophyte (leaf P increased from 0.14 to 0.51% of dry weight within the plot). Although the markedly increased epiphytic algal growth easily accommodated the increased nutrient loading in this case, that capacity is often exceeded simply by excessive nutrient loading or because the duration of nutrient-laden water movement along the gradients from the land to the lake *per se* is too short (short residence times). Excessive nutrient loading can lead to dense epiphytic communities that compete with the supporting submersed macrophyte to the point of restriction of light and suppression of photosynthesis of the macrophyte (Phillips *et al.,* 1978). As noted earlier, such responses occur similarly on rock, sand, and other substrata in response to enhanced nutrient loadings.

Macroinvertebrate grazing upon attached microbial communities is well known (e.g., Allanson, 1973; Cattaneo, 1983; Cuker, 1983; Bronmark, 1989; see Steinman, Chapter 12). Additionally, it is clear that enhanced growth of epiphytic algae can markedly increase the growth and development of grazers. For example, in the studies by Moeller *et al.* (1995) just discussed in which epiphytic algae were experimentally enriched with continuous nutrient releases, epiphytic biomass increased at Site A in spite of an increase in the intensity of grazing by gastropods. Snail biomass per unit area of lake bottom increased 73% [115 versus 66 mg dry weight per 0.05 m^2; $P<0.05$; Moeller *et al.* (1996)]. Snails were collected from surficial sediment with overlying macrophytes ($N = 4$ enriched quadrats <1 m from the center of Site A and 4 control quadrats >2 m from the center). Snail dry mass was based on regressions of weight on shell length, which excluded shell and operculum. Snail biomass per unit of macrophyte biomass increased 46% [34 versus 23 mg dry wt per g organic wt (ash-free dry mass, AFDM); $P<0.1$]. The biomass response was attributable to two dominant snails (cf. Fig. 2) that constitute more than 80% of the total snail biomass and feed on epiphytic microflora and detritus. Both species, which live only one year, had reproduced before the enrichment experiments began. Snail densities per unit area of lake bottom or unit of macrophyte biomass did not change, which indicated that neither immigration nor reproduction contributed significantly to the response. Individuals of both species were significantly larger ($P<0.005$) within the zone of increased epiphyte biomass (Fig. 2). Such increased snail growth rate in response to an increased food supply has been shown often (cf. Bronmark, 1989). What is not clear in this and most similar studies is the effect on nutrient recycling by grazing of the invertebrates or by feces-associated microflora.

Along the wetland and littoral gradients of lentic waters, the emergent macrophytes are the most productive of all of the plant communities. Although the emergent plants function as major exporters of dissolved organic matter, nutrients and particularly limiting nutrients tend to be efficiently acquired, retained, and reutilized. The macroprocesses can be illustrated with phosphorus.

Among actively growing emergent macrophytes, bacteria and algae associated with dissolved and particulate detrital organic matter of the detrital litter understory and sediments incorporate, retain, and recycle (see Section III, B) much of the imported phosphorus, in both inorganic and organic forms (Fig. 3). Essentially all phosphorus of the macrophyte is assimilated from the interstitial water of the sediments by emergent, rooted floating-leaved, and most submersed macrophytes (e.g., Moeller *et al.*, 1988). Phosphorus within the microbes and detritus is actively recycled and some is interred with senescing cells and organic deposits. Uptake and incorporation of P of surrounding water are moderately efficient; relatively little phosphorus leaves the emergent wetland during the active growing season, particularly for the attached microbes, except during high water flows through the wetland community when water residence times are too brief or severe flushing actions occur.

Much of the plant P is translocated back to the rooting tissues at the time of maturation of the emergent macrophytes, as well as during continual phased leaf senescence and turnover of leaf cohorts of these common clonal herbaceous perennial plants (Fig. 3). As plant foliage senesces and collapses to the litter and detrital sediments, much of the released phosphorus is retained in the particulate organic matter and attendant bacterial–algal microbial consortia (Fig. 3). The macrophyte and microbial communities are functionally coupled (see the following) to maximize retention and recycling of this nutrient both in inorganic form as well as from organic compounds after enzymatic hydrolysis.

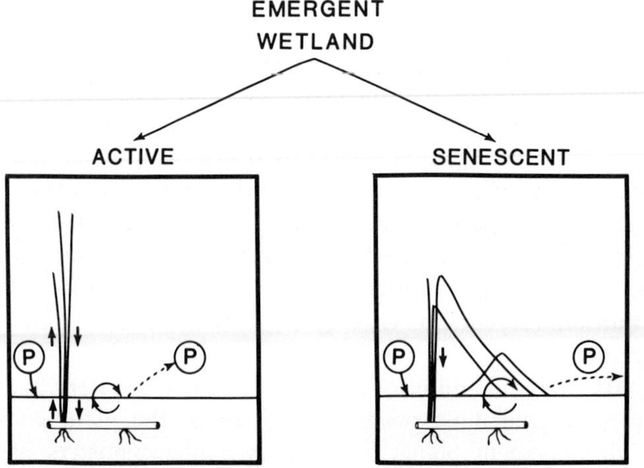

FIGURE 3 Diagrammatic representation of phosphorus (P) fluxes among emergent macrophyte communities. (Reproduced with permission from Wetzel, 1990.)

For many years the epiphytic microflora on submersed portions of macrophytes has been treated largely as a curiosity and ignored as being quantitatively insignificant in most lentic ecosystems. A comprehensive model of the major metabolic pathways and interactions of the attached algae and bacteria within the littoral zone gradients was set forth by Wetzel and Allen (1970; cf. also confirmations by Allanson 1973) to correct these fallacious promulgations. Evidence now exists that the epiphytic algae and coupled microheterotrophs are not only productive but are often a major or the dominant regulator of nutrient fluxes in fresh waters.

The productivity of epiphytic algae is much greater than previously believed. For example, even though the littoral zone occupies only 15% of Lawrence Lake, Michigan, the epiphytic algae contributed between 70 and 85% of the lake primary productivity (Burkholder and Wetzel, 1989). In every situation where epiphytic algal productivity is evaluated accurately, very high rates are found.

Reasons for the high productivity of epiphytic algae are coupled to both the physical and nutrient characteristics of the macrophytic substrata. Submersed macrophytes are dominated by perennials, and many species grow or physically persist in a dormant or "evergreen" condition for much of the year, including under ice cover. Above-sediment biomass of submersed macrophytes commonly reaches a maximum in spring or early summer and then gradually declines as resources are diverted to rooting tissues (Fig. 4). An appreciable macrophytic leaf biomass and associated

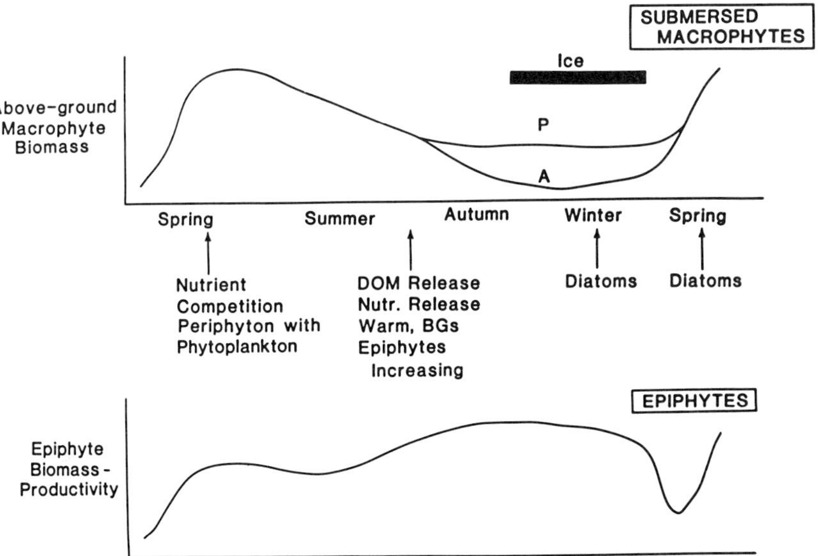

FIGURE 4 Generalized relationships of epiphytic algal biomass and productivity in relation to variations in above-sediment biomass of submersed macrophytes (see text). P = perennial plants; A = annual plants.

surface area persist all year in subtemperate and tropical regions and even under ice in temperate regions. As a result, epiphytic algal biomass and productivity tend to be high and relatively constant throughout the year. Competition among epiphytic algae and phytoplankton for nutrients can reduce epiphytic development in the midyear period, but epiphytic growth, particularly by cyanobacteria, often increases markedly in late summer and autumn (e.g., Burkholder and Wetzel, 1989). Diatom productivity is often especially high during winter and spring (Fig. 4).

The epiphytic algal community is exposed to nutrients from the water within and passing through the littoral zone and to those released from the supporting host macrophyte tissues. Studies have demonstrated complex nutrient interactions between the macrophyte tissues and epiphytic algae and other microbes (see the following). Even when the nutrient concentrations of the water are very high, some nutrients are obtained from the macrophytes, simply because diffusion into and within the complex algal–bacterial community is too slow to meet their metabolic demands (e.g., Riber and Wetzel, 1987; Moeller *et al.*, 1988; Fairchild and Sherman, 1993).

As discussed for the emergent macrophytes, most of the nutrients of actively growing submersed macrophytes are obtained from the sediments. A number of studies have demonstrated that most (95–99%) phosphorus, for example, is obtained from sediment interstitial water and is retained within the plant and recycled repeatedly. Phosphorus from littoral water is actively assimilated by the loosely attached epiphytic algae and bacteria, incorporated into the periphyton, and intensively recycled (Fig. 5). Little

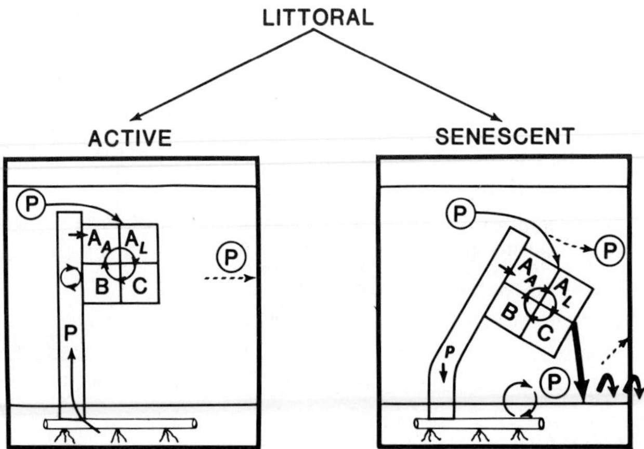

FIGURE 5 Diagrammatic representation of fluxes of phosphorus (P) from the sediments to submersed littoral macrophytes and among the epiphytic microflora of the periphyton. A_A = adnate algae; A_L = loosely attached algae; B = bacteria; C = inorganic or organic particulate detritus, such as calcium carbonate. (Reproduced with permission from Wetzel, 1990.)

phosphorus from the water passes to the macrophyte (Moeller *et al.*, 1988). The epiphytic algae and bacteria, rather than the submersed macrophytes, function as the primary scavengers for limiting nutrients such as phosphorus from the water. The same uptake abilities occur within the microgradients among the epilithic, epipelic, and epipsammic algae attached to nonliving surfaces.

B. Microgradients

Algal and associated microbial community development on surfaces is entirely different in community attributes and environmental constraints than those of planktonic microbiota. The dynamics of colonization and growth of bacterial films have been thoroughly studied, particularly in regard to alteration of water flow in pipes and along similar confined surfaces (e.g., Characklis and Wilderer, 1989; Characklis and Marshall, 1990). Successional development of microbial consortia follows a typical logarithmic sigmoid growth curve until the depth of the community is achieved where portions of the most recent community development slough from the main strata and are exported from the community by gravity or water movements (Fig. 6).

Rates of diffusion of solutes and gases from the ambient medium into the microbial communities are influenced by water movements, concentration gradients, metabolism within the communities, and other factors (e.g., Riber and Wetzel, 1987). Water movements reduce boundary-layer thicknesses and distances for molecular diffusion of gases and nutrients between microbiota and the overlying water (cf. Stevenson, Chapter 11, this volume). It must be recognized that among standing waters, turbulent flow is very rare at interfaces between the substrata and water. Earlier estimates of nonturbulent boundary thicknesses of ca. 10 μm (Raven, 1970; Smith and Walker, 1980; Koch, 1990) are much too small. Hydrodynamic boundary-layer thicknesses for attached communities are much larger and in the range of 10^2 to 10^4 μm (Losee and Wetzel, 1993). In addition to the boundary-layer-associated resistance to diffusion across the water/attached microbial interface, the thicknesses of the microbial communities of standing waters are very large, often to several millimeters or even centimeters, and much more complexly developed than is the case in flowing waters.

In addition, attached benthic algal and microbial communities are not porous, but rather are relatively impervious to throughflow (Wetzel *et al.*, 1995). Most attached communities consist of a tight integration of living and dead algal and bacterial cells, particulate detritus, and inorganic particles bound within an organic matrix of largely extracellular polysaccharide secretions (EPS) (e.g., Lange, 1976; Lock *et al.*, 1984; Costerton *et al.*, 1987; Lock, 1990). Large filamentous cyanobacterial-dominated commu-

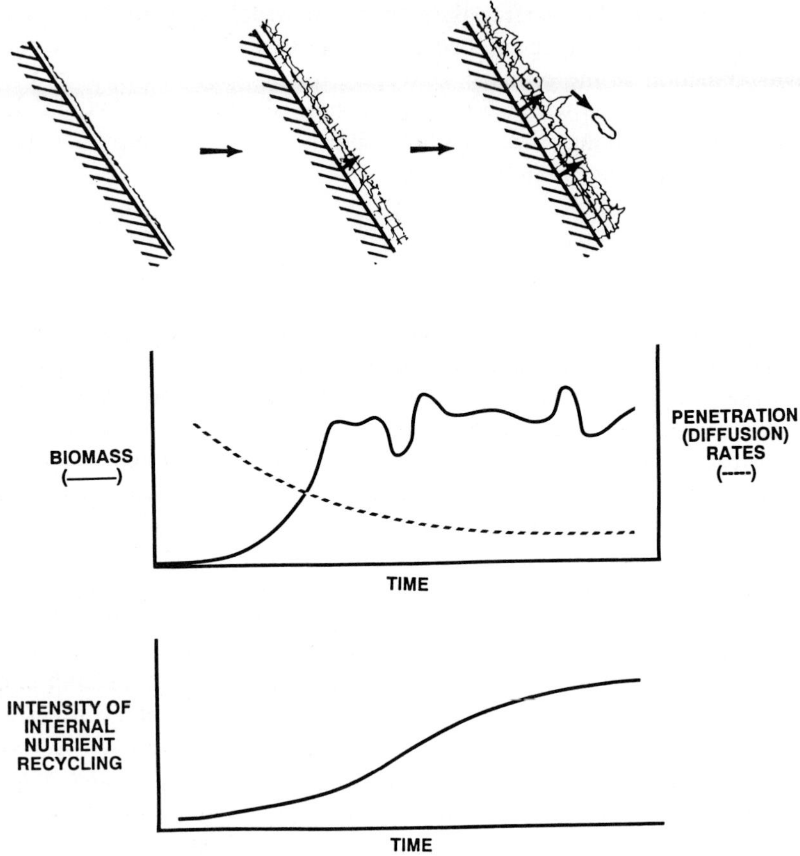

FIGURE 6 Successional development of and increased nutrient recycling within microbial communities on substrata. (After Wetzel, 1993.)

nities are particularly nonporous. Stalked, nonmucilaginous diatoms and filamentous green algae can presumably increase the porosity of the community (J. Stevenson, personal communication).

In contrast to the microbial communities, many of the substrata upon which the algae grow, such as sand and organic particles, are very porous to intrusions from surface flows and from groundwater sources. The scale of this movement, however, is much larger than that of the metabolic environment of individual cells of the attached microbial communities. An important distinction is that although the substrata may be porous, the microbial biofilms are not; within biofilms, transport of nutrients is strictly by diffusion along concentration-mediated gradients. Those gradients are highly dynamic as mediated by the photosynthetic and respiratory metabolism of the microbiota, and in the case of macrophytes the living substrata. It should be emphasized that water currents in lentic environments are slow in comparison to streams and rivers. Even wave-swept areas of littoral areas are only intermittently active and the macrocurrents of the bulkwater bear no relationship to those at the boundary layers of the periphyton. Within macrophyte beds, flows are very slight, usually much less than 0.5 cm s^{-1} (Losee and Wetzel, 1993). Most flow rates result in large laminar flow boundary-layer thicknesses of 1 to 10 mm, usually much larger than the communities themselves.

Therefore, in most natural environments of lentic waters, attached communities are not exposed to appreciable flows, and boundary layers are often, if not usually, manifold thicker than the cells themselves. As a result, diffusion of nutrients is regulated by internal concentration gradients within the microbial complexes as well as diffusive boundary-layer thickness. These gradients can change very rapidly (seconds) and are regulated by the coupled metabolism of algal and macrophytic photosynthesis and of bacterial, fungal, protistan, and plant respiration and photorespiration. As a result, the microbial algal and bacterial communities have adapted most efficiently to a microenvironment in which they have coupled their physiological processes mutualistically by intensive, rapid recycling of carbon and nutrients (Wetzel, 1993).

The general physiological couplings of algal and bacterial metabolism within attached periphytic communities have been a common observation (Haack and McFeters, 1982; Hamilton, 1987; Paerl and Carlton, 1988; Paerl *et al.,* 1993; Neely, 1994). Some of the couplings within attached communities are known with considerable specificity. For example, the diurnal interactions of oxygenic photosynthesis and N_2 fixation, as well as other nitrogen metabolic processes, are known for microbial mats (e.g., Bebout *et al.,* 1987; Paerl *et al.,* 1989, 1993). Most of these interactions are at a general level rather than in regard to specific movements of an element and their rates of nutrient cycling. For example, epiphytic algal and bacterial productivities were directly coupled over a photon flux density range of 20–400 µmol m^{-2} s^{-1} (Neely and Wetzel, 1995). Application of a photosystem II inhibitor of photosynthesis resulted in an immediate reduction in bacterial production by 46%, but had no such effect on the productivity of bacterial isolates. Thus, although bacterial production in periphyton is closely coupled to algal photosynthesis and products of bacterial metabolism (CO_2, inorganic and organic micronutrients) are clearly affecting epiphytic algal photosynthesis and growth, the rates of nutrient cycling at these microlevels are not well defined and quantified. Quantification of these rates certainly represents a most challenging and potentially rewarding area of study.

All attached communities are exposed to nutrient sources from the surrounding water as well as from the substrata upon which they grow. In lentic waters, four major nutrient sources from the substrata are now known qualitatively; some quantitative data indicate major if not dominant importance of these sources, particularly in oligotrophic waters.

1. The microdistribution of species and groups of epilithic algae is clearly correlated directly with differences in microdistribution of rock facies and differences in solubility of specific elements from the rock (Smith *et al.*, 1992). Mineral nutrients (Fe, Si, trace elements) can leach from the rock substratum and be utilized by the attached microflora with alterations in microdistributions of algal species. The distribution and productivity of epipsammic algae are influenced by microscale differences in diffusion and microflows from interstitial waters through sandy sediments, as well as sand grain morphology and topography (Krejci and Lowe, 1986, 1987). The bacteria of interstitial waters degrade organic substrates and recycle nutrients that subsequently move with flows into surficial sandy sediments of photic regions. These evasions from the substrata directly affect the distribution and growth of adnate microbiota; it is unclear how far this influence extends into the overlying loosely attached communities of rocks and sand. Greater insight into the importance of nutrient recycling movements is available from study of epiphytic algae and bacteria than is the case for epilithic and epipsammic algae.

2. The epiphytic algae of lentic waters, if present, are clearly the most productive among planktonic and other attached algae, and maintain large collective biomasses throughout the year. In addition to the very large surface areas for colonization and the retention of algae by supporting macrophytes into overlying lighted zones, host plants also provide significant sources of nutrients from both living as well as senescing tissues (reviewed by Wetzel, 1983a, 1990; Moeller *et al.*, 1988; Burkholder and Wetzel, 1990; Burkholder *et al.*, 1990). These nutrients internally supplement those diffusing into the attached communities from the water within and passing through the littoral zone. Complex nutrient interactions exist among algal species. For example, the development of littoral diatoms can be coupled to availability of silica in the water and from macrophyte tissues. In Furesø, Denmark, the spring maximum of the planktonic diatoms, primarily *Stephanodiscus,* reduced silica concentrations (< 40 μg liter^{-1}) to levels that were shown experimentally to severely inhibit the growth of these algae (E. G. Jørgensen, 1957). An immediate increase in epiphytic diatoms growing on the submersed portions of the emergent macrophyte *Phragmites* accompanied this decline in silica. The *Phragmites* stems were shown to possess large quantities of easily dissolved silica, and their silica content decreased during the development of the epiphytic diatoms.

In another example, certain algal species growing adnate to the macrophyte can obtain over 60% of their phosphorus from the macrophytes (Fig. 7). Algae near the outer portion of the community obtained less phosphorus from the macrophyte, but nonetheless significant quantities emanated from the plant. Even when phosphorus concentrations are very

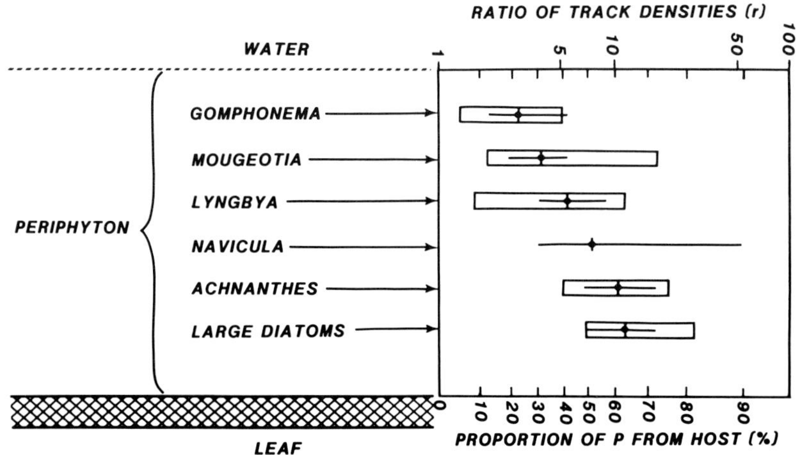

FIGURE 7 Sources of phosphorus within an epiphytic community on *Najas flexilis*. Epiphytic algal taxa are arranged according to increasing proportion of phosphorus received from the host plant. The relation between *r* (upper) and the proportion (lower) is implicit in the dual scaling. A 90% C.I. was calculated from raw data (open box) and corrected data (bar). [Reprinted with permission from Wetzel (1990), based on data from Moeller *et al.* (1988).]

high in the water, some phosphorus, and presumably other nutrients, is obtained from the macrophyte, simply because diffusion rates from the water into and within the complex epiphytic community are too slow to meet metabolic demands.

Although nutrients from the surrounding water may not be adequate to meet the collective demands of extant epiphytic algae and associated microflora, phosphorus and many other nutrients are actively assimilated by the loosely attached epiphytic periphyton, incorporated into the complex of algae, bacteria, and inorganic and organic detritus, and intensively recycled. Phosphorus uptake kinetics of periphyton, for example, were found to be acutely limited by boundary-layer mass transfer and a power function of flow velocity (Riber and Wetzel, 1987). Kinetic calculations based on turnover measurements indicated that internal recycling of phosphate and recycling from the boundary layer, rather than external uptake, accounted for most phosphate turnover within intact periphyton biofilms. Under optimal conditions, it was estimated that the turnover time of phosphorus was so rapid that phosphorus was recycling between algae and bacteria every 15 seconds. Much variation was found in the rates within periphyton, however, and was related to the effects of community patchiness. Little phosphorus incorporated into the periphyton passed to the macrophyte (Moeller *et al.,* 1988). The periphyton functions as the primary scavenger of limiting nutrients, such as phosphorus, from the water. With reasonable exposure times of the periphyton to the water, nutrient removal and retention by periphytic algae and bacteria are very efficient (Fig. 20.6). As macrophytic tissue senesces, much of the phosphorus is translocated to rooting tissues. Loss of cellular integrity of the leaves results in leaching from the leaves. Nutrients and dissolved organic compounds are readily utilized by periphytic microflora, which tends to develop profusely during autumn and winter periods (Fig. 4). As the senescing macrophyte tissue with their epiphytic communities collapse to the detrital mass near the sediments, much of the phosphorus is incorporated into the detritus of the sediments and actively retained and recycled by the detrital and sediment epipelic microflora (e.g., Federle and Vestal, 1980; Carlton and Wetzel, 1988).

Among all periphytic algal communities of lentic waters, not only is the environment radically changed within the periphyton aggregates *per se*, but the products of the metabolic activities of these highly productive communities can extend quite a distance into the surrounding waters. These altered conditions can radically change the nutrient availabilities within the macroenvironments external to the periphyton within and beyond the boundary layers, as well as influence the retention and release of nutrients from one part of the ecosystem to another.

The combined photosynthesis of epiphytic algae and submersed macrophytes can lead to marked, abrupt decreases in the concentrations of inorganic carbon and increases in pH both in close proximity to the communities as well as within the littoral zones (e.g., Goulder, 1969; Brammer and Wetzel, 1984). As these communities photosynthesize, steep microgradients also occur within them and immediately adjacent to them (e.g., Carlton and Wetzel, 1988). The high pH, often exceeding a pH of 9, can induce the precipitation of calcium carbonate with the simultaneous adsorption and/or coprecipitation of phosphate (Otsuki and Wetzel, 1972) and metabolically important organic compounds, such as vitamin B_{12} (White and Wetzel, 1985) and organic acids (Otsuki and Wetzel, 1973). Some of these nutrients can be stored in the complexed forms with the carbonates and resolubilized subsequently; however, many are permanently inactivated with the precipitated carbonates.

Although major alterations in the distributions of inorganic carbon and pH in littoral areas of submersed plants have been attributed to the macrophytic metabolism (e.g., Goulder, 1969), it is likely that the very active photosynthetic activity of the epiphytic algae was the primary driver of these changes. Indeed, epiphytic algae have been shown experimentally to induce carbonate precipitation and a cascade of associated alterations in nutrient losses and altered availabilities of both nutrients and dissolved organic carbon (Mickle and Wetzel, 1978a,b, 1979).

3. Epipelic algae growing upon organic sediments often develop into dense communities, both loosely attached and in dense matlike aggregations. Often these communities will be several millimeters in thickness. But

even in relatively sparsely developed communities on sediments, photosynthesis of the epipelic algae can markedly affect nutrient fluxes from the sediments to the overlying water. Metabolic activities affect fluxes by (1) modification of redox gradients, which in turn influence chemical mobilities, and (2) direct assimilation and utilization. The recent development and application of microelectrodes to these communities under natural and experimental conditions have enhanced our understanding of distributions, fluxes, and cycling of oxygen, pH, phosphorus, combined nitrogen, and other chemical constituents of the interstitial water of sediments and sediment–water interface algal–microbial communities (Revsbech and Jørgensen, 1986).

Photosynthetic oxygen production can rapidly exceed that of respiratory consumption and extend appreciably into the interstitial waters of the sediments (Fig. 8). This production of oxygen and penetration well into the sediments result in supersaturated oxygen conditions of interstitial water and rapid shifts, in a matter of minutes, in the redox potentials

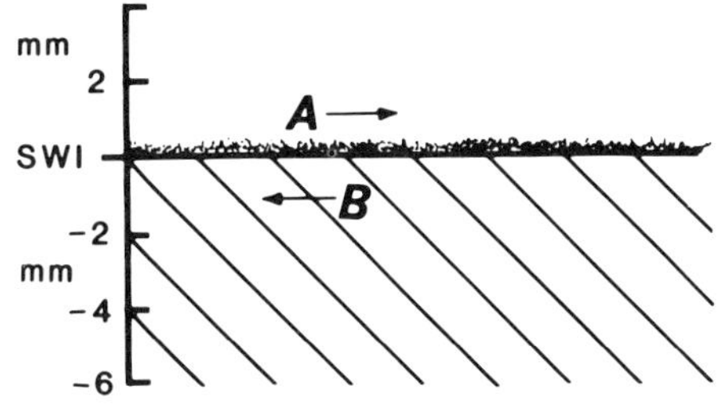

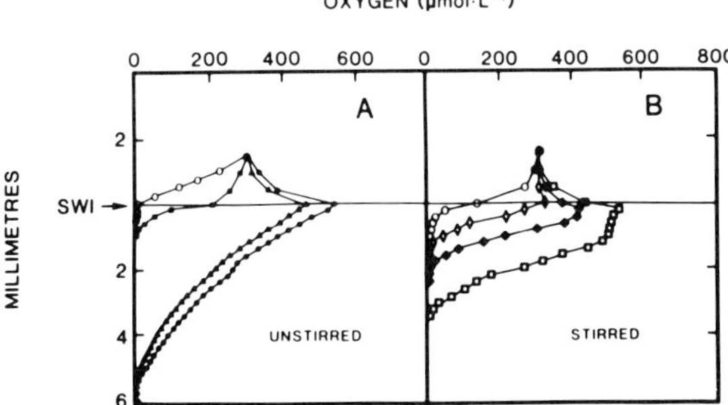

FIGURE 8 (Upper) Sediment–water microzone, with microprofiles of oxygen in the sediments colonized by microalgae (A →), which tend to increase dissolved oxygen, and redox and bacteria (← B), which rapidly decrease oxygen and redox. (Lower) (A) During darkness (○) and after 1 h (■), 8 h (▲) and 10 h (●) of illumination with 10 μE m^{-2} s^{-1}; overlying water not stirred. (B) During darkness (○) and after 8 h of illumination with 10 (◊), 30 (♦), and 45 μE m^{-2} s–1 (□); overlying water stirred. Vertical scale is distance from sediment–water interface. Temperature = 16°C; 100% saturation = 308 μmol $liter^{-1}$ O_2. (Modified from Carlton and Wetzel, 1987.)

from highly reducing to highly oxidizing conditions. Fluxes of phosphorus from the sediments to the overlying water are nearly totally suppressed under these conditions (Carlton and Wetzel, 1988). Upon cessation of photosynthesis on a diurnal basis, reducing conditions return rapidly and phosphorus mobility increases and diffuses to the epipelic algae and overlying water.

Such marked diurnal fluctuations in oxygen and other chemical microgradients have been shown in various epipelic loosely attached algal and mat-type algal communities (e.g., Jørgensen *et al.,* 1979, 1983; Carlton and Wetzel, 1987, 1988; Hansson, 1989). Although algal and cyanobacterial mat communities tend to be very compressed and relatively inefficient because of steep chemical and biotic gradients (cf. Wetzel, 1993), carbon and nutrient cycling tend to be very rapid (seconds to minutes) within and between the cells and the immediate interstitial water of the sediments. Chemical microelectrodes, such as for nitrous oxide and oxygen (Nielsen *et al.,* 1990), as well as ^{15}N dilution methods (Blackburn, 1979; Jansson, 1980; Thybo-Christesen and Blackburn, 1993), hold much potential for effective evaluation of rates of denitrification, ammonia assimilation and cycling, and other nutrient cycling measurements *within* these important microcommunities of benthic algae.

In addition to the indirect effects of benthic algae on nutrient fluxes from the sediments, the algae themselves are effective scavengers of interstitial nutrients migrating from the sediments. Both phosphorus (Carlton and Wetzel, 1988) and nitrogen as NH_4-N (Jansson, 1980) evasion to the water is suppressed by active assimilation into the algae. The benthic algae likely release appreciable quantities of dissolved organic nitrogen into the water above the sediments. Some of the combined nitrogen is denitrified internally by the microbial consortium, as noted earlier, but relatively little is known of the specific rates of internal recycling in algae of these communities.

4. Nutrient cycling within algal communities of the metaphyton is largely unstudied. Metaphytic algae are frequently filamentous and often develop initially on surfaces, particularly the sediments (Hillebrand, 1983). After a period of submersed growth, formation of oxygen bubbles within the community leads to ascent of the algal clusters. During flotation, these algae are exposed to extreme conditions of intense radiation, dissolved oxygen, pH, and localized nutrient and carbon depletion (e.g., Spencer *et al.,* 1994).

Mats of *Cladophora* frequently form large algal aggregations floating above the sediments in shallow areas. Large diurnal fluctuations in concentrations of NH_4^+, NO_3^-, PO_4^{3-}, and dissolved oxygen and in pH and temperature were found in the metaphytic communities (Thybo-Christesen *et al.,* 1993). Much recycling of nitrogen and phosphorus occurred in the mat. Highest NH_4^+ uptake rates were in the lower portion of the mat community, probably because of the proximity of enriched nutrients in and near the sediments. A large release of NH_4^+ at the onset of darkness, followed by low uptake rates in the dark, was evidence for release of loosely bound rather than intracellular NH_4^+. A budget for June showed the importance of recycling of N within the mat. Consumption by the filamentous algae accounted for up to 95% of the available N and 85% of the available P. The metaphyton was nearly a closed system, with 63% of its N requirements being supplied by internal recycling (2.6 mmol m^{-2} d^{-1}) and the remainder coming from the sediment (0.83 mmol m^{-2} d^{-1}), lateral water advection (0.3 mmol m^{-2} d^{-1}), and atmospheric precipitation (0.4 mmol m^{-2} d^{-1}). The algal mat had a relatively high biomass (June; 87 g DW m^{-2}) with an active breakdown (2.6 mmol N m^{-2} d^{-1}) and resynthesis (3.9 mmol N m^{-2} d^{-1}).

In shallow littoral areas, particularly in wetlands, loosely attached metaphytic algal communities often form massive aggregates of algae and bacteria in various dynamics of photosynthesis and decomposition. In the vast Everglades wetland ecosystem of southern Florida, for example, metaphytic algae are a dominant component of the productivity and nutrient assimilation and sequestering, particularly by direct phosphorus acquisition and further by immobilization with photosynthetically induced precipitation of calcium carbonate (Browder *et al.*, 1994). Intimate, coupled nutrient and carbon exchanges are clearly occurring at very rapid rates, as has been demonstrated among epiphytic microbial communities. On the basis of studies of other attached algal communities, nutrient microgradients must be steep and fluxes within and in close proximity to the metaphytic algal communities must be rapid and highly dynamic. Despite the complexities of heterogeneous development of these aggregates, studies of nutrient fluxes to, within, and from metaphytic algae and other microbes are fertile areas of inquiry of major limnological importance.

IV. SUMMARY AND CONCLUSIONS

Attached algal communities constitute the major or dominant primary producer and source of autochthonous organic matter of most standing waters because most lakes and reservoirs are predominantly small and shallow. These very high rates of photosynthesis are only possible because the efficiencies of nutrient retention and recycling are much greater among closely aggregated attached algal–microbial–substratum communities than is the case among the plankton.

Four major attached algal community habitats and their nutrient gradient characteristics were summarized: emergent macrophyte–detrital aggregates and hydrosoils; submersed macrophytes and attendant epiphytic microfloral communities; epilithic, epipsammic, and epipelic benthic algal communities; and metaphyton of littoral areas. These habitats differ in internal nutrient microgradients but are collectively integrated in nutrient cycling along macrogradients from the land to the open water. Much of the extant information on nutrient cycling in these wetland and littoral communities is at the macroscale level, which yields relatively few insights into the regulating parameters of the microbial communities within the complex communities. Studies at the microscale levels during the past decade, however, have examined nutrient fluxes into, within, and from the attached algal–microbial communities. The periphytic microbial cells are highly efficient metabolically because the microbiota are in close juxtaposition, sometimes in direct contact adnate to each other and to the living or dead supporting substrata. Diffusion distances are very short and concentration gradients can be kept steep by constant metabolic utilization (sinks) within the attached communities.

The importance of recycling of both nutrients and gases cannot be overemphasized (Wetzel, 1993). Only by means of intensive recycling of essential nutrients, particularly phosphorus, nitrogen, and inorganic carbon, can the problems of slow diffusion transport across boundary layers be overcome sufficiently to permit the extremely high levels of growth and productivity observed to occur in periphytic algae of both eutrophic and particularly of oligotrophic waters. As external nutrient loading and availability increase in the surrounding water, as for example with eutrophication, one would anticipate rates of internal recycling to decrease. This relation is not robust, however, because of the rapidly increasing physical thicknesses of the periphytic algal communities with nutrient enrichment and attendant problems of diffusion and light reduction. Although rates of internal nutrient recycling with the greatest levels of nutrient conservation occur among attached algal communities in oligotrophic waters, the physical constraints allow only moderate shifts to greater reliance upon external nutrient sources among attached algae of eutrophic waters. Every selective physical and biotic process among the species must be operating to maximize resource retention and conservative recycling in order to maintain maximal growth and reproductive potential among benthic algae.

ACKNOWLEDGMENTS

The support of the National Science Foundation (DEB-9220822 and OSR-91-08761) and the Department of Energy for research leading to this synthesis is gratefully acknowledged. The critical comments of Drs. Robert P. Neely, Rex Lowe, R. G. Carlton, and R. J. Stevenson are particularly appreciated.

REFERENCES

Allanson, B. R. (1973). The fine structure of the periphyton of *Chara* sp. and *Potamogeton natans* from Wytham Pond, Oxford, and its significance to the macrophyte–periphyton model of R. G. Wetzel and H. L. Allen. *Freshwater Biol.* **3,** 535–541.

Bebout, B. M., Paerl, H. W., Crocker, K. M., and Prufert, L. E. (1987). Diel interactions of oxygenic photosynthesis and N_2 fixation (acetylene reduction) in marine microbial mat communities. *Appl. Environ. Microbiol.* **53,** 2353–2362.

Behre, K. (1956). Die Algenbesiedlung einiger Seen um Bremen and Bremerhaven. *Veroeff. Inst. Meeresforsch. Bremerhaven* **4,** 221–283.

Blackburn, T. H. (1979). Method for measuring rates of NH_4^+ turnover in anoxic marine sediments, using a ^{15}N-NH_4^+ dilution technique. *Appl. Environ. Microbiol.* **37,** 760–765.

Bowden, W. B. (1987). The biogeochemistry of nitrogen in freshwater wetlands. *Biogeochemistry* **4,** 313–348.

Brammer, E. S., and Wetzel, R. G. (1984). Uptake and release of K^+, Na^+, and Ca^{++} by the water soldier, *Stratiotes aloides* L. *Aquat. Bot.* **19,** 119–130.

Bronmark, C. (1989). Interactions between epiphytes, macrophytes and freshwater snails: A review. *J. Molluscan Stud.* **55,** 299–311.

Browder, J. A., Gleason, P. J., and Swift, D. R. (1994). Periphyton in the Everglades: Spatial variation, environmental correlates, and ecological implications. *In* "Everglades: The Ecosystem and Its Restoration" (S. M. Davis and J. C. Ogden, eds.), pp. 379–418. St. Lucie Press, Delray Beach, FL.

Burkholder, J. M., and Wetzel, R. G. (1989). Epiphytic microalgae on natural substrata in a hardwater lake: Seasonal dynamics of community structure, biomass and ATP content. *Arch. Hydrobiol., Suppl.* **83,** 1–56.

Burkholder, J. M., and Wetzel, R. G. (1990). Epiphytic alkaline phosphatase on natural and artificial plants in an oligotrophic lake: Re-evaluation of the role of macrophytes as a phosphorus source for epiphytes. *Limnol. Oceanogr.* **35,** 736–747.

Burkholder, J. M., Wetzel, R. G., and Klomparens, K. L. (1990). Direct comparison of phosphate uptake by adnate and loosely attached microalgae within an intact biofilm matrix. *Appl. Environ. Microbiol.* **56,** 2882–2890.

Carignan, R., and Kalff, J. (1982). Phosphorus release by submerged macrophytes: Significance to epiphyton and phytoplankton. *Limnol. Oceanogr.* **27,** 419–427.

Carlton, R. G., and Wetzel, R. G. (1987). Distributions and fates of oxygen in periphyton communities. *Can. J. Bot.* **65,** 1031–1037.

Carlton, R. G., and Wetzel, R. G. (1988). Phosphorus flux from lake sediments: Effect of epipelic algal oxygen production. *Limnol. Oceanogr.* **33,** 562–570.

Cattaneo, A. (1983). Grazing on epiphytes. *Limnol. Oceanogr.* **28,** 124–132.

Characklis, W. G., and Marshall, K. C., eds. (1990). "Biofilms." Wiley, New York.

Characklis, W. G., and Wilderer, P. A., eds. (1989). "Structure and Function of Biofilms." Wiley, Chichester.

Coffin, C. C., Hayes, F. R., Jodrey, L. H., and Whiteway, S. G. (1949). Exchange of materials in a lake as studied by the addition of radioactive phosphorus. *Can. J. Res., Sect. D* **27,** 207–222.

Costerton, J. W., Cheng, K. J., Geesey, G. G., Ladd, T. I., Nickel, J. C., Dasgupta, D., and Marrie, T. J. (1987). Bacterial biofilms in nature and disease. *Ann. Rev. Microbiol.* **41,** 435.

Coveney, M. F., and Wetzel, R. G. (1995). Biomass, production, and specific growth rate of bacterioplankton and coupling to phytoplankton in an oligotrophic lake. *Limnol. Oceanogr.* **40,** 1187–1200.

Cuker, B. E. (1983). Competition and coexistence among the grazing snail *Lymnaea,* Chironomidae, and microcrustacea in an arctic epilithic lacustrine community. *Ecology* **64,** 10–15.

DeAngelis, D. L. (1992). "Dynamics of Nutrient Cycling and Food Webs." Chapman & Hall, New York.

Dickerman, J. A., and Wetzel, R. G. (1985). Clonal growth in *Typha latifolia:* Population dynamics and demography of the ramets. *J. Ecol.* **73**, 535–552.

Fairchild, G. W., and Sherman, J. W. (1993). Algal periphyton response to acidity and nutrients in softwater lakes: Lake comparison vs. nutrient enrichment approaches. *J. North Am. Benthol. Soc.* **12**, 157–167.

Federle, T. W., and Vestal, J. R. (1980). Microbial colonization and decomposition of *Carex* litter in an arctic lake. *Appl. Environ. Microbiol.* **39**, 888–893.

Godshalk, G. L., and Wetzel, R. G. (1978a). Decomposition of aquatic angiosperms. I. Dissolved components. *Aquat. Bot.* **5**, 281–300.

Godshalk, G. L., and Wetzel, R. G. (1978b). Decomposition of aquatic angiosperms. II. Particulate components. *Aquat. Bot.* **5**, 301–328.

Goulder, R. (1969). Interactions between the rates of production of a freshwater macrophyte and phytoplankton in a pond. *Oikos* **20**, 300–309.

Haack, T. K., and McFeters, G. A. (1982). Nutritional relationships among microorganisms in an epilithic biofilm community. *Microb. Ecol.* **8**, 115–126.

Hamilton, W. A. (1987). Biofilms: Microbial interactions and metabolic activities. *In* "Ecology of Microbial Communities" (M. Fletcher, T. R. G. Gray, and J. G. Jones, eds.), pp. 361–385. Cambridge Univ. Press, New York.

Hansson, L.-A. (1989). The influence of a periphytic biolayer on phosphorus exchange between substrate and water. *Arch. Hydrobiol.* **115**, 21–26.

Hawes, I., and Smith, R. (1993). Effect of localised nutrient enrichment on the shallow epilithic periphyton of oligotrophic Lake Taupo, New Zealand. *N. Z. J. Mar. Freshwater Res.* **27**, 365–372.

Hayes, F. R., and Coffin, C. C. (1951). Radioactive phosphorus and the exchange of lake nutrients. *Endeavour* **10**, 78–81.

Hillebrand, H. (1983). Development and dynamics of floating clusters of filamentous algae. *In* "Periphyton of Freshwater Ecosystems" pp. 31–39. (R. G. Wetzel, ed.), Dr. W. Junk Publishers, The Hague.

Howard-Williams, C. (1981). Studies on the ability of a *Potamogeton pectinatus* community to remove dissolved nitrogen and phosphorus compounds from lake water. *J. Appl. Ecol.* **18**, 619–637.

Imberger, J., and Patterson, J. C. (1990). Physical limnology. *Adv. Appl. Mech.* **27**, 303–475.

Jansson, M. (1980). Role of benthic algae in transport of nitrogen from sediment to lake water in a shallow clearwater lake. *Arch. Hydrobiol.* **89**, 101–109.

Jørgensen, B. B., Revsbech, N. P., Blackburn, T. H., and Cohen, Y. (1979). Diurnal cycle of oxygen and sulfide microgradients and microbial photosynthesis in a cyanobacterial mat sediment. *Appl. Environ. Microbiol.* **38**, 46–58.

Jørgensen, B. B., Revsbech, N. P., and Cohen, Y. (1983). Photosynthesis and structure of benthic microbial mats: Microelectrode and SEM studies of four cyanobacterial communities. *Limnol. Oceanogr.* **28**, 1075–1093.

Jørgensen, E. G. (1957). Diatom periodicity and silicon assimilation. *Dans. Bot. Ark.* **18**(1), 1–54.

Koch, A. L. (1990). Diffusion: The crucial process in many aspects of the biology of bacteria. *Adv. Microb. Ecol.* **11**, 37–70.

Krejci, M. E., and Lowe, R. L. (1986). Importance of sand grain mineralogy and topography in determining micro-spatial distribution of epipsammic diatoms. *J. North Am. Benthol. Soc.* **5**, 211–220.

Krejci, M. E., and Lowe, R. L. (1987). Spatial and temporal variation of epipsammic diatoms in a spring-fed brook. *J. Phycol.* **23**, 585–590.

Kuznetsov, S. I. (1968). Recent studies on the role of microorganisms in the cycling of substances in lakes. *Limnol. Oceanogr.* **13**, 211–224.

Kuznetsov, S. I. (1970). "Mikroflora ozer i ee geokhimicheskaya deyatel'nost' (Microflora of Lakes and Their Geochemical Activities)" (in Russian). Izdatal'stvo Nauka, Leningrad.

Lange, W. (1976). Speculations on a possible essential function of the gelatinous sheath of blue-green algae. *Can. J. Microbiol.* **22,** 1181–1185.

Lock, M. A. (1990). The dynamics of dissolved and particulate organic material over the substratum of water bodies. *In* R. S. Wotton, (ed.), "The Biology of Particles in Aquatic Systems" pp. 117–144. CRC Press, Roca Baton, FL.

Lock, M. A., Wallace, R. R., Costerton, J. W., Ventullo, R. M., and Charlton, S. E. (1984). River epilithon (biofilm): Toward a structural functional model. *Oikos* **42,** 10–22.

Losee, R. F., and Wetzel, R. G. (1993). Littoral flow rates within and around submersed macrophyte communities. *Freshwater Biol.* **29,** 7–17.

Meadows, P. S., and Anderson, G. C. (1966). Micro-organisms attached to marine and freshwater sand grains. *Nature (London)* **212,** 1059–1060.

Mickle, A. M., and Wetzel, R. G. (1978a). Effectiveness of submersed angiosperm epiphyte complexes on exchange of nutrients and organic carbon in littoral systems. I. Inorganic nutrients. *Aquat. Bot.* **4,** 303–316.

Mickle, A. M., and Wetzel, R. G. (1978b). Effectiveness of submersed angiosperm epiphyte complexes on exchange of nutrients and organic carbon in littoral systems. II. Dissolved organic carbon. *Aquat. Bot.* **4,** 317–329.

Mickle, A. M., and Wetzel, R. G. (1979). Effectiveness of submersed angiosperm epiphyte complexes on exchange of nutrients and organic carbon in littoral systems. III. Refractory organic carbon. *Aquat. Bot.* **6,** 329–335.

Moeller, R. E., Burkholder, J. M., and Wetzel, R. G. (1988). Significance of sedimentary phosphorus to a rooted submersed macrophyte (*Najas flexilis* (Willd.) Rostk. and Schmidt) and its algal epiphytes. *Aquat. Bot.* **32,** 261–281.

Moeller, R. E., Wetzel, R. G., and Osenberg, C. W. (1996). Concordance of phosphorus limitation in lakes: Phytoplankton, epiphytes, and rooted macrophytes. (Submitted; unpublished manuscript).

Moshiri, G. A., ed. (1993). "Constructed Wetlands for Water Quality Improvement." Lewis Publishers, Boca Raton, FL.

Moss, B., and Round, F. E. (1967). Observations on standing crops of epipelic and epipsammic algal communities in Shear Water, Wilts. *Br. Phycol. Bull.* **3,** 241–248.

Naumann, E. (1931). "Limnologische Terminologie. Handbuch der biologischen Arbeitsmethoden," Sect. IX, Part 8. Urban & Schwarzenberg, Berlin.

Neely, R. K. (1994). Evidence for positive interactions between epiphytic algae and heterotrophic decomposers during the decomposition of *Typha latifolia. Arch. Hydrobiol.* **129,** 443–457.

Neely, R. K., and Wetzel, R. G. (1995). Simultaneous use of ^{14}C and ^{3}H to determine autotrophic production and bacterial protein production in periphyton. *Microb. Ecol.* **30,** 227–237.

Niederhauser, P., and Schanz, F. (1993). Effects of nutrient (N, P, C) enrichment upon the littoral diatom community of an oligotrophic high-mountain lake. *Hydrobiologia* **269/270,** 453–462.

Nielsen, L. P., Christensen, P. B., Revsbech, N. P., and Sorensen, J. (1990). Denitrification and oxygen respiration in biofilms studied with a microsensor for nitrous oxide and oxygen. *Microb. Ecol.* **19,** 63–72.

Otsuki, A., and Wetzel, R. G. (1972). Coprecipitation of phosphate with carbonates in a marl lake. *Limnol. Oceanogr.* **17,** 763–767.

Otsuki, A., and Wetzel, R. G. (1973). Interaction of yellow organic acids with calcium carbonate in fresh water. *Limnol. Oceanogr.* **18,** 490–494.

Paerl, H. W. (1985). Microzone formation: Its role in the enhancement of aquatic N_2 fixation. *Limnol. Oceanogr.* **30,** 1246–1252.

Paerl, H. W. (1990). Physiological ecology and regulation of N_2 fixation in natural waters. *Adv. Microb. Ecol.* **11,** 305–344.

Paerl, H. W., and Carlton, R. G. (1988). Control of nitrogen fixation by oxygen depletion in surface-associated microzones. *Nature (London)* **332,** 260–262.

Paerl, H. W., Bebout, B. M., and Prufert, L. E. (1989). Naturally occurring patterns of oxygenic photosynthesis and N_2 fixation in a marine microbial mat: Physiological and ecological ramifications. *In* "Microbial Mats: Physiological Ecology of Benthic Microbial Communities" (Y. Cohen and E. Rosenberg, eds.), pp. 326–341. Am. Soc. Microbiol., Washington, DC.

Paerl, H. W., Bebout, B. M., Joye, S. B., and Des Marais, D. J. (1993). Microscale characterization of dissolved organic matter production and uptake in marine microbial mat communities. *Limnol. Oceanogr.* **38,** 1150–1161.

Phillips, G. L., Eminson, D., and Moss, B. (1978). A mechanism to account for macrophyte decline in progressively eutrophicated fresh waters. *Aquat. Bot.* **4,** 103–126.

Pieczyńska, E. (1972). Production and decomposition in the eulittoral zone of lakes. *In* "Productivity Problems of Freshwaters" (Z. Kajak and A. Hillbricht-Ilkowska, eds.), pp. 271–285. IBP-UNESCO, Warszawa-Krakow.

Pringle, C. M. (1987). Effects of water and substratum nutrient supplies on lotic periphyton growth: An integrated bioassay. *Can. J. Fish. Aquat. Sci.* **44,** 619–629.

Raven, J. A. (1970). Exogenous inorganic carbon sources in plant photosynthesis. *Biol. Rev. Cambridge Philos. Soc.* **44,** 167–221.

Reckhow, K. H., and Chapra, S. C. (1983). "Engineering Approaches for Lake Management. I. Data Analyses and Empirical Modeling." Butterworth, Boston.

Reuter, J. E., Loeb, S. L., and Goldman, C. R. (1983). Nitrogen fixation in periphyton of oligotrophic Lake Tahoe. *In* "Periphyton of Freshwater Ecosystems" (R. G. Wetzel, ed.), pp. 101–109. Dr. W. Junk Publishers, The Hague.

Revsbech, N. P., and Jørgensen, B. B. (1986). Microelectrodes: Their use in microbial ecology. *Adv. Microb. Ecol.* **9,** 293–352.

Riber, H. H., and Wetzel, R. G. (1987). Boundary-layer and internal diffusion effects on phosphorus fluxes in lake periphyton. *Limnol. Oceanogr.* **32,** 1181–1194.

Rigler, F. H. (1956). A tracer study of the phosphorus cycle in lake water. *Ecology* **37,** 550–562.

Rigler, F. H. (1964). The phosphorus fractions and the turnover time of inorganic phosphorus in different types of lakes. *Limnol. Oceanogr.* **9,** 511–518.

Rigler, F. H. (1973). A dynamic view of the phosphorus cycle in lakes. *In* "Environmental Phosphorus Handbook" (E. J. Griffith, A. Beeton, J. M. Spencer, and D. T. Mitchell, eds.), pp. 539–572. Wiley, New York.

Round, F. E. (1981). "The Ecology of Algae." Cambridge Univ. Press, Cambridge, UK.

Sculthorpe, C. D. (1967). "The Biology of Aquatic Vascular Plants." St. Martin's Press, New York.

Sládečková, A. (1960). Limnological study of the Reservoir Sedlice near Zeliv. XI. Periphyton stratification during the first year-long period (June 1957–July 1958). *Sci. Pap. Inst. Chem. Technol. Prague, Fac. Technol. Fuel Water* **4,** 143–261.

Smith, F. A., and Walker, N. A. (1980). Photosynthesis by aquatic plants: Effects of unstirred layers in relation to assimilation of CO_2 and HCO_3 and to carbon isotopic discrimination. *New Phytol.* **86,** 245–259.

Smith, J., Ward, A. K., and Stock, M. S. (1992). Quantitative estimation of epilithic algal patchiness caused by microtopographical irregularities of different rock types. *Bull. North Am. Benthol. Soc.* **9,** 148.

Spencer, W. E., Teeri, J., and Wetzel, R. G. (1994). Acclimation of photosynthetic phenotype to environmental heterogeneity. *Ecology* **75,** 301–314.

Thybo-Christesen, M., and Blackburn, T. H. (1993). Internal N-cycling, measured by $^{15}NH_4^+$ dilution, in *Cladophora sericea* in a shallow Danish bay. *Mar. Ecol.: Prog. Ser.* **100,** 283–286.

Thybo-Christesen, M., Rasmussen, M. B., and Blackburn, T. H. (1993). Nutrient fluxes and growth of *Cladophora sericea* in a shallow Danish bay. *Mar. Ecol.: Prog. Ser.* **100,** 273–281.

Triska, F. J., and Oremland, R. S. (1981). Denitrification associated with periphyton communities. *Appl. Environ. Microbiol.* **42,** 745–748.

Wetzel, R. G. (1964). A comparative study of the primary productivity of higher aquatic plants, periphyton, and phytoplankton in a large, shallow lake. *Int. Rev. gesamten Hydrobiol.* **49,** 1–61.

Wetzel, R. G. (1983a). Attached algal–substrata interactions: Fact or myth, and when and how? *In* "Periphyton of Freshwater Ecosystems" pp. 207–215. (R. G. Wetzel, ed.), Dr. W. Junk Publishers, The Hague.

Wetzel, R. G. (1983b). "Limnology," 2nd ed. Saunders College Publishing, Philadelphia.

Wetzel, R. G., ed. (1983c). "Periphyton of Aquatic Ecosystems," Dev. Hydrobiol. Vol. 17. Dr. W. Junk Publishers, The Hague.

Wetzel, R. G. (1990). Land–water interfaces: Metabolic and limnological regulators. *Verh.—Int. Ver. Theor. Angew. Limnol.* **24,** 6–24.

Wetzel, R. G. (1993). Microcommunities and microgradients: Linking nutrient regeneration, microbial mutualism, and high sustained aquatic primary production. *Neth. J. Aquat. Ecol.* **27,** 3–9.

Wetzel, R. G. (1996). In preparation.

Wetzel, R. G., and Allen, H. L. (1970). Functions and interactions of dissolved organic matter and the littoral zone in lake metabolism and eutrophication. *In* "Productivity Problems of Freshwaters" (Z. Kajak and A. Hillbricht-Ilkowska, eds.), pp. 333–347. PWN Polish Scientific Publishers, Warsaw.

Wetzel, R. G., and Ward, A. K. (1992). Primary production. *In* "Rivers Handbook" (P. Calow and G. E. Petts, eds.), pp. 354–369. Blackwell, Oxford.

Wetzel, R. G., Rich, P. H., Miller, M. C., and Allen, H. L. (1972). Metabolism of dissolved and particulate detrital carbon in a temperate hard-water lake. *Mem. Ist. Ital. Idrobiol. Dott. Marco de Marchi* **29**(Suppl.), 185–243.

Wetzel, R. G., Ward, A. K., and Stock, M. (1995). Reduction in mucilaginous matrices of biofilm communities in response to natural dissolved organic matter. *Limnol. Oceanogr.* (submitted for publication).

White, W. S., and Wetzel, R. G. (1985). Association of vitamin B_{12} with calcium carbonate in hardwater lakes. *Arch. Hydrobiol.* **104,** 305–309.

21

Modeling Benthic Algal Communities: An Example from Stream Ecology

C. David McIntire,* Stanley V. Gregory,† Alan D. Steinman,‡ and Gary A. Lamberti§

*Department of Botany and Plant Pathology, Oregon State University, Corvallis, Oregon 97331

†Department of Fisheries and Wildlife, Oregon State University, Corvallis, Oregon 97331

‡Research Department, South Florida Water Management District, West Palm Beach, Florida 33416

§Department of Biological Sciences, University of Notre Dame, Notre Dame, Indiana 46556

Algal Ecology

I. INTRODUCTION

A. Modeling and Models

From a scientific perspective, modeling is the process of putting structure on knowledge, and a model is some kind of statement of relationships. Therefore, all research scientists are modelers in the sense that they are involved in generating and updating conceptual models that evolve from field and laboratory studies. In some cases, it is useful to transform conceptual models into integrated numerical systems by mathematical formalization. In ecology, mathematical modeling is the translation of an ecological system into mathematical form and the subsequent investigation of the mathematical system, usually by computer simulation.

The overall goals of model building are description and prediction. In particular, mathematical models can be used for simple forecasting (e.g., the weather) or for scientific purposes: (1) for hypothesis generation; (2) to synthesize the results of field and laboratory studies; (3) to evaluate a data base; and (4) to set priorities for future research. The process of model building usually includes the selection and classification of variables, equation writing and parameterization, simulation, and the comparison of model output with the behavior of the natural system under consideration (model testing). In this chapter, we discuss the use of mathematical modeling for the scientific investigation of benthic algal assemblages and related variables in lotic ecosystems.

B. Examples of Benthic Algal Models from Stream Ecology

Most mathematical models that include algae as a biological component focus on the water column of marine or freshwater ecosystems and simulate patterns of primary production rather than successional trajectories of individual taxa. However, models that represent benthic algae in some way often are similar in mathematical form to models that simulate the production dynamics of planktonic algae (Straškraba and Gnauck, 1985).

One of the first models related to the production dynamics of benthic algae was based on experimental data from laboratory streams (McIntire, 1973). This model represented benthic algae and associated heterotrophic microorganisms as a quasi-species, and there was no attempt to partition the assemblage into individual algal taxa. Later, the model became the primary production module of a stream ecosystem model (McIntire *et al.*, 1975; McIntire and Colby, 1978). In 1974, the Desert Biome (I.B.P.) issued a progress report (Wlosinski, 1974) that described the mathematical details of another stream ecosystem model. In this model, the benthic plant assemblage was divided into four taxonomic groups: two algal species [*Cladophora glomerata* (L.) Kütz. and *Chara vulgaris* L.], all diatom species collectively, and a vascular plant (*Potamogeton pectinatis* L.). The desert stream model was eventually used to examine compatibility between model behavior and changes in a natural stream at different levels of taxonomic resolution (Wlosinski and Minshall, 1983). Another modeling approach to the study of benthic algae in streams partitioned the dynamics of attached diatom assemblages into the processes of immigration, reproduction, growth, and emigration (Stevenson, 1986; Stevenson and Peterson, 1991). In this case, models were used to examine alternative mechanisms that could account for changes in cell density.

C. Objectives

In small streams, benthic algae may provide the only significant source of autochthonous organic material, particularly when conditions are unsuitable for the establishment of bryophytes or aquatic vascular plants. Experimental work with benthic algae in lotic ecosystems usually has involved the isolation of periphyton assemblages from other components of the ecosystem and the subsequent manipulation of selected environmental variables in some kind of controlled system (see Lamberti and Steinman, 1993, for review). Although such assemblages consist of complex aggregations of microorganisms, the research approach is frequently autecological in the sense that the periphyton assemblage is the focus of the study and is viewed as a single unit or quasi-species, while all other variables are treated as part of the environment. Individual experiments of this design are more meaningful when they are an integral part of a sequence of related studies that systematically examine relationships between benthic algae and other components of the ecosystem. This chapter is primarily concerned with the role of mathematical modeling in the design, evaluation, interpretation, and synthesis of such studies during an ongoing research program.

Specifically, the objectives of this chapter are (1) to describe a modeling approach for the integration of experimental and observational studies of benthic algae in flowing water; (2) to illustrate the approach by examples of output from an existing model; (3) to present some hypotheses that correspond to the model output; and (4) to summarize the kinds of insights and research directions that models of benthic algal assemblages can provide.

II. A MODELING APPROACH

Benthic algae, in their natural surroundings, do not live in isolation. Individual taxa are subjected to complex interactions with abiotic and biotic components within an algal assemblage, and the assemblage as a whole changes its structural and functional attributes in response to direct and indirect relationships with components outside the boundaries of the assemblage. In the field or laboratory, an individual, replicated experiment is usually designed to examine effects of only one or at most a few variables in systems that maintain the same environment with respect to the variables not under investigation. The role of modeling, as presented in this chapter, is to provide a conceptual and structural basis for a series of experiments that couple together in a way that will help optimize the relevancy of each individual experiment to the objective questions under consideration.

First, we propose a research approach that includes: (1) identification of research goals and specific objectives; (2) system conceptualization, a process that involves the definition of system variables and their coupling structure, and the determination of appropriate levels of resolution relative to time, space, and biological organization; (3) translation of the biological concepts into mathematical form and the subsequent investigation of the mathematical model in relation to the objective questions under consideration; (4) model testing by comparisons of model behavior with observational data from the field; (5) generation of new hypotheses that are based on priorities revealed by modeling and field observations; (6) the design and performance of experiments to examine new hypotheses; (7) modification of the mathematical model and system conceptualization based on the latest experimental results; and (8) reevaluation of specific objectives and research progress in relation to the level of understanding generated by the most recent round of experimental work and modeling. In summary, this approach is iterative and synthetic, and involves the careful interplay between modeling, field observations, experimentation, and a periodic update of specific objectives in relation to an overall research goal.

In the following sections, we emphasize steps (5) and (7) by presenting an example of an ongoing study of plant–herbivore interactions in lotic ecosystems. In the example, an existing model of biological processes in small streams is updated (step 7) to represent and synthesize the results of recent experiments with benthic algal assemblages. Next, behavior of the modified stream model is used to generate new hypotheses (i.e., step 5 in the next iteration of studies) that relate to direct and indirect relationships between benthic algae and other components of lotic ecosystems.

III. THE MCINTIRE AND COLBY STREAM MODEL

Results of experimental studies at Oregon State University provided the basis for an updated version of an existing lotic ecosystem model (McIntire and Colby, 1978; McIntire, 1983). The original version of the model, referred to here as the McIntire and Colby stream model or M & C model, was developed to generate hypotheses, to synthesize the results of field and laboratory research, and to set priorities for future research. The model has also been used for the integration and evaluation of our latest laboratory stream research with benthic algae and selected herbivores (Steinman and McIntire, 1986, 1987; Steinman *et al.,* 1987a; Lamberti *et al.,* 1987, 1989; DeNicola and McIntire, 1990a,b, 1991; DeNicola *et al.,* 1990) and for the generation of new hypotheses related to the processes of primary production, grazing, shredding, collecting, and predation in lotic ecosystems.

The M & C model is an expansion of the lotic periphyton model described by McIntire (1973). Briefly, the M & C model represents biological processes that are usually active in most streams. From this perspective, stream ecosystems are conceptualized as two coupled subsystems, the processes of primary consumption and predation (Fig. 1). The Primary Consumption subsystem is composed of all processes associated with the direct consumption and decomposition of both autotrophic organisms and detritus, including the production dynamics of the autotrophic organisms collectively. Primary Consumption has two subsystems that represent the processes of herbivory and detritivory. The Herbivory subsystem consists of all processes associated with the production and consumption of benthic algae within the system, that is, the processes of primary production and grazing; whereas Detritivory includes processes related to the consumption and decomposition of detrital inputs, namely, shredding, collecting, and microbial decomposition. The Predation subsystem is composed of the processes of invertebrate and vertebrate predation, processes concerned with the transfer of energy among primary, secondary, and tertiary macroconsumers. In this chapter, we focus on the dynamics of benthic algal assemblages, a component of the Herbivory subsystem, and model output is investigated in relation to changes in parameters, inputs, and the internal structure of the Primary Production and Grazing subsystems.

IV. AN UPDATED HERBIVORY SUBSYSTEM MODEL

A new Herbivory subsystem model was developed by the isolation and modification of the Herbivory subsystem of the M & C model. The structure of the isolated subsystem model was changed and expanded to help synthesize and interpret the results of some recent experiments with benthic algae in laboratory streams.

In the first version of the M & C model, the Herbivory subsystem contains subsystems that represent the processes of primary production and grazing (Fig. 1). The state variable in each of these subsystems represents the biomass that is involved in the corresponding process at any time. New data from the experimental work (Steinman and McIntire, 1986, 1987; Steinman *et al.,* 1987a; Lamberti *et al.,* 1987, 1989) allow the state vari-

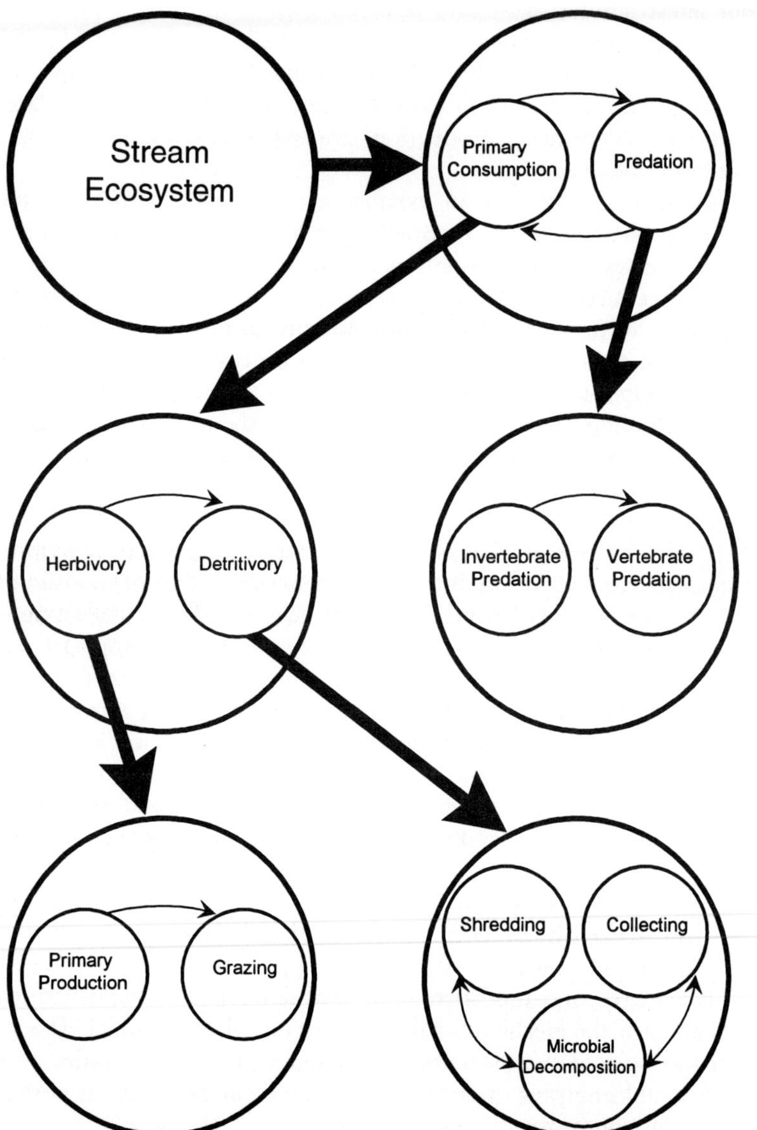

FIGURE 1 Schematic representation of a lotic ecosystem showing the hierarchical decomposition of the Primary Consumption and Predation subsystems and the structure of the Herbivory and Detritivory subsystems. Circles represent biological processes, and small arrows between processes indicate directions of energy flow.

able inside the Primary Production subsystem to be partitioned into three new state variables that are related to the taxonomic composition and successional stage of the algal assemblage. In this case, the state variables represent the collective biomass of filamentous and coenobic chlorophytes, diatoms, and cyanobacteria, with each functional group including associated heterotrophic microorganisms. In addition, feeding experiments by Lamberti *et al.* (1989, and unpublished data) provide a preliminary basis for establishing a mathematical relationship between the relative abundance of the three algal functional groups and the food consumption rates and assimilation efficiencies associated with the process of grazing.

In the updated Herbivory subsystem model, primary production is modeled according to the mathematical relationships described by McIntire and Colby (1978). This means that calculations of photosynthetic rates, respiratory expenditures, and export losses are based on the total periphyton biomass (i.e., rates are not calculated separately for individual algal functional groups). The algorithm that incorporates the new information into the Herbivory subsystem model has the following characteristics:

1. The new primary production calculated for each day is partitioned among the algal functional groups according to these rules:
   (a) If the irradiance is <30 μmol quanta m^{-2} s^{-1} or the periphyton biomass (ash-free dry weight) is <2 g m^{-2}, the new production is 100% diatoms (Fig. 2A);
   (b) If irradiance is >30 and <150 μmol quanta m^{-2} s^{-1}, the new production is partitioned by a linear relationship between light energy and the proportions of diatoms and cyanobacteria, reaching a maximum of 19% cyanobacteria at 50 μmol quanta m^{-2} s^{-1} when the algal biomass is >2 g m^{-2} (Figs. 2A and 2B);
   (c) If the irradiance is >150 μmol quanta m^{-2} s^{-1}, chlorophytes, diatoms, and cyanobacteria are all part of the new production in proportions that are determined by linear relationships with irradiance and algal biomass (Figs. 2A, 2B and 2C); and
   (d) If irradiance is >300 μmol quanta m^{-2} s^{-1}, the algal assemblage eventually will assume a composition of 48% diatoms, 48% chlorophytes, and 4% cyanobacteria when the algal biomass is >25 g m^{-2} (Figs. 2A, 2B and 2C).
2. The assimilation efficiency and a food quality limiting factor associated with the process of grazing are a function of the proportion of diatoms in the algal assemblage according to these rules:
   (a) Assimilation efficiency is a linear function of the proportion of diatoms in the assemblage, varying between 0.53 (48% diatoms) and 0.73 (100% diatoms); and

(b) A food quality limiting factor expressed as a proportional adjustment of the food demand (i.e., the food consumption rate with an optimum diet and unlimited food supply) is a linear function of the proportion of diatoms in the assemblage, varying between 0.28 (48% diatoms) and 1.00 (100% diatoms). The rate of grazing is adjusted to the composition of the algal assemblage by multiplying the food demand by the food quality limiting factor (for more detail, see Section V,C).

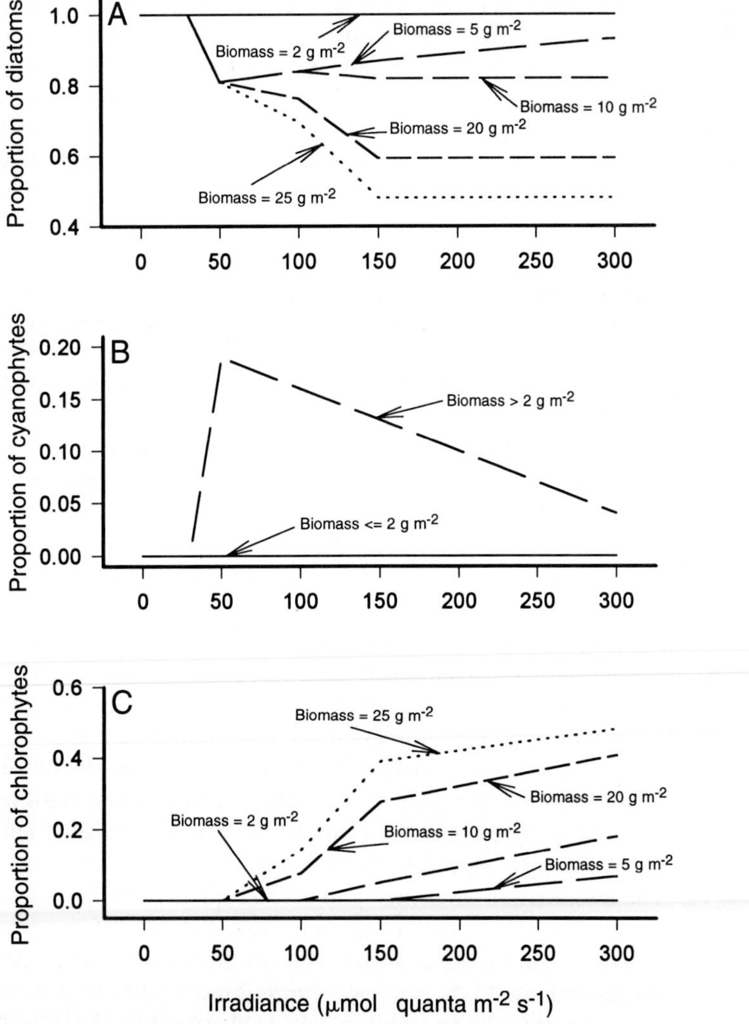

FIGURE 2 Relationships between irradiance and the proportions of diatoms (A), cyanophytes (B), and chlorophytes (C) in the daily increment of biomass generated by primary production. Curves represent five different values of periphyton biomass.

In summary, the new version of the Herbivory subsystem model tracks the successional trajectory and production dynamics of the algal assemblage, as well as the response of grazers to corresponding changes in food quality and quantity. This representation also expresses the feedback control that the process of grazing has on successional changes within the algal assemblage.

V. BEHAVIOR OF THE HERBIVORY SUBSYSTEM MODEL

Behavior of the Herbivory subsystem model in isolation was examined first by obtaining output from a standard run, with and without the process of grazing. This output then was compared to runs designed to investigate the sensitivity of selected variables to changes in the light energy input schedule and to parameters that control the rate of food consumption. Input tables for a standard run were the same as input for the standard run of the M & C model, that is, the Berry Creek light schedule and allochthonous inputs in Fig. 3 of McIntire and Colby (1978). Such tables provide for the simulation of a small, low-order stream receiving annual allochthonous organic inputs of 473 g m^{-2}. The corresponding light schedule generates maximum energy inputs in the spring, with very low inputs during the summer months when the stream is assumed to be shaded by a dense canopy of riparian vegetation. The parameters under investigation, explained in the following section, are a multiplier that controls the light input schedule, a parameter that provides the algal assemblage with varying degrees of protection from the effects of grazing, and five parameters that control the food consumption rates for the process of grazing. For each change in parameters or input tables, the model is allowed to run at a daily time resolution until it exhibits a new steady-state behavior. The simulation period required for the system to reach a steady state may vary between a few years to as long as 40 years, depending on how the changes affect system dynamics. Since input tables correspond to a period of one year, a steady state usually means that state variables have repeatable annual trajectories. However, in some cases the system exhibits repeatable cycles of 2, 3, or even 4 years. Output from each simulation run is usually displayed as a plot of state variables for a 1-year period or is reported as values that are part of an annual energy budget (e.g., an annual production rate or annual mean biomass for functional groups of organisms).

A. Standard Run

Output from a standard run indicates that diatoms dominate the algal assemblage when the system is in a steady state and the process of grazing is in equilibrium with available food resources (Fig. 3). In this case, the model predicts that the algal biomass turns over about 63 times each year,

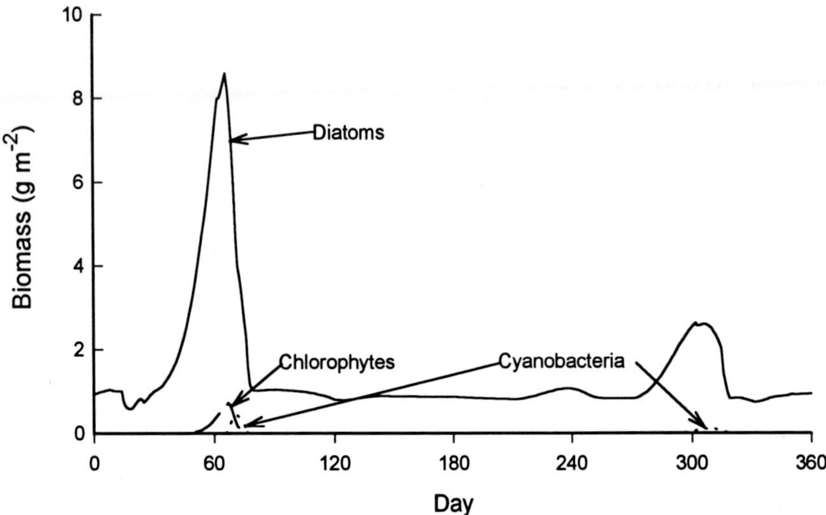

FIGURE 3 Steady-state, seasonal dynamics of state variables representing algal functional groups of the updated version of the Herbivory subsystem model. The graph depicts a run with standard input tables (see text) with grazing. Model is based on 360-day years and 30-day months with the beginning of January as Day 1 and the end of December as Day 360.

and that annual gross primary production, expressed as organic matter, is 106 g m^{-2}, of which green algae and cyanobacteria contribute only 3%. With the standard set of inputs, annual production of herbivore biomass is 6.6 g m^{-2} (organic matter), with a corresponding turnover of 3.4 times per year. If light energy inputs are increased from shaded conditions to full sunlight, the annual mean biomass of grazers increases, whereas annual mean periphyton biomass decreases slightly (Fig. 4A). Also, annual primary production, annual turnover of periphyton biomass, and annual grazer production increase with corresponding increases in light energy (Figs. 4B and 4C), whereas annual turnover of grazer biomass exhibits relatively little change (Fig. 4C). Although the algorithm allows the green algae to reach high biomasses at high inputs of light energy, this does not happen in the presence of grazing, as high rates of algal consumption restrict the flora to a diatom assemblage that is more characteristic of an early stage of succession.

In the absence of grazing (i.e., grazer biomass remains zero), the standard run predicts that the annual mean algal biomass is 20 g m^{-2}, and that all three algal groups are prominent in the spring and fall of the year (Fig. 5). Without grazing, annual gross primary production is 530 g m^{-2}, of which the diatoms, chlorophytes, and cyanobacteria account for 73.3%, 19.3%, and 7.4% of this total, respectively. Corresponding annual turnover

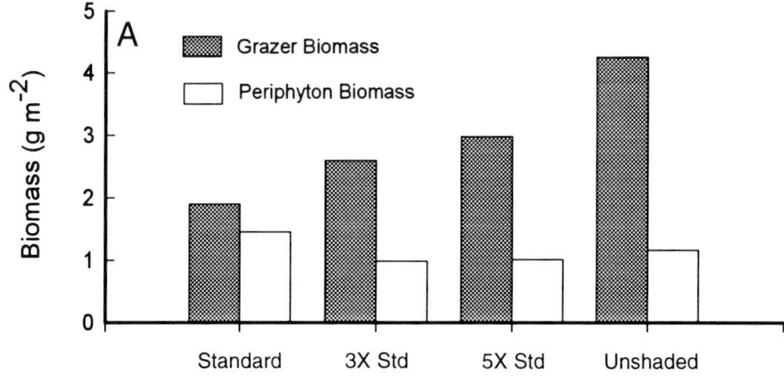

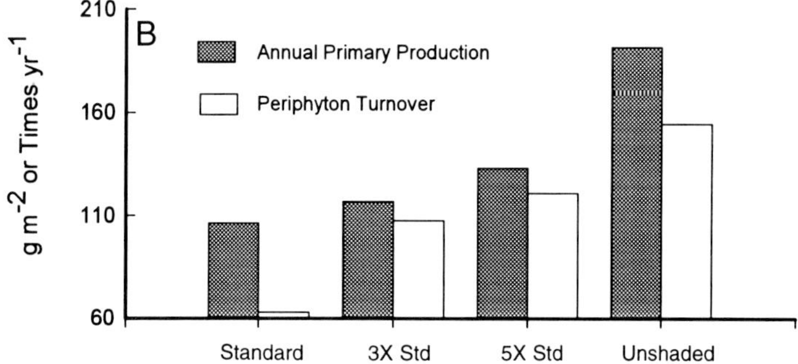

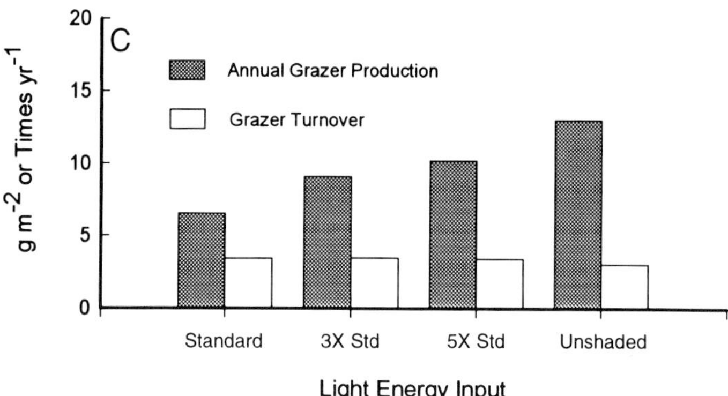

FIGURE 4 Mean grazer and periphyton biomass (A), annual gross primary production and periphyton turnover (B), and annual grazer production and turnover (C) associated with the updated Herbivory subsystem at four levels of light energy input. Levels of irradiance are a standard table of inputs (see text), 3× each value in the standard table, 5× each value in the standard table, and a constant value above the light saturation intensity for photosynthesis (unshaded).

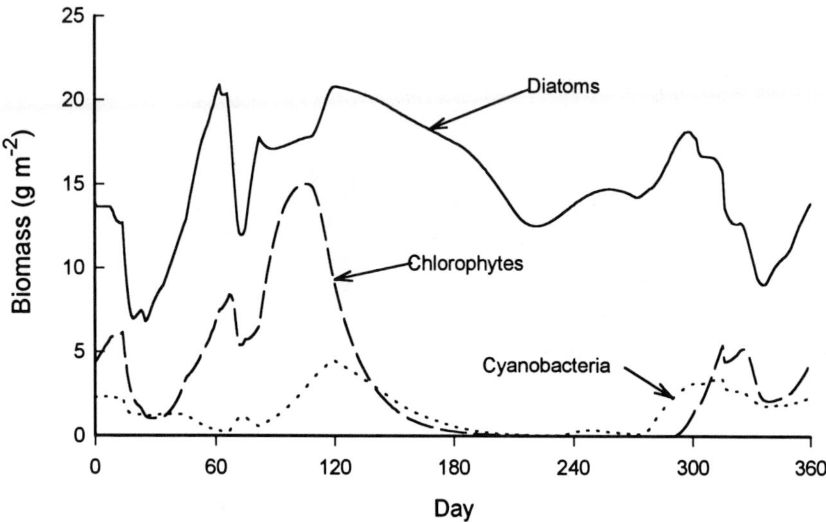

FIGURE 5 Steady-state, seasonal dynamics of state variables representing algal functional groups of the updated version of the Herbivory subsystem model. The graph depicts a run with standard input tables (see text) without grazing. The time scale is the same as in Fig. 3.

numbers for these groups are 14.4, 16.1, and 19.4 times per year. Furthermore, annual energy losses from the algal assemblage, without grazing, partition into 41.8% respiration, 41.9% particulate export, and 16.3% dissolved organic matter (DOM) leakage. In contrast, the standard run with grazing indicates that 62.6% of annual gross primary production and 73.1% of annual net primary production are consumed by herbivores; corresponding losses from respiration, particulate export, and DOM leakage are 14.4%, 20.4%, and 2.6% of gross primary production, respectively. In the latter case, diatoms lose 62.8% to grazing, whereas chlorophytes lose only 49.5%, a manifestation of the effects of the food quality limiting factor on consumption rates.

B. Algal Refuge

Definition: The algal refuge is the algal biomass, expressed as g m^{-2} organic matter, below which the consumption rate by the process of grazing is equal to zero.

A parameter controls the algal refuge level for the Herbivory subsystem. The ecological justification for this parameter is related to (1) differences in feeding efficiencies among consumers with different patterns of behavior and mouthpart morphologies (Steinman *et al.*, 1987a); (2) differences in food availability that result from substrate heterogeneity (DeNicola and McIntire, 1991); and (3) different susceptibilities of algal growth forms to grazing (Steinman *et al.*, 1992). For the standard run, the refuge parameter is set at 0.7 g m^{-2}, a value roughly compatible with algal biomasses observed on flat tile substrata subjected to heavy grazing pressure by snails *(Juga silicula)*.

Relationships between the algal refuge parameter and the production dynamics of the Herbivory subsystem were investigated by a series of 11 simulation runs. With the standard set of inputs, annual grazer production and annual mean biomass of grazers are greatest at an algal refuge between 5 and 7 g m^{-2} after the system reaches steady-state behavior (Fig. 6A). These results suggest that secondary production can be limited by

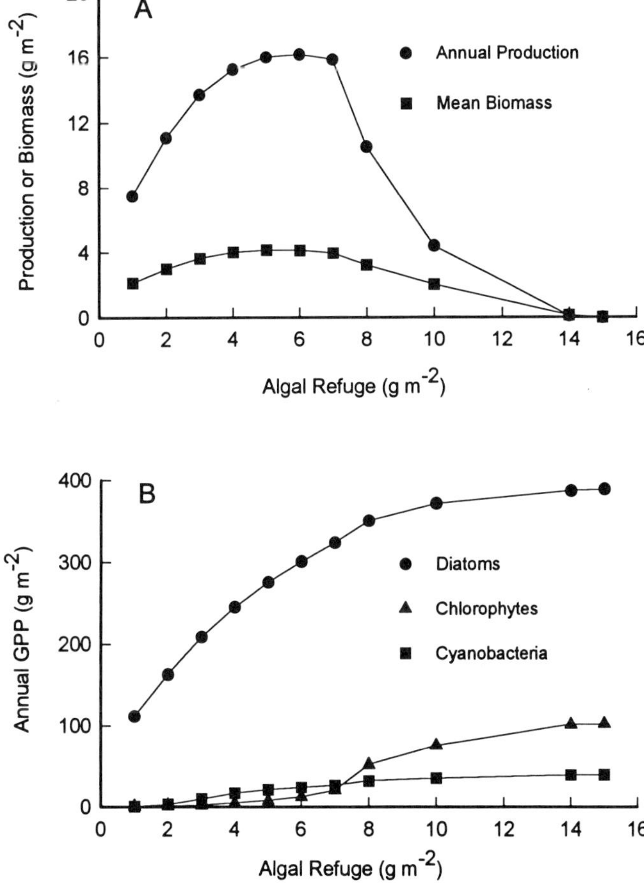

FIGURE 6 Relationships between algal refuge and annual mean grazer biomass and production (A) and between algal refuge and annual gross primary production partitioned by algal functional group (B), as indicated by the updated version of the Herbivory subsystem model. Algal refuge is defined in the text.

overexploitation of food resources under some circumstances. The model also predicts that at refuge values above 7 g m^{-2}, green algae account for a larger proportion of gross primary production (Fig. 6B), a factor that lowers food quality and further contributes to a decline in secondary production (Fig. 6A). As the algal refuge approaches 15 g m^{-2}, secondary production in the Herbivory subsystem goes to zero and algal production reaches its maximum annual rate because grazer losses to emergence and predation exceed the gains through assimilation of algal biomass at this refuge level.

C. Food Consumption and Demand

Model behavior was also investigated in relation to the interaction between the rate of food consumption and inputs of light energy . Food consumption for the process of grazing is a function of food demand and food density (i.e., the biomass of benthic algae and associated microorganisms).

Definition: Food demand is the consumption rate when food is in unlimited supply and the quality of the resource is optimal.

In the model, food demand is a function of temperature and the biomass of the consumer functional group, grazers in this case. Food demand has a maximum value at 18°C and goes to zero as the temperature approaches a low of 0°C and a high of 30°C. In natural streams, food demand also would be expected to vary with the physiological state, life-history stage, and genetic composition of the functional group of grazers. After food demand is calculated, the model determines the realized food consumption rate by multiplying the demand by food quality and food density limiting factors.

Definition: The food quality limiting factor is the proportion of the demand that is allowed by the quality of the food resource.

If the quality of the food resource is optimal, the food quality limiting factor is equal to 1, whereas if the food resource is inedible, the value is zero. The food density limiting factor also ranges from 0 to 1, and is a nonlinear function of the biomass of the food resource [see Eq. (21) in McIntire and Colby, 1978].

The simulation runs generated output for the standard demand (i.e., the demand set up for the standard run), and for a series that included 90%, 80%, 70%, 60%, and 50% of the standard demand. Annual patterns of light energy input were controlled by a multiplier that adjusted the standard table of inputs to the desired level. For the simulation runs reported here, the light schedules included the standard table of inputs, 3× each value in the standard table, 5× each value in the standard table, and a constant input above the saturation intensity for photosynthesis. Irradiance values in all tables were less than 2000 μmol quanta m^{-2} s^{-1}.

At the standard demand and 90% of the standard demand, the model predicts that increases in the inputs of light energy are accompanied by corresponding increases in gross primary production and the biomass and production associated with the process of grazing (Figs. 7A, 7B and 8B); annual mean algal biomass remains low and virtually constant under these conditions, between 1.0 and 1.8 g m^{-2} (Fig. 8A). When demand is reduced to 80% of the standard or below, there is a pronounced increase

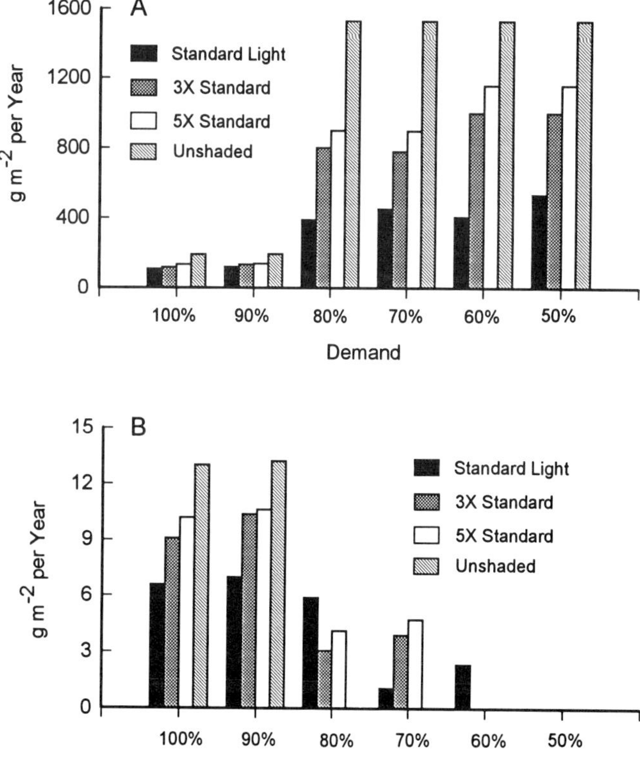

FIGURE 7 Relationships between annual gross primary production and grazer food demand (A) and between annual grazer production and grazer food demand (B) at different levels of light energy input, as indicated by the updated version of the Herbivory subsystem model. Levels of irradiance are the same as in Fig. 4. The concept of food demand is defined in the text.

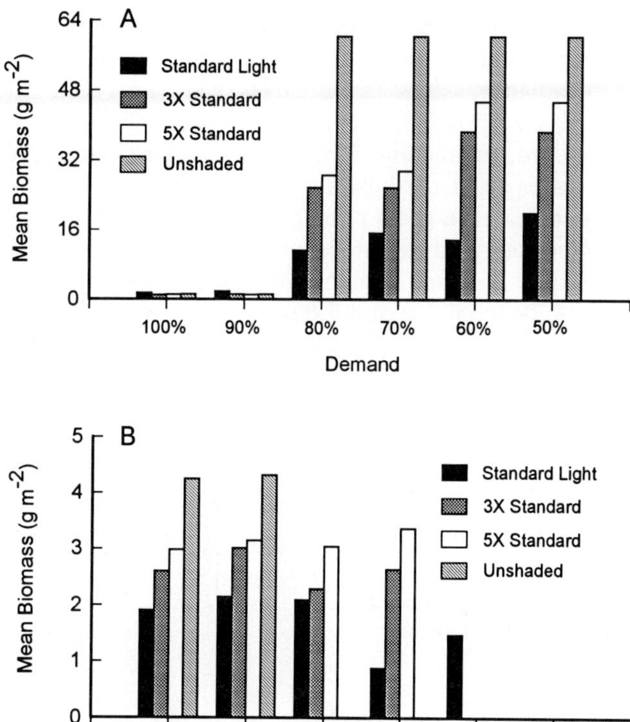

FIGURE 8 Relationship between annual mean periphyton biomass and grazer food demand (A) and between annual mean grazer biomass and grazer food demand (B) at different levels of light energy input, as indicated by the updated version of the Herbivory subsystem model. Levels of irradiance are the same as in Fig. 4. The concept of food demand is defined in the text.

in algal primary production and annual mean algal biomass, and increases in light energy inputs bring about corresponding increases in both of these variables. At 80% demand or below, the system does not support the process of grazing at the highest level of light energy inputs (5× the standard table). On the basis of strictly bioenergetic considerations, these predictions are counterintuitive, as the lowest level of light energy (1× the standard table) supports some grazing at 60% of the standard demand. The explanation for this behavior is related to a model structure that can generate changes in the taxonomic composition of the algal assemblage (Fig. 9). At the highest light energy input and a consumption rate of 80% of the standard demand or below, the model predicts that green and blue-green algae become prominent members of the

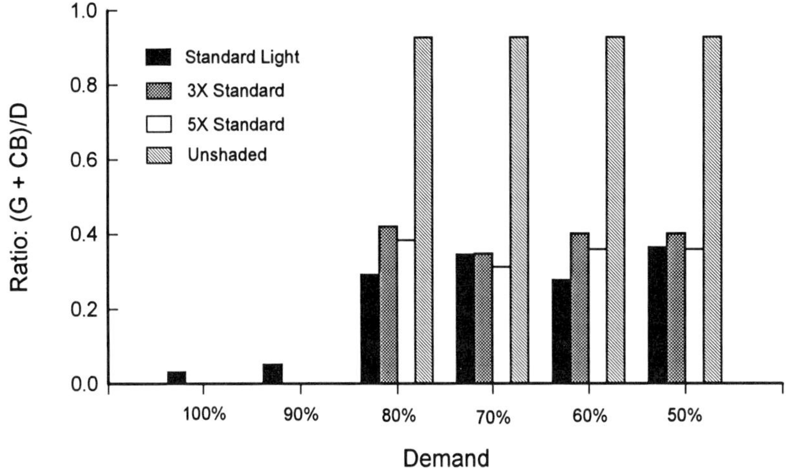

FIGURE 9 Relationship between the taxonomic structure of the algal assemblage and grazer food demand at different levels of light energy input, as indicated by the updated version of the Herbivory subsystem model. Taxonomic structure is defined as the ratio of chlorophyte and cyanobacteria production to diatom production. Levels of irradiance are the same as in Fig. 4. The concept of food demand is defined in the text.

assemblage, a change that has a negative effect on food quality. When consumption is reduced to 50% of the standard demand, the process of grazing is not supported at any of the light energy levels. Although these predictions have not been examined or tested in the field by experimentation, they do suggest that the impact of food quality on trophic relations between benthic algae and grazers may be significant enough to affect patterns of energy flow in streams.

VI. BEHAVIOR OF THE UPDATED M & C MODEL

A. Irradiance and Algal Refuge

After the investigation of the updated Herbivory subsystem in isolation, the new subsystem model was reinserted into the M & C ecosystem model as a replacement for the original representation of the Herbivory subsystem. Behavior of the updated M & C model was examined by manipulating parameters controlling light energy inputs and algal refuge. For these simulations, the corresponding output was structured to demonstrate hypothetical relationships between the process of herbivory in streams and associated consumer processes of shredding, collecting, invertebrate predation, and vertebrate predation. Particular emphasis was placed on the exploration of indirect relationships among biological processes when model behavior indicated that analogous dynamics in natural streams may be misinterpreted or remain unexplained. Results from the simulations are presented in six graphs that illustrate the relationship between annual production associated with a particular process and algal refuge at four levels of light energy (Figs. 10A–10C and 11A–11C). For these runs the light schedules included the standard table of inputs (1×) and 2×, 2.5×, and 3× the values in the standard table; algal refuge varied between 0.7 g m^{-2}, the value for the standard run, and 30 g m^{-2}, a value above which the grazer biomass remained zero regardless of the level of irradiance.

The updated M & C model predicts that secondary production associated with the process of grazing increases from 11 g m^{-2} yr^{-1} at an algal refuge of 0.7 g m^{-2} to a maximum of 53 g m^{-2} yr^{-1} at a refuge of 3 g m^{-2} when the light level is 2.5× the standard schedule (Fig. 10A). With the unaltered standard light schedule (1×), production also is maximized at a refuge level of 3 g m^{-2}, but at a value of about 18 g m^{-2} yr^{-1}. In comparison, grazer production is maximized at an algal refuge of 6 g m^{-2} when the Herbivory subsystem model is run in isolation without processes associated with the Predation and Detritivory subsystems (Fig. 6A), a pattern that indicates regulatory effects of other consumer processes, particularly invertebrate and vertebrate predation. At 3× the standard schedule, grazer production is relatively low (<4 g m^{-2} yr^{-1}) and annual gross primary production is at a maximum at all algal refuge levels (Fig. 11A). In the latter case, the model predicts a release from the regulatory effects of grazing on the algal assemblage when light energy inputs are increased from 2.5× to 3× the standard schedule, a change that generates a pronounced increase in the production and biomass of chlorophytes and a corresponding decrease in food quality. Therefore, model behavior suggests that the combination of low food quality and predation, which is supported by resources generated by shredding and collecting, can account for relatively low grazer production at a high level of irradiance.

Model simulations also indicate that input variables that directly affect the Herbivory subsystem (e.g., the algal refuge parameter and the schedule of light energy inputs) have indirect effects on the production dynamics of shredding and collecting, the primary consumer processes that utilize detrital materials as a food resource (Figs. 10B and 10C). At algal refuge levels from 3 to 10 g m^{-2}, both shredder and collector production are maximized at 3× the standard light schedule and are minimum at the intermediate light levels (2× and 2.5×), a pattern that is indirectly related to low grazer production with the 3× schedule and the corresponding decreases in the production of vertebrate and invertebrate predators. In other words, the Herbivory subsystem indirectly controls the process of shredding, in part, by its direct effects on invertebrate and vertebrate predation. At algal refuges greater than 10 g m^{-2}, the light schedule has relatively little effect on shredder production because the concentration of the food resource

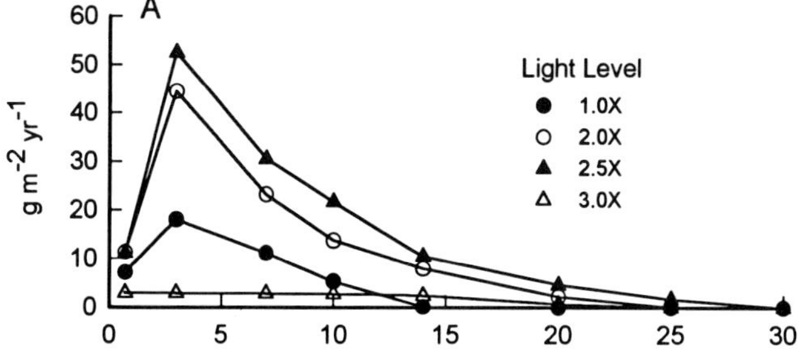

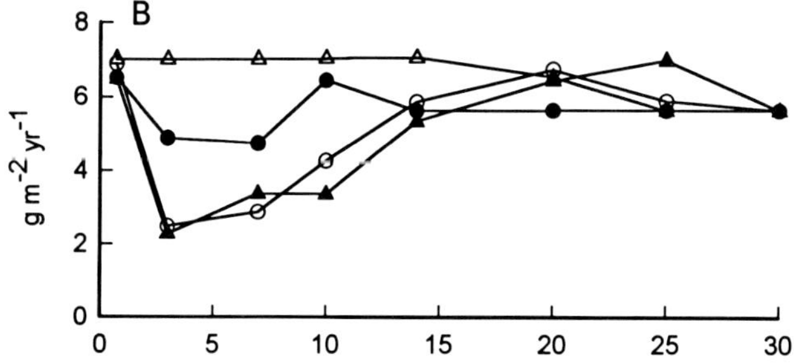

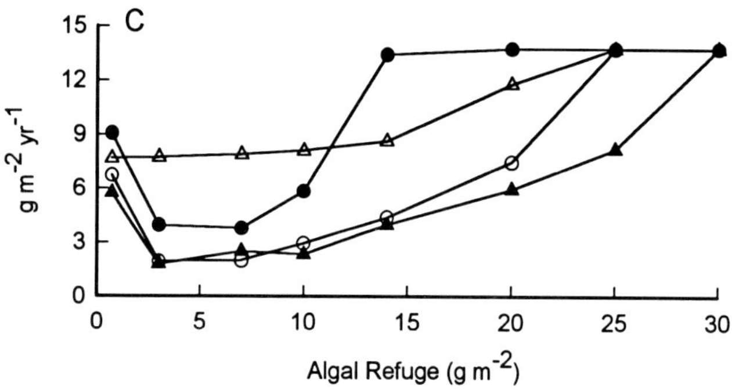

FIGURE 10 Relationships between annual grazer production (A), annual shredder production (B), and annual collector production (C) and algal refuge at different levels of light energy input, as indicated by the updated version of the M & C stream ecosystem model. Levels of irradiance are the standard table of inputs (1×), and 2×, 2.5×, and 3× each value in the standard table.

(i.e., allochthonous particulate organic matter) is independent of the other biological processes and the production of vertebrate and invertebrate predators is relatively low. In contrast, collector production exhibits pronounced light-related differences at algal refuges of 14, 20, and 25 g m^{-2} (Fig. 10C), differences that represent indirect responses to the mechanisms

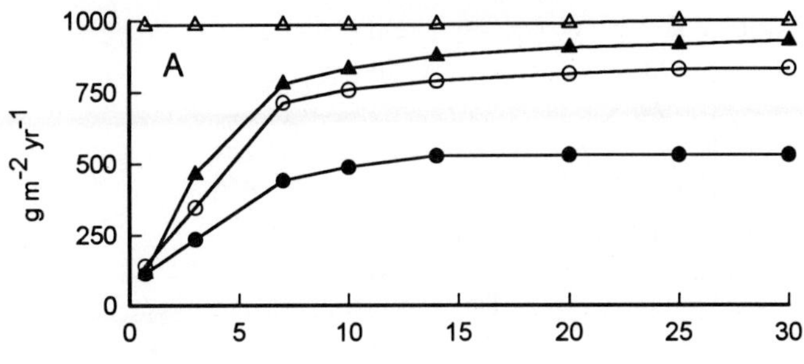

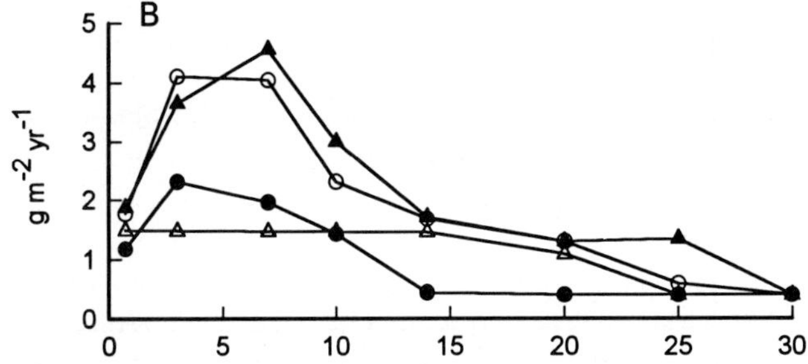

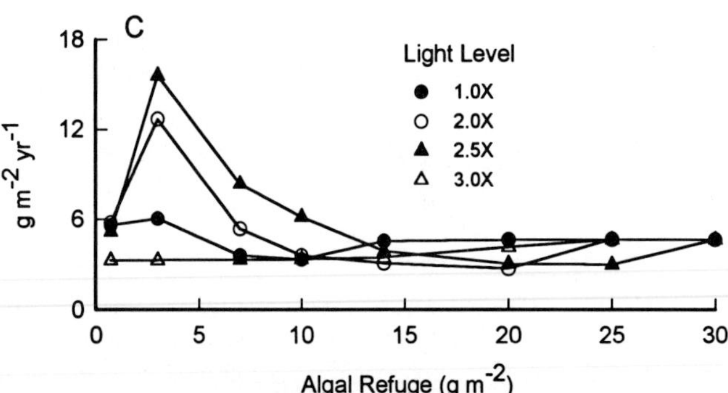

FIGURE 11 Relationships between annual gross primary production (A), annual invertebrate predator production (B), and annual vertebrate predator production (C) and algal refuge at different levels of light energy input, as indicated by the updated version of the M & C stream ecosystem model. Levels of irradiance are the same as in Fig. 10.

that limit shredder production. For most of the selected combinations of inputs, the model predicts that the process of shredding is limited by predation and emergence losses, and that resources are in unlimited supply during most seasons of the year. However, with the standard schedule of allochthonous inputs (Fig. 3 of McIntire and Colby, 1978), shredding becomes food resource limited for a short period in the late spring and early summer in certain cases (e.g., at an algal refuge of 14 g m^{-2} and the 1× light schedule). When this occurs, production and biomass of the functional groups of predators decrease, a change that allows a concurrent increase in collector production. At an algal refuge of 30 g m^{-2}, the system does not support grazing, and the processes of shredding and collecting become independent of the light input schedule because invertebrate and vertebrate predation are no longer affected by changes in the production and biomass of grazers.

Output from the updated M & C model indicates that the processes of vertebrate and invertebrate predation are tightly coupled to the dynamics of the Herbivory subsystem. In its present form this version of the model allows the assimilation efficiency for grazing to vary between 53 and 73%, and food demand to vary between 28 and 100% of maximum (G. A. Lamberti, unpublished data), depending on the composition of the algal assemblage. Corresponding assimilation efficiencies for shredding and collecting are 18 and 21%, respectively. As a result of this representation, manipulation of variables that affect algal production and composition have a pronounced effect on the production dynamics of the Predation subsystem of the model. Although direct relationships between algae, grazers, and predators are relatively easy to interpret, indirect relationships between the Herbivory subsystem and the processes of shredding and collecting are much less intuitive. Moreover, mechanisms accounting for differences in patterns exhibited by the processes of vertebrate and invertebrate predation are not obvious. In the model, invertebrate predators also serve as a food resource for vertebrate predators. Consequently, the model predicts that the processes of vertebrate and invertebrate predation reach a production maximum at different algal refuge levels: 3 and 7 g m^{-2} at the 2.5× light level, respectively (Figs. 11B and 11C).

B. Allochthonous Inputs

Most model simulations were run with an annual input of allochthonous organic matter of 473 g m^{-2}, a value derived from measured litterfall and lateral movement into a small stream at the H. J. Andrews Experimental Forest in western Oregon (McIntire and Colby, 1978). During the standard run, that is, with an algal refuge of 0.7 g m^{-2} and the 1× light schedule, the model predicts that losses of this material to microbial decomposition, the process of shredding, export, and mechanical conversion to fine particulate organic matter are 112 (23.7%), 233 (49.3%), 73 (15.4%), and 55 g m^{-2} (11.6%), respectively. To examine the sensitivity of the Predation subsystem to an increase in detrital inputs, the annual allochthonous input was doubled to 946 g m^{-2}, an input that is higher than values usually reported for natural streams. With this increase, losses to shredding increase to only 281 g m^{-2} (29.7%), and the fate of most of the additional inputs is microbial decomposition (41.6%) and export (21.6%). Patterns of grazer and collector production with all light schedules and at all algal refuge levels were relatively unaffected by the increase in allochthonous inputs. However, with the input of 946 g m^{-2}, the process of shredding was never limited by food resources and its corresponding pattern of production was similar to that found for the collectors (i.e., similar to patterns in Fig. 9C). The increase in allochthonous material had relatively little impact on the general patterns of invertebrate and vertebrate predation because the process of shredding was limited much more by predation and emergence losses than by food resources, and the dynamics of both functional groups of predators were more tightly coupled to the Herbivory subsystem than to the processes of shredding and collecting. The model also predicts that doubling allochthonous inputs has very little or no effect on annual primary production or mean algal biomass. At low algal refuge levels (<10 g m^{-2}), the process of shredding is not limited by food resources, and therefore the dynamics of this process are not affected by the addition of more allochthonous material. As the algal refuge increases to values above 10 g m^{-2}, effects of grazing on the algal assemblage are minimal because of a decrease in food quality, and therefore indirect relationships between grazing and shredding, through the process of predation, have no significant effects on primary production.

C. Food Quality and Nutrients

The behavior of the updated M & C ecosystem model indicates that the system is sensitive to changes in taxonomic composition of the benthic algal assemblage. This sensitivity is related to direct effects of species composition on food consumption by grazers and indirect effects of grazer production on the processes of vertebrate predation, invertebrate predation, shredding, and collecting. Moreover, in the simulations presented in the foregoing, it is assumed that grazer food demand varies between 28% of the maximum, when the algal assemblage is 48% diatoms, 48% chlorophytes, and 4% cyanobacteria, to maximum demand when the assemblage is 100% diatoms. Because the representation of food quality effects is based on limited experimental data, there is a possibility that such effects are overstated in the model when the productive capacity of the system is relatively high (i.e., when high irradiance and nutrient concentrations generate algal assemblages with high proportions of chlorophytes and cyanophytes).

To investigate algal composition and food quality in more detail, the model was modified to allow control over the effects of food quality on grazer food demand. This was accomplished by introducing a new parameter that sets the minimum value for the food quality limiting factor (see Section V,C for definition). For example, when the value of this parameter is 0.28, food demand is 28% of its maximum when the algal assemblage has its lowest percentage of diatoms (48%) and highest percentage of chlorophytes (48%); when the value is 0.75, demand is 75% of the maximum with this taxonomic composition. Therefore, as the parameter increases in value, the effect of food quality on grazer demand decreases, and at a value of 1.0, food quality has no effect on demand. At a value of 0.28, model behavior is identical to patterns illustrated in Figs. 10 and 11.

Simulations that were set for a study of the new parameter also were designed to investigate effects of a limiting nutrient at high inputs of irradiance (i.e., when light energy is not limiting). Such conditions are more typical of larger rivers than of smaller, lower-order streams. It was assumed that the limiting nutrient was nitrate nitrogen and that the range of concentrations of interest was between 0.01 and 0.5 mg liter^{-1}. Annual allochthonous input for these simulations was 210 g m^{-2}, whereas irradiance varied only in relation to daylength and was set at a constant value greater than the saturation intensity for photosynthesis. The schedule of allochthonous inputs was derived from data for the Willamette River (Oregon). In this case, approximately 75% of the detrital inputs were introduced at a time corresponding to a period from the beginning of September to the end of December.

If irradiance is not limiting photosynthesis, the model predicts that the dynamics of the algal assemblage are particularly sensitive to the availability of a limiting nutrient, and that primary and secondary consumers are indirectly affected by nutrient changes, in this case, especially when the nitrate concentration is below 0.1 mg liter^{-1} (Figs. 12A–12C, 13A, and 13B). At a nitrate concentration of 0.03 mg liter^{-1} or less, grazers are able to persist in the system when the food quality parameter is low (<0.6) because the algal biomass and primary production are never high enough to allow the growth of chlorophytes (Fig. 13C). In other words, when nutrient supply is low, food quality remains high regardless of the value of the parameter, because diatoms dominate the assemblage at low biomasses. At nitrate concentrations of 0.07 mg liter^{-1} or greater, grazer production is much more sensitive to changes in the food quality limiting factor. A threshold response value for the food quality parameter is between 0.7 and 0.8. From an ecological perspective, this means that when grazer food demand can be reduced to 70% of the maximum or below by changes in food quality, grazer production actually decreases when increases in nutrient supply and primary production bring about corresponding increases in the proportional abundances of taxa that decrease food quality. Although these taxa are chlorophytes and cyanophytes in the current version of the model, the model can be reparameterized for other relevant functional groups of benthic algae when research dictates a change in mathematical structure. When the effect of food quality is minimal (i.e., the food quality

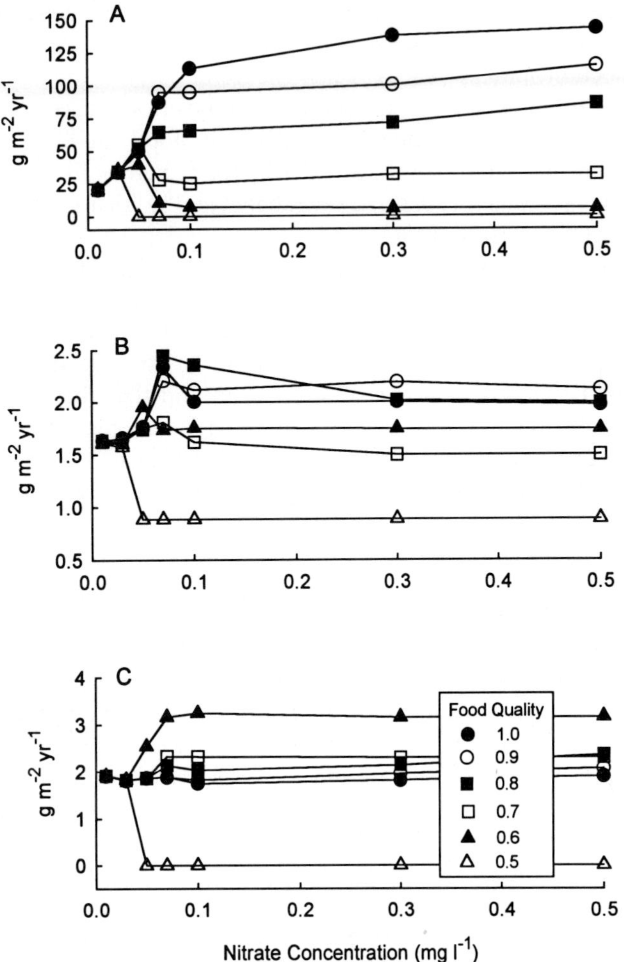

FIGURE 12 Relationships between nitrate concentration (mg liter^{-1}) and annual grazer production (A), annual shredder production (B), and annual collector production (C) at different levels of food quality (see text for explanation), as indicated by the updated version of the M & C stream ecosystem model. In these simulations, irradiance is always above the light saturation value for photosynthesis.

parameter is 0.8 or greater), grazer production increases or is relatively unaffected when nitrate concentration increases to values above 0.07 mg liter^{-1}.

The model predicts that shredders, collectors, and predators have different responses to changes in nitrate concentration and the food quality limiting factor. The response of vertebrate predators is similar to the pattern exhibited by the grazers, indicating that these functional groups are tightly coupled bioenergetically (Figs. 12A and 13A). With a relatively high

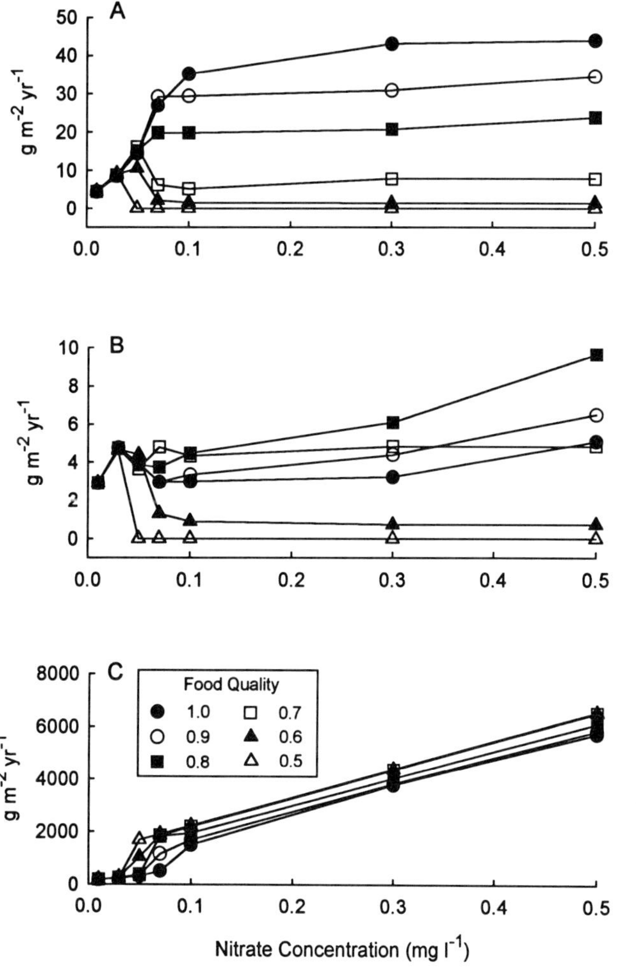

FIGURE 13 Relationships between nitrate concentration (mg liter^{-1}) and annual vertebrate predator production (A), annual invertebrate predator production (B), and annual gross primary production (C) at different levels of food quality (see text for explanation), as indicated by the updated version of the M & C stream ecosystem model. In these simulations, irradiance is always above the light saturation value for photosynthesis.

nitrate concentration (0.1 mg liter^{-1} or greater), invertebrate predators reach maximum production when the food quality parameter is 0.8 (Fig. 13B), a pattern that is a manifestation of the trade-offs between availability of food resources (grazer, shredder, and collector biomasses) and biomass losses to vertebrate predators. At nitrate concentrations of 0.07 mg liter^{-1} and greater, collector production is inversely related to the food quality parameter except when grazer production is zero (parameter =

0.5). This pattern suggests that the process of collecting is controlled by vertebrate predation and becomes resource-limited only when the process of grazing fails to produce enough detrital particles. In reality, there are other sources of fine particulate organic matter in natural streams, and the process of collecting is probably not as tightly coupled to the processes of grazing and shredding as the model suggests. The relationship between shredder production and the food quality limiting factor is complex (Fig. 12B) and is directly related to the seasonal pattern of allochthonous inputs and indirectly related to the response of the functional groups of predators to changes in grazer biomass. However, for any given value of the food quality parameter, shredder production and collector production are relatively unaffected by changes in nitrate supply when concentrations are above 0.07 mg liter^{-1}.

VII. HYPOTHESIS GENERATION

Ongoing laboratory and field studies can provide the basis for a periodic update of the mathematical representation of stream ecosystem dynamics. The end product is a new or modified version of an existing model that is used for synthesis and hypothesis generation. Therefore, modeling can make valuable contributions to a research program by providing new questions and directions for experimental studies. If modeling is used as an iterative approach to the synthesis of past and present research, new hypotheses that emerge from the study of model behavior tend to be highly relevant to the goals of the research program and to the understanding of the corresponding system under investigation. Moreover, hypotheses generated by modeling often are related to questions that are not obvious from the results of individual laboratory or field studies. Examples of hypotheses that follow from some of the simulations described earlier in the chapter are presented in this section.

Hypothesis I: The process of grazing can affect succession in benthic algal assemblages by preventing the development of seral stages with relatively high biomasses of filamentous and colonial chlorophytes.

The structure and parameterization of the updated Herbivory subsystem model was based on experimental work by Steinman *et al.* (1987a) and Lamberti *et al.* (1989). In the study by Steinman *et al.* (1987a), successional trajectories of benthic algal assemblages in laboratory streams depended on the degree of grazing pressure and the kind of grazer (snail or caddis fly) introduced into the system. In addition, results of other studies suggest that grazers can prevent the dominance of filamentous and large, erect unicellular or colonial taxa in benthic algal assemblages, because such taxa are more easily removed during feeding than the smaller, prostrate taxa (Eichenberger and Schlatter, 1978; Sumner and McIntire, 1982; Gregory, 1983; Perrin *et al.,* 1987; DeNicola *et al.,* 1990; Steinman, 1992). Hypothesis I is worded in terms of taxonomic classes when, in fact, it may be more appropriate to classify according to growth form (e.g., filamentous, colonial, large erect unicellular, and small prostrate growth forms). Moreover, some chlorophytes are heterotrichous (e.g., *Stigeoclonium*) and have the potential to expand from an assemblage of basal cells, under heavy grazing, to a filamentous form, in the absence of grazing. In any case, the current version of the model presents a crude representation of relationships between grazing and algal community structure, and simulates direct and indirect effects that these relationships can have on other components of lotic ecosystems.

Hypothesis II: If the productive capacity of a stream ecosystem is increased by inputs that enhance benthic algal production (e.g., an increase in irradiance or nutrient supply), there is a corresponding increase in the annual mean biomass of primary and secondary consumers while the mean algal biomass may change very little or actually decrease slightly.

The relative importance of allochthonous detrital inputs and autochthonous primary production to the bioenergetics of lotic ecosystems is controversial and has been the subject of considerable research for the past 30 years. Hypothesis II implies that the biomass of benthic algae at any particular time may be a poor indicator of the relative contribution of benthic autotrophs to the food resources of macroconsumers in streams; or stated more specifically, benthic algal biomass may be a poor predictor of the capacity of a stream to support grazing. Model simulations indicate that an increase in primary production can be expressed as an increase in grazer production and biomass rather than a conspicuous increase in algal biomass, because of an increase in the number of times that the algal biomass will turn over during a given period of time.

Most field and laboratory experiments have not been run for a long enough period of time or have concurrent measurements of the necessary variables to provide a satisfactory test for Hypothesis II. Lamberti *et al.* (1989) found that a herbivorous snail *(Juga silicula)* in laboratory streams exhibited a relatively low growth rate at a photon flux density of 20 μmol quanta m^{-2} s^{-1} and relatively high growth rates at irradiances of 100 and 400 μmol quanta m^{-2} s^{-1}. In this experiment, the snails were able to delay the accumulation of algal biomass, but animal densities were not high enough to prevent algal biomass from reaching levels comparable to those in streams without grazers. However, Steinman *et al.* (1987a) clearly demonstrated that in laboratory streams stocked with high densities of either snails or caddis flies, algal biomasses were less than 3 g m^{-2} when irradiance was 400 μmol quanta m^{-2} s^{-1}. In a field experiment, Steinman (1992) also found that biomass-specific algal photosynthesis was enhanced by an increase in irradiance, whereas algal biomass was controlled by grazing pressure. Unfortunately, grazer production was not measured during that experiment.

Hypothesis III: When inputs into a stream ecosystem remain unchanged (i.e., the productive capacity does not change), secondary production is maximized when benthic algal assemblages are protected by mechanisms that prevent overgrazing.

In this chapter, the concept of algal refuge is used in a broad sense to mean any mechanism that prevents harvest of an algal food resource by a consumer when the algal biomass falls below a lower threshold level. Mechanisms associated with such a threshold may relate to substrate heterogeneity (DeNicola and McIntire, 1990a,b, 1991) or to morphological and behavioral characteristics of the consumer organisms (Wiley and Kohler, 1981, 1984; Hart and Resh, 1980; Hart, 1981; Lamberti and Moore, 1984). In the M & C model, the focus is on a spatial scale of one square meter, and it is assumed that the distribution of the algal biomass in relation to the degree of substrate heterogeneity within that area can affect the rate of food consumption by grazers. Steinman *et al.* (1987a) found that in laboratory streams with a smooth, uniform substrate and stocked with high densities of snails (500 m^{-2}) or caddis flies (200 m^{-2}), grazers still were not able to consume all of their algal food resources, and with heavy grazing pressure, algal assemblages consisted of a monolayer of diatoms and *Stigeoclonium* basal cells. As the attachment substrate becomes more irregular, access to an algal food supply presumably becomes more difficult, and the algal biomass below which consumption is zero increases to a level that is determined by complex interactions between the morphological and behavioral characteristics of the grazer and the spatial distribution and microhabitat of individual algal taxa. Therefore, tests of Hypothesis III require measurements of primary production, secondary production, and algal biomass in experimental systems within which resource accessibility can be controlled.

Hypothesis IV: An increase in the productive capacity of a stream ecosystem concurrent with a decrease in the quality of the algal food resource may have a negative effect on grazer production even though algal biomass and primary productivity of the system increase.

Very little information is available concerning effects of the quality of algal food resources on secondary production. In the M & C stream model, food quality is conceptualized as any property of the algal food resource that affects rates of consumption and assimilation by consumers when the resource is in unlimited supply. Examples of properties that could affect food quality are community physiognomy, the size and shape of individual taxa, and the biochemical composition of the algal food resource.

McIntire *et al.* (1969) demonstrated differences in the fatty acid composition of algal assemblages subjected to different irradiance levels and current velocities in laboratory streams, and Steinman *et al.* (1987b) found that the introduction of herbivores (snails and caddis flies) into laboratory streams altered the fatty acid and species composition of benthic algae, but had less effect on the relative concentrations of amino acids. However, relationships between biochemical composition of algae and grazer production are less clear because secondary production usually is not measured concurrently with studies of algal chemical composition. Lamberti and Moore (1984) suggested that, because of their thick cellulose cells walls and mucous coating, chlorophytes and cyanophytes are digested less easily by grazers than are diatoms. Also, feeding preference studies (Cargill *et al.*, 1985) revealed that the caddis fly *Clistoronia magnifica* preferentially ingested lipid-coated detritus during the last larval instar, indicating that at least some aquatic insects are sensitive to changes in the biochemical components of the food resource during certain periods of their life cycle.

Hypothesis V: When environmental conditions are favorable for the production of benthic algae in streams, predator production is more tightly coupled to the dynamics of grazer populations than to the production of shredders and collectors, even in the presence of relatively high inputs of allochthonous detritus.

Indirect effects of the composition and production of benthic algae on the production of secondary consumers have received relatively little attention from stream ecologists. Some studies of the effects of artificial fertilization on stream communities indicate that nutrient enrichment often is accompanied by an increase in fish production, suggesting that secondary consumers are sensitive to changes in the food supply of insects that feed on benthic algae (Perrin *et al.*, 1987; Deegan and Peterson, 1992; Peterson *et al.*, 1993). Warren *et al.* (1964) found that trout production in Berry Creek, near Corvallis, Oregon, was 21 times greater in riffles enriched with sucrose than in unenriched riffles. In this case, enrichment stimulated the production of *Sphaerotilus natans,* a filamentous bacterium that blanketed the streambed in the enriched section, and both herbivorous and carnivorous insects, the primary food resources for the trout population.

Model simulations suggest that mechanisms accounting for the close association between vertebrate predation and the process of grazing are related to the short generation times for algal assemblages (McIntire and

Colby, 1978) and relatively high assimilation efficiencies found for organisms that consume living algae (Lamberti *et al.*, 1989; McCullough and Minshall, 1979). Consequently, when allochthonous inputs of organic matter are relatively high, the process of grazing tends to be food resource limited, whereas processes of shredding and collecting are limited primarily by predation (see Figs. 6 and 7 in McIntire and Colby, 1978). Therefore, the model implies that when algal food resources are abundant, the capacity of the grazer functional group to support predator production is greater than the capacity of functional groups of detrital feeders, irrespective of the abundance of the detrital food supply. This conclusion is based entirely on bioenergetic considerations and does not consider differences in the behavioral ecology of the different functional groups or negative effects of changes in algal food quality.

Hypothesis VI: When the production of benthic algae in streams is limited by nutrient supply, changes in the nutrient concentration directly affect the quantity and quality of the algal food resource and indirectly affect shredder, collector, and predator production.

Corollary: If an increase in a limiting nutrient generates a decrease in algal food quality below a threshold value, grazer production decreases with an increase in algal productivity, a response that has indirect effects on the processes of shredding, collecting, and predation.

Indirect effects of nutrient enrichment on detritivores and functional groups of predators have received little attention from stream ecologists. Hypothesis VI is similar to Hypothesis IV in that it focuses on indirect relationships between variables that control productive capacity, nutrients in this case, and macroconsumer processes. The proposed mechanisms of interaction relate to trade-offs between algal quantity and quality in relation to associated effects on grazers and indirect effects on shredders, collectors, and predators. The model predicts that indirect effects on shredders and collectors operate through the process of predation, which is tightly coupled to changes in grazer production and biomass.

VIII. DISCUSSION AND CONCLUSIONS

Simulations and hypotheses presented in earlier sections of this chapter are examples of the kinds of insights and research directions that modeling can provide. It is interesting to note that we often learn more when model output is inconsistent with reality than when trajectories of state variables are similar to what we observe in nature. In some cases, nothing succeeds like failure, because when the model does not exhibit the expected or desired behavior, its current structure represents an explicit expression of ignorance that can be analyzed and evaluated for the purpose of setting priorities for future research. Often, reevaluation of model structure in relation to its current behavior generates new ways of thinking about the system under investigation. In the examples presented for benthic algae in streams, model behavior suggested that we can learn a great deal from studies that examine direct, and particularly the indirect, relationships between the algal assemblage and the primary and secondary consumers in the system. These kinds of studies are much more difficult to design than studies that focus on individual algal taxa or assemblages of taxa in isolation.

Modeling also provides a basis for partitioning ecological processes into their component parts. Simulation runs from the M & C stream model are performed by the FLEX model processor (Overton, 1972, 1975) and are based on a discrete time increment of one day. The update algorithm is a simple difference equation,

$$\mathbf{x}(k + 1) = \mathbf{x}(k) + \Delta(k) \tag{1}$$

where $\mathbf{x}$ is a vector of state variable values at time k, $\mathbf{x}(k + 1)$ is a vector of values for the same variables one day later, and $\Delta(k)$ is a vector of the net changes in $\mathbf{x}$ between time k and $k + 1$ estimated at time k. In the case of primary consumers (grazers, shredders, and collectors),

$$\Delta_i = a_i C_i - R_i - E_i - M_i - P_i \tag{2}$$

where C is the food consumed between k and $k + 1$; R, E, M, and P are corresponding losses to respiration, emergence and export, natural mortality, and predation, respectively; and a is the assimilation efficiency. To understand how the system works, each of the components of Δ_i must be investigated. Moreover, components at this level are functions of other variables and can be partitioned into sets of subcomponents. In the examples presented earlier in the chapter, consumption of algal biomass by grazers is a function of food demand and a food density limiting factor, which itself is a function of the algal biomass minus the algal refuge level. Food demand is a function of temperature and is adjusted by the food quality limiting factor.

The value of partitioning ecological processes into their component parts goes beyond the exercise of creating a mathematical model. The identification of process components requires a fundamental understanding of the process and provides an explicit set of variables for research purposes and review. Furthermore, the definition of process components can lead to useful ecological concepts that can serve as a basis for experimental design and hypothesis testing. Examples of such concepts from the M & C stream model include: (1) food demand, the consumption of a food resource when the supply is unlimited; (2) algal refuge, the algal biomass below which consumption by macroconsumers is zero; and (3) the food quality limiting factor, a value that adjusts the food demand to the quality of the food resource. All three of these concepts can be incorporated into hypotheses and the design of future experiments with benthic algal assemblages.

Output from the M & C stream model clearly demonstrates that links between resource production and consumption are altered by access to the resource. Availability of algal resources in the model is controlled by both physical and biological factors. Substrate heterogeneity and elevation of algal growth forms above the substrate surface are physical characteristics that modify the outcome of grazer–periphyton interactions, whereas biological features that alter the access of herbivores to food resources include food quality, morphology of mouthparts and food-gathering structures, and behavioral patterns. In the M & C model, the algal refuge parameter and a parameter that controls the food quality limiting factor affect food availability, consumption, and assimilation. Both of these parameters have a strong effect on the behavior of the Herbivory subsystem of the model and, as a result, have the capacity to change the production of other components of the system that are indirectly linked to the process of herbivory. Studies of herbivory in streams usually are based on an unstated assumption that 100% of the plant biomass is available to herbivores. However, it is unlikely that this assumption is consistent with the structural and functional attributes of most natural streams. The stream model predicts that biological components in natural streams are sensitive to resource availability and indicates that different patterns of herbivory could be observed in seemingly similar systems.

One of the more interesting hypotheses presented in the previous section indicates that dynamics of vertebrate predator populations may be tightly coupled to patterns of benthic primary production when conditions are favorable for the growth of attached algae (Hypothesis V). If this is really true, indirect relationships between vertebrate predators and benthic algae have management implications in fisheries. If the hypothesis is false, or when it is false, it would be interesting to know why the natural system exhibits behavior that is counter to the outcome predicted by bioenergetic considerations. In streams, periodic dominance of physical factors in interaction with peculiarities of the life-history characteristics of individual taxa may cause deviations from patterns predicted by models in which such details are not represented at the process level of organization. Therefore, model output can sometimes indicate when it is appropriate to do the research necessary to elaborate the structure of the model subsystems in greater detail. The expansion of the Herbivory subsystem of the M & C stream model illustrated how a new set of research objectives required the development of new model structures and concepts at a finer level of resolution.

Mathematical modeling also can be used to address some of the broader, more theoretical aspects of benthic algal ecology. As an example, we consider the question of whether stream ecosystems are controlled by "bottom-up" or "top-down" processes and how the dynamics of benthic algal assemblages relate to this question. In stream ecology, "top-down" control usually refers to a case when an increase in a resource that limits primary production (e.g., light energy or nutrients) has no effect on algal biomass, because autotrophic biomass is controlled by grazers (Steinman, 1992; Rosemond *et al.*, 1993). In contrast, "bottom-up" control means that algal biomass increases significantly with an increase in the input of some limiting factor. Some of the ambiguities about "top-down" and "bottom-up" mechanisms relate to what is actually meant by control and whether the focus is on an individual population, a functional group, or the ecosystem as a whole. For example, the M & C stream model predicts that under some conditions, an increase in the level of a limiting factor can enhance primary production without a conspicuous change in algal biomass, because the biomass turns over more rapidly in response to the increase in resources and concurrent increases in macroconsumer production and biomass. Consequently, "bottom-up" control is achieved without much change in the mean algal biomass. This indicates that it might be less ambiguous to define limitation or control in terms of production instead of biomass. However, in the case of streams, which often obtain their resources from both autochthonous and allochthonous sources, an increase in detrital inputs from the surrounding terrestrial environment will always result in an increase in energy flux through the ecosystem ("bottom-up" control) regardless of the effects of predators on primary consumers. In other words, if shredders and collectors do not process the new material, it will ultimately be processed by the microbial flora. The pronounced seasonality of allochthonous and autochthonous inputs and the frequent disturbance regimes in stream ecosystems make it unlikely that simple "top-down" or "bottom-up" effects would occur throughout a food web. Instead, controls are likely to be transient, and the complex array of life histories and generation times characteristic of lotic ecosystems tends to obscure mechanisms of control and patterns of resource limitation and exploitation.

Experience with the M & C stream model suggests a more direct approach to the understanding of process limitation and control. Modeling for research purposes often requires that each process be partitioned into its component parts [see Eqs. (1) and (2)], each part of which represents either a gain or loss to the associated state variable. Therefore, mechanisms of regulation and control are revealed by the relative importance of the positive or negative effects of each part on the process. McIntire and Colby (1978) and McIntire (1983) defined a new set of variables that allow a graphic display of the factors that prevent a state variable from reaching its maximum potential specific growth rate. For example, model output predicted that in a shaded stream receiving relatively high allochthonous inputs (473 g m^{-2} yr^{-1}), the process of grazing is controlled by the algal food resource, whereas the processes of shredding and collecting are affected more by predation than by resource limitation (see Figs. 5 and 6 in McIntire, 1983). Model output also indicated that it is possible for such control to vary seasonally and that at certain times physical processes or losses relating to life-history characteristics (e.g., emergence) may have much greater effects on process dynamics than trophic interactions.

In summary, theoretical generalizations can evolve from a systematic investigation of different model structures, while varying inputs and parameters. Experimental and observational studies of benthic algae in streams provide the data base necessary for a modeling approach to the synthesis of existing information and concepts into an integrated theory of how the structure and function of benthic algal assemblages relate to physical processes and to other biological components of ecological systems. In particular, modeling is a powerful research tool when it is used in close association with related laboratory and field studies.

ACKNOWLEDGMENTS

The senior author wishes to thank W. S. Overton for providing the theory upon which the M & C stream ecosystem model was based and for his leadership in the development of FLEX4, a general model processor available at Oregon State University. The authors are also indebted to Brad Smith for programming the latest version of the FLEX model processor. The work reported in this chapter was supported in part by National Science Foundation Grants BSR-8318386, BSR-8907968, and BSR-90-11663 (LTER).

REFERENCES

Cargill, A. S., Cummins, K. W., Hanson, B. J., and Lowry, R. R. (1985). The role of lipids, fungi, and temperature in the nutrition of a shredder caddisfly, *Clistoronia magnifica*. *Freshwater Invertebr. Biol.* **4**, 64–78.

Deegan, L. A., and Peterson, B. J. (1992). Whole-river fertilization stimulates fish production in an arctic tundra river. *Can. J. Fish. Aquat. Sci.* **49**, 1890–1901.

DeNicola, D. M., and McIntire, C. D. (1990a). Effects of substrate relief on the distribution of periphyton in laboratory streams. I. Hydrology. *J. Phycol.* **26**, 625–633.

DeNicola, D. M., and McIntire, C. D. (1990b). Effects of substrate relief on the distribution of periphyton in laboratory streams. II. Interactions with irradiance. *J. Phycol.* **26**, 634–641.

DeNicola, D. M., and McIntire, C. D. (1991). Effects of hydraulic refuge and irradiance on grazer–periphyton interactions in laboratory streams. *J. North Am. Benthol. Soc.* **10**, 251–262.

DeNicola, D. M., McIntire, C. D., Lamberti, G. A., Gregory, S. V., and Ashkenas, L. R. (1990). Temporal patterns of grazer–periphyton interactions in laboratory streams. *Freshwater Biol.* **23**, 475–489.

Eichenberger, E., and Schlatter, A. (1978). Effect of herbivorous insects on the production of benthic algal vegetation in outdoor channels. *Verh.—Int. Ver. Theor. Angew. Limnol.* **20,** 1806–1810.

Gregory, S. V. (1983). Plant–herbivore interactions in stream systems. *In* "Stream Ecology" (J. R. Barnes and G. W. Minshall, eds.), pp. 157–189. Plenum, New York.

Hart, D. D. (1981). Foraging and resource patchiness: Field experiments with a grazing stream insect. *Oikos* **37,** 46–52.

Hart, D. D., and Resh, V. H. (1980). Movement patterns and foraging ecology of a stream caddisfly larva. *Can. J. Zool.* **58,** 1174–1185.

Lamberti, G. A., and Moore, J. W. (1984). Aquatic insects as primary consumers. *In* "The Ecology of Aquatic Insects" (V. H. Resh and D. M. Rosenberg, eds.), pp. 164–195. Praeger, New York.

Lamberti, G. A., and Steinman, A. D. (1993). Research in artificial streams: Applications, uses, and abuses. *J. North Am. Benthol. Soc.* **12,** 313–384.

Lamberti, G. A., Ashkenas, L. R., Gregory, S. V., and Steinman, A. D. (1987). Effects of three herbivores on periphyton communities in laboratory streams. *J. North Am. Benthol. Soc.* **6,** 92–104.

Lamberti, G. A., Gregory, S. V., Ashkenas, L. R., Steinman, A. D., and McIntire, C. D. (1989). Productive capacity of periphyton as a determinant of plant–herbivore interactions in streams. *Ecology* **70,** 1840–1856.

McCullough, D. A., and Minshall, G. W. (1979). Bioenergetics of a stream "collector" organism *Tricorythodes minutus* (Insecta: Ephemeroptera). *Limnol. Oceanogr.* **24,** 45–58.

McIntire, C. D. (1973). Periphyton dynamics in laboratory streams: A simulation model and its implications. *Ecol. Monogr.* **43,** 399–420.

McIntire, C. D. (1983). A conceptual framework for process studies in lotic ecosystems. *In* "Dynamics of Lotic Ecosystems" (T. D. Fontaine and S. M. Bartell, eds.). pp. 43–68. Ann Arbor Sci. Publ., Ann Arbor, MI.

McIntire, C. D., and Colby, J. A. (1978). A hierarchical model of lotic ecosytems. *Ecol. Monogr.* **48,** 167–190.

McIntire, C. D., Tinsley, I. J., and Lowry, R. R. (1969). Fatty acids in lotic periphyton: Another measure of community structure. *J. Phycol.* **5,** 26–32.

McIntire, C. D., Colby, J. A., and Hall, J. D. (1975). The dynamics of small lotic ecosystems: A modeling approach. *Verh.—Int. Ver. Theor. Angew. Limnol.* **19,** 1599–1609.

Overton, W. S. (1972). Toward a general model structure for a forest ecosystem. *In* "Proceedings: Research on Coniferous Forest Ecosystems, a Symposium" (J. F. Franklin, L. J. Dempster, and R. H. Waring, eds.), pp. 37–47. Pacific Northwest Forest and Range Experiment Station, Portland, OR.

Overton, W. S. (1975). The ecosystem modeling approach in the coniferous forest biome. *In* "Systems Analysis and Simulation in Ecology" (B. C. Patten, ed.). Vol. 3, pp. 117–138. Academic Press, New York.

Perrin, C. J., Bothwell, M. L., and Slaney, P. A. (1987). Experimental enrichment of a coastal stream in British Columbia: Effects of organic and inorganic additions on autotrophic periphyton production. *Can. J. Fish. Aquat. Sci.* **44,** 1247–1256.

Peterson, B. J., Deegan, L., Helfrich, J., Hobbie, J. E., Hullar, M., Moller, B., Ford, T. E., Hershey, A., Hiltner, A., Kipphut, G., Lock, M. A., Fiebig, D. M., McKinley, V., Miller, M. C., Vestal, J. R., Ventullo, R., and Volk, G. (1993). Biological responses of a tundra river to fertilization. *Ecology* **74,** 653–672.

Rosemond, A. D., Mulholland, P. J., and Elwood, J. W. (1993). Top-down and bottom-up control of stream periphyton: Effects of nutrients and herbivores. *Ecology* **74,** 1264–1280.

Steinman, A. D. (1992). Does an increase in irradiance influence periphyton in a heavily-grazed woodland stream? *Oecologia* **91,** 163–170.

Steinman, A. D., and McIntire, C. D. (1986). Effects of current velocity and light energy on the structure of periphyton assemblages in laboratory streams. *J. Phycol.* **22,** 352–361.

Steinman, A. D., and McIntire, C. D. (1987). Effects of irradiance on the community structure and biomass of algal assemblages in laboratory streams. *Can. J. Fish. Aquat. Sci.* **44,** 1640–1648.

Steinman, A. D., McIntire, C. D., Gregory, S. V., Lamberti, G. A., and Ashkenas, L. R. (1987a). Effects of herbivore type and density on taxonomic structure and physiognomy of algal assemblages in laboratory streams. *J. North Am. Benthol. Soc.* **6,** 175–188.

Steinman, A. D., McIntire, C. D., and Lowry, R. R. (1987b). Effects of herbivore type and density on chemical composition of algal assemblages in laboratory streams. *J. North Am. Benthol. Soc.* **6,** 189–197.

Steinman, A. D., Mulholland, P. J., and Hill, W. R. (1992). Functional response associated with growth form in stream algae. *J. North Am. Benthol. Soc.* **11,** 229–243.

Stevenson, R. J. (1986). Mathematical model of epilithic diatom accumulation. *In* "Proceedings of the Eighth International Diatom Symposium" (M. Ricard, ed.), pp. 323–335. Koeltz Scientific Books, Koenigstein, Germany.

Stevenson, R. J., and Peterson, C. G. (1991). Emigration and immigration can be important determinants of benthic diatom assemblages in streams. *Freshwater Biol.* **26,** 279–294.

Straškraba, M., and Gnauck, A. H. (1985). "Freshwater Ecosystems, Modelling and Simulation," Dev. Environ. Modell., Vol. 8. Elsevier, New York.

Sumner, W. T., and McIntire, C. D. (1982). Grazer–periphyton interactions in laboratory streams. *Arch. Hydrobiol.* **93,** 135–157.

Warren, C. E., Wales, J. H., Davis, G. E., and Doudoroff, P. (1964). Trout production in an experimental stream enriched with sucrose. *J. Wildl. Manage.* **28,** 617–660.

Wiley, M. J., and Kohler, S. L. (1981). An assessment of biological interactions in an epilithic stream community using time-lapse cinematography. *Hydrobiologia.* **78,** 183–188.

Wiley, M. J., and Kohler, S. L. (1984). Behavioral adaptations of aquatic insects. *In* "The Ecology of Aquatic Insects" (V. H. Resh and D. M. Rosenberg, eds.), pp. 101–133. Praeger, New York.

Wlosinski, J. H. (1974). A description and preliminary user's guide to the desert biome stream ecosystem model. *US/IBP Desert Biome Res. Memo.* **74–60.**

Wlosinski, J. H., and Minshall, G. W. (1983). Predictability of stream ecosystem models of various levels of resolution. *In* "Dynamics of Lotic Ecosystems" (T. D. Fontaine and S. M. Bartell, eds.), pp. 69–86. Ann Arbor Sci. Publ., Ann Arbor, MI.

22

Benthic Algal Communities as Biological Monitors

Rex L. Lowe* and Yangdong Pan†

*Department of Biological Sciences,
Bowling Green State University,
Bowling Green, Ohio 43403,
and
The University of Michigan Biological Station,
Pellston, Michigan 49769

†Water Resources Laboratory,
University of Louisville,
Louisville, Kentucky 40292

I. INTRODUCTION

Access to abundant supplies of water has always been essential for the development of human civilizations. Water is not only a requisite for human nutrition but also for agriculture and domestic livestock. In addition, water is important for transportation and also for manufacturing. Thus, centers of human population are often found associated with coastal or riparian zones. Water is also used to export the waste products of human populations. By-products of modern civilization often include a diverse and bewildering assemblage of organic and inorganic chemicals. Some may be beneficial in small quantities (plant nutrients) whereas others may be extremely toxic (heavy metals, chlorinated hydrocarbons). This often complex assemblage of chemicals has the potential to severely damage the aquatic ecosystems that receive it; therefore, it is important that potentially impacted aquatic ecosystems be regularly monitored for biotic integrity if we are to protect them.

Biologists are often asked "Why monitor aquatic habitats biologically? Why not just monitor aquatic ecosystems by measuring physical and chemical parameters directly?" Instrumentation, including ion-specific chemical probes and continuous data loggers has been developed to the point where dozens of potentially influential chemicals can be inexpensively monitored on a continuous basis. However, the accessibility of chemical monitoring equipment might be difficult in many situations (e.g., small industries and municipalities), and leaving such equipment in the field puts it at risk for vandalization or theft. Furthermore, some aquatic ecosystems that are being polluted from diffuse sources, such as contamination from the atmosphere, are in remote areas where the transport and placement of continuous chemical monitoring equipment is not practical. In addition, even continuous chemical monitoring and data logging can miss events that might seriously impact key members of the biological community. It is also difficult to predict the interactive or synergistic influences of combinations of chemicals on aquatic biota.

Therefore, it is important that aquatic ecosystems are monitored biologically. Life is the ultimate monitor of environmental quality. If anthropogenic changes in the water quality of aquatic habitats fall outside the tolerance range for a species or a set of species, those taxa will decline and ultimately disappear. Aquatic organisms integrate all of the influential biotic and abiotic parameters in their habitat. They provide a continuous record of environmental quality. The mechanisms responsible for decline or local extinction of populations may be difficult to determine, but the populations' responses produce unambiguous signals concerning the quality of the habitat. These responses can provide regulators with information about where and when they should investigate potential pollution sources more intensively.

Benthic algae possess many attributes that make them ideal organisms to employ in water quality monitoring investigations:

1. Benthic algae, because they are primarily autotrophic, occupy a pivotal position in aquatic ecosystems at the interface of the chemical-physical and biotic components of the food web. This is a critical link in aquatic ecosystems and disruptions at this link can profoundly influence the rest of the aquatic community. Pollution indices based on diatom composition have been shown to yield more precise and valid predictions than

those based on protozoa or benthic macroinvertebrates because the former respond more directly to organic pollutants (Stewart et al. 1985, Leclercq and Maquet 1987).

2. Because benthic algae are sessile, they cannot avoid potential pollutants through migration or other means. They must either tolerate their surrounding abiotic environment or perish.

3. Benthic algal communities are usually species-rich relative to other aquatic groups. A few square centimeters of substratum may support in excess of 100 different species of algae (Kingston *et al.,* 1983). Each species, of course, has its own set of environmental tolerances and preferences (Lowe, 1974; Beaver, 1981; VanLandingham, 1982). Thus, the entire assemblage represents an information-rich system for environmental monitoring.

4. Benthic algae have relatively short life cycles. Cells of some species may divide more than twice daily (Eppley, 1977), which allows a rapid response to shifts in environmental conditions. Extant benthic algal communities are typically quite representative of current environmental conditions since they are among the first organisms to respond to environmental stress and among the first to recover from stress.

5. Benthic algal communities are spatially compact, and representative natural communities can be collected from a few square centimeters of substratum. Artificial substrata can be simple and relatively small.

6. Samples are easy to handle and curate. Benthic algal collections can be curated and stored in small spaces, and this facilitates long-term storage of organisms with little concern for space allocation. Soft benthic algae can be permanently stored in suspension or in glucose mounts (Stevenson, 1984a). Diatoms can be stored on permanent diatom mounts providing long-term access to the benthic algal assemblages.

7. Identification is not exceedingly difficult. Taxonomy of benthic algae is usually based on cell or thallus morphology, which is easily discernible through the light microscope. Excellent taxonomic keys exist for identification of benthic algae in most parts of the world.

II. APPLICATIONS

The most common application of biological monitoring with benthic algae involves the investigation of the impact of point-source disturbances of lotic systems (Hansmann and Phinney, 1973; Lowe and McCullough, 1974; Cooper and Wilhm, 1975; VanLandingham, 1976; Dickman and Gochnauer, 1978; Stevenson and Lowe, 1986; Morgan, 1987; Biggs, 1989). In this case, the suspected pollutant enters a stream at a point and the benthic algal community is employed to assess the impact of the suspected pollutant on the stream community. The benthic algal assemblage is normally analyzed upstream and downstream from the point source of interest. Sampling sites are located strategically to examine communities above, within, and below the influence of the point source (Fig. 1).

Benthic algae can also be employed to monitor long-term changes in aquatic communities that result from changes in diffuse (nonpoint) pollution sources that change through time. For example, one might want to biologically monitor the changes that occur through time in a watershed where agricultural practices are changing from full tillage to conservation tillage. Such changes in agricultural methods should result in a gradual decline in surface erosion of soil and phosphorus loading to the adjacent aquatic ecosystem but with increased runoff of agricultural herbicides (Fig. 2).

Finally, some components of the benthic algal assemblage, particularly diatoms, lend themselves to monitoring prehistoric conditions in aquatic ecosystems (Smol *et al.*, 1986, Dixit *et al.*, 1992). This science, called paleolimnology, has been most often employed in lakes but may also be useful in rivers with stable pools, in reservoirs, and in wetlands.

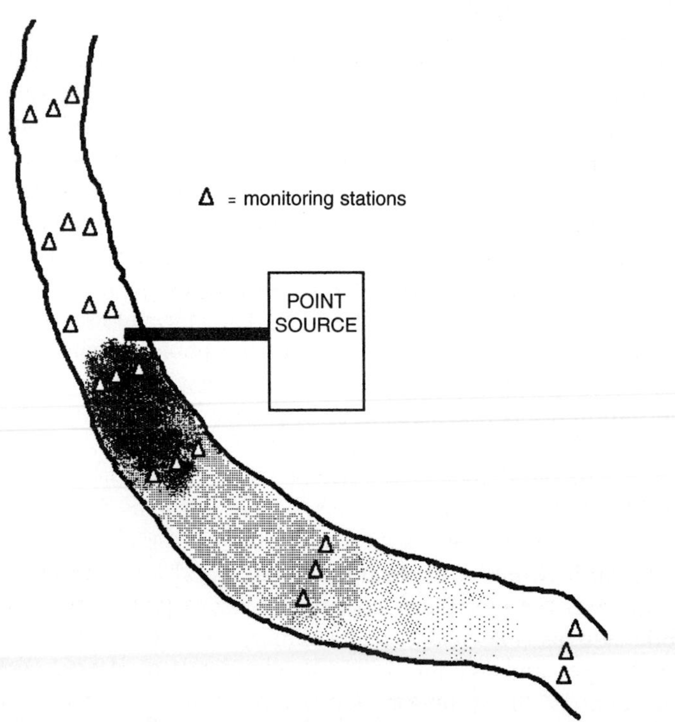

FIGURE 1 Monitoring a point-source perturbation in a lotic ecosystem. There are seven monitoring sites with three replicate substrata at each site. Shading represents the intensity of the point-source disturbance.

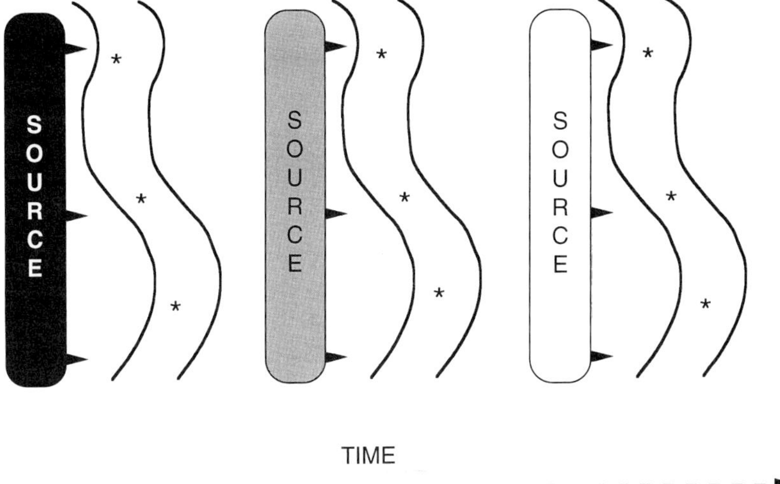

FIGURE 2 Monitoring a diffuse (nonpoint) source perturbation in a lotic ecosystem through time. There are three replicate substrata (artificial or natural) in the stream which are indicated by asterisks. Shading represents the intensity of the disturbance. The time scale may represent years as the non-point-source changes (is reduced or intensified).

III. METHODOLOGY

The methodology employed to collect and process benthic algae and generate data is partially dictated by the objectives of the monitoring investigation. The following methodological description is directed toward monitoring a point-source disturbance in a lotic ecosystem, although benthic algae are also useful in the biological monitoring of lentic systems (Ennis, 1977; Kettunen, 1983; Kann and Falter, 1989; Guzkowska and Gasse, 1990a,b; Hawes and Smith, 1993).

A. Site Selection

An initial visual and physicochemical reconnaissance of the site should be completed prior to the selection of benthic algal collection sites. The reconnaissance can help determine the entry site of the point-source perturbation in the stream, the stream distance where the perturbation persists, and the location of the potential recovery zone. The nature of data collected on this initial reconnaissance is a function of the nature of the perturbation to be monitored. If the disturbance is an organic load (sewage treatment plant, etc.), dissolved organic carbon or biochemical oxygen demand should be measured. Specific conductance and chemical-specific probes (such as dissolved oxygen) are also helpful in identifying the plume. If heavy metals or synthetic pesticides are suspected of being released, concentrations should be measured. The objective of the initial reconnaissance is not to generate an encyclopedic list of the concentrations of all chemicals through the zone of disturbance, but simply to identify the zone of the river that should be monitored biologically. Once the site of point-source loading and the probable recovery zone have been tentatively identified, benthic algae can be collected for biologically monitoring of the site.

B. Collection of Benthic Algal Samples

Data collection is frequently influenced by budgetary concerns; it is always desirable to optimize data collection by generating the necessary quantity and quality of data for the minimum cost. Undercollecting and overcollecting of data should both be avoided. If the data are to be scrutinized by inferential statistical tests (ANOVA, etc.), a minimum of three replicates per treatment or station is desirable. Further replication will provide increased degrees of freedom and facilitate the discovery of significant differences between treatments in a "noisy" data set. Preliminary studies or review of the literature may provide measures of variance that can then be used in power analyses (Gerrodette, 1987) to determine optimal sample sizes.

After determining the number of replicate samples to be collected from each sampling station ("treatment"), the number and location of stations should be chosen. The absolute minimum is a control treatment immediately above the entry of the point source disturbance and a disturbance treatment within the plume of the point source to be monitored. With three replicates collected at each station, this yields a total of six samples. If the objectives of monitoring include an analysis of the magnitude of the impact downstream and the location of the recovery zone, then more stations must be located downstream from the plume. The number and distance of stations from the point-source should be a function of the objectives of monitoring, the budget, and the data generated from the initial physicochemical reconnaissance. Figure 1 illustrates a monitoring design with seven stations of three replicate samples each.

Benthic algae should be collected by employing a technique that eliminates confounding environmental variables. It is essential that the primary variable between sampling stations is the point-source variable. Biggs (1985) has shown that triplicate samplers deployed in a stream in seemingly identical microhabitats can develop algal densities that are significantly different from each other. Thus, all other potentially influential variables such as light, current, and the nature of the substratum should be fastidiously standardized between stations. In small streams, it is necessary to examine the canopy of riparian vegetation and locate collecting stations in similar light regimes (Keithan and Lowe, 1985; Robinson and Rushforth, 1987; Duncan and Blinn, 1989). This leads to the question of how to collect the benthic algae and the nature of the substratum.

As just mentioned, it is imperative to collect benthic algae from the same kind of substratum at all sampling locations in the river in order to establish the impact of the point-source disturbance on the community (Meier *et al.,* 1983; Stevenson and Lowe, 1986). The nature of the substratum is strongly influential on the benthic algal community structure (Siver, 1977; Gale *et al.,* 1979; Tuchman and Blinn, 1979; Tuchman and Stevenson, 1980; Austin *et al.,* 1981; Antoine and Benson-Evans, 1985; Burkholder and Wetzel, 1989; Burkholder *et al.,* 1990). One cannot expect to make conclusions about the impact of point-source loading on the stream biota through benthic algal monitoring if benthic algal communities are collected from rocks at one station, macrophytes at another, and mud at yet another! It is often difficult to find replicate natural substrata at all stream segments of interest for sampling. The point source itself may have influenced the distribution, and thus the availability, of natural substrata through siltation or scouring. Thus, it is usually most expeditious to employ artificial substrata for periphyton collection.

Artificial benthic algal substrata were popularized by the research at the Philadelphia Academy of Natural Sciences and the development of the Catherwood Diatometer (Patrick *et al.,* 1954). This largely Plexiglas device floats in the water and supports glass microscope slides as artificial substrata (Fig. 3). Benthic algae colonize the microscope slides, which sit a few centimeters below the water surface and are oriented with their long axes parallel to the river channel. A curved Plexiglas shield on the front of the diatometer protects the glass slides from floating debris and also helps orient the diatometer in the current. Thus, if deployment sites are matched for riparian canopy cover, the diatometer standardizes variables such as light,

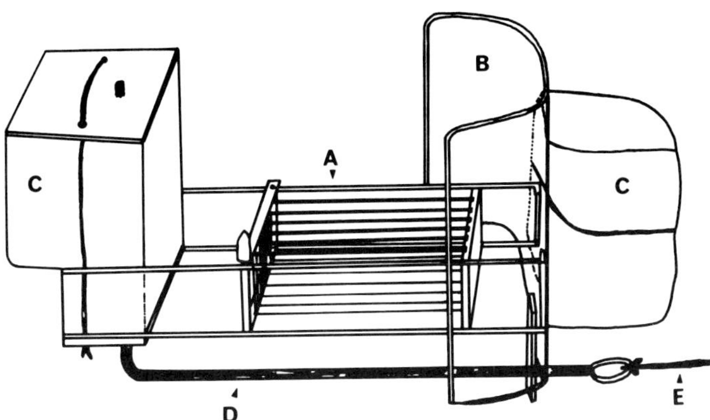

FIGURE 3 Floating periphyton sampler modeled after the Patrick et al. (1954) diatometer. Glass slides in a Plexiglas box (A) are behind a Plexiglas shield (B). Styrofoam floats (C) and a brass rod (D) provide buoyance and balance. A nylon cord (E) is tied to a submerged anchor.

current, and substratum among sites. Many other replicable artificial substrates, in addition to glass slides in a diatometer, have been employed to collect benthic algae (Aloi, 1990).

One potentially confounding variable that may impact benthic algal community structure differently at each station is the quantity and quality of invertebrate grazers. If, for example, the point source reduces or eliminates benthic algal grazers, increased algal biomass at the point source might be an indirect rather than direct effect of the disturbance. Careful analysis of the benthic algal community (discussed later) can help detect these relationships. The Catherwood Diatometer and other similar floating artificial substrata based on its design have been used extensively in biological monitoring projects (Hohn and Hellerman, 1963; Lowe and McCullough, 1974; Bahls and Bahls, 1974; Lowe and Gale, 1980; Austin *et al.,* 1981; Cairns *et al.,* 1983; Chessman, 1985; Gill *et al.,* 1993). Floating benthic algal substrata work well if the investigator is aware of the limitations of artificial substrata and the objectives of the monitoring project are clear. It must be kept in mind, however, that the benthic algae accruing on the samplers may be quite different from the community growing on the stream bottom or margins, with green and blue-green algae often underrepresented (Cattaneo and Amireault, 1992). The objective is to collect communities on comparable substrata through the riverine zone of interest, not to investigate naturally occurring communities.

Although these devices work well and are widely employed in the field, they are subject to frequent vandalization. Floating benthic algal artificial substrata are likely to be stolen, smashed, shot at, and generally abused by the public. Most people with a few years of experience can wax for hours on novel ways they have tried (and usually failed) to protect artificial substrata from destruction. Informative signs sometimes help, but often only attract further attention from "limboids" bent on destroying anything that they do not understand. The best course of action for deployment and protection of artificial substrata is probably to make them as obscure and inaccessible as possible. This can be accomplished by avoiding bright colors on floats and by deploying them, if possible, in low-traffic areas.

One popular and effective method of protecting artificial substrata is to place them under water (Gale *et al.,* 1979; Austin *et al.,* 1981; Lowe *et al.,* 1986; Biggs, 1988a). In this case, great care must be taken not to introduce extraneous and confounding variables. Submerged substrata should be deployed at similar depths and current speeds. Orientation of the substratum with respect to flow is also important (Stevenson, 1984b). In deep clear rivers (Gale *et al.,* 1979) or in lakes (Carrick *et al.,* 1988), the deployment and collection of substrata may require SCUBA technology, which will add some expense to a monitoring project but may be well worth the cost for data integrity. In shallow streams, the substrata can be handled by wading (Biggs, 1988a). If submerged substrata are used, care must also be taken not to dislodge benthic algae from the substratum when it is retrieved. Dislodgment can be prevented by placing the substratum in a plastic bag or other container while still under water before it is removed from the river. As with floating substrata, a variety of materials and designs have been employed for submerged substrata (Austin *et al.*, 1981; Pringle and Bowers, 1984; Fairchild *et al.*, 1985; Biggs, 1988a; Aloi, 1990).

Once the sampling sites and number of replicate samples have been chosen, the artificial substrata can be deployed (or, if using natural substrata, they can be collected directly). The length of exposure often varies between rivers and seasons as a function of light, temperature, and invertebrate grazing intensity (Biggs, 1988b; Gale *et al.*, 1979; Horner and Welch, 1981). Patrick *et al.* (1954) asserted that 2 weeks of exposure of their diatometer was sufficient in the summer to obtain a representative benthic algal community. Gale *et al.* (1979) demonstrated that in the Susquehanna River there was minimal new accrual on benthic samplers deployed in the winter. It is prudent to inspect the substrata after 2 to 3 weeks to look for the development of color (biofilm). If a noticeable biofilm, is present the substrata can be collected. The thickness of this biofilm may vary among stations, which should not be surprising because the stations were selected to monitor the impact of a point source. There is some danger in allowing the substrata to remain in the field too long, since they are continuously vulnerable to vandalization or storm events. In addition, thick biofilms developing on the substrata have an increased liability to sloughing relative to thin biofilms (Stevenson, 1990; Biggs and Close, 1989).

Under many circumstances, it may be more advantageous to use natural substrata rather than artificial substrata for collecting benthic algae (Bahls, 1973). For example, if one is interested in determining the impact of point-source perturbation on the indigenous benthic algal community, then the indigenous community should be sampled on the substrata present in the stream. The use of natural substrata is also more cost-effective since only one trip into the field is required rather than a trip to deploy plus a trip to retrieve artificial substrata. In addition, artificial substrata are frequently lost or vandalized, resulting in the loss of the entire data set. When using natural substrata, an effort should be made to collect from the same substratum throughout the zone of interest in the river if the impact of a point source is being monitored. Also, when natural substrata are used, extra care must be taken to ensure that all the potentially confounding parameters not associated with the point source vary minimally across sites (depth, current velocity, angle of substratum, etc.). There are several relevant examples in the literature of the employment of natural substrata in lotic benthic algal monitoring investigations on sand (Krejci and Lowe, 1987), stones (Round, 1991), and plants (van Dam, 1975).

C. Bioassays and Artificial Stream Systems

If the primary objective of biological monitoring is to determine the impact of suspected pollutants on aquatic communities, it is feasible in some situations to satisfy this objective in more controlled experiments using artificial stream mesocosms (Genter *et al.,* 1987; Bélanger *et al.,* 1994). The use of artificial streams makes it possible to control many of the extraneous influential variables such as light, current velocity, grazers, and so on that have the potential to contribute "noise" to experiments in which a single chemical variable is being manipulated. Artificial stream experiments have contributed greatly to our knowledge of the impact of metals (Genter *et al.,* 1987), pesticides (Rose and McIntire, 1970; Kosinski, 1984), nutrients (Bothwell, 1985; Pan and Lowe, 1994), and physical variables (Bothwell, 1988; Maurice *et al.,* 1987) on aquatic ecosystems, as well as to our general knowledge of benthic algal biology (McIntire, 1968; Steinman and McIntire, 1986; Steinman *et al.,* 1987, 1989; Mulholland *et al.,* 1991). Benthic algae lend themselves particularly well to artificial stream mesocosms because of the compact nature of the community and the ease of manipulation. Many ecologically realistic replicate mesocosms can be constructed inexpensively in a relatively small space (Pan and Lowe, 1994), thus enhancing the statistical predictive power of such bioassays. Artificial stream mesocosms have the additional advantage of determining the impact of potential environmental toxins in a controlled environment rather than risking environmental damage in a natural stream. An example of the application of artificial stream mesocosms to aquatic toxicology investigations utilizing benthic algae is presented in Bélanger *et al.* (1994).

D. Analysis of the Benthic Algal Community

The type of benthic algal analyses to be performed and the nature of the data to be generated should be determined well in advance of beginning the monitoring project and should be a function of the nature and the objectives of the project. For example, if an increase in algal nutrients is being monitored, one might expect the impact to include an increase in algal growth rates (Biggs, 1990). If the impact of toxic substances is being monitored, a decline in sensitive species and a change in community structure might be hypothesized. Data generation should be planned carefully to maximize relevant information and minimize effort and costs. A good general scheme to follow, if one is not certain about how many data are necessary, is to prioritize data generation based on an information/effort ratio. Data that maximize this ratio should be generated first as time, budget, and objectives allow. For example, biomass and chlorophyll are more easily generated than detailed taxonomic data. The following methodology begins with generation of community-level data and proceeds to the population level.

1. Estimations of Productivity

Biomass changes through time have been used as indirect estimates of productivity and are appropriate in situations where biomass loss through sloughing or grazing is minimal. In systems where eukaryotic filamentous algae dominate, wet weights provide a suitable estimate for benthic algal biomass across treatments (Power *et al.,* 1985). But when community composition varies among treatments, dry weight (DW) or ash-free dry weight (AFDW) are better estimators of benthic algal community biomass. Methodologies for data generation have been standardized (American Public Health Association *et al.,* 1985). Benthic algal biomass can also be estimated using chlorophyll as a proxy (American Public Health Association *et al.,* 1985). This method has been used extensively in phytoplankton community measurements, but chlorophyll concentrations in algal cells can be quite variable and may be unreliable estimators of algal biomass. This is particularly problematic when comparing chlorophyll data between habitats that vary widely in physical (light) or chemical (nutrients) parameters. A direct measure (DW, AFDW) is a much more reliable means of determining periphyton biomass. However, chlorophyll data are essential if the autotrophic index (see Section III,F) of the periphyton community is to be measured.

Benthic algal productivity can be measured directly by measuring changes in biomass through time in controlled experiments where grazing and sloughing are not a concern (Stevenson, 1990; Bothwell, 1988; Pan and Lowe, 1994) or by measuring rates of carbon fixation or oxygen evolution (Loeb, 1981; Turner *et al.*, 1983; Fairchild and Everett, 1988). These techniques are not suitable for standard monitoring applications because of their labor- and equipment-intensive nature, but may be employed in bioassay applications where specific chemical or physical parameters such as nutrients (Fairchild and Everett, 1988), commercial chemicals (Bélanger *et al.*, 1994), or light (Turner *et al.*, 1983) are targeted. Direct measures of productivity have most often been made in laboratory or artificial stream systems.

2. Community Enumeration

One of the unique advantages of employing benthic algae in water quality monitoring is the species-rich nature of benthic algal communities. Changes in the community structure represent the sum of changes in potentially hundreds of populations. There are two approaches for using benthic algal data in water quality monitoring investigations: assessing the presence of indicator species and analyzing the community structure. These will be discussed in detail later, but in both approaches the community must be enumerated.

The benthic algal community is likely to contain populations of species from several different algal divisions that require different methodologies of preparation for enumeration. The first step is to preserve the community as quickly as possible in the field. Once the benthic algal community is removed from its habitat it should be scraped from the substratum and preserved. Cells may continue to divide, and division rates of each population may vary; thus, immediate preservation is desirable. Waiting for only 24 hours before preservation could allow some populations to double in size. A preservative such as formalin or glutaraldehyde in a final concentration of 3–5% works well. Preserved samples should be kept out of direct sunlight. If portions of the collection are to be analyzed for chlorophyll or AFDW, they should be subsampled and processed prior to preservation.

Several methodologies are available for enumeration of benthic algae (Weber, 1973) and the technique of choice is often a matter of availability of equipment and the nature of the collection. A research-grade light microscope equipped with an ocular micrometer is essential. In general, it is best to scan a wet mount of the sample under low (60–100×) and high (400–500×) magnification to become comfortable and familiar with the community and to note the general "health" of the community. This is also the time to generate a list of taxa present to ease the subsequent analysis of community structure. It may not be possible to identify diatoms beyond genus or form (e.g., naviculoid, gomphonemoid, cymbelloid) during this initial scan, but species of other divisions of "soft algae" (greens, blue-greens, reds, etc.) can be identified and listed. The selection of taxonomic keys will vary depending on the region of the world from which the samples are collected, but a basic set of taxonomic literature should include Bourrelly (1966, 1968, 1970); Krammer and Lange-Bertalot (1986, 1988, 1991a,b); Hustedt (1930); Germain (1981); Prescott (1962); Prescott *et al.* (1972, 1975, 1977, 1981); Patrick and Reimer (1966, 1975); and Whitford and Schumacher (1973).

Enumeration of benthic algal samples is often done with a Palmer–Maloney Nannoplankton Counting Chamber (PMC). This device holds 0.1 ml of algal suspension and is shallow enough (400 μm) that samples can be analyzed with high-power objective (40–45×), which allows a total magnification (400–450×) that facilitates accurate identification of algae. Similar counting chambers can be inexpensively constructed by punching a hole (10 to 20 mm diameter) in a plastic coverslip with a cork borer and bonding the modified coverslip to a glass microscope slide. It is not necessary to count all the algae in the chamber. Volumes of individual microscope fields or strips across the PMC are easily calculated. During enumeration, algal taxa are identified to species if possible. Species of some genera require sexually mature specimens for identification (e.g., *Oedogonium, Spirogyra, Zygnema*) and can only be identified accurately to genus. Diatoms should be counted and identified to the lowest taxonomic level with which the microscopist is comfortable (genus or form). Species names of diatoms can be generated later from permanent diatom mounts (Patrick and Reimer, 1966) and these data subsequently applied to the PMC counts. For example, if ten cymbelloid diatom cells were observed in the PMC, then ten cymbelloid diatoms encountered randomly on the permanent diatom mounts *(Amphora, Cymbella, Encyonema)* would be identified to species and those names applied to the count. In making the diatom species identifications one should try to identify similarly sized specimens in both the PMC and the diatom mount.

Some microscopists prefer to enumerate benthic algae from semipermanent glucose mounts (Taft, 1978; Stevenson, 1984b). This technique has the advantage of establishing a "permanent" microscope slide of the benthic algal community, allowing the application of oil immersion objectives for identification of soft algae and diatoms from the same preparation. Yet it has the disadvantage of mounting diatoms in a medium of relatively low refractive index with their cytoplasmic contents intact, often making them difficult to identify to species. Glucose mounts, on the other hand, have the advantage of allowing the microscopist to determine if the diatom was living (with cytoplasm intact) at the time of collection.

One of the most commonly asked questions about enumeration of algal samples is "How many cells should I count?" The answer depends on the type of statistical analyses to which the data are to be subjected. If a log-normal curve is to be fit to the data (Patrick *et al.*, 1954), it may be necessary to enumerate thousands of specimens. However, if interest is restricted to shifts in population densities of dominant and common taxa, then it has been suggested (Round, 1991) that it is necessary to enumerate only 50 to 100 cells per sample. Most investigators enumerate from 300 to 500 organisms.

Increased speed and accuracy of the generation of benthic algal data can be achieved by interfacing the enumeration process with a computer (Lowe and Johnson, 1992). The computer keyboard becomes the data-entry tool, with each algal taxon assigned a number. Data generated in this way need not be reentered manually into a computer for statistical analyses, thus reducing the chance of error during data entry. In addition, the counting software can quickly provide periodic outputs of community metrics such as species diversity (Johnson and Lowe, 1993), allowing the analyst to determine when enough individuals have been examined to adequately describe the community.

It has been proposed that the computerization of benthic algal community analyses might even include computer identification of diatoms by matching diatoms with a library of diatom holograms (Cairns *et al.*, 1977). This appears to be technologically feasible but is not currently available.

Algal taxa from a single sample can vary widely in biovolume (Table I), and many investigators feel that densities of algal populations are best described on a biovolume basis rather than as cell number. This is particu-

TABLE I Algal Cell Biovolume Estimates (μm^3) from Big Darby Creek and Little Miami River, Ohio

DIATOMS					
Achnanthidium clevei	170	*Gomphonema sphaerophorum*	600	*Nitzschia hungarica*	600
Achnanthidium detha	90	*Gomphonema subclavatum*	400	*Nitzschia kutzingiana*	350
Achnanthidium lanceolata	180	*Gomphonema truncatum*	400	*Nitzschia linearis*	850
Achnanthidium linearis	70	*Gyrosigma acuminatum*	3500	*Nitzschia palea*	200
Achnanthidium minutissimum	70	*Gyrosigma scalproides*	800	*Nitzschia parvula*	350
Actinocyclus rothii	520	*Gyrosigma sciotoense*	1600	*Nitzschia reversa*	280
Amphipleura pellucida	600	*Gyrosigma spencerii*	1200	*Nitzschia sigma*	520
Amphora ovalis	350	*Hantzschia amphioxys*	750	*Nitzschia sigmoidea*	1200
Amphora ovalis v. *pediculus*	130	*Luticola mutica*	400	*Nitzschia triblionella* v. *debilis*	500
Amphora submontana	100	*Martyana martyi*	150	*Nitzschia triblionella*	700
Amphora veneta	150	*Mastogloia grevillei*	1700	*Nitzschia vitrea*	1200
Amphora perpusilla	100	*Meridon circulare*	450	*Pinnularia borealis*	400
Anomoeoneis serians	240	*Navicula accomoda*	400	*Pinnularia brebissonii*	700
Anomoeoneis vitrea	180	*Navicula anglica*	450	*Pinnularia obscura*	500
Asterionella formosa	550	*Navicula atomus*	80	*Pleurosira laevis*	10,600
Aulacosira distans	200	*Navicula auriculata*	120	*Reimeria sinuata*	130
Aulacosira granulata	550	*Navicula bacillum*	350	*Rhoicosphenia curvata*	500
Bacillaria paradoxa	800	*Navicula capitata*	200	*Rhopalodia gibba*	1300
Caloneis bacillum	250	*Navicula cryptocephala*	350	*Sellaphora pupula*	400
Caloneis hyalina	100	*Navicula decussis*	300	*Skeletonema potamus*	60
Caloneis ventricosa	400	*Navicula elginensis*	300	*Stauroneis anceps*	650
Cocconeis pediculus	450	*Navicula exigua*	475	*Stephanodiscus invisitatus*	150
Cocconeis placentula	160	*Navicula exigua* v. *capitata*	460	*Stephanodiscus niagarae*	700
Craticula cuspidata	1100	*Navicula gregaria*	300	*Surirella angustata*	300
Cyclotella atomus	80	*Navicula halophila*	500	*Surirella ovata*	360
Cyclotella menenghiniana	150	*Navicula hambergii*	240	*Synedra rumpens*	560
Cyclotella stelligera	120	*Navicula heufleri*	450	*Synedra ulna*	1200
Cymatopleura solea	3500	*Navicula lanceolata*	700	*Thalassiosira weisflogii*	200

Cymbella affinis	350	*Navicula luzonensis*	100	GREEN ALGAE	
Cymbella cistula	600	*Navicula menisculus*	300	*Ankistrodesmus falcatus*	80
Cymbella prostrata	800	*Navicula minima*	130	*Cosmarium* sp.	780
Cymbella tumida	800	*Navicula minuscula*	80	*Cladophora glomerata*	2200
Diadesmus contenta	90	*Navicula pelliculosa*	70	*Closterium* sp.	750
Diatoma vulgare	900	*Navicula radiosa*	3500	*Coelastrum sphaericum*	900
Diploneis puella	250	*Navicula latans*	250	*Crucigenia tetrapedia*	120
Epithemia adnata	2800	*Navicula radiosa* v. *tenella*	300	*Dictyosphaerium pulchellum*	50
Eucocconeis flexella	500	*Navicula rynchocephala*	650	*Mougeotia* sp.	2200
Encyonema minuta	150	*Navicula seminulum*	60	*Oedogonium* sp.	200
Eunotia pectinalis	450	*Navicula viridula*	750	*Pandorina morum*	500
Fallacia pygmea	280	*Neidium affine*	1500	*Scenedesmus quadracauda*	320
Fragilaria construens	100	*Neidium iridis*	1500	*Spirogyra* sp.	3300
Fragilaria crotonensis	700	*Nitzschia accomodata*	300	*Stigeoclonium tenue*	200
Fragilaria vaucheriae	410	*Nitzschia acicularis*	280	*Tetraedron minimum*	90
Frustulia rhomboides	750	*Nitzschia amphibia*	200	*Ulothrix variabilis*	130
Frustulia vulgaris	600	*Nitzschia angustata*	300		
Gomphoneis herculeana	3200	*Nitzschia bacata*	300	BLUE-GREEN ALGAE	
Gomphoneis olivacea	320	*Nitzschia dissipata*	300		
Gomphonema acuminatum	600	*Nitzschia dubia*	350	*Chamaesiphon incrustans*	40
Gomphonema angustatum	300	*Nitzschia filiformis*	300	*Microcystis incerta*	10
Gomphonema brasiliense	300	*Nitzschia fonticola*	250	*Schizothrix calcicola*	40
Gomphonema dichotomum	280	*Nitzschia frustulum*	250		
Gomphonema gracile	450	*Nitzschia gandersheimensis*	300		
Gomphonema intricatum	250	*Nitzschia gracilis*	400		
Gomphonema parvulum	300				

larly critical when comparing the relative response of algal populations to some environmental parameter. For example, if a population of *Achnanthidium lanceolata* contains 100 cells mm^{-2} and a population of *Pleurosira laevis* has only 5 cells mm^{-2}, one might conclude that *A. lanceolata* is overwhelmingly dominant. If one considers, however, that cells of *A. lanceolata* have a volume of about 180 μm^3, whereas cells of *P. laevis* contain about 10,600 μm^3 (Fig. 4), then *P. laevis* would be considered the dominant population of this pair of taxa. Biovolume may not be the best measure of the importance of a taxon in a mixed community, but intuitively it is better than cell number when comparing population-level responses among algal taxa. Often taxa with relatively large cellular biovolumes such as *P. laevis* and *Cladophora glomerata* are not numerically common and are likely to be missed if a limited number of cells are enumerated. If such large, but uncommon taxa are present in the community it is advisable that some sort of stratified counting procedure be employed. In this procedure, the community is reexamined at low magnification after the PMC or semipermanent mount has been analyzed at high magnification. In the reexamination, the lower magnification utilized is normally 100×, and small taxa are ignored while large, less common taxa are quantified. The data from this second low-magnification count are added to the data from the initial high-magnification count while keeping track of volumes analyzed. Finally, the number of cells mm^{-2} of substratum can be calculated by the formula

$$C_i = n_i V_t V_e^{-1} A^{-1}$$

Where

> $\boldsymbol{C_i}$ is the density of taxon i in cells mm^{-2};
> $\boldsymbol{n_i}$ is the number of cells of taxon i counted in the PMC;
> $\boldsymbol{V_t}$ is the total volume of the benthic algal suspension;
> $\boldsymbol{V_e}$ is the volume of sample enumerated; and
> $\boldsymbol{A}$ is the area in mm^2 of the substratum scraped for benthic algae.

E. Quality Assurance, Quality Control, and Standard Operating Procedure

Many pollution monitoring investigations have the potential to generate data that may be used in a court of law. A standard example might be a regional or federal regulatory agency in litigation with a private corporation or individual over environmental damage. When large sums of money are involved in litigation, the integrity of benthic algal monitoring data will be assiduously scrutinized. If the accuracy of the data is equivocal, those data might be useless to the prosecution or defense. It is thus important to establish good laboratory practices (GLP). GLP involve the

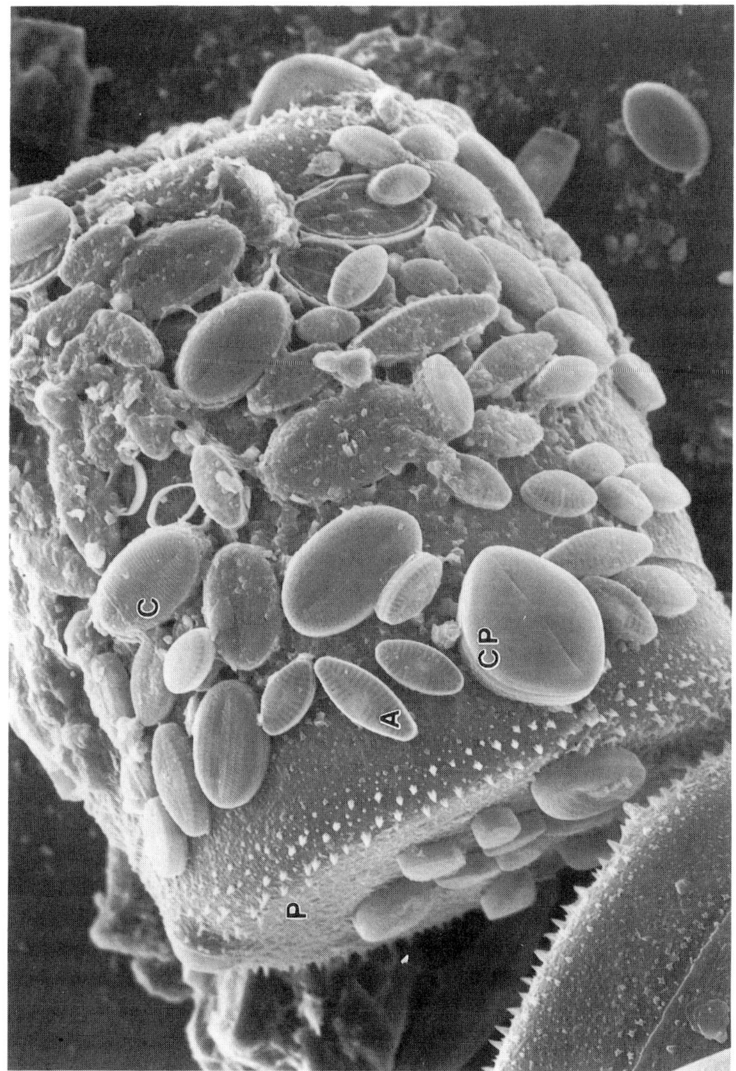

FIGURE 4 Scanning electron micrograph of sympatric diatom taxa from the Little Miami River, Ohio, with different biovolumes. P = *Pleurosira laevis* (10,600 μm^3), A = *Achnanthidium lanceolata* (180 μm^3), Cp = *Cocconeis pediculus* (450 μm^3), C = *Cocconeis placentula* (160 μm^3).

utilization of standard operating procedure (SOP) in the field and in the laboratory. SOP should be documented in detail, with the documentation conserved in a safe place. The documentation includes the operating manuals of any instrumentation employed (microscopes, spectrophotometers, etc.), as well as the curriculum vitae of all scientists and technicians associated with the project. The SOP documentation should also present all the details of data generation and analysis. As samples are retrieved from the field and transported to the laboratory (or perhaps posted to another laboratory), a chain of custody document must travel with the sample so that one can later reconstruct the record of sample control at any given time. To maintain quality assurance, a percentage of the samples (usually a minimum of 10–15%) should be analyzed in duplicate to assure that the same sample consistently yields the same information. Quality control of data is demonstrated by the ability of two independent investigators to arrive at similar end point values (AFDM, chlorophyll, community structure, etc.). A minimum of 1% of the samples should be submitted to an independent laboratory for analysis of quality control of the data. Samples should be conserved for at least five years (longer if possible) to permit the reevaluation of questionable data. The small size of algal samples expedites this procedure.

F. Data Analysis and Statistical Procedures

In the simplest analysis, the response of benthic algal community biomass to point-source perturbation can be measured. For example, if the point source is rich in algal nutrients, and the benthic algal biomass in the stream is indeed nutrient-limited, an increase in algal biomass might be expected (Stockner and Shortreed, 1978; Freeman, 1986; Wharfe *et al.,* 1984). Assays such as relative specific growth rates (Biggs, 1990) can be useful in making predictions about the impact of nutrient loading. If the point source is rich in organic substances, then an increase in bacterial or fungal biomass might occur. Such changes could be detected by applying the autotrophic index, which is calculated as total ash-free dry weight (mg m^{-2}) per chlorophyll *a* (mg m^{-2}). The greater the concentration of organic matter (BOD), the higher the autotrophic index of the benthic algal community. Autotrophic index values greater than 100 may result from organic pollution (Weber, 1973), however, this index must be applied cautiously since high autotrophic index values may result from other circumstances.

Two complementary approaches have been employed to analyze the response of benthic algal populations to point-source disturbance. The first approach focuses on the response of periphyton populations or sets of populations that are affected (or unaffected) by the perturbation. This approach, known as the autecological or indicator organism approach, relies on a preexisting body of knowledge about the ecological profiles of target species. The densities of dominant or sensitive taxa can be subjected to an ANOVA (Zar, 1983) above and below the point source. Significant differences in patterns of distribution indicate an impact of the point source. For example, if the population of *Diatoma hiemale* significantly declines below a suspected point source of chemical perturbation, it could be concluded that some component of the point source is adversely affecting this population. *D. hiemale* is "indicating" something about the point source. By accessing the preexisting body of knowledge on the ecological profile of *D. hiemale,* it is possible to pinpoint the nature of the perturbation.

Catalogs of ecological profiles have been generated for hundreds of common species of benthic algae (Sládeček, 1973; Lowe, 1974; Beaver, 1981; VanLandingham, 1982). In addition, many other taxa have well-known ecological profiles but have not been formally cataloged (Lange-Bertalot, 1978, 1979b). Often sets, or guilds (Carrick *et al.*, 1988), of taxa are affected similarly by prevailing ecological conditions, resulting in information redundancy and strengthening the conclusions about the nature of the perturbation. The indicator organism approach to water quality monitoring has received considerable attention and has undergone considerable refinement in the investigation of European rivers (Whitton *et al.*, 1991b), although almost all the benthic algal indices devised to date are used to assess organic pollution (Wantanabe *et al.*, 1988; Lange-Bertalot, 1979a; Sumita and Wantanabe, 1983). Algal taxa within watersheds have been assigned not only numerical values based on their ecological profiles but also an "indicating value" based on the amplitude of their response (Descy, 1979). An exception to the use of benthic algae for monitoring organic pollution is the Whitton *et al.* (1991a) proposal of a standard "package" of ten plant species, which includes three algal species for monitoring heavy metals in British waterways.

Community ecology represents a second approach to the interpretation of population data and is simply an extension of the autecological or indicator organism approach. In the community ecology or "synecology" approach, the sum of the responses of all taxa is converted into a new metric such as species diversity. In general, environmental disturbances or perturbations that deviate greatly from the "environmental norm" of a community eliminate sensitive species or reduce their populations significantly. More tolerant populations may expand, utilizing resources relinquished by sensitive species. The result of this shift in populations is a reduction in species diversity (Archibold, 1972; van Dam, 1982). There are many choices of diversity indices, from the Sequential Comparison Index (Cairns *et al.*, 1968) that requires no taxonomic expertise to the log-normal analysis of distribution (Patrick *et al.*, 1954) to the Shannon and Wiener (Shannon, 1948) diversity index that is widely used in both benthic algal and benthic macroinvertebrate monitoring applications. In general, the more severe the perturbation, the greater the reduction in species diversity. However, Stevenson (1984c) has shown that in some instances minor disturbances may increase diversity. As with population-level data, diversity indices can be subjected to an ANOVA both upstream and downstream from a point source to test for significant differences.

The response of algal assemblages to perturbation has also been described using sophisticated data summarization procedures (Guzkowska and Gasse, 1990a,b) that exploit the information-rich autecological profiles of algal species. With the widespread access to increasing computer power, more computer-intensive procedures can be adopted for numerical ecology (see review by Birks, 1993). These autecological inference procedures are an excellent means of detecting patterns of distribution of algal species related to physical and chemical parameters.

Cluster analysis arranges samples (collection sites) into a small number of groups or clusters and generates a hierarchical diagram of relatedness based on the taxonomic structural similarity among the samples. The upstream and downstream sites are expected to be located in different clusters if effects of point-source pollution on algal composition are severe. Cluster analysis can be divided into divisive and agglomerative methods. The major difference between the two methods is that divisive methods emphasize global differences and agglomerative methods are heavily influenced by local differences (van Tongeren, 1987). The interpretation of results of cluster analysis can be facilitated by additional analysis such as discriminant function analysis (terBraak, 1986a). Divisive methods start from the top: one group is divided into two subgroups and so on. One method of divisive cluster analysis commonly used in community ecology is TWo INdicator SPecies ANalysis (TWINSPAN) (Hill *et al.*, 1975). TWINSPAN is a two-step process. First, sites are ordinated by correspondence analysis (CA). Sites along the first CA axis are divided at the middle to form two groups. The resulting division is refined using variables with maximum indicator value, so-called iterative character weighing. The process is repeated on each of the resulting divisions. Agglomerative cluster methods, on the other hand, start from the individual site, combining it with other sites into larger groups. There are many types of agglomerative cluster methods, but they differ only in distance measures among sites and in fusing rules for clustering sites. Readers are referred to Everitt (1993), Gordon (1981), and Digby and Kempton (1987) for detailed coverage on the choices of distance measures and criterion for fusing sites.

The assessment of algal response to pollution, especially to non-point-source pollution, may require more sophisticated multivariate statistical approaches such as ordination. Ordination, such as principal component analysis (PCA) and correspondence analysis, arranges sites along axes according to species composition. The end product of ordination is a low-dimension (often two or three) graphical summary of species data. The sites with similar species composition are close together in the ordination diagram. The axes are the unknown theoretical variables that can best explain species distribution. The interpretation of the axes requires additional analysis such as linear regression or correlation analysis if environmental variables are available. Specifically, the axes can be interpreted according to the strength of linear association between the axes (vectors of sites or cores) and environmental variables. An alternative approach is to interpret axes based on autecological information of species that heavily load to the axis indicated by their high species scores. Therefore, ordination is a two-step approach that can be used to detect underlying structure and relate it to external environmental variables.

One of the most commonly used approaches to constrained ordination is canonical correspondence analysis (CCA) (Birks *et al.,* 1994). CCA incorporates ordination and multiple regression into one technique (terBraak, 1986b, 1987a). CCA does exactly the same thing as ordination, but with the axes constrained to be linear combinations of environmental variables. Such analysis is also referred to as direct gradient analysis, in contrast to PCA and CA, which are two-step indirect gradient analyses. CCA is extremely powerful for detecting patterns of species distribution related to associated physical and chemical parameters. Since terBraak introduced CCA in 1986, and especially after the implementation of his computer program CANOCO (terBraak, 1987b), a total of 379 research papers using CCA and its close relatives have been published in different areas ranging from community ecology, biogeography, and paleolimnology to management (Birks *et al.,* 1994). Detailed coverage on ordination can be found in Pielou (1984), Manly (1991), Digby and Kempton (1987), Jongman *et al.* (1987), Jackson (1991), and Reyment (1991). A chapter by terBraak (Jongman *et al.,* 1987) on ordination including CCA is highly recommended. Readers who are interested in CCA and other gradient analysis methods should refer to a review article by terBraak and Prentice (1988).

The assumption for using CCA is that species respond to environmental parameters in a unimodal manner. This assumption is commonly evaluated by running detrended correspondence analysis (DCA), in which the gradient length is calculated. If the gradient length is greater than 2 units of standard deviation, CCA is an appropriate method. Otherwise, redundancy analysis, a multivariate technique based on linear response model, should be used.

Like ordination, CCA can reduce the dimensionality of a complex data set and graphically summarize data in low-dimensional diagrams (Fig. 5). Unlike an ordination diagram, environmental variables are also displayed as arrows with species and sites as points in a CCA diagram. Arrow length represents the strength of the correlation between the environmental variable and the ordination axes. In addition to graphical display of data in

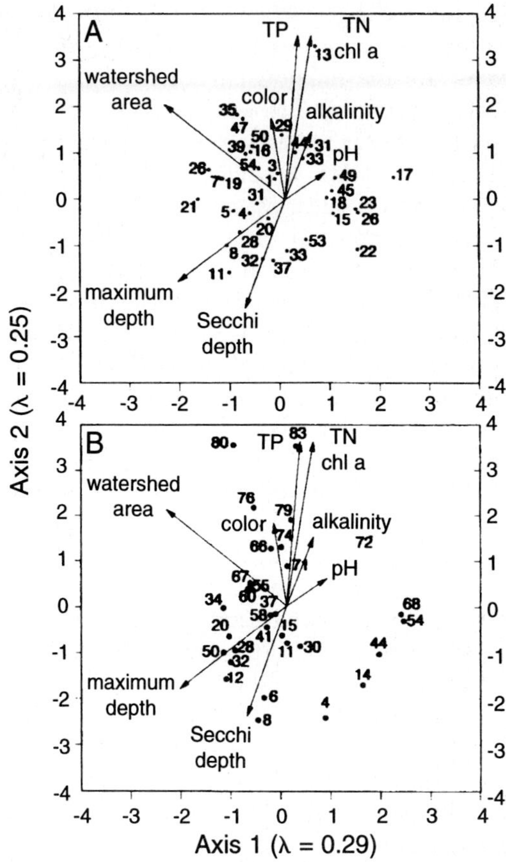

FIGURE 5 CCA ordination showing (A) sample scores generated from the linear combination of environmental variables with numbers corresponding to lakes found in Table III and (B) diatom taxa biplot with numbers corresponding to diatom taxa found in Table V. Environmental arrows are magnified (4×) in both figures. λ = eigenvalue. (From Christie and Smol (1993), with permission of the *Journal of Phycology*.)

CCA diagrams, CANOCO also generates a summary table (Table II). The importance of each canonical axis can be evaluated by associated eigenvalues. The larger the eigenvalue, the better sites are dispersed along the axis and therefore the more important that axis is as an explanation of physical as well as biological variation in measured ecosystem parameters. Total inertia is the sum of all unconstrained eigenvalues, in other words, the total amount of variance of species data. The sum of all canonical eigenvalues is the amount of variance explained by all measured environmental variables. The proportion of variance accounted for by all environmental variables can be calculated as follows: (sum of all canonical eigenvalues * 100)/total inertia. One can then calculate the proportion of

TABLE II Summary of Canonical Correspondence Analysis (CCA)[a]

Axes	*1*	*2*	*3*	*4*	*Total inertia*
Eigenvalues	0.445	0.359	0.283	0.166	5.574
Species–environment correlations	0.868	0.869	0.824	0.798	
Cumulative percentage variance					
Of species data	8.0	14.4	19.5	22.5	
Of species–environment relation	26.6	48.0	64.9	74.8	
Sum of all unconstrained eigenvalues					5.574
Sum of all canonical eigenvalues					1.675

[a]From Y. Pan and R. J. Stevenson (unpublished data).

variance explained by each axis as follows: (eigenvalue of the axis * 100)/ sum of all canonical eigenvalues. This is referred to as the percentage variance of the species–environment relation by CANOCO and is the measure of the strength of the relationship between species distribution and environmental variables. Canonical coefficients represent the weight that each environmental variable contributes to the axes, which can be tested for significance by approximate *t*-test (Table III). The intraset correlations reflect

TABLE III Canonical Coefficients, Their Approximate *t*-Test Scores, and Intraset Correlations of the Environment Variables[a]

	Canonical coefficients		t-*values of canonical coefficients*		*Intraset correlations*	
Environmental variable	*Axis 1*	*Axis 2*	*Axis 1*	*Axis 2*	*Axis 1*	*Axis 2*
Watershed area	−0.62	0.03	−4.87*	0.40	0.10	0.88
Maximum depth	−0.43	−0.03	−2.91*	−0.37	−0.48	−0.46
TN	−0.50	−0.01	−1.28	−0.60	0.10	0.88
TP	0.19	0.30	0.68	2.05*	0.04	0.90
Chlorophyll *a*	0.33	0.40	1.30	2.99*	0.10	0.87
Secchi depth	0.12	0.27	0.58	2.40*	−0.13	−0.60
pH	0.00	0.27	−0.02	2.18*	0.14	0.33
Alkalinity	0.40	−0.21	1.90	−1.85	0.22	0.13
Color	0.28	0.09	1.75	1.09	0.10	−0.60

[a]From Christie and Smol (1993), with permission of the *Journal of Phycology*. Asterisk indicates approximate *t*-test significant at $P = 0.05$.

the correlation between the environmental variables and the axes. Significant canonical coefficients and high intraset correlations of the environmental variables indicate which environmental gradient the axis represents. For example, axis 2 may represent a gradient that is related to lake trophic status (Table III and Fig. 5).

Once important gradients are identified, CANOCO permits testing the significance of the gradients by performing a Monte Carlo permutation test. This test can assess a specific statistical hypothesis by comparing an observed value (in this case an eigenvalue) with a significant level. The significant level is determined by an empirical distribution of values for the statistic (e.g., eigenvalue) obtained by numerous permutations of the data under some assumed model (Birks *et al.,* 1994). Since many environmental variables are likely to be correlated, tests for independence of important variables are necessary. A series of partial CCA can remove the effects of some environmental variables that are likely to be statistically related (terBraak, 1988). For example, a partial CCA with its first axis constrained to TN and correlated variables, such as TP, as a covariable generates an eigenvalue that can be used to calculate the amount of variance solely explained by TN (Table IV). After a series of partial CCAs, one can partition explained variance into several components as with an ANOVA table. In conjunction with Monte Carlo permutation tests, environmental variables that account for a certain species distribution pattern can be identified and tested for significance. Similar approaches can be used to remove the nuisance from the data set. The nuisance is the variance that is present that is not of primary interest and sometimes confounds the analysis of the variable of primary interest (terBraak, 1988). For example, species data

TABLE IV Results of the Monte Carlo Permutation Test (from CCA in CANOCO) When the First Axis Is Constrained to be One of the Significant Variables (Chosen Using Forward Selection in CANOCO v.3.1) and the Correlated Variables (from Table II) Were Used as Covariables[a]

Significant variant ($P \le 0.05$)	*Covariable(s)*	I_1	I_2	*Probability*
TN	Chlorophyll *a*	0.07	0.54	0.83*
	TP	0.06	0.54	0.82*
	Secchi depth	0.13	0.54	0.09*
Alkalinity	pH	0.10	0.53	0.25*
Maximum depth	TN	0.17	0.48	0.02
	Secchi depth	0.15	0.49	0.05

[a]From Christie and Smol (1993), with permission of the *Journal of Phycology.* Asterisk indicates when the results are not significant ($P > 0.05$) and the effects of the significant variable and covariable on the distribution of the diatom assemblages cannot be separated.

may be collected in different seasons or years. In some cases, seasonal variation can be substantial but is not of primary interest. Temporal effects can be removed statistically by creating dummy variables for sampling dates, declaring them as covariables, and running a series of partial CCAs.

Species optima and tolerances with respect to the important environmental variables can be calculated with weighted average regression once gradients are identified by CCA (Table V). Weighted average regression is based on the assumption that the species–response curve, with respect to the environmental variable, is unimodal (terBraak and Barendregt, 1986; terBraak and Looman, 1986; terBraak and van Dam, 1989). In a lake with a certain TN concentration, the most abundant taxon is assumed to have an optimum close to the lake TN. Once species optima and tolerances (standard deviation of optima) are estimated from a reference set, a simple estimate of an unknown lake TN can be inferred by taking a sum of products of each taxon's TN optimum and its relative abundance, which can be done using computer software such as WACALIB (v.2.1) by Line and Birks (1990). Taxa with a small TN tolerance are good TN indicators and are given more weight in inferring TN (terBraak and van Dam, 1989). It is recommended that sites be sampled from a variety of habitats that cover the gradient of interest as completely as possible. Sites should be distributed homogeneously along the gradient. It might be necessary to build up a large data base or library of estimated species optima across different geographic regions and ecoregions to increase the accuracy of the weighted averaging predictive model.

Species optima and tolerances can be classified using a cluster analysis. Each cluster can be assigned a rank based on species preference with respect to the environmental variable, with the rank bring ordinal or nominal. If species optima with respect to all important environmental variables are calculated and assigned as ranks, multidimensional biotic indices similar to the fish IBI (Karr, 1981) can be developed. The weighted average approach can extract autecological information from species-rich (=information-rich) communities. In conjunction with other important parameters such as algal biomass, primary productivity, and total tissue nutrients (TP, TN), biotic indices based on multidimensional matrices can be more robust and accurate in reflecting ecological conditions.

Meanwhile, other commonly used biotic indices such as the Shannon diversity index (1948), autotrophic index (Weber, 1973), and the pollution index (Lange-Bertalot, 1979a) can be reevaluated by comparing them with algal IBI based on a weighted average approach and incorporated into multimetric indices.

The accuracy of indices can be tested by the procedure called split-sampling, which is less computer-intensive than other procedures. For example, if 100 sites are sampled for benthic algae, we can randomly divide them into two subsets: 75 sites as a training set, and 25 sites as a

TABLE V Weighted Averages (Optima) and Tolerances of Diatom Taxa to TN Estimated Using WACALIB. Sorted by Ascending TN Optimum[a]

	Diatom taxon	*Optima* (*μg liter*$^{-1}$)	*Tolerance* (*μg liter*$^{-1}$)	*Number*
1.	*Nitzschia romana* Grun.	303.4	61.4	5
2.	*Cyclotella ocellata* Pant.	327.3	68.9	4
3.	*Fragilaria pinnata* v. *intercedens* (Grun.) Hust.	347.5	122.4	11
4.	*Cyclotella* sp. #2	364.0	77.7	10
5.	*Cyclotella* sp. #3	365.6	80.1	26
6.	*Cyclotella bodanica* v. *affinis* Grun.	369.8	99.0	43
7.	*Stephanodiscus medius* Hak.	371.5	68.0	5
8.	*Cyclotella* sp. #1	374.1	74.6	19
9.	*Cymbella minuta* f. *latens* (Krass.) Rabh.	374.1	119.3	19
10.	*Synedra ulna* v. *chasiana* Thomas	382.8	125.3	22
11.	*Cyclotella* sp. #4.	383.7	102.7	14
12.	*Synedra filiformia* v. *exilis* Cl.-Eul.	388.2	106.2	28
13.	*Achnanthes exigua* Grun.	400.9	123.9	23
14.	*Fragilaria construens* v. *venter* (Ehrenb.) Grun	400.9	121.5	36
15.	*Cyclotella stelligera* (Cl. & Grun.) V. H.	406.4	107.6	36
16.	*Synedra radians* Kütz.	407.4	106.7	9
17.	*Aulacoseira* cf. *distans* (Ehrenb.)	408.3	88.3	5
18.	*Synedra delicatissima* W. Sm.	412.1	137.4	15
19.	*Synedra amphicephala* (Kütz.)	412.1	57.8	7
20.	*Asterionella formosa* Hass.	413.1	108.1	39
21.	*Tabellaria flocculosa* (Roth) Kütz. v. *linearis* Koppen	414.0	54.8	10
22.	*Cyclotella* sp. #5	415.0	120.8	14
23.	*Tabellaria flocculosa* (Roth) Kütz. Strain III *sensu* Koppen	415.9	88.0	12
24.	*Tabellaria flocculosa* (Roth) Kütz. Strain IV *sensu* Koppen	419.8	26.9	7
25.	*Amphora ovalis* Kütz.	420.7	100.5	10
26.	*Cyclotella meneghiniana* Kütz.	421.7	87.6	6
27.	*Fragilaria brevistriata* v. *inflata* (Pant.) Hust	422.7	139.7	27
28.	*Tabellaria flocculosa* (Roth) Kütz. Strain IIIp *sensu* Koppen	415.9	88.0	12
29.	*Cocconeis dimunuta* Pant.	426.6	104.3	25
30.	*Achnanthes pinnata* Hust.	426.6	97.0	27
31.	*Navicula seminuloides* Hust.	428.6	58.8	12
32.	*Aulacoseira subarctica* (O. Mull.) Haworth	436.5	105.5	34
33.	*Nitzschia gracilis* Hust.	436.5	128.4	23
34.	*Fragilaria crotonensis* Kitt.	438.5	131.6	50
35.	*Syndera parasitica* (W. Sm.) Hust.	439.5	108.7	16
36.	*Anomoeoneis vitrea* (Grun.) Ross	446.7	100.3	10
37.	*Tabellaria fenestrata* (Lyngb.) Kütz.	447.7	115.9	30
38.	*Navicula minima* Grun.	447.7	104.4	31
39.	*Achnanthes microcephala* (Kütz.) Grun	451.0	96.2	11

(Continues)

TABLE V (continued)

Diatom taxon	Optima (μg liter^{-1})	Tolerance (μg liter^{-1})	Number
40. *Navicula arvensis* Hust.	451.0	107.7	15
41. *Stephanodiscus niagarae* Ehrenb.	451.9	126.2	36
42. *Amphora perpusilla* Grun.	455.0	154.6	37
43. *Achnanthes* sp. #1	457.1	87.4	12
44. *Fragilaria brevistriata* Grun.	458.1	118.6	38
45. *Achnanthes lanceolata* (Bréb. ex Kütz.) Grun.	458.1	129.3	25
46. *Fragilaria brevistriata* v. *capitata* Herib.	460.3	80.5	2
47. *Fragilaria pinnata* v. *acuminata* May.	463.5	55.4	6
48. *Achnanthes linearis* (W. Sm.) Grun.	463.5	102.8	25
49. *Navicula minima* v. *okamurae* Skvort.	466.7	100.9	28
50. *Stephanodiscus alpinus* Hust.	466.7	118.1	28
51. *Fragilaria vaucheriae* Kütz.	469.9	127.1	15
52. *Achnanthes kryophila* Peter	471.0	141.2	10
53. *Achnanthes biasolettiana* (Kütz) Grun.	471.0	90.1	18
54. *Fragilaria pinnata* Ehrenb.	472.1	100.7	39
55. *Cyclotella michiganiana* Skvort.	472.1	118.1	29
56. *Fragilaria* sp. 2 PIRLA	473.2	86.6	25
57. *Synedra rumpens* Kütz.	477.5	80.9	11

[a]From Christie and Smol (1993), with permission of the *Journal of Phycology*. WACALIB is discussed in Line and Birks (1990). These data can be used to describe the distribution of diatom taxa along a trophic gradient. "Number" represents the number of occurrences of each taxon.

testing set. The training set is used to develop the model and ultimately the indices. The accuracy of the biotic indices can be assessed with the remaining sites in the testing set by correlation analysis. If samples are taken for multiple years, newly sampled sites can be used to test the model developed from last year's data set provided temporal variation is removed. In computer resampling procedures, 100 original sites from a reference area are used to develop indices. Then the same 100 sites form a data pool. Each time, 100 sites are randomly selected from the data pool and used to develop indices. The same procedure is repeated *n* times (e.g., 999 times). A distribution of the biotic indices can then be created and therefore a standard error or a 95% confidence interval can be estimated for observed indices. This procedure is commonly called bootstrapping (Efron, 1979). A similar procedure called "jackknifing" is an approximation of bootstrapping (Tukey, 1958). In jackknifing, one sample is excluded to calculate a parameter. In our example, one site would be excluded (99 sites are used to calculate indices each time). In bootstrapping, resampling is done with replacement and the same sample can be selected more than once. The foregoing procedures can provide robust, reliable, and relatively unbiased standard errors for indices. Therefore, accuracy and sensitivity of biotic indices can be evaluated.

An automated monitoring system can be developed to facilitate utilization of biotic indices in monitoring ecosystems. The system should be programmed to a level so that users only need to input simple biological parameters such as species relative abundance and algal biomass (AFDM or chlorophyll *a*), which results in the production of a score indicating ecological conditions based on built-in biotic indices. With such automated and user-friendly systems, people who receive only very brief biological training can be involved in monitoring programs. For example, the Mississippi River is monitored by a number of high school and elementary school students through an NSF-funded program called The River Program (R. Williams, personal communication). With the development by researchers of a series of biotic indices for large ecosystems, a monitoring network could be established around the nation involving state agencies and, more importantly, students.

IV. SUMMARY AND CONCLUSIONS

Benthic algal communities abound in most aquatic habitats containing sunlit substrata. The species richness and physiological diversity within this community render it an excellent choice for water quality monitoring. As with any monitoring program, care must be exercised to collect data of high quality that are truly reflective of the environment to be monitored. It is also important to remember that the interface of the abiotic environment and the benthic algae is not isolated from the remainder of the aquatic community and that changes in the benthic algal community induced from chemical changes in the habitat may be modified or masked by other factors. For example, point sources of nutrients may result in increased benthic algal productivity with little increase in benthic algal biomass because of rapid transfer of organic matter to grazers. These types of top-down forces that mask benthic algal response have been well documented in several aquatic ecosystems (Peterson *et al.*, 1985; Colletti *et al.*, 1987, Marks and Lowe, 1989). Thus, benthic algal monitoring data are most valuable when combined with a suite of monitoring data including physical and chemical measurements and analysis of other biota of aquatic communities such as invertebrates.

ACKNOWLEDGMENTS

We thank Stephen Porter, Jan Stevenson, and Gina LaLiberte for helpful reviews of this contribution and David Johnson for assistance with electron microscopy.

REFERENCES

Aloi, J. E. (1990). A critical review of recent freshwater periphyton field methods. *Can. J. Fish. Aquat. Sci.* **47**, 656–670.

American Public Health Association, American Water Works Association and Water Pollution Control Federation (1985). *In* "Standard Methods for the Examination of Water and Wastewater" (A. E. Greenburg, R. R. Trussell, and L. S. Clesceri, eds.). APHA, Washington, DC.

Antoine, S. E., and Benson-Evans, K. (1985). Colonisation rates of benthic algae on four different rock substrata in the River Ithon, Mid Wales, U. K. *Limnologica* **16**, 307–313.

Archibold, R. E. M. (1972). Diversity in some South African diatom associations and its relation to water quality. *Water Res.* **6**, 1229–1238.

Austin, A., Lang, S., and Pomeroy, M. (1981). Simple methods for sampling periphyton with observations on sampler design criteria. *Hydrobiologia* **85**, 33–47.

Bahls, L. L. (1973). Diatom community response to primary wastewater effluent. *J. Water Pollut. Control Fed.* **45**, 134–144.

Bahls, P. A., and Bahls, L. L. (1974). Trophic response to a hatchery effluent. *Proc. Mont. Acad. Sci.* **34**, 5–11.

Beaver, J. (1981). "Apparent Ecological Characteristics of Some Common Freshwater Diatoms." Ontario Ministry of the Environment, Technical Support Section, Don Mills, Ontario, Can.

Bélanger, S. E., Barnum, J. B., Woltering, D. M., Bowling, J. W., Ventullo, R. M., Schermerhorn, S. D., and Lowe, R. L. (1994). Algal periphyton structure and function in response to consumer chemicals in stream mesocosms. *In* "Aquatic Mesocosm Studies in Ecological Risk Assessment" (R. L. Graney, J. H. Kennedy, and J. H. Rogers, eds.), SETAC Spec. Publ. Ser., Lewis Publishers, Ann Arbor, MI.

Biggs, B. J. F. (1985). The use of periphyton in the monitoring of water quality. *In* "Biological Monitoring in Freshwaters: Proceedings of a Seminar" (R. D. Pridmore and A. D. Cooper, eds.), Water and Soil Misc. Publ. No. 82. Ministry of Works and Development for the National Water and Soil Conservation Authority, Wellington, NZ.

Biggs, B. J. F. (1988a). A periphyton sampler for shallow, swift rivers. *N. Z. J. Mar. Freshwater Res.* **22**, 189–199.

Biggs, B. J. F. (1988b). Artificial substrate exposure times for periphyton biomass estimates in rivers. *N. Z. J. Mar. Freshwater Res.* **22**, 507–515.

Biggs B. J. F. (1989). Biomonitoring of organic pollution using periphyton, South Branch, Canterbury, New Zealand. *N. Z. J. Mar. Freshwater Res.* **23**, 263–274.

Biggs, B. J. F. (1990). Use of relative specific growth rates of periphytic diatoms to assess enrichment of a stream. *N. Z. J. Mar. Freshwater Res.* **24**, 9–18.

Biggs, B. J. F., and Close, M. E. (1989). Periphyton biomass dynamics in gravel bed rivers: The relative effects of flow and nutrients. *Freshwater Biol.* **22**, 209–231.

Birks, H. J. B. (1993). "Impact of Computer-intensive Procedures in Testing Palaeoecological Hypothesis," Newsletter No. 9. Inqua-Commission for the Study of the Holocene: Working Group on Data-handling Methods.

Birks, H. J. B., Peglar, S. M., and Austin, H. A. (1994). "An Annotated Bibliography of Canonical Correspondence Analysis and Related Constrained Ordination Methods 1986–1993. Botanical Institute, University of Bergen, Bergen, Norway.

Bothwell, M. L. (1985). Phosphorus limitation of lotic periphyton growth rates: An intersite comparison using continuous-flow troughs (Thompson River System, British Columbia). *Limnol. Oceanogr.* **30**, 527–542.

Bothwell, M. L. (1988). Growth rate responses of lotic periphyton diatoms to experimental phosphorus enrichment: The influence of temperature and light. *Can. J, Fish. Aquat. Sci.* **45**, 261–270.

Bourrelly, P. (1966). "Les Algues d'eau Douce. Vol. 1. Les Algues Vertes." Boubee, Paris.

Bourrelly, P. (1968). "Les Algues D'eau Douce. Vol. 2. Les Algues Jaunes et Brunes." Boubee, Paris.

Bourrelly, P. (1970). "Les Algues D'eau Douce. Vol. 2. Les Algues Bleues et Rouges." Boubee, Paris.

Burkholder, J. M., and Wetzel, R. G. (1989). Microbial colonization on natural and artificial macrophytes in a phosphorus-limited hardwater lake. *J. Phycol.* **25,** 55–65.

Burkholder, J. M., Wetzel, R. G., and Klomparens, K. L. (1990). Direct comparison of phosphate uptake by adnate and loosely attached microalgae within an intact biofilm matrix. *Appl. Environ. Microb.* **56,** 2882–2890.

Cairns, J., Jr., Albough, D. W., Busey, F., and Chanay, M. D. (1968). The sequential comparison index—A simplified method for nonbiologists to estimate relative differences in biological diversity in stream pollution studies. *J. Water Pollut. Control Fed.* **40,** 1607–1613.

Cairns, J., Jr., Dickson, K. L., and Slocomb, J. (1977). The ABC's of diatom identification using laser holography. *Hydrobiologia* **54,** 7–16.

Cairns, J., Jr., Plafkin, J. L., Kaesler, R. L., and Lowe, R. L. (1983). Early colonization patterns of diatoms and protozoa in fourteen fresh-water lakes. *J. Protozool.* **30,** 47–51.

Carrick, H. J., Lowe, R. L., and Rotenberry, J. T. (1988). Guilds of benthic algae along nutrient-gradients: Relationships with algal community diversity. *J. North Am. Benthol. Soc.* **2,** 117–128.

Cattaneo, A., and Amireault, M. C. (1992). How artificial are artificial substrata for periphyton? *J. North Am. Benthol. Soc.* **11,** 244–256.

Chessman, B. C. (1985). Artificial-substratum periphyton and water quality in the lower La Trobe River, Victoria. *Aust. J. Mar. Freshwater Res.* **36,** 855–871.

Christie, C. E., and Smol, J. P. (1993). Diatom assemblages as indicators of lake trophic status in southeastern Ontario lakes. *J. Phycol.* **29,** 575–586.

Colletti, P.J., Blinn, D. W., Pickart, A., and Wagner, V. T. (1987). Influence of different densities of the mayfly grazer *Heptogenia criddlei* on lotic diatom communities. *J. North Am. Benthol. Soc.* **6,** 270–280.

Cooper, J. M., and Wilhm, J. (1975). Spatial and temporal variation in productivity species diversity and pigment diversity of periphyton in a stream receiving domestic oil refinery effluents. *Southwest Nat.,* pp. 413–428.

Descy, J. P. (1979). A new approach to water quality estimation using diatoms. *Nova Hedwi., Beih.* **64,** 305–323.

Dickman, M. D., and Gochnauer, M. B. (1978). Impact of sodium chloride on the microbiota of a small stream. *Environ. Pollut.* **17,** 109–126.

Digby, P. G. N., and Kempton, R. A. (1987). "Multivariate Analysis of Ecological Communities." Chapman & Hall, London.

Dixit, S. S., Cumming, B. F., Smol, J. P., and Kingston, J. C. (1992). Monitoring environmental changes in lakes using algal microfossils. *In* "Ecological Indicators" (D. H. McKenzie, D. E. Hyatt, and V. J. MacDonald, eds.), pp. 1135–1155. Elsevier, Amsterdam.

Duncan, S. W., and Blinn, D. W. (1989). Importance of physical variables on the seasonal dynamics of epilithic algae in a highly shaded canyon stream. *J. Phycol.* **25,** 455–461.

Efron, B. (1979). Bootstrap methods: Another look at the Jackknife. *Ann. Stat.* **7,** 1–26.

Ennis, G. L. (1977). Attached algae as water quality indicators in phosphorous enriched Kootenay Lake, British Columbia. *J. Phycol., Suppl.* **13,** 20.

Eppley, R. W. (1977). The growth and culture of diatoms. *In* "The Biology of Diatoms" (D. Werner, ed.). Univ. of California Press, Berkeley.

Everitt, B. (1993). "Cluster Analysis." Arnold, London.

Fairchild, G. W., and Everett, A. C. (1988). Effects of nutrient (N, P, C) enrichment upon periphyton standing crop, species composition and primary production in an oligotrophic softwater lake. *Freshwater Biol.* **19,** 57–70.

Fairchild, G. W., Lowe, R. L. and Richardson, W. T. (1985). Algal periphyton growth on nutrient-diffusing substrates: An *in situ* bioassay. *Ecology* **66**, 465–472.

Freeman, M. C. (1986). The role of nitrogen and phosphorus in the development of *Cladophora glomerata* in the Manawatu River, New Zealand. *Hydrobiologia* **131**, 23–30

Gale, W. F., Gurzynski, A. and Lowe, R. L. (1979). Colonization rates and standing crops of periphytic algae in a large Pennsylvania river. *J. Phycol.* **15**, 117–123.

Genter, R. B., Cherry, D. S., Smith, E. P., and Cairns, J., Jr. (1987). Algal–periphyton population and community changes from zinc stress in stream mesocosms. *Hydrobiologia* **153**, 261–275.

Germain H. (1981). "Flore des Diatomés. Diatomophyees eaux douces et saumâtres du Massif Armoricain et des contrées voisines d'Europe occidental." Boubee, Paris.

Gerrodette, T. (1987). A power analysis for detecting trends. *Ecology* **68**, 1364–1372.

Gill, M. A., Lowe, R. L., and Sferra, J. C. (1993). Use of river benthic community structure for water quality assessment: A case study from a consultant's perspective. *Bull. Soc. Chem. Environ. Tox.* **14**, 175.

Gordon, A. D. (1981). "Classification. Method for the Exploratory Analysis of Multivariate Data." Chapman & Hall, London.

Guzkowska, M. A. J., and Gasse, F. (1990a). Diatoms as indicators of water quality in some English urban lakes. *Freshwater Biol.* **23**, 233–250.

Guzkowska, M. A. J., and Gasse, F. (1990b). The seasonal response of diatom communities to variable water quality in some English urban lakes. *Freshwater Biol.* **23**, 251–264.

Hansmann, E. E., and Phinney, H. K. (1973). Effects of logging on periphyton in coastal streams of Oregon. *Ecology* **54**, 194–199.

Hawes, I., and Smith, R. (1993). Effect of localised nutrient enrichment on the shallow epilithic periphyton of oligotrophic Lake Taupo, New Zealand. *N. Z. J. Mar. Freshwater Res.* **27**, 365–372.

Hill, M. O., Bunce, R. G. H., and Shaw, M. W. (1975). Indicator species analysis, a divisive polythetic method of classification, and its application to a survey of native pine woods in Scotland. *J. Ecol.* **63**, 597–613.

Hohn, M. H., and Hellerman, J. (1963). The taxonomy and structure of diatom populations from three eastern North American rivers using three sampling methods. *Trans. Am. Microsc. Soc.* **87**, 250–329.

Horner, R. R., and Welch, E. B. (1981). Stream periphyton development in relation to current velocity and nutrients. *Can. J. Fish. Aquat. Sci.* **38**, 449–457.

Hustedt, F. (1930). Bacillariophyta (Diatomeae). *In* "Die Susswasser-Flora Mitteleuropas" (A. Pascher ed.), No. 10. Fischer, Jena.

Jackson, J. E. (1991). "A User's Guide to Principal Components." Wiley, Chichester, UK.

Johnson, D. W., and Lowe, R. L. (1993). Advances in quality control of biological monitoring in streams: Interfacing periphyton analyses with the computer. *Bull. North Am. Benthol. Soc.* **10**, 176–177.

Jongman, R. H. G., terBraak, C. J. F. and vanTongeren, O. F. R. (1987). "Data Analysis in Community and Landscape Ecology." Pudoc, Wageningen, The Netherlands.

Kann, J., and Falter, C. M. (1989). Periphyton as indicators of enrichment in Lake Pend Oreille, Idaho. *Lake Reserv. Manage.* **5**, 39–48.

Karr, J. R. (1981). Assessment of biotic integrity using fish communities. *Fisheries* **6**, 21–27.

Keithan, E. D., and Lowe, R. L. (1985). Primary productivity and structure of phytolithic communities in streams in the Great Smoky Mountains National Park. *Hydrobiologia* **123**, 59–67.

Kettunen, I. (1983). A study of the periphyton of Lake Saimaa, polluted by waste waters of the pulp industry. *In* "Periphyton of Freshwater Ecosystems" (R. G. Wetzel, ed.), pp. 331–335. Dr. W. Junk Publishers, The Hague.

Kingston, J. C., Lowe, R. L., Stoermer, E. F., and Ludewski, T. (1983). Spatial and temporal distribution of benthic diatoms in northern Lake Michigan. *Ecology* **64,** 1566–1580.

Kosinski, R. J. (1984). The effect of terrestrial herbicides on the community structure of stream periphyton. *Environ. Pollut.* **36,** 165–189.

Krammer, K., and Lange-Bertalot, H. (1986). "Süsswasserflora von Mitteleuropa. Bacillariophyceae 1. Teil, Naviculaceae," p. 876. Fischer, New York.

Krammer, K., and Lange-Bertalot, H. (1988). "Süsswasserflora von Mitteleuropa. Bacillariophyceae 2. Teil, Bacillariaceae, Epithemiaceae, Surirellaceae," p. 596. Fischer, New York.

Krammer, K., and Lange-Bertalot, H. (1991a). "Süsswasserflora von Mitteleuropa. Bacillariophyceae 3. Teil, Centrales, Fragilariaceae, Eunotiaceae," p. 576. Fischer, New York.

Krammer, K., and Lange-Bertalot, H. (1991b). "Süsswasserflora von Mitteleuropa. Bacillariophyceae 4. Teil, Achnanthaceae, Kritische Ergängzugen zu *Navicula* (Lineolatae) und *Gomphonema,*" p. 437. Fischer, New York.

Krejci, M. E., and Lowe, R. L. (1987). Spatial and temporal variation of epipsammic diatoms in a spring-fed brook. *J. Phycol.* **23,** 585–590.

Lange-Bertalot, H. (1978). Diatomeen-Differentialarten an Stelle von Leitformen: Ein geeigneteres Kriterium des Gewässerbelastung. *Arch. Hydrobiol., Suppl.* **51,** 393–427.

Lange-Bertalot, H. (1979a). Pollution tolerance of diatoms as a criterion for water quality estimation. *Nova Hedw. Beih.* **64,** 285–304.

Lange-Bertalot, H. (1979b). Toleranzgrenzen und Populationsdynamik benthischer Diatomeen bei unterschiedlich starker Abwasserbelastung. *Arch. Hydrobiol., Suppl.* **56,** 184–219.

Leclercq, L., and Maquet, B. (1987). Deux nouveux indicies diatomique et de qualité existants. *Cah. Biol. Mar.* **28,** 303–310.

Line, J. M., and Birks, H. J. B. (1990). WACALIB version 2.1—A computer program to reconstruct environmental variables from fossil assemblages by weighted averaging. *J. Paleolimnol.* **3,** 170–173.

Loeb, S. L. (1981). An *in situ* method for measuring the primary productivity and standing crop of the epilithic periphyton community in lentic systems. *Limnol. Oceanogr.* **26,** 394–399.

Lowe, R. L. (1974). "Environmental Requirements and Pollution tolerance of Freshwater Diatoms." Environ. Monit. Ser. 670/4-74-005. USEPA, Washington, DC.

Lowe, R. L., and Gale, W.F. (1980). Monitoring periphyton with artificial benthic substrates. *Hydrobiologia* **69,** 235–244.

Lowe, R. L., and Johnson, D. W. (1992). Advances in quality control of biological monitoring in streams: Interfacing periphyton analyses with the computer. *Bull. Soc. Chem. Environ. Toxicol.* **13,** 272.

Lowe, R. L., and McCullough, J. M. (1974). The effect of sewage-treatment-plant effluent on diatom communities of the Portage River, Wood County, Ohio. *Ohio J. Sci.* **74,** 154–161.

Lowe, R. L., Golladay, S., and Webster, J. (1986). Periphyton response to nutrient manipulation in a clearcut and forested watershed. *J. North Am. Benthol. Soc.* **3,** 211–220.

Manly, B. F. J. (1991). "Multivariate Statistical Methods. A Primer." Chapman & Hall, London.

Marks, J. C., and Lowe, R. L. (1989). The independent and interactive effects of snail grazing and nutrient enrichment on structuring periphyton communities. *Hydrobiologia* **185,** 9–17.

Maurice, C. G., Lowe, R. L., Burton, T. M., and Stanford, R. M. (1987). Biomass and compositional changes in the periphytic community of an artificial stream in response to lowered pH. *Water, Air, Soil Pollut.* **33,** 165–177.

McIntire, C. D. (1968). Structural characteristics of benthic algal communities in laboratory streams. *Ecol.* **49,** 520–537.

Meier, P. G., O'Conner, D., and Dilks, D. (1983). Artificial substrata for reducing periphytic variability on replicated samples. *In* "Periphyton of Freshwater Ecosystems" (R. G. Wetzel ed.), pp. 283–286. Dr. W. Junk Publishers, The Hague.

Morgan, M. D. (1987). Impact of nutrient enrichment and alkalinization on periphyton communities in the New Jersey Pine Barrens. *Hydrobiol.* **144**, 233–241.

Mulholland, P. J., Steinman, A. D., Palumbo, A. V., and Elwood, J. W. (1991). Role of nutrient cycling and herbivory in regulating periphyton communities in laboratory streams. *Ecology* **72**, 966–982.

Pan, Y., and Lowe, R. L. (1994). Independent and interactive effects of nutrients and grazers on benthic algal community structure. *Hydrobiologia* **291**, 201–209.

Patrick, R., and Reimer, C. W. (1966). "The Diatoms of the United States Exclusive of Alaska and Hawaii," Monogr. 13, Vol. I. Academy of Natural Science, Philadelphia.

Patrick, R., and Reimer, C. W. (1975). "The Diatoms of the United States Exclusive of Alaska and Hawaii," Monogr. 13, Vol. 2, Part 1. Academy of Natural Science, Philadelphia.

Patrick, R., Hohn, M. H., and Wallace, J. (1954). A new method for determining the pattern of the diatom flora. *Not. Nat. Acad. Nat. Sci. Philadelphia* **259**, 1–12.

Peterson, B. J., Hobbie, J. E., Hershey, A. E., Lock, M. A., Ford, T. E., Vestal, J. R., McKinley, V. L., M. Hullan, A. J., Miller, M. C., Ventullo, R. M., and Volk, G. S. (1985). Transformation of a tundra river from heterotrophy to autotrophy by addition of phosphorus. *Science* **229**, 1383–1386.

Piclou, E. C. (1984). "The Interpretation of Ecological Data. A Primer on Classification and Ordination." Wiley, New York.

Power, M. E., Matthews, W. J., and Stewart, A. J. (1985). Grazing minnows, piscivorous bass and stream algae: Dynamics of a strong interaction. *Ecology* **66**, 1448–1456.

Prescott, G. W. (1962). "Algae of the Western Great Lakes Area." Wm. C. Brown, Dubuque, IA.

Prescott, G. W., Croasdale, H. T., and Vinyard, W. C. (1972). "North American Flora. Series II, Part 6, Desmidiales. Part I. Saccodermae, Mesotaeniaceae." New York Botanical Garden, Bronx, N.Y.

Prescott, G. W., Croasdale, H. T., and Vinyard, W. C. (1975). "A Synopsis of North American Desmids. Part II. Desmidiaceae: Placodermae Section 1." Univ. of Nebraska Press, Lincoln.

Prescott, G. W., Croasdale, H. T., and Vinyard, W. C. (1977). "A Synopsis of North American Desmids. Part II. Desmidiaceae: Placodermae Section 2." Univ. of Nebraska Press, Lincoln.

Prescott, G. W., Croasdale, H. T., Vinyard, W. C. and Bicudo, C. E. M. (1981). "A Synopsis of North American Desmids. Part II. Desmidiaceae: Placodermae Section 3." Univ. of Nebraska Press, Lincoln.

Pringle, C. M., and Bowers, J. A. (1984). An *in situ* substratum fertilization technique: Diatom colonization on nutrient-enriched sand substrata. *Can. J. Fish. Aquat. Sci.* **41**, 1247–1251.

Reyment, R. A. (1991). "Multidimensional Palaeobiology." Pergamon, Oxford.

Robinson, C. T., and Rushforth, S. R. (1987). Effect of physical disturbance and canopy cover on attached diatom community structure in an Idaho stream. *Hydrobiologia* **154**, 49–59.

Rose, F. L., and McIntire, C. D. (1970). Accumulation of dieldrin by benthic algae in laboratory streams. *Hydrobiologia* **35**, 481–493.

Round, F. E. (1991). Use of diatoms for monitoring rivers. *In* "Use of Algae for Monitoring Rivers" (B. A. Whitton, E. Rott, and G. Friedrich, eds.). Publ. Inst. Bot., AG Hydrobotanik, Universität Innsbruck, Innsbruck, Austria.

Shannon, C. E. (1948). A mathematical theory of communication. *Bell Syst. Tech. J.* **27**, 379–423.

Siver, P. A. (1977). Comparison of attached diatom communities on natural and artificial substrates. *J. Phycol.* **13**, 402–406.

Sládeček, V. (1973). System of water quality from the biological point of view. *Arch. Hydrobiol./Ergeb. Limnol.* **7**, 1–218.

Smol, J. P., Battarbee, R. W., Davis, R. B., and Meriläinen, J., eds. (1986). "Diatoms and Lake Acidity." Junk Publ., Dordrecht; The Netherlands.

Steinman, A. D., and McIntire, C. D. (1986). Effects of current velocity and light on the structure of periphyton assemblages in laboratory streams. *J. Phycol.* **22,** 352–361.

Steinman, A. D., McIntire, C. D., Gregory, S. V., Lamberti, G. A., and Ashkenas, L. R. (1987). Effects of herbivore type and density on taxonomic structure and physiognomy of algal assemblages in laboratory streams. *J. North Am. Benthol. Soc.* **6,** 175–188.

Steinman, A. D., McIntire, C. D., Gregory, S. V., and Lamberti, G. A. (1989). Effects of irradiance and grazing on lotic algal assemblages. *J. Phycol.* **25,** 478–485.

Stevenson, R. J. (1984a). Procedures for mounting algae in syrup medium. *Trans. Am. Microsc. Soc.* **107,** 320–321.

Stevenson, R. J. (1984b). How currents on different sides of substrates in streams affect mechanisms of benthic algal accumulation. *Int. Rev. Gesamten Hydrobiol.* **69,** 241–262.

Stevenson, R. J. (1984c). Eplithic and epipelic diatoms in the Sandusky River, with emphasis on species diversity and water pollution. *Hydrobiologia* **114,** 161–175.

Stevenson, R. J. (1990). Benthic community dynamics in a stream during and after a spate. *J. North Am. Benthol. Soc.* **9,** 277–288.

Stevenson, R. J., and Lowe, R. L. (1986). Sampling and interpretation of algal patterns for water quality assessments. *In* "Rationale for Sampling and Interpretation of Ecological Data in the Assessment of Freshwater Ecosystems" (B. Isom, ed.), ASTM STP 894. Am. Soc. Test. Mater., Philadelphia.

Stewart, P. M., Pratt, J. R., Cairns, J., Jr., and Lowe, R. L. (1985). Diatom and protozoan accrual on artificial substrates in lentic habitats. *Trans. Am. Microsc. Soc.* **104,** 369–377.

Stockner, J. G., and Shortreed, K. R. S. (1978). Enhancement of autotrophic production by nutrient addition in a coastal rainforest stream on Vancouver Island. *J. Fish. Res. Board Can.* **35,** 28–34.

Sumita, N., and Wantanabe, T. (1983). New general estimation of river pollution using new diatom community index (NDCI) as biological indicator based on specific composition of epilithic diatom communities. Applied to the Asana-gawa and the Sai-gawa rivers in Ishikawa Prefecture. *Jpn. J. Limnol.* **44,** 329–340.

Taft, C. E. (1978). A mounting medium for fresh-water plankton. *Trans. Am. Microsc. Soc.* **97,** 263–264.

terBraak, C. J. F. (1986a). Interpreting a hierarchical classification with simple discriminant functions: An ecological example. *In* "Data Analysis and Informatatics IV" (Diday *et al.,* eds.) Elsevier/North-Holland, Amsterdam.

terBraak, C. J. F. (1986b). Canonical correspondence analysis: A new eigenvector technique for multivariate direct gradient analysis. *Ecology* **67,** 1667–1679.

terBraak, C. J. F. (1987a). The analysis of vegetation–environment relationships by canonical correspondence analysis. *Vegetatio* **69,** 69–77.

terBraak, C. J. F. (1987b). "CANOCO—A FORTRAN Program for Canonical Community Ordination by [Partial][Detrended][Canonical] Correspondence Analysis, Principle Components Analysis and Redundancy Analysis (Version 2.1)." TNO Institute of Applied Computer Science, Wageningen, The Netherlands.

terBraak, C. J. F. (1988). Partial canonical correspondence analysis. *In* "Classification and Related Methods of Data Analysis" (H. H. Lock, ed.). Elsevier/North-Holland, Amsterdam.

terBraak, C. J. F., and Barendregt, L. G. (1986). Weighted averaging of species indicator values: Its efficiency in environmental calibration. *Math. Biosci.* **78,** 57–72.

terBraak, C. J. F., and Looman, W. N. (1986). Weighted averaging, logistic regression and the Gaussian response model. *Vegetatio* **65,** 3–11.

terBraak, C. J. F., and Prentice, I. C. (1988). A theory of gradient analysis. *Adv. Ecol. Res.* **18,** 271–317.

terBraak, C. J. F., and van Dam, H. (1989). Inferring pH from diatoms: A comparison of old and new calibration methods. *Hydrobiologia* **178**, 209–223.

Tuchman, M. L., and Blinn, D. W. (1979). Comparison of attached algal communities on natural and artificial substrata along a thermal gradient. *Br. Phycol. J.* **14**, 243–254.

Tuchman, M. L., and Stevenson, R. J. (1980). A comparison of clay tile, sterilized rock and natural substrate diatom communities in a small stream in southeastern Michigan, USA. *Hydrobiologia* **75**, 873–79.

Tukey, J. W. (1958). Bias and confidence in not quite large samples. *Anns. Math. Stat.* **29**, 614.

Turner, M. A., Schindler, D. W., and Graham, R. W. (1983). Photosynthesis–irradiance relationships of epilithic algae measured in the laboratory and *in situ*. *In* "Periphyton of Freshwater Ecosystems" (R. G. Wetzel ed.), pp. 73–87. Dr. W. Junk Publishers, The Hague.

van Dam, H. (1975). De invloed van vervuiling, speciaal op epifytische diatomeeëngemeenschappen, in het plassengebied rond Ankeveen. *Oeverdruk Levende Nat.* **78**, 37–48.

van Dam, H. (1982). On the use of measures of structure and diversity in applied diatom ecology. *Nova Hedw., Beih.* **73**, 97–113.

VanLandingham, S. L. (1976). Comparative evaluation of water quality on the St. Joseph River (Michigan and Indiana, U.S.A.) by three methods of algal analysis. *Hydrobiologia* **48**, 145–173.

VanLandingham, S. L. (1982). "Guide to the Identification, Environmental Requirements and Pollution Tolerance of Freshwater Bluegreen Algae (Cyanophyta)," Environ. Monit. Ser. 600/3-82-072. USEPA Washington, DC.

van Tongeren, O. F. R. (1987). Cluster analysis. *In* "Data Analysis in Community and Landscape Ecology" (R. H. G. Jongman, C. J. F. terBraak, and O. F. R. van Tongeren, eds.), pp. 174–212. Pudoc, Wageningen, The Netherlands.

Wantanabe, T., Asai, K. and Houki, A. (1988). Numerical water quality monitoring of organic pollution using diatom assemblages. *In* "Proceedings of the 9th International Diatom Symposium" (F. E. Round, ed.), pp 123–141. Biopress Ltd., Bristol, UK.

Weber, C. I. (1973). Recent developments in the measurement of the response of plankton and periphyton to changes in their environment. *In* "Bioassay Techniques and Environmental Chemistry" (G. E. Glass, ed.), pp. 119–138. Ann Arbor Sci. Publi., Ann Arbor, MI.

Wharfe, J. R., Taylor, K. S., and Montgomery, H. A. (1984). The growth of *Cladophora glomerata* in a river receiving sewage effluent. *Water Res.* **18**, 971–979.

Whitford, L. A., and Schumacher, G. J. (1973). "A Manual of Fresh-Water Algae." Sparks Press, Raleigh, NC.

Whitton, B. A., Kelly, M. G., Harding, J. P. C., and Say, P. J. (1991a). "Use of Plants to Monitor Heavy Metals in Fresh Waters 1991." Standing Committee of Analysts, H. M. Stationery Office, London.

Whitton, B. A., Rott, E., and Friedrich, G. (1991b). "Use of Algae for Monitoring Rivers." E. Rott, Publ. Inst. Bot., AG Hydrobotanik, Universität Innsbruck, Innsbruck, Austria.

Zar, J. H. (1983). "Biostatistical Analysis." Prentice-Hall, Englewood Cliffs, NJ.

Taxonomic Index

Subject Index

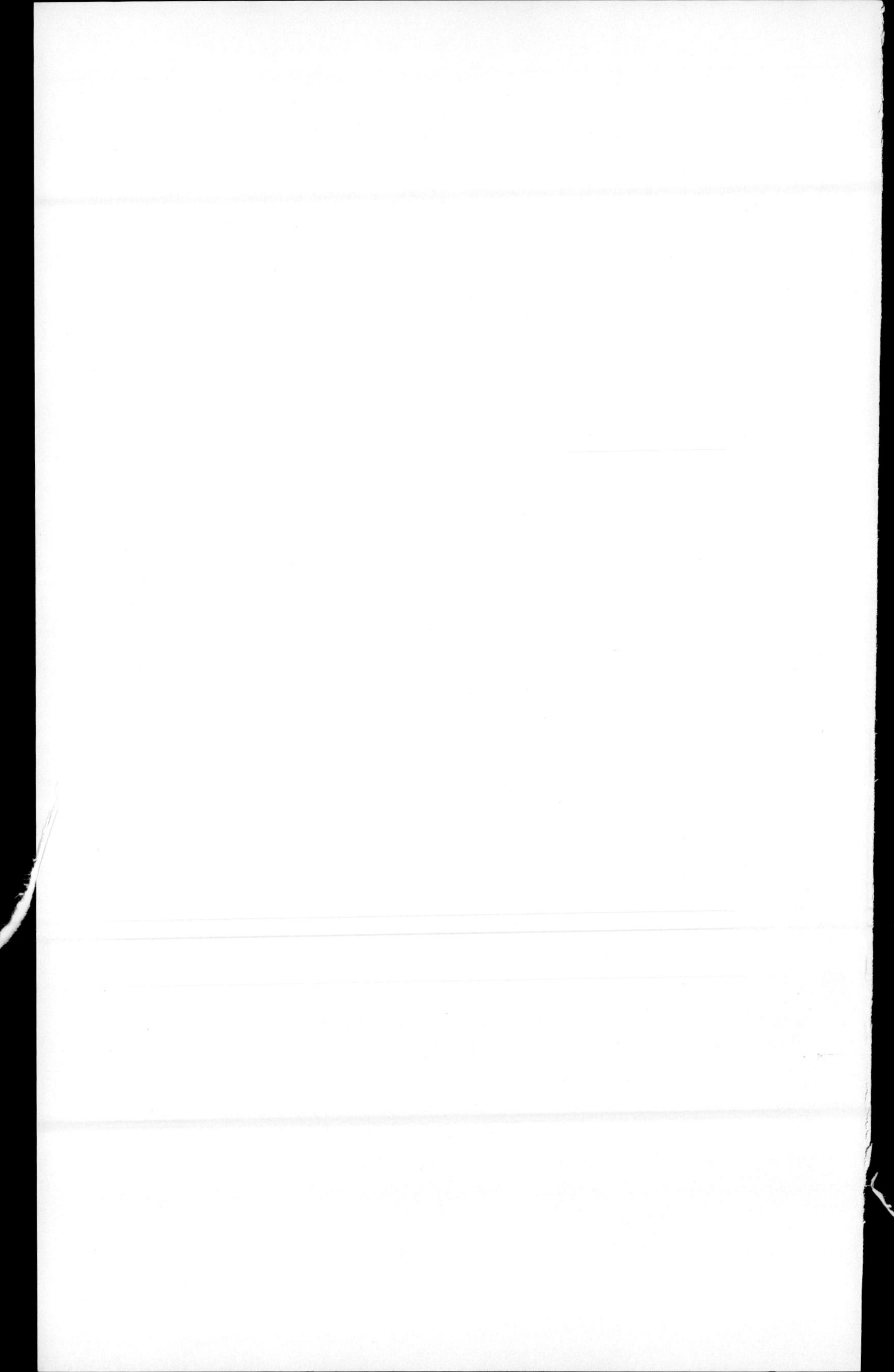